Baccharis

Geraldo Wilson Fernandes • Yumi Oki
Milton Barbosa
Editors

Baccharis

From Evolutionary and Ecological Aspects to Social Uses and Medicinal Applications

Springer

Editors
Geraldo Wilson Fernandes
Universidade Federal de Minas Gerais
Belo Horizonte, Minas Gerais, Brazil

Yumi Oki
Universidade Federal de Minas Gerais
Belo Horizonte, Minas Gerais, Brazil

Milton Barbosa
Universidade Federal de Minas Gerais
Belo Horizonte, Minas Gerais, Brazil

ISBN 978-3-030-83510-1 ISBN 978-3-030-83511-8 (eBook)
https://doi.org/10.1007/978-3-030-83511-8

This Springer imprint is published by the registered company Springer Nature Switzerland AG
The registered company address is: Gewerbestrasse 11, 6330 Cham, Switzerland

We dedicate this book to the memory of Graziela Maciel Barroso, known as the first lady of botany in Brazil. Graziela was born in the then small town of Corumbá near the border with Paraguay in 1912 and passed away in Rio de Janeiro in 2003. She started studying botany at the age of 30 to become the most iconic botanical taxonomist in Brazil. She was the first female naturalist to work in the Botanical Garden of Rio de Janeiro in 1946. When she was 47 years old, she started biology at the State University of Guanabara, today State University of Rio de Janeiro. Graziela was 60 years old when she received her PhD from the University of Campinas, working on the systematics of Brazilian Baccharis. She passed away when she was 91 years old, one month before entering the Brazilian Academy of Sciences. Dedicating this book to Graziela Barroso is fair and appropriate as her attitudes and way of life resemble that of Baccharis: conquer and thrive in unlikely places, under the most difficult pressures to circumvent.

Foreword

Imagine that you want to find a plant model system to study ecology, evolutionary biology, phytochemistry, ethnobotany, plant-animal interactions, biogeography, plant development, and morphology. Imagine, also, that you want to use this model system for any of these subjects, or combinations thereof, in a variety of ecological settings—from drylands to rainy areas, from lowlands to relatively high elevations. Imagine, further, that you want to have access to a model system that, being of ethnobotanical importance—resulting from its use for medicinal purposes by traditional indigenous and rural cultures– is promising for bioprospecting and holds potential for industrial development. Your first option would be to compile a large group of plants that encompasses this diversity of ecological, evolutionary, ethnobotanical, and pharmacological traits. To be sure, the exuberant plant kingdom—composed of an estimated contingent of some 380,000 species of vascular plants—that currently populates the planet will allow you to find some species of some lineage to study one or two of the aspects of your research program; another species or group of species from another lineage would be convenient for you to examine other aspects, and so on, across multiple lineages of the plant kingdom.

Beyond the practical complexities that such an interspecific approach would bring, you will have to deal with the challenge of the correct interpretation of several aspects of your research. Consider, by way of example, a hypothetical case of studies on the ecology and evolution of anti-herbivore plant compounds under contrasting edaphic environmental settings—poor-nutrient soils, compared to nutrient-rich soils, for instance. You could predict that species adapted to poor-nutrient soils will be more defended than those adapted to nutrient-rich soils under the Resource Availability Hypothesis. Suppose that you uncover differences in the profile of defensive compounds in your study plants, consistent with your theoretical expectations. However, your interpretations of such exciting differences will have to factor out the fact that the differences may have arisen due to the species' different evolutionary histories represented by their different lineages—a sort of phylogenetic inertia—instead of, or in addition to, the species adaptations to thrive in different soil environments. This challenge regarding the correct interpretation of your findings will be applicable to many of the topics of your research program (although we now have analytical tools to at least partly take this into account).

A more direct, "cleaner" alternative would be to find a plant lineage that is rich in taxa that collectively represents a diversity of species and, yet, they are phylogenetically related, such that several of your fundamental and applied interrogations can be framed in the absence, or minimal influence of the plant's phylogeny. If you had such an aspiration, then you would do well in selecting the genus *Baccharis* as a model system. On par with the remarkable features of the hyper cosmopolitan and specious Asteraceae mega-family—the lineage to which this genus belongs—*Baccharis* exhibits an explosive diversification, with 440 known species. Likely an underestimate, given the incomplete botanical exploration across its range, and the still limited application of molecular tools to fine tune the group's taxonomy.

With an impressive geographical range, going from southern Chile to Nova Scotia, Canada, this cosmopolitan generic plant lineage includes species that thrive from sea level to elevations that surpass 5000 m, thriving under climatic envelopes that encompass a wide range from drylands to mesic forests. Naturally, the plant life forms (i.e., functional groups) that represent the adaptations necessary to thrive in such a diversity of ecological sceneries is equally diverse, ranging from leafless shrubs (in water-deficient locations) to evergreen shrubs (e.g., in chaparral systems), small herbs to lianas and treelets (in more climate-benign forests). In some of these ecological settings, some members of the genus are endemic, while others feature a much broader geographic distribution. Throughout their range, some species are very generalists (e.g., occurring in open and somewhat shaded sites), while others are much more stenoic (e.g., occurring exclusively in disturbed sites). Additionally, some are present at high local abundances and others thrive at low densities. Thus, *Baccharis* conforms an assemblage of closely related species that encompass the entire gradient of rarity—from stenoic species of narrow geographic distribution and low local density to sister species of broad geographic distribution, habitat-generalists, and locally abundant.

The ecological and evolutionary versatility of this unique plant lineage is manifested, as could be expected, in a plethora of traits that add to its attractiveness as model study system, prominent among which is their phytochemical diversity represented in the (at least) 150 isolated and characterized secondary metabolites. Such a chemical repertoire underscores the importance of *Baccharis* in chemical prospecting leading to the multiplicity of promising medical applications to deal with multiple maladies—including several types of cancer—as well as its importance in the ethnobotanical sphere. In my own experience as a field biologist, I have come across the case of *Baccharis conferta*, used by several Mexican indigenous and rural communities for diarrhea, or *B. tricuneata*, the local name of which, *sanalotodo* (rough translation from Spanish to English: "good cure for anything"), reflects the multi-faceted ethnomedical value of this species and the genus as a whole, and its promising potential in bioprospecting. From a very different perspective, it is worth noting that its ecological and evolutionary diversification that has allowed *Baccharis* to make presence in a significant portion of the Western Hemisphere also allows it to be present, as an accidentally or deliberately introduced species, in other parts of the world—in several of which it represents a management challenge. In sum, this versatile genus offers opportunities of study on basically any field of enquiry for plant biologists.

To this summarized account of the biological and chemical aspects that make *Baccharis* a convenient cluster of closely related species, one can mention that, from a practical, logistical perspective, the genus is also friendly to researchers: the plants are frequently abundant, of reachable size, typically generous in seed production, and a manageable plant to examine in terms of the multiple hosts it harbors. Indeed, several *Baccharis* species can be regarded as foundation species given their role as habitat or trophic resource for phytophagous insects (including an astonishing diversity of galling taxa) and fungal associates (both mutualistic endophytes and pathogenic ones), as well as pollinators. Related to this aspect of *Baccharis* serving as a foundation species, I have personally benefited from having *Baccharis pilularis* at hand to study its role as a facilitating mechanism for the establishment and recruitment of oak seedlings and saplings in the biotically and abiotically stressing grassland ecosystems of Northern California. And not only that, for *B. pilularis* has proven to be a convenient teaching tool for my field ecology classes in the grassland areas of Stanford's Jasper Ridge Biological Preserve. Using this friendly plant, students can learn about facilitation ecology, habitat colonization, plant responses to global change factors, intersexual resource allocation (given the dioecious character of the species), and gall-plant interactions.

In this prologue, I have endeavored in showcasing the tremendous value of *Baccharis* as a model study system, given what we know, so far, of this genus. Although we know that there is much that we do not know yet, what we know is precious and is splendidly compiled in this volume that brings together, for the first time, the most thorough synthesis of this amazing plant lineage. The volume also aptly includes an articulation of the most critical remaining knowledge lacunae and spells out the promising avenues for the understanding of how the plant kingdom works, seen here through the lens of this unique cluster of sister species. Such botanical understanding is increasingly critical in this era of daunting anthropogenic global environmental change.

Stanford University
Stanford, CA, USA

Rodolfo Dirzo

Acknowledgements

We want to express our special gratitude to all authors for accepting the challenge of contributing chapters to this book, without whom this project would not have been possible. We are also indebted to many of them who also invested additional time and patience to review other chapters. We are thankful for the reviews kindly done by Carolina Siniscalchi, Fernão Braga, Gari Vidal Cana Capatinta, Germán Ávila, Gustavo Heiden, Lorena Ashworth, Paulete Romoff, Rosy Isaias, Sandra Ludwig, and Takanori Tsuda. We are grateful to Barbara Rubia Silveira, Barbara Araujo Dias, and Natalia Moia for helping with the formatting and editing of some chapters and to Patricia Angrisano for the colour pencil illustrations of the *Baccharis* species. We also thank the Conselho Nacional de Pesquisas – CNPq, Coordenação de Aperfeiçoamento de Pessoal de Nível Superior – CAPES/Print, Fundação de Desenvolvimento da Pesquisa do Estado de Minas Gerais – FAPEMIG, Ministério da Ciência, Tecnologia e Inovações – MCTI, Planta Ltda, Reserva Vellozia, Instituto Brasileiro do Meio Ambiente e dos Recursos Naturais Renováveis (IBAMA), Associação Brasileira de Estudos das Abelhas – A. B. E. L. H. A. / Ministério do Meio Ambiente, Instituto Estadual de Florestas – IEF, Instituto Chico Mendes de Conservação da Biodiversidade – ICMBio, Programa de Pesquisa Ecológica de Longa Duração – PELD/CNPq, Programa de Pesquisa em Biodiversidade (PPBio/MCTI), and all other supporting agencies that contributed to the development of this book.

We dedicate this book to the memory of Graziela Maciel Barroso, known as the first lady of botany in Brazil. Graziela was born in the then small town of Corumbá near the border with Paraguay in 1912 and passed away in Rio de Janeiro in 2003. She began studying botany at the age of 30 to become the most iconic botanical taxonomist in Brazil. She was the first female naturalist to work in the Botanical Garden of Rio de Janeiro in 1946. When she was 47 years old, she began her study of biology at the State University of Guanabara, today State University of Rio de Janeiro. Graziela was 60 years old when she received her PhD from the University of Campinas, working on the systematics of Brazilian *Baccharis*. She passed away when she was 91 years old, one month before entering the Brazilian Academy of Sciences. Dedicating this book to Graziela Barroso is fair and appropriate as her attitudes and way of life resemble that of *Baccharis*: conquer and thrive in unlikely places, under the most difficult pressures to circumvent.

About the Book

This unique book provides a comprehensive overview of the most up-to-date knowledge of the plant genus *Baccharis*. The chapters cover a broad scope and are written by multiple authors who are experts in different subjects. The content of this book is organised into four major topics encompassing the evolution, ecology, chemistry, as well as environmental and medical applications of the genus. The genus *Baccharis* comprises more than 440 species, many of which are of primary ecological, economic, and cultural importance. Several species play crucial roles in biodiversity maintenance as foundation species, while others are invasive species with economic implications around the world. *Baccharis* is also well-known for being the source of innumerable chemical compounds widely used in folk medicine, by the cosmetics and pharmaceutical industries. As a result, *Baccharis* is the most studied Neotropical plant genus, and it has been studied by researchers around the world within fields spanning from ecology to pharmacology, chemistry, medicine, molecular biology, and economy. Even countries where the genus is not found, such as Lithuania, Denmark, France, and Canada, have produced articles on *Baccharis*. This book dedicates a whole part to the chemistry and the unique pharmacological potential of the green propolis produced by bees using resins collected from *Baccharis dracunculifolia*, and used worldwide to treat various diseases. This book brings to light many new data and provides updated and new syntheses in many of its chapters. This publication is a major reference for an audience of practising researchers, academics, PhD students, and other scientists in a wide-ranging collection of fields, from sociology to medicine to bioeconomy.

Belo Horizonte, MG, Brazil

Geraldo Wilson Fernandes
Yumi Oki
Milton Barbosa

Contents

About the Editors

Geraldo Wilson Fernandes PhD in evolutionary ecology (Northern Arizona University, USA, 1992), Tinker Professor at Stanford University (2015/2016), visiting professor at Universidad de Seville (2019/2020). Full member of the Brazilian Academy of Sciences. Has experience in plant-animal interactions, biomonitoring, restoration ecology, climate change, and bioeconomy. Six books published (three of them with Springer), more than five hundred peer-reviewed articles published on various subjects in the field of environmental sciences, and more than hundred graduate students supervised. Currently, full professor at the Universidade Federal de Minas Gerais, Brazil.

Yumi Oki PhD in entomology (University of São Paulo, Brazil, 2005). Postdoctoral fellow in the Department of Earth Sciences at the University of Alberta, Canada. She has experience in the fields of ecology, with emphasis on ecosystem ecology, applied ecology, environmental restoration, environmental services, environmental microbiology, and bioprospecting. Current research subjects include diversity, insect-plant interactions, biological activity, environmental services, symbiotic microorganisms and endophytic fungi (insect-plant mediation and bioprospecting), and environmental restoration. Currently, a postdoctoral researcher at the Universidade Federal de Minas Gerais, Brazil

Milton Barbosa PhD in zoology (University of Oxford, England, 2017). Worked in the Sea Around Us project, University of British Columbia, Canada. Was a fellow in the "Emerging Leader in the Americas" programme at the University of Alberta, Canada. Participated in the IBISCA (Investigating Biodiversity of Insects in Soil and Canopy) international expeditions to Australia and Vanuatu, and in the Tropi-Dry project, studying tropical dry forests in Serra do Cipó, Brazil. He is currently a postdoctoral fellow at the Universidade Federal de Minas Gerais Brazil. He has experience in the field of ecology, working mainly on community assembly, indirect interaction networks, and insect-plant interactions.

Part I
Ecological and Evolutionary Pathways

Part I presents an evaluation of the ecological and evolutionary pathways of *Baccharis* and includes the most updated information on this genus taxonomy, ecology and evolution.

Baccharis is a genus comprising more than 440 species distributed from the United States to Southern South America. The genus has enormous ecological importance since its species represent a year-round resource for a large number of organisms, including herbivores and pollinators. A diverse fauna of herbivorous insects feeds on these genus species, including the largest diversity of galling insects in the Neotropics. In addition, *Baccharis* species are mostly pioneers, with adaptations to resist extreme environmental conditions and, therefore, have great potential for the restoration and maintenance of biodiversity in several habitats. Due to biological features, diverse associated fauna, and wide distribution—usually at high frequency and across gradients (altitudinal, hygrothermal, salinity, and habitat disturbance)—the species of *Baccharis* have been extensively used as study models in ecological research; for instance, to monitor the impacts of climate change. On the other hand, several species of *Baccharis* are considered invasive species of difficult management that occupy ruderal areas, pasturelands, and crops in Australia, the USA, and Europe. This part covers it all, bringing to light plenty of new information on *Baccharis* ecological and evolutionary pathways, as well as opening space for more studies.

***Baccharis magnifica*. Illustration by Patrícia Angrisano**

Chapter 1
The Ecological and Applied Potential of *Baccharis*

G. Wilson Fernandes, Yumi Oki, and Milton Barbosa

Abstract The genus *Baccharis* is composed of ca. 440 species, distributed primarily in South and Central America, many of which are of great ecological, economic, and cultural importance. *Baccharis* species are mostly dioecious and highly diverse in chemistry, ecology, architecture, and phenology, occupying many different niches and habitats across several gradients of light, temperature, humidity, altitude, and succession. Its species are found in natural, urban, and highly polluted environ ments. Many species host a large number of associated organisms, including the largest fauna of gall-inducing insects in the Neotropics, and play crucial roles in biodiversity maintenance as foundation species or ecosystem engineers, while others are invasive species with economic implications around the world. Many species are geographically restricted or endemic. *Baccharis* is also well known for being the source of innumerable chemical compounds widely used in folk medicine and in the cosmetics and pharmaceutical industries. It is one of the most studied genera in the world, owing to these multiple factors that have captured the attention of the scientific community.

Keywords Asteraceae · Bioeconomy · Ecosystem engineer · Foundation species · Nurse species · Plant-animal interactions

1 The Universe of *Baccharis*

The genus *Baccharis* is relatively large with ca. of 440 species so far described in the New World. Its distribution is broad, ranging from 55 degrees South (*Baccharis magellanica*: Isla Hornos, Chile: Lat −55.9416629, Lon −67.26916559) to 43 degrees North (*Baccharis halimifolia*: Nova Scotia, Canada: Lat 43.99676, Lon −65.869709); therefore, with an estimated 11,187 km distance between the most extreme species populations. *Baccharis pilularis* has the northwestern-most

G. W. Fernandes (✉) · Y. Oki · M. Barbosa
Ecologia Evolutiva & Biodiversidade/DGEE, Universidade Federal de Minas Gerais, Belo Horizonte, Minas Gerais, Brazil

G. W. Fernandes et al. (eds.), *Baccharis*,
https://doi.org/10.1007/978-3-030-83511-8_1

distribution (Seattle, EUA: Lat 47.660917, Lon −122.42093). The species of *Baccharis* are found from sea level at both the Pacific and Atlantic oceans up to 5050 meters above sea level in the Andes. They inhabit forests, savanna, grassland, peat bogs, rocky outcrops, and desert ecosystems. They survive in mesic and xeric habitats, under saline conditions, in the shade and the sun, in extremely nutrient-impoverished environments, and even in polluted or contaminated areas. Many are early succession species (Westman et al. 1975), thriving in habitat conditions where nutrients and light are abundant. Some are rare and endemic to very specific habitats. *Baccharis* range from small herbaceous to treelet species and lianas, while some species are aphyllous. They are mostly dioecious and evergreen and provide the basis for the assembly of many animal communities and, in some cases, function as nurse species. The genus likely underwent true adaptive radiation in the New World, an aspect yet to be explored.

Several species of *Baccharis* are important whether for their beneficial or harmful effects (Palmer 1986; Boldt and Robbins 1987, 1990; Boldt 1989; Palmer and Haseler 1992a, b; Palmer and Tomley 1993; Palmer et al. 1993; Torres et al. 2000; Park et al. 2004; Oliveira et al. 2005; Abad and Bermejo 2007; Morales et al. 2008; Resende et al. 2012; Rabelo and Costa 2018; Cock and Hierro 2020; Schild et al. 2020). Beneficial effects include their use in controlling erosion, as ornamental plants or as a hedge (Thompson et al. 1995), and their medicinal properties (e.g., Budel et al. 2005; Verdi et al. 2005; Rabelo and Costa 2018). *Baccharis* species are important producers of hundreds of bioactive compounds used in several industries (e.g., fiber sensors coupled with an anti-theft alarm, food, chemical, cosmetics) or popularly used to treat many and different illnesses. The diversity of compounds produced by the species is probably a result of their wide range distribution in the most stressful conditions of the globe, which demands biochemical and physiological adaptations to survive. Furthermore, many species of this genus show a high cultural relevance. On the other hand, several species of *Baccharis* are considered invasive species of difficult management that occupy ruderal areas, pasturelands, and crops in Australia, the United States, and Europe (Sims-Chilton et al. 2010; Caño et al. 2016; Calleja et al. 2019). Other *Baccharis* species (e.g., *B. megapotamica* and *B. coridifolia*) produce toxic compounds that can kill livestock (Habermehl et al. 1985; Jarvis et al. 1987, 1988, 1996; Rizzo et al. 1997; Rissi et al. 2005).

Although this book brings to light the most recent update on the scientific studies on the use and relevance of *Baccharis* in the wild, as sources of compounds with potential industrial or commercial application, or as a model system in science, the gaps in the knowledge of the genus are still enormous. For most species, basic knowledge is still lacking, such as information on life span, interactions with other organisms including pollinators and herbivores, population genetics, and propagation, among others. Only a few species (e.g., *B. halimifolia*, *B. trimera*, and *B. dracunculifolia*) have been studied in more detail (Gene et al. 1996; Caño et al. 2016; Fernandes et al. 2018; Rabelo and Costa 2018; Barbosa et al. 2017, 2019; Calleja et al. 2019; Monteiro et al. 2020; Rodrigues et al. 2020). There are few works on niche modeling developed for *Baccharis* (Gonzáles et al. 2019).

While we have often argued that *Baccharis* is highly successful in habitat colonization due to its large seed production and long-distance dispersion, this has only been studied for very few species (e.g., *B. pilularis*, *B. halimifolia*, *B. dracunculifolia*). Most species are not that abundant in nature and inhabit harsh habitats in deserts and mountaintops. The ability to germinate under dark conditions, seed tolerance to shade, wide adaptability to soil nutrient conditions and salinity, survival under high soil humidity, and resprouting capabilities after fire (Westman et al. 1975; Gomes and Fernandes 2002) have also been listed as important causes of *Baccharis* success and widespread distribution. Boldt (1989) argue that these characteristics associated with high root growth capacity, intense sprouting after damage, high carbohydrate storage in the root system, and efficient water uptake and water use suggest several mechanisms responsible for the widespread occurrence and diversity of the genus. But these studies were mostly done for common species in North America, while most species are unknown beyond their taxonomy (e.g., Fernandes et al. 2007).

Baccharis represents a relatively large genus with wide distribution in the New World and presents high scientific and cultural relevance and huge economic potential, such as in producing important goods and services for human well-being. Among the countries that have published scientific literature about *Baccharis* species in the last 40 years are Brazil, United States, Argentina, Chile, Canada, and Mexico. More recently, even countries where the genus does not naturally occur, such as Spain, Japan, Lithuania, Denmark, and France, have produced articles on *Baccharis*.

2 The Ecological Path of *Baccharis*

Baccharis species are widely distributed from their origin, likely the mountaintop areas in eastern Brazil (Barroso 1976; Heiden et al. 2019). In the process of expansion and adaptation of *Baccharis* species, colonization by many species of insects and pathogens occurred, originating several types of associations. Some *Baccharis* species offer excellent resources for herbivorous and pollinating insects since they remain green and in bloom throughout the year (Boldt 1989; Espírito-Santo and Fernandes 1998; Espírito-Santo et al. 2004, 2007, 2012; Marques and Fernandes 2016; Watts et al. 2016; Fernandes et al. 2018; Moreira et al. 2018; Matilde-Silva et al. 2019; Monteiro et al. 2020). Some species bloom in autumn, which makes them very attractive to honey bees in a period when other flowers are absent. For example, *B. salicifolia, B. pilularis* and *B. sarothroides* are late summer and autumn honey plants (Boldt 1989). *Baccharis dracunculifolia* is a very important honey plant in Brazil and the main source of substances for the production of green propolis (Bastos and Oliveira 1999; Santos et al. 2011; Fernandes et al. 2018), while *B. concinna* produces flowers throughout the year (Madeira and Fernandes 1999; Espírito Santo and Fernandes 1998; Espírito Santo et al. 2012; Marques et al. 2002). These characteristics provide a unique scenario where problems of central relevance in ecology and biodiversity can be studied in detail (see Fagundes et al. 2005; Silva et al. 2007) and across large biogeographical regions.

Baccharis species are, to a large extent, primary colonizers of disturbed habitats (eg, *B. dracunculifolia, B. concinna, B. pseudomyriocephala, B. halimifolia, B. pilularis*) and, thus, are very important for the recovery, functioning and maintenance of biodiversity in various ecosystems, including those under natural succession (Boldt 1989; Araújo et al. 2003). Due to their biological features, diverse associated fauna, and wide distribution – usually in high frequency and across gradients (altitudinal, hygrothermal, and of habitat disturbance) – the species of *Baccharis* have been extensively used as study models in ecological research; such as in monitoring the impacts of climate change. *Baccharis pilularis,* for instance, a major facilitating species in the chaparral of California (Pelaez et al. 2019), has been used as a model for studies of climate change due to its biological characteristics and ease of experimental manipulation (see Zavaleta 2006; Zavaleta and Kettley 2006). In Brazil, several aspects of the species *B. dracunculifolia* have also been studied, including its invasiveness, environmental recovery capacity (see Julião et al. 2005; Fernandes et al. 2016; Adenesky-Filho et al. 2017), function as a nurse species (Perea et al. 2019) and experimental species for testing the effects of climate change (e.g., Sá et al. 2014; Oki et al. 2020).

Several species of *Baccharis* are, on the other hand, considered pests that are difficult to manage in pastures, growing in recreational areas. These invasions usually occur after changes in the environment and, due to their rapid growth, dense stands are formed (Boldt 1989). In central Chile, the formation of degraded vegetation resulted in optimal conditions for the establishment of hybrids and backcross progenies for some species of *Baccharis* (Faini et al. 1991). Some species interfere with the use of soil water and the maintenance of irrigation and drainage channels (Timmons 1959; Parker 1972; Ellis 2001; Caño et al. 2016; Fried et al. 2016). At least one species is invasive, *Baccharis halimifolia,* which was introduced in Australia (Bailey 1900), France, Spain (Dupont 1966) and Italy (Boldt 1989), being the only species occurring outside the Americas. On the other hand, recent studies point to other *Baccharis* invasions in Europe, such as that of *B. spicata*, and may represent a worrying threat (Verloove et al. 2018). In Brazil, pastures are completely unviable when the invasion by *B. dracunculifolia* is intense (Lorenzi 1992; Kissmann and Groth 1992; Altesor et al. 2005). However, it is a plant that colonizes degraded or abandoned areas and an abandoned pasture can be considered a degraded area compared to the natural environment.

Many studies have been carried out in the United States, Mexico, Brazil, and Australia to verify the richness and importance of insects on some native and introduced *Baccharis* species. In Australia, these studies focus on potential agents for the biological control of *B. halimifolia*, which has reached high population densities, replacing native vegetation (Palmer 1986; Boldt and Robbins 1987, 1990; Boldt 1989; Palmer and Haseler 1992a, b; Palmer et al. 1993; Palmer and Tomley 1993; Donders et al. 2005, Sims-Chilton et al. 2010; Green et al. 2012). This same species is becoming one of the most troublesome invaders in the European continent (Caño et al. 2013; Calleja et al. 2019). Among the organisms that cause damage to the host plant, the most significant are Chrysomelidae, Curculionidae, Tephritidae, and Cecidomyiidae (Tilden 1953; Palmer 1986; Boldt 1989; Cordo et al. 1999; Oki et al. 2009; Fagundes and Fernandes 2011; Espirito-Santo et al. 2012; Fernandes et al. 2014; Monteiro et al. 2020). Chrysomelidae species consume large amounts of *Baccharis* spp. in South America (Blackwelder 1946), and the Brazilian species *Lioplacis elliptica* was introduced into Australia for biological control of *B. halimifolia* (Buzzi 1977; McFadyen 1978). Although only distributed in the North American Southwest, *B. pilularis* is another species that has been widely studied due to its importance as an invader of urban areas and water sources in the United States (Ellis 2001; Laris et al. 2017).

Gall-inducing insects can reach large population densities on some hosts and hence could be of importance in the biological control of *Baccharis* species (see reviews in Fernandes and Santos 2014). The potential for using the gall inducer *Baccharopelma dracunculifoliae* (Homoptera: Psyllidae) to control *B. dracunculifolia* where it represents potential problems due to its invasibility (e.g., Cochabamba, Bolivia) can be high due to its high frequency, impact, and wide distribution (Lara and Fernandes 1996; Espírito Santo and Fernandes 1998; Burkhadt et al. 2004; Araujo et al. 2006). Seed predator and borer insects are also of great relevance in studies of *Baccharis* biological control (Brailovsky 1982; Palmer 1986; McFadyen 1978) but have not been studied with the detail it deserves in recent years.

Thirty-three species of insects cause parasitic diseases on *Baccharis* in the United States (Cummings 1978). However, very little is known about the herbivorous insect fauna that attacks the hundreds of other native *Baccharis* species in the Americas, despite some timid advances made in recent years (e.g., Collevatti and Sperber 1997; Hudson and Stiling 1997; Espírito Santo and Fernandes 1998; Burkhardt et al. 2004; Carneiro et al. 2005, 2006, 2009a, b; Fagundes et al. 2005; Fagundes and Fernandes 2011; Neves et al. 2011; Oki et al. 2009; Espirito-Santo et al. 2012; Monteiro et al. 2020). Less well known are the most attacked species and circumstances or factors that influence the resistance and/or susceptibility to attack by natural enemies, although some initial progress has also been made (e.g., Espírito Santo et al. 2007, 2012).

3 *Baccharis* Interactions and Community Structuring

In the wilderness, *Baccharis* plays a key role in creating opportunities for community assembly and maintenance. A few other genera or species have been reported to be superhosts of gall-inducing insects in the Nearctic and Palearctic regions: *Quercus* (Felt 1940; Abrahamson et al. 1998; Manos et al. 1999; Maldonado-Lopez et al. 2016; Pérez-Lopez et al. 2016), *Larrea tridentata* (Waring and Price 1990), *Salix* (Price et al. 1995), *Populus* (Floate and Whitham 1995), *Rosa* (Shorthouse and Rohfritsch 1992; Stone et al. 2002), *Chrysothamnus* in southwestern North America (Fernandes et al. 2000), and *Solidago* (Abrahamson and Weis 1997). These widely colonized host species have served as laboratories to test for generalities of ecological interactions (see Fernandes and Barbosa 2014). In the Neotropics and southern temperate region, some host plant genera have the same role, such as *Copaifera* (Leguminosae) (Costa et al. 2010, 2011), *Nothofagus* (Quintero et al. 2014), *Protium* (Maia 2011; Julião et al. 2014), and *Baccharis* (Fernandes et al. 1996; Fernandes and Barbosa 2014; Formiga et al. 2015; Barbosa et al. 2017, 2019). In the reviews by Fernandes et al. (1996) and Fernandes and Barbosa (2014), the *Baccharis* hosts that supported the highest numbers of galling insects were *B. dracunculifolia* (17 spp.), *B. concinna* (15 spp.), *B. salicifolia* (13 spp.), and *Baccharis* sp. 1 (11 spp.). In the southeastern mountains of Brazil, the Mantiqueira and Espinhaço Mountains, Coelho et al. (2018) reported 106 galling species on 17 *Baccharis* species. The highest richness of galling insects (13 galling species) was recorded on *B. dracunculifolia*, confirming the previous literature surveys for the species. The study also recorded a high richness of galling insects on *B. minutiflora* (12 spp.), *B. cognata* (10 spp.), *B. reticularia* (9 spp.), *B. intermixta* (8 spp.), and *B. concinna* (7 spp.). The hosts *B. ramosissima*, *B. helychrysoides*, and *B. truncata* supported six galling species each, while *B. serrulata*, *B. ligustrina*, and *B. glutinosa* each had three galling species recorded. A remarkable feature of *Baccharis* is that its galling organisms are from many different orders; e.g., Diptera, Lepidoptera, and Hemiptera. Based on a Web of Science search with the words "insect galls, galls, cecidia, galling insects, galhas, gallmucken, and agallas," we were able to record at least 47 studies on galling insects on 8 species of *Baccharis* in the last 75 years (1945–2020).

While no one has yet listed the number of insects attracted to the flowers of *Baccharis* (but see Ferracini et al. 1995), our own experience indicates it is large. In a short observation on the number of insects attracted to the flowers of *Baccharis dracunculifolia* during a very limited number of days (2–3 days), we have been able to list more than 30 different species (in review). These data confirm that some *Baccharis* species are extremely important in providing resources to pollinators. They also confirm that this species' effects on ecosystem functioning must be even higher where these plants are abundant or are key strategic resources for the community, such in mountaintop regions and deserts (e.g., Boldt 1989; Griffin 1997). Hence, *Baccharis* species could be used in programs to attract pollinators.

On the other hand, the understanding of some ecological and evolutionary paths in *Baccharis* is incipient. For instance, genetic studies are in their infancy, and more applied aspects such as propagation for several uses are not well developed. Genetic studies could be of great relevance in promoting other sorts of ecological and evolutionary studies in the future.

4 The Chemistry of *Baccharis*

Baccharis species are known in traditional culture for the treatment of diseases such as gastrointestinal and liver disorders, anemia, diabetes, diarrhea, infections, cancer, gout, rheumatism, ulcers, and skin problems, among others (Vidari et al. 2003; Abad and Bermejo 2007; Hocayen et al. 2016; Rabelo and Costa 2018;

Romero-Benavides et al. 2018; Ascari et al. 2019; Basso et al. 2019; Costa et al. 2019; Bonin et al. 2020; Paniagua-Zambrana et al. 2020; Souza et al. 2020). Several studies present the most updated information on the production of phytochemicals for pharmaceutical, cosmetic, and other applications, and therefore we will not review these here (Verdi et al. 2005; Grecco et al. 2010; Galvão et al. 2012; Vannini et al. 2012; González et al. 2018; Jaramillo-García et al. 2018; Ueno et al. 2018).

This high usage in folk medicine and interest by the pharmaceutical industry has its origin in the rich chemical properties of the genus. Several species of *Baccharis* produce chemical compounds that are under investigation by many institutes and laboratories around the world (e.g., Jarvis et al. 1988; Brown 1994; Fournet et al. 1994; Verdi et al. 2005; Pereira et al. 2017; Romero-Benavides et al. 2018; Bonin et al. 2020). In folk medicine, tea made from *B. douglasii* is used to treat ulcerations and wounds. Other teas are used to treat headaches and as emetics (Boldt 1989). *B. trimera* ethyl acetate extracts have been used against *Schistosoma* infections in Brazil (Herz et al. 1977). In Argentina, about 50 species of *Baccharis* are used in folk medicine (Boldt 1989). Bandoni et al. (1978) found that two flavonoids extracted from *B. crispa* and *B. notosergila* have antimicrobial activity, but since then this number of studies has continued to grow, showing the relevance of this genus.

Various chemical compounds in *Baccharis* are potentially effective in fighting cancer. Baccharin trichothecene extracted from the leaves, buds, and dried flowers of *B. megapotamica* acts against leukemia and tumors of the colon of mice (Kupchan et al. 1976, 1977; Arcamone et al. 1980; Carvalho et al. 2016; Rodrigues et al. 2020). Two additional groups of trichothecenes, roridins, and verrucarins, found in *B. coridifolia*, are active against nasopharyngeal tumor cells (Jarvis et al. 1988, see also Budel et al. 2005; Verdi et al. 2005). About 180 species have already been analyzed chemically, leaving about 260 species to be prospected for their chemical constituents and efficacy. On the other hand, the potential for discovering new chemicals is greatly expanded when long-term studies are carried out. Collections of botanical material for phytochemical studies are generally gathered in a single opportunity, thus losing the enormous variability and seasonality of the production of compounds (Gershenzon 1984). Hence, the genus has an enormous potential to contribute a large number of chemical substances, some of which might be new to science.

According to Abad and Bermejo (2007), over 150 compounds have been isolated and identified from the *Baccharis* genus. Many substances isolated from this genus have been used as medicine (e.g., *B. trinervis*, used as anti-HIV), perfumes (essential oils of *B. dracunculifolia*, *B. uncinella*, *B. genistelloides*, *B. trimera*), and repellents (terpenoids and flavonoids found in many species), among other products (Jarvis et al. 1988; Argandoña and Faini 1993; Ferracini et al. 1995; Palomino et al. 2002; Agostini et al. 2005; Verdi et al. 2005; Wollenweber et al. 2006). The *Baccharis* species more deeply studied chemically are *B. megapotamica*, *B. incarum*, *B. trimera*, *B. trinervis*, *B. salicifolia*, *B. crispa*, *B. coridifolia*, *B. dracunculifolia*, *B. uncinella*, *B. retusa*, *B. linearis*, *B. grisebachii*, *B. obtusifolia*, and *B. tricuneata* (Bohlmann et al. 1982; Verdi et al. 2005; Budel et al. 2005; Besten et al. 2012;

Campos et al. 2016; Moraes Neto et al. 2019). This arsenal of applicability has led to the filing of 226 *Baccharis* patents. Half of these patents are aimed at the pharmacological area (53.5% of patents), mainly in the treatment of cancer (40 patents). Among the patents, another highlight is the application of *B. gaudichaudiana* in the treatment of coronavirus (Junxing et al. 2003). This information emphasizes the pharmacological potentialities around this genus. In southeastern Brazil, oils extracted from leaves and stems of *B. dracunculifolia* and *B. genistelloides* are used as fragrances (Chialva and Doglia 1990; Suttisri et al. 1994; Fabiane et al. 2008; Frizzo et al. 2001, 2008; Queiroga et al. 1990, 2014), but the potential for new findings is enormous as most of the species were not yet screened for the production of them.

5 Other Applications

Despite their low nutritional content and known unpalatability (Pelaez et al. 2019; Cock and Hierro 2020), some species of *Baccharis*, considered to be non-toxic, have been used as fodder for cattle (Benson and Darrow 1981). *B. sarothroides* has been used as vegetation cover and to correct soils degraded by copper mining in Arizona, USA (Day and Ludeke 1980; Norem et al. 1982; Haque et al. 2008; Haque et al. 2009). In Brazil, mainly in Minas Gerais, where mineral exploration is carried out in the open, studies on the use of plants of this genus for the recovery of degraded areas were initiated, and the species *B. dracunculifolia* and *B. concinna* are suggested as viable alternatives to the introduction of exotic species to habitat and region (e.g., Fernandes et al. 2007; Negreiros et al. 2014; Gomes et al. 2015; Fernandes et al. 2016).

A serious problem is cattle poisoning by *Baccharis*. Animal death cases have been recorded in Brazil (Occhioni 1944; Tokarnia and Dobereiner 1976), Uruguay, and Argentina (Schang 1929). The vegetative parts are toxic throughout the year, while the flowers are 4–8 times more toxic (Tokarnia and Dobereiner 1976). The most common toxic compounds found in some *Baccharis* species studied were macrocyclic trichothecenes (Kupchan et al. 1976, 1977; Habermehl et al. 1985; Jarvis et al. 1988, 1996; Rizzo et al. 1997; Driemeier et al. 2000; Varaschin and Alessi 2003). The symptoms exhibited by the poisoned cattle are anorexia, lack of coordination and direction in walking, tremors, and convulsions. Postmortem analyses reveal lesions in the rumen, necrosis, and detachment of the intestinal mucosa (Tokarnia and Dobereiner 1976; see also Jarvis et al. 1996; Driemeier et al. 2000; Budel et al. 2005; Verdi et al. 2005; Oliveira-Filho et al. 2011; Panziera et al. 2015).

An important applied aspect is that of *B. dracunculifolia* and Africanized honey bee *Apis mellifera* (Kumazawa et al. 2003). This bee collects resins from apical buds of *B. dracunculifolia* and uses it to produce a resinous layer inside the hive, known as green propolis (Teixeira et al. 2005; Fernandes et al. 2018). This resinous mass has antiseptic, anti-inflammatory, anticancer, and healing properties and thus has been widely studied, commercialized, and used, primarily by the pharmaceutical and cosmetics industry (Banskota et al. 2001; Chan et al. 2012; Veiga et al. 2017; Endo et al. 2018). Among the chemicals isolated from propolis, it is worth mentioning the presence of flavonoids, phenylpropanoids, phenolic acids, and essential oils (Kumazawa et al. 2003; Teixeira et al. 2005; Takashima et al. 2019).

Another application of *Baccharis* is associated with its high diversity of symbiotic organisms (bacteria, endophytic fungi, and mycorrhizae), which not only helps in plant survival and development but generates a potential for use in the field of bioprospecting (Oki et al. 2009, 2016; Cuzzi et al. 2012; Vieira et al. 2014; Coutinho et al. 2019). *Baccharis* endophytic fungi have been shown to be effective in antimicrobial and antifungal (Oki et al. 2016; Vieira et al. 2014), as well as anti-herbivory, activities (Oki et al. 2016, 2021).

6 The Content of the Book

This book is arranged into four main parts. Chapter 1 focuses on the main ecological and evolutionary aspects of the genus. Chapter 2 presents the most current understanding of the taxonomy and distribution of the 442 species of *Baccharis*, discussing the genus origins and diversification. Chapter 3 offers a historical overview of genetic studies on *Baccharis* species, including recent method developments. Chapter 4 addresses the relationship between intersexual differences in resource allocation and herbivory in dioecious *Baccharis* species. Chapter 5 provides a detailed description of the network of direct and indirect interactions among arthropods in the well-studied *B. dracunculifolia* system. Chapter 6 brings to light the world of endophytic fungi associated with *Baccharis* and their importance in helping plants cope with environmental stresses and natural enemies and as a source of bioactive compounds. Chapter 7 reveals the crucial role of *Baccharis* species as nurse plants and in community assembly, particularly in stressful and herbivore-dominated environments in the Americas. Then, Chap. 8 closes the section discussing the causes and consequences of biological invasion by *Baccharis* in the world.

The second part concentrates on the structural and chemical particularities of the *Baccharis* species. Chapter 9 provides a comprehensive review of the morphology and anatomy of *Baccharis*, including morphological and anatomical features of particular taxonomic relevance. Chapter 10 reviews the chemical composition of essential oils of *Baccharis* species and their wide range of biological activities. Chapter 11 shows a comprehensive overview of the wide variety of flavonoids present in *Baccharis* species. Chapter 12 presents ethnopharmacological uses of phenolic compounds, focusing on folk medicine, and it also discusses the toxicity of some *Baccharis* species. The main volatile terpenes, which play key roles in the biological activities of *Baccharis* species, are presented in Chap. 13. Trichothecenes are covered in detail in Chap. 14, while in Chap. 15 livestock poisoning by some species of *Baccharis* is reviewed.

Part three explores the social and economic importance of *Baccharis*. Chapter 16 reviews the wide variety of popular uses of several species of *Baccharis* in South

America. Chapter 17 describes the development of the cultivar of *B. trimera*, the first Brazilian medicinal plant to be registered and patent-protected. Chapter 18 exposes the potential of *Baccharis* secondary metabolites in the development of new drugs to fight cancer. The last chapter in this part, Chap. 19, portrays the current status of scientific and technological innovations involving species of *Baccharis*.

This book's final part is devoted to green propolis, whose chief plant source is *B. dracunculifolia*. Chapter 20 reviews the chemical constituents and antioxidant properties of Brazilian green propolis. Chapter 21 examines the potential of green propolis components in the prevention and treatment of obesity and diabetes. Chapter 22 discusses the effects of the green propolis on the immune response. Finally, Chap. 23 presents the current status of innovation and markets of propolis, emphasizing green propolis.

Acknowledgments We thank the funding agencies CAPES, CNPq, Planta Ltda, and FAPEMIG for the field and laboratory supports granted to the authors involved.

References

Abad MJ, Bermejo P (2007) *Baccharis* (Compositae): a review update. ARKIVOC 7:76–96

Abrahamson WG, Weis AE (1997) Evolutionary ecology across three trophic levels: goldenrods, gallmakers and natural enemies. Princeton University, Princeton

Abrahamson WG, Melika G, ScraVord R, Csóka G (1998) Gall-inducing insects provide insights into plant systematic relationships. Am J Bot 85:1159–1165

Adenesky-Filho E, Maçaneiro JP, Vitorino MD (2017) How to select potential species for ecological restoration of rain forest–Southern Brazil. Appl Ecol Environ Sci 15:1671–1684

Agostini F, Santos ACA, Rossato M, Pansera MR, Zattera F, Wasum R, Serafini LA (2005) Studies on the essential oils from several *Baccharis* (Asteraceae) from Southern Brazil. Rev Bras Farmacogn 15:215–219

Altesor A, Oesterheld M, Leoni E, Lezama F, Rodríguez C (2005) Effect of grazing on community structure and productivity of a Uruguayan grassland. Plant Ecol 179:83–91

Araújo APA, Carneiro MAA, Fernandes GW (2003) Efeitos do sexo, do vigor e do tamanho da planta hospedeira sobre a distribuição de insetos indutores de galhas em *Baccharis pseudo-myriocephala* Teodoro (Asteraceae). Rev Bras Entomol 47:483–490

Araújo APA, Paula JD, Carneiro MAA, Schoereder JH (2006) Effects of host plant architecture on colonization by galling insects. Austral Ecol 31:343–348

Arcamone R, Cassinelli G, Casazza AM (1980) New antitumor drugs from plants. J Ethnopharmacol 2:41–46

Argandoña VH, Faini F (1993) Oleanolic acid content in *Baccharis linearis* and its effects on *Heliothis zea* larvae. Phytochemistry 33:1377–1379

Ascari J, Oliveira MS, Nunes DS, Granato D, Scharf DR, Simionatto E, Otuki M, Soley B, Heiden G (2019) Chemical composition, antioxidant and anti-inflammatory activities of the essential oils from male and female specimens of *Baccharis punctulata* (Asteraceae). J Ethnopharmacol 234:1–7. https://doi.org/10.1016/j.jep.2019.01.005

Bailey L (1900) The Queensland flora, vol 3. A. Diddams, Brisbane

Bandoni A, Medina J, Rondina R, Coussio J (1978) Genus *Baccharis* L. 1. Phytochemical analysis of a non polar fraction from *B. crispa* Sprengel. Planta Med 34:328–331

Banskota AH, Tezuka Y, Kadota S (2001) Recent progress in pharmacological research of propolis. Phytother Res 15:561–571

Barbosa M, Fernandes GW, Lewis OT, Morris RJ (2017) Experimentally reducing species abundance indirectly affects food web structure and robustness. J Anim Ecol 86:327–336

Barbosa M, Fernandes GW, Morris RJ (2019) Interaction engineering: non-trophic effects modify interactions in an insect galler community. J Anim Ecol 88:1168–1177

Barroso GM (1976) Compositae-subtribo Baccharidinae-Hoffman: estudo das espécies ocorrentes no Brasil. Rodriguesia 40:3–273

Basso BS, Mesquita FC, Dias HB, Krause GC, Scherer M, Santarém ER, Oliveira JR (2019) Therapeutic effect of *Baccharis anomala* DC. extracts on activated hepatic stellate cells. EXCLI J 18:91–105

Bastos EM, Oliveira VC (1999) Aspectos morfo-anatômicos da folha de *Baccharis dracunculifolia* DC. (Asteraceae) visando à identificação da origem botânica da própolis. Acta Bot Bras 12:431–439

Benson L, Darrow R (1981) Trees and shrubs of the southwestern desert. University of Arizona Press, Tucson

Besten MA, Jasinski VC, Costa ÂDG, Nunes DS, Sens SL, Wisniewski A Jr, Simionatto EL, Riva D, Dalmarco JB, Granato D (2012) Chemical composition similarity between the essential oils isolated from male and female specimens of each five *Baccharis* species. J Braz Chem Soc 23:1041–1047

Blackwelder R (1946) Checklist of the coleopterous insects of Mexico, Central America, the West Indies and South America. Bull Am Mus Nat Hist 185:733–757

Bohlmann F, Kramp W, Jakupovic J, Robinson H, King RM (1982) Diterpenes from *Baccharis* species. Phytochemistry 21:399–403

Boldt PE (1989) *Baccharis* (Asteraceae), a review of its taxonomy, phytochemistry, ecology, economic status, natural enemies and the potential for its biological control in the United States. USDA, Agricultural Research Service. Grassland, soil and water research laboratory. Temple, Texas

Boldt PE, Robbins TO (1987) Phytophagous and pollinating insect fauna of *Baccharis neglecta* (Compositae) in Texas. Environ Entomol 16:887–895

Boldt PE, Robbins TO (1990) Phytophagous and flower-visiting insect fauna of *Baccharis salicifolia* (Asteraceae) in the southwestern United States and northern Mexico. Environ Entomol 19:515–523

Bonin E, Carvalho VM, Avila VD, Santos NCA, Benassi-Zanqueta É, Lancheros CAC, Previdelli ITS, Ueda-Nakamura T, Abreu Filho BA, Prado IN (2020) *Baccharis dracunculifolia*: chemical constituents, cytotoxicity and antimicrobial activity. LWT 120:1–10. https://doi.org/10.1016/j.lwt.2019.108920

Brailovsky H (1982) Revision del complejo *Ochrimnus* con descripcion de nuevas especies y nuevos generos (Hemiptera-Heteroptera-Lygaeidae-Lygaeinae). Folia Entomol Mex 51:1–163

Brown GD (1994) Phenylpropanoids and other secondary metabolites from *Baccharis linearis*. Phytochemistry 35:1037–1042

Budel JM, Duarte MR, Santos CAM, Farago PV, Matzenbacher N (2005) O progresso da pesquisa sobre o gênero *Baccharis*, Asteraceae: I-Estudos botânicos. Rev Bras Farmacogn 15:268–271

Burckhardt D, Espírito-Santo MM, Fernandes GW, Malenovský I (2004) Gall-inducing jumping plant-lice of the Neotropical genus *Baccharopelma* (Hemiptera, Psylloidea) associated with *Baccharis* (Asteraceae). J Nat Hist 38:2051–2071

Buzzi Z (1977) Uma nova espécie de *Lioplacis* (Coleoptera: Chrysomelidae) do sul do Brasil. Dusenia 10:229–232

Calleja F, Ondiviela B, Juanes JA (2019) Invasive potential of *Baccharis halimifolia*: experimental characterization of its establishment capacity. Environ Exp Bot 162:444–454

Campos FR, Bressan J, Jasinski VCG, Zuccolotto T, Silva LE, Cerqueira, LB (2016) *Baccharis* (Asteraceae): chemical constituents and biological activities. Chem Biodivers 13:1–17

Caño L, Campos JA, García-Magro D, Herrera M (2013) Replacement of estuarine communities by an exotic shrub: distribution and invasion history of *Baccharis halimifolia* in Europe. Biol Invasions 15:1183–1188. https://doi.org/10.1007/s10530-012-0360-4

Caño L, Fuertes-Mendizabal T, García-Baquero HM, González-Moro MB (2016) Plasticity to salinity and transgenerational effects in the nonnative shrub *Baccharis halimifolia*: insights into an estuarine invasion. Am J Bot 103:808–820

Carneiro MAA, Fernandes GW, De Souza OFF (2005) Convergence in the variation of local and regional galling species richness. Neotrop Entomol 34:547–553

Carneiro MAA, Fernandes GW, De Souza OFF, Souza WVM (2006) Sex-mediated herbivory by galling insects on *Baccharis concinna* (Asteraceae). Rev Bras Entomol 50:394–398

Carneiro MAA, Branco CSA, Braga CED, Almada ED, Costa MBM, Maia VC, Fernandes GW (2009a) Are gall midge species (Diptera, Cecidomyiidae) host-plant specialists? Rev Bras Entomol 53:365–378

Carneiro MAA, Borges RAX, Araújo APA, Fernandes GW (2009b) Insetos indutores de galhas da porção sul da Cadeia do Espinhaço, MG. Rev Bras Entomol 53:570–592

Carvalho MP, Weich H, Abraham WR (2016) Macrocyclic trichothecenes as antifungal and anticancer compounds. Curr Med Chem 23:23–35

Chan GC, Cheung K, Sze DM (2012) The immuno modulatory and anticancer properties of propolis. Clin Rev Allergy Immunol. https://doi.org/10.1007/s12016-012-8322-2

Chialva F, Doglia G (1990) Essential oil from Carqueja (Baccharis genistelloides Pers.). J Essent Oil Res 2:173–177

Cock MC, Hierro JL (2020) Native weed protects species that sustain cattle raising in semi-arid natural grasslands. J Arid Environ 175:1–8. https://doi.org/10.1016/j.jaridenv.2019.104088

Coelho MS, Carneiro MAA, Branco CA, Borges RAX, Fernandes GW (2018) Species turnover drives β-diversity patterns across multiple spatial scales of plant-galling interactions in mountaintop grasslands. PLoS One 13:e0195565

Collevatti RG, Sperber CF (1997) The gall maker *Neopelma baccharidis* Burck. (Homoptera: Psyllidae) on *Baccharis dracunculifolia* DC. (Asteraceae): individual, local, and regional patterns. Ann Soc Entomol Brasil 26:45–53

Cordo HA, DeLoach CJ, Habeck DH (1999) Biology of *Heilipodus ventralis* (Coleoptera: Curculionidae), an Argentine weevil for biological control of snakeweeds (*Gutierrezia* spp.) in the United States. Biol Control 15:210–227

Costa FV, Fagundes MF, Neves FS (2010) Arquitetura da planta e diversidade de galhas associadas à *Copaifera langsdorffii* (Fabaceae). Ecol Aust 20:9–17

Costa FV, Neves FS, Silva JO, Fagundes M (2011) Relationship between plant development, tannin concentration and insects associated with *Copaifera langsdorffii* (Fabaceae). Arthropod Plant Interact 5:9–18

Costa P, Boeing T, Somensi LB, Cury BJ, Espíndola VL, França TCS, Almeida MO, Arruda C, Bastos JK, Silva LM, Andrade SF (2019) Hydroalcoholic extract from *Baccharis dracunculifolia* recovers the gastric ulcerated tissue, and p-coumaric acid is a pivotal bioactive compound to this action. Biofactors 45:479–489

Coutinho ES, Beiroz W, Barbosa M, Xavier JHA, Fernandes GW (2019) Arbuscular mycorrhizal fungi in the rhizosphere of saplings used in the restoration of the rupestrian grassland. Ecol Restor 37:152–162

Cummings G (1978) Rust fungi on legumes and composites in North America. Arizona University Press, Tucson

Cuzzi C, Link S, Vilani A, Sartori C, Onofre SB (2012) Endophytic fungi of the "vassourinha" (*Baccharis dracunculifolia* D. *C.* – Asteraceae*)*. Rev Bras Bioci 10:135–139

Day A, Ludeke K (1980) Reclamation of copper mine wastes with shrubs in the southwestern U.S.A. J Arid Environ 3:107–112

Donders TH, Wagner F, Visscher H (2005) Quantification strategies for human-induced and natural hydrological changes in wetland vegetation, southern Florida, USA. Quat Res 64:333–342

Driemeier D, Cruz C, Loretti AP (2000) *Baccharis megapotamica* var *Weirii* poisoning in Brazilian cattle. Vet Hum Toxicol 42:220–221

Dupont P (1966) L' extension de *Baccharis halimifolia* entre Loire et Gironde. Bull Soc Sci Bretagne 41:141–144

Ellis LM (2001) Short-term response of woody plants to fire in a Rio Grande riparian forest, Central New Mexico, USA. Biol Conserv 97:159–170

Endo S, Hoshi M, Matsunaga T, Inoue T, Ichihara K, Ikari A (2018) Autophagy inhibition enhances anticancer efficacy of artepillin C, a cinnamic acid derivative in Brazilian green propolis. Biochem Biophys Res Commun 497:437–443

Espírito-Santo MM, Fernandes GW (1998) Abundance of *Neopelma baccharidis* (Homoptera: Psyllidae) galls on the dioecious shrub *Baccharis dracunculifolia* (Asteraceae). Environ Entomol 27:870–876

Espírito-Santo MM, Faria ML, Fernandes GW (2004) Parasitoid attack and its consequences to the development of the galling psyllid *Baccharopelma dracunculifoliae*. Basic Appl Ecol 5:475–484

Espírito-Santo MM, Neves FS, Andrade-Neto FR, Fernandes GW (2007) Plant architecture and meristem dynamics as the mechanism determining the diversity of gall-inducing insects. Oecologia 153:353–364

Espírito-Santo MM, Neves FS, Fernandes GW, Silva JO, Andrade-Neto FR (2012) Plant phenology and absence of sex-biased gall attack on three species of *Baccharis*. PLoS One 7:e46896

Fabiane KC, Ferronatto R, Santos ACD, Onofre SB (2008) Physicochemical characteristics of the essential oils of *Baccharis dracunculifolia* and *Baccharis uncinella* DC (Asteraceae). Rev Bras Farmacogn 18:197–203

Fagundes M, Fernandes GW (2011) Insect herbivores associated with *Baccharis dracunculifolia* (Asteraceae): responses of gall-forming and free-feeding insects to latitudinal variation. Rev Biol Trop 59:1419–1432

Fagundes M, Neves FS, Fernandes GW (2005) Direct and indirect interactions involving ants, insect herbivores, parasitoids, and the host plant *Baccharis dracunculifolia* (Asteraceae). Ecol Entomol 30:28–35

Faini F, Hellwig F, Labbe C, Castillo M (1991) Hybridization in the genus *Baccharis*: *Baccharis linearis* X *B. macraei*. Biochem Syst Ecol 1:53–57

Felt EP (1940) Plant galls and gall makers. Comstock, Ithaca

Fernandes GW, Barbosa M (2014) Bottom-up effects on gall distribution. In: Fernandes G, Santos J (eds) Neotropical insect galls. Springer, Dordrecht, pp 99–113

Fernandes GW, Carneiro MAA, Lara ACF, Allain LA, Andrade GI, Julião G, Reis TC, Silva IM (1996) Galling insects on neotropical species of *Baccharis* (Asteraceae). Trop Zool 9:315–332

Fernandes GW, Santos JC (eds) (2014) Neotropical insect galls. Springer, Dordrecht

Fernandes GW, Saraiva C, Cornelissen TG, Price PW (2000) Diversity and morphology of insect galls on *Chrysothamnus nauseous* (Asteraceae) in North Arizona. Bios 8:39–48

Fernandes GW, Rodarte LH, Negreiros D, Franco AC (2007) Aspectos nutricionais em *Baccharis concinna* (Asteraceae), espécie endêmica e ameaçada da Serra do Espinhaço, Brasil. Lundiana 8:83–88

Fernandes GW, Silva JO, Espírito-Santo MM, Fagundes M, Oki Y, Carneiro MAA (2014) *Baccharis*: a neotropical model system to study insect plant interactions. In: Fernandes GW, Santos JC (eds) Neotropical insect galls. Springer, Dordrecht, pp 193–219

Fernandes GW, Toma TSP, Angrisano P, Overbeck G (2016) Challenges in the restoration of quartzitic and ironstone rupestrian grasslands. In: Fernandes GW (ed) Ecology and conservation of mountaintop grasslands in Brazil. Springer, Switzerland, pp 449–477

Fernandes GW, Oki Y, Belmiro MS, Resende FM, Correa Junior AC, Azevedo JL (2018) Multitrophic interactions among fungal endophytes, bees, and *Baccharis dracunculifolia*: resin tapering for propolis production leads to endophyte infection. Arthropod Plant Interact 12:329–337

Ferracini VL, Paraiba LC, Leitão-Filho HF, Silva AGD, Nascimento LR, Marsaioli AJ (1995) Essential oils of seven Brazilian *Baccharis* species. J Essent Oil Res 7:355–367

Floate KD, Whitham TG (1995) Insects as traits in plant systematics: their use in discriminating between hybrid cottonwoods. Can J Bot 73:1–13

Formiga AT, Silveira FAO, Fernandes GW, Isaias RMS (2015) Phenotypic plasticity and similarity among gall morphotypes on a superhost, *Baccharis reticularia* (Asteraceae). Plant Biol J 17:512–521. https://doi.org/10.1111/plb.12232

Fournet A, Barrios A, Muñoz V (1994) Leishmanicidal and trypanocidal activities of Bolivian medicinal plants. J Ethnopharmacol 41:19–37

Fried G, Caño L, Brunel S, Beteta E, Charpentier A, Herrera M, Starfinger U, Panetta FD (2016) Monographs on Invasive Plants in Europe: *Baccharis halimifolia* L. Bot Letters 163:127–153. https://doi.org/10.1080/23818107.2016.1168315

Frizzo CD, Serafini LA, Dellacassa E, Lorenzo D, Moyna P (2001) Essential oil of *Baccharis uncinella* DC. from Southern Brazil. Flavour Fragr J 16:286–288

Frizzo CD, Atti-Serafini L, Laguna SE, Cassel E, Lorenzo D, Dellacassa E (2008) Essential oil variability in Baccharis uncinella DC and *Baccharis dracunculifolia* DC growing wild in southern Brazil, Bolivia and Uruguay. Flavour Fragr J 23:99–106

Galvão LCC, Furletti VF, Bersan SMF, Cunha MG, Ruiz ALTC, Carvalho JE, Santoratto A, Rehder VLG, Figueira GM, Duarte MCT, Ikegaki M, Alencar SM, Rosalen PL (2012) Antimicrobial activity of essential oils against *Streptococcus mutans* and their antiproliferative effects. Evid Based Complement Alternat Med 2012:1–15

Gene RM, Cartañá C, Adzet T, Marin E, Parella T, Canigueral S (1996) Anti-inflammatory and analgesic activity of *Baccharis trimera*: identification of its active constituents. Planta Med 62:232–235

Gershenzon J (1984) Changes in the levels of plant secondary metabolites under water and nutrient stress. Recent Adv Phytochem 18:273–320

Gomes V, Fernandes GW (2002) Germinação de aquênios de *Baccharis dracunculifolia* D. C. (Asteraceae). Acta Bot Bras 16:421–427

Gomes VM, Negreiros D, Carvalho V, Fernandes GW (2015) Growth and performance of rupestrian grasslands native species in quartzitic degraded areas. Neotropical Biol Conserv 10:159–168

Gonzáles P, Caño A, Müller J (2019) An unusual new record of *Baccharis* (Asteraceae) from the Peruvian Andes and its relation with the northern limit of the dry puna. Acta Bot Mex 126:e1393

González ML, Joray MB, Laiolo J, Crespo MI, Palacios SM, Ruiz GM, Carpinella MC (2018) Cytotoxic activity of extracts from plants of central Argentina on sensitive and multidrug-resistant leukemia cells: isolation of an active principle from *Gaillardia megapotamica*. Evid Based Complement Alternat Med 2018:1–13. https://doi.org/10.1155/2018/9185935

Grecco SS, Reimão JQ, Tempone AG, Sartorelli P, Romoff P, Ferreira MJ, Fávero OA, Lago JH (2010) Isolation of an antileishmanial and antitrypanosomal flavanone from the leaves of *Baccharis retusa* DC.(Asteraceae). Parasitol Res 106:1245–1248

Green BJ, Simpson RW, Dettmann ME (2012) Assessment of airborne Asteraceae pollen in Brisbane, Australia. Aerobiologia 28:295–301. https://doi.org/10.1007/s10453-011-9224-0

Griffin GP (1997) Pollination in the genus *Baccharis* (Asteraceae): the role of wind and insects. Doctoral dissertation, Texas A&M University

Habermehl GG, Busam L, Heydel P, Mebs D, Tokarnia CH, Döbereiner J, Spraul M (1985) Macrocyclic trichothecenes: cause of livestock poisoning by the Brazilian plant *Baccharis coridifolia*. Toxicon 23:731–745

Haque N, Peralta-Videa JR, Jones GL, Gill TE, Gardea-Torresdey JL (2008) Screening the phytoremediation potential of desert broom (*Baccharis sarothroides* Gray) growing on mine tailings in Arizona, USA. Environ Pollut 153:362–368

Haque N, Peralta-Videa JR, Jones GL, Gill TE, Gardea-Torresdey JL (2009) Differential effect of metals/metalloids on the growth and element uptake of mesquite plants obtained from plants grown at a copper mine tailing and commercial seeds. Bioresour Technol 100:6177–6182

Heiden G, Antonelli A, Pirani JR (2019) A novel phylogenetic infrageneric classification of *Baccharis* (Asteraceae: Astereae), a highly diversified American genus. Taxon 68:1048–1081

Herz W, Pilotti A, Soderholm A, Shuhama I, Vichnewsy W (1977) New ent-clerodane-type diterpenoids from *Baccharis trimera*. J Organomet Chem 42:3913–3917

Hocayen PDA, Grassiolli S, Leite NC, Pochapski MT, Pereira RA, Silva LA, Snack A, Michel RG, Kagimura FY, Cunha MAA, Malfatti CR (2016) *Baccharis dracunculifolia* methanol extract enhances glucose-stimulated insulin secretion in pancreatic islets of monosodium glutamate induced-obesity model rats. Pharm Biol 54: 1263–1271

Hudson EE, Stiling P (1997) Exploitative competition strongly affects the herbivorous insect community on *Baccharis halimifolia*. Oikos 79:521–528

Jaramillo-García V, Trindade C, Lima E, Guecheva TN, Villela I, Martinez-Lopez W, Corrêa DS, Ferraz ABF, Moura S, Sosa MQ, da Silva J, Henriques JAP (2018) Chemical characterization and cytotoxic, genotoxic, and mutagenic properties of *Baccharis trinervis* (Lam, Persoon) from Colombia and Brazil. J Ethnopharmacol 213:210–220

Jarvis BB, Wells KM, Lee YW, Bean GA, Kommedahl T, Barros CS, Barros SS (1987) Macrocyclic trichothecene mycotoxins in Brazil species of *Baccharis*. Phytopathology 77:980–984

Jarvis BB, Midiwo JO, Bean GA, Abdoul-Nasr MB, Barras CS (1988) The mystery of trichothecene antibiotics in *Baccharis* species. J Nat Prod 51:736–744

Jarvis BB, Wang S, Cox C, Rao MM, Philip V, Varaschin MS, Barros CS (1996) Brazilian *Baccharis* toxins: livestock poisoning and the isolation of macrocyclic trichothecene glucosides. Nat Toxins 4:58–71

Julião GR, Fernandes GW, Negreiros D, Bedê L, Araújo RC (2005) Insetos galhadores associados a duas espécies de plantas invasoras de áreas urbanas e peri-urbanas. Rev Bras Entomol 49:97–106

Julião GR, Almada ED, Fernandes GW (2014) Galling insects in the pantanal wetland and Amazonian rainforest. In: Fernandes G, Santos J (eds) Neotropical insect galls. Springer, Dordrecht. https://doi.org/10.1007/978-94-017-8783-3_19

Junxing D, Huijuan Z, Aiping T (2003) Use of dicaffeoylquinic acid derivative and analogs in treating disease related to coronavirus infection. Applicant: Institute of Radiation Medicine, Chinese Academy of Military Medical Sciences. CN1449751A. https://patents.google.com/patent/CN1449751A/en#citedBy

Kissmann KG, Groth D (1992) Plantas infestantes e nocivas. Tomo I, II e III. BASF, São Paulo

Kumazawa S, Yoneda M, Shibata I, Kanaeda J, Hamasaka T, Nakayama T (2003) Direct evidence for the plant origin of Brazilian propolis by the observation of honeybee behavior and phytochemical analysis. Chem Pharm Bull 51:740–742

Kupchan S, Jarvis B, Dailey R Jr, Bright W, Bryan R, Shizuri Y (1976) Baccharin, a novel potent antileukemic tricothecene triepoxide from *Baccharis megapotamica*. J Am Chem Soc 98:7092–7093

Kupchan SM, Streelman DR, Jarvis BB, Dailey RG Jr, Sneden AT (1977) Isolation of potent new antileukemic trichothecenes from *Baccharis megapotamica*. J Organomet Chem 22:4221–4225

Lara ACF, Fernandes GW (1996) The highest diversity of galling insects: Serra do Cipó, Brazil. Biodivers Lett 3:111–114

Laris P, Brennan S, Engelberg K (2017) The coyote brush invasion of southern California grasslands and the legacy of mechanical disturbance. Geogr Rev 107:640–659

Lorenzi H (1992) Plantas daninhas do Brasil, 2nd edn. Plantarum, São Paulo

Madeira JA, Fernandes GW (1999) Reproductive phenology of sympatric taxa of *Chamaecrista* (Leguminosae) in Serra do Cipó. Br J Trop Ecol 15:463–479

Maia VC (2011) Characterization of insect galls, gall makers, and associated fauna of Platô Bacaba (Porto de Trombetas, Pará, Brazil). Biota Neotrop 11:1–18. https://doi.org/10.1590/S1676-06032011000400003

Maldonado-López Y, Cuevas-Reyes P, Oyama K (2016) Diversity of gall wasps (Hymenoptera: Cynipidae) associated with oak trees (Fagaceae: *Quercus*) in a fragmented landscape in Mexico. Arthropod Plant Interact 10:29–39

Manos PS, Doyle JJ, Nixon KC (1999) Phylogeny, biogeography, and processes of molecular differentiation in *Quercus* subgenus *Quercus* (Fagaceae). Mol Phylogenet Evol 12:333–349

Marques ES, Fernandes GW (2016) The gall inducing insect community on *Baccharis concinna* (Asteraceae): the role of shoot growth rates and seasonal variations. Lundiana 12:17–26

Marques AR, Fernandes GW, Reis IA, Assunção RM (2002) Distribution of adult male and female *Baccharis concinna* (Asteraceae) in the rupestrian fields of Serra do Cipó, Brazil. Plant Biol 4:94–103

Matilde-Silva M, Boeger MRT, Melo Júnior JCFD (2019) O vigor da planta altera a densidade de galhas em populações de *Baccharis longiattenuata* (Asteraceae) sob distintas condições de solo? Rodriguésia 70:e02752017. https://doi.org/10.1590/2175-7860201970018

McFadyen P (1978) A review of the biocontrol of groundsel-bush (*Baccharis halimifolia* L.). In: Proceedings of the first conference of the council Australia Weed Science Society of Melbourne, Australia, pp 123–125

Monteiro GF, Macedo-Reis LE, Dáttilo W, Fernandes GW, Siqueira de Castro F, Neves FS (2020) Ecological interactions among insect herbivores, ants and the host plant *Baccharis dracunculifolia* in a Brazilian mountain ecosystem. Austral Ecol 45:158–167

Moraes Neto RN, Setúbal RFB, Higino TMM, Brelaz-de-Castro MCA, Silva LCN, Aliança ASDS (2019) Asteraceae plants as sources of compounds against Leishmaniasis and Chagas disease. Front Pharmacol 10:1–20. https://doi.org/10.3389/fphar.2019.00477

Morales G, Paredes A, Sierra P, Loyola LA (2008) Antimicrobial activity of three *Baccharis* species used in the traditional medicine of Northern Chile. Molecules 13:790–794

Moreira X, Nell CS, Katsanis A, Rasmann S, Mooney KA (2018) Herbivore specificity and the chemical basis of plant–plant communication in *Baccharis salicifolia* (Asteraceae). New Phytol 220:703–713

Negreiros D, Esteves D, Fernandes GW, Berbara RLL, Oki Y, Vichiato M, Chalub C (2014) Growth-survival tradeoff in the widespread tropical shrub *Baccharis dracunculifolia* (Asteraceae) in response to a nutrient gradient. Trop Ecol 55:167–176

Neves FS, Fagundes M, Sperber CF, Fernandes GW (2011) Tri-trophic level interactions affect host plant development and abundance of insect herbivores. Arthropod Plant Interact 5:351–357

Norem MA, Day AD, Ludeke KL (1982) An evaluation of shrub and tree species used for revegetating copper mine wastes in the South-Western United States. J Arid Environ 5:299–304

Occhioni P (1944) Contribuição ao estudo do "mio-mio" *Baccharis coridifolia* DC. Rev Bras Med Vet 13:193–209

Oki Y, Soares NR, Belmiro MS, Correa-Junior A, Fernandes GW (2009) The influence of the endophytic fungi on the herbivores from *Baccharis dracunculifolia* (Asteraceae). Neotrop Biol Conserv 4:83–88. https://doi.org/10.4013/5119

Oki Y, Goto BT, Jobim K, Rosa LH, Ferreira MC, Coutinho ES, Xavier JH de A, Carvalho F, de Souza MFM, Berbara RLL, Fernandes GW (2016) Arbuscular mycorrhiza and endophytic fungi in ruspestrian grasslands. In: Fernandes GW (ed) Ecology and conservation of mountaintop grasslands in Brazil. Springer, Switzerland, pp. 157–179

Oki Y, Arantes-Garcia L, Costa MB, Nunes BC, Silveira BR, Gélvez-Zúñiga I, Franco A, Fernandes GW (2020) CO_2 fertilizer effect on growth, polyphenols, and endophytes in two *Baccharis* species. Braz Arch Biol Technol 63:e20190302. https://doi.org/10.1590/1678-4324-2020190302

Oki Y, Nascimento IM, Costa NB, Maia RA, Takahashi JA, Ferraz V, Fernandes GW (2021) Effectiveness of endophytic fungi from *Baccharis dracunculifolia* against sucking insects and fungal pathogens. In: Rosa LH (ed) Neotropical endophytic fungi. Springer, Cham, pp 337–349

Oliveira ACP, Endringer DC, Amorim LAS, Graças LBM, Coelho MM (2005) Effect of the extracts and fractions of *Baccharis trimera* and *Syzygium cumini* on glycaemia of diabetic and non-diabetic mice. J Ethnopharmacol 102:465–469

Oliveira-Filho JC, Carmo PM, Lucena RB, Pierezan F, Barros CS (2011) *Baccharis megapotamica* var. *weirii* poisoning in water buffalo (*Bubalus bubalis*). Vet Diagn Invest 23:610–614

Palmer WA (1986) The host range of *Trirhabda flavolimbata* (Mannnerheim) (Coleoptera: Chrysomelidae) and its suitability as a biological control agent for *Baccharis* spp. (Asteraceae: Asterae). Coleopt Bull 40:149–153

Palmer WA, Haseler WH (1992a) The host specificity and biology of *Trirhabda bacharidis* (Weber) (Coleoptera: Chrysomelidae), a species introduced into Australia for the biological control of *Baccharis halimifolia* L. Coleopt Bull 46:61–66

Palmer WA, Haseler WH (1992b) Food plant specificity and biology of *Oidaematophorus balanotes* (Pterophoridae): a North American moth introduced into Australia for the biological control of *Baccharis halimifolia*. J Lepid Soc 46:195–202

Palmer WA, Tomley AJ (1993) The host range and biology of *Amniscus perplexus* Haldeman (Coleoptera: Cerambycidae), a candidate evaluated for the biological control of *Baccharis halimifolia* in Australia. Coleopt Bull 47:27–34

Palmer WA, Diatloff G, Melkshan J (1993) The host specificity of *Rhopalomyia californica* Felt (Diptera: Cecidomyiidae) and its importation into Australia as a biological control agent for *Baccharis halimifolia* L. Proc Entomol Soc Wash 95:1–6

Palomino SS, Abad MJ, Bedoya LM, García J, Gonzales E, Chiriboga X, Bermejo P, Alcami J (2002) Screening of South American plants against human immunodeficiency virus: preliminary fractionation of aqueous extract from *Baccharis trinervis*. Biol Pharm Bull 25:1147–1150

Paniagua-Zambrana NY, Bussmann RW, Romero C, Echeverría J (2020) *Baccharis genistelloides* (Lam.) Pers. Asteraceae. In: Paniagua-Zambrana N, Bussmann R (eds) Ethnobotany of the Andes. Ethnobotany of mountain regions. Springer, Cham, pp 291–296. https://doi.org/10.1007/978-3-030-28933-1_304

Panziera W, Gonçalves MA, Lorenzett MP, Damboriarena P, Argenta FF, Laisse CJ, Pavarini SP, Driemeier D (2015) Intoxicação natural por *Baccharis megapotamica* var. weirii em caprinos. Pesqui Vet Bras 35:360–364

Park YK, Paredes-Guzman JF, Aguiar CL, Alencar SM, Fujiwara FY (2004) Chemical constituents in *Baccharis dracunculifolia* as the main botanical origin of southeastern Brazilian propolis. J Agric Food Chem 52:1100–1103

Parker K (1972) An illustrated guide to Arizona weeds. University of Arizona Press, Tucson

Peláez M, Dirzo R, Fernandes GW, Perea R (2019) Nurse plant size and biotic stress determine quantity and quality of plant facilitation in oak savannas. Forest Ecol Manag 437:435–442

Perea R, Cunha JS, Spadeto C, Gomes VM, Moura AL, Silveira BS, Fernandes GW (2019) Nurse shrubs to mitigate plant invasion along roads of montane Neotropics. Ecol Eng 136:193–196

Pereira CB, Kanunfre CC, Farago PV, Borsato DM, Budel JM, Maia BHLNS, Campesatto EA, Sartoratto A, Miguel MD, Miguel OG (2017) Cytotoxic mechanism of *Baccharis milleflora* (Less.) DC. essential oil. Toxicol in Vitro 42:214–221

Pérez-López G, González-Rodríguez A, Oyama K, Cuevas-Reyes P (2016) Effects of plant hybridization on the structure and composition of a highly rich community of cynipid gall wasps: the case of the oak hybrid complex *Quercus magnoliifolia* x *Quercus resinosa* in Mexico. Biodivers Conserv 25:633–651

Price PW, Craig TP, Roininen H (1995) Working toward theory on galling sawfly population dynamics. In: Cappuccino N, Price PW (eds) Population dynamics: new approaches and synthesis. Academic Press, San Diego, pp 321–338

Queiroga CL, Fukai A, Marsaioli A (1990) Composition of the essential oil of vassoura. J Braz Chem Soc 1:105–109

Queiroga CL, Cavalcante MQ, Ferraz PC, Coser RN, Sartoratto A, Magalhães PM (2014) High-speed countercurrent chromatography as a tool to isolate nerolidol from the *Baccharis dracunculifolia* volatile oil. J Essent Oil Res 26:334–337

Quintero C, Garibaldi LA, Grez A, Polidori C, Nieves-Aldrey JL (2014) Galls of the Temperate Forest of Southern South America: Argentina and Chile. In: Fernandes G, Santos J (eds) Neotropical insect galls. Springer, Dordrecht, pp 429–463

Rabelo ACS, Costa DC (2018) A review of biological and pharmacological activities of *Baccharis trimera*. Chem Biol Interact 296:65–75

Resende FA, Munari CC, Monteiro Neto MAB, Tavares DC, Bastos JK, Silva Filho AA, Varanda EA (2012) Comparative studies of the (anti) mutagenicity of *Baccharis dracunculifolia* and artepillin C by the bacterial reverse mutation test. Molecules 17:2335–2350

Rissi DR, Rech RR, Fighera RA, Cagnini DQ, Kommers GD, Barros CS (2005) Intoxicação espontânea por *Baccharis coridifolia* em bovinos. PesqVet Bras 25:111–114

Rizzo I, Varsavky E, Haidukowski M, Frade H (1997) Macrocyclic trichothecenes in *Baccharis coridifolia* plants and endophytes and *Baccharis artemisioides* plants. Toxicon 35:753–757

Rodrigues DM, De Souza MC, Arruda C, Pereira RAS, Bastos JK (2020) The role of *Baccharis dracunculifolia* and its chemical profile on green propolis production by *Apis mellifera*. J Chem Ecol 46:150–162

Romero-Benavides JC, Ortega-Torres GC, Villacis J, Vivanco-Jaramillo SL, Galarza-Urgilés KI, Bailon-Moscoso N (2018) Phytochemical study and evaluation of the cytotoxic properties of methanolic extract from *Baccharis obtusifolia*. Int J Med Chem 2018:1–5. https://doi.org/10.1155/2018/8908435

Sá CEM, Negreiros D, Fernandes GW, Dias MC, Franco AC (2014) Carbon dioxide-enriched atmosphere enhances biomass accumulation and meristem production in the pioneer shrub *Baccharis dracunculifolia* (Asteraceae). Acta Bot Bras 28:646–650. https://doi.org/10.1590/0102-33062014abb3329

Santos JC, Almeida-Cortez JS, Fernandes GW (2011) Richness of gall-inducing insects in the tropical dry forest (Caatinga) of Pernambuco. Rev Bras Entomol 55:45–54

Schang R (1929) Acción tóxica del romerillo o mio-mio (*Baccharis coridifolia*) algunos conceptos nuevos. Rev Bras Med Vet 11:151–181

Schild CO, Oliveira LGS, Miraballes C, Giannitti F, Casaux ML, Aráoz V, Silveira CS, Boabaid FM, Riet-Correa F (2020) *Baccharis coridifolia* poisoning in livestock in Uruguay. Toxicon 188:5–10

Shorthouse JD, Rohfritsch O (1992) Biology of insect-induced galls. Oxford University, New York

Silva RMD, Fernandes GW, Lovato MB (2007) Genetic variation in two *Chamaecrista* species (Leguminosae), one endangered and narrowly distributed and another widespread in the Serra do Espinhaço, Brazil. Can J Bot 85:629–636

Sims-Chilton NM, Zalucki MP, Buckley YM (2010) Long term climate effects are confounded with the biological control programme against the invasive weed *Baccharis halimifolia* in Australia. Biol Invasions 12:3145–3155

Souza MMQ, Silva GRD, Cola IM, Silva AO, Schaedler MI, Guarnier LP, Palozi RAC, Barboza LN, Menetrier JV, Froelich DL, Auth PA, Veiga AA, Souza LM, Lovato ECW, Ribeiro-Paes JT, Gasparotto Junior A, Lívero FADR (2020) *Baccharis trimera* (Less.) DC: an innovative cardioprotective herbal medicine against multiple risk factors for cardiovascular disease. J Med Food 23:676–684

Stone GN, Schönrogge K, Atkinson RJ, Bellido D, Pujade-Villar J (2002) The population biology of oak gall wasps (Hymenoptera: Cynipidae). Annu Rev Entomol 47:633–668

Suttisri R, Kinghorn AD, Wright AD, Stichert O (1994) Neo-clerodane diterpenoids and other constituents from *Baccharis genistelloides*. Phytochemistry 35:443–446

Takashima M, Ichihara K, Hirata Y (2019) Neuroprotective effects of Brazilian green propolis on oxytosis/ferroptosis in mouse hippocampal HT22 cells. Food Chem Toxicol 132:1–10. https://doi.org/10.1016/j.fct.2019.110669

Teixeira EW, Negri G, Meira RM, Message D, Salatino A (2005) Plant origin of green propolis: bee behavior, plant anatomy and chemistry. Evid Based Complement Alternat Med 2:85–92

Thompson AE, Lee CW, Gass RE (1995) Development of hybrid *Baccharis* plants for desert landscaping. HortScience 30:1357–1362

Tilden J (1953) Biological notes on *Trirhabda flavolimbata*. Coleopt Bull 7:43–53

Timmons F (1959) Phreatophytes – water wasters – a menace in the arid west. Reclam Era 45:85–88

Tokarnia C, Dobereiner J (1976) Intoxicação experimental em ovinos por "mio-mio", *Baccharis coridifolia*. Pesqui Agropecu Bras 11:19–26

Torres LMB, Gamberini MT, Roque NF, Lima-Landman MT, Souccar C, Lapa AJ (2000) Diterpene from *Baccharis trimera* with a relaxant effect on rat vascular smooth muscle. Phytochemistry 55:617–619

Ueno AK, Barcellos AF, Costa-Silva TA, Mesquita JT, Ferreira DD, Tempone AG, Romoff P, Antar GM, Lago JHG (2018) Antitrypanosomal activity and evaluation of the mechanism of action of diterpenes from aerial parts of *Baccharis retusa* (Asteraceae). Fitoterapia 125:55–58

Vannini AB, Santos TG, Fleming AC, Purnhagen LRP, Lourenço LA, Butzke ETB, Kempt M, Begnini IM, Rebelo RA, Dalmarco EM, Cruz AB, Schmit AP, Cruz RCB, Yamanaka CN, Steindel M (2012) Chemical characterization and antimicrobial evaluation of the essential oils from *Baccharis uncinella* DC and *Baccharis semiserrata* DC (Asteraceae). J Essent Oil Res 24:547–554

Varaschin MS, Alessi AC (2003) Poisoning of mice by *Baccharis coridifolia*: an experimental model. Vet Hum Toxicol 45:42–44

Veiga RS, Mendonça S, Mendes PB, Paulino N, Mimica MJ, Lagareiro Netto AA, Lira IS, López BGC, Negrão V, Marcucci MC (2017) Artepillin C and phenolic compounds responsible for antimicrobial and antioxidant activity of green propolis and *Baccharis dracunculifolia* DC. J Appl Microbiol 122:911–920

Verdi LG, Brighente MC, Pizzolatti MG (2005) Gênero *Baccharis* (Asteraceae): Aspectos químicos, econômicos biológicos. Quim Nova 28:85–94

Verloove F, Dana ED, Alves P (2018) *Baccharis spicata* (Asteraceae), a new potentially invasive species to Europe. Plant Biosyst 152:416–426. https://doi.org/10.1080/11263504.2017.1303001

Vidari G, Finzi PV, Zarzuelo A, Gálvez, Zafra C, Chiriboga X, Berenguer B, La Casa C, Lastra CA, Motilva V, Martin MJ (2003) Antiulcer and antidiarrhoeic effect of *Baccharis teindalensis*. Pharm Biol 41:405–411

Vieira ML, Johann S, Hughes FM, Rosa CA, Rosa LH (2014) The diversity and antimicrobial activity of endophytic fungi associated with medicinal plant *Baccharis trimera* (Asteraceae) from the Brazilian savannah. Can J Microbiol 60:847–856. https://doi.org/10.1139/cjm-2014-0449

Waring GL, Price PW (1990) Plant water stress and gall formation (Cecidomyiidae: *Asphondylia* spp.) on creosote bush. Ecol Entomol 15:87–95

Watts S, Dormann CF, González AMM, Ollerton J (2016) The influence of floral traits on specialization and modularity of plant–pollinator networks in a biodiversity hotspot in the Peruvian Andes. Ann Bot 118:415–429

Westman WE, Panetta FD, Stanely TD (1975) Ecological studies on reproduction and establishment of the woody weed, groundsel bush (*Baccharis halimifolia* L.: Asteraceae). Aust J Agric Res 26:855–870

Wollenweber E, Valantvetschera KM, Fernandes GW (2006) Chemodiversity of exudate flavonoids in *Baccharis concinna* and three further South-American *Baccharis* species. Nat Prod Commun 1:627–632

Zavaleta ES (2006) Shrub establishment under experimental global changes in a California grassland. Plant Ecol 184:53–63

Zavaleta ES, Kettley LS (2006) Ecosystem change along a woody invasion chronosequence in a California grassland. J Arid Environ 66:290–306. https://doi.org/10.1016/j.jaridenv.2005.11.008

Chapter 2
Baccharis: Diversity and Distribution

Gustavo Heiden

Abstract *Baccharis* is a monophyletic genus characterized by functionally unisexual florets, generally distributed in distinct individuals (dioecy), but also including monoecious, gynodioecious, and polygamous species. The genus has not been revised taxonomically as a whole for nearly two centuries. Recent country- or dependent territory-level checklists are hardly comparable and mostly outdated. A comprehensive checklist on the diversity and distributions of *Baccharis* at generic, infrageneric, and specific levels, including putative hybrids, and adventitious occurrences is provided. *Baccharis* comprises 442 species classified into 47 sections and 7 subgenera. The genus is native in the Americas, from southeastern Canada and northwestern USA to Tierra del Fuego, with species native to the Falkland/Malvinas Islands, across most of the Caribbean islands and the Galápagos archipelago. Complete lists of species per country and territory are provided. Brazil (185 species, 114 endemics), Argentina (110 species, 25 endemics), and Bolivia (76 species, 22 endemic) are the richest countries for the genus. Four species are highlighted for occurring in more than ten countries or territories within their native range (*B. dioica*, *B. pedunculata*, *B. trinervis*, *B. salicifolia*), while at least 218 species are endemic to a single country or territory. The role of hybridization in the genus diversity, ecology, and evolution is still a neglected subject, and 38 putative hybrid taxa were described so far. Some species were spread outside the American continent by anthropogenic dispersals, and at least two have established naturalized alien populations: *B. halimifolia* in Europe and Oceania and *B. spicata* in Europe.

Keywords Endemism · Infrageneric classification · Phylogeny · Plant distribution · Taxonomy

---

G. Heiden (✉)
Embrapa Clima Temperado, Pelotas, Rio Grande do Sul, Brazil
e-mail: gustavo.heiden@embrapa.br

G. W. Fernandes et al. (eds.), *Baccharis*,
https://doi.org/10.1007/978-3-030-83511-8_2

1 What Is *Baccharis*?

Baccharis L. (Asteraceae, Astereae, Baccharidinae) is a monophyletic American genus mostly characterized by a tufted indumentum of trichomes with several adjoining basal cells on leaves and stems and functionally unisexual florets (Heiden et al. 2019). The broad circumscription of *Baccharis* proposed by Nesom (1988) and Müller (2006), including monoecious (*Baccharidastrum* Cabrera), gynodioecious (*Heterothalamus* Less.), and polygamous taxa (*Baccharidiopsis* G.M.Barroso), is supported by molecular data and phylogenetic reconstructions (Heiden et al. 2019).

Nesom (2020) presented an updated subtribal classification of Astereae and proposed a new expanded concept of Baccharidinae including hypothesis of relationships within the subtribe influenced by the phylogenetic molecular data published by Vargas et al. (2017), Vargas (2018) and Heiden et al. (2019). *Baccharis* and several genera formerly placed in the now synonymized subtribes Podocominae and Hinterhuberinae are currently included in Baccharidinae. So, subtribe Baccharidinae encompasses 24 genera and is advocated to represent a monophyletic group sister to the North American lineages (Nesom 2020).

The best available evidence points that *Baccharis* is sister to *Exostigma* G.Sancho (Vargas 2018). These two genera are sisters of a clade composed by the Central Andean genera *Floscaldasia* Cuatrec., *Parastrephia* Nutt., and *Diplostephium* Kunth. This clade containing *Baccharis*, *Exostigma*, and the Central Andean genera is then nested in a clade with Central and South American *Archibaccharis* Heering, *Aztecaster* G.L.Nesom, *Laennecia* Cass., *Lagenophora* Cass., and *Westoniella* Cuatrec., which is sister of a mostly North Andean clade composed by *Laestadia* Kunth ex Less, *Hinterhubera* Sch.Bip., *Blakiella* Cuatrec., and *Linochilus* Benth. (Vargas et al. 2017).

The available phylogenies support a South American origin of *Baccharis*. The most recent common ancestor of *Baccharis* and *Exostigma* has split about 6.93 million years ago (Vargas et al. 2017) in the Miocene. After an early diversification in South America (Heiden et al. 2019), some derived lineages entered the North America subcontinent several times by multiple events of dispersal followed by subsequent radiations as represented by the occurrence of independent but not closely related lineages (Heiden et al. 2019) such as *Baccharis* sect. *Aristidentes* G.L.Nesom, *B.* sect. *Baccharis*, *B.* sect. *Bogotenses* Cuatrec., *B.* sect. *Corymbosae* Heering, and *B.* sect. *Punctatae* Giuliano & G.L.Nesom in North America, Central America, and the Caribbean. The most recent common ancestor of *B.* subgen. *Heterothalamus* (Less.) G.Heiden and *B.* subgen. *Baccharis* has split 4.02 million years ago in the Pliocene, and the first divergence event within *B.* subgen. *Baccharis* (*B.* sect. *Caulopterae* and the remaining sections within the subgenus) occurred 2.44 million years ago (Vargas et al. 2017) in the Pliocene/Pleistocene transition.

Functionally unisexual florets are present in all lineages of *Baccharis*, while its disposition in distinct plants (dioecy) is known in all the subgenera and most of the sections (Fig. 2.1). Variations from perfect dioecy are scattered in some lineages as

Fig. 2.1 Representative of a functionally dioecious species: *Baccharis tarchonanthoides*. (**a**) Habit. (**b**) Male capitula with only tubulose staminate florets. (**c**) Female capitula with only filiform pistillate florets. (Photos **a**, **b**, **c**: G.Heiden)

secondarily derived states. Monoecious taxa are represented by the sister species *B. breviseta* DC. and *B. vulneraria* Baker (Fig. 2.2), as well as *B. monoica* G.L.Nesom. Gynodioecious species are represented by *B. aliena* (Spreng.) Joch. Müll. (Fig. 2.3), *B. hyemalis* Deble, and *B. psiadioides* (Less.) Joch.Müll., while *B. pohlii* (Baker) Deble & A.S.Oliveira, *B. polygama* Ariza, and *B. punctulata* DC. (Fig. 2.4) are polygamous taxa sister to dioecious species.

The tufted indumentum cannot be considered as a synapomorphy of *Baccharis*, as suggested by Müller (2006), since it is absent in *B.* subgen. *Tarchonanthoides* Heering, which is the first diverging lineage within the genus. This subgenus shares the functionally unisexual florets with its sister clade composed of the remaining species of *Baccharis* but lacks the tufted indumentum. So, the tufted indumentum occurs only in the sister clade composed of the remaining species of *Baccharis*, with a posterior reversal (loss) in *B.* subgen. *Coridifoliae* (Heiden et al. 2019).

Baccharis has never been revised taxonomically as a whole since Candolle (1836). Recently published checklists are hardly comparable and mostly outdated due to differences in the genus circumscription, nomenclatural changes, description of new species, and adoption of conflicting taxa circumscriptions by different authors. Heiden et al. (2019) published a revised checklist at infrageneric and species level, but data on distribution was not encompassing species distributions at country, dependent territory, and first political division level. In this chapter, an updated checklist on the diversity and distributions of *Baccharis* at generic, infrageneric, and specific levels, including putative hybrids, and adventitious occurrences for countries and dependent territories is provided.

2 How Was This List Built?

The taxonomic checklist and geographical distributional data provided here are the results of the collation of published and original information. An updated checklist was obtained by means of literature species citations revisited based on nomenclatural updating, author circumscription of the accepted species, and double-checking of vouchers cited in references. This checklist was then complemented by fieldwork data, revision of herbaria specimens personally or remotely, and punctual additions of occurrence data from databases with verifiable and confident information, hence representing the most updated and solid information on this genus.

The baseline for the new checklist construction is the infrageneric classification published by Heiden et al. (2019), complemented with data from Malagarriga (1976), Müller (2013), Plants of the World Online (POWO 2019), and Vascular Plants of the Americas (Ulloa et al. 2018). These sources resulted in the addition of ten neglected names missing from the previous list by Heiden et al. (2019) (*B. argentina* Heering, *B. concava* (Ruiz. & Pav.) Pers., *B. cordobensis* Heering, *B. douglasii* DC., *B. emoryi* A.Gray, *B. glomerata* Joch.Müll., *B. isabelae* Soria & Zardini,

Fig. 2.2 Representative of a gynodioecious species: *Baccharis aliena*. (**a**) Habit. (**b**) Hermaphrodite capitula with pistillate ray florets and tubulose staminate disk florets. (**c**) Female capitula with only short-rayed pistillate florets. (Photos **a**, **b**, **c**: G.Heiden)

Fig. 2.3 Two representatives of monoecious species. *Baccharis breviseta*: (**a**) Habit. (**b**) Hermaphrodite capitula with marginal filiform pistillate florets and one to few central tubulose staminate florets. *B. vulneraria*: (**c**) Habit. (**d**) Hermaphrodite capitula with marginal filiform pistillate florets and few central tubulose staminate florets. (Photos **a**, **b**, **c**, **d**: G.Heiden)

Fig. 2.4 Representative of a polygamous species: *Baccharis punctulata*. (**a**) Habit. (**b**) Male capitula with only tubulose staminate florets. (**c**) Female capitula with only filiform pistillate florets. Hermaphrodites are rare and not shown here; their capitula have marginal filiform pistillate florets and few central tubulose staminate florets. (Photos **a**, **b**, **c**: G.Heiden)

B. lancifolia DC., *B. polygama* Ariza, and *B. vitis-idaea* Oliver. ex Thurn.), plus the acceptance of the recircumscription of *B. trimera* (Less.) DC. by Valtierra (2018), and other two new species described later (*B. funkiae* Bonif. et al. and *B. rectialata* V.Valtierra et al.).

Subsequently, regional floras and checklists available for North America (Sundberg and Bogler 2006), Mesoamerica (Pruski and Robinson 2018), the Caribbean (Acevedo-Rodríguez and Strong 2007), the United Kingdom Overseas Territories (UKOT 2020), the Guiana Shield (Funk et al. 2007), and the Southern Cone of South America (Zuloaga et al. 2008) were consulted for country and first political order geographical subdivision distribution records. Country-level checklists, floras, and new occurrences were then consulted to Argentina (Giuliano and Plos 2014), Bolivia (Müller 2006; Hind 2011), Brazil (Heiden 2020), Canada (Fielding 2001), Chile (Heering 1903; Hellwig 1990; Moreira-Muñoz et al. 2016), Colombia (Cuatrecasas 1967; Ávila et al. 2020), Cuba (Alain 1962; Greuter and Rodriguez 2017), Ecuador (Robinson et al. 1999), Mexico (Villaseñor 2016), Peru (Brako and Zarucchi 1993; Beltrán et al. 2006; Gonzáles et al. 2019), Uruguay (Valtierra and Bonifacino 2014; Valtierra 2018), and Venezuela (Hocke et al. 2008). All the species lists and voucher specimens, when cited, were nomenclaturally updated and taxonomically checked for identification. The information on this regard is comprehensive for most cases; however, it is clearly not perfect and reflects current knowledge, especially for subdivisions of countries lacking recent taxonomic studies focusing on *Baccharis*, especially the Northern Andean countries of Colombia, Ecuador, and Peru.

The data from literature were checked and complemented with information from herbaria specimens consulted personally (B, BAA, BAB, BHCB, BHZB, BM, C, CDS, CEN, CEPEC, CESJ, COL, CORD, CTES, E, ECT, ESA, F, FCAB, FLOR, FUEL, FURB, G, GB, GUA, HAC, HAJB, HAS, HB, HBG, HBR, HEPH, HPL, HRJ, HUCS, HUEFS, HURG, HVAT, IAC, IBGE, ICN, INPA, IPA, JPB, K, LIL, LP, MBM, MBML, MEDEL, MEXU, MG, MO, MOL, MT, MVFA, MVFQ, MVJB, MVM, NY, O, OUPR, P, PACA, PAMG, QCEN, PEL, R, RB, RBR, RFFP, RUSU, S, SI, SP, SPF, SPSF, TOLI, UB, UEC, UFP, UFRN, UPCB, US, USZ) or remotely (BR, GH, GOET, HCSM, HDCF, JE, LE, M, MA, OBI, OXF, W, WU).

GBIF (2020) data were used for building the genus overview distribution map (Fig. 2.5). This database was also used when species records contributed for occurrence knowledge in some states, provinces, or departments and some Caribbean Islands that were missing from previous publications. However, GBIF data were included only when these specimens' records were determined by specialists who have published confident taxonomic studies on the genus along the last hundred years (Barroso, Bonifacino, Cuatrecasas, Giuliano, Heiden, Hellwig, Hind, Malagarriga Heras, Müller, Nesom, Pruski, Robinson, Soria, and Valtierra).

Modern taxonomic revisions independent of regional constraints and revising monophyletic groups are scarce, and the work of Heiden & Pirani (2016) is still the only recent revision under this framework. Doubtlessly, revisions of other clades

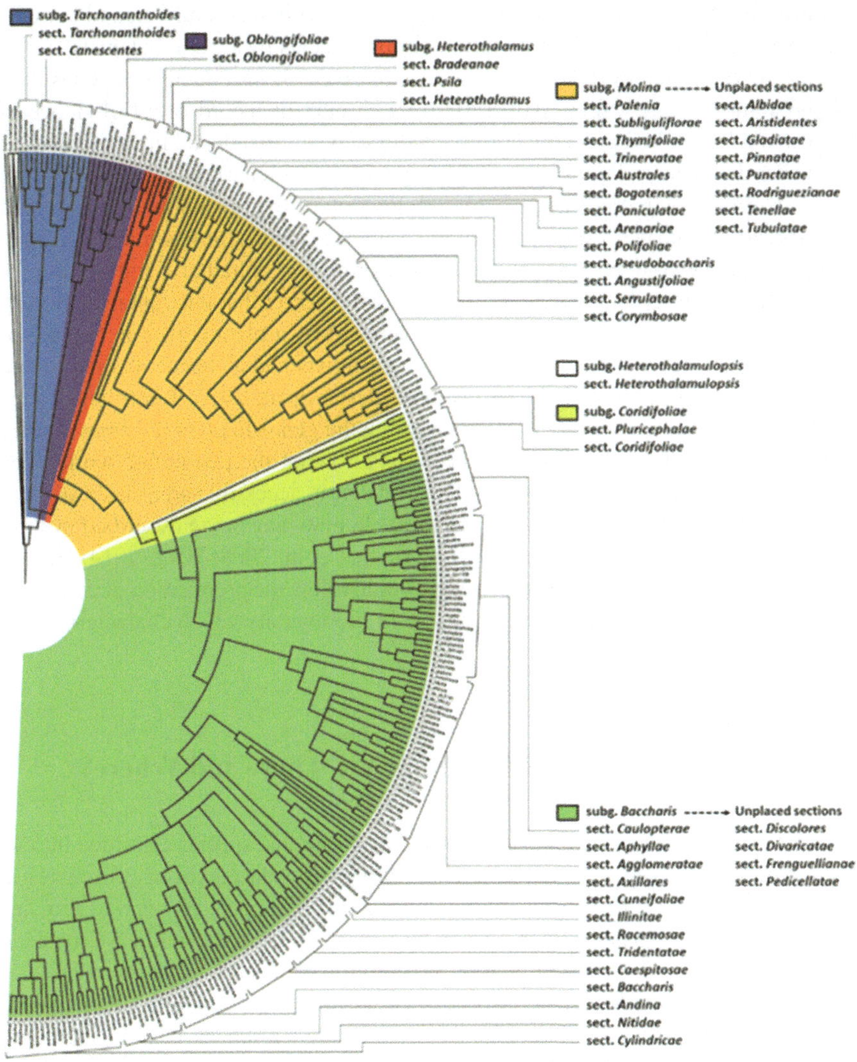

Fig. 2.5 Schematic phylogenetic infrageneric classification of *Baccharis* (Asteraceae) comprising 7 subgenera, 47 sections, and 248 sampled taxa out of 442 accepted species. (Based on Heiden et al. 2019)

currently underway and fully or almost fully sampled in the phylogeny (*B.* subgen. *Oblongifolia*, *B.* subgen. *Heterothalamus*, *B.* subgen. *Coridifoliae*, and *B.* subgen. *Heterothalamulopsis*) will greatly improve the understanding of these subgenera and its sections, series, and species composition and relationships. Whereas deeper knowledge for *B.* subgen. *Molina* and *B.* subgen. *Baccharis* are still far from being achieved completely, progress continues to be done with the reevaluation of complicated species complexes and description of new species by several botanists.

Despite the great efforts for herbaria revision, 105,952 specimens were consulted by the author so far, while GBIF (2020) has a current record of 147,141 occurrences, and lots of herbaria collections are still to be digitized, and most of the digitized ones do not have digital images of the specimens which would allow confirming more identification records and permit acceleration of the taxonomy of a large and widespread group such *Baccharis*. Doubtlessly, the genus still has many new species never collected to be described to science or housed at herbaria and waiting for taxonomists to describe them, as well as it is not a secret that uncountable areas of the Americas has never been stepped and sampled by a plant collector. Several of these areas are being destroyed even before scientists had a chance to study them, and conservation actions could be taken (Tedesco et al. 2012; Pimm and Joppa 2015). Time urges for a great effort on sampling the genus in potential areas and for funding towards digitizing all the collections allowing the taxonomic work to be sped up. Moreover, even well-explored and relatively accessible areas keep surprising researchers with noteworthy and remarkable new taxa being described even in well-known and well-sampled areas (e.g., Heiden et al. 2009, 2012, 2014; Heiden and Pirani 2014). A reasonable knowledge would allow the proposition of conservation measures and avoid the irreversible loss of biodiversity which characterizes the Anthropocene.

3 How Many Species of *Baccharis* Are There and Where?

Baccharis comprises 442 formally described and accepted species classified into 47 sections and 7 subgenera (Fig. 2.5). The infrageneric placement of 33 species at sectional level, 17 in *B.* subgen. *Molina* and 16 in *B.* subgen. *Baccharis*, remains pending. The genus is native in the Americas (Fig. 2.6), from the southeastern island of Nova Scotia in Canada and coastal Washington State in northwestern USA to southern Isla Hornos in Tierra del Fuego in the Chilean Patagonia, with species native to the Falkland/Malvinas Islands, most of the Caribbean islands, including several endemics, and the Galápagos archipelago, including one endemic species too. *Baccharis halimifolia* L. from North America was introduced as an ornamental in Europe and Oceania and became naturalized in coastal environments (Fried et al. 2016), while *B. spicata* (Lam.) Baill. was accidentally introduced in Europe and became naturalized (Verloove et al. 2018).

The highlands of eastern Brazil are the main area for the early radiations of *Baccharis* with the likely highest phylogenetic and species diversity. This assumption is attested by the presence of all the seven recognized subgenera and by the distribution of several of the earliest-divergent lineages of the genus.

Most of the infrageneric taxa accepted here are based on the results of the Bayesian analyses of combined molecular data (nuclear ETS and ITS and

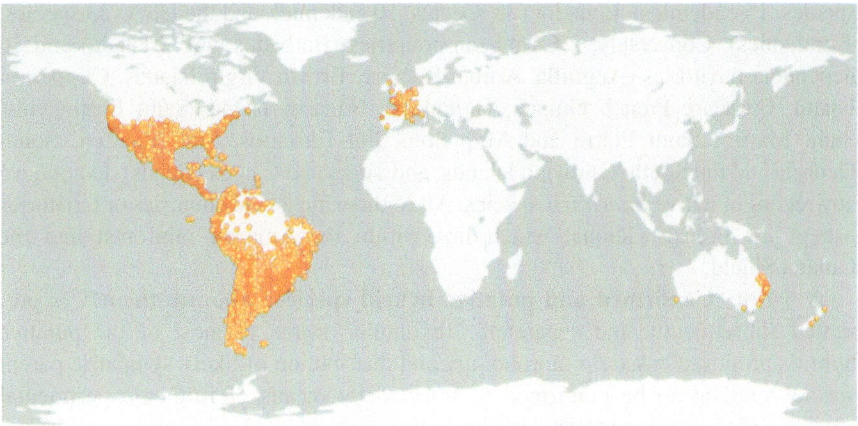

Fig. 2.6 World distribution of *Baccharis* within its native range in the Americas and naturalized introduced populations in Europe and Oceania. Orange spots and dots represent specimens and observation records distributions based on cleaned GBIF occurrence data

chloroplastidial *trn*H-*psb*A and *trn*L-F DNA) and are mostly strongly supported by statistical posterior probability (PP) values, as published by Heiden et al. (2019). The complete infrageneric classification of *Baccharis* comprising the subgenera, sections, and taxa accepted at species level is presented in a simplified schematic tree (Fig. 2.5) based on the cladograms as an index/guideline to reflect the hierarchical divergence events and relationships among subgenera and sections. Therefore, subgenera and sections are presented in the text following the divergence events reconstructed in the trees and from less-diversified to more-diversified/sampled groups when no hierarchical order (polytomies) was recovered. Species placed within each section, sections with unknown relationships within a subgenus (not sampled in the phylogeny, but confident taxonomic position in a subgenus proposed based on morphology), and species with unknown relationships within a subgenus (not sampled in the phylogeny, but confident taxonomic position possible based on morphology in a subgenus but not within a section) are presented alphabetically by the end of *Baccharis* subgen. *Molina* and *Baccharis* subgen. *Baccharis*.

Synthetic data on *Baccharis* diversity and distributions are further recorded and detailed in this chapter. The complete list of subgenera, sections, and species distributions across the 35 American sovereign countries and the 24 dependent territories from the Americas along with the introduced potentially naturalized occurrences in Europe and Oceania are presented in the paragraph: "***Baccharis* diversity and distribution: who they are and where to find them?**" section. Countries with the higher number of species richness are Brazil (185 species, 114 endemics), Argentina (110 species, 25 endemics), Bolivia (76 species, 22 endemic), Peru (61 species, 21 endemics), Uruguay (54 species, 5 endemic), Chile (48 species, 15 endemics), Paraguay (47 species, 2 endemics), Mexico (46 species, 22 endemic), Colombia (39 species, 15 endemics), Ecuador (38 species, 10 endemic), and the USA (23 species, 4 endemics). Conversely, 2 Caribbean countries (Barbados and Grenada) and 16 dependent territories (Anguilla, Aruba, Bonaire, British Virgin Islands, Clipperton Island, Curaçao, French Guiana, Greenland, Navassa Island, Saint Barthélemy, Saint Martin, Saint Pierre and Miquelon, Sint Eustatius, Sint Maarten, South Georgia and the South Sandwich Islands, and Turks and Caicos Islands) do not have any record of native *Baccharis* species. All of them are island countries or territories except for French Guiana, which lies within the Amazon rainforest and the Guiana Shield.

A list of "**Confirmed and putative hybrid species: who are them?**" is presented subsequently and separately. Taxonomic status of most of the putative hybrids proposed based on morphology and distribution of likely sympatric parent species remains to be confirmed or rejected by means of further experimental crosses and in situ population genetic studies.

Complete checklists of recorded species and endemics for the 35 American sovereign countries and the 24 dependent territories from the Americas are presented in the "**Distribution of *Baccharis* by sovereign countries or dependent territories: how many species are out there?**" section followed by the "**Going worldwide? List of introduced naturalized adventitious distributions outside the Americas**" section, where records of established self-sustained populations (historically recorded but currently eradicated or still thriving and under no efficient human intervention for extirpation) are presented. Furthermore, there are sparse records in herbaria and literature of historical or recent exotic species occurrences, beyond the listed areas of Europe and Oceania, of species of *Baccharis* out of their native range in Africa, Asia, Europe, and Oceania, but these other punctual records representing cultivate or waifs (not naturalized, not persistent introductions) are not presented here, since they do not represent sustainable alien introductions.

Several *Baccharis* species have the potential to become invasive outside the Americas, as have happened with *B. halimifolia* (Caño et al. 2013; Fried et al. 2016) and *B. spicata* (Verloove et al. 2018). Thus, on purpose introductions should be avoided as well as preventive measures and monitoring to suppress any new record of introduction outside the native range should always be considered priorities.

The genus is commonly seen as weedy or invasive because some species of *Baccharis* are widespread within the Americas. Twenty-eight species occur in more than five countries or territories within their native ranges. Four species of *Baccharis* are highlighted since they are recorded to more than ten countries and territories (Fig. 2.7). *B. dioica* Vahl (Fig. 2.7a) grows in coastal vegetations from 12 Caribbean countries and territories, many of them occupying smallish lands, including the Caribbean coast of Mexico and USA, although it is locally extinct in the later and currently recorded for this country only under cultivation. *B. pedunculata* (Mill.) Cabrera (Fig. 2.7b) is a common forest species occurring in the understory and forest edges and climbing up to the canopy in tropical forests, being widespread in 19

Fig. 2.7 Four species of *Baccharis* recorded to more than ten countries and territories (**a**) *Baccharis dioica* in the USA. (**b**) *B. pedunculata* in Bolivia. (**c**) *B. salicifolia* in Argentina. (**d**) *B. trinervis* in Brazil. (Photos **a**, **b**, **c**, **d**: G.Heiden)

countries and territories from Central America, the Caribbean islands, and Northern Andean South America countries. *B. salicifolia* (Ruiz & Pav.) Pers. (Fig. 2.7b) is by far the most widespread species of the genus occurring in a wide range of biomes from deserts to tropical forests and high-elevation vegetations. It is considered a phreatophyte and always occurs in sunny sites around springs and along riverine vegetation from North America, Central America, and South America, but it is not recorded in the Caribbean Islands. *B. trinervis* Pers. (Fig. 2.7d) is also a common forest species occurring in the understory and forest edges and climbing up to the canopy in tropical forests, but it also grows as a shrub in savannas and anthropogenic areas, being widespread in 18 countries and territories in Central America, South America, and Trinidad and Tobago.

There are other widely distributed species occurring in 7 (*B. glutinosa* Pers., *B. linearifolia* (Lam.) Pers., *B. nitida* (Ruiz & Pav.) Pers., *B. oblongifolia* (Ruiz & Pav.) Pers., and *B. sagittalis* (Less.) DC.) or 6 countries (*B. alpina* Kunth, *B. chilco* Kunth, *B. latifolia* (Ruiz & Pav.) Pers., and *B. trimera* (Less.) DC.), while 47 species are recorded occurring in 3–5 countries and 105 recorded from 2 countries or territories.

Some of the widespread species are highly variable morphologically, which could be a result of the wide distribution and/or the fate caused by the merging of distinct species into a single broad species concept, such as *B. linearifolia*, *B. oblongifolia*, *B. sagittalis*, *B. salicifolia*, and *B. trinervis*. These taxa have an extensive list of synonyms, some segregated as distinct species by different authors, and the circumscription of them and their segregates or synonyms is far from consensus, demanding further taxonomic research to elucidate narrower or broader hypothesis for these entities.

Conversely, the genus is prolific in species restricted to just one country or dependent territory with 263 species in this situation, although several of them could be common and widely distributed within countries as big as Argentina, Brazil, Mexico, and the USA. Countries with the higher absolute number of endemic species are Brazil (114 endemics), Argentina (25 endemics), Bolivia (22 endemics), Mexico (22 endemics), Peru (21 endemics), Chile (15 endemics), Colombia (15 endemics), and Ecuador (10 endemics). However, it is important to keep in mind that political borders are generally not a precise way to count endemics, due to the incomparable size of political entities, distinct levels of geographical isolation, and frequent geographical meaninglessness of several political frontiers. Moreover, some really narrow distributed species could occur exactly along geographical barriers that divide political territories giving the false idea that such species are largely spread. Anyway, counting of endemics under political geographical boundaries gives an idea of the importance of endemism within the genus and allows us to recognize the responsibility of each political entity in protecting its endemic biota.

Narrow endemics are found from the USA (e.g., *B. glabrata* (Hoover) G.Heiden) to Chilean Patagonia (e.g., *B. mylodontis* F.H.Hellw.), including endemics in the Caribbean (e.g., *B. mornicola* (Urb.) G.Heiden) and Galápagos (e.g., *B. steetzii* Andersson) islands. All the seven recognized subgenera have narrow endemic species, and some of them are depicted (Figs. 2.8 and 2.9) and exemplified here,

Fig. 2.8 Narrow endemic species of *Baccharis*. (**a**) *B.* (subgen. *Tarchonanthoides*) *chionolaenoides* in Santa Catarina, Brazil. (**b**) *B.* (subgen. *Oblongifoliae*) *friburgensis* in Rio de Janeiro, Brazil. (**c**). *B.* (subgen. *Heterothalamus*) *magnifica* in Espírito Santo, Brazil. (**d**). *B.* (subgen. *Molina*) *bifrons* in Rio de Janeiro, Brazil. (Photos **a**, **b**, **d**: G.Heiden; **c**: M.A.T.Silva)

Fig. 2.9 Narrow endemic species of *Baccharis*. (**a**) *B.* (subgen. *Heterothalamulopsis*) *wagenitzii* in Santa Catarina, Brazil. (**b**) *B.* (subgen. *Coridifoliae*) *pluricapitulata* in Rio Grande do Sul, Brazil. (**c**). *B.* (subgen. *Baccharis*) *scopulorum* in Santa Catarina, Brazil. (**d**). *B.* (subgen. *Baccharis*) *obdeltata* in Minas Gerais, Brazil. (Photos **a**, **b**, **c**: G.Heiden; **d**: C.M.Siniscalchi)

although the number of narrow endemics is much higher than the eight illustrative examples shown.

For example, *B. chionolaenoides* D.B.Falkenb. & Deble (Fig. 2.8a), belonging to *B.* subgen. *Tarchonanthoides* Heering, is one of the narrowest endemic species of the genus occurring only adhered to steep cliffs in Morro da Igreja, Parque Nacional de São Joaquim, in Urubici, Santa Catarina, Brazil (Heiden and Pirani 2016). *B. friburgensis* G.Heiden & al. (Fig. 2.8b), belonging to *B.* subgen. *Oblongifoliae* (DC.) G.Heiden, is only known for a few inselbergs within the Atlantic Rainforest in a small area of Macaé de Cima District, in Nova Friburgo, Rio de Janeiro, Brazil (Heiden et al. 2009). *B. magnifica* G.Heiden & al. (Fig. 2.8c), belonging to *B.* subgen. *Heterothalamus* (Less.) G.Heiden, is a rare endemic from the escarpments of Pico da Bandeira massif, at Parque Nacional do Caparaó, in the border between Espírito Santo and Minas Gerais, Brazil (Heiden et al. 2014). *B. bifrons* Baker, belonging to *B.* subgen. *Molina* (Pers.) Heering (Fig. 2.8d), has a restricted area of occurrence and is only known for the seasonally dry tropical forests and coastal vegetation in the Cabo Frio peninsula in Rio de Janeiro, Brazil (Heiden et al. 2012).

Another example is *B. wagenitzii* (F.H.Hellw.) Joch.Müll. (Fig. 2.9a), the only species belonging to *B.* subgen. *Heterothalamulopsis* (Deble, A.S.Oliveira & Marchiori) G.Heiden that only occurs amid basaltic and sandstone rock outcrops or edges of cloud forests on the canyon cliffs of Parque Nacional Aparados da Serra and Serra Geral, along the borders of Rio Grande do Sul and Santa Catarina, Brazil (Hellwig 2003). *B. pluricapitulata* (Deble) G.Heiden (Fig. 2.9b), belonging to *B.* subgen. *Coridifoliae* (DC.) G.Heiden, is known only for few localities in swamps and peat bogs in the hilltops of Serra do Sudeste, amid the pampas temperate grasslands, in the municipalities of Pedras Altas and Pinheiro Machado, Rio Grande do Sul, Brazil (Heiden 2013). *B. scopulorum* A.A.Schneid. & G.Heiden (Fig. 2.9c), belonging to the first early divergent lineage from the remaining *B.* subgen. *Baccharis*, is restricted to a single locality growing on sandstone shaded cliffs in Serra do Corvo Branco, the boundary between Grão Pará and Urubici municipalities at Santa Catarina, Brazil (Schneider et al. 2011), while *B. obdeltata* G.Heiden (Fig. 2.9d), belonging to the core of *B.* subgen. *Baccharis*, is endemic to the mountaintop quartzite tropical grasslands (*campos rupestres*) on the slopes of Pico do Breu, a massif in the Espinhaço Range at Santana do Riacho, Minas Gerais, Brazil (Heiden and Pirani 2014). *Baccharis* has plenty of other examples of narrow endemic species, and mapping their occupied area and studying their species population demography and ecology would be worthwhile for conservation measures to be proposed and put into action.

4 What Do We Know (or Not) About the Role of Hybridization on *Baccharis* Diversity?

The role of hybridization on *Baccharis* diversity, distribution, evolution, and ecology remains a neglected subject deserving more attention. Several authors were puzzled by complexes of similar species and morphologically intermediate specimens between not morphologically closely related taxa. Natural interspecific hybrids are scarcely properly documented in the literature and herbaria, although it is commonly advocated as a widespread phenomenon within the genus. Alboff and Kurtz (1896) and Spegazzini (1896), as cited by Hellwig (1990), were among the first ones to highlight the hybridization of *B. magellanica* (Lam.) Pers. and *B. patagonica* Hook. & Arn., on studies of the Tierra del Fuego flora.

Malagarriga (1949, under his priest name Irmão Teodoro Luis) was the first to formally recognize a putative hybrid (*B.* × *paulopolitana* L.Teodoro & W.Hoehne). Later, he described other specimens as additional natural hybrids (*B.* × *fraudulenta* L.Teodoro, *B.* × *heeringiana* L.Teodoro, *B.* × *hoehneana* L.Teodoro, *B.* × *wilsoniana* L.Teodoro) suggesting the likely parental taxa (Malagarriga 1954, also as Irmão Teodoro Luis). This author defended the legitimacy of naming and describing hybrid taxonomical entities and the indication of putative parental species towards the recognition of the role of hybridization for the understanding of hybrid complexes represented by the segregation of an offspring between not closely related species. Malagarriga (1954) also briefly discussed the role of dioecy in hybridization and introgression and their consequences, for example, the hybrids' occupancy of intermediate ecological niches, not suitable for the parental taxa.

Hellwig (1990) described 18 hybrid species and 15 hybrid subspecies and hypothesized the hybrid origin of 3 taxa (*B.* × *concava* (Ruiz & Pav.) Pers., *B.* × *intermedia* DC., and *B.* × *volckmannii* Phil.) published by former authors, discussing their likely origins, geography, and ecology in Chile, intending to draw attention from future collectors and deeper research on this subject. He highlighted the fact that *B. intermedia* behaves as a complex of hybridization, with the parental *B. macraei* Hook. & Arn. occurring along sandy soils by the coast, while *B. linearis* (Ruiz & Pav.) Pers., the other parent species, is widespread but restricted to inland, never found close to the sea. The gap of ecological distribution, as well as the morphological spectrum between the two parental species, is filled by *B.* × *intermedia*. Besides the case of *B. intermedia*, whose individuals can be easily spotted as hybrids, Hellwig (1990) also explained other cases as the hybrids between *B. obovata* and *B. magellanica*, yielding large and easily recognizable swarms of intermediate forms. Meanwhile, on the opposite side, he also gave the example of other hybrids more difficult to recognize, as when hybridization takes part between morphologically similar parent species as the cases of crosses of *B. neaei* versus *B. linearis*, whose offspring is hard to tell apart from the parent species.

The case of *B.* × *intermedia* was later investigated by Faini et al. (1991), who examined a zone of sympatric occurrence between *B. linearis* and *B. macraei*, to determine the hybrid nature of *B.* × *intermedia* using morphological and chemical data. The authors found that hybrids show intermediate branching pattern, all intermediate shapes of leaves, and intermediate peduncle size and capitula arrangement. For the chemical profile, it was found that while *B. linearis* has five species-specific components, predominantly aromatic derivatives, *B. macraei* metabolizes mainly furanoclerodane diterpenes and, apart from the triterpenoids of common occurrence in the genus, the chemical composition of *B.* × *intermedia* is a clear combination of the metabolites present in the two parent species. The main conclusion of the work was that these abovementioned taxa form part of a homogamic complex which comprises almost the whole genus in Chile with the species adapted to conditions which characterize their habitats and connected by hybrids that inhabit transitional zones between the ranges of the parental species. They highlight that in some cases, alteration of the natural vegetation by human impact resulted in the formation of degraded vegetation types where contact zones create conditions for the establishment of hybrids and backcross progeny which are favorable.

Zanowiak (1991) investigated, by the mean of cpDNA, nrDNA, and pollen fertility data, a region of overlapping distribution for *B. halimifolia* L. and *B. neglecta* Britton, where a continuum of leaf morphology from broader to narrower, characterizing the first and the last species respectively, occurs. The results confirmed that hybridization does indeed exist between the two species and that hybrids retain a level of fertility sufficient for introgression to occur. By the way, according to Sundberg and Bogler (2006), not formally named natural hybrids between *B. halimifolia* and *B. angustifolia* Michx. or *B. neglecta* have been recorded in Arkansas, Louisiana, and East Texas, while hybrids between *B. halimifolia* and *B. angustifolia* are also known in Florida.

Meanwhile, many doubts remain concerning the patterns and processes of hybridization in *Baccharis*; this mechanism is being successfully explored by the applied sciences since a while, and it gives a lot of clues of what could be happening under natural circumstances. Artificial hybrids, between the allopatric *B. pilularis* (coyote bush), from California and Oregon, and *B. sarothroides* (desert broom), from Arizona, California, New Mexico, and Texas in the USA and Baja California, Baja California Sur, Sinaloa, and Sonora in Mexico, were successfully obtained by Thompson et al. (1976). The aim of the author was to apply the desirable hybrid forms as an ornamental shrub for desert landscaping combining desirable features of arid-land adaptability and drought tolerance of the desert broom with the leafiness and compact, procumbent growth habit of the coyote bush. A total of 301 hybrid plants within the progeny of this cross were evaluated by Lee et al. (1984) over a period of 6 years, and individual plant selections were successfully propagated asexually by means of rooted cuttings. Thompson et al. (1995) later described in detail the whole process of obtaining these hybrids, and the reading is worthwhile as the most thorough case study of hybridization and backcrossing in *Baccharis*. The hybrid plants started to flower during the second year of growth. Morphologically, the size and shape of the leaves of the fully grown hybrid plants were intermediate between the two parental types at the same stage of maturity. The pistillate capitula of the hybrid plants were identical to those of the desert broom maternal parent, but they were variable in size and reduced in length and diameter, and the bristles of the pappus were noticeably shorter. In general, hybrid plants produced fewer capitula and smaller and lighter seeds. The high degree of variation in height, width, width/height ratio, leaf shapes, branching habit, and number of florets per plant suggests a high level of genetic heterogeneity within the hybrid population. Although the hybrid plants have not been subjected to rigorous testing, according to Thompson (1995), they appeared to have a high level of tolerance to various abiotic stresses, especially drought and low humidity. The plants survived summer heat of 45 °C without wilting and winter cold as low as −10 °C. After evaluations (Lee et al. 1984), one plant was selected as having the best combination of desirable horticultural characteristics and was released as a new cultivar called "Centennial" to commemorate the 100th year of the University of Arizona. All the interspecific hybrid plants were pistillate, and the reason for this type of sex expression has not been determined by Thompson (1985), but he proposed that maternal cytoplasmic factor could be operating. Because excessive production of pappus on female plants can be a nuisance and fire hazard, in an area naturally prone for this phenomena, male or staminate flowering would be preferred, and later Thompson (1995) selected hybrid plants that have been backcrossed to the parental *B. sarothroides*, and seedlings of the resulting progenies were evaluated and utilized to clarify the mechanism and inheritance of sex expression with the aim to obtain superior selections from these progenies to combine drought tolerance, procumbent and compact growth habit, and staminate sex expression. The second-generation seedling plants displayed a wide range of leaf types. As the plants matured, most tended to exhibit either the oblong, serrated leaf shape of "Centennial" or the smooth, linear shape of the desert broom parent. Backcrossed hybrids revealed the presence of several staminate plants. Because some genetic segregation for sex expression was evident in the second-generation hybrid seedlings, propagation focused on an all-staminate hybrid that would eliminate the dispersion of unwanted seedlings, although potentially creating backcross hybrid progeny via pollen dispersal. Thompson (1995) stated that the most positive aspect would be eliminating dispersal of unwanted pappus, and after growing the selections in field plots for 3 years, a staminate, second-generation hybrid plant, which had the desirable degree of procumbency and vigor, was chosen and propagated resulting in a plant which has a leaf shape intermediate between "Centennial" and the male backcross parent. This new cultivar was named "Thompson" and represents a second-generation interspecific hybrid.

Although it was not the aim of the plant breeders, the development of the two cultivars demonstrated the feasibility of hybridization and backcrossing in *Baccharis*, resulting in a wide range of intermediate forms and factual evidence of introgression. Conversely, if it is proven possible for different *Baccharis* species to hybridize, it is still not understood what are the genetic and ecological barriers that keep species boundaries working at a high degree in nature, and research on this field would raise lots of interesting questions and answers on the evolution and origins of diversity in the genus.

5 *Baccharis* Diversity and Distribution: Who They Are and Where to Find Them?

The complete list of subgenera, sections, and species distributions across the 35 American sovereign countries and the 24 dependent territories from the Americas along with the introduced potentially naturalized occurrences in Europe and Oceania is presented in the following section.

I. ***Baccharis*** subgen. ***Tarchonanthoides*** Heering: Two sections and thirteen species. The group has diversified mainly in open vegetation of southeastern South America, occurring in Brazil, Paraguay, Uruguay, and Argentina. The main center of species richness is found in the southern Brazilian highlands along the states of Paraná, Santa Catarina, and Rio Grande do Sul. Nonetheless, endemic taxa also occur in the highland rocky grasslands (*campos rupestres*) of Minas Gerais, in southeastern Brazil, and in the lowland temperate grasslands from the pampas of Argentina, Uruguay, and southern Brazil (Heiden et al. 2019).

I.I. ***Baccharis*** sect. ***Tarchonanthoides*** (Heering) Cuatrec.: Six species occurring in savannas and grasslands from southeastern Brazil to southern Uruguay.

1. ***B. chionolaenoides*** D.B.Falkenb. & Deble. **Brazil** (Santa Catarina).
2. ***B. curitybensis*** Heering ex Malme. **Brazil** (Minas Gerais, Paraná, Rio de Janeiro, Rio Grande do Sul, Santa Catarina, São Paulo).
3. ***B. lychnophora*** Gardner. **Brazil** (Minas Gerais).
4. ***B. nebularis*** G.Heiden. **Brazil** (Paraná, Santa Catarina).
5. ***B. patens*** Baker. **Brazil** (Rio Grande do Sul). **Uruguay** (Canelones, Lavalleja, Maldonado, Montevideo, Rocha, San José).
6. ***B. tarchonanthoides*** DC. **Brazil** (Bahia, Minas Gerais, Paraná, Rio de Janeiro, Santa Catarina, São Paulo).

I.II. ***Baccharis*** sect. ***Canescentes*** Giuliano: Seven species occurring in a wide array of ecosystems in southeastern South America, from eastern Paraguay and southeastern Brazil south to the province of Buenos Aires in Argentina.

7. ***B. gibertii*** Baker. **Brazil** (Rio Grande do Sul). **Uruguay** (Canelones, Cerro Largo, Maldonado, Montevideo, Rocha, San José).
8. ***B. gnaphalioides*** Spreng. **Argentina** (Buenos Aires). **Brazil** (Rio Grande do Sul, Santa Catarina). **Uruguay** (Canelones, Maldonado, Montevideo, Rivera, Rocha, Tacuarembó, San José).
9. ***B. helichrysoides*** DC. **Argentina** (Misiones). **Brazil** (Mato Grosso do Sul, Minas Gerais, Paraná, Rio de Janeiro, Rio Grande do Sul, Santa Catarina, São Paulo). **Paraguay** (Alto Paraná, Amambay, Caaguazú, Caazapá, Guairá).
10. ***B. leucocephala*** Dusén. **Brazil** (Paraná, Rio Grande do Sul, Santa Catarina).
11. ***B. leucopappa*** DC. **Brazil** (Rio Grande do Sul, Santa Catarina).
12. ***B. phylicifolia*** DC. **Brazil** (Minas Gerais, Rio de Janeiro, Rio Grande do Sul, Paraná, Santa Catarina, São Paulo).

13. ***B. uleana*** Malag. **Brazil** (Rio Grande do Sul, Santa Catarina, São Paulo).

II. ***Baccharis*** subgen. ***Oblongifoliae*** (DC.) G.Heiden: 18 species mostly in the summits of the main mountainous ranges of South America, the highest diversity found in southeastern Brazil. Species such as *B. oblongifolia* (Ruiz & Pav.) Pers. and *B. rufidula* (Spreng.) Joch.Müll. occur in submontane forests. *Baccharis* sect. *Oblongifoliae* is the only known section to this subgenus (Heiden et al. 2019).

II.I. ***Baccharis*** sect. ***Oblongifoliae*** DC.: Same composition and distribution of the subgenus.

14. ***B. alpestris*** Gardner. **Brazil** (Rio de Janeiro).
15. ***B. antioquensis*** Killip & Cuatrec. **Colombia** (Antioquia, Valle de Cauca).
16. ***B. ciliata*** Gardner. **Brazil** (Rio de Janeiro).
17. ***B. coronata*** Giuliano. **Brazil** (Rio Grande do Sul, Santa Catarina).
18. ***B. crassipappa*** Deble & A.S.Oliveira. **Brazil** (Minas Gerais).
19. ***B. cutervensis*** Hieron. **Ecuador** (Azuay). **Peru** (Amazonas, Cajamarca).
20. ***B. densa*** (N.E.Br.) V.M.Badillo. **Guyana**. **Brazil** (Roraima). **Venezuela** (Amazonas, Bolívar).
21. ***B. dichotoma*** G.Heiden & L.D.Meireles. **Brazil** (Minas Gerais, São Paulo).
22. ***B. friburgensis*** G.Heiden & al. **Brazil** (Rio de Janeiro).
23. ***B. grandimucronata*** L.Teodoro & J.Vidal. **Brazil** (Espírito Santo, Minas Gerais, Paraná, Rio de Janeiro, Santa Catarina, São Paulo).
24. ***B. ligustrina*** DC. **Brazil** (Bahia, Distrito Federal, Goiás, Minas Gerais, Paraná, São Paulo).
25. ***B. macrophylla*** Dusén. **Brazil** (Minas Gerais, Rio de Janeiro, São Paulo).
26. ***B. meridensis*** Steyerm. **Venezuela** (Mérida).
27. ***B. myricifolia*** DC. **Brazil** (Distrito Federal, Goiás, Minas Gerais, Paraná, Santa Catarina, São Paulo).
28. ***B. oblongifolia*** (Ruiz & Pav.) Pers. **Argentina** (Misiones). **Bolivia** (Cochabamba, La Paz, Santa Cruz). **Brazil** (Amazonas, Bahia, Espírito Santo, Minas Gerais, Paraná, Rio de Janeiro, Rio Grande do Sul, Roraima, Santa Catarina, São Paulo). **Colombia** (Antioquia, Boyacá, Caquetá, Cauca, Cesar, Chocó, Cundinamarca, La Guajira, Huila, Magdalena, Meta, Nariño, Norte de Santander, Putumayo, Santander, Valle de Cauca). **Ecuador** (Azuay, Carchi, Loja, Morona-Santiago, Napo, Pichincha, Sucumbios, Tungurahua, Zamora-Chinchipe). **Peru** (Amazonas, Cajamarca, Cuzco, Junín, Huánuco, Loreto, Pasco, San Martín, Ucayali). **Venezuela** (Amazonas, Bolívar, Lara, Mérida, Táchira, Trujillo, Zulia).
29. ***B. rufidula*** (Spreng.) Joch.Müll. **Brazil** (Espírito Santo, Minas Gerais, Rio de Janeiro).
30. ***B. vismioides*** DC. **Brazil** (Minas Gerais, Paraná, São Paulo).
31. ***B. vitis-idaea*** Oliver. ex Thurn. **Brazil** (Amazonas, Roraima). **Guyana**. **Venezuela** (Amazonas, Bolívar).

III. ***Baccharis*** subgen. ***Heterothalamus*** (Less.) G.Heiden: Three sections and seven species. The three lineages are found in distinct regions, in Serra do Caparaó, southeastern Brazil, along the Andes from northern Argentina to Peru, and in pampean mountains and hilly ranges of Argentina, southern Brazil, and Uruguay (Heiden et al. 2019).

III.I. ***Baccharis*** sect. ***Bradeanae*** G.Heiden: Two species, both endemic to Serra do Caparaó in southeastern Brazil.

32. ***B. dubia*** Deble & A.S.Oliveira. **Brazil** (Espírito Santo, Minas Gerais).
33. ***B. magnifica*** G.Heiden & al. **Brazil** (Espírito Santo, Minas Gerais).

III.II. ***Baccharis*** sect. ***Psila*** (Phil.) Cuatrec.: Two species restricted to salt plains, marshes, and streams in high-elevation areas along the Andes.

34. ***B. acaulis*** (Wedd. ex R.E.Fr.) Cabrera. **Argentina** (Catamarca, Jujuy, Salta). **Bolivia** (Cochabamba, La Paz, Oruro, Potosí, Tarija). **Chile** (Arica y Parinacota, Tarapacá). **Peru** (Cusco, Puno).
35. ***B. davidsonii*** Cuatrec. **Peru** (Junín).

III.III. ***Baccharis*** sect. ***Heterothalamus*** (Less.) Giuliano: Three species occurring in central Argentina, southern Brazil, and Uruguay.

36. ***B. aliena*** (Spreng.) Joch.Müll. **Argentina** (Córdoba, La Rioja, San Luis). **Brazil** (Rio Grande do Sul, Santa Catarina). **Uruguay** (Canelones, Cerro Largo, Durazno, Florida, Lavalleja, Maldonado, Montevideo, Río Negro, Rivera, Rocha, Tacuarembó, Treinta y Tres).
37. ***B. hyemalis*** Deble. **Brazil** (Rio Grande do Sul).
38. ***B. psiadioides*** (Less.) Joch.Müll. **Brazil** (Rio Grande do Sul, Santa Catarina). **Uruguay** (Cerro Largo, Rivera, Tacuarembó, Treinta y Tres).

IV. ***Baccharis*** subgen. ***Molina*** (Pers.) Heering: 21 sections and 150 species, from southwestern USA to southern South America. Mexico, the Andes, and southeastern Brazil are the main centers of species richness. 27 species are still placed in 8 sections with unknown relationships, and 18 species still lack a hypothesis of relationships with any of the remaining sections and species of the subgenus (Heiden et al. 2019).

IV.I. ***Baccharis*** sect. ***Palenia*** Giuliano: *Baccharis nivalis* is the only species placed in the section, occurring in Argentinean and Chilean Patagonia.

39. ***B. nivalis*** (Wedd.) Sch.Bip. ex Phil. **Argentina** (Chubut, Neuquén, Río Negro, Santa Cruz, Tierra del Fuego). **Chile** (Aysén, Araucanía, Los Lagos, Los Ríos, Magallanes y Antártica, Maule).

IV.II. ***Baccharis*** sect. ***Subliguliflorae*** Giuliano: Seven species occurring mostly in the monte dry thorn scrub and prepuna and puna mountain vegetations from northwestern Argentina and Bolivia.

40. ***B. beckii*** Joch.Müll. **Bolivia** (Chuquisaca, Potosí, Tarija).

41. ***B. cabrerae*** Ariza. **Argentina** (Catamarca, Salta, Tucumán).
42. ***B. niederleinii*** Heering. **Argentina** (La Rioja, San Juan).
43. ***B. potosiensis*** H.Rob. **Bolivia** (Chuquisaca, Potosí).
44. ***B. potrerillana*** (Ariza) G.Heiden. **Argentina** (La Rioja, San Juan).
45. ***B. torricoi*** Joch.Müll. **Bolivia** (Cochabamba, Potosí, Tarija).
46. ***B. woodii*** Joch.Müll. **Bolivia** (Potosí).

IV.III. ***Baccharis*** sect. ***Thymifoliae*** Giuliano: Two species from the prepuna and puna vegetations from northwestern Argentina and Bolivia.

47. ***B. grisebachii*** Hieron. **Argentina** (Catamarca, Jujuy, La Rioja, Mendoza, Salta, San Juan, Tucumán). **Bolivia** (Potosí, Tarija).
48. ***B. thymifolia*** Hook. & Arn. **Argentina** (Mendoza).

IV.IV. ***Baccharis*** sect. ***Trinervatae*** DC.: 13 species, most of them are widespread and distributed in forest edges of tropical Central and South America. *Baccharis pedunculata* (Mill.) Cabrera and *B. trinervis* Pers. are widespread species, while *B. steetzii* Andersson is endemic to the Galápagos, and *B. acutata* (Alain) Borhidi, *B. nipensis* Urb., and *B. orientalis* Alain are endemic to Cuba. *Baccharis retamoides* Phil. is a desert subshrub from central Argentina.

49. ***B. acutata*** (Alain) Borhidi. **Cuba**.
50. ***B. bifrons*** Baker. **Brazil** (Rio de Janeiro).
51. ***B. cinerea*** DC. **Brazil** (Alagoas, Bahia, Ceará, Distrito Federal, Espírito Santo, Minas Gerais, Paraíba, Pernambuco, Rio de Janeiro, Rio Grande do Norte, Sergipe, São Paulo).
52. ***B. debilis*** Rusby. **Argentina** (Chaco, Corrientes, Formosa, Jujuy, Misiones, Salta). **Bolivia** (Beni, Chuquisaca, Cochabamba, La Paz, Pando, Santa Cruz, Tarija). **Paraguay** (Concepción). **Peru** (Cajamarca, Cusco, Huánuco, San Martín). **Venezuela** (Aragua, Mérida).
53. ***B. nervosa*** DC. **Guadeloupe**. **Martinique**. **Trinidad and Tobago**.
54. ***B. nipensis*** Urb. **Cuba**.
55. ***B. orientalis*** Alain. **Cuba**.
56. ***B. pedunculata*** (Mill.) Cabrera. **Bolivia** (Beni, Cochabamba, La Paz, Santa Cruz). **Colombia** (Antioquia, Boyacá, Caldas, Cauca, Chocó, Cundinamarca, Huila, Nariño, Norte de Santander, Putumayo, Quindío, Santander, Tolima, Valle de Cauca). **Costa Rica**. **Dominica**. **Ecuador** (Azuay, Cañar, Chimborazo, Cotopaxi, Imbabura, Morona-Santiago, Napo, Pastaza, Pichincha, Sucumbíos, Tungurahua, Zamora-Chinchipe). **El Salvador**. **Guatemala**. **Honduras**. **Martinique**. **Mexico** (Chiapas). **Montserrat**. **Nicaragua**. **Panama**. **Peru** (Cajamarca, Cusco, Huánuco, Junín, Madre de Dios, Pasco, Piura, San Martín). **Saba**. **Saint Kitts and Nevis**. **Saint Lucia**. **Saint Vincent and the Grenadines**. **Venezuela** (Distrito Federal, Mérida, Miranda, Portuguesa, Táchira, Trujillo).
57. ***B. quitensis*** Kunth. **Argentina** (Catamarca, Jujuy, La Rioja, Salta, Tucumán). **Bolivia** (Chuquisaca, Cochabamba, La Paz, Santa Cruz, Tarija). **Brazil** (Mato Grosso, Mato Grosso do Sul, Minas Gerais, Rondônia, São Paulo).

Ecuador (Guayas, Loja, Pichincha). **Peru** (Amazonas, Apurimac, Cajamarca, La Libertad, Loreto, Piura, Puno).

58. ***B. retamoides*** Phil. **Argentina** (Catamarca, La Rioja, Mendoza, Salta, San Juan, Tucumán).
59. ***B. spartea*** Benth. **Peru** (Ancash, Ayacucho, Cajamarca, La Libertad, Lima).
60. ***B. steetzii*** Andersson. **Ecuador** (Galápagos).
61. ***B. trinervis*** Pers. **Argentina** (Chaco, Corrientes, Formosa, Jujuy, Misiones, Salta, Santiago del Estero). **Belize**. **Bolivia** (Beni, Chuquisaca, Cochabamba, La Paz, Pando, Santa Cruz, Tarija). **Brazil** (Acre, Alagoas, Bahia, Distrito Federal, Espírito Santo, Minas Gerais, Pará, Paraná, Pernambuco, Rio de Janeiro, Rio Grande do Sul, Rondônia, Roraima, Santa Catarina, São Paulo). **Colombia** (Antioquia, Atlántico, Bolívar, Boyacá, Caldas, Caquetá, Cauca, Cesar, Chocó, Cundinamarca, La Guajira, Guaviare, Huila, Magdalena, Meta, Nariño, Norte de Santander, Putumayo, Quindío, Risaralda, Santander, Sucre, Tolima, Valle de Cauca). **Costa Rica**. **Ecuador** (Azuay, Bolívar, Carchi, Chimborazo, Cotopaxi, El Oro, Esmeraldas, Guayas, Imbabura, Loja, Los Ríos, Manabí, Morona-Santiago, Napo, Orellana, Pastaza, Pichincha, Sucumbíos, Tungurahua, Zamora-Chinchipe). **El Salvador**. **Guatemala**. **Guyana**. **Honduras**. **Mexico** (Campeche, Chiapas, Colima, Estado de México, Guerrero, Hidalgo, Jalisco, Michoacán, Morelos, Nayarit, Oaxaca, Puebla, Querétaro, Quintana Roo, San Luis Potosí, Sinaloa, Tabasco, Tamaulipas, Veracruz, Yucatán). **Nicaragua**. **Panama**. **Paraguay** (Alto Paraguay, Amambay, Boquerón, Canindeyú, Central, Concepción, Cordillera, Guairá, Paraguarí, Presidente Hayes, San Pedro). **Peru** (Amazonas, Cajamarca, Cuzco, Huánuco, Junín, Lambayeque, La Libertad, Loreto, Madre de Dios, Pasco, Piura, Puno, San Martín, Tumbes, Ucayali). **Trinidad and Tobago**. **Venezuela** (Amazonas, Anzoátegui, Aragua, Barinas, Bolívar, Carabobo, Distrito Federal, Lara, Mérida, Miranda, Monagas, Nueva Esparta, Portuguesa, Sucre, Táchira, Trujillo, Zulia).

IV.V. ***Baccharis*** sect. ***Australes*** Giuliano: *Baccharis racemosa* is the only species and occurs in Patagonia.

62. ***B. racemosa*** (Ruiz & Pav.) DC. **Argentina** (Chubut, Neuquén, Río Negro). **Chile** (Araucanía, Aysén, Bío Bío, Coquimbo, Los Lagos, Los Ríos, Maule, Ñuble, O'Higgins, Valparaíso).

IV.VI. ***Baccharis*** sect. ***Bogotenses*** Cuatrec.: 18 species occurring from Argentina and Uruguay to North America, where it is slightly more diverse.

63. ***B. bogotensis*** Kunth. **Colombia** (Boyacá, Cesar, Cundinamarca, Meta, Norte de Santander, Santander).
64. ***B. brachyphylla*** A.Gray. **Mexico** (Baja California, Baja California Sur, Chihuahua, Sonora). **USA** (Arizona, California, Nevada, New Mexico, Texas).
65. ***B. charucoensis*** G.L.Nesom. **Mexico** (Michoacán).
66. ***B. erosoricola*** Rzed. **Mexico** (Estado de México, Hidalgo, Quintana Roo).
67. ***B. glabrata*** (Hoover) G.Heiden. **USA** (California).

68. ***B. gracilis*** DC. **Brazil** (Distrito Federal, Goiás, Minas Gerais, Paraná, Rio de Janeiro, São Paulo). **Paraguay** (Caazapá, Guairá).
69. ***B. macrocephala*** Sch.Bip. ex Greenm. **Mexico** (Estado de México, Hidalgo, Morelos, Puebla, Veracruz).
70. ***B. maxima*** Baker. **Brazil** (Minas Gerais, Rio de Janeiro, São Paulo).
71. ***B. mutisiana*** Cuatrec. **Colombia** (Boyacá, Cundinamarca, Meta, Norte de Santander, Santander). **Venezuela** (Mérida, Táchira, Trujillo).
72. ***B. occidentalis*** S.F.Blake. **Mexico** (Guanajuato, Guerrero, Jalisco, Nayarit, Sinaloa, San Luis Potosí, Zacatecas).
73. ***B. plummerae*** A.Gray. **USA** (California).
74. ***B. pteronioides*** DC. **Mexico** (Aguascalientes, Chihuahua, Coahuila, Colima, Durango, Estado de México, Guanajuato, Guerrero, Hidalgo, Jalisco, Michoacán, Morelos, Nayarit, Nuevo León, Oaxaca, Puebla, Querétaro, San Luis Potosí, Sinaloa, Sonora, Tamaulipas, Tlaxcala, Veracruz, Zacatecas). **USA** (Arizona, New Mexico, Texas).
75. ***B. ramiflora*** A.Gray. **Mexico** (Guanajuato, Guerrero, Hidalgo, Querétaro, San Luis Potosí, Veracruz, Yucatán).
76. ***B. saliens*** Rusby. **Bolivia** (Cochabamba, La Paz, Santa Cruz). **Peru** (Cusco).
77. ***B. serranoi*** H.Rob. **Bolivia** (Chuquisaca, Cochabamba, La Paz, Santa Cruz).
78. ***B. solomonii*** H.Rob. **Bolivia** (La Paz).
79. ***B. texana*** (Torr. & A.Gray) A.Gray. **Mexico** (Chihuahua, Coahuila, Durango, Nuevo León, Tamaulipas). **USA** (New Mexico, Oklahoma, Texas).
80. ***B. wrightii*** A.Gray. **Mexico** (Chihuahua, Durango, Sonora). **USA** (Arizona, Colorado, Kansas, New Mexico, Oklahoma, Texas, Utah).

IV.VII. ***Baccharis*** sect. ***Paniculatae*** Heering: Two species, *B. paniculata*, from Chile, and *B. effusa*, from northwestern Argentina and southwestern Bolivia.

81. ***B. effusa*** Griseb. **Argentina** (Catamarca, Jujuy, Salta, Tucumán).
82. ***B. paniculata*** DC. **Chile** (Atacama, Bío Bío, Coquimbo, Maule, Metropolitana, Ñuble, O'Higgins, Valparaíso).

IV.VIII. ***Baccharis*** sect. ***Arenariae*** Giuliano: *Baccharis arenaria* is the only species and occurs in riverine and lake shore vegetation in northeastern Argentina, southern Brazil, and Uruguay.

83. ***B. arenaria*** Baker. **Argentina** (Corrientes, Entre Ríos). **Brazil** (Rio Grande do Sul). **Uruguay** (Canelones, Cerro Largo, Colonia, Florida, Lavalleja, Maldonado, Montevideo, Rocha, San José, Tacuarembó).

IV.IX. ***Baccharis*** sect. ***Polifoliae*** G.Heiden: *Baccharis polifolia* is the only species and occurs in the Andes from northwestern Argentina to southwestern Bolivia.

84. ***B. polifolia*** Griseb. **Argentina** (Catamarca, Jujuy, La Rioja, Mendoza, Salta, San Juan, Tucumán). **Bolivia** (Tarija).

IV.X. ***Baccharis*** sect. ***Pseudobaccharis*** (Cabrera) Cuatrec.: Three species, occurring in the puna, prepuna, and altoandina vegetations in northwestern Argentina and Bolivia.

85. ***B. boliviensis*** (Wedd.) Cabrera. **Argentina** (Catamarca, Jujuy, Salta, Tucumán). **Bolivia** (Cochabamba, La Paz, Oruro, Potosí, Tarija). **Chile** (Arica y Parinacota, Antofagasta, Tarapacá). **Peru** (Arequipa, Cusco, La Libertad, Puno, Tacna).
86. ***B. spartioides*** (Hook. & Arn. ex DC.) Remy. **Argentina** (Buenos Aires, Catamarca, Chubut, Córdoba, Jujuy, La Pampa, La Rioja, Mendoza, Neuquén, Río Negro, Salta, Santa Cruz, San Juan, San Luis, Tucumán). **Chile** (Atacama).

IV.XI. ***Baccharis*** sect. ***Angustifoliae*** Baker: Seven species from the northern Andes (*B. arguta*) to Patagonia, where most of the species are found, with one species (*B. orbignyana* Klatt) in the eastern Bolivian and central Brazilian tropical savannas (cerrado).

87. ***B. arguta*** Gillies ex Hook. & Arn. **Argentina** (Catamarca, Córdoba, Jujuy, La Rioja, Mendoza, Salta, San Juan, San Luis, Tucumán). **Bolivia** (Chuquisaca, Cochabamba, La Paz, Santa Cruz). **Peru** (Apurímac, Ayacucho, Cusco, Puno).
88. ***B. darwinii*** Hook. & Arn. **Argentina** (Buenos Aires, Catamarca, Chaco, Chubut, Córdoba, Corrientes, Entre Ríos, Formosa, La Pampa, La Rioja, Mendoza, Neuquén, Río Negro, Salta, Santa Cruz, Tucuman). **Bolivia** (Cochabamba, La Paz, Potosí). **Uruguay** (Río Negro).
89. ***B. gilliesii*** A.Gray. **Argentina** (Buenos Aires, Catamarca, Chaco, Chubut, Córdoba, Corrientes, Entre Ríos, Jujuy, La Pampa, La Rioja, Mendoza, Río Negro).
90. ***B. melanopotamica*** Speg. **Argentina** (Buenos Aires, Chubut, Córdoba, La Pampa, Río Negro, San Luis).
91. ***B. orbignyana*** Klatt. **Bolivia** (Santa Cruz). **Brazil** (Bahia, Distrito Federal, Goiás, Minas Gerais).
92. ***B. petrophila*** R.E.Fr. **Argentina** (Jujuy, Salta).
93. ***B. ulicina*** Hook. & Arn. **Argentina** (Buenos Aires, Catamarca, Chaco, Córdoba, Corrientes, Entre Ríos, Jujuy, La Pampa, La Rioja, Río Negro, Salta, Santa Cruz, Santiago del Estero, Santa Fe, San Luis, Tucumán). **Bolivia** (Chuquisaca, Cochabamba, Potosí, Santa Cruz, Tarija).

IV.XII. ***Baccharis*** sect. ***Serrulatae*** Cuatrec.: Eight species from the páramos vegetation of the northern Andes, mainly Colombia and Ecuador.

94. ***B. arbutifolia*** Vahl. **Ecuador** (Azuay, Bolívar, Carchi, Chimborazo, Cotopaxi, Imbabura, Morona-Santiago, Napo, Pichincha, Tungurahua).
95. ***B. caldasiana*** Cuatrec. **Colombia** (Caldas, Risaralda, Tolima).
96. ***B. ledifolia*** Kunth. **Colombia** (Amazonas, Cajamarca).

97. ***B. padifolia*** Hieron. **Colombia** (Cauca, Magdalena, Nariño). **Ecuador** (Azuay, Carchi, Chimborazo, Cotopaxi, Imbabura, Napo, Pichincha, Tungurahua).
98. ***B. prunifolia*** Kunth. **Colombia** (Antioquia, Arauca, Boyacá, Caldas, Cauca, Cesar, Cundinamarca, Magdalena, Meta, Nariño, Norte de Santander, Santander, Valle de Cauca). **Venezuela** (Lara, Mérida, Táchira, Trujillo).
99. ***B. revoluta*** Kunth. **Colombia** (Boyacá, Caldas, Cundinamarca, Tolima, Santander, Valle de Cauca).
100. ***B. rupicola*** Kunth. **Colombia** (Arauca, Boyacá, Caldas, Cauca, Cundinamarca, Meta, Norte de Santander, Risaralda, Santander, Tolima, Valle de Cauca).
101. ***B. vacciniifolia*** Cuatrec. **Colombia** (Boyacá, Cundinamarca, Tolima, Valle de Cauca).

IV.XIII. ***Baccharis*** sect. ***Corymbosae*** Heering: 37 species occurring from the southwestern USA to central Argentina and Chile. Two centers of diversity: one along the eastern slope of the Andes and the other along the mountain ranges of southeastern Brazil.

102. ***B. alnifolia*** Meyen & Walp. **Chile** (Arica y Parinacota, Antofagasta, Tarapacá). **Peru** (Arequipa, La Libertad, Moquegua, Tacna).
103. ***B. anomala*** DC. **Argentina** (Entre Ríos, Misiones). **Brazil** (Minas Gerais, Paraná, Rio Grande do Sul, Santa Catarina, São Paulo). **Paraguay** (Alto Paraná, Caaguazú, Caazapá, Guairá). **Uruguay** (Artigas, Cerro Largo, Durazno, Maldonado, Río Negro, Rivera, Rocha, Tacuarembó, Treinta y Tres).
104. ***B. auriculigera*** Hieron. **Ecuador** (Chimborazo, Cañar, Loja). **Peru** (Amazonas, Cajamarca, Cusco, La Libertad, Lambayeque, Piura).
105. ***B. breviseta*** DC. **Argentina** (Buenos Aires, Chaco, Corrientes, Entre Ríos, Misiones). **Brazil** (Bahia, Distrito Federal, Mato Grosso, Mato Grosso do Sul, Minas Gerais, Paraná, Rio de Janeiro, Rio Grande do Sul, Santa Catarina, São Paulo). **Colombia** (Antioquia, Boyacá, Cundinamarca). **Paraguay** (Caazapá, Central, Concepción, Cordillera, Guairá, Itapúa, Paraguarí). **Uruguay** (Canelones, Cerro Largo, Colonia, Florida, Montevideo, Río Negro, Rocha, San José, Tacuarembó).
106. ***B. calliprinos*** Griseb. **Argentina** (Catamarca, La Rioja, Mendoza, Salta, San Juan, Tucumán).
107. ***B. cana*** Joch.Müll. **Bolivia** (Chuquisaca).
108. ***B. capitalensis*** Heering. **Argentina** (Jujuy, Salta, Tucumán). **Bolivia** (Santa Cruz, Tarija).
109. ***B. clavata*** (Joch.Müll.) G.Heiden. **Argentina** (Catamarca, Jujuy, La Rioja, Salta). **Bolivia** (Cochabamba, La Paz, Oruro, Potosí, Tarija).
110. ***B. conyzoides*** (Less.) DC. **Brazil** (Paraná, Rio Grande do Sul, Santa Catarina, São Paulo).

111. ***B. decussata*** (Klatt) Hieron. **Colombia** (Antioquia, Boyacá, Caldas, Cauca, Cundinamarca, Huila, Nariño, Norte de Santander, Putumayo, Santander, Tolima, Valle de Cauca). **Ecuador** (Cotopaxi, Loja, Morona-Santiago, Pichincha). **Peru** (Amazonas, Cajamarca, Huánuco). **Venezuela** (Mérida, Táchira, Trujillo).
112. ***B. douglasii*** DC. **Mexico** (Baja California). **USA** (California, Oregon).
113. ***B. densiflora*** Wedd. **Bolivia** (Chuquisaca, Cochabamba, La Paz).
114. ***B. famatinensis*** Ariza. **Argentina** (La Rioja).
115. ***B. floribundoides*** Cuatrec. **Colombia** (Valle de Cauca).
116. ***B. glutinosa*** Pers. **Argentina** (Buenos Aires, Catamarca, Chaco, Chubut, Córdoba, Corrientes, Distrito Federal, Entre Ríos, Formosa, Jujuy, La Pampa, La Rioja, Mendoza, Misiones, Neuquén, Río Negro, Salta, Santiago del Estero, Santa Fe, San Juan, San Luis, Tucuman). **Bolivia** (Chuquisaca, Cochabamba, La Paz, Santa Cruz, Tarija). **Brazil** (Bahia, Espírito Santo, Mato Grosso, Mato Grosso do Sul, Minas Gerais, Paraná, Pernambuco, Rio de Janeiro, Rio Grande do Sul, Santa Catarina, São Paulo). **Chile** (Bío Bío, Los Lagos, Los Ríos, Maule, Metropolitana, Ñuble, O'Higgins, Valparaíso). **Paraguay** (Alto Paraguay, Amambay, Caaguazú, Canindeyú, Central, Chaco, Concepción, Cordillera, Guairá, Misiones, Nueva Asunción, Paraguarí, Presidente Hayes). **Peru** (Cusco). **Uruguay** (Canelones, Cerro Largo, Colonia, Maldonado, Montevideo, Paysandú, Río Negro, Salto, San José, Soriano).
117. ***B. haitiensis*** Heering. **Dominican Republic**. **Haiti**.
118. ***B. imbricata*** L.Teodoro & J.Vidal. **Brazil** (Espírito Santo, Minas Gerais).
119. ***B. jelskii*** Hieron. **Bolivia** (Cochabamba, La Paz). **Colombia** (Antioquia, Boyacá, Cauca, Cundinamarca, Huila, Meta, Nariño, Tolima, Valle de Cauca). **Ecuador** (Azuay, Bolívar, Chimborazo, Loja, Napo, Pastaza, Pichincha, Tungurahua, Zamora-Chinchipe). **Peru** (Amazonas, Cajamarca, Pasco).
120. ***B. juncea*** (Lehm.) Desf. **Argentina** (Buenos Aires, Catamarca, Chubut, Córdoba, Entre Ríos, Jujuy, La Pampa, La Rioja, Mendoza, Neuquén, Río Negro, Salta, Santa Cruz, Santiago del Estero, Santa Fe, San Juan, San Luis, Tucumán). **Chile** (Arica y Parinacota, Antofagasta, Atacama, Coquimbo, Maule, Tarapacá). **Uruguay** (Montevideo).
121. ***B. latifolia*** (Ruiz & Pav.) Pers. **Argentina** (Catamarca, Jujuy, La Rioja, Salta, Tucumán). **Bolivia** (Chuquisaca, Cochabamba, La Paz, Santa Cruz, Tarija). **Colombia** (Antioquia, Bolívar, Boyacá, Caldas, Cauca, Chocó, Cundinamarca, La Guajira, Huila, Magdalena, Meta, Nariño, Norte de Santander, Putumayo, Quindío, Risaralda, Santander, Tolima, Valle de Cauca). **Ecuador** (Azuay, Bolívar, Cañar, Carchi, Chimborazo, Cotopaxi, Imbabura, Loja, Morona-Santiago, Napo, Orellana, Pastaza, Pichincha, Sucumbíos, Tungurahua, Zamora-Chinchipe). **Peru** (Amazonas, Ancash, Apurimac, Arequipa, Ayacucho, Cajamarca, Callao, Cuzco, Huánuco, Huancavelica, Junín, La Libertad, Lima, Moquegua, Pasco, Puno). **Venezuela** (Mérida, Táchira, Trujillo).

122. ***B. lewisii*** (H.Rob.) Joch.Müll. **Bolivia** (La Paz).
123. ***B. lilloi*** Heering. **Argentina** (Catamarca, Jujuy, Salta, Tucumán). **Bolivia** (Tarija).
124. ***B. monoica*** G.L.Nesom. **El Salvador**. **Guatemala**. **Honduras**. **Nicaragua**.
125. ***B. multibracteata*** (Joch.Müll.) G.Heiden. **Peru** (Apurimac, Cusco, Junín).
126. ***B. multiflosculosa*** Heering. **Argentina** (Catamarca, Salta Tucumán). **Bolivia** (Tarija).
127. ***B. oxyodonta*** DC. **Argentina** (Misiones). **Brazil** (Espírito Santo, Minas Gerais, Paraná, Rio de Janeiro, Rio Grande do Sul, Santa Catarina, São Paulo). **Paraguay** (Alto Paraná, Alto Paraguay, Amambay, Caazapá, Central, Guairá, Paraguarí). **Uruguay** (Canelones, Colonia, Florida, Maldonado, Montevideo).
128. ***B. pentlandii*** DC. **Bolivia** (Cochabamba, La Paz). **Peru** (Puno).
129. ***B. perulata*** Kuntze. **Argentina** (Catamarca, Jujuy, La Rioja, Salta, Tucumán).
130. ***B. pingraea*** DC. **Argentina** (Buenos Aires, Catamarca, Chaco, Chubut, Córdoba, Corrientes, Distrito Federal, Entre Ríos, Formosa, Jujuy, La Pampa, La Rioja, Mendoza, Misiones, Neuquén, Río Negro, Salta, San Juan, San Luis, Santa Fé, Santiago del Estero). **Chile** (Araucanía, Atacama, Bío Bío, Coquimbo, Maule, Metropolitana, Ñuble, O'Higgins, Tarapacá, Valparaíso). **Brazil** (Rio Grande do Sul). **Paraguay** (Misiones). **Uruguay** (Colonia, Lavalleja, Montevideo, Salto, San José, Soriano).
131. ***B. polygama*** Ariza. **Argentina** (Tucumán).
132. ***B. punctulata*** DC. **Argentina** (Buenos Aires, Catamarca, Chaco, Corrientes, Entre Ríos, Formosa, Jujuy, Misiones, Santa Fe, Tucumán, Salta). **Bolivia** (Chuquisaca, Santa Cruz). **Brazil** (Espírito Santo, Mato Grosso do Sul, Minas Gerais, Paraná, Rio de Janeiro, Rio Grande do Sul, Santa Catarina, São Paulo). **Paraguay** (Alto Paraná, Amambay, Caaguazú, Caazapá, Canindeyú, Central, Cordillera, Guairá, Paraguarí, San Pedro). **Uruguay** (Artigas, Cerro Largo, Colonia, Durazno, Florida, Maldonado, Montevideo, Paysandú, Salto, San José, Soriano, Tacuarembó, Treinta y Tres).
133. ***B. pycnantha*** Phil. **Chile** (Atacama, Tarapacá).
134. ***B. salicifolia*** (Ruiz & Pav.) Pers. **Argentina** (Buenos Aires, Catamarca, Chaco, Chubut, Córdoba, Corrientes, Distrito Federal, Entre Ríos, Formosa, Jujuy, La Pampa, La Rioja, Mendoza, Misiones, Neuquén, Río Negro, Salta, Santa Cruz, Santiago del Estero, Santa Fe, San Juan, San Luis, Tucumán). **Bolivia** (Beni, Chuquisaca, Cochabamba, La Paz, Oruro, Potosí, Santa Cruz, Tarija). **Brazil** (Acre, Mato Grosso, Minas Gerais). **Chile** (Antofagasta, Araucanía, Arica y Parinacota, Atacama, Bío Bío, Coquimbo, Los Lagos, Los Ríos, Maule, Metropolitana, Ñuble, O'Higgins, Tarapacá, Valparaíso). **Colombia** (Casanare, Cauca, Huila, Meta, Putumayo, Tolima). **Ecuador** (Chimborazo, Loja, Napo, Zamora-Chinchipe). **El Salvador**. **Guatemala**. **Honduras**. **Mexico** (Aguascalientes, Baja California, Baja California Sur, Chiapas, Chihuahua, Coahuila,

Colima, Durango, Estado de México, Guanajuato, Guerrero, Hidalgo, Jalisco, Michoacán, Morelos, Nayarit, Nuevo León, Oaxaca, Puebla, Querétaro, San Luis Potosí, Sinaloa, Sonora, Tamaulipas, Tlaxcala, Veracruz, Zacatecas). **Paraguay** (Alto Paraguay, Boquerón, Central, Nueva Asunción, Presidente Hayes). **Peru** (Amazonas, Ancash, Arequipa, Apurímac, Cajamarca, Callao, Cusco, Huánuco, Ica, Junín, La Libertad, Lambayeque, Lima, Loreto, Madre Dios, Pasco, San Martín, Tacna, Tumbes, Ucayali). **USA** (Arizona, California, Colorado, Nevada, New Mexico, Texas, Utah). **Uruguay** (Colonia). **Venezuela** (Portuguesa, Trujillo).

135. ***B. scandens*** (Ruiz & Pav.) Pers. **Chile** (Antofagasta, Arica y Parinacota, Atacama, Tarapacá). **Peru** (Arequipa, Cajamarca, Lambayeque, La Libertad, Lima, Tacna).
136. ***B. sculpta*** Griseb. **Argentina** (Catamarca, Jujuy, La Rioja, Salta, Tucumán). **Bolivia** (Chuquisaca, Cochabamba, Oruro, Potosí, Tarija).
137. ***B. serrulata*** (Lam.) Pers. **Brazil** (Alagoas, Bahia, Espírito Santo, Minas Gerais, Paraíba, Pernambuco, Rio de Janeiro, São Paulo).
138. ***B. sphaerocephala*** Hook. & Arn. **Chile** (Araucanía, Bío Bío, Los Lagos, Los Ríos).
139. ***B. stenophylla*** Ariza. **Argentina** (Buenos Aires, Catamarca, Córdoba, La Pampa, La Rioja, Santa Fe, San Luis). **Uruguay** (Colonia, Flores, Florida, Montevideo, Río Negro).
140. ***B. stylosa*** Gardner. **Brazil** (Minas Gerais, Rio de Janeiro, São Paulo).
141. ***B. vulneraria*** Baker. **Argentina** (Buenos Aires, Chaco, Corrientes, Entre Ríos, Misiones, Santa Fe). **Brazil** (Espírito Santo, Mato Grosso, Mato Grosso do Sul, Minas Gerais, Paraná, Rio de Janeiro, Rio Grande do Sul, Santa Catarina, São Paulo). **Paraguay** (Caaguazú, Central, Cordillera, Guairá, Paraguarí). **Uruguay** (Artigas, Canelones, Colonia, Florida, Maldonado, Montevideo, Río Negro, Rivera, Rocha, Salto, San José, Tacuarembó, Treinta y Tres).

Sections of unknown relationships within *Baccharis* subgen. *Molina*.

IV.XIV. ***Baccharis*** sect. ***Albidae*** Giuliano: *Baccharis albida* is the only species and occurs in marshes in northeastern Argentina.

142. ***B. albida*** Hook. & Arn. **Argentina** (Buenos Aires, Chaco, Corrientes, Entre Ríos, Santa Fe).

IV.XV. ***Baccharis*** sect. ***Aristidentes*** G.L.Nesom: Nine species, mostly from North America, except for *B. hirta* DC. Native to southeastern Brazil and Uruguay.

143. ***B. brevipappa*** (McVaugh) G.L.Nesom. **Mexico** (Aguascalientes, Colima, Durango, Guerrero, Jalisco, Michoacán, Zacatecas).
144. ***B. herbacea*** (McVaugh) G.L.Nesom. **Mexico** (Michoacán).
145. ***B. hirta*** DC. **Brazil** (Espírito Santo, Minas Gerais, Paraná, Rio Grande do Sul, Santa Catarina, São Paulo). **Uruguay** (Maldonado, Montevideo, Rivera).

146. ***B. horizontalis*** G.L.Nesom. **Mexico** (Durango, Sinaloa).
147. ***B. malibuensis*** R.M.Beauch. & Henrickson. **USA** (California).
148. ***B. multiflora*** Kunth. **Mexico** (Chiapas, Durango, Estado de México, Guanajuato, Guerrero, Hidalgo, Jalisco, Morelos, Oaxaca, Puebla, Querétaro, San Luis Potosí, Sinaloa, Tlaxcala, Veracruz, Zacatecas).
149. ***B. praetermissa*** G.L.Nesom. **Mexico** (Baja California Sur).
150. ***B. serrifolia*** DC. **El Salvador**. **Guatemala**. **Honduras**. **Mexico** (Chiapas, Estado de México, Guerrero, Hidalgo, Oaxaca, Puebla, San Luis Potosí, Tlaxcala, Veracruz). **Nicaragua**.
151. ***B. sordescens*** DC. **Mexico** (Chiapas, Estado de México, Guanajuato, Guerrero, Hidalgo, Michoacán, Morelos, Oaxaca, Puebla, Querétaro, San Luis Potosí, Tamaulipas, Tlaxcala, Veracruz).

IV.XVI. ***Baccharis*** sect. ***Gladiatae*** Cuatrec.: *Baccharis marcetiifolia* is the only pecies and occurs in the Andes from southern Colombia to Ecuador.

152. ***B. marcetiifolia*** Benth. **Ecuador** (Chimborazo, Napo, Pichincha). **Colombia** (Nariño).

IV.XVII. ***Baccharis*** sect. ***Pinnatae*** Cuatrec.: Three species that occur in the northern Andes from Colombia to Ecuador.

153. ***B. ayacuchensis*** Cuatrec. **Peru** (Ayacucho).
154. ***B. buddlejoides*** Kunth. **Colombia** (Antioquia, Cauca, Nariño, Tolima, Valle de Cauca). **Ecuador** (Imbabura, Morona-Santiago, Napo, Pastaza, Pichincha, Sucumbíos, Tungurahua, Zamora-Chinchipe).
155. ***B. raulii*** S.Díaz & Cuatrec. **Colombia** (Quindío).

IV.XVIII. ***Baccharis*** sect. ***Punctatae*** Giuliano & G.L.Nesom: Eight North American species, from Mexico and southwestern USA.

156. ***B. bigelovii*** A.Gray. **Mexico** (Chihuahua, Coahuila, Durango, Nuevo León, San Luis Potosí, Sonora, Tamaulipas). **USA** (Arizona, New Mexico, Texas).
157. ***B. crassicuneata*** G.L.Nesom. **Mexico** (Coahuila, Nuevo León, Tamaulipas).
158. ***B. mexicana*** Cuatrec. **Mexico** (Guerrero, Oaxaca, Puebla, Tamaulipas, Tlaxcala, Veracruz).
159. ***B. sulcata*** DC. **Mexico** (Aguascalientes, Chihuahua, Coahuila, Durango, Guanajuato, Hidalgo, Jalisco, Michoacán, Nuevo León, Querétaro, San Luis Potosí, Tamaulipas, Zacatecas). **USA** (New Mexico, Texas).
160. ***B. supplex*** G.L.Nesom. **Mexico** (Durango).
161. ***B. thesioides*** Kunth. **Mexico** (Aguascalientes, Chihuahua, Durango, Estado de México, Guanajuato, Hidalgo, Jalisco, Michoacán, Morelos, Nayarit, Nuevo León, Querétaro, San Luis Potosí, Sinaloa, Sonora, Tamaulipas, Tlaxcala, Veracruz, Zacatecas). **USA** (Arizona, New Mexico).
162. ***B. zamoranensis*** Rzed. **Mexico** (Guanajuato, Queretaro).
163. ***B. zamudiorum*** Rzed. **Mexico** (Queretaro).

IV.IX. ***Baccharis*** sect. ***Rodriguezianae*** Giuliano: *Baccharis rodriguezii* is the only species and occurs in the prepuna vegetation in northwestern Argentina.

164. ***B. rodriguezii*** Ariza. **Argentina** (Salta, Tucumán).

IV.XX. ***Baccharis*** sect. ***Tenellae*** Giuliano: *Baccharis tenella* is the only species placed in the section and occurs in central and southern Argentina.

165. ***B. tenella*** Hook. & Arn. **Argentina** (Buenos Aires, Catamarca, Chubut, Córdoba, Corrientes, La Pampa, La Rioja, Mendoza, Néuquen, Río Negro, San Luis, Santa Cruz).

IV.XXI. ***Baccharis*** sect. ***Tubulatae*** Cuatrec.: Seven species, occurring mostly in the northern Andes, mainly between Peru and Venezuela.

166. ***B. farallonensis*** (Cuatrec.) G.Heiden. **Colombia** (Valle de Cauca).
167. ***B. fraterna*** Cuatrec. **Colombia** (Antioquia, Quindío).
168. ***B. grandiflora*** Kunth. **Ecuador** (Azuay, Carchi, Chimborazo, Cotopaxi, Imbabura, Napo, Pichincha, Sucumbíos). **Colombia** (Caldas, Cauca, Cundinamarca, Nariño, Tolima, Valle de Cauca).
169. ***B. huairacajensis*** Hieron. **Ecuador** (Azuay, Chimborazo, Cañar, Cotopaxi, Loja).
170. ***B. klattii*** Benoist. **Ecuador** (Azuay, Carchi, Chimborazo, Imbabura, Napo, Pichincha). **Colombia** (Nariño, Putumayo).
171. ***B. volubilis*** Kunth. **Ecuador** (Azuay, Cañar, Loja, Napo, Pichincha). **Peru** (Amazonas, Lambayeque, Piura, San Martín).
172. ***B. zumbadorensis*** V.M.Badillo. **Venezuela** (Táchira).

IV.XXX. **Species with unknown relationships in *B.* subgen. *Molina*.**

173. ***B. alamosana*** S.F.Blake. **Mexico** (Sonora).
174. ***B. cymosa*** Phil. **Chile** (Los Lagos, Los Ríos).
175. ***B. gnidiifolia*** Kunth. **Bolivia** (La Paz). **Chile** (Tarapacá). **Ecuador** (Chimborazo, Loja, Orellana). **Peru** (Amazonas, Ancash, Ayacucho, Cajamarca, La Libertad, Lambayeque, Lima, Piura, Tacna).
176. ***B. hambatensis*** Kunth. **Ecuador** (Chimborazo, Tungurahua).
177. ***B. hutchisonii*** Cuatrec. **Peru** (Cajamarca, La Libertad, Lambayeque, Piura).
178. ***B. johnwurdackiana*** H.Rob. **Peru** (Cusco).
179. ***B. libertadensis*** (S.B.Jones) H.Rob. **Peru** (La Libertad).
180. ***B. mandonii*** Klatt. **Bolivia** (Cochabamba, La Paz).
181. ***B. mollis*** Kunth. **Ecuador** (Pichincha).
182. ***B. palmeri*** Greenm. **Mexico** (Aguascalientes, Durango, Guanajuato, Jalisco, Zacatecas).
183. ***B. pohlii*** (Baker) Deble & A.S.Oliveira. **Brazil** (Distrito Federal, Goiás, Minas Gerais).
184. ***B. seemannii*** A.Gray. **Mexico** (Aguascalientes, Durango, Guanajuato, Jalisco, Nayarit, San Luis Potosí, Zacatecas).
185. ***B. taltalensis*** I.M.Johnst. **Chile** (Antofagasta, Atacama).

186. ***B. tarmensis*** Cuatrec. **Peru** (Amazonas, Cajamarca, Junín).
187. ***B. uniflora*** (Ruiz & Pav.) Pers. **Peru** (Ancash, Cusco, Ica, Junín, La Libertad, Lambayeque).
188. ***B. vanessae*** R.M.Beauch. **USA** (California).
189. ***B. zongoensis*** Joch.Müll. **Bolivia** (Cochabamba, La Paz).

V. ***Baccharis*** subgen. ***Heterothalamulopsis*** (Deble, A.S.Oliveira & Marchiori) G. Heiden: *Baccharis* sect. *Heterothalamulopsis* is the only known section to belong to the subgenus, and *B. wagenitzii*, a rupicolous shrub found in cloud forest edges on basaltic cliffs in southern Brazil, is the only species known to belong to this depauperate lineage (Heiden et al. 2019).

V.I. ***Baccharis*** sect. ***Heterothalamulopsis*** (Deble, A.S.Oliveira & Marchiori) G. Heiden: A monospecific section, as stated above.

190. ***B. wagenitzii*** (F.H.Hellw.) Joch.Müll. **Brazil** (Santa Catarina).

VI. ***Baccharis*** subgen. ***Coridifoliae*** (DC.) G.Heiden: Two sections and ten species occurring in moist or dry grasslands and savannas from Bolivia and central Brazil, south to central Argentina (Heiden et al. 2019).

VI.I. ***Baccharis*** sect. ***Pluricephalae*** (Deble) G.Heiden: Two narrowly endemic species from southern Brazil, occurring in highland marshes and peat bogs.

191. ***B. pluricapitulata*** (Deble) G.Heiden. **Brazil** (Rio Grande do Sul).
192. ***B. scabrifolia*** G.Heiden. **Brazil** (Rio Grande do Sul, Santa Catarina).

VI.II. ***Baccharis*** sect. ***Coridifoliae*** Giuliano: Eight species occurring in grasslands and savannas from Bolivia and central Brazil, south to central Argentina.

193. ***B. albilanosa*** A.S.Oliveira & Deble. **Brazil** (Rio Grande do Sul).
194. ***B. artemisioides*** Hook. & Arn. **Argentina** (Buenos Aires, Catamarca, Córdoba, Corrientes, Entre Ríos, La Pampa, Río Negro, Santa Fe, San Luis, Tucumán). **Uruguay** (Soriano).
195. ***B. bicolor*** (Joch.Müll.) G.Heiden. **Bolivia** (Cochabamba, La Paz).
196. ***B. coridifolia*** DC. **Argentina** (Buenos Aires, Catamarca, Chaco, Córdoba, Corrientes, Entre Ríos, Formosa, Jujuy, La Pampa, La Rioja, Mendoza, Misiones). **Bolivia** (Chuquisaca, Santa Cruz, Tarija). **Brazil** (Paraná, Rio Grande do Sul, Santa Catarina, São Paulo). **Paraguay** (Misiones). **Uruguay** (Artigas, Canelones, Cerro Largo, Colonia, Durazno, Flores, Florida, Lavalleja, Maldonado, Montevideo, Paysandú, Río Negro, Rivera, Rocha, Salto, San José, Soriano, Tacuarembó, Treinta y Tres).
197. ***B. erigeroides*** DC. **Argentina** (Misiones). **Brazil** (Distrito Federal, Goiás, Mato Grosso do Sul, Minas Gerais, Paraná, Rio Grande do Sul, Santa Catarina, São Paulo).
198. ***B. napaea*** G.Heiden. **Brazil** (Paraná, Rio Grande do Sul, Santa Catarina).
199. ***B. ochracea*** Spreng. **Argentina** (Entre Ríos). **Brazil** (Paraná, Rio Grande do Sul, Santa Catarina). **Uruguay** (Cerro Largo, Lavalleja, Maldonado, Montevideo, Rivera, Rocha, Tacuarembó, Treinta y Tres).
200. ***B. suberectifolia*** A.S.Oliveira & Deble. **Brazil** (Paraná).

VII. ***Baccharis*** subgen. ***Baccharis***: 17 sections and 241 species, occurring from the northeastern USA to southern South America.

VII.I. ***Baccharis*** sect. ***Caulopterae*** DC.: 15 species occurring from Colombia to central Argentina, with the greatest diversity in southeastern and southern Brazil (Heiden et al. 2019).

201. ***B. alpina*** Kunth. **Argentina** (Catamarca, Jujuy, Salta, Tucumán). **Bolivia** (Chuquisaca, Cochabamba, La Paz, Oruro, Potosí, Tarija). **Chile** (Arica y Parinacota, Tarapacá). **Colombia** (Caldas, Cauca, Cundinamarca, La Guajira, Risaralda). **Ecuador** (Azuay, Carchi, Chimborazo, Cotopaxi, Imbabura, Napo, Pichincha, Tungurahua). **Peru** (Apurimac, Arequipa, Cusco, Huancavelica, Junín, Pasco, Puno, Tacna).
202. ***B. altimontana*** G.Heiden & al. **Brazil** (Minas Gerais, Rio de Janeiro, São Paulo).
203. ***B. crispa*** Spreng. **Uruguay** (Lavalleja, Maldonado).
204. ***B. decurrens*** (Vell.) Stellf. **Brazil** (Rio de Janeiro),
205. ***B. dunensis*** A.A.Schneid. & G.Heiden. **Brazil** (Rio Grande do Sul).
206. ***B. genistelloides*** (Lam.) Pers. **Bolivia** (Cochabamba, La Paz, Oruro). **Chile** (Arica y Parinacota, Tarapacá). **Colombia** (Antioquia, Caldas, Cauca, Cundinamarca, Huila, Nariño, Putumayo, Risaralda, Tolima, Valle de Cauca). **Ecuador** (Azuay, Bolívar, Cañar, Carchi, Chimborazo, Cotopaxi, El Oro, Imbabura, Loja, Morona-Santiago, Napo, Orellana, Pastaza, Pichincha, Sucumbíos, Tungurahua, Zamora-Chinchipe). **Peru** (Amazonas, Ancash, Arequipa, Cajamarca, Cuzco, Junín, Huancavelica, Huánuco, La Libertad, Lambayeque, Lima, Loreto, Moquegua, Pasco, Piura, Puno, San Martín, Tacna).
207. ***B. jocheniana*** G.Heiden & Macias. **Brazil** (Rio Grande do Sul). **Uruguay** (Canelones, Lavalleja, Maldonado, Rocha).
208. ***B. lorentzii*** (Joch.Müll.) Deble. **Argentina** (Buenos Aires, Córdoba).
209. ***B. microcephala*** (Less.) DC. **Argentina** (Chaco, Corrientes, Distrito Federal, Entre Ríos, Formosa, Misiones, Santa Fe). **Brazil** (Bahia, Minas Gerais, Paraná, Rio Grande do Sul, Santa Catarina, São Paulo). **Paraguay** (Caaguazú, Caazapá, Central, Concepción, Cordillera, Guairá, Ñeembucú, Paraguarí, Presidente Hayes, San Pedro). **Uruguay** (Artigas, Canelones, Cerro Largo, Florida, Lavalleja, Maldonado, Montevideo, Paysandú, Río Negro, Rocha, Salto, San José, Soriano, Tacuarembó, Treinta y Tres).
210. ***B. myriocephala*** DC. **Brazil** (Bahia, Ceará, Distrito Federal, Espírito Santo, Goiás, Minas Gerais, Paraná, Pernambuco, Rio de Janeiro, Rio Grande do Sul, Santa Catarina, São Paulo). **Paraguay** (Caazapá).
211. ***B. opuntioides*** Mart. ex Baker. **Brazil** (Espírito Santo, Minas Gerais).
212. ***B. polygona*** Baker. **Brazil** (Rio Grande do Sul).
213. ***B. rectialata*** V.Valtierra et al. **Uruguay** (Cerro Largo, Durazno, Treinta y Tres).
214. ***B. riograndensis*** L.Teodoro & J.Vidal. **Brazil** (Rio Grande do Sul). **Uruguay** (Cerro Largo, Rivera, Treinta y Tres).

215. ***B. scopulorum*** A.A.Schneid. & G.Heiden. **Brazil** (Santa Catarina).
216. ***B. triangularis*** Hauman. **Argentina** (Buenos Aires, Chubut, La Pampa, Río Negro, San Luis).
217. ***B. trimera*** (Less.) DC. **Argentina** (Buenos Aires, Catamarca, Chubut, Córdoba, Corrientes, Entre Ríos, Jujuy, La Pampa, La Rioja, Mendoza, Misiones, Río Negro, Salta, Santa Fe, San Juan, San Luis, Tucumán). **Bolivia** (Chuquisaca, Cochabamba, La Paz, Santa Cruz, Tarija). **Brazil** (Bahia, Distrito Federal, Espírito Santo, Goiás, Mato Grosso, Mato Grosso do Sul, Minas Gerais, Paraná, Pernambuco, Rio de Janeiro, Rio Grande do Sul, Santa Catarina, São Paulo). **Paraguay** (Alto Paraná, Amambay, Caaguazú, Caazapá, Canindeyú, Central, Concepción, Guairá, Itapúa, Misiones, Paraguarí, San Pedro). **Peru** (Cusco, Puno). **Uruguay** (Artigas, Canelones, Cerro Largo, Colonia, Durazno, Lavalleja, Maldonado, Montevideo, Paysandú, Río Negro, Rivera, Rocha, Salto, San José, Soriano, Tacuarembó, Treinta y Tres).

VII.II. ***Baccharis*** sect. ***Aphyllae*** Baker: 33 species found along a diverse array of habitats, from sand dunes and dry grasslands to edges of forests and swampy environments, reaching its greatest diversity in southern Brazil, but with many endemic taxa in marginal areas of distribution such as Bolivia, Peru, and the Argentinean Patagonia.

218. ***B. aphylla*** (Vell.) DC. **Bolivia** (La Paz, Santa Cruz). **Brazil** (Bahia, Minas Gerais, Paraná, São Paulo).
219. ***B. apicifoliosa*** A.A.Schneid. & Boldrini. **Brazil** (Rio Grande do Sul, Santa Catarina).
220. ***B. articulata*** (Lam.) Pers. **Argentina** (Buenos Aires, Catamarca, Chaco, Córdoba, Corrientes, Entre Ríos, Jujuy, La Pampa, La Rioja, Misiones, Salta, Santiago del Estero, Santa Fe, San Luis, Tucumán). **Bolivia** (Chuquisaca, Santa Cruz). **Brazil** (Paraná, Rio Grande do Sul, Santa Catarina, São Paulo). **Paraguay** (Alto Paraná, Amambay, Caaguazú, Central, Cordillera, Guairá, Itapúa, Paraguarí, San Pedro). **Uruguay** (Artigas, Canelones, Cerro Largo, Colonia, Durazno, Florida, Lavalleja, Maldonado, Montevideo, Paysandú, Río Negro, Rivera, Rocha, Salto, San José, Soriano, Treinta y Tres).
221. ***B. burchellii*** Baker. **Brazil** (Minas Gerais, Rio de Janeiro, São Paulo).
222. ***B. campos-portoana*** Malag. **Brazil** (São Paulo).
223. ***B. chubutensis*** Speg. **Argentina** (Chubut).
224. ***B. deblei*** A.S.Oliveira & Marchiori. **Brazil** (Rio Grande do Sul, Santa Catarina).
225. ***B. flexuosiramosa*** A.A.Schneid. & Boldrini. **Brazil** (Rio Grande do Sul, Santa Catarina).
226. ***B. genistifolia*** DC. **Argentina** (Buenos Aires, Chubut, Córdoba, La Pampa, Santa Fe). **Uruguay** (Canelones, Flores, Maldonado, Montevideo, Rocha, San José).

227. ***B. glaziovii*** Baker. **Brazil** (Mato Grosso do Sul, Minas Gerais, Paraná, Rio de Janeiro, Rio Grande do Sul, Santa Catarina, São Paulo). **Paraguay** (Alto Paraná).
228. ***B. hemiptera*** G.Heiden & A.A.Schneid. **Brazil** (Espírito Santo, Minas Gerais).
229. ***B. junciformis*** DC. **Brazil** (Espírito Santo, Minas Gerais, Paraná, Rio de Janeiro, Rio Grande do Sul, Santa Catarina, São Paulo). **Uruguay** (Cerro Largo, Rivera).
230. ***B. megapotamica*** Spreng. **Argentina** (Misiones). **Brazil** (Espírito Santo, Minas Gerais, Paraná, Rio de Janeiro, Rio Grande do Sul, Santa Catarina, São Paulo). **Paraguay** (Caaguazú, Caazapá, Guairá). **Uruguay** (Rivera, Rocha, Tacuarembó, Treinta y Tres).
231. ***B. milleflora*** (Less.) DC. **Brazil** (Minas Gerais, Paraná, Rio Grande do Sul, Santa Catarina, São Paulo).
232. ***B. organensis*** Baker. **Brazil** (Minas Gerais, Paraná, Rio de Janeiro, Rio Grande do Sul, Santa Catarina, São Paulo).
233. ***B. palustris*** Heering. **Brazil** (Minas Gerais, Rio Grande do Sul, Santa Catarina). **Uruguay** (Canelones, Florida).
234. ***B. paranensis*** Heering & Dusén. **Brazil** (Paraná, Rio Grande do Sul, Santa Catarina, São Paulo).
235. ***B. penningtonii*** Heering. **Argentina** (Buenos Aires, Corrientes, Distrito Federal, Entre Ríos, Santa Fe). **Brazil** (Rio Grande do Sul). **Uruguay** (Canelones, Cerro Largo, Colonia, Lavalleja, Rocha, San José, Treinta y Tres).
236. ***B. pentaptera*** (Less.) DC. **Brazil** (Minas Gerais, Paraná, Rio Grande do Sul, Santa Catarina, São Paulo).
237. ***B. phyteuma*** Heering. **Argentina** (Buenos Aires, Entre Ríos, Santa Fe).
238. ***B. phyteumoides*** (Less.) DC. **Argentina** (Buenos Aires, Chaco, Corrientes, Entre Ríos, Misiones, Santa Fe). **Brazil** (Rio Grande do Sul). **Paraguay** (Ñeembucú). **Uruguay** (Artigas, Canelones, Cerro Largo, Colonia, Flores, Florida, Lavalleja, Maldonado, Montevideo, Paysandú, Río Negro, Rivera, Rocha, Salto, San José, Soriano, Tacuarembó).
239. ***B. pseudovillosa*** L.Teodoro & J.Vidal. **Brazil** (Rio Grande do Sul, Santa Catarina).
240. ***B. ramboi*** G. Heiden & Macias. **Brazil** (Paraná, Rio Grande do Sul, Santa Catarina).
241. ***B. regnellii*** Sch.Bip. ex Baker. **Brazil** (Espírito Santo, Minas Gerais, São Paulo).
242. ***B. reticulata*** (Ruiz & Pav.) Pers. **Ecuador** (Azuay, Loja, Morona-Santiago, Napo, Zamora-Chinchipe). **Peru** (Amazonas, Cajamarca, Huánuco, San Martín).
243. ***B. sagittalis*** (Less.) DC. **Argentina** (Chubut, Mendoza, Neuquén, Río Negro, San Juan). **Bolivia** (Cochabamba, La Paz, Santa Cruz). **Brazil** (Bahia, Espírito Santo, Minas Gerais, Paraná, Rio Grande do Sul, Santa Catarina, São Paulo). **Chile** (Aysén, Araucanía, Atacama, Bío Bío,

Coquimbo, Los Lagos, Los Ríos, Maule, Metropolitana, Ñuble, O'Higgins, Valparaíso). **Ecuador** (Azuay, Loja). **Paraguay** (Alto Paraná, Caaguazú, Caazapá). **Peru** (Amazonas, Cajamarca, Junín, La Libertad, Pasco, Piura).

244. ***B. sphagnophila*** A.A.Schneid. & G.Heiden. **Brazil** (Paraná, Rio Grande do Sul, Santa Catarina).
245. ***B. subalata*** Wedd. **Bolivia** (Chuquisaca, Cochabamba, La Paz, Santa Cruz). **Peru** (Cusco, Puno).
246. ***B. subbimera*** Hieron. **Ecuador** (Zamora-Chinchipe). **Peru** (Amazonas, Cajamarca, Loreto, San Martín).
247. ***B. subtropicalis*** G.Heiden. **Uruguay** (Florida, Maldonado, Montevideo, Rocha, San José, Soriano).
248. ***B. vargasii*** (Joch.Müll.) G.Heiden. **Bolivia** (Cochabamba, Santa Cruz).
249. ***B. vincifolia*** Baker. **Brazil** (Paraná, Rio Grande do Sul, Santa Catarina, São Paulo).
250. ***B. weirii*** Baker. **Brazil** (Paraná, Rio Grande do Sul, Santa Catarina, São Paulo). **Uruguay** (Rivera).
251. ***B. woytkowskii*** Joch.Müll. **Peru** (Amazonas, Cajamarca).

VII.III. ***Baccharis*** sect. ***Agglomeratae*** Giuliano: 34 species with a high level of diversity and endemism in southeastern Brazil, mainly in the tropical savannas and tropical high-elevation grasslands, some species reaching subtropical highland grasslands from southern Brazil, while *B. platypoda* has a disjunct distribution occurring in the eastern slope of the Andes close to the Bolivian and Peruvian border and along the main mountain chains of eastern Brazil.

252. ***B. alleluia*** A.S.Oliveira & Deble. **Brazil** (Bahia).
253. ***B. angusticeps*** Heering ex Malme. **Brazil** (Paraná, Santa Catarina).
254. ***B. claussenii*** Baker. **Brazil** (Minas Gerais).
255. ***B. concinna*** G.M.Barroso. **Brazil** (Minas Gerais).
256. ***B. elliptica*** Gardner. **Brazil** (Bahia, Minas Gerais).
257. ***B. intermixta*** Gardner. **Brazil** (Bahia, Minas Gerais, Espírito Santo, Paraná, Rio de Janeiro, Rio Grande do Sul, Santa Catarina, São Paulo).
258. ***B. itatiaiae*** Wawra. **Brazil** (Minas Gerais, Rio de Janeiro).
259. ***B. lateralis*** Baker. **Brazil** (Minas Gerais, Paraná, Rio de Janeiro, Rio Grande do Sul, Santa Catarina, São Paulo).
260. ***B. mesoneura*** DC. **Brazil** (Paraná, Rio Grande do Sul, Santa Catarina, São Paulo).
261. ***B. obdeltata*** G.Heiden. **Brazil** (Minas Gerais).
262. ***B. orbiculata*** Deble & A.S.Oliveira. **Brazil** (Bahia).
263. ***B. oreophila*** Malme. **Brazil** (Minas Gerais, Paraná, Rio de Janeiro, Santa Catarina, São Paulo).
264. ***B. parvidentata*** Malag. **Brazil** (Espírito Santo, Minas Gerais, Rio de Janeiro, São Paulo).
265. ***B. pauciflosculosa*** DC. **Brazil** (Paraná, Minas Gerais, Santa Catarina, São Paulo).
266. ***B. perlata*** Sch.Bip. ex Baker. **Brazil** (Minas Gerais).

267. ***B. platypoda*** DC. **Bolivia** (La Paz). **Brazil** (Bahia, Espírito Santo, Minas Gerais, Rio de Janeiro, São Paulo). **Peru** (Puno).
268. ***B. polyphylla*** Gardner. **Brazil** (Bahia, Minas Gerais).
269. ***B. pseudomyriocephala*** Malag. **Brazil** (Minas Gerais, Paraná, Rio de Janeiro, São Paulo).
270. ***B. pseudovaccinioides*** L.Teodoro. **Brazil** (Rio de Janeiro).
271. ***B. ramosissima*** Gardner. **Brazil** (Bahia, Distrito Federal, Goiás, Mato Grosso, Mato Grosso do Sul, Minas Gerais, Paraná, Pernambuco, Rio de Janeiro, São Paulo).
272. ***B. reticularia*** DC. **Brazil** (Bahia, Distrito Federal, Espírito Santo, Goiás, Mato Grosso do Sul, Minas Gerais, Paraná, Rio de Janeiro, Santa Catarina, São Paulo).
273. ***B. reticularioides*** Deble & A.S.Oliveira. **Brazil** (Paraná, São Paulo).
274. ***B. retusa*** DC. **Brazil** (Rio Grande do Sul, Santa Catarina).
275. ***B. salzmannii*** DC. **Brazil** (Bahia, Minas Gerais).
276. ***B. schultzii*** Baker. **Brazil** (Minas Gerais, Rio de Janeiro, São Paulo).
277. ***B. serrula*** Sch.Bip. ex Baker. **Brazil** (Minas Gerais).
278. ***B. simplex*** G.Heiden. **Brazil** (Minas Gerais).
279. ***B. truncata*** Gardner. **Brazil** (Bahia, Minas Gerais).

VII.IV. ***Baccharis*** sect. ***Axillares*** (Giuliano) G.Heiden: 14 species diversified on the high-elevation tropical and subtropical grasslands from eastern Brazil and on the low-elevation temperate grasslands from southern Brazil and Uruguay.

280. ***B. aracatubaensis*** Malag. **Brazil** (Paraná, Santa Catarina).
281. ***B. axillaris*** DC. **Brazil** (Paraná, Rio Grande do Sul, Santa Catarina, São Paulo).
282. ***B. cultrata*** Baker. **Brazil** (Rio Grande do Sul, Santa Catarina). **Uruguay** (Lavalleja, Maldonado, Rivera, Tacuarembó).
283. ***B. floccosa*** Deble & A.S.Oliveira. **Brazil** (Rio Grande do Sul, Santa Catarina).
284. ***B. gaucha*** G.Heiden. **Brazil** (Rio Grande do Sul).
285. ***B. hypericifolia*** Baker. **Brazil** (Rio Grande do Sul, Santa Catarina).
286. ***B. incisa*** Hook. & Arn. **Brazil** (Paraná, Rio Grande do Sul, Santa Catarina, São Paulo).
287. ***B. leptospermoides*** DC. **Brazil** (Minas Gerais, Paraná, Rio Grande do Sul, Santa Catarina, São Paulo).
288. ***B. lymanii*** G.M.Barroso ex G.Heiden. **Brazil** (Paraná, Rio Grande do Sul, Santa Catarina).
289. ***B. minutiflora*** Mart. ex Baker. **Brazil** (Minas Gerais).
290. ***B. sphenophylla*** Dusén ex Malme. **Brazil** (Minas Gerais, Paraná, Santa Catarina, São Paulo).
291. ***B. trilobata*** A.S.Oliveira & Marchiori. **Brazil** (Paraná, Santa Catarina).
292. ***B. trineura*** Soria & Zardini. **Brazil** (Minas Gerais, São Paulo).
293. ***B. umbellata*** G.Heiden & Ribas. **Brazil** (Paraná).

VII.V. ***Baccharis*** sect. ***Cuneifoliae*** DC.: 14 species, all from Argentinean and Chilean Patagonia.

294. ***B. concava*** (Ruiz. & Pav.) Pers. **Chile** (Bío Bío).
295. ***B. elaeoides*** Remy. **Argentina** (Chubut, Río Negro). **Chile** (Los Lagos).
296. ***B. macraei*** Hook. & Arn. **Chile** (Coquimbo, Maule, Metropolitana, Ñuble, O'Higgins, Valparaíso).
297. ***B. magellanica*** (Lam.) Pers. **Argentina** (Chubut, Mendoza, Neuquén, Río Negro, Santa Cruz, Tierra del Fuego). **Chile** (Aysén, Araucanía, Bío Bío, Los Lagos, Los Ríos, Magallanes y Antártica, Maule, Metropolitana, Ñuble, O'Higgins). **Falkland/Malvinas Islands**.
298. ***B. minor*** (F.H.Hellw.) G.Heiden. **Chile** (Magallanes y Antártica).
299. ***B. mylodontis*** F.H.Hellw. **Chile** (Magallanes y Antártica).
300. ***B. neaei*** DC. **Argentina** (Chubut, Mendoza, Neuquén, Río Negro). **Chile** (Aysén, Bío Bío, Coquimbo, Los Lagos, Los Ríos, Maule, Metropolitana, Ñuble, O'Higgins, Valparaíso).
301. ***B. neotruncata*** G.Heiden. **Chile** (Bío Bío, Maule, O'Higgins, Valparaíso).
302. ***B. palenae*** Phil. **Chile** (Aysén, Araucanía, Bío Bío, Los Lagos, Los Ríos).
303. ***B. patagonica*** Hook. & Arn. **Argentina** (Chubut, Neuquén, Río Negro, Santa Cruz, Tierra del Fuego). **Chile** (Araucanía, Aysén, Los Lagos, Los Ríos, Magallanes).
304. ***B. pilcensis*** F.H.Hellw. **Chile** (Bío Bío, Maule).
305. ***B. rhomboidalis*** Remy. **Chile** (Araucanía, Bío Bío, Los Lagos, Los Ríos).
306. ***B. umbelliformis*** DC. **Argentina** (Neuquén). **Chile** (Araucanía, Bío Bío).
307. ***B. vernalis*** F.H.Hellw. **Chile** (Bío Bío, Coquimbo, Maule, Ñuble, O'Higgins, Valparaíso).
308. ***B. zoellneri*** F.H.Hellw. **Chile** (Aysén, Los Lagos, Magallanes y Antártica). **Falkland/Malvinas Islands**.

VII.VI. *Baccharis* sect. ***Illinitae*** G.Heiden: Two species of moist or flooded grasslands from southern and central Brazil and eastern Paraguay.

309. ***B. illinita*** DC. **Brazil** (Distrito Federal, Goiás, Mato Grosso, Mato Grosso do Sul, Minas Gerais, Paraná, Rio Grande do Sul, Santa Catarina, São Paulo). **Paraguay** (Amambay, Caaguazú).
310. ***B. illinitoides*** Malag. **Brazil** (Mato Grosso do Sul). **Paraguay** (Amambay, Caaguazú).

VII.VII. ***Baccharis*** sect. ***Racemosae*** Ariza: Nine species, most of them from northeastern Argentina, southeastern Brazil, and eastern Paraguay and Uruguay in grasslands, forests, and savannas. *Baccharis dracunculifolia* has a wider distribution and is also recorded from the Bolivian Andes.

311. ***B. amambayensis*** Zardini & Soria. **Paraguay** (Amambay).
312. ***B. calvescens*** DC. **Brazil** (Bahia, Espírito Santo, Minas Gerais, Paraná, Pernambuco, Rio Grande do Sul, Rio de Janeiro, Santa Catarina, São Paulo).

313. ***B. dracunculifolia*** DC. **Argentina** (Buenos Aires, Catamarca, Chaco, Corrientes, Entre Ríos, Formosa, Jujuy, Misiones, Salta, Santiago del Estero, Santa Fe, Tucumán). **Bolivia** (Chuquisaca, Cochabamba, La Paz, Oruro, Potosí, Santa Cruz, Tarija). **Brazil** (Bahia, Distrito Federal, Espírito Santo, Goiás, Mato Grosso, Mato Grosso do Sul, Minas Gerais, Paraná, Rio Grande do Sul, Rio de Janeiro, Santa Catarina, São Paulo). **Paraguay** (Amambay, Caaguazú, Caazapá, Canindeyú, Central, Cordillera, Guairá, Misiones, Paraguarí). **Uruguay** (Artigas, Canelones, Cerro Largo, Florida, Lavalleja, Maldonado, Montevideo, Paysandú, Río Negro, Rivera, Rocha, Salto, Tacuarembó, Treinta y Tres).
314. ***B. erioclada*** DC. **Brazil** (Minas Gerais, Paraná, Rio de Janeiro, Rio Grande do Sul, Santa Catarina, São Paulo). **Paraguay** (Amambay).
315. ***B. longiattenuata*** A.S.Oliveira & Deble. **Brazil** (Rio Grande do Sul, Santa Catarina). **Uruguay** (Cerro Largo, Treinta y Tres).
316. ***B. montana*** DC. **Argentina** (Misiones). **Brazil** (Espírito Santo, Minas Gerais, Paraná, Rio de Janeiro, Rio Grande do Sul, Santa Catarina, São Paulo). **Paraguay** (Alto Paraná, Caaguazú, Canindeyú, Guairá, Itapúa, Paraguarí).
317. ***B. rivularis*** Gardner. **Brazil** (Distrito Federal, Goiás, Mato Grosso do Sul, Minas Gerais, São Paulo, Tocantins).
318. ***B. semiserrata*** DC. **Argentina** (Misiones). **Brazil** (Minas Gerais, Paraná, Rio de Janeiro, Rio Grande do Sul, Santa Catarina, São Paulo). **Paraguay** (Guairá, Itapúa, Paraguarí).
319. ***B. uncinella*** DC. **Brazil** (Espírito Santo, Minas Gerais, Paraná, Rio de Janeiro, Rio Grande do Sul, Santa Catarina, São Paulo).

VII.VIII. ***Baccharis*** sect. ***Tridentatae*** Giuliano: Six species, from Bolivia and southeastern Brazil south to eastern Argentina, most of the species occurring in southern Brazil.

320. ***B. caprariifolia*** DC. **Argentina** (Buenos Aires, Corrientes, Entre Ríos, Misiones). **Brazil** (Minas Gerais, Paraná, Rio de Janeiro, Rio Grande do Sul, Santa Catarina, São Paulo). **Paraguay** (Alto Paraná, Caaguazú, Canindeyú, Guairá, San Pedro). **Uruguay** (Rivera, Treinta y Tres).
321. ***B. deltoidea*** Baker. **Brazil** (Rio Grande do Sul).
322. ***B. isabelae*** Soria & Zardini. **Paraguay** (Central, Cordillera, Paraguarí).
323. ***B. nummularia*** Heering ex Malme. **Brazil** (Paraná, Rio Grande do Sul, Santa Catarina).
324. ***B. spicata*** (Lam.) Baill. **Argentina** (Buenos Aires, Chaco, Córdoba, Corrientes, Distrito Federal, Entre Ríos, Formosa, La Pampa, Santiago del Estero, Santa Fe). **Brazil** (Paraná, Rio Grande do Sul, Santa Catarina, São Paulo). **Paraguay** (Central, Cordillera, Paraguarí, Presidente Hayes). **Uruguay** (Canelones, Cerro Largo, Colonia, Florida, Lavalleja, Maldonado, Montevideo, Río Negro, Rocha, Salto, San José, Soriano, Tacuarembó, Treinta y Tres). ***Introduced in Portugal***.

325. ***B. tridentata*** Vahl. **Argentina** (Buenos Aires, Chaco, Córdoba, Corrientes, Entre Ríos, Formosa, Jujuy, La Pampa, Misiones, Salta, Santa Fe). **Bolivia** (Chuquisaca, Cochabamba, Santa Cruz, Tarija). **Brazil** (Distrito Federal, Goiás, Mato Grosso, Mato Grosso do Sul, Minas Gerais, Paraná, Rio de Janeiro, Rio Grande do Sul, Santa Catarina, São Paulo). **Paraguay** (Amambay, Central, Caaguazú, Caazapá, Canindeyú, Concepción, Cordillera, Guairá, Misiones). **Uruguay** (Canelones, Lavalleja, Maldonado, Montevideo, Rivera, Tacuarembó).
326. ***B. urvilleana*** Brongn. **Brazil** (Rio Grande do Sul, Santa Catarina).

VII.IX. ***Baccharis*** sect. ***Caespitosae*** Giuliano: 20 Andean species, occurring from Colombia to Argentina and Chile.

327. ***B. buchtienii*** H.Rob. **Bolivia** (Cochabamba, La Paz, Santa Cruz). **Peru** (Cusco).
328. ***B. caespitosa*** (Ruiz & Pav.) Pers. **Bolivia** (Cochabamba, La Paz, Oruro, Potosí, Tarija). **Peru** (Ancash, Apurímac, Arequipa, Ayacucho, Cajamarca, Cusco, Huánuco, Huancavelica, Junín, La Libertad, Lambayeque, Lima, Moquegua, Pasco, Puno).
329. ***B. chaparensis*** (Joch.Müll.) G.Heiden. **Bolivia** (Cochabamba).
330. ***B. chrysophylla*** (F.H.Hellw.) G.Heiden. **Bolivia** (Oruro). **Chile** (Arica y Parinacota, Tarapacá).
331. ***B. corymbosa*** (Ruiz & Pav.) Pers. **Peru** (Huánuco, Pasco).
332. ***B. fimbriata*** (Joch.Müll.) G.Heiden. **Argentina** (Jujuy, Salta). **Bolivia** (Potosí, Oruro, Tarija).
333. ***B. incarum*** (Wedd.) Perkins. **Argentina** (Jujuy). **Bolivia** (Cochabamba, La Paz, Oruro, Potosí, Tarija). **Chile** (Tarapacá). **Peru** (Arequipa, Puno, Tacna).
334. ***B. integrifolia*** (Joch.Müll.) G.Heiden. **Bolivia** (Cochabamba, La Paz, Santa Cruz).
335. ***B. kessleri*** (Joch.Müll.) G.Heiden. **Bolivia** (Cochabamba).
336. ***B. lapidosa*** Deble & A.S.Oliveira. **Bolivia** (Chuquisaca, Santa Cruz).
337. ***B. longipedicellata*** (Joch.Müll.) G.Heiden. **Bolivia** (La Paz). **Peru** (Puno).
338. ***B. neoaustralis*** G.Heiden. **Bolivia** (Chuquisaca, Cochabamba, Potosí).
339. ***B. neorupestris*** Deble & A.S.Oliveira. **Argentina** (Catamarca, Jujuy, Salta, Tucumán).
340. ***B. odorata*** Kunth. **Colombia** (Nariño).
341. ***B. papillosa*** Rusby. **Bolivia** (Chuquisaca, Cochabamba, La Paz, Potosí).
342. ***B. procumbens*** Hieron. **Peru** (Cajamarca, La Libertad).
343. ***B. santelicis*** Phil. **Argentina** (Mendoza, San Juan). **Chile** (Arica y Parinacota).
344. ***B. tola*** Phil. **Argentina** (Catamarca, Jujuy, La Rioja, Mendoza, Salta, San Juan, Tucumán). **Bolivia** (Chuquisaca, Oruro, Potosí, Tarija). **Chile** (Atacama, Tarapacá).
345. ***B. viscosissima*** (Kuntze) G.Heiden. **Argentina** (Salta). **Bolivia** (Potosí). **Chile** (Antofagasta).

346. ***B. yungensis*** (Joch.Müll.) G.Heiden. **Bolivia** (Cochabamba, La Paz).

VII.X. ***Baccharis*** sect. ***Baccharis***: 24 species, most from North America and the Caribbean. Species in the section show a wide array of variation, from broom-like shrubs with leaves reduced to scales to leafy treelets.

347. ***B. angustifolia*** Michx. **Bahamas. USA** (Alabama, Massachusetts, Florida, Georgia, Louisiana, Mississippi, North Carolina, South Carolina).
348. ***B. angustior*** (DC.) Britton ex Malag. **Cuba**.
349. ***B. buxifolia*** (Lam.) DC. **Colombia** (Nariño). **Ecuador** (Azuay, Bolívar, Cañar, Carchi, Chimborazo, Cotopaxi, Imbabura, Loja, Morona-Santiago, Napo, Pichincha, Tungurahua). **Peru** (Ancash, Ayacucho, Apurímac, Cajamarca, Cusco, Huánuco, Junín, La Libertad, Lima, San Martín).
350. ***B. conferta*** Kunth. **Mexico** (Chiapas, Durango, Estado de México, Guanajuato, Guerrero, Hidalgo, Jalisco, Michoacán, Morelos, Oaxaca, Puebla, Querétaro, San Luis Potosí, Tlaxcala, Veracruz).
351. ***B. confertoides*** G.L.Nesom. **El Salvador. Guatemala. Honduras. Mexico** (Chiapas).
352. ***B. dioica*** Vahl. **Antigua and Barbuda. Bahamas. Cayman Islands. Cuba. Dominican Republic. Haiti. Jamaica. México** (Campeche, Quintana Roo, Yucatán). **Montserrat. Puerto Rico. United States Virgin Islands. USA** (Florida).
353. ***B. emoryi*** A.Gray. **Mexico** (Baja California, Sonora). **USA** (Arizona, California, Nevada, Utah).
354. ***B. glandulifera*** G.L.Nesom. **Guatemala. Mexico** (Chiapas, Oaxaca).
355. ***B. glomeruliflora*** Pers. **Bahamas. Bermuda. Cuba. USA** (Alabama, Massachusetts, Florida, Georgia, Mississippi, North Carolina, South Carolina).
356. ***B. halimifolia*** L. **Bahamas. Canada** (Nova Scotia). **Mexico** (Nuevo León, San Luis Potosí, Tamaulipas, Veracruz). **USA** (Alabama, Arkansas, Connecticut, Delaware, Florida, Georgia, Louisiana, Massachusetts, Maryland, Mississippi, North Carolina, New Jersey, New York, Oklahoma, Pennsylvania, Rhode Island, South Carolina, Texas, Virginia). ***Introduced in Australia*** *(New South Wales, Queensland, Victoria, Western Australia).* ***Belgium. France. Georgia. Italy. Netherlands. New Zealand. Spain. United Kingdom*** *(England, Scotland).*
357. ***B. heterophylla*** Kunth. **Guatemala. Mexico** (Aguascalientes, Chiapas, Chihuahua, Colima, Durango, Estado de México, Guanajuato, Guerrero, Hidalgo, Jalisco, Michoacán, Morelos, Nayarit, Nuevo León, Oaxaca, Puebla, Querétaro, San Luis Potosí, Sinaloa, Sonora, Tamaulipas, Tlaxcala, Veracruz, Zacatecas).
358. ***B. kurtziana*** Ariza. **Argentina** (La Rioja, Salta, San Juan).
359. ***B. lancifolia*** DC. **Mexico** (Hidalgo, Queretaro, Veracruz).
360. ***B. mornicola*** (Urb.) G.Heiden. **Dominican Republic. Haiti.**
361. ***B. myrsinites*** (Lam.) Pers. **Dominican Republic. Haiti. Puerto Rico.**

362. ***B. neglecta*** Britton. **Mexico** (Chihuahua, Coahuila, Durango, Nuevo León, San Luis Potosí, Sonora, Tamaulipas, Zacatecas). **USA** (Texas).
363. ***B. nesomiana*** Rzed. & Zamudio. **Mexico** (Tamaulipas).
364. ***B. notosergila*** Griseb. **Argentina** (Buenos Aires, Chaco, Córdoba, Corrientes, Entre Ríos, Formosa, Santiago del Estero, Santa Fe). **Brazil** (Mato Grosso do Sul, Rio Grande do Sul). **Paraguay** (Caaguazú, Central, Cordillera, Presidente Hayes). **Uruguay** (Canelones, Flores, Florida, Maldonado, Montevideo, Paysandú, Río Negro, Salto, San José, Soriano, Tacuarembó).
365. ***B. pilularis*** DC. **Mexico** (Baja California). **USA** (California, New Mexico, Oregon).
366. ***B. salicina*** Torr. & A.Gray. **Mexico** (Baja California, Chihuahua, Coahuila, Durango, Nuevo León, Sonora, Tamaulipas). **USA** (Arizona, California, Colorado, Kansas, Nevada, New Mexico, Oklahoma, Texas, Utah).
367. ***B. sarothroides*** A.Gray. **Mexico** (Baja California, Baja California Sur, Chihuahua, Sinaloa, Sonora). **USA** (Arizona, California, New Mexico, Texas).
368. ***B. scoparia*** (L.) Sw. **Cuba**. **Jamaica**.
369. ***B. scoparioides*** Griseb. **Cuba**.
370. ***B. sergiloides*** A.Gray. **Mexico** (Baja California, Baja California Sur, Sonora). **USA** (Arizona, California, Nevada, Utah).
371. ***B. shaferi*** Britton. **Cuba**.
372. ***B. tucumanensis*** Hook. & Arn. **Argentina** (Catamarca, Córdoba, Jujuy, La Rioja, Salta, San Luis, Tucumán). **Bolivia** (Tarija).

VII.XI. ***Baccharis*** sect. ***Andina*** G.Heiden: 15 species of shrubs and treelets mostly from subpáramo and páramos of the northern Andes, from Venezuela to Ecuador.

373. ***B. angelica*** Benoist. **Ecuador** (Carchi).
374. ***B. balnearia*** Benoist. **Ecuador** (Bolívar, Chañar, Cotopaxi, Loja, Morona-Santiago).
375. ***B. boyacensis*** Cuatrec. **Colombia** (Boyacá, Cundinamarca).
376. ***B. cochensis*** Hieron. **Colombia** (Nariño).
377. ***B. emarginata*** (Ruiz & Pav.) Pers. **Peru** (Amazonas, Ancash, Ayacucho, Cajamarca, Huánuco, Junín, Lambayeque, Piura).
378. ***B. grandicapitulata*** Hieron. **Peru** (Amazonas, Ayacucho, Ancash, Cajamarca, Huánuco, Piura).
379. ***B. lehmannii*** Klatt. **Colombia** (Antioquia, Cauca, Cundinamarca, Meta, Nariño, Tolima).
380. ***B. lloensis*** Hieron. **Colombia** (Arauca, Boyacá, Caldas, Cauca, Cesar, Chocó, Cundinamarca, La Guajira, Huila, Meta, Nariño, Putumayo, Tolima). **Ecuador** (Azuay, Bolívar, Carchi, Chañar, Chimborazo, Cotopaxi, Imbabura, Loja, Morona-Santiago, Napo, Pichincha, Tungurahua).
381. ***B. pachycephala*** Hieron. **Peru** (Cajamarca).
382. ***B. paramicola*** Cuatrec. **Colombia** (Caldas, Cauca, Tolima).

383. ***B. sinuata*** Kunth. **Ecuador** (Azuay, Chañar, Loja, Orellana, Zamora-Chinchipe). **Peru** (Amazonas, Cajamarca, Piura).
384. ***B. teindalensis*** Kunth. **Colombia** (Cauca, Cundinamarca, Nariño). **Ecuador** (Azuay, Cañar, Carchi, Cotopaxi, Imbabura, Morona-Santiago, Napo, Pichincha).
385. ***B. tetroica*** G.Heiden. **Colombia** (Boyacá, Cundinamarca, Norte de Santander, Santander). **Venezuela** (Mérida, Táchira, Trujillo).
386. ***B. tricuneata*** (L.f.) Pers. **Colombia** (Antioquia, Arauca, Boyacá, Caldas, Cundinamarca, Magdalena, Meta, Norte de Santander, Quindío, Risaralda, Santander, Tolima, Valle de Cauca). **Venezuela** (Mérida, Táchira, Trujillo).

VII.XII. ***Baccharis*** sect. ***Nitidae*** Cuatrec.: Four species, inhabiting the edges of tropical rainforests, in the Atlantic Rainforest of Brazil or in the Yungas from the northern Andes.

387. ***B. dentata*** (Vell.) G.M.Barroso. **Argentina** (Misiones). **Brazil** (Espírito Santo, Minas Gerais, Paraná, Pernambuco, Rio de Janeiro, Rio Grande do Sul, Santa Catarina, São Paulo). **Paraguay** (Amambay, Guairá).
388. ***B. malmei*** Joch.Müll. **Brazil** (Bahia, Distrito Federal, Espírito Santo, Goiás, Minas Gerais, Paraná, Pernambuco, Rio de Janeiro, Santa Catarina, São Paulo).
389. ***B. nitida*** (Ruiz & Pav.) Pers. **Argentina** (Salta). **Bolivia** (Chuquisaca, Cochabamba, La Paz, Santa Cruz, Tarija). **Colombia** (Antioquia, Bolívar, Boyacá, Caldas, Caquetá, Cauca, Cesar, Chocó, Cundinamarca, La Guajira, Huila, Magdalena, Meta, Nariño, Norte de Santander, Putumayo, Quindío, Risaralda, Santander, Tolima, Valle de Cauca). **Ecuador** (Bolívar, Carchi, Chimborazo, Cotopaxi, Loja, Napo, Pastaza, Pichincha, Tungurahua). **Guyana**. **Peru** (Amazonas, Ayacucho, Cajamarca, Cuzco, Huánuco, Junín, La Libertad, Lambayeque, Pasco, Puno, San Martín, Ucayali). **Venezuela** (Aragua, Distrito Federal, Lara, Mérida, Miranda, Táchira, Trujillo).
390. ***B. singularis*** (Vell.) G.M.Barroso. **Brazil** (Bahia, Espírito Santo, Paraná, Rio de Janeiro, Rio Grande do Sul, Santa Catarina, São Paulo).

VII.XIII. ***Baccharis*** sect. ***Cylindricae*** Heering: 24 species from South America, the highest diversity in the grasslands and savannas from Argentina, Brazil, and Uruguay.

391. ***B. argentina*** Heering. **Argentina** (Catamarca, La Rioja, Salta).
392. ***B. brevifolia*** DC. **Brazil** (Bahia, Mato Grosso, Mato Grosso do Sul, Minas Gerais, Paraná, Rio de Janeiro, Rio Grande do Sul, Santa Catarina, São Paulo). **Uruguay** (Treinta y Tres).
393. ***B. chilco*** Kunth. **Bolivia** (Beni, La Paz, Santa Cruz). **Brazil** (Mato Grosso). **Colombia** (Boyacá, Caldas, Caquetá, Cauca, Cundinamarca, Huila, Meta, Nariño, Tolima, Valle de Cauca). **Ecuador** (Chimborazo, Imbabura, Loja, Pichincha, Zamora-Chinchipe). **Paraguay** (Presidente Hayes). **Peru** (Amazonas, Ancash, Cajamarca, Cusco, Huánuco, Junín, Puno, San Martín).

394. ***B. cognata*** DC. **Argentina** (Corrientes, Misiones). **Brazil** (Bahia, Distrito Federal, Goiás, Mato Grosso, Mato Grosso do Sul, Minas Gerais, Paraná, Rio Grande do Sul, Santa Catarina, São Paulo). **Paraguay** (Amambay, Caaguazú, Caazapá, Central, Concepción, Cordillera, Guairá, Misiones, Paraguarí). **Uruguay** (Tacuarembó, Treinta y Tres).
395. ***B. cordobensis*** Heering. **Argentina** (Córdoba).
396. ***B. flabellata*** Hook. & Arn. **Argentina** (Córdoba, La Pampa, La Rioja, Santiago del Estero, San Juan, San Luis).
397. ***B. funkiae*** Bonif. et al. **Uruguay** (Treinta y Tres).
398. ***B. gracillima*** Heering & Dusén. **Brazil** (Paraná).
399. ***B. humilis*** Sch.Bip. ex Baker. **Brazil** (Distrito Federal, Goiás, Mato Grosso, Mato Grosso do Sul, Minas Gerais, São Paulo).
400. ***B. inexspectata*** Deble & A.S.Oliveira. **Brazil** (Rio Grande do Sul).
401. ***B. linearifolia*** (Lam.) Pers. **Argentina** (Buenos Aires, Catamarca, Chaco, Córdoba, Corrientes, Entre Ríos, Jujuy, La Pampa, La Rioja, Mendoza, Misiones, Río Negro, Salta, Santa Fe, San Juan, San Luis). **Brazil** (Amazonas, Bahia, Distrito Federal, Goiás, Maranhão, Mato Grosso, Mato Grosso do Sul, Minas Gerais, Pará, Paraná, Pernambuco, Piauí, Rio de Janeiro, Rio Grande do Sul, Rondônia, Roraima, Santa Catarina, São Paulo, Tocantins). **Guyana**. **Paraguay** (Alto Paraná, Amambay, Caaguazú, Caazapá, Canindeyú, Central, Cordillera, Guairá, Paraguarí, Presidente Hayes, San Pedro). **Suriname**. **Uruguay** (Canelones, Colonia, Florida, Maldonado, Montevideo, Paysandú, Río Negro, Rivera, Rocha, Tacuarembó, Treinta y Tres). **Venezuela** (Bolívar).
402. ***B. maritima*** Baker. **Brazil** (Rio Grande do Sul, Santa Catarina). **Uruguay** (Maldonado, Montevideo, Rocha, San José).
403. ***B. microdonta*** DC. **Argentina** (Catamarca, Corrientes, Jujuy, Misiones, Salta, Tucumán). **Bolivia** (Chuquisaca, Cochabamba, Santa Cruz, Tarija). **Brazil** (Minas Gerais, Paraná, Rio de Janeiro, Rio Grande do Sul, Santa Catarina, São Paulo). **Paraguay** (Cordillera). **Uruguay** (Canelones, Florida, Lavalleja, Maldonado, Montevideo, Rocha, Salto).
404. ***B. multifolia*** A.S.Oliveira & al. **Brazil** (Rio Grande do Sul).
405. ***B. pampeana*** A.S.Oliveira & al. **Brazil** (Rio Grande do Sul).
406. ***B. pedersenii*** Cabrera. **Argentina** (Corrientes, Entre Ríos, Santa Fe). **Brazil** (Rio Grande do Sul). **Paraguay** (Cordillera, Ñeembucú). **Uruguay** (Colonia).
407. ***B. pentodonta*** Malme. **Brazil** (Paraná, Rio Grande do Sul, Santa Catarina, São Paulo). **Paraguay** (Caaguazú, Guairá).
408. ***B. petraea*** Heering. **Argentina** (Misiones). **Brazil** (Rio Grande do Sul, Santa Catarina, São Paulo). **Paraguay** (Central, Cordillera, Paraguarí). **Uruguay** (Flores, Florida).
409. ***B. polycephala*** Wedd. **Argentina** (Jujuy, Salta). **Bolivia** (Chuquisaca, Cochabamba, La Paz, Potosí, Santa Cruz, Tarija). **Peru** (Puno).
410. ***B. saltensis*** Baker. **Uruguay** (Salto).
411. ***B. santiagensis*** Heering. **Chile** (Metropolitana).

412. ***B. sessiliflora*** Vahl. **Argentina** (Corrientes, Misiones, Salta). **Bolivia** (Santa Cruz). **Brazil** (Bahia, Distrito Federal, Goiás, Mato Grosso, Mato Grosso do Sul, Minas Gerais, Paraná, Pernambuco, Rio Grande do Sul, Santa Catarina, São Paulo, Sergipe). **Paraguay** (Alto Paraná, Alto Paraguay, Amambay, Caaguazú, Canindeyú, Central, Concepción, Cordillera, Presidente Hayes, San Pedro). **Uruguay** (Maldonado, Montevideo, Paysandú, Rivera).
413. ***B. subdentata*** DC. **Brazil** (Bahia, Distrito Federal, Goiás, Mato Grosso, Mato Grosso do Sul, Maranhão, Minas Gerais, Paraná, Rio Grande do Sul, Santa Catarina, São Paulo). **Paraguay** (Caaguazú, Caazapá).
414. ***B. subopposita*** DC. **Brazil** (Rio Grande do Sul). **Paraguay** (Caaguazú, Caazapá, Canindeyú, Cordillera, Guairá, Misiones).
415. ***B. variabiliflora*** Deble & A.S.Oliveira. **Brazil** (Minas Gerais).
416. ***B. vernicosa*** Hook. & Arn. **Argentina** (Misiones). **Brazil** (Rio Grande do Sul). **Uruguay** (Artigas, Paysandú, Rivera, Salto).

Sections of unknown relationships within *Baccharis* subgen. *Baccharis*.

VII.XIV. ***Baccharis*** sect. ***Discolores*** DC.: *Baccharis phylicoides* is the only species and occurs along the Andes from Ecuador to Bolivia.

417. ***B. phylicoides*** Kunth. **Bolivia** (Amazonas, Ancash, Ayacucho, Cajamarca, Cusco, Huánuco, Junín, La Libertad, Lambayeque, Piura). **Peru** (La Paz).

VII.XV. ***Baccharis*** sect. ***Divaricatae*** Giuliano: *Baccharis divaricata* is the only species and occurs in coastal dunes of the Argentinean Patagonia.

418. ***B. divaricata*** Hauman. **Argentina** (Buenos Aires, Chubut, Río Negro).

VII.XVI. ***Baccharis*** sect. ***Frenguellianae*** Giuliano: *Baccharis frenguelli* is the only species and occurs in riverine vegetation in northeastern Argentina and Paraguay.

419. ***B. frenguellii*** Cabrera. **Argentina** (Chaco, Corrientes, Entre Ríos, Santa Fe). **Paraguay** (Central, Cordillera, Paraguarí).

VII.XVII. ***Baccharis*** sect. ***Pedicellatae*** Heering: Seven species from Argentina and Chile.

420. ***B. austropedicellata*** (F.H.Hellw.) G.Heiden. **Argentina** (Neuquén). **Chile** (Bío Bío, Maule, Ñuble, O'Higgins).
421. ***B. linearis*** (Ruiz & Pav.) Pers. **Argentina** (Chubut, Mendoza, Neuquén, Río Negro, San Juan). **Chile** (Araucanía, Atacama, Bío bío, Coquimbo, Los Lagos, Los Ríos, Maule, Metropolitana, Ñuble, O'Higgins, Valparaíso).
422. ***B. lycioides*** Remy. **Argentina** (Neuquén). **Chile** (Araucanía, Bío Bío, Los Lagos, Los Ríos).
423. ***B. obovata*** Hook. & Arn. **Argentina** (Chubut, Neuquén, Río Negro, Santa Cruz). **Chile** (Aysén, Araucanía, Bío Bío, Los Lagos, Los Ríos, Maule, Ñuble, O'Higgins, Valparaíso).
424. ***B. ocellata*** Phil. **Argentina** (Neuquén). **Chile** (Bío Bío, Maule, Ñuble, O'Higgins).

425. ***B. poeppigiana*** DC. **Argentina** (Neuquén). **Chile** (Coquimbo, Metropolitana, Valparaíso).
426. ***B. pycnocephala*** (F.H.Hellw.) G.Heiden. **Argentina** (Chubut, Neuquén, Río Negro). **Chile** (Araucanía, Bío Bío, Los Lagos, Los Ríos, Maule, Ñuble, Valparaíso).

Species of unknown relationships within *Baccharis* subgen. *Baccharis.*

427. ***B. alaternoides*** Kunth. **Peru** (Amazonas, Áncash, Cajamarca, Cuzco, La Libertad, Lima).
428. ***B. chachapoyasensis*** Cuatrec. **Peru** (Amazonas, Cajamarca).
429. ***B. dependens*** (Ruiz & Pav.) Pers. **Peru** (Huánuco).
430. ***B. erectifolia*** Steyerm. **Venezuela** (Anzoátegui, Sucre, Trujillo).
431. ***B. glomerata*** Joch.Müll. **Bolivia** (La Paz).
432. ***B. hieronymi*** Heering. **Ecuador** (Azuay, Cañar, Morona-Santiago).
433. ***B. humifusa*** Kunth. **Ecuador** (Azuay, Bolivar, Chañar, Chimborazo, Cotopaxi, Imbabura, Napo, Pastaza, Pichincha, Tungurahua).
434. ***B. itapirocensis*** A.S.Oliveira & Deble. **Brazil** (Paraná).
435. ***B. paucicostata*** Joch.Müll. & Giuliano. **Argentina** (Mendoza, San Juan).
436. ***B. pseudoalpestris*** L.Teodoro. **Brazil** (Minas Gerais).
437. ***B. pumila*** Joch.Müll. **Bolivia** (Tarija).
438. ***B. samensis*** Joch.Müll. **Argentina** (Salta). **Bolivia** (Tarija).
439. ***B. scabra*** (Ruiz & Pav.) Pers. **Peru** (Junín).
440. ***B. schomburgkii*** Baker. **Brazil** (Roraima). **Guyana**. **Venezuela** (Amazonas, Bolívar, Táchira, Trujillo).
441. ***B. tenuicapitulata*** Joch.Müll. **Ecuador** (Azuay, Chimborazo, Chañar, Loja).
442. ***B. tomentosa*** (Ruiz & Pav.) Pers. **Peru** (Ancash).

6 Confirmed and Putative Hybrid Species: Who Are Them?

Putative hybrid species and subspecies were formally described 38 times for *Baccharis* (Malagarriga 1949, 1954; Hellwig 1990). They are presented here ordered alphabetically. Likely parental species are within bracts, followed by geographic distribution and, when pertinent, comments on current taxonomic status. Taxonomic status of most of the putative hybrids proposed based on morphology and distribution of likely sympatric parent species remains to be confirmed or rejected by means of further experimental crosses and in situ population genetic studies.

×1. ***B.* × *alboffii*** F.H.Hellw. (*B. patagonica* × *B. zoellneri*). **Chile** (Los Lagos).
×2. ***B.* × *antucensis*** F.H.Hellw. (*B. umbelliformis* × *B. neaei*). **Chile** (Bío Bío).
×3. ***B.* × *arcuata*** F.H.Hellw. (*B. magellanica* × *B. obovata*). **Chile** (Aysén).
×4. ***B.* × *arcuata*** F.H.Hellw. nothosubsp. ***arcuata***. (*B. magellanica* × *B. obovata*). **Chile** (Aysén).

×5. **B. × *arcuata*** F.H.Hellw. nothosubsp. ***chamaearcuata***. (*B. magellanica* × *B. obovata* × B. *umbelliformis*). **Argentina** (Neuquén). **Chile** (Araucanía).
×6. ***B.* × *australis*** F.H.Hellw. (*B. magellanica* × *B. zoellneri*). **Chile** (Los Lagos, Magallanes y Antárctica). **Falkland/Malvinas Islands**.
×7. ***B.* × *australis*** F.H.Hellw. nothosubsp. ***australis***. (*B. magellanica* × *B. zoellneri*). **Chile** (Los Lagos, Magallanes y Antárctica). **Falkland Islands**.
×8. ***B.* × *australis*** F.H.Hellw. nothosubsp. ***minutifolia***. (*B. magellanica* × *B. minor*). **Chile**. (Araucanía).
×9. ***B.* × *caramavidensis*** F.H.Hellw. (*B. minor* × *B. obovata*). **Chile** (Bío Bío).
×10. ***B.* × *caramavidensis*** F.H.Hellw. nothosubsp. ***caramavidensis***. (*B. minor* × *B. obovata*). **Chile** (Bío Bío).
×11. ***B.* × *caramavidensis*** F.H.Hellw. nothosubsp. ***maior***. (*B. zoellneri* × *B. obovata*). **Chile** (Los Lagos).
×12. ***B.* × *chillanensis*** F.H.Hellw. (*B. magellanica* × *B. ocellata*). **Chile** (Bío Bío).
×13. ***B.* × *chillanensis*** F.H.Hellw. nothosubsp. ***chillanensis***. (*B. magellanica* × *B. ocellata*). **Chile** (Bío Bío).
×14. ***B.* × *chillanensis*** F.H.Hellw. nothosubsp. ***procumbens***. (*B. magellanica* × *B. austropedicellata*). **Chile** (Bío Bío, Maule).
×15. ***B.* × *concava*** (Ruiz & Pav.) Pers. (*B. rhomboidalis* × *B. obovata*). **Chile** (Bío Bío). =Currently accepted as ***B. concava***.
×16. ***B.* × *concavoides*** F.H.Hellw. (*B. rhomboidalis* × *B. elaeoides* × *B. obovata*). **Chile** (Los Lagos).
×17. ***B.* × *crenatolycioides*** F.H.Hellw. (*B. obovata* × *B. lycioides*). **Chile** (Bío Bío).
×18. ***B.* × *demissa*** F.H.Hellw. (*B. magellanica* × *B. mylodontis*). **Chile** (Magallanes y Antáctica).
×19. ***B.* × *exspectata*** F.H.Hellw. (*B. obovata* × *B. patagonica*). **Chile** (Los Lagos).
×20. ***B.* × *exspectata*** F.H.Hellw. nothosubsp. ***expectata***. (*B. obovata* × *B. patagonica*). **Chile** (Los Lagos).
×21. ***B.* × *exspectata*** F.H.Hellw. nothosubsp. ***crenatopalenae***. (*B. palenae* × *B. patagonica*). **Chile** (Los Lagos).
×22. ***B.* × *fraudulenta*** L.Teodoro. (*B. subopposita* × *B. subdentata*). **Brazil** (São Paulo). =Currently accepted as ***B. linearifolia***.
×23. ***B.* × *heeringiana*** L.Teodoro (*B. junciformis* × *B. milleflora*). **Brazil** (São Paulo). =Currently accepted as ***B. sagittalis***.
×24. ***B.* × *hoehneana*** L.Teodoro. (*B. linearifolia* × *B. cognata*). **Brazil** (São Paulo). =Currently accepted as ***B. linearifolia***.
×25. ***B.* × *intermedia*** DC. (*B. macraei* × *B. linearis*). **Chile** (Coquimbo, O'Higgins, Valparaíso). Hybrid status supported by morphological and chemical data (Faini et al. 1991).
×26. ***B.* × *paulopolitana*** L.Teodoro & W.Hoehne. (*B. dracunculifolia* × *B. linearifolia*). **Brazil** (São Paulo). Currently accepted as ***B. linearifolia***.
×27. ***B.* × *pseudolycioides*** F.H.Hellw. (*B. neaei* × *B. lycioides*). **Chile** (Bío Bío, Araucanía).

×28. ***B.* × *pseudoneaei*** F.H.Hellw. (*B. neaei* × *B. magellanica*). **Chile**. (Maule).

×29. ***B.* × *pseudopalenae*** F.H.Hellw. (*B. elaeoides* × *B. palenae*). **Chile** (Los Lagos).

×30. ***B.* × *pseudopilcensis*** F.H.Hellw. (*B. pilcensis* × *B. neaei*). **Chile** (Bío Bío).

×31. ***B.* × *septentrionalis*** F.H.Hellw. (*B. macraei* × *B. vernalis*). **Chile** (Coquimbo, Valparaíso).

×32. ***B.* × *spegazzinii*** F.H.Hellw. (*B. magellanica* × *B. patagonica*). **Chile** (Magallanes y Antáctica).

×33. ***B.* × *subaequalis*** F.H.Hellw. (*B. elaeoides* × *B. rhomboidalis*). **Chile** (Los Lagos).

×34. ***B.* × *tarapacana*** F.H.Hellw. (*B. viscosissima* × *B. santelicis* or *B. chrysophylla*). **Chile** (Tarapacá).

×35. ***B.* × *volckmannii*** Phil. (*B. linearis* × *B. neaei*). **Chile** (Coquimbo, Valparaíso, Metropolitana).

×36. ***B.* × *volckmannii*** Phil. nothosubsp. ***volckmannii***. (*B. linearis* × *B. neaei*). **Chile** (Coquimbo, Metropolitana, Valparaíso).

×37. ***B.* × *volckmannii*** Phil. nothosubsp. ***hybrida***. (*B. pycnocephala* × *B. neaei*). **Chile** (Bío Bío, Maule).

×38. ***B.* × *wilsoniana*** L.Teodoro. (*B. linearifolia* × *B. pentodonta*). **Brazil** (São Paulo). Currently accepted as ***B. microdonta***.

7 Distribution of *Baccharis* by Sovereign Countries or Dependent Territories: How Many Species Are Out There?

A checklist of recorded species and endemics for the 35 American sovereign countries and the 24 dependent territories from the Americas is presented in the following section.

Sovereign Countries

Argentina. 110 species, 25 of them endemic*: *B. acaulis, B. albida*, B. aliena, B. alpina, B. anomala, B. arenaria, B. argentina*, B. arguta, B. artemisioides, B. articulata, B. austropedicellata, B. boliviensis, B. breviseta, B. cabrerae*, B. calliprinos*, B. capitalensis, B. caprariifolia, B. chubutensis*, B. clavata, B. cognata, B. cordobensis*, B. coridifolia, B. darwinii, B. debilis, B. dentata, B. divaricata*, B. dracunculifolia, B. effusa, B. elaeoides, B. famatinensis*, B. fimbriata, B. flabellata*, B. frenguellii, B. genistifolia, B. gilliesii*, B. glutinosa, B. gnaphalioides, B. grisebachii, B. helichrysoides, B. incarum, B. juncea, B. kurtziana*, B. latifolia, B. lilloi, B. linearifolia, B. linearis, B. lorentzii*, B. lycioides, B. magellanica, B. megapotamica, B. melanopotamica*, B. microcephala, B. microdonta, B. montana, B. multiflosculosa, B. neaei, B. neorupestris*, B. niederleinii*, B. nitida, B. nivalis, B. notosergila, B. oblongifolia, B. obovata, B. ocellata, B. ochracea, B. oxyodonta, B. patagonica, B. paucicostata*, B. peder-*

senii, B. penningtonii, B. perulata, B. petraea, B. petrophila*, B. phyteuma*, B. phyteumoides, B. pingraea, B. poeppigiana, B. polifolia, B. polycephala, B. polygama*, B. potrerillana*, B. punctulata, B. pycnocephala, B. quitensis, B. racemosa, B. retamoides*, B. rodriguezii*, B. sagittalis, B. salicifolia, B. samensis, B. santelicis, B. sculpta, B. semiserrata, B. sessiliflora, B. spartioides, B. spicata, B. stenophylla, B. tenella*, B. thymifolia*, B. tola, B. triangularis, B. tridentata, B. trimera, B. trinervis, B. tucumanensis, B. ulicina, B. umbelliformis, B. vernicosa, B. viscosissima, B. vulneraria.*

Antigua and Barbuda. 1 species: *B. dioica.*

Bahamas. 4 species. *B. angustifolia, B. dioica, B. glomeruliflora, B. halimifolia.*

Barbados. 0 species native, 1 cultivated: *B. halimifolia.*

Belize. 1 species: *B. trinervis.*

Bolivia. 76 species, 22 endemic*: *B. acaulis, B. alpina, B. aphylla, B. arguta, B. articulata, B. beckii*, B. bicolor*, B. boliviensis, B. buchtienii, B. caespitosa, B. cana*, B. capitalensis, B. chaparensis*, B. chilco, B. chrysophylla, B. clavata, B. coridifolia, B. darwinii, B. debilis, B. densiflora*, B. dracunculifolia, B. effusa, B. fimbriata, B. genistelloides, B. glomerata*, B. glutinosa, B. gnidiifolia, B. grisebachii, B. incarum, B. integrifolia*, B. jelskii, B. kessleri*, B. lapidosa*, B. latifolia, B. lewisii*, B. lilloi, B. longipedicellata, B. mandonii*, B. microdonta, B. multiflosculosa, B. neoaustralis*, B. nitida, B. oblongifolia, B. orbignyana, B. papillosa*, B. pedunculata, B. pentlandii, B. phylicoides, B. platypoda, B. polifolia, B. polycephala, B. potosiensis*, B. pumila*, B. punctulata, B. quitensis, B. sagittalis, B. salicifolia, B. saliens, B. samensis, B. sculpta, B. serranoi*, B. sessiliflora, B. solomonii*, B. subalata, B. tola, B. torricoi*, B. tridentata, B. trimera, B. trinervis, B. tucumanensis, B. ulicina, B. vargasii*, B. viscosissima, B. woodii*, B. yungensis*, B. zongoensis*.*

Brazil. 185 species, 114 of them endemic*: *B. albilanosa*, B. aliena, B. alleluia*, B. alpestris*, B. altimontana*, B. angusticeps*, B. anomala, B. aphylla, B. apicifoliosa*, B. aracatubaensis*, B. arenaria, B. articulata, B. axillaris*, B. bifrons*, B. brevifolia, B. breviseta, B. burchellii*, B. calvescens*, B. campos-portoana*, B. caprariifolia, B. chilco, B. chionolaenoides*, B. ciliata*, B. cinerea*, B. claussenii*, B. cognata, B. concinna*, B. conyzoides*, B. coridifolia, B. coronata*, B. crassipappa*, B. cultrata, B. curitybensis*, B. deblei*, B. decurrens*, B. deltoidea*, B. densa, B. dentata, B. dichotoma*, B. dracunculifolia, B. dubia*, B. dunensis*, B. elliptica*, B. erigeroides, B. erioclada, B. flexuosiramosa*, B. floccosa*, B. friburgensis*, B. gaucha*, B. gibertii, B. glaziovii, B. glutinosa, B. gnaphalioides, B. gracilis, B. gracillima*, B. grandimucronata*, B. helichrysoides, B. hemiptera*, B. hirta, B. humilis*, B. hyemalis*, B. hypericifolia*, B. illinita, B. illinitoides, B. imbricata*, B. incisa*, B. inexspectata*, B. intermixta*, B. itapirocensis*, B. itatiaiae*, B. jocheniana, B. junciformis, B. lateralis, B. leptospermoides*, B. leucocephala*, B. leucopappa*, B. ligustrina*, B. linearifolia, B. longiattenuata, B. lychnophora*, B. lymanii*, B. macrophylla*, B. magnifica*, B. malmei*, B. maritima, B. maxima*, B. megapotamica, B. mesoneura*, B. microcephala, B. microdonta, B. milleflora*, B. minutiflora*,*

B. montana, B. multifolia, B. myricifolia*, B. myriocephala, B. napaea*, B. nebularis*, B. notosergila, B. nummularia*, B. obdeltata*, B. oblongifolia, B. ochracea, B. opuntioides*, B. orbiculata*, B. orbignyana, B. oreophila*, B. organensis*, B. oxyodonta, B. palustris, B. pampeana*, B. paranensis*, B. parvidentata*, B. patens, B. pauciflosculosa*, B. pedersenii, B. penningtonii, B. pentaptera*, B. pentodonta, B. perlata*, B. petraea, B. phylicifolia*, B. phyteumoides, B. pingraea, B. platypoda, B. pluricapitulata*, B. pohlii*, B. polygona*, B. polyphylla*, B. pseudoalpestris*, B. pseudomyriocephala*, B. pseudovaccinioides*, B. pseudovillosa*, B. psiadioides, B. punctulata, B. quitensis, B. ramboi*, B. ramosissima*, B. regnellii*, B. reticularia*, B. reticularioides*, B. retusa, B. riograndensis, B. rivularis*, B. rufidula*, B. sagittalis, B. salicifolia, B. salzmannii*, B. scabrifolia*, B. schomburgkii, B. schultzii*, B. scopulorum*, B. semiserrata, B. serrula*, B. serrulata*, B. sessiliflora, B. simplex*, B. singularis*, B. sphagnophila*, B. sphenophylla*, B. spicata, B. stylosa*, B. subdentata, B. suberectifolia*, B. subopposita, B. tarchonanthoides*, B. triangularis, B. tridentata, B. trilobata*, B. trimera, B. trinervis, B. trineura*, B. truncata*, B. uleana*, B. umbellata*, B. uncinella*, B. urvilleana*, B. variabiliflora*, B. vernicosa, B. vincifolia*, B. vismioides*, B. vitis-idaea, B. vulneraria, B. wagenitzii*, B. weirii.*

Canada. 1 species: *B. halimifolia.*

Chile. 48 species, 15 endemics*: *B. acaulis, B. alnifolia, B. alpina, B. austropedicellata, B. boliviensis, B. chrysophylla, B. concava*, B. cymosa*, B. elaeoides, B. genistelloides, B. glutinosa, B. gnidiifolia, B. incarum, B. juncea, B. linearis, B. lycioides, B. macraei*, B. magellanica, B. minor*, B. mylodontis*, B. neaei, B. neotruncata*, B. nivalis, B. obovata, B. ocellata, B. palenae*, B. paniculata*, B. patagonica, B. pilcensis*, B. pingraea, B. poeppigiana, B. pycnantha*, B. pycnocephala, B. racemosa, B. rhomboidalis*, B. sagittalis, B. salicifolia, B. santelicis, B. santiagensis*, B. scandens, B. spartioides, B. sphaerocephala*, B. taltalensis*, B. tola, B. umbelliformis, B. vernalis*, B. viscosissima, B. zoellneri.*

Colombia. 39 species, 15 endemic*: *B. alpina, B. antioquensis*, B. bogotensis*, B. boyacensis*, B. breviseta, B. buddlejoides, B. buxifolia, B. caldasiana*, B. chilco, B. cochensis*, B. decussata, B. farallonensis*, B. floribundoides*, B. fraterna*, B. genistelloides, B. grandiflora, B. jelskii, B. klattii, B. latifolia, B. lehmannii*, B. lloensis, B. marcetiifolia, B. mutisiana, B. nitida, B. oblongifolia, B. odorata*, B. padifolia, B. paramicola*, B. pedunculata, B. prunifolia, B. raulii*, B. revoluta*, B. rupicola*, B. salicifolia, B. teindalensis, B. tetroica, B. tricuneata, B. trinervis, B. vacciniifolia*.*

Costa Rica. 2 species: *B. pedunculata, B. trinervis.*

Cuba. 9 species, 6 endemic*: *B. acutata*, B. angustior*, B. dioica, B. glomeruliflora, B. nipensis*, B. orientalis*, B. scoparia, B. scoparioides*, B. shaferi*.*

Dominica. 1 species: *B. pedunculata.*

Dominican Republic. 4 species: *B. dioica, B. haitiensis, B. mornicola, B. myrsinites.*

Ecuador. 38 species, 10 endemic*: *B. alpina, B. angelica*, B. arbutifolia*, B. auriculigera, B. balnearia*, B. buddlejoides, B. buxifolia, B. chilco, B. cutervensis, B. decussata, B. genistelloides, B. gnidiifolia, B. grandiflora, B. hambatensis*, B. hieronymi*, B. huairacajensis*, B. humifusa*, B. jelskii, B. klattii, B. latifolia, B. lloensis, B. marcetiifolia, B. mollis*, B. nitida, B. oblongifolia, B. padifolia, B. pedunculata, B. quitensis, B. reticulata, B. sagittalis, B. salicifolia, B. sinuata, B. steetzii*, B. subbimera, B. teindalensis, B. tenuicapitulata*, B. trinervis, B. volubilis.*

El Salvador. 6 species: *B. confertoides, B. monoica, B. pedunculata, B. salicifolia, B. serrifolia, B. trinervis.*

Grenada. 0 species.

Guatemala. 8 species: *B. confertoides, B. glandulifera, B. heterophylla, B. monoica, B. pedunculata, B. salicifolia, B. serrifolia, B. trinervis.*

Guyana. 7 species: *B. densa, B. linearifolia, B. nitida, B. oblongifolia, B. schomburgkii, B. trinervis, B. vitis-idaea.*

Haiti. 4 species: *B. dioica, B. haitiensis, B. mornicola, B. myrsinites.*

Honduras. 6 species: *B. confertoides, B. monoica, B. pedunculata, B. salicifolia, B. serrifolia, B. trinervis.*

Jamaica. 2 species: *B. dioica, B. scoparia.*

Mexico. 46 species, 22 endemic*: *B. alamosana*, B. bigelovii, B. brachyphylla, B. brevipappa*, B. charucoensis*, B. conferta*, B. confertoides, B. crassicuneata*, B. dioica, B. douglasii, B. emoryi, B. erosoricola*, B. glandulifera, B. halimifolia, B. herbacea*, B. heterophylla, B. horizontalis*, B. lancifolia, B. macrocephala*, B. mexicana*, B. monoica, B. multiflora*, B. neglecta, B. nesomiana*, B. occidentalis*, B. palmeri*, B. pedunculata, B. pilularis, B. praetermissa*, B. pteronioides, B. ramiflora*, B. salicifolia, B. salicina, B. sarothroides, B. seemannii*, B. sergiloides, B. serrifolia, B. sordescens*, B. sulcata, B. supplex*, B. texana, B. thesioides, B. trinervis, B. wrightii, B. zamoranensis*, B. zamudiorum*.*

Nicaragua. 4 species: *B. monoica, B. pedunculata, B. serrifolia, B. trinervis.*

Panama. 2 species: *B. pedunculata, B. trinervis.*

Paraguay. 47 species, 2 endemic*: *B. amambayensis*, B. anomala, B. articulata, B. breviseta, B. caprariifolia, B. chilco, B. cognata, B. coridifolia, B. debilis, B. dentata, B. dracunculifolia, B. erigeroides, B. erioclada, B. frenguellii, B. glaziovii, B. glutinosa, B. gracilis, B. helichrysoides, B. illinita, B. illinitoides, B. isabelae*, B. linearifolia, B. megapotamica, B. microcephala, B. microdonta, B. montana, B. myriocephala, B. notosergila, B. oxyodonta, B. pedersenii, B. pentodonta, B. petraea, B. phyteumoides, B. pingraea, B. punctulata, B. retusa, B. sagittalis, B. salicifolia, B. semiserrata, B. sessiliflora, B. spicata, B. subdentata, B. subopposita, B. tridentata, B. trimera, B. trinervis, B. vulneraria.*

Peru. 61 species, 21 endemic*: *B. acaulis, B. alaternoides*, B. alnifolia, B. alpina, B. arguta, B. auriculigera, B. ayacuchensis*, B. boliviensis, B. buchtienii, B. buxifolia, B. caespitosa, B. chachapoyasensis*, B. chilco, B. clavata, B. corymbosa*, B. cutervensis, B. davidsonii*, B. debilis, B. decussata, B. dependens*, B.*

*emarginata**, *B. genistelloides, B. glutinosa, B. gnidiifolia, B. grandicapitulata**, *B. hutchisonii**, *B. incarum, B. jelskii, B. johnwurdackiana**, *B. latifolia, B. ledifolia**, *B. libertadensis**, *B. longipedicellata, B. multibracteata**, *B. nitida, B. oblongifolia, B. pachycephala**, *B. pedunculata, B. pentlandii, B. phylicoides, B. platypoda, B. polycephala, B. procumbens**, *B. quitensis, B. reticulata, B. sagittalis, B. salicifolia, B. saliens, B. scabra**, *B. scandens, B. sinuata, B. spartea**, *B. subalata, B. subbimera, B. tarmensis**, *B. tomentosa**, *B. trimera, B. trinervis, B. uniflora**, *B. volubilis, B. woytkowskii**.

Saint Kitts and Nevis. 1 species: *B. pedunculata*.

Saint Lucia. 1 species: *B. pedunculata*.

Saint Vincent and the Grenadines. 1 species: *B. pedunculata*.

Suriname. 1 species: *B. linearifolia*.

Trinidad and Tobago. 2 species: *B. nervosa, B. trinervis*.

United States of America. 23 species, 4 endemic*: *B. angustifolia, B. bigelovii, B. brachyphylla, B. dioica, B. douglasii, B. emoryi, B. glabrata**, *B. glomeruliflora, B. halimifolia, B. malibuensis**, *B. neglecta, B. pilularis, B. plummerae**, *B. pteronioides, B. salicifolia, B. salicina, B. sarothroides, B. sergiloides, B. sulcata, B. texana, B. thesioides, B. vanessae**, *B. wrightii*.

Uruguay. 54 species, 5 endemic*: *B. aliena, B. anomala, B. arenaria, B. artemisioides, B. articulata, B. brevifolia, B. breviseta, B. caprariifolia, B. cognata, B. coridifolia, B. crispa**, *B. cultrata, B. darwinii, B. dracunculifolia, B. funkiae**, *B. genistifolia, B. gibertii, B. glutinosa, B. gnaphalioides, B. hirta, B. jocheniana, B. juncea, B. junciformis, B. linearifolia, B. longiattenuata, B. maritima, B. megapotamica, B. microcephala, B. microdonta, B. notosergila, B. ochracea, B. oxyodonta, B. palustris, B. patens, B. pedersenii, B. penningtonii, B. petraea, B. phyteumoides, B. pingraea, B. psiadioides, B. punctulata, B. rectialata**, *B. riograndensis, B. salicifolia, B. saltensis**, *B. sessiliflora, B. spicata, B. stenophylla, B. subtropicalis**, *B. tridentata, B. trimera, B. vernicosa, B. vulneraria, B. weirii*.

Venezuela. 19 species, 3 endemic*: *B. debilis, B. decussata, B. densa, B. erectifolia**, *B. latifolia, B. linearifolia, B. meridensis**, *B. mutisiana, B. nitida, B. oblongifolia, B. pedunculata, B. prunifolia, B. salicifolia, B. schomburgkii, B. tetroica, B. tricuneata, B. trinervis, B. vitis-idaea, B. zumbadorensis**.

Dependent Territories

Anguilla. 0 species.

Aruba. 0 species.

Bermuda. 1 species: *B. glomeruliflora*.

Bonaire. 0 species.

British Virgin Islands. 0 species.

Cayman Islands. 1 species: *B. dioica*.

Clipperton Island. 0 species.

Curaçao. 0 species.

Falkland/Malvinas Islands. 2 species: *B. magellanica, B. zoellneri*.

French Guiana. 0 species.

Greenland. 0 species.
Guadeloupe. 1 species: *B. nervosa.*
Martinique. 2 species: *B. nervosa, B. pedunculata.*
Montserrat. 2 species: *B. dioica, B. pedunculata.*
Navassa Island. 0 species.
Puerto Rico. 2 species: *B. dioica, B. myrsinites.*
Saba. 1 species: *B. pedunculata.*
Saint Barthélemy. 0 species.
Saint Martin. 0 species.
Saint Pierre and Miquelon. 0 species.
Sint Eustatius. 0 species.
Sint Maarten. 0 species.
South Georgia and the South Sandwich Islands. 0 species.
Turks and Caicos Islands. 0 species.
United States Virgin Islands. 1 species: *B. dioica.*

Going Worldwide? List of Introduced Naturalized Adventitious Distributions Outside the Americas

Records of established self-sustained populations of *Baccharis* species, historically recorded outside their native range but currently eradicated or still thriving and under no efficient human intervention for extirpation, are presented as follows.

EUROPE.
Belgium. France. Georgia. Great Britain. The Netherlands. Spain. *B. halimifolia.*
Portugal. *B. spicata.*

OCEANIA.
Australia (New South Wales, Queensland, Victoria, Western Australia).
New Zealand. *B. halimifolia.*

Acknowledgments I would like to acknowledge financial support from CNPq/ Brazil (314590/2020-0), FAPERGS (17/2551-0000333-0), IAPT Research Grants in Plant Systematics 2010, the Smithsonian Institution's 2011 Cuatrecasas Fellowship Award, the Systematics Research Fund 2015/2016, and LinnéSys 2020/2021. I also would like to thank Carolina Moriani Siniscalchi and Benoît Patrice Francis Loeuille for their invaluable comments and contributions for the improvement of this chapter.

References

Acevedo-Rodríguez P, Strong MT (2007) Catalogue of the seed plants of the West Indies. Smithsonian Institution. https://naturalhistory2.si.edu/botany/WestIndies/. Accessed 30 May 2020

Alain H (1962) Flora de Cuba, 5. Rubiales – Valerianales – Cucurbitales – Campanulales – Asterales. Universidad de Puerto Rico, Río Piedras

Alboff N, Kurtz F (1896) Contributions a la Flore de la Terre de Feu. Revista Mus. La Plata 7:1–277

Ávila F, Funk VA, Diazgranados M, Díaz-Piedrahíta S, Vargas O (2020) *Baccharis* (Asteraceae). In: Bernal R, Gradstein SR, Celis M (eds) Catálogo de plantas y líquenes de Colombia. Universidad Nacional de Colombia. http://catalogoplantasdecolombia.unal.edu.co/en/resultados/genero/baccharis/ Accessed 30 May 2020

Beltrán H, Granda A, León B, Sagástegui A, Sánchez I, Zapata M (2006) Asteraceae endémicas del Perú. Rev Peru Biol 13:64s–164s

Brako L, Zarucchi JL (1993) Catalog of the flowering plants and gymnosperms of Peru. Monogr Syst Bot Miss Bot Gard 45:1–1286

Candolle AP de (1836) Asteroideae. In: Candolle AP de (ed) Prodromus systematis naturalis regni vegetabilis, vol 5. sumptibus sociorum Treuttel et Würtz, Parisiis [Paris], pp 201–497. https://doi.org/10.5962/bhl.title.286

Caño L, Campos JA, García-Magro D, Herrera M (2013) Replacement of estuarine communities by an exotic shrub: distribution and invasion history of *Baccharis halimifolia* in Europe. Biol Invasions 15:1183–1188

Cuatrecasas J (1967) Revisión de las especies Colombianas del género *Baccharis*. Revista Acad Colomb Ci Exact 13:5–102

Faini F, Hellwig FH, Labbé C, Castillo M (1991) Hybridization in the genus *Baccharis*: *Baccharis linearis* × *B. macraei*. Biochem Syst Ecol 19:53–57

Fielding RR (2001) *Baccharis*: a genus of the Asteraceae new to Canada. Proc Trans Nova Scotian Inst Nat Sci 41:214–215

Fried G, Caño L, Brunel S, Beteta E, Charpentier A, Herrera M, Starfinger U, Panetta FD (2016) Monographs on invasive plants in Europe: *Baccharis halimifolia* L. Bot Lett 163:127–153

Funk VA, Hollowell TH, Berry PE, Kelloff CL, Alexander SN (2007) Checklist of the plants of the Guiana Shield (Venezuela: Amazonas, Bolivar, Delta Amacuro; Guayana, Surinam, French Guiana). Contr US Natl Herb 55:1–584

GBIF (Global Biodiversity Information Facility) (2020) GBIF.org – GBIF Occurrence Download. https://doi.org/10.15468/dl.dsbcpc. Accessed 29 June 2020

Giuliano DA, Plos A (2014) Género *Baccharis*. In: Zuloaga FO, Belgrano MJ, Anton AM (eds) Flora Argentina: flora vascular de la República Argentina, vol 7. Instituto de Botánica Darwinion, San Isidro, pp 43–123

Gonzáles P, Cano A, Müller J (2019) An unusual new record of *Baccharis* (Asteraceae) from the Peruvian Andes andits relation with the northern limit of the dry puna. Acta Bot Mex 126(e1393):1–11

Greuter W, Rodriguez RR (2017) Vascular plants of Cuba. A preliminary checklist. Second, updated edition of The Spermatophyta of Cuba, with Pteridophyta added. Botanischer Garten & Botanisches Museum Berlin, Jardín Botánico Nacional, Universidad de La Habana

Heering W (1903) *Baccharis* L. In: Reiche C (ed) Estudios críticos sobre la Flora de Chile. Anales Univ Chile 111:151–196

Heiden G (2013) Two new combinations in *Baccharis* (Asteraceae: Astereae). Phyton 78:1–2

Heiden G (2020) *Baccharis* (Asteraceae: Astereae). In. Flora do Brasil 2020 under construction. Jardim Botânico do Rio de Janeiro. http://floradobrasil.jbrj.gov.br/reflora/floradobrasil/FB5151. Accessed 30 May 2020

Heiden G, Pirani JR (2014) Two new species of *Baccharis* subgen. *Baccharis* (Asteraceae, Astereae) with single-flowered female capitula from the Serra do Cipó, Minas Gerais. Brazil Phytotaxa 164:141–148. https://doi.org/10.11646/phytotaxa.164.2.8

Heiden G, Pirani JR (2016) Taxonomy of *Baccharis* subgen. *Tarchonanthoides* (Asteraceae: Astereae: Baccharidinae), a group from the southeastern South American grasslands and savannas. Phytotaxa 241:1–70. https://doi.org/10.11646/phytotaxa.241.1.1

Heiden G, Baumgratz JFA, Esteves RL (2009) New species of *Baccharis* (Asteraceae: Astereae) from Rio de Janeiro State, Southeastern Brazil. J Bot Res Inst TX 3:139–145

Heiden G, Baumgratz JFA, Esteves RL (2012) *Baccharis* subgen. *Molina* (Asteraceae) no estado do Rio de Janeiro. Brasil Rodriguésia 63:649–687. https://doi.org/10.1590/S2175-78602012000300013

Heiden G, Leoni LS, Nakajima JN (2014) *Baccharis magnifica* (Asteraceae, Astereae): a striking new species endemic to the summits of Serra do Caparaó, southeastern Brazil. Phytotaxa 162:211–216. https://doi.org/10.11646/phytotaxa.162.4.3

Heiden G, Antonelli A, Pirani JR (2019) A novel phylogenetic infrageneric classification of *Baccharis* (Asteraceae: Astereae), a highly diversified American genus. Taxon 68:1048–1081. https://doi.org/10.1002/tax.12128

Hellwig FH (1990) Die Gattung *Baccharis* L. (Compositae-Astereae) in Chile. Mitt Bot Staatssamml München 29:1–456

Hellwig FH (2003) *Heterothalamus wagenitzii* (Compositae, Astereae), a new species from Brazil. Haussknechtia 9:135–139

Hind DJN (2011) An annotated preliminary checklist of the Compositae of Bolivia, ver. 2. Royal Botanic Gardens, Kew. http://www.kew.org/science/tropamerica/boliviacompositae. Accessed 30 May 2020

Hokche O, Berry PE, Huber O (2008) Nuevo Catálogo de la Flora Vascular de Venezuela. Fundación Instituto Botánico de Venezuela Dr. Tobias Lasser, Caracas

Lee CW, Thompson AE, Jones WD, Hogan L (1984) 'Centennial' *Baccharis* interspecific hybrid. HortScience 19:903

Malagarriga Heras RP [1976] (1977) Nomenclator baccharidinarum omnium. Mem Soc Ci Nat La Salle 37:129–224

Malagarriga Heras RP [Luis IT] (1949) De “Re-Botanica”. Bol Inform Inst Geobiol La Salle 2:9–11

Malagarriga Heras RP [Luis IT] (1954) Exsiccatae Baccharidinarum. I. Plantae Wilson-Hoehneana. Contrib Inst Geobiol La Salle Bras 3:3–20

Moreira-Muñoz A, Muñoz-Schick M, Marticorena A, Morales V (2016) Catálogo de Asteraceae (Compositae) de la Región de Arica y Parinacota, Chile. Gayana 73(2):226–267

Müller J (2006) Systematics of *Baccharis* (Compositae-Astereae) in Bolivia, including an overview of the genus. Syst Bot Monogr 76:1–341

Müller J (2013) World checklist of *Baccharis* L. (Compositae–Astereae). Universität Jena. http://www.spezbot.uni-jena.de/blog/publikationen/world-checklist-of-baccharis-l/. Accessed 3 May 2018

Nesom GL (1988) *Baccharis* sect. *Baccharidastrum* (Compositae: Astereae), including two monoecious and one dioecious species. Phytologia 65:169–173. https://doi.org/10.5962/bhl.part.7254

Nesom GL (2020) Revised subtribal classification of Astereae. Phyton 53:1–39

Pimm SL, Joppa LN (2015) How many plant species are there, where are they, and at what rate are they going extinct? Ann Missouri Bot Gard 100:170–176

POWO (2019) Plants of the World Online. Royal Botanic Gardens, Kew. http://www.plantsoftheworldonline.org/ Accessed 30 May 2020

Pruski J, Robinson H (2018) Asteraceae. In: Davidse G, Sousa-Sánchez M, Knapp S, Chiang-Cabrera F (eds) Fl. Mesoamer. 5:i–xix, 1–608

Robinson H, Nordenstam B, Lundin R, Jørgensen PM (1999) Asteraceae. In: Jørgensen PM, León-Yánez S (eds) Catalogue of the vascular plants of Ecuador. Monogr Syst Bot Missouri Bot Gard 75:260–314

Schneider AA, Heiden G, Boldrini II (2011) *Baccharis scopulorum*, a new species of section *Caulopterae* (Asteraceae: Astereae) from rocky cliffs of southern Brazil. Phytotaxa 15:9–14

Spegazzini C (1896) Plantae per Fuegiam a. c. s., anno 1882 collectae. Anales Mus Buenos Aires V:39–103

Sundberg SD, Bogler DJ (2006) *Baccharis*. In: Flora of North America. Editorial Committee. http://www.efloras.org/flora_page.aspx?flora_id=1. Accessed 30 May 2020

Tedesco PA, Bigorne R, Bogan AE, Giam X, Jézéquel C, Hugueny B (2012) Estimating how many undescribed species have gone extinct. Conserv Biol 28:1360–1370

Thompson AE (1985) New native plants for the arid southwest. Econ Bot 39:436–453

Thompson AE, Jones WD, Hogan L (1976) Breeding and selection of native landscape plant materials for arid regions. HortScience 11:330

Thompson AE, Lee CW, Gass RE (1995) Development of hybrid *Baccharis* plants for desert landscaping. HortScience 30:1357–1362

UKOT (United Kingdom Overseas Territories) (2020) UKOTs Online Herbarium. Royal Botanic Gardens. http://brahmsonline.kew.org/UKOT. Accessed 30 May 2020

Ulloa CM, Acevedo-Rodríguez P, Beck S, Belgrano MJ, Bernal R, Berry PE, Brako L, Celis M, Davidse G, Forzza RC, Gradstein SR, Hokche O, León B, León-Yánez S, Magill RE, Neill DA, Nee M, Raven PH, Stimmel H, Strong MT, Villaseñor JL, Zarucchi JL, Zuloaga FO, Jørgensen PM (2018 onwards) Vascular Plants of the Americas (VPA) Website. Missouri Botanical Garden. http://www.tropicos.org/Project/VPA. Accessed 30 May 2020

Valtierra V (2018) Revisión taxonómica de *Baccharis* L. sect. *Caulopterae* DC. (Asteraceae: Astereae) en Uruguay. Dissertation, Universidad de la República

Valtierra V, Bonifacino JM (2014) Revisión taxonómica de *Baccharis* sect. *Heterothalamus* (Asteraceae: Astereae) en Uruguay. Bol Soc Argent Bot 49:613–620

Vargas OM (2018) Reinstatement of the genus *Piofontia*: a phylogenomic-based study reveals the biphyletic nature of *Diplostephium* (Asteraceae: Astereae). Syst Bot 43:485–496. https://doi.org/10.1600/036364418X697210

Vargas OM, Ortiz EM, Simpson BB (2017) Conflicting phylogenomic signals reveal a pattern of reticulate evolution in a recent high-Andean diversification (Asteraceae: Astereae: *Diplostephium*). New Phytol 214:1736–1750. https://doi.org/10.1111/nph.14530

Verloove F, Dana ED, Alves P (2018) *Baccharis spicata* (Asteraceae), a new potentially invasive species to Europe. Plant Biosyst 152:416–426

Villaseñor JL (2016) Checklist of the native vascular plants of Mexico. Revista Mex Biodivers 87:559–902

Zanowiak DJ (1991) An analysis of systematic and phyletic relationships within the Baccharidinae (Asteraceae: Astereae). Dissertation, Texas A&M University, 82p

Zuloaga FO, Morrone O, Belgrano MJ (2008) Catálogo de las plantas vasculares del Cono Sur (Argentina, Sur de Brasil, Chile, Paraguay y Uruguay). Monogr Syst Bot Missouri Bot Gard 107:i-xcvi + 1–3348

Chapter 3
The Evolution of Genetic Studies on *Baccharis*

Patrícia de Abreu Moreira, Heloiza Navarro de Novaes, and G. Wilson Fernandes

Abstract The genetic studies on the genus *Baccharis* started in 1945 and accompanied the development of tools for genetics investigation. First of all, the karyotype of some *Baccharis* species was determined, followed by reports on the chromosome number for some species. The majority of the information on the molecular biology of the *Baccharis* genus was generated to clarify the taxonomic identity of the taxon. In the 2000s, an intraspecific genetic study with the rare and endemic *Baccharis concinna* using randomly amplified polymorphic DNA markers was performed in an altitudinal gradient in Southeastern Brazil. Despite the high genetic variability within populations of *B. concinna*, the populations studied were very similar, and genetic variability was not related to variation in altitude. It was an important study that marked the population genetic investigation on the genus *Baccharis*. Then, next-generation sequencing technology was used to develop microsatellite markers for *B. dracunculifolia*. This set of microsatellite markers was efficient in kinship and gene flow analyses, and a low combined probability of genetic identity was attained when the six loci were included in the analysis. Two other species, *B. concinna* and *B. aphylla*, were evaluated for the transferability of microsatellite markers developed for *B. dracunculifolia*. Five microsatellite markers that successfully amplified fragments were obtained both in *B. concinna* and *B. aphylla*. Otherwise, more genetic studies on *Baccharis* genus are called for as the importance of its species in community assembly and ecosystem services is increasing.

Keywords Genetic variability · Heterozygosity · Molecular biology · Microsatellite markers · Phylogeny

P. de Abreu Moreira (✉)
Laboratório de Genética Evolutiva e de Populações, Universidade Federal de Ouro Preto, Ouro Preto, Brasil

H. N. de Novaes · G. W. Fernandes
Ecologia Evolutiva & Biodiversidade, Departamento de Genética, Ecologia & Evolução / ICB, Universidade Federal de Minas Gerais, Belo Horizonte, Brasil

G. W. Fernandes et al. (eds.), *Baccharis*,
https://doi.org/10.1007/978-3-030-83511-8_3

Until the mid-1960s, the genetic diversity of populations was accessed by morphological traits as sizes, shapes, or color patterns. These kinds of markers contributed to broaden our knowledge about population genetics. However, there are innumerous limitations to morphological traits; for instance, genetic variation could be overestimated because of phenotypic plasticity (Freeland 2005). Later, the genetic diversity was accessed by the sizes, shapes, and numbers of chromosomes, which was used to reconstruct the evolutionary history of *Drosophila pseudoobscura* by Sturtevant and Dobzhansky (1936). Chromosomal variation was studied between species and populations, but there was no consistent relationship between morphological and chromosomal variation (Rowe et al 2004). Hence, the development of molecular markers revolutionized this scenario, and genetic variation could be accessed from polymorphic proteins or DNA sequences.

A very large amount of information about different species could be performed, which allowed quantifying genetic diversity, population subdivision, gene flow, effective population size, breeding structure, inbreeding depression, natural selection, and genetic drift. All these studies are predominantly intraspecific (within a particular species). Nevertheless, molecular markers as DNA sequences are used in both systematics and phylogenetic studies which focus on the species level of classification (Rowe et al 2004). All this genetic knowledge has become extremely important for the effective conservation of many species.

The studies on the genus *Baccharis* accompanied the development of tools for genetics investigation. Genetic studies of *Baccharis* started in 1945 when chromosome numbers of 28 families of angiosperms were described by Bowden (1945) and included *Baccharis pingraea*, *B. genistelloides*, *B. genistifolia*, *B. halimifolia*, and *B. phyteumoides*. This report contained polar views of meiotic stages and mitotic metaphases of different species. The same number of chromosomes was registered for all these species ($n = 9$ or $2n = 18$). On the other hand, reports on the chromosome number for species in the genus have increased dramatically (see Table 3.1), mainly to better understand the systematics and phylogeny of the group. The chromosome number of *B. dracunculifolia* ($n = 9$) was first described in 1970 using botanical materials collected in the state of Minas Gerais, Brazil (Coleman 1970). It was suggested that the ancestral chromosome number of the family Asteraceae is $n = 9$ (Solbrig et al. 1969; Solbrig 1977; Nesom 2000; Mota et al. 2016). While the majority of *Baccharis* species has a basic chromosome number of $x = 9$, some exceptions were detected as in *B. latifolia* (Turner et al. 1967; Powell and King 1969; Spooner et al. 1995), *B. glutinosa* (Ariza-Espinar 1974), and *B. salicifolia* (Solbrig et al. 1969), all of them with $n = 18$ or in *B. nitida* where $n = 25$ (Powell and King 1969). There is a description of $n = 10$ for *B. tricuneata* (Turner et al. 1967), but it is a questionable count since authors pointed that it is possible that a supernumerary or "B" chromosome was mistaken as a bivalent. In fact, B chromosome and some fragments were described in species of the genus as in *B. decussata* (Turner et al. 1967; Powell and King 1969), *B. flabellata* (Wulff et al. 1996), *B. prunifolia* (Powell and Cuatrecasas 1970, 1975), *B. punctulata* (Rozenblum et al. 1985), *B. serrifolia* (Anderson et al. 1974), and *B. thesioides* (Spellenberg and Ward 1988). However, as the genus *Baccharis* appears to exhibit chromosomal stability

Table 3.1 Gametic (*n*) and somatic (2*n*) chromosome numbers in *Baccharis*

Species	*n*	2*n*	References
B. acaulis (Wedd. ex R.E.Fr.) Cabrera	–	18 + 1B	Hellwig (1990)
B. aliena (Spreng.) Joch.Müll. (as *Heterothalamus alienus* (Spreng.) O. Kuntze)	9	–	Bernardello (1986)
B. anomala DC.	–	18	Ruas et al. (1989)
B. articulata (Lam.) Pers.	9	–	Rozenblum et al. (1985)
B. boliviensis (Wedd.) Cabrera (as *Heterothalamus boliviensis* Wedd.)	9	–	Rozenblum et al. (1985)
B. brachyphylla A.Gray	9	–	Keil and Pinkava (1976)
	9	–	Spellenberg and Ward (1988)
B. braunii (Polak.) Standl.	9	–	Anderson et al. (1974)
B. brevifolia DC.	9	–	Coleman (1970)
	9	–	Solbrig et al. (1969)
B. breviseta DC. (as *Baccharidastrum argutum* (Less.) Cabrera	9	–	Coleman (1968)
B. burchellii Baker	9	–	Coleman (1970)
B. buxifolia (Lam.) Pers. (as *B. revoluta* Kunth)	9	–	Hunziker et al. (1989)
B. chachapoyasensis Cuatrec.	ca. 9	–	Turner et al. (1967)
B. chilco Kunth	9	–	Turner et al. (1967)
B. coridifolia DC.	9	–	Ariza-Espinar (1974)
	9	18	Hunziker et al. (1990)
B. cutervensis Hieron. (as *B. spathulata* Klatt)	9	–	Turner et al. (1967)
B. darwinii Hook. & Arn. (as *B. heterothalamoides Britton*)	9	–	Turner et al. (1979)
B. decussata (Klatt) Hieron.	9 + 3–5 fragments	–	Powell and King (1969)
	9 + 1B	–	Turner et al. (1967)
B. dracunculifolia DC.	9	–	Casas (1981)
	9	–	Coleman (1970)
B. effusa Griseb.	9 II	–	Wulff et al. (1996)
B. elaeoides Remy	–	18	Hellwig (1990)
B. flabellata Hook. & Arn.	9II	–	Wulff et al. (1996)
B. flabellata Hook. & Arn. var. *argentina* (Heering) Ariza	9II	–	Wulff et al. (1996)
B. genistelloides (Lam.) Pers.	9	–	Coleman (1968)
	9	–	Bowden (1945)
B. genistifolia DC.	9	–	Bowden (1945)

(continued)

Table 3.1 (continued)

Species	*n*	2*n*	References
B. glutinosa Pers.	ca. 9	–	Anderson et al. (1974)
B. glutinosa Pers. (as *B. douglasii* DC.)	9	–	Keil and Pinkava (1976)
	9	–	Solbrig et al. (1964)
B. glutinosa Pers. (as *B. pingraea* DC.)	9	–	Bowden (1945)
	9	–	Covas and Schnack (1946)
	18	–	Ariza-Espinar (1974)
	9	–	Hunziker et al. (1989)
	9	–	Turner et al. (1979)
B. gnidiifolia Kunth (as *B. sternbergiana* Steud.)	9 II	–	Sundberg et al. (1986)
B. grandicapitulata Hieron.	9	–	Turner et al. (1967)
	9II	–	Sundberg et al. (1986)
B. halimifolia L.	9	–	Bowden (1945)
	9	–	Westman et al. (1975)
B. helichrysoides DC.	9	–	Coleman (1968)
B. heterophylla Kunth	9	–	Keil and Stuessy (1977)
B. latifolia (Ruiz & Pav.) Pers.	ca. 18	–	Powell and King (1969)
	18	–	Spooner et al. (1995)
	ca. 18	–	Turner et al. (1967)
	–	18	Müller (2006)
B. linearifolia (Lam.) Pers. (as *B. leptophylla* DC.)	9	–	Hunziker et al. (1989)
B. linearifolia (Lam.) Pers. (as *B. rufescens* Spreng.)	9	–	Ariza-Espinar (1974)
B. linearifolia (Lam.) Pers. (as *B. subcapitata* Gardner)	ca. 9		Turner et al. (1967)
B. ligustrina DC.	9	–	Coleman (1970)
B. linearis (Ruiz & Pav.) Pers.	9	–	Jansen and Stuessy (1980)
	9	–	Hunziker et al. (1990)
	–	18	Hellwig (1990)
B. mexicana Cuatrec.	9	–	Jackson (1970)

(continued)

Table 3.1 (continued)

Species	*n*	2*n*	References
B. mesoneura DC.	9	–	Coleman (1970)
B. microdonta DC.	–	18	Müller (2006)
B. montana DC. (as *B. elaegnoides* Steud. ex Baker)	9	–	Coleman (1970)
B. multiflora Kunth	9	–	Jackson (1970)
B. nitida (Ruiz & Pav.) Pers.	25	–	Powell and King (1969)
B. nivalis (Wedd.) Sch. Bip. ex Phil.	9	–	Solbrig et al. (1964)
B. notosergila Griseb.	9	–	Hunziker et al. (1989)
B. oblongifolia (Ruiz & Pav.) Pers.	9	–	Turner et al. (1967)
B. obovata (Ruiz & Pav.) DC.	–	18	Hellwig (1990)
B. oxyodonta DC. (as *B. melastomifolia* Hook. & Arn.)	9	–	Turner and Irwin (1960)
B. oxyodonta DC.	9	–	Coleman (1968)
B. patagonica Hook. & Arn.	–	18	Dollenz (1976)
B. petiolata DC.	9	–	Hunziker et al. (1989)
B. phylicoides Kunth	9 II	–	Sundberg et al. (1986)
	ca. 9	–	Turner et al. (1967)
B. phyteumoides (Less.) DC.	9	–	Bowden (1945)
B. pilularis DC. subsp. *consanguinea* (DC.) C.B.Wolf	9	–	Raven et al. (1960)
B. pilularis DC. subsp. *pilularis*	9	–	DeJong and Montgomery (1963)
B. plummerae A.Gray	9	–	Raven et al. (1960)
B. poeppigiana DC. subsp. *ocellata* F.H. Hellwig	–	18 + 1B	Hellwig (1990)
B. polifolia Griseb.	9	–	Turner et al. (1979)
B. prunifolia Kunth	9 + 2B	–	Powell and Cuatrecasas (1970)
B. punctulata DC.	9	–	Carr et al. (1999)
	9	–	Rozenblum et al. (1985)
	9 + 4B	–	Rozenblum et al. (1985)
	9	–	Turner et al. (1979)
B. racemosa (Ruiz & Pav.) DC.	9	–	Jansen and Stuessy (1980)
B. rhomboidalis Remy var. *rhomboidalis*	–	18	Hellwig (1990)
B. rhomboidalis Remy subsp. *truncata* (Phil.) F.H. Hellwig	–	18	Hellwig (1990)

(continued)

Table 3.1 (continued)

Species	*n*	2*n*	References
B. riograndensis Teodoro & J.E.Vidal		18	Heiden et al. (2006)
B. rupicola Kunth	9	–	Jansen et al. (1984)
B. salicifolia (Rui & Pav.) Pers.	9	–	DeJong and Montgomery (1963)
	9	–	Jackson (1970)
	9	–	Keil and Stuessy (1975)
	9	–	Keil and Stuessy (1977)
	9	–	Pinkava and Keil (1977)
	9	–	Rozenblum et al. (1985)
	9	–	Solbrig et al. (1964)
	18	–	Solbrig et al. (1969)
	9	–	Turner and Irwin (1960
	ca. 9	–	Turner et al. (1979)
	–	18	Müller (2006)
B. salicifolia Pers. (as *B. lanceolata* Kunth)	9	–	Covas and Schnak (1946)
B. sarothroides A.Gray	9	–	DeJong and Montgomery (1963)
	9	–	Keil and Pinkava (1976)
B. scandens (Ruiz & Pav.) Pers. (as *B. alnifolia* Meyen & Walp.)	–	18	Dillon and Turner (1982)
B. sergiloides A.Gray	9	–	DeJong and Montgomery (1963)
B. serranoi H.Rob.	–	18	H. Rob.
B. serrifolia DC.	9 + 1 fragment + 2B	–	Anderson et al. (1974)
	–	18	Sundberg et al. (1986)
B. sordescens DC.	9 + 1B	–	Keil and Stuessy (1977)
	9	–	Powell and Turner (1963)
B. tarchonanthoides Baker	9	–	Coleman (1970)

(continued)

Table 3.1 (continued)

Species	*n*	2*n*	References
B. thesioides Kunth	9 + 3–4 s	–	Spellenberg and Ward (1988)
Baccharis tola Phil. subsp. *sanctelicis* (Phil.) Joch.Müll. (published as *B. santelicis* Phil.)	–	18	Hellwig (1990)
B. tricuneata (L. f.) Pers. (as *B. magellanica* (Lam.) Pers.)	9	–	Wulff (1984)
	–	18 + 1B	Hellwig (1990)
B. tricuneata Pers. (as *B. tricuneata* (L. f.) Pers. var. *callaenis* Cuatrec.	10	–	Turner et al. (1967)
B. tricuneata Pers. (as *B. tricuneata* (L. f.) Pers. var. *paramorum* Cuatrec.)	9	–	Powell and King (1969)
B. trinervis (lam.) Pers.	Ca. 9 + fragments	–	Anderson et al. (1974)
	9	–	Jackson (1970)
	9	–	Solbrig et al. (1969)
	9	–	Turner and Irwin (1960)
B. trinervis (Lam.) Pers. (as *B. trinervis* (Lam.) Pers. var. *rhexioides* (Kunth) Baker)	9	–	Powell and Cuatrecasas (1970)
	9 II	–	Wulff et al. (1996)
B. vulneraria Baker (as *Baccharidastrum triplinervium* (Less.) Cabrera))	9	–	Coleman (1970)
B. tucumanensis Hook. & Arn.	9 II	–	Wulff et al. (1996)
B. ulicina Hook. & Arn.	9	–	Ariza-Espinar (1974)
	9	–	Turner et al. (1979)
B. vanessae R.M.Beauch.	9	–	Beauchamp (1980)
Baccharis ventanicola (Cabrera) Soria & Zardini (published as *B. rufescens* var. *ventanicola* Cabrera)	9	–	Hunziker et al. (1989)
B. vernalis F.H.Hellw.	–	18	Hellwig (1990)
B. wrightii A.Gray	9 II + 1I		Powell and Powell (1977)
	9 II + 2B		Weedin and Powell (1978)

Adapted from Heiden et al. (2006)

(Solbrig et al. 1969; Solbrig 1977), molecular markers are needed to provide information about genetic diversity within and between populations.

The majority of the information on the molecular biology of the *Baccharis* genus was generated to clarify the taxonomic identity of this taxon. Despite being the fourth largest genus of the family Asteraceae and the most specious within the tribe Astereae (Heiden 2014), the evolutionary relationship of the genus is still in discussion. Zanowiak (1991) studied the systematic and phyletic relationships within the subtribe Baccharidinae. This chloroplast DNA study suggests that South American

Conyza spp. should be included in the subtribe Baccharidinae, that the Baccharidinae consists of some species of *Baccharis* (published as *Heterothalamus*) and *Archibaccharis* clades, while another clade includes South American *Exostigma notebellidiastrum* (published as *Conyza notebellidiastrum*), *Baccharis*, and *Baccharidastrum*. In this same study, Zanowiak (1991) verified that *B. neglecta* and *B. halimifolia* hybridize, with *B. neglecta* being the maternal parent. Some novelties towards a phylogenetic infrageneric classification of *Baccharis* were published by Heiden and Pirani (2016) which includes names of new taxa, new combinations, and names at new rank for subgenera and sections of the genus. Later, Heiden et al. (2019), based mostly on phylogenetic grounds, proposed that *Baccharis* should comprise 440 species classified into 7 subgenera and 47 sections.

In the 2000s, Gomes et al. (2004), for the first time, conducted an intraspecific genetic study in this genus. These authors investigated the genetic variability in *Baccharis concinna* using randomly amplified polymorphic DNA (RAPD) markers. This species is a rare, dioecious, and threatened shrub, endemic to Serra do Espinhaço, Southeastern Brazil. The authors studied 335 individuals belonging to 6 populations along an altitudinal gradient. Despite the high genetic variability within populations of *B. concinna*, the populations studied were very similar, and genetic variability was not related to the altitudinal gradient. The authors argued that their findings could be explained by the *B. concinna* mating system. This shrub is pollinated and dispersed by wind, which may promote an intense gene flow among the studied species patches, independent of elevation. The authors also emphasized the absence of a physical barrier to gene flow by pollen and seed dispersal among the studied patches of individuals in the landscape.

The RAPD technique was developed in the 1990s (Welsh and Mcclelland 1990; Williams et al. 1990); it is quick and easily generated by PCR and requires no prior DNA sequence information. Although RAPD markers were commonly used for genetic diversity in plants (e.g., Wachira et al. 1995; Iqbal et al. 1997; Ram et al. 2008), they are subject to some limitations. Due to the dominance of the RAPD markers, it is not possible to distinguish between homozygotes (one copy of allele) and heterozygotes (two copies of allele) individuals. Furthermore, the RAPD markers do not allow the investigation of direct gene flow using paternity analysis. In addition, RAPD markers are of limited reproducibility because the segments of DNA are amplified by PCR using arbitrary primers that copy genome regions according to the annealing temperature of user selection. Nowadays, the use of other molecular markers is needed to investigate the genetic diversity of plant species with confidence.

Alternatively, the microsatellite markers, also known as simple sequence repeats (SSRs) and short tandem repeats (STR) (Jacob et al. 1991; Edwards et al. 1991), have been used in population and conservation genetics studies (Guichoux et al. 2011). The microsatellite markers are repeating motifs in tandem that are found at high frequency in most taxa genomes and exhibit high levels of polymorphism due to the high mutation rate that make them more informative than other molecular markers (i.e., single nucleotide polymorphism – SNP) (Bhargava and Fuentes 2010). The microsatellite markers are relatively uniformly distributed in the genomes of species, and due to their co-dominance, the distinction between homozygote and heterozygote individuals is possible. Traditionally, microsatellite development was slow, costly, and labor-intensive and required the construction of genomic libraries using recombinant DNA enriched for a few targeted SSR motifs. The repeating motifs can be mono-, di-, tri-, tetra-, penta-, and hexanucleotide repeats (Litt and Luty 1989; Zane et al. 2002).

Dinucleotide microsatellite repeats are commonly produced by the genomic library technique. However, dinucleotide repeats are prone to polymerase slippage during the PCR amplification (slipped-strand mispairing) and suffer from stutter bands (PCR products from the same fragment that are shorter by one or a few repeats) (Chambers and MacAvoy 2000). Unfortunately, this feature of dinucleotide repeats may lead to genotype scoring errors (Clarke et al. 2001; McDowell et al. 2002) making allele definition difficult (Levinson and Gutman 1987; Meldgaard and Morling 1997), especially for heterozygotes with adjacent alleles (Guichoux et al. 2011). Besides that, tri-, tetra-, and pentanucleotide loci perform better than dinucleotides because they are less prone to enzyme slippage (Edwards et al. 1991; Acharige et al. 2012).

The conservation of the sequence in the primer sites flanking the microsatellite loci and the stability of those sequences during evolution (Dayanandan et al. 1997; Ciampi et al. 2008; Feres et al. 2009) allows the use of SSR markers developed from one species to another. The transfer of polymorphic markers in plants is mainly successful within genera, and it has been successfully applied to the genetic analysis of tropical species (Zucchi et al. 2002; Cota et al. 2012; Moreira et al. 2012). Thus, in the last two decades, microsatellite markers have been used to assess gene flow at the population level and recent demographic events and aided in phylogenetic inferences (Braga et al. 2007; Ciampi et al. 2008; Moreira et al. 2008; Cruz et al. 2012; Muñoz-Pajares et al. 2017; Larranaga et al. 2017; Morris and Shaw 2018).

Despite the great number of species in the genus *Baccharis*, there are microsatellite markers developed for *B. dracunculifolia* only (Belini et al. 2016). A set of 17 markers was developed for *B. dracunculifolia* using a genomic microsatellite library (Belini et al. 2016), but out of 17, 12 are dinucleotide microsatellite that implies genotype scoring errors detailed above. Besides, six of them were monomorphic for three *B. dracunculifolia* populations ($N = 315$ individuals), which reinforced the need to advance in the development of new microsatellite markers for this species.

1 Development of Microsatellite Markers for *Baccharis dracunculifolia* Using NGS Technology

A good strategy to develop microsatellite markers with better performance is using next-generation sequencing (NGS) technology (Zalapa et al. 2012; Ambreen et al. 2015; Bonatelli et al. 2015; Hodel et al. 2016). The NGS allows the rapid and efficient development of microsatellite markers for non-model organisms for ecological and evolutionary studies. Moreover, the advent of NGS provided a cheaper and faster microsatellite development (Guichoux et al. 2011). We followed this approach to develop microsatellite markers for *B. dracunculifolia*, as described next.

A genomic library was built from 100 µg of one individual of *B. dracunculifolia*, in which DNA was extracted from leaves, through paired-end strategy that was sequenced using MiSeq® platform (Illumina©, San Diego, CA) to produce paired-end 250 base reads. A total of 21.4 million reads were obtained, and we used the Perl script PAL_FINDER_v0.02.04 (see Castoe et al. 2012) to identify 11,296 potentially amplifiable locus (PAL) (Table 3.2). We extracted reads that contained perfect dinucleotide, trinucleotide, tetranucleotide, pentanucleotide, and hexanucleotide tandem SSRs, totaling 7,277 PALs.

The SSR repeat motifs consisted of 24.08% dinucleotide, 58.55% trinucleotide, 6.95% tetranucleotide, 5.35% pentanucleotide, and 5.04% hexanucleotide repeat units (Fig. 3.1a). The AT/TA motif repeats were the most abundant dinucleotide SSR, accounting for 76.32% of all dinucleotide repeats (Fig. 3.1b). The trinucleotide AAT/TAA motif repeats were the most abundant type, accounting for 27.20% of all trinucleotide repeat motifs, while ATA/TAT and ATT/TAA repeats accounted for 17.10% and 14.90%, respectively (Fig. 3.1c).

Then, a manual filtering step was performed to select exclusively SSR loci with long and perfect repeats motifs since they tended to be more polymorphic (Zalapa et al. 2012). Thus, we chose tri-, tetra-, penta-, or hexanucleotide repeats present in long reads, larger than 274 bp, and obtained 1356 microsatellite loci candidates for microsatellite markers. To ensure that SSR loci chosen for *B. dracunculifolia* could follow as much as possible the stepwise mutation model used in coalescent-based methods to infer demographic events as proposed by Estoup et al. (2001), we selected just perfect motifs. Thus, 36 primer pairs flanking the SSR loci were designed using Primer3 software (Untergasser et al. 2012). Following, a set of 17 perfect microsatellite markers were chosen for amplification screening using 15 *B. dracunculifolia* individuals from the rupestrian grassland vegetation in Serra do Cipó, Brazil (Fig. 3.2). The PCR products were viewed on a polyacrylamide gel electrophoresis (PAGE) 6% and stained with silver nitrate (Sanguinetti et al. 1994).

Table 3.2 Simple sequence repeat types in *Baccharis dracunculifolia* contigs sequences

Motif length	Number of SSR	Frequency (%)
Mononucleotide	523	4.6
Dinucleotide	1753	15.5
Trinucleotide	4261	37.7
Tetranucleotide	506	4.5
Pentanucleotide	390	3.5
Hexanucleotide	367	3.2
Compound	3240	28.7
Broken	256	2.3
Total	11,296	

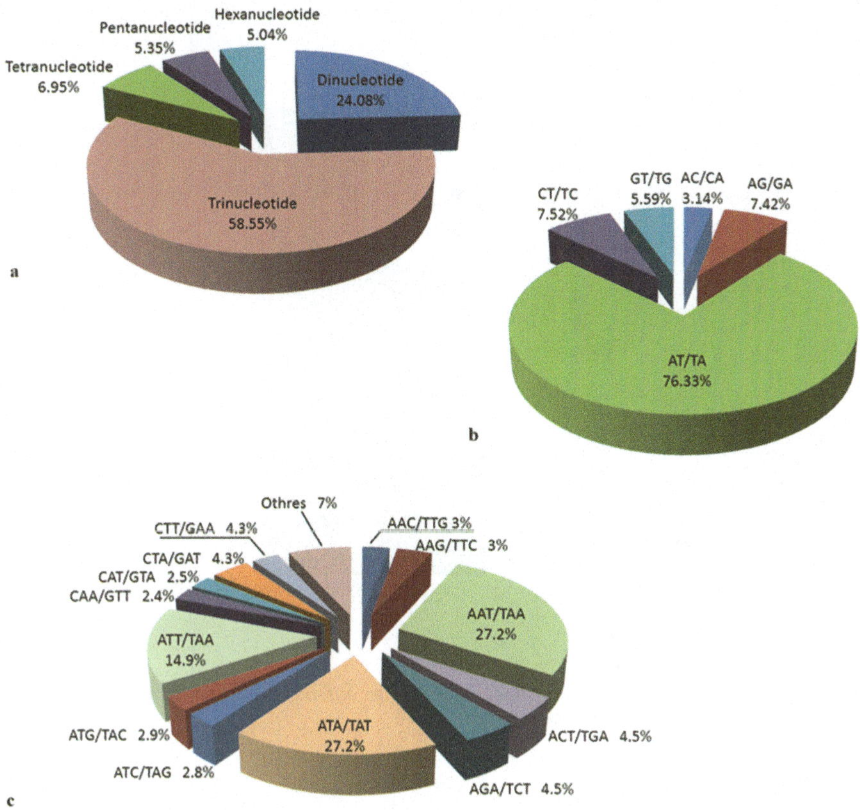

Fig. 3.1 Characteristics of SSR motifs identified in *Baccharis dracunculifolia* using NGS sequencing. (**a**) Frequency of di-, tri-, penta-, tetra-, and hexanucleotide SSR motif repeats. (**b**) Frequency of different dinucleotide SSR motifs. (**c**) Frequency of different trinucleotide SSR motifs

A total of 12 microsatellite loci (Bdr6, Bdr7, Bdr9, Bdr11, Bdr13, Bdr20, Bdr21, Bdr22, Bdr25, Bdr26, Bdr31, Bdr34) produced clear amplicons with expected size in the acrylamide gel (Fig. 3.3). Then, we designed all these 12 primer pairs, and the forward primers were marked with 4 dyes: VIC®, 6-FAM™, PET®, and NED™.

To assess the polymorphism and population genetic parameters with these microsatellite markers, we genotyped 60 individuals of *Baccharis dracunculifolia* from the Serra do Cipó region: 20 individuals between 760 and 839 m, 20 between 1026 and 1040 m, and 20 from 1348 to 1356 m altitudes (Fig. 3.2). The DNA was extracted from leaves using CTAB 2% protocol (Doyle and Doyle 1990). DNA purity and concentration were checked using NanoDrop 2000 Spectrophotometer (Thermo Fisher Scientific, Waltham, Massachusetts, USA).

All *B. dracunculifolia* individuals were genotyped with six microsatellite loci (Bdr6, Bdr7, Bdr22, Bdr25, Bdr26, Bdr31). PCR amplifications were performed in a 13 μL volume containing 10.0 μM of each primer, 1.5 μL of 5X special IVB PCR buffer, 1 unit of Taq DNA polymerase (Phoneutria, BR), 0.25 mM of each dNTP,

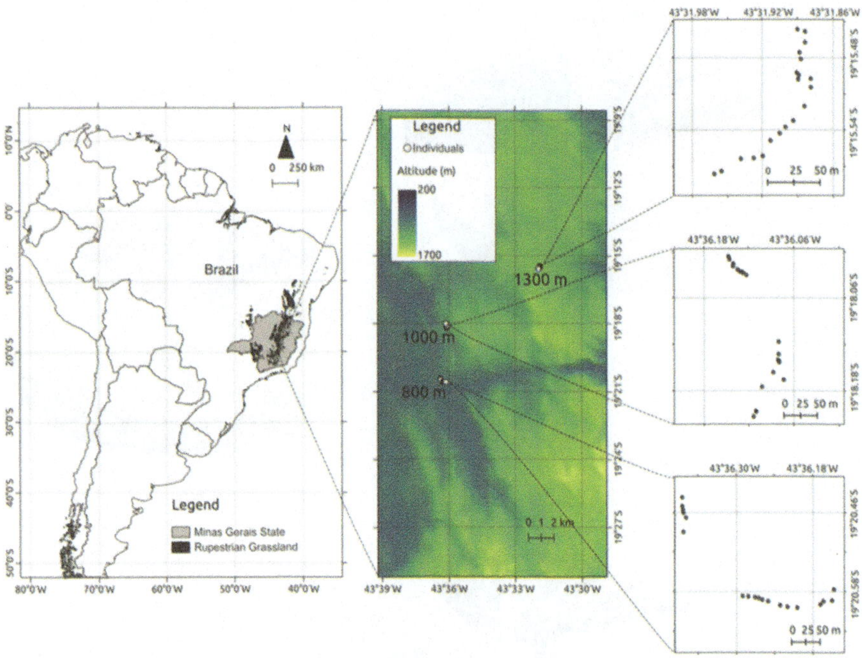

Fig. 3.2 Sites where individuals of *Baccharis dracunculifolia* were sampled in the rupestrian grassland vegetation in Serra do Cipó, Brazil

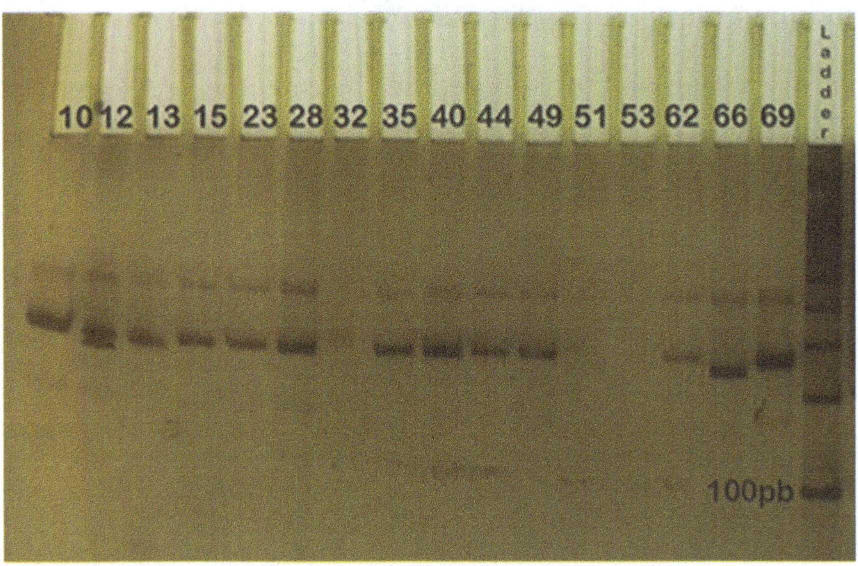

Fig. 3.3 An acrylamide gel stained with silver nitrate to verify polymorphism in Bdr6 microsatellite marker developed for *Baccharis dracunculifolia* before labeling them with fluorophores. The numbers above correspond to *B. dracunculifolia* individuals

Table 3.3 Characterization of 12 polymorphic microsatellite markers developed in *Baccharis dracunculifolia*

Locus	Primer sequence (5′–3′)	Repeat motif	Expected allele size (bp)	T_a (°C)	[$MgCl_2$]
Bdr6[a]	F: ACAGGCGGGAATACTTACCA R: CCCTAATGAAACCAGGACCA	$(AATT)_6$	231	60	2.15 mM
Bdr7[a]	F: GAGAAGGGGGAGAGGCTTTA R: CCCATTTTAAGGCTGTTTGA	$(AGAA)_6$	245	52	2.15 mM
Bdr9	F:GGAGCCGAAAGTGAAAAACA R: TGTTCAGCGGGTGTTGTAAA	(TGA)7	272	52	2.15 mM
Bdr11	F: TCCTTCATCTTGTTGCTCCA R: TGTCCGCCATTTTCTTCTCT	(GGAT)6	213	60	2.15 mM
Bdr13	F: GATGGTGGTTCGGGTAAGAA R: CGCCATTGAAATTGTTGTTG	(TATC)6	200	62	0.43 mM
Bdr20	F: CCCAAAGAAATGGATGAAGC R: TGGAATGGAGTTGTGTGTTGA	(TCTT)6	195	60	2.15 mM
Bdr21	F: TGCCACCATCTCTCTCTCTCT R: AATTAGCACCCACGCCATT	(TTTA)6	197	56	0.86 mM
Bdr22[a]	F: CCAATTTGAAACGACATGACTC R: CGGCTACGTCAACGACTATG	$(ATTT)_6$	157	58	0.43 mM
Bdr25[a]	F: GGAGCCGAAAGTGAAAAACA R: TGTTCAGCGGGTGTTGTAAA	$(TGA)_7$	272	52	2.15 mM
Bdr26[a]	F: AGCTGTTGTTGTGCCTGAGA R: GGATCGTCATCTCGTGTCCT	$(ATG)_8$	171	60	0.215 mM
Bdr31[a]	F: CCTGCATATTGAAAGCTCGTC R: GCTTGAATGACCCACGAAC	$(GCTCG)_5$	246	60	2.15 mM
Bdr34	F: CCGAGGCCAAATGAAATCT R: CTTGTCGAATGCCGAAAAAT	(TATTT)7	221	52	2.15 mM

[a]Microsatellite markers used to genotype a *Baccharis dracunculifolia* population

0.215–2.15 mM of $MgCl_2$ (according to each primer – Table 3.3), and 10.0 ng of template DNA. DNA amplification was accomplished in a PCR system (Veriti™ 96-Well Thermal Cycler; Applied Biosystems, California, USA) under the following conditions: 94 °C for 5 min (one cycle); 94 °C for 1 min, 52–62 °C for 1 min (according to each primer – Table 3.3); 72 °C for 1 min (35 cycles); and 72 °C for 50 min. The PCR products were electrophoresed on an ABI Prism 3730 automated DNA sequencer (Fig. 3.4) (Applied Biosystems, California, USA) and were sized by comparison to a *GeneScan* 500 LIZ dye Size Standard (Applied Biosystems, California, USA). Fluorescent PCR products were automatically sized using Geneious 10.2.3 (Kearse et al. 2012).

The number of alleles (*A*) ranged from 2 to 8 per locus, and the average number of alleles in this population was 5.33 (Table 3.4). Despite the use of just 6 microsatellite markers, we found higher allelic richness than Belini et al. (2016) using 11 microsatellite markers based on 315 samples from 3 populations. Besides, Belini et al. (2016) developed six other microsatellite markers which were monomorphic in all these individuals. All these monomorphic markers were dinucleotide repeats,

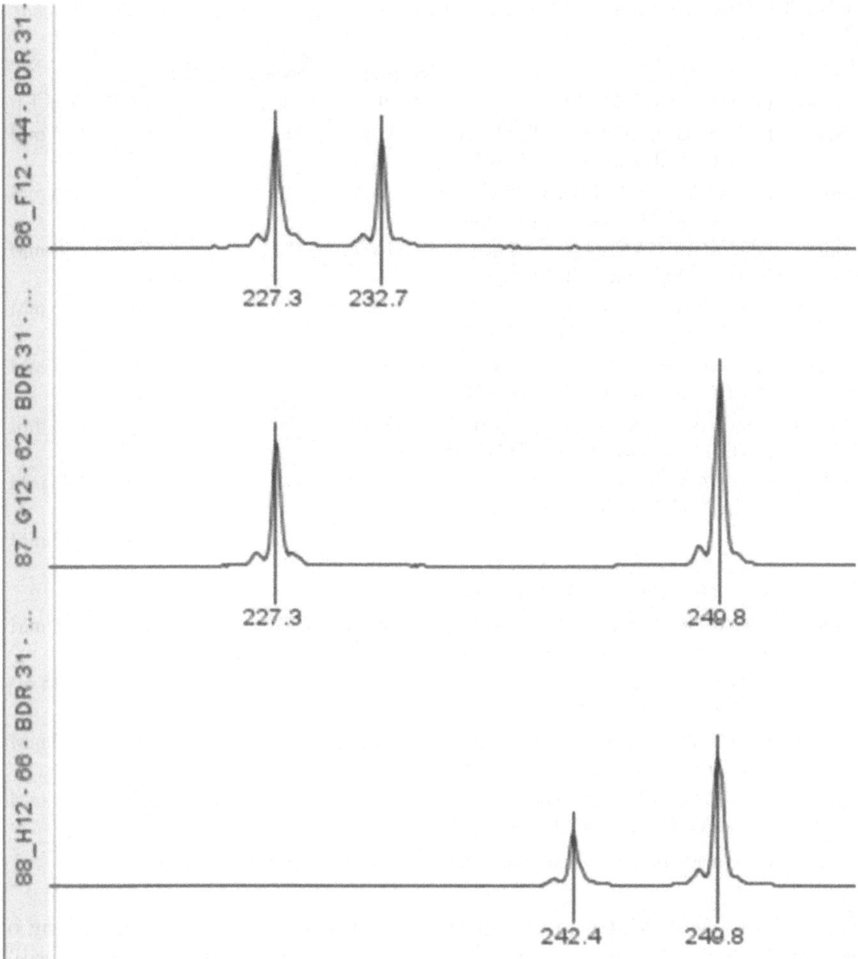

Fig. 3.4 Microsatellite profile of Bdr31 marker developed for *Baccharis dracunculifolia* detecting different heterozygote peaks

and the polymorphic markers were composed of six dinucleotide and five compound markers. Also, these authors used the traditional genomic microsatellite library to develop microsatellite markers. This result reinforces the better outcomes when using NGS to identify microsatellites that could enable the selection of tri-, tetra-, or pentanucleotide motif repeats to avoid "stutter" bands and genotype scoring errors because of dinucleotide repeat. All microsatellite markers developed by us have tri- or more motif repeats, and all of them are perfect microsatellite markers, which are considered more polymorphic. These characteristics highlight the potential of our markers.

Table 3.4 Characterization of 6 microsatellite loci based on a sample of 60 adult individuals of *Baccharis dracunculifolia* from Serra do Cipó, Brazil

Locus	Alleles range (bp)	A	Ho	He	Fis	Q	I
Bdr6	131–234	8	0.231	0.808	0.714***	0.743	0.061
Bdr7	225–253	8	0.545	0.594	0.082***	0.743	0.187
Bdr22	152–168	5	0.333	0.605	0.449***	0.595	0.210
Bdr25	270–273	2	0.286	0.408	0.300ns	0.187	0.433
Bdr26	164–170	3	0.789	0.547	−0.443*	0.370	0.270
Bdr31	225–249	6	0.583	0.659	0.115**	0.659	0.178
Over all loci	194–224	5.33	0.461	0.603	0.202	QC = 0.995	IC = 5.1×10^{-05}

A number of alleles, *He* expected heterozygosity, *Ho* observed heterozygosity, *Fis* fixation index, *Q* probability of paternity exclusion, *QC* combined probability of paternity exclusion, *I* probability of genetic identity, *IC* combined probability of genetic identity

*Loci deviating from HWE equilibrium after Bonferroni corrections (values followed by ns did not statistically differ from zero, $^{*}p < 0.05$, $^{**}p < 0.01$, $^{***}p < 0.001$)

The observed heterozygosity (*Ho*) ranged from 0.231 to 0.789 per locus, and the average was 0.461. The expected heterozygosity (*He*) ranged from 0.408 to 0.807 per locus, and the average was 0.603. For most loci (except Bdr25), the observed heterozygosity was lower than expected under the Hardy-Weinberg equilibrium (HWE), with fixation indexes (*Fis*) significantly different from zero (Table 3.4). These HWE deviations may be the presence of null alleles or due to the low number of analyzed individuals, which must be insufficient to reveal all possible genotypic combinations (López-Márquez et al. 2016), hence leading to a possible underestimation of allele frequencies and heterozygosity (McInerney et al. 2011). In addition, the observed heterozygosity was lower than expected under HWE which may be due to excess homozygotes since all evaluated individuals belong to only one population and inbreeding can occur in this population.

The probability of excluding two individuals as related when they are not was 74.3% with the Bdr6 locus. However, the combined probability of paternity exclusion increased to 99.5% when the six loci were included in the analysis (QC = 0.995 – Table 3.4), indicating that this set of microsatellite markers is efficient in kinship and gene flow analyses. The probability of genetic identity (I) ranged from 0.187 to 0.934 per locus (Table 3.3), and a low combined probability of genetic identity (IC) was attained when the six loci were included in the analysis (IC = 5.1×10^{-05} – Table 3.3). Although some loci presented a significant excess of homozygotes, the higher combined probability of paternity exclusion and lower combined probability of genetic identity show that this battery of microsatellite markers is suitable for population genetic analyses.

2 Cross-Amplification in *Baccharis*

We sampled other two species, *B. concinna* and *B. aphylla*, to evaluate the transferability of microsatellite markers developed for *B. dracunculifolia*. All of these three species occur in sympatry in the rupestrian grasslands of Serra do Cipó. We collected leaves from eight individuals of *B. concinna* and nine individuals of *B. aphylla* and extracted their DNA using the CTAB 2% protocol (Doyle and Doyle 1990). DNA purity and concentration were checked using NanoDrop 2000 Spectrophotometer (Thermo Fisher Scientific, Waltham, Massachusetts, USA).

The cross-amplification analysis was realized with the six microsatellite markers characterized in *B. dracunculifolia* (see Table 3.3) under the same PCR conditions used to amplify *B. dracunculifolia*. The PCR fragments were viewed on a 6% polyacrylamide gel electrophoresis (PAGE) stained with silver nitrate (Sanguinetti et al. 1994). Five microsatellite markers that successfully amplified fragments were obtained both in *B. concinna* (Bdr6, Bdr7, Bdr25, Bdr26, Bdr31) (Fig. 3.5a) and in *B. aphylla* (Bdr6, Bdr7, Bdr22, Bdr25, Bdr31) (Fig. 3.5b).

Modifications of the tested PCR conditions (mainly annealing temperature, DNA and $MgCl_2$ concentration) may have increased this preliminary success of cross-species amplification in *Baccharis*. In addition, other microsatellites developed for *B. dracunculifolia*, but not yet characterized for this species, can be used in future cross-amplification in this genus. These microsatellite markers must provide new information about the population genetic structure of *B. dracunculifolia* and related species and may help elucidate more details on the evolutionary relationships in this genus. Besides, this new molecular tool may help in the management and conservation of *B. dracunculifolia* as well as other species in the genus. We argue that urgent genetic studies on the genus *Baccharis* are called for as the importance of its species in community assembly, ecosystem services, and potential invisibility of disturbed communities is increasing in the recent decade.

a b
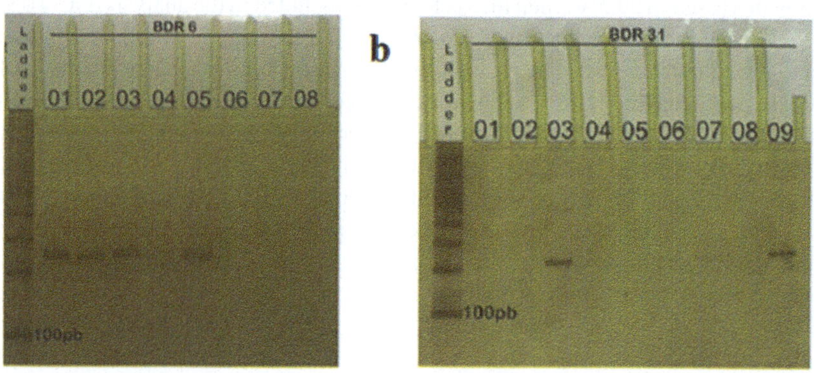

Fig. 3.5 Microsatellite profile of microsatellite markers developed for *Baccharis dracunculifolia* amplified in congeneric species, (**a**) Bdr6 profile in *Baccharis concinna*, PCR fragments were detected in individuals 1–5, (**b**) Bdr31 profile in *Baccharis aphylla*, PCR fragments were detected in individuals 3, 6, 7, and 9

Acknowledgements We thank the Universidade Federal de Ouro Preto (UFOP) for logistical support, Coordenação de Aperfeiçoamento de Pessoalde Nível Superior (CAPES), Fundação de Amparo à Pesquisa do Estado de Minas Gerais (FAPEMIG) and Conselho Nacional deDesenvolvimento Científico e Tecnológico (CNPq) for financial support.

References

Acharige SD, Chand V, Mather P (2012) Development and characterization of tri- and tetranucleotide polymorphic microsatellite markers for skipjack tuna (*Katsuwonus pelamis*). Ceylon J Sci Biol Sci 41:11–17

Ambreen H et al (2015) Development of genomic microsatellite markers in *Carthamus tinctorius* L. (Safflower) using next generation sequencing and assessment of their cross-species transferability and utility for diversity analysis. PLoS One 10:e0135443

Anderson LC, Kyhos DW, Mosquin T, Powell AM, Raven PH (1974) Chromosome numbers in Compositae. IX. Haplopappus and other Astereae. Am J Bot 61:665–671

Ariza Espinar L (1974) Las especies de *Baccharis* (Compositae) de Argentina Central. Museo Botánico, Universidad Nacional de Córdoba, Córdoba

Beauchamp RM (1980) *Baccharis vanessae*, a new species from San Diego County, California. Phytologia 46:216–222

Belini CM et al (2016) Characterization of microsatellite markers for *Baccharis dracunculifolia* (Asteraceae). Appl Plant Sci 4:apps.1500093

Bernardello LM (1986) Números cromosómicos en Asteraceae de Córdoba. Darwin 27:169–178

Bhargava A, Fuentes FF (2010) Mutational dynamics of microsatellites. Mol Biotechnol 44:250–266

Bonatelli IA, Carstens BC, Moraes EM (2015) Using next generation RAD sequencing to isolate multispecies microsatellites for *Pilosocereus* (Cactaceae). PLoS One 10:e0142602

Bowden WM (1945) A list of chromosome numbers in higher plants. I. Acanthaceae to Myrtaceae. Am J Bot 32:81–92

Braga AC, Reis AMM, Leoi LT, Pereira RW, Collevatti RG (2007) Development and characterization of microsatellite markers for the tropical tree species *Tabebuia aurea* (Bignoniaceae). Mol Ecol Notes 7:53–56

Carr GD, King RM, Powell AM, Robinson H (1999) Chromosome numbers in Compositae. XVIII. Am J Bot 86:1003–1013

Castoe TA et al (2012) Rapid microsatellite identification from Illumina paired-end genomic sequencing in two birds and a snake. PLoS One 7:e30953

Chambers GK, MacAvoy ES (2000) Microsatellites: consensus and controversy. Comp Biochem Physiol B Biochem Mol Biol 126:455–476

Ciampi AY, Azevedo VCR, Gaiotto FA, Ramos ACS, Lovato MB (2008) Isolation and characterization of microsatellite loci for *Hymenaea courbaril* and transferability to *Hymenaea stigonocarpa*, two tropical timber species. Mol Ecol Res 8:1074–1077

Clarke LA, Rebelo CS, Goncalves J, Boavida MG, Jordan P (2001) PCR amplification introduces errors into mononucleotide and dinucleotide repeat sequences. Mol Pathol 54:351–353

Coleman JR (1968) Chromosome numbers in some Brazilian Compositae. Rhodora 70:228–240

Coleman JR (1970) Additional chromosome numbers in Brazilian Compositae. Rhodora 72:94–99

Cota LG, Moreira PA, Menezes EV, Gomes AS, Ericsson AR, Oliveira DA, Melo AF Jr (2012) Transferability and characterization of simple sequence repeat markers from *Anacardium occidentale* to *A. humile* (Anacardiaceae). Genet Mol Res 11:4609–4616

Covas G, Schnak B (1946) Número de chromosomas enantofites de la region de Cuyo (Republica Argentina). Rev Argent Agron 13:153–166

Cruz MV, Rodrigues JG, Souza HA, Lovato MB (2012) Isolation and characterization of microsatellite markers for *Plathymenia reticulata* (Fabaceae). Am J Bot 99:e210–e212

Dayanandan S, Bawa K, Kesseli R (1997) Conservation of microsatellites among tropical trees (Leguminosae). Am J Bot 84:1658

Dejong D, Montgomery F (1963) Chromosome numbers in some Californian Compositae: Astereae. Aliso 5:255–256

Dillon M, Turner BL (1982) Chromosome numbers of Peruvian Compositae. Rhodora 84:131–137

Dollenz AO (1976) Números cromósomicos de *Verbena tridens* Lag., *Baccharis patagonica* Hook. & Arn. Y *Adesmia boronoides* Hook. f. Anal Inst Patagonia 7:163–167

Doyle JJ, Doyle JL (1990) Isolation of plant DNA from fresh tissue. Focus 12:13–15

Edwards A, Civitello A, Hammond HA, Caskey CT (1991) DNA typing and genetic mapping with trimeric and tetrameric tandem repeats. Am J Hum Genet 49:746–756

Estoup A, Wilson IJ, Sullivan C, Cornuet JM, Moritz C (2001) Inferring population history from microsatellite and enzyme data in serially introduced cane toads, *Bufo marinus*. Genetics 159:1671–1687

Feres JM, Martinez ML, Martinez CA, Mestriner MA, Alzate-Marin AL (2009) Transferability and characterization of nine microsatellite markers for the tropical tree *species Tabebuia roseo-alba*. Mol Ecol Res 9:434–437

Fernandez Casas J (1981) Recuentos cromosomaticos de algunas angiospermas de Bolivia y Peru. Saussurea 12:157–164

Freeland JR (2005) Molecular ecology. Wiley-Blackwell, Hoboken

Gomes V, Collevatti R, Silveira F, Fernandes G (2004) The distribution of genetic variability in *Baccharis concinna* (Asteraceae), an endemic, dioecious and threatened shrub of rupestrian fields of Brazil. Conserv Genet 5:157–165

Guichoux E, Lagache L, Wagner S, Chaumeil P, Léger P, Lepais O, Lepoittevin C, Malausa T, Revardel E, Salin F, Petit RJ (2011) Current trends in microsatellite genotyping. Mol Ecol Res 11:591–611

Heiden G (2014) Systematics of *Baccharis* (Asteraceae: Astereae). Thesis, Universidade de São Paulo

Heiden G, Pirani JR (2016) Novelties towards a phylogenetic infrageneric classification of *Baccharis* (Asteraceae, Astereae). Phytotaxa 289:285–290

Heiden G, Iganci J, Stein V, Bobrowski V (2006) Número cromossômico de *Baccharis riograndensis* Malag. & J.E.Vidal (Asteraceae). Pesq Bot 57:121–136

Heiden G, Antonelli A, Pirani JR (2019) A novel phylogenetic infrageneric classification of *Baccharis* (Asteraceae: Astereae), a highly diversified American genus. Taxon 68:1048-1081

Hellwig FH (1990) Die Gattung *Baccharis* L. (Compositae – Asteraceae) in Chile. Mitt Bot Staatssamml München 29:1–456

Hodel RGJS-SMC, Landis JB, Crowl AA, Sun M, Liu X, Gitzendanner MA, Douglas NA, Germain-Aubrey CC, Chen S, Soltis DE, Soltis PS (2016) The report of my death was an exaggeration: a review for researchers using microsatellites in the 21st century. Appl Plant Sci 4:1–13

Hunziker J, Escobar A, Xifreda CC, Gamerro J (1990) Estudios cariológicos en Compositae VI. Darwin 30:115–121

Hunziker JH, Wulff AF, Wulf AF, Xifreda CC, Escobar A (1989) Estudios cariologicos en Compositae V. Darwiniana 29:25–39

Iqbal MJ, Aziz N, Saeed NA, Zafar Y, Malik KA (1997) Genetic diversity evaluation of some elite cotton varieties by RAPD analysis. Theor Appl Genet 94:139–144

Jackson J (1970) IOPB chromosome numbers reports XXV. Taxon 19:102–113

Jacob HJ, Lindpaintner K, Lincoln SE, Kusumi K, Bunker RK, Mao YP, Ganten D, Dzau VJ, Lander ES (1991) Genetic mapping of a gene causing hypertension in the stroke-prone spontaneously hypertensive rat. Cell 67:213–224

Jansen RK, Stuessy TF (1980) Chromosome counts of Compositae from Latin America. Am J Bot 67:585–594

Jansen RK, Stuessy TF, Díaz-Piedrahíta S, Funk VA (1984) Recuentos cromosómicos en Compositae de Colombia. Caldasia 14:7–20

Kearse M, Moir R, Wilson A, Stones-Havas S, Cheung M, Sturrock S, Buxton S, Cooper A, Markowitz S, Duran C, Thierer T, Ashton B, Meintjes P, Drummond A (2012) Geneious basic: an integrated and extendable desktop software platform for the organization and analysis of sequence data. Bioinformatics 28:1647–1649

Keil DJ, Pinkava DJ (1976) Chromosome counts and taxonomic notes for Compositae from the United States and Mexico. Am J Bot 63:1393–1403

Keil DJ, Stuessy TF (1975) Chromosome counts of Compositae from the United States, Mexico and Guatemala. Rhodora 77:171–195

Keil DJ, Stuessy TF (1977) Chromosome counts of Compositae from Mexico and the United States. Am J Bot 64:791–798

Larranaga N, Albertazzi FJ, Fontecha G, Palmieri M, Rainer H, van Zonneveld M, Hormaza JI (2017) A Mesoamerican origin of cherimoya (*Annona cherimola* Mill.): implications for the conservation of plant genetic resources. Mol Ecol 26:4116–4130

Levinson G, Gutman GA (1987) Slipped-strand mispairing: a major mechanism for DNA sequence evolution. Mol Biol Evol 4:203–221

Litt M, Luty JA (1989) A hypervariable microsatellite revealed by in vitro amplification of a dinucleotide repeat within the cardiac muscle actin gene. Am J Hum Genet 44:397–401

López-Márquez V, García-Jiménez R, Templado J, Machordom A (2016) Development and characterization of 26 novel microsatellite loci for the trochid gastropod *Gibbula divaricata* (Linnaeus, 1758), using Illumina MiSeq next generation sequencing technology. PeerJ 4:e1789

McDowell JR, Diaz-Jaimes P, Graves JE (2002) Isolation and characterization of seven tetranucleotide microsatellite loci from Atlantic northern bluefin tuna *Thunnus thynnus thynnus*. Mol Ecol Notes 2:214–216

McInerney CE, Allcock AL, Johnson MP, Bailie DA, Prodohl PA (2011) Comparative genomic analysis reveals species-dependent complexities that explain difficulties with microsatellite marker development in molluscs. Heredity 106:78–87

Meldgaard M, Morling N (1997) Detection and quantitative characterization of artificial extra peaks following polymerase chain reaction amplification of 14 short tandem repeat systems used in forensic investigations. Electrophoresis 18:1928–1935

Moreira RG, McCauley RA, Cortés-Palomec AC, Lovato MB, Fernandes GW, Oyama K (2008) Isolation and characterization of microsatellite loci in *Coccoloba cereifera* (Polygonaceae), an endangered species endemic to the Serra do Cipó, Brazil. Mol Ecol Resour 8:854–856

Moreira PA, Sousa SA, Oliveira FA, Araujo NH, Fernandes GW, Oliveira DA (2012) Characterization of nine transferred SSR markers in the tropical tree species *Enterolobium contortisiliquum* (Fabaceae). Genet Mol Res 11:3729–3734

Morris AB, Shaw J (2018) Markers in time and space: a review of the last decade of plant phylogeographic approaches. Mol Ecol 27:2317–2333

Mota L, Torices R, Loureiro J (2016) The evolution of haploid chromosome numbers in the sunflower family. Genome Biol Evol 8:3516–3528

Müller J (2006) Systematics of *Baccharis* (Compositae-Astereae) in Bolivia, including an overview of the genus. Syst Bot Monogr 76:1–341

Muñoz-Pajares AJ, García C, Abdelaziz M, Bosch J, Perfectti F, Gómez JM (2017) Drivers of genetic differentiation in a generalist insect-pollinated herb across spatial scales. Mol Ecol 26:1576–1585

Nesom GL (2000) Generic conspectus of the tribe Astereae (Asteraceae) in North America and Central America, the Antilles, and Hawai. Botanical Research Institute of Texas, Texas

Pinkava D, Keil D (1977) Chromosome counts of Compositae from the United States and Mexico. Am J Bot 64:680–686

Powell AM, Cuatrecasas J (1970) Chromosome numbers in Compositae: Colombian and Venezuelan Species. Ann Mo Bot Gard 57:374–379

Powell AM, Cuatrecasas J (1975) In IOPB chromosome number reports L. Taxon 24:671–678

Powell AM, King RM (1969) Chromosome numbers in the Compositae: Colombian species. Am J Bot 56:116–121
Powell AM, Powell SA (1977) Chromosome numbers of gypsophilic plant species of the Chihuahuan desert. Sida, Contrib Bot (Online) 7:80–90
Powell A, Turner B (1963) Chromosome numbers in the Compositae VII. Additional species from the southwestern United States and Mexico. Madrono 17:128–140
Ram SG, Parthiban K, Kumar RS, Thiruvengadam V, Paramathma M (2008) Genetic diversity among *Jatropha* species as revealed by RAPD markers. Genet Resour Crop Evol 55:803–809
Raven PH, Solbrig OT, Kyhos DW, Snow R (1960) Chromosome numbers in Compositae. I. Astereae. Am J Bot 47:124–132
Rowe G, Sweet M, Beebee TJC (2004) An introduction to molecular ecology. Oxford University Press, New York
Rozenblum E, Waisman CE, Hunziker JH (1985) Estudios cariolό́gicos en Compositae. II Darwiniana 26:15–25
Ruas PM, Bertillacchi SRA, Vieira AOS, Matzenbacher NI (1989) Citogenética do gênero *Baccharis* (Compositae). Ciên Cult 41:702
Sanguinetti CJ, Dias Neto E, Simpson AJ (1994) Rapid silver staining and recovery of PCR products separated on polyacrylamide gels. BioTechniques 17:914–921
Solbrig OT (1977) Chromosomal cytology and evolution in the family Compositae. In: Heywood VH, Harborne JB, Turner BL (eds) The biology and chemistry of the Compositae. Academic Press, London, pp 267–281
Solbrig OT, Anderson LC, Kyhos DW, Raven PH, Rudenberg L (1964) Chromosome numbers in Compositae V. Astereae II. Am J Bot 51:513–519
Solbrig OT, Anderson LC, Kyhos DW, Raven PH (1969) Chromosome numbers in Compositae VII: Astereae III. Am J Bot 56:348–353
Spellenberg R, Ward DE (1988) Chromosome counts of angiosperms from New Mexico and adjacent areas. Phytologia 64:390–398
Spooner DM, De Jong DCD, Byung-Yun S, Stuessy TF, Gengler KM, Nesom GL, Berry PE (1995) Chromosome counts of Compositae from Ecuador and Venezuela. Ann Mo Bot Gard 82:596–602
Sturtevant AH, Dobzhansky T (1936) Inversions in the third chromosome of wild races of *Drosophila pseudoobscura*, and their use in the study of the history of the species. Proc Natl Acad Sci U S A 22:448–450
Sundberg S, Cowan CP, Turner BL (1986) Chromosome counts of Latin American Compositae. Am J Bot 73:33–38
Turner BL, Irwin HS (1960) Chromosome numbers in the Compositae II. Meiotic counts for 14 species of Brazilian Compositae. Rhodora 62:122–126
Turner BL, Powell AM, Cuatrecasas J (1967) Chromosome numbers in Compositae. XI. Peruvian Species. Ann Mo Bot Gard 54:172–177
Turner BL, Bacon J, Urbatsch L, Simpson B (1979) Chromosome numbers in South American Compositae. Am J Bot 66:173–178
Untergasser A, Cutcutache I, Koressaar T, Ye J, Faircloth BC, Remm M, Rozen SG (2012) Primer3 – new capabilities and interfaces. Nucleic Acids Res 40:e115–e115
Wachira FN, Waugh R, Powell W, Hackett CA (1995) Detection of genetic diversity in tea (*Camellia sinensis*) using RAPD markers. Genome 38:201–210
Weedin J, Powell A (1978) IOPB chromosome number reports LX. Taxon 27:223–231
Welsh J, McClelland M (1990) Fingerprinting genomes using PCR with arbitrary primers. Nucleic Acids Res 18:7213–7218
Westman WE, Panetta FD, Stanley TD (1975) Ecological studies on reproduction and establishment of the woody weed, groundsel bush (*Baccharis halimifolia* L.: Asteraceae). Aust J Agric Res 26:855–870
Williams JG, Kubelik AR, Livak KJ, Rafalski JA, Tingey SV (1990) DNA polymorphisms amplified by arbitrary primers are useful as genetic markers. Nucleic Acids Res 18:6531–6535

Wulff AF (1984) Estudios cromosómicos en Compuestas de las floras Patagonica y Bonaerense. Darwin 25:17–26

Wulff A, Hunziker JH, Escobar A (1996) Estudios cariológicos en Compositae. Darwin 34:213–231

Zalapa JE, Cuevas H, Zhu H, Steffan S, Senalik D, Zeldin E, McCown B, Harbut R, Simon P (2012) Using next-generation sequencing approaches to isolate simple sequence repeat (SSR) loci in the plant sciences Am J Bot 99:193–208

Zane L, Bargelloni L, Patarnello T (2002) Strategies for microsatellite isolation: a review. Mol Ecol 11:1–16

Zanowiak DJ (1991) An analysis of systematic and phyletic relationships within Baccharidinae (Asteraceae: Astereae). PhD Thesis, Texas A&M University.

Zucchi MI, Brondani RPV, Pinheiro JB, Brondani C, Vencovsky R (2002) Transferability of microsatellite markers from *Eucalyptus* spp. a *Eugenia dysenterica* (família Myrtaceae). Mol Ecol Notes 2:512–513

Chapter 4
Intersexual Differences in Demography, Resource Investment, and Herbivory in *Baccharis*

Evelyn M. Frazier, Jhonathan O. Silva, Mário M. Espírito-Santo, Tatiana Cornelissen, and G. Wilson Fernandes

Abstract Intersexual differences in resource allocation between host plant sexes have been the most proposed mechanism to explain variation in herbivory in dioecious plants. In this sense, the sex-biased herbivory hypothesis predicts that male plants have higher growth rates and should be more susceptible to herbivores than females. Most studies testing this hypothesis were conducted in temperate regions and focused on a few host plant genera. Currently, the male-biased herbivory as a rule for dioecious species has been questioned. In this study, we reviewed the aforementioned hypotheses for *Baccharis*, performed a meta-analysis contrasting herbivory and resource allocation in *Baccharis* versus other dioecious systems, and addressed two case studies: (1) intersexual comparisons in plant architectural traits and herbivory for 12 species of *Baccharis* and (2) a fine-scale analysis for the mountaintop endemic *Baccharis concinna* in long- and short-term studies. In general, most *Baccharis* species showed no intersexual differences for vegetative resource allocation, except plant biomass that exhibited a positive trend to increase in male individuals. Most *Baccharis* species do not support the sex-biased herbivory

E. M. Frazier (✉)
Department of Biological Sciences, Florida Atlantic University, Boca Raton, FL, USA
e-mail: efrazier@fau.edu

J. O. Silva
Colegiado de Ecologia, Universidade Federal do Vale do São Francisco – Univasf, Senhor do Bonfim, Bahia, Brazil
e-mail: jhonathan.oliveira@univasf.edu.br

M. M. Espírito-Santo
Departamento de Biologia Geral, Universidade Estadual de Montes Claros – Unimontes, Montes Claros, Minas Gerais, Brazil

Department of Geography, University of Exeter, Exeter, UK

T. Cornelissen · G. W. Fernandes
Departamento de Genética, Ecologia & Evolução, Universidade Federal de Minas Gerais, Belo Horizonte, Minas Gerais, Brazil
e-mail: taticornelissen@ufmg.br

G. W. Fernandes et al. (eds.), *Baccharis*,
https://doi.org/10.1007/978-3-030-83511-8_4

hypothesis, likely due to an absence of intersexual differences in resource allocation. Although host plant sex was a weak predictor explaining galling insect abundance in this host plant genus, the plant traits evaluated here were important drivers of gall attack. Due to the lack of differential resource allocation and sex-biased herbivory in both short- and long-term sampling events, we have consistent support to reject the male-biased attack hypothesis on *Baccharis*.

Keywords Dioecy · Plant gender · Plant-herbivore interactions · Resource allocation · Sex-biased herbivory

1 Introduction

Selection Theory: Male and Female Functions

The theory of sexual selection was originally proposed by Charles Darwin (1871) to explain the evolution of different secondary sexual characteristics between male and female animals, and he argued that any trait that differed between the sexes would be the result of selection by one of the sexes and such trait would have evolved as a result of differential mating success. The theory of sexual selection predicts different energetic investments between the sexes (see Emlen and Oring 1977; Arnold and Wade 1984; Jaenicke and Morrow 2018). Males would have a lower energetic investment in reproduction because they only produce sperm, whereas females would have a higher energetic investment having to produce the larger eggs as well as, in the case of plants, investing in the development and protection of fruits and seeds (Stanton et al. 1986). This differential investment in reproduction would result in different reproductive success limitations between the sexes. A male's reproductive success would be limited by how much sperm a male could spread to females, while a female's reproductive success would be limited by the number of eggs that a female could produce (Jaenicke and Morrow 2018). Although the theory of sexual selection was originally developed to explain reproductive patterns observed in animals, it also applies to plants, and the selection of secondary traits in plants can be mediated via pollinators by selecting flowers that offer more resources (Stanton et al. 1986; Knauer and Schiesti 2017; Delph 2019).

The ancestral condition for plants is the presence of hermaphroditic flowers, with female and male reproductive structures within the same flower (Bawa 1980, Charlesworth 2018; Henry et al. 2018). In hermaphroditic flowers, pollinators are attracted to food resources such as nectar and pollen grains, but these monoecious flowers incur greater inbreeding reducing the plant's fitness (Barrett 2002). Plants evolved dioecy, or male and female flowers in different plants, probably as a result of pressures to promote outcrossing and reduce inbreeding depression (Barret 2002), but herbivory has also been suggested as a potential selective pressure (Herms and Mattson 1992; Ashman 2002; Cornelissen and Stiling 2005; Moritz et al. 2017; Moreira et al. 2019; LeRoy et al. 2020). Although dioecy is not as common as hermaphroditism, approximately 37 plant families out of a total of 51 families encompass dioecious species (Bawa 1980; Henry et al. 2018), with some large genera containing almost exclusive dioecious species, such as *Baccharis* (Asteraceae).

Dioecy created challenges to pollination as male flowers provide resources to pollinators as pollen or nectaries, but female flowers compete with male flowers for pollinators and provide no resources in exchange for the pollinator's visit. Female flowers already have a high energetic cost of producing ovaries containing eggs, which after fertilization become fruits containing seeds. Also, in female flowers, seed production (fitness) is frequently limited by resources and not by pollinator visits (Sutherland and Delph 1984; Vaughton and Ramsey 1998; Barrett 2002). Therefore, many female flowers mimic male flowers in color, shape, and size as closely as possible to trick pollinators into visiting them (Sutherland and Delph 1984; Vaughton and Ramsey 1998; Vega-Frutis et al. 2013).

Trade-Offs in Growth, Reproduction, and Defense Against Herbivores

In general, female plants allocate more resources into sexual reproduction, while male plants allocate more resources into vegetative growth (e.g., Loyd and Webb 1977; Wallace and Rundel 1979; Ågreen 1988; Popp and Reinartz 1988; Obeso 2002; Harris and Pannel 2008; Pfeiffer et al. 2019). The greater allocation of resources to growth by male plants results in the exhaustion of the pool of carbon molecules available to form secondary defensive compounds (e.g., Herms and Mattson 1992; Cornelissen and Fernandes 2001; Imaji and Seiwa 2010). Consequently, male plants should represent a food resource of high nutritional quality and low chemical defenses to insect herbivores.

This differential allocation of resources within the host plant represents an important source of variation in plant quality for insect herbivores (e.g., Boecklen et al. 1990; Boecklen and Hoffman 1993; Multikainen and Delph 1996; Espírito-Santo et al. 2003; Barret and Hough 2013; Erb 2018). Indeed, several studies have recorded male-biased herbivory, in which male plants sustain higher herbivore densities or greater levels of herbivory when compared with female plants (e.g., Lovett-Doust et al. 1987; Boecklen et al. 1990; Jing and Coley 1990; Pearson et al. 1990; Åhman 1997; Cornelissen and Stiling 2005; Rivkin et al. 2018). Nevertheless, exceptions to this pattern exist where insect herbivores preferentially attack female plants in the genus *Baccharis* (e.g., Espírito-Santo and Fernandes 1998; Faria and Fernandes 2001) or do not show attack preferences driven by plant gender (e.g., Boecklen et al. 1994; Åhman 1997; Espírito-Santo et al. 1999, 2012; Araújo et al. 2006; Marques and Fernandes 2016) (see Table 4.1). Boecklen et al. (1994) did not detect intersexual variation in herbivory in three herbivorous species on *Salix*

Table 4.1 Sex-biased herbivory studies on several species of *Baccharis*

References	Species	Hypotheses tested	Measures	Length of study	Findings
Krischik and Denno (1990)	*B. halimifolia*	1. Differential resource allocation between host plant sex 2. Sex-biased herbivory	Shoot growth, feeding trials, nutrients in leaf	6 months	No clear pattern for herbivory
Espírito-Santo and Fernandes (1998)	*B. dracunculifolia*	1. Differential resource allocation between host plant sex 2. Sex-biased herbivory	Abundance of one galling species	17 months	Plant sex did not affect gall abundance, tannin production, or gall mortality rates
Espírito-Santo et al. (1999)	*B. dracunculifolia*	Male plants are better defended than female plants	Tannin concentration	12 months	No difference in tannin concentration between sexes, no support for differential resource allocation hypothesis
Faria and Fernandes (2001)	*B. dracunculifolia*	1. Differential resource allocation between host plant sex 2. Sex-biased herbivory 3. Plant vigor	Shoot length, number of leaves per shoot, galling attack and survival of *Neopelma baccharidis*	Single destructive sampling event (1994)	1. No difference in shoot length between male and female plants 2. No differences on number of galls or survival between the sexes 3. Longer shoots harbored more galls

(continued)

Table 4.1 (continued)

References	Species	Hypotheses tested	Measures	Length of study	Findings
Ribeiro-Mendes et al. (2002)	*B. dracunculifolia*	Host plant sex and galling herbivore survivorship	Survivorship of galls over geographic range of plant	Single destructive collection	No support for sex-biased herbivory
Araújo et al. (2003)	*B. pseudomyriocephala*	1. Increase of plant architecture increases galling species richness and abundance 2. Plant architecture increases galling insect survival and decreases parasitism	1.Number of shoots, branches, and biomass (240 plants) 2. Galling species richness and abundance	Single destructive sampling event in 2000 and 2001	1. Higher galling species richness and abundance on plants with greater biomass 2. Higher galling species richness and abundance on architecturally more complex plants 3. Plant architectural complexity was not related with parasitism or gall survival
Carneiro et al. (2006)	*B. concinna*	1. Altitudinal gradient in galling species richness 2. Sex-biased herbivory 3. Habitat-mediated herbivory	Galling species and Cecidomyiidae richness on 425 plants across four sites along an altitudinal gradient	One time sampling event (1993)	1. Altitude had a significant effect on number of galling and Cecidomyiidae species richness, with mid-elevation peak 2. Males plants had greater galling and Cecidomyiidae species richness 3. Xeric habitats harbored greater galling and Cecidomyiidae species richness

(continued)

Table 4.1 (continued)

References	Species	Hypotheses tested	Measures	Length of study	Findings
Espírito-Santo et al. (2012)	*B. dracunculifolia*, *B. concinna*, *B. ramosissima*	1. Differential resource allocation between host plant sex 2. Sex-biased herbivory	Shoot growth rates, number of inflorescences, and number of species and gall abundance were sampled on 15 male and 15 female plants for each plant species	12 months (2001–2002)	1. No difference in resource allocation between the sexes was found 2. No intersexual difference was found on galling species richness or abundance
Marques and Fernandes (2016)	*B. concinna*	1. Differential resource allocation between host plant sex 2. Sex-biased herbivory	Lateral shoot growth rates and insect galling community	12 months (1998–1999)	1.Male plants have greater lateral shoot growth rates 2. No intersexual difference was found on galling species richness or abundance

lasiolepis (Salicaceae) and argued that the herbivore species were probably utilizing resources that were not sexually dimorphic in this host plant, such as leaves. On the other hand, Avila-Sakar and Romanow (2012) argued that the lack of sex preference by insect herbivores may be common and indicate that male and female plants could be allocating resources in a similar manner. These authors have pointed some methodological bias in male-biased herbivory studies, such as: (a) taxonomic bias, e.g., research efforts have focused on few species within a few orders and families, and (b) failure to make the connection between sex-biased herbivore damage and intersexual differences in growth rate and reproduction. Finally, Avila-Sakar and Romanow (2012) recommend a standard protocol for evolutionary-ecological studies, suggesting an increase in the taxonomic breadth of the studies of herbivory in dioecious species because only 2% of the dioecious species have been studied, mostly from the Salicaceae family. Moreover, other studies have focused on herbaceous dioecious species.

In temperate plants, resource allocation is concentrated in the Spring and Summer months, so herbivores have a small window of time to attack the host plant. Therefore, sampling insects only in a couple of months may be adequate to obtain a realistic picture of herbivory patterns in temperate areas. Furthermore, most of these studies only measured herbivore damage inflicted by one or a few herbivore species.

This is not the scenario found in tropical regions where plants may grow for a much longer period of time and in which they are challenged by a much richer herbivore fauna. One study conducted in a tropical plant species by Wolfe (1997) found higher herbivore attack rates on male plants of *Neea psychotrioides* (Nyctaginaceae) in Costa Rica. This study showed that male plants had more and larger flowers when compared to female plants and that two galling insect species and a free-living lepidopteran species attacked flowers on male plants more often. But these data related to flower herbivory which is a resource with a differential allocation between the sexes, and the study did not address the sex-biased herbivory hypothesis which predicted greater herbivory on vegetative portions of the male plants. In this way, more studies in tropical regions are necessary to provide a realistic global pattern on intersexual differences in herbivore attack.

The Genus Baccharis *as a Study System for Sex-Biased Herbivory*

Baccharis represents an excellent model system to the study of plant-animal interactions because of the tremendously high number of insects of different guilds that feed upon them and its occurrence in temperate and tropical regions along gradients of salinity, humidity, altitude, and temperature (see Fernandes et al. 2014). Furthermore, this genus has more than 440 species, with most of them being dioecious (Heiden et al. 2019). Most of the studies on herbivory in dioecious tropical systems have been conducted by our research group on the component communities of galling insects mainly on *B. concinna* and *B. dracunculifolia* but also in other species in this genus (e.g., Marques 1997; Madeira et al. 1997; Wolfe 1997; Espírito-Santo and Fernandes 1998; Marques et al. 2002; Espírito-Santo et al. 2007, 2012; Marques and Fernandes 2016). So far, no clear patterns of differential resource allocation between plants of different sexes and sex-biased attack by herbivores have been detected in this system (see Table 4.1 and Box 4.1). Therefore, we set out to design studies that included several measures of resource allocation such as shoot growth, height, number of meristems, plant architecture, and aboveground biomass, as well as intensive samples of their associated herbivore communities. Sex-biased herbivory and intersexual differences in resource allocation have been studied mostly on *B. concinna*, *B. dracunculifolia*, and *B. ramosissima* (Espírito-Santo et al. 2012; Fernandes et al. 2014; see Table 4.1 for more details). In the present study, we bring new data involving intersexual comparisons for several species of *Baccharis*, aiming to detect a clearer picture of how the communities of insect herbivores respond to resource allocation patterns of their host plants.

An analysis of the available literature on the sex-biased herbivore hypothesis suggests that most studies were conducted in a few plant species (see Box 4.1), in short sampling periods, often involving one species of insect herbivore, and mostly in temperate latitudes. We set out to address these weaknesses by designing studies

Box 4.1: The Influence of Plant Gender on Plant-Insect Interactions and Baccharis as a Study System: An Integrative Review

We quantitatively reviewed the effects of plant gender on insect abundance and damage on dioecious plants, by systematically reviewing the published literature. Searches were conducted on Web of Science and Scopus databases, using as keywords "plant dioecy," "dioecy," "plant gender," "gender," "insect*," "herbivor*," and their combinations. We also used the database of Cornelissen and Stiling (2005), which meta-analytically reviewed the evidence for sex-biased herbivory. From a total of 127 studies found, 47 papers (Appendix 1) met the criteria of language (English) and statistics clearly reported (data of means and a measurement of variance reported separately for males and females) and were included in our review. To compare insect abundance and plant damage (i.e., herbivory) on male and female plants of dioecious systems, response mean values (X_{male}, X_{female}), standard deviations (S_{male}, S_{female}), and sample size (N_{male}, N_{female}) were gathered from the text, tables, and/or figures in each study. Insect abundance encompassed data on counts, density, number of eggs laid, number of galls and damage including leaf area removed, and number of feeding holes and/or leaf scars. Insect survivorship on both sexes was excluded from the analyses due to the low number of comparisons ($n = 3$). To address the effects of plant gender on insect data, we used the standardized mean difference between male plants and females, calculating Hedge's d, and the cumulative Hedge's d was calculated using a weighing method with the reciprocal of the sampling variance. For statistical purposes only, male plants were considered an experiment group (male-biased herbivory has been previously reported in the literature; see Cornelissen and Stiling 2005), and female plants were used as control. We ran analyses for each parameter (abundance, damage) separately using a random model and ran further analyses using only *Baccharis* species. All analyses were conducted on OPEN MEE (Wallace et al. 2017).

Forty-seven published studies, between 1978 and 2018, were included in our analysis, enabling 120 independent comparisons. These studies evaluated the effects of plant sex on 49 different plant species in 21 botanical families. Asteraceae and Salicaceae were the most represented families, encompassing almost half of the plants studied. Almost 50 species of insects were studied, and gall-formers and folivores were the most common guilds on the plants (almost 78% of all independent comparisons), and other guilds less represented were florivores, stem borers, and suckers.

Our analysis for all plants indicates male-biased herbivory, with male plants supporting almost 50% more insects than female plants (insect abundance, E++ = 0.4946, CI = 0.2271 to 0.7622, df = 69) and 20% higher damage (E++ = 0.203, CI = 0.1657 to 0.4718, df = 4) than female plants (Fig. 4.1). These results were strongly influenced by gall-formers for data on abundance (E++ = 0.3369) and for folivores for data on damage (E++ = 0.4430), and these guilds were significantly different from each other (QB = 12.09, $P = 0.05$).

Box 4.1 (continued)

Species of *Baccharis* represented almost 30% of the species found in our database (14 out of 49 species, 39 independent comparisons), and we ran the same analyses using *Baccharis* only. Most studies with *Baccharis* addressed gall abundance on leaves and twigs (85% of the comparisons), and a few studies evaluated abundance of folivores (beetles), florivores (thrips), and sap-suckers (aphids).

Differently from the entire dataset, *Baccharis* species do not show evidence for male-biased herbivory (E++ = 0.1261, CI = −0.05 to 0.309, df = 33, Fig. 4.1), as shown by the confidence intervals around the mean effect size that encompasses the zero value. Therefore, although our dataset shows evidence for male-biased herbivory in terms of insect abundance and plant damage, the same does not hold true for *Baccharis* species here evaluated. Reasons for the absence of male-biased herbivory for these plants are discussed along the text.

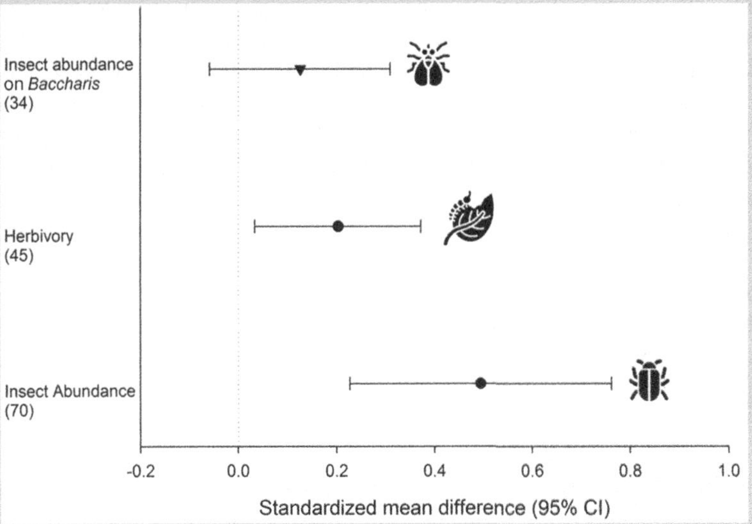

Fig. 4.1 Standardized mean difference in insect abundance and herbivory (Hedge's d) between male and female plants. Circles represent the cumulative effect size with associated confidence intervals, and the triangle represents the effect size for *Baccharis* species only. Numbers in parentheses indicate the number of comparisons for each response variable, and effects are significant if confidence intervals do not encompass zero

that addressed several host plant species, all herbivore species of a guild present on the host plant, and long-term studies up to 11 months, over a broad geographical range. Below, we address two cases of sex-biased herbivory in the genus *Baccharis*. The first one relates to a study across the genus in which the intersexual differences in resource investment and herbivory are evaluated in 12 *Baccharis* species. The second case evaluates resource allocation between plant sexes in *B. concinna* in a short-term destructive sampling and a long-term (12 months) study while also recording herbivory by galling insect species.

2 Study 1. Intersexual Differences on Resource Investment and Herbivory for 12 *Baccharis* Species

We investigated intersexual differences in resource allocation and gall attack in 12 species of *Baccharis* along the southern portion of the Espinhaço Mountain Range in Minas Gerais, Brazil, (see Table 4.2) during one rainy season in 2002–2003 (see Espírito-Santo et al. 2007 for further details on locations and sampling design). We sampled 30 individuals (15 males and 15 females) of each *Baccharis* species that were arbitrarily marked in the field. Plant sex was determined by analysis of floral morphology. Five traits related to plant architecture were evaluated in the study: plant height, number of fourth-level shoots, average ramification level, number of ramifications, and biomass (Table 4.2). For each plant, we recorded gall species and their abundance on leaves, flowers, and stems. Details on sampling design and statistics conducted and study sites can be found in Espírito-Santo et al. (2007). Fourth-level shoots were considered an optimal indicator of the amount of active meristems in a given individual (see Espírito-Santo et al. 2012; Fernandes et al. 2014).

In relation to resource investment, only one species showed difference in number of ramifications (*B. minutiflora*), where resource investment was found to be higher in female plants, contrary to the prediction of the differential resource allocation hypothesis (Ågreen 1987; Rivkin et al. 2018) (Table 4.2). No *Baccharis* species exhibited intersexual differences in the average ramification level. Only two species showed intersexual difference in height (*B. minutiflora* and *B. medullosa*), both being higher in female plants. Five *Baccharis* species showed intersexual differences in the number of fourth-level shoots, with *B. aphylla*, *B. ramosissima*, and *B. dracunculifolia* being female-biased and *B. cognata* and *B. trinervis* male-biased (see Table 4.2). Nine species showed intersexual differences in biomass, with three being female-biased (*B. aphylla, B. minutiflora, and B. medullosa*) and six male-biased (*B. ramosissima*, *B. cognata*, *B. serrulata*, *B. dracunculifolia*, *B. concinna*, and *B. medullosa*). In general, most *Baccharis* species showed no intersexual differences for vegetative resource allocation (Tables 4.2 and 4.3), except for plant biomass that showed a trend to be greater in male individuals (see Table 4.3). The biomass was the best predictor variable to verify the differential sex resource investment in *Baccharis* (Table 4.3). Other studies also have not detected differential allocation of resources between the sexes, even when total plant biomass was considered (Delph et al. 1993; Hemborg and Karlsson 1999; Espírito-Santo et al. 2012).

The 12 species of *Baccharis* were attacked by 55 different gall morphotypes (see Espírito-Santo et al. 2007 for a complete species list). We detected intersexual variations in herbivory for only five species (Tables 4.2 and 4.3), four male-biased (*B. aphylla*, *B. truncata*, *B. minutiflora*, and *B. dracunculifolia*) and one female-biased (*B. ramosissima*). Hence, most *Baccharis* species seem not to support the

Table 4.2 Intersexual comparison of architectural traits and gall abundance (mean ± SE) on 12 *Baccharis* species distributed along the southern portion of the Espinhaço Mountain Range, in Minas Gerais, Brazil

Species	Sex	Number of ramifications	Level of ramification	Height (m)	Number of fourth-level shoots	Biomass (g)	Gall abundance
B. aphylla	Female	1.42 ± 0.11a	3.41 ± 0.25a	0.56 ± 0.03a	30.00 ± 16.16a	58.30 ± 24.48a	3.60 ± 1.29a
	Male	1.91 ± 0.29a	3.17 ± 0.20a	0.50 ± 0.02a	13.33 ± 3.90b	26.42 ± 5.58b	6.60 ± 2.02b
B. ramosissima	Female	21.09 ± 2.76a	4.98 ± 0.19a	2.15 ± 0.10a	122.67 ± 14.50a	199.78 ± 27.67a	16.73 ± 2.94a
	Male	20.17 ± 2.72a	4.60 ± 0.33a	1.86 ± 0.11a	75.67 ± 13.75b	261.24 ± 81.91b	11.40 ± 2.91b
B. cognata	Female	21.07 ± 2.07a	4.12 ± 0.10a	0.94 ± 0.04a	70.80 ± 15.56a	30.06 ± 7.40a	0.53 ± 0.21a
	Male	22.08 ± 3.99a	4.10 ± 0.15a	0.90 ± 0.05a	117.33 ± 26.15b	60.92 ± 10.08b	1.27 ± 0.44a
B. helychrysoides	Female	1.00 ± 0.01a	2.12 ± 0.17a	1.51 ± 0.06a	0.13 ± 0.13a	27.82 ± 7.63a	0.67 ± 0.23a
	Male	1.44 ± 0.21a	2.43 ± 0.21a	1.47 ± 0.04a	2.07 ± 1.68a	27.87 ± 11.44a	0.40 ± 0.13a
B. truncata	Female	6.91 ± 1.04a	4.87 ± 0.22a	0.71 ± 0.06a	35.73 ± 6.70a	57.31 ± 10.71a	1.20 ± 0.42a
	Male	5.27 ± 0.68a	4.36 ± 0.34a	0.74 ± 0.07a	38.33 ± 11.29a	53.38 ± 18.24a	5.07 ± 1.36b
B. minutiflora	Female	42.02 ± 4.10a	5.58 ± 0.33a	1.05 ± 0.07a	123.53 ± 18.86a	106.63 ± 22.85a	34.13 ± 7.84a
	Male	30.67 ± 3.39b	5.67 ± 0.21a	0.85 ± 0.06b	116.87 ± 20.53a	42.15 ± 8.72b	50.53 ± 6.08b
B. serrulata	Female	2.49 ± 0.49a	2.93 ± 0.23a	0.64 ± 0.09a	2.73 ± 1.03a	15.46 ± 3.17a	2.13 ± 0.82a
	Male	2.80 ± 0.50a	3.23 ± 0.20a	0.80 ± 0.11a	4.53 ± 1.00a	28.78 ± 11.44b	2.20 ± 1.17a
B. dracunculifolia	Female	10.38 ± 0.95a	5.51 ± 0.29a	1.78 ± 0.13a	140.67 ± 21.52a	304.84 ± 40.31a	37.80 ± 9.11a
	Male	10.96 ± 1.33a	5.40 ± 0.25a	1.72 ± 0.14a	100.00 ± 17.31b	358.92 ± 43.80b	51.00 ± 12.40b
B. concinna	Female	14.73 ± 1.44a	6.07 ± 0.22a	1.21 ± 0.11a	87.80 ± 11.41a	48.16 ± 13.67a	10.27 ± 3.88a
	Male	11.42 ± 1.11a	6.47 ± 0.29a	1.06 ± 0.05a	87.93 ± 21.72a	90.27 ± 15.68b	9.67 ± 2.62a
B. trinervis	Female	4.49 ± 0.51a	3.71 ± 0.22a	1.00 ± 0.07a	30.00 ± 12.80a	117.99 ± 42.3a	3.13 ± 0.93a
	Male	3.60 ± 0.44a	3.89 ± 0.37a	1.22 ± 0.09a	59.60 ± 28.65b	152.52 ± 89.19b	1.33 ± 0.60a
B. medullosa	Female	3.62 ± 0.64a	3.33 ± 0.18a	2.17 ± 0.11a	16.00 ± 5.01a	122.00 ± 15.23a	3.73 ± 1.30a
	Male	3.69 ± 0.55a	3.31 ± 0.12a	1.88 ± 0.08b	14.60 ± 4.86a	68.25 ± 10.17b	2.47 ± 0.69a

(continued)

Table 4.2 (continued)

Species	Sex	Number of ramifications	Level of ramification	Height (m)	Number of fourth-level shoots	Biomass (g)	Gall abundance
Baccharis sp. 1	Female	3.62 ± 1.40a	2.69 ± 0.13a	1.84 ± 0.07a	3.73 ± 1.80a	78.84 ± 19.25a	0.07 ± 0.07a
	Male	2.16 ± 0.30a	2.42 ± 0.11a	1.81 ± 0.09a	0.40 ± 0.28a	59.63 ± 19.55a	0.20 ± 0.14a

The differences among the structural characteristics and gall attack were tested using generalized linear models. Identical letters between sexes of the same species indicate averages that are not statistically significantly different ($P > 0.05$)

Table 4.3 Intersexual trends for each architectural trait and gall abundance on 12 *Baccharis* species during 6 months of sampling during the rainy season

	Number of species in each category			
Variables	Male < Female	Male = Female	Male > Female	Final trend
Number of ramifications	1	11	0	Male = Female
Level of ramification	0	12	0	Male = Female
Height (m)	2	10	0	Male = Female
Number of fourth-level shoots	4	7	1	Male = Female
Biomass (g)	3	3	6	Male > Female
Gall abundance	1	7	4	Male = Female

These results were obtained on statistical significance from Table 4.2

sex-biased herbivory hypotheses (see Table 4.3), likely due to lack of differences in resource allocation between the host sexes (Espírito-Santo et al. 2012; Fernandes et al. 2014; Marques and Fernandes 2016). Indeed, the total reproductive investment may not differ between male and female individuals for most *Baccharis* species. Although we did not quantify reproductive effort here, Espírito-Santo et al. (2012) found no differences in inflorescence number for two out of three *Baccharis* species in 1 year. For the species considered in the present study, inflorescences have a similar size in both sexes, and female flowers do not invest in pollinator rewards (authors' personal observations). Seeds are tiny and wind-dispersed, and although their maturation may represent a high relative cost to female individuals, this investment may be compensated by male's pollen production. Considering that we did not observe any marked difference in spatial distribution among sexes, it is likely that the lack of differences in reproductive investment is reflected in similar vegetative characteristics that regulate gall densities. In this way, both sexes of *Baccharis* would be equally susceptible to herbivore attack.

Relationship Between Plant Architectural Traits and Gall Attack

Although there was no clear intersexual difference in resource allocation and herbivory, we tested the effects of plant architectural traits on gall abundance for each of the 12 *Baccharis* species studied here. To summarize all architectural traits into a unique variable, we performed a principal component analysis (PCA) for each *Baccharis* species. We used PCA scores instead of the raw data because they were not orthogonal (i.e., not independent); therefore, the use of multiple regressions would not be recommended. The scores of the first axes from each PCA were chosen to indicate architectural gradients because they summarized the patterns observed in the data and explained most of the data variation, being suitable for use in the regression models described below.

We found a significant statistical relationship between gall insect abundance and the first axis of the PCA in 6 out of 12 *Baccharis* species (Fig. 4.2). For all significant relations, the first axis of the PCA was positively correlated with the number of

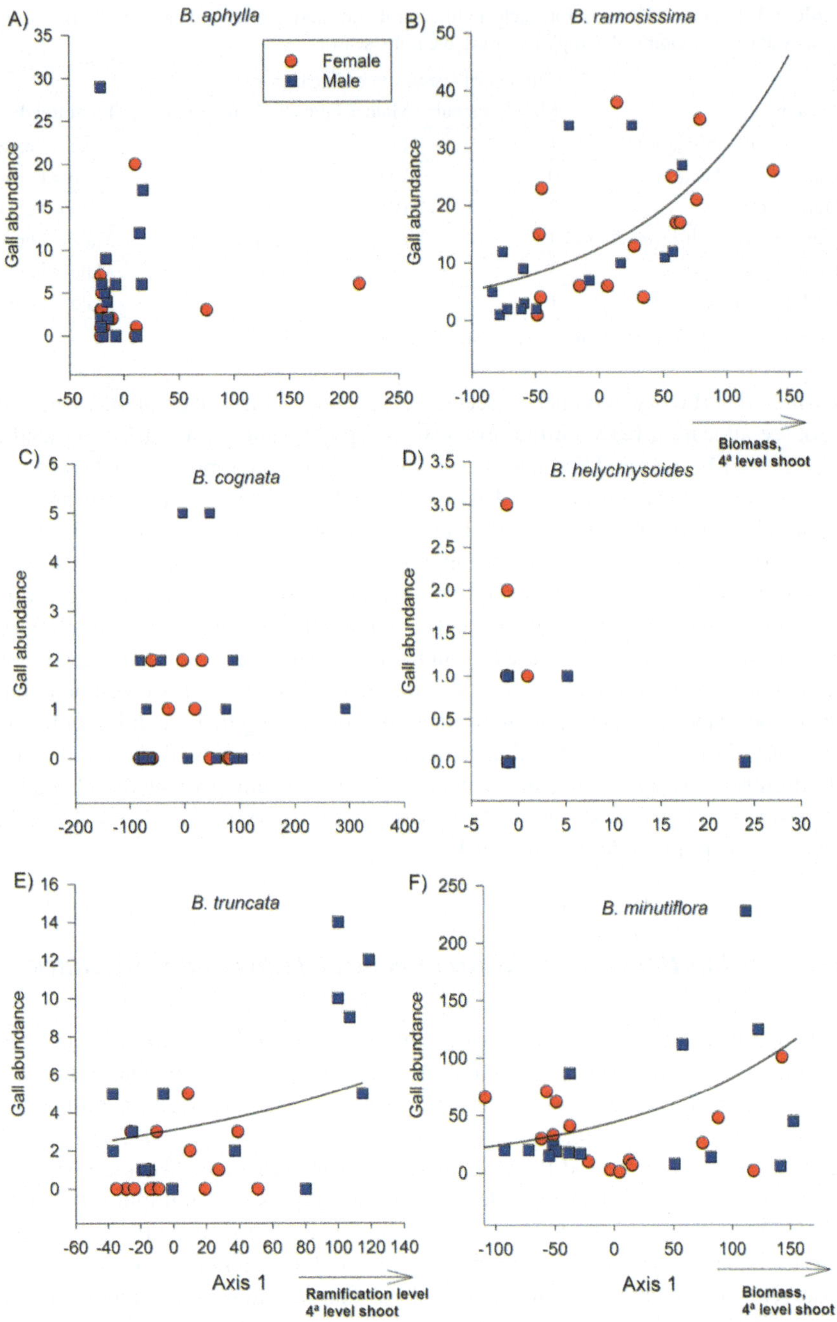

Fig. 4.2 Galling insect attack in relation to host architectural traits (summarized by the scores of the first axes of principal component analysis – PCA) for 12 species of *Baccharis*. The curves were adjusted based on parameters estimated from the analysis of generalized linear models (n = 30). Arrow below the x axis indicates the variables that positively correlated with the PCA axis-1

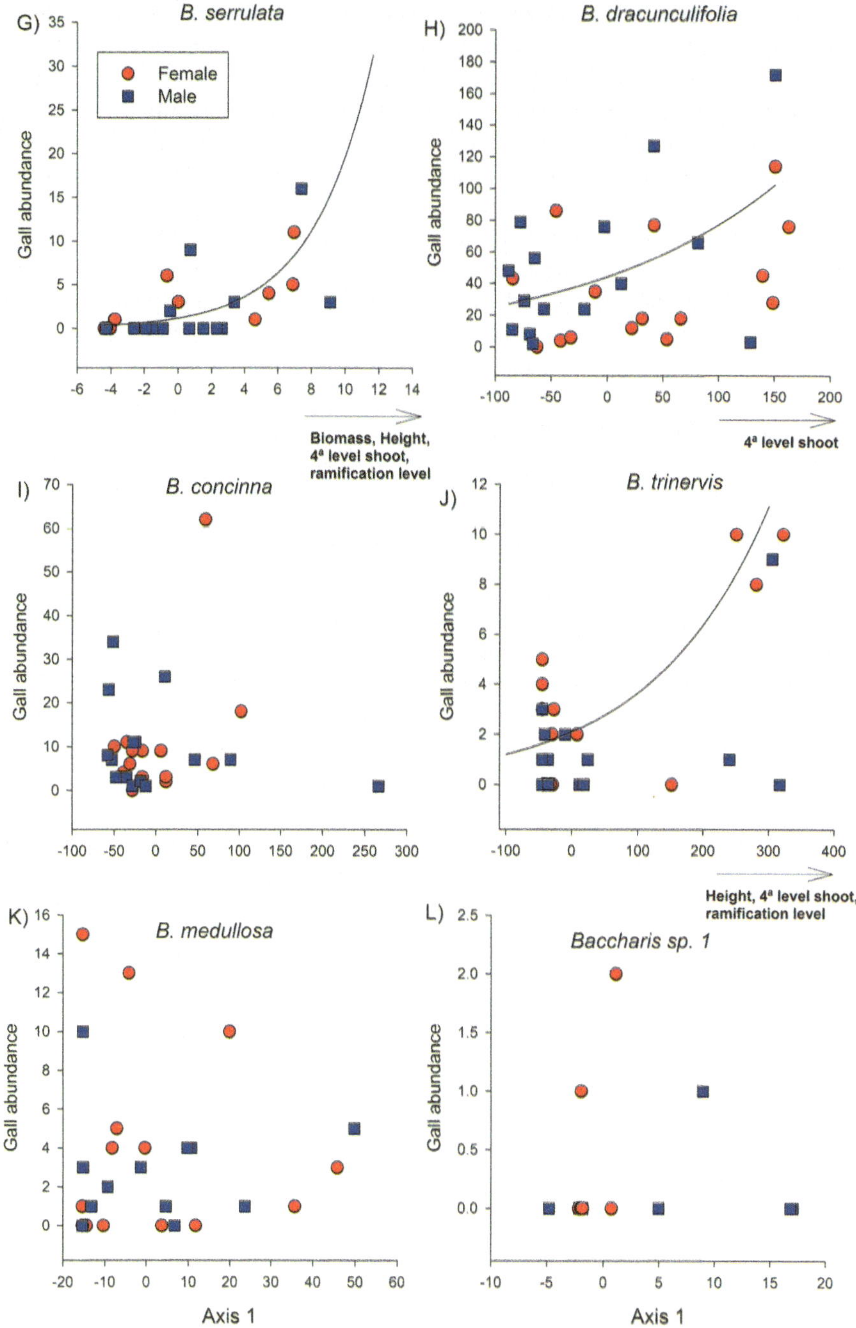

Fig. 4.2 (continued)

fourth-level shoots, followed by three species which had the first axis of the PCA positively correlated with biomass and ramification level. Finally, only two species had the first axis of the PCA positively correlated with height (Fig. 4.2). These results reinforce the role of fourth-level shoots as the best proxy for architectural traits leading to gall attack in *Baccharis* species, as this variable is a good indicator of plant meristem number (Espírito-Santo et al. 2007, 2012). Indeed, meristematic activity influences the plant's availability of young tissue, which is key to gall development (see Weis et al. 1988; Rohfritsch 1992; Carneiro et al. 2017) and highly attractive even to free-feeding herbivores (Boege 2005; Silva et al. 2012). In our study, intersexual differences in the number of fourth-level shoots were found for five *Baccharis* species, but no clear trend was observed in the gall abundance for these species (Table 4.1). Considering all species together, the lack of sex-related differences in gall attack is consistent with the general architectural similarity among male and female individuals of *Baccharis*.

3 Case Study 2. A Fine-Scale Study for *Baccharis concinna*

The sex-biased herbivory hypothesis has not been supported in any of the studies conducted on several species of *Baccharis* in tropical areas of Brazil (Box 4.1), although differential resource allocation may have been detected within the host plant (see Table 4.1). Therefore, we set out to test the differential resource allocation and sex-biased herbivory hypotheses at both a short- and a long-term sampling event on *B. concinna* to observe whether these would show consistent results. We included several measures of resource allocation such as shoot growth and total aboveground biomass, and we also sampled entire galling insect communities over 11 months. The hope was that we would be able to detect a clearer picture of how galling insect communities are impacted by the resource allocation patterns of this host plant. *Baccharis concinna* is a perennial shrub species with continuous production of very small flowers and growth meristems throughout the year. It harbors 15 species of gall-inducing insects that attack leaves, flowers, and stems as described by Fernandes et al. (1996, 2014).

Long-Term Study

In the long-term study, one population of *Baccharis concinna* occurring along Geraldinho creek at 1100 m in elevation at the Reserva Vellozia, in Minas Gerais, Brazil, was studied for 12 months. Forty individuals of each sex were haphazardly chosen and marked. Plant size was estimated by measuring plant height and two measures of crown width. We measured the growth of lateral shoots as they became available for colonization by gall-inducing insect species throughout the 11 months. Lateral shoots are here defined as shoots growing along a main stem and with determinate growth. To assess the lateral shoot growth rates of the 40 previously marked plants, 10 haphazardly chosen lateral shoots were marked with bird tags and labeled 1 through 10 (see Marques and Fernandes 2016).

Our findings revealed that male and female plants did not differ in plant height and were 1.15 m tall (mean female height = 115.25 ± 4.05 cm; mean male height = 114.9 ± 3.89 cm; $P > 0.05$). Nevertheless, when total crown area was considered (crown width*crown length), male plants had, on average, twice the crown area of female plants (mean male area = 6726.37 ± 628.4 cm^2; mean female area = 3220.25 ± 335.13 cm^2; $P < 0.01$), suggesting that male plants have a greater number of shorter branches when compared to female plants. Overall, monthly growth rates of lateral shoots were greater on male plants with a slow but steady increase from March (the end of the rainy season) throughout May and September (dry season) and showing a greater increase in November, December, and January (rainy season) (Fig. 4.3). Cumulative growth rates of lateral shoots varied between sex and months with a non-significant interaction term – plant sex and months (LN) (sex: $F = 4.16$, df = 1, $P < 0.042$; months: $F = 498.12$, df = 9, $P < 0.01$; sex*month: $F = 2.82$, df = 9, $P < 0.01$, $N = 5175$; Fig. 4.3). The lateral shoots marked in February of 1998 were 1 year old and on average 0.6 cm longer in male plants. By the end of the second growing season, lateral shoots in male plants were on average almost 2 cm longer than those in female plants. The differential resource allocation

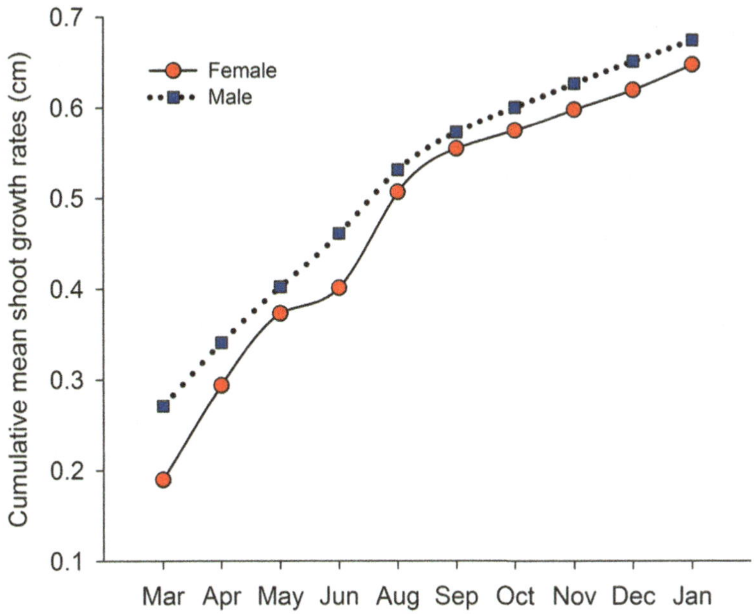

Fig. 4.3 Cumulative mean lateral shoot growth rates (average of 10 shoots per plant) on 40 female (solid line and circles) and 40 male plants (dashed line and squares) of *Baccharis concinna*, monthly measured between 1998 and 1999, at the Reserva Vellozia, in Minas Gerais, Brazil (standard error for all points was below 0.002 cm)

hypothesis that predicted male plants would allocate more resources towards growth was corroborated in this study.

Sex-biased herbivory – The community of galling insects in this population over 11 months consisted of 11 species out of the 15 known species to be associated with *Baccharis concinna* (Fernandes et al. 1996). Only five of these species had abundances above the hundreds, and only four species were common throughout the study period (Table 4.4). Male plants accumulated 11 gall-inducing insect species, while female plants had 10 species over the 12 months of study. Otherwise, gall-inducing insect species accumulated at a faster rate on male plants as opposed to female plants. Male plants had accumulated all 11 species by August (after 6 months of sampling), while female plants had all 10 species by December (after 10 months). Four galling insect morphospecies (A, B, F, and G) were present and abundant year-round, while most other species were common in a few months or were rare (low abundances) throughout the study. Gall-inducing insect richness showed an initial peak in March (the end of the rainy season) through July (dry season), with a sharp decrease in November (beginning of the rainy season). The difference between galling insect species between male and female plants was driven by rare species, which took longer to be detected on female plants. Mean gall-inducing insect abundance on male and female plants showed a very similar seasonal pattern to that of insect richness. Mean galling insect abundance differed between months but not between plant sex (sex: $F = 0.057$, df = 1, $P > 0.813$; months: $F = 5.5$, df = 8, $P < 0.01$; sex*month: $F = 1.88$, df = 8, $P > 0.064$, $N = 355$). Similarly, cumulative galling insect abundance differed between months but not between plant sex (natural logarithm +1) (sex: $F = 0.17$, df = 1, $N = 48$, $P > 0.17$; months: $F = 178.9$, df = 8, $N = 368$, $P < 0.01$; sex*month: $F = 0.83$, df = 8, $N = 368$, $P > 0.05$). In some months

Table 4.4 Composition of the galling insect community on male and female plants (M/F) of *Baccharis concinna*, monthly measured between 1998 and 1999, in a population in Serra do Cipó, Brazil

	1998								1999
Sp.	Mar	Apr	May	Jun	Aug	Sept	Nov	Dec	Jan
A	14/25	17/17	18/12	9/14	0/2	9/10	18/6	13/18	5/8
B	44/55	13/20	27/10	47/50	16/11	20/13	0/8	6/12	0/7
C	9/2	7/1	0/1	0/1	0/0	0/0	0/0	0/0	0/0
D	0/0	1/0	3/0	1/0	0/0	0/0	0/0	0/1	0/0
E	0/0	0/0	5/0	1/0	13/6	0/1	0/2	0/0	0/0
F	110/90	76/71	70/60	37/31	0/9	17/29	6/3	10/9	14/21
G	60/70	22/17	27/70	24/28	7/3	15/32	9/17	162/111	71/108
H	0/1	9/2	2/4	2/11	4/3	0/3	0/0	0/0	0/0
K	0/0	0/0	0/0	2/1	0/0	0/0	0/0	0/1	0/0
M	0/0	0/0	0/0	2/0	0/0	0/0	0/0	0/1	0/0
N	0/0	0/0	0/0	0/0	7/25	359/487	0/0	0/1	0/0

Coding for galling morphospecies according to Fernandes et al. (1996). Galling insect species attacked new shoots (B, D, H, K, and L), apical buds (A, C, F, and G), leaves (E and N), and flowers (M) (Marques and Fernandes 2016)

male plants would have a greater abundance of galling insects associated, while the opposite pattern was seen in other months where female plants had a greater abundance of galling insects. A few abundant species directed the patterns observed (Table 4.4). The only clear pattern observed was the seasonal distribution of gall-inducing insects on *B. concinna*. Gall-inducing insect species richness and abundance were higher in the drier months decreasing towards the rainy season. Therefore, the sex-biased herbivory hypothesis was not supported in this study.

Short-Term Study

We performed a more precise evaluation of intersexual differences in resource allocation by assessing the biomass of resources and shoot length on several populations of *B. concinna*. We further tested the sex-biased herbivory hypothesis by recording the galling insect species richness and abundance on the entire plant in one destructive sampling event at several locations and in the same region where the long-term study was conducted. All mature plants from both sexes were collected from one population at each of seven dry sites. The numbers of male and female plants collected at each site were I (28 M and 31 F), II (25 M and 35 F), III (18 M and 29 F), IV (24 M and 25 F), V (25 M and 25 F), VI (24 M and 25 F), and VII (22 M and 26 F), totalling 362 plants. Plants were cut at ground level and frozen at −10 °C until the analyses were performed. Plant sex was determined by analyzing the external morphology of flowers, which are always present in mature plants. To evaluate resource allocation to growth between plants of different sexes, three parameters were measured: (i) the current year's growth of lateral shoots (cm), (ii) total plant dry biomass (g), and (iii) proportion of soft tissues (leaves and flowers) and hard tissues (stems) (%). Current year's growth was evaluated by measuring the green portion of ten haphazardly chosen lateral shoots per plant. The number and abundance of new (green) as well as old (brown) galls per plant were counted under stereoscopic microscopes in the laboratory and were identified following Fernandes et al. (1996).

The data on plant biomass, shoot length, richness, and abundance of galling insects were log-transformed to meet the assumptions of normality of the tests utilized (Sokal and Rohlf 2012). Statistical analyses were conducted in two stages. First, two-way analyses of variance (ANOVAs) were used to check for differences in current year's growth of lateral shoots and plant biomass between male and female plants. Second, multiple linear regressions were used to test the relationship between plant biomass, shoot length, plant sex, and galling species richness. Multiple linear regressions with backward stepwise procedure were also used to test the relationship between the variables mentioned above and the abundance of insect galls on *B. concinna*. The model included the variables site, sex, shoot length, biomass, and respective interactions on the richness and abundance of galling insect species.

Differential resource allocation for lateral shoot length varied between plant sex and sites with a significant interaction between site and plant sex (Table 4.5). Male plants produced longer lateral shoots when compared to female plants at three different sites, but at the other three sites no differences in shoot length were evident (Tables 4.6 and 4.7). Total plant biomass did not differ between plant sexes in *B. concinna*, but there were differences in plant biomass between sites (Tables 4.5 and 4.7). Similar results were observed when the resources were broken down into leaves and flowers and stems. No differences were observed on leaf and flower (soft tissues) and stem (woody tissue) dry biomass between male and female plants, while differences were evident between sites. The significance of the interaction terms suggests a sex-by-site interaction (leaf and flower site: $F = 15.11$, df = 1, $n = 359$, $P < 0.05$; sex: $F = 2.44$, df = 1, $n = 359$, $P > 0.119$; site*sex: $F = 2.22$, df = 1, $n = 359$, $P < 0.05$) (site*stem: $F = 11.85$, df = 1, $n = 357$, $P < 0.05$; sex: $F = 1.861$, df = 1, $n = 357$, $P > 0.18$; site*sex: $F = 2.92$, df = 1, $n = 357$, $P < 0.05$).

Resource allocation and plant growth varied between sites suggesting that *B. concinna* is strongly affected by environmental conditions. This becomes evident due to the significant interaction between site and sex suggesting a sex-by-site interaction where male plants growing in certain sites – maybe less stressful environmental conditions – would be able to produce longer shoots. Although all sites were dry and contained the same soil types, it is possible that soil conditions could differ between sites (see Marques et al. 2002). The differential growth of lateral shoots observed on male plants did not result in greater biomass of leaves and flowers or stems on male plants (data not shown). Since resource allocation for growth did not

Table 4.5 Effects of plant sex, growth rate, and month on galling insect species richness and abundance on *Baccharis concinna*, in Serra do Cipó, Brazil (Marques and Fernandes 2016)

Multiple regression		df	F	P
Galling species richness				
Months		1	77.97	<0.05
Plant sex			0.26	0.61
Mean shoot growth rate			0.12	0.73
Error		353		
Total		355		
Minimum adequate model	n	R^2	F	P
Lnsp = 1.397–0.06 month	355	0.406	69.63	<0.05
Galling abundance				
Months		1	29.07	0.05
Plant sex			0.119	<0.73
Mean shoot growth rate		1	4.7	<0.032
Error		326		
Total		329		
Minimum adequate model	n	R^2	F	P
Lnab = 2.62–5.94 shoot, 0.074 month	329	0.294	15.5	<0.05

Table 4.6 Mean gall abundance and SE (standard error) of morphospecies on male and female plants of *Baccharis concinna* in Serra do Cipó, Brazil

		Galling morphospecies												
		A	B	C	D	E	F	G	H	I	K	L	M	N
SITE I														
Female	Mean	1.5	6.4	2.42	0	0	4.5	20	1	0	0	0	0	0
	SE	0.75	1.02	0.47	0	0	0.74	0	0	0	0	0	0	0
Male	Mean	2	3.1	3.1	1	1	4.5	2.12	1	0	1	0	0	0
	SE	0	0.74	0.7	0	0	1.28	0.74	0	0	0	0	0	0
SITE II														
Female	Mean	1.33	24	3	1	2.4	12.6	0	1.3	0	1	1	0	1
	SE	0.33	4.02	1.27	0	0.77	1.60	0	0.33	0	0	0	0	0
Male	Mean	1.43	8.54	1.54	0	1	11.39	0	1	0	0	0	0	0
	SE	0.41	1.51	0.31	0	0	1.71	0	0	0	0	0	0	0
SITE III														
Female	Mean	1.25	4.5	4.6	1	4.28	20	0	0	0	0	0	0	0
	SE	0.22	1.78	1.25	0	0.74	4.75	0	0	0	0	0	0	0
Male	Mean	1.5	1.66	2.25	0	1	7.75	0	0	0	0	0	0	0
	SE	0.49	0.49	0.41	0	0	1.83	0	0	0	0	0	0	0
SITE IV														
Female	Mean	1.33	2.4	2.8	1	3	5.33	1	1	0	0	0	0	0
	SE	0.26	0.47	0.62	0	0	0.85	0	0	0	0	0	0	0
Male	Mean	1.25	2.4	2.8	1	3	5.33	1	1	0	0	0	0	0
	SE	0.29	0.74	0.59	0	0	1.0	0	0	0	0	0	0	0
SITE V														
Female	Mean	1.4	2	0	0	0	2.36	0	0	0	0	0	0	0
	SE	0.19	0.44	0	0	0	0.42	0	0	0	0	0	0	0
Male	Mean	2.25	1.4	0	0	0	1.86	0	0	0	0	0	0	0
	SE	1.1	0.22	0	0	0	0.22	0	0	0	0	0	0	0
SITE VI														
Female	Mean	1	1.63	2	1	11.93	5.31	4.75	1.25	4.33	4.25	1	1	0
	SE	0	0.29	0.45	0	9.17	1.08	1.07	0.25	1.45	1.12	0	0	0
Male	Mean	1.25	1.66	1.84	1	2.57	2.23	2.16	1	2	2	1	1	1
	SE	0.25	0.42	0.36	0	0.61	0.45	0.57	0	0.99	0.7	0	0	0
SITE VII														
Female	Mean	1.715	1.25	1	0	0	24.5	2	1.71	0	0	1.66	0	0
	SE	0.24	0.14	0	0	0	4.3	0	0.36	0	0	0.44	0	0
Male	Mean	2.711	1.25	1.5	1	1	17.28	0	5.62	0	0	2.77	0	1
	SE	0.84	0.22	0.5	0	0	3.21	0	1.63	0	0	0.93	0	0

Letters stand for: A = Curculionidae, B = Lepidoptera, C = Cecidomyiidae, D = Cecidomyiidae, E = Psyllidae, F = Cecidomyiidae, G = Cecidomyiidae, H = Lepidoptera, I = Cecidomyiidae, K = Cecidomyiidae, L = Cecidomyiidae, M = Cecidomyiidae, N = Cecidomyiidae. See Fernandes et al. (1996) for details

Table 4.7 Effects of site and plant sex on current year's lateral shoot length (cm), biomass (g), and species richness and abundance on *Baccharis concinna* in Serra do Cipó, Brazil

Analysis of variance	df	SS	*F*	*P*
Lateral shoots				
Site	5	112.88	81.24	<0.05
Plant sex	1	6.93	24.95	<0.05
Plant sex*site	5	8.56	6.16	<0.05
Error	2409	669.45		
Total	2420	797.82		
Plant biomass				
Site	6	30.520	13.844	<0.05
Plant sex	1	0.671	1.826	0.177
Plant sex*site	6	4.580	2.077	0.055
Error	344	126.400		
Total	357			
Insect galling richness				
Site	6	13.13	8.51	<0.05
Plant sex	1	0.106	0.413	0.521
Plant sex*site	6	2.26	1.47	0.189
Error	285	73.33		
Total	298			
Insect galling abundance				
Site	6	121.44	23.33	<0.05
Plant sex	1	19.38	22.35	<0.05
Plant sex*site	6	5.67	1.09	0.368
Error	302	408.38		
Total	315			

differ between the sexes, the differential resource allocation hypothesis was not corroborated in this study.

Sex-biased herbivory hypothesis was not corroborated in this study when the totality of all galling insect species was considered. A total of 11 galling insect morphospecies were found associated with *B. concinna* in this study. Galling morphospecies B, D, H, K, and L attacked new stems, whereas the morphospecies A, C, F, G, and I attacked apical buds; morphospecies E and N attacked leaves, and morphospecies M attacked flowers (Table 4.6). Galling insect richness differed between sites, but not between male and female plants (Table 4.7). Also, galling insect richness was not evenly distributed across sites. A total of 9 morphospecies were found at sites I and II, 6 morphospecies at site III, 8 morphospecies at site IV, 3 morphospecies at site V, 11 morphospecies at site VI, and 10 morphospecies at site VII (Table 4.6). Three morphospecies were common to all seven sites, three morphospecies occurred at six sites, and the five remaining morphospecies were rare (Table 4.6). Common galling morphospecies, those that occurred at more sites, were also more abundant. Mean gall-inducing insect richness differed between sites but not between plant sexes, with a non-significant interaction term – plant sex and sites (Tables 4.7 and 4.8). Male and female plants of *B. concinna* showed similar richness of galling insect species although richness varied between sites. Galling insect abundance differed between sites and plant sexes (Tables 4.6 and 4.7). Female plants supported greater galling insect abundance at site II, while male plants supported greater galling insect abundance at sites III and VI (Tables 4.6 and 4.7). Shoot length did not show a clear pattern with galling insect species abundance (Table 4.8), and neither did plant total biomass and species richness (Table 4.9).

In conclusion, lateral shoot length varied between plant sex and sites with a significant interaction between site and plant sex. Male plants produced longer lateral shoots when compared to female plants at three different sites, but at the other three sites no differences in shoot length were evident (Tables 4.7 and 4.8). Total plant biomass did not differ between plant sexes, but there were differences in plant biomass between sites (Tables 4.7, 4.8, and 4.9).

Resource allocation in *B. concinna* has been studied by measures of size and biomass for the entire plant (this study) and for apical and lateral shoots (Marques 1997; Madeira et al. 1997; Carneiro et al. 2005, 2006). Although in general no differential resource allocation pattern was seen for most of the resources studied in *B. concinna*, male plants had on average longer lateral shoots at some of the populations studied when compared to female plants (short-term study). This suggests the existence of architectural differences between male and female plants, where male plants probably compensate for the shorter apical shoots by investing on longer lateral shoots as shown in this study and that of Madeira et al. (1997). Longer apical shoots in female plants and lateral shoots in male plants did not translate into greater total plant, stem, or leaf biomass on plants of different sexes at these sites, as would be expected if any sex were allocating more resources into growth. Nevertheless, the end result is that there is no difference in leaf, stem, or total plant biomass between male and female plants as shown by this study.

In the long-term study involving one site, we found evidence to suggest that male plants of *B. concinna* had greater average lateral shoot length over 11 months. But these findings again suggest that plants at different sites reflect different growth patterns. The long-term study of gall-inducing insect richness and abundance between male and female plants differed only in 3 months but with no clear trends (Table 4.4). Galling richness was greater on female plants in March but greater on male plants in May and August. Although male plants accumulated morphospecies at a faster rate when compared with female plants, the most abundant morphospecies were present on both plant sexes, and the rare species, with very few occurrences, were responsible for driving this pattern. Gall-inducing insect abundance was greater on male plants only for the month of August with no difference observed between sexes for the remaining 10 months.

Nevertheless, an equivalent short-term sampling study conducted in the second case study across seven sites and which assessed total plant biomass in *B. concinna* did not detect any intersexual differences in this species but reflected differences in plant biomass at different sites. When plant biomass was broken down into its separate components of soft tissues and hard tissues, still no difference was detected

Table 4.8 Mean shoot length (cm) and galling insect abundance (SE – standard error) on male (M) and female (F) plants of *Baccharis concinna* in Serra do Cipó, Brazil

		Shoot length	Significance		Abundance		Significance
Site	Statistics	F	M	*P*	F	M	*P*
I							
	Mean	6.71	6.9	ns	10.2	6.0	ns
	S.E.	0.27	0.37		1.9	1.4	
II							
	Mean	8.0	9.37	<0.05	36.5	17.7	<0.05
	S.E.	0.4	0.63		4.9	2.4	
III							
	Mean	7.51	9.78	<0.05	8.9	21.5	<0.05
	S.E.	0.4	0.3		1.9	4.6	
IV							
	Mean	5.5	5.7	ns	5.3	7.7	ns
	S.E.	0.23	0.13		1.6	1.17	
V							
	Mean	10.18	12.6	<0.05	2.5	1.6	ns
	S.E.	0.32	0.42		0.49	0.3	
VI							
	Mean	–	–	–	5.8	16.0	<0.05
	S.E.	–	–		0.87	5.2	
VII							
	Mean	8.5	8.1	ns	24.8	18.7	ns
	S.E.	0.33	0.36		5.4	4.1	

The significance level (*P*) corresponds to t tests comparing means between male and female plants for each site separately

Table 4.9 Result of post-doc test on mean plant biomass (g) and galling insect richness (SE – standard error) on male and female plants of *Baccharis concinna* in Serra do Cipó, Brazil

Site	Plant biomass	Richness
I	26.64 ± 2.46a	2.1 ± 0.19a
II	47.4 ± 3.6b	2.6 ± 0.14b
III	36.26 ± 4.02c	2.0 ± 0.18a
IV	43.89 ± 4.3c	2.2 ± 0.26a
V	43.12 ± 4.2c	1.8 ± 0.31a
VI	63.13 ± 4.14d	2.3 ± 0.22a
VII	43.5 ± 3.9c	3.77 ± 0.60c

The letters group the means according to the Tukey test

between the host plant sexes. These findings further support other studies conducted by our laboratory which did not detect differential resource allocation in *Baccharis*. Delph et al. (1993) did not detect differential allocation of resources for growth or reproduction in *Carex picta* (Cyperaceae) and argued that the energetic cost of reproduction did not differ between the sexes because *C. picta* had dry, energetically inexpensive fruits much like *B. concinna*.

Plant quality in this tropical host species might also be affected by the extremely nutrient-poor soils of the rupestrian grasslands, at Serra do Cipó (Marques 1997), suggesting the existence of sex-by-site interaction where under such stressful environmental conditions male and female plants will perform differently (Boecklen and Hoffman 1993). For example, we know that male plants of *B. concinna* are more susceptible to aluminum in acidic soil conditions, such as those common to soils of rupestrian grasslands, when compared to female plants (Marques 1997).

4 Concluding Remarks

The two case studies outlined in this chapter provide a broad picture of all aspects of resource allocation towards growth over 12 species of *Baccharis*, over several sites, and a closer look at plant growth in *B. concinna* over 11 months. The first case study detailed the single sampling event of 12 species of *Baccharis* in Brazil, with no clear pattern of differential resource allocation being detected. When lateral shoot growth was considered in *B. concinna* in the second case study, the single destructive sampling event at seven sites revealed that male plants produced longer lateral shoots at three different sites, but at three other sites no differences were observed. In the long-term study involving one site, we found evidence to suggest that male plants of *B. concinna* had greater average lateral shoot length over 11 months. But these findings again suggest that plants at different sites reflect different growth patterns. These findings further support other studies conducted by our laboratory which did not detect differential resource allocation in *Baccharis* (Table 4.2).

The lack of consistency in differential resource allocation patterns suggests that resource allocation in *Baccharis* varies greatly with abiotic conditions including soil type, aluminum content, pH levels, precipitation, and temperature (Marques et al. 2002). *Baccharis* species are pioneer and colonizing plant species known to grow in disturbed habitats such as the sites of roads and clearings (Marques 1997). Their male and female flowers are abundant throughout the year in tropical species but are similar in size and small suggesting very similar energetic investment; their dry fruits suggest that female plants do not allocate much energy to the fruit and there is no investment in attracting dispersers because seeds are wind-dispersed.

In spite of the many studies done on the differential allocation of resources and resulting species distribution here reviewed, the understanding of the evolutionary ecology of *Baccharis* is still rudimentary. There is an enormous gap in the knowledge on *Baccharis* phenological trends, ecological niche, population, and reproductive ecology and genetics, which altogether builds the basis to understand its relationship with the associated fauna and biodiversity. Since male and female plants allocated most resources equally, we did not expect to find differences in the richness and abundance of the galling species associated with *Baccharis* species.

Indeed, we conducted meta-analyses to determine the occurrence of sex-biased herbivory in general dioecious systems and for the genus *Baccharis* in particular (Box 4.1). Unlike the general pattern of higher herbivory on male plants (Boecklen et al. 1990; Åhman 1997; Cornelissen and Stiling 2005; Rivkin et al. 2018), no intersexual differences were found in *Baccharis*.

We selected to study gall-inducing insects because they utilize meristematic tissues that are still growing and differentiating (Mani 1964). That was indeed the case for most studies conducted by our group (Table 4.1), with one exception (Carneiro et al. 2005). However, these studies did not corroborate the differential resource allocation and the plant sex-biased herbivory hypotheses. We suggest that both sexes of these species, which grow slowly and produce vegetative and reproductive meristems continuously throughout the year, might be better chemically defended against herbivores when compared to temperate plants that experience one flush of growth in a year (see Herms and Mattson 1992; Sagers and Coley 1995). The second point was raised by Boecklen et al. (1994), who inquired if these studies were actually measuring the resources used by the insects or if, in reality, the resources utilized by these species were not sexually dimorphic. From the perspective of galling insects, it would be more realistic if studies measured the allocation of the resources which are actually utilized by the insects.

Furthermore, true patterns in nature should become stronger when more populations of the host plant are studied. That did not seem to be the case in this study, since as more populations were studied the less clear the patterns became. As seen, patterns of abundance could be attributed to one sex if only a few populations were studied, but by adding more populations the pattern became weaker. In addition, we must be very careful when selecting the scale (apical meristems, lateral meristems, or whole plant) used to search for patterns in galling species diversity/density for they can address different adaptive strategies of both host plants and galling insects. To further clarify the relationship between galling species richness, abundance, and plant sex in this system, future studies should consider the points mentioned above as well as measures of availability of resources throughout the year as meristems become available and galls colonize their host plants.

Herbivores are affected by multiple top-down and bottom-up forces that vary both spatially and temporally. As such, detecting general patterns through field studies in complex tropical environments is a quite complicated task. In the case of *Baccharis*, our results indicate that galling insects do not select for plant sexes, but long-term studies under controlled environmental conditions are necessary to confirm such a pattern. The genus *Baccharis* has the greatest number of galling insect species in any genus so far studied (Fernandes et al. 1996) and presents a great model to study these mainstream hypotheses on plant-animal interactions.

References

Ågreen J (1988) Sexual differences in biomass and nutrient allocation in the dioecious *Rubus chamemorus*. Ecology 69:962–973

Ågren J (1987) Intersexual differences in phenology and damage by herbivores and pathogens in dioecious *Rubus chamaemorus* L. Oecologia 72:161–169

Åhman I (1997) Growth, herbivory, and disease in relation to gender in *Salix viminalis* L. Oecologia 111:61–68

Araújo APA, Carneiro MAA, Fernandes GW (2003) Efeitos do sexo, do vigor e do tamanho da planta hospedeira sobre a distribuição de insetos indutores de galhas em *Baccharis pseudomyriocephala* Teodoro (Asteraceae). Rev Bras Entomol 47:483–490

Araújo APA, De Paula JD, Carneiro MAA, Schoereder JH (2006) Effects of host plant architecture on colonization by galling insects. Austral Ecol 31:343–348

Arnold SJ, Wade MJ (1984) On the measurement of natural and sexual selection: theory. Evolution 38:709–719

Ashman TL (2002) The role of herbivores in the evolution of separate sexes from hermaphroditism. Ecology 83:1175–1184

Avila-Sakar G, Romanow CA (2012) Divergence in defence against herbivores between males and females of dioecious plant species. Int J Evol Biol 12. ID 897157. https://doi.org/10.1155/2012/897157

Barrett SCH (2002) The evolution of plant sexual diversity. Nature 3:274–284

Barrett SCH, Hough J (2013) Sexual dimorphism in flowering plants. J Exp Bot 64:67–82. https://doi.org/10.1093/jxb/ers308

Bawa KS (1980) Evolution of dioecy in flowering plants. Annu Rev Ecol Syst 11:15–39

Boecklen WJ, Hoffman MT (1993) Sex-biased herbivory in *Ephedra trifurca*: the importance of sex-by-environment interactions. Oecologia 96:49–55

Boecklen WJ, Price PW, Mopper S (1990) Sex drugs and herbivores: sex-biased herbivory in arroyo willow (*Salix lasiolepis*). Ecology 71:581–588

Boecklen WJ, Mopper S, Price PW (1994) Sex-biased herbivory in arroyo willow: are there general patterns among herbivores? Oikos 71:267–272

Boege K (2005) Herbivore attack in *Casearia nitida* influenced by plant ontogenetic variation in foliage quality and plant architecture. Oecologia 143:117–125

Carneiro MAA, Fernandes GW, De Souza OFF (2005) Convergence in the variation of local and regional galling species richness. Neotrop Entomol 34:547–553

Carneiro MAA, Fernandes GW, De Souza OFF, Souza WVM (2006) Sex-mediated herbivory by galling insects on *Baccharis concinna* (Asteraceae). Rev Bras Entomol 50:394–398

Carneiro RGS, Isaias RMS, Moreira ASFP, Oliveira DC (2017) Reacquisition of new meristematic sites determines the development of a new organ, the cecidomyiidae gall on *Copaifera langsdorffii* Desf. (Fabaceae). Front Plant Sci 8:1622. https://doi.org/10.3389/fpls.2017.01622

Charlesworth D (2018) Does sexual dimorphism in plants promote sex chromosome evolution? Environ Exp Bot 146:5–12. https://doi.org/10.1016/j.envexpbot.2017.11.005

Cornelissen TG, Fernandes GW (2001) Defence, growth and nutrient allocation in the tropical shrub *Bauhinia brevipes* (Leguminosae). Austral Ecol 26:246–253

Cornelissen T, Stiling P (2005) Sex-biased herbivory: a meta-analysis of the effects of gender on plant-herbivore interactions. Oikos 111:488–500

Darwin C (1871) The descent of man, and selection in relation to sex. Murray, London

Delph LF (2019) Pollen competition is the mechanism underlying a variety of evolutionary phenomena in dioecious plants. New Phytol 224:1075–1079. https://doi.org/10.1111/nph.15868

Delph LF, Lu Y, Jayne LD (1993) Patterns of resource allocation in a dioecious *Carex* (Cyperaceae). Am J Bot 80:607–615

Emlen ST, Oring LW (1977) Ecology, sexual selection and the evolution of mating systems. Science 197:215–223

Erb M (2018) Plant defenses against herbivory: closing the fitness gap. Trends Plant Sci 23:187–194. https://doi.org/10.1016/j.tplants.2017.11.005

Espírito-Santo MM, Fernandes GW (1998) Abundance of *Neopelma baccharidis* (Homoptera:Psyllidae) galls on the dioecious shrub *Baccharis dracunculifolia* (Asteraceae). Environ Entomol 27:870–876

Espírito-Santo MM, Fernandes GW, Allain LR, Reis TRF (1999) Tannins in *Baccharis dracunculifolia*: effects of seasonality, water availability and plant sex. Acta Bot Bras 13:167–174

Espírito-Santo MM, Madeira BG, Neves FS, Faria ML, Fagundes M, Fernades GW (2003) Sexual differences in reproductive phenology and their consequences for the demography of *Baccharis dracunculifolia* (Asteraceae), a dioecious tropical shrub. Ann Bot 91:13–19

Espírito-Santo MM, Neves FS, Andrade-Neto FR, Fernandes GW (2007) Plant architecture and meristem dynamics as the mechanisms determining the diversity of gall-inducing insects. Oecologia 153:353–364

Espírito-Santo MM, Neves FS, Fernandes GW, Silva JO (2012) Plant phenology and absence of sex-biased gall attack on three species of *Baccharis*. PLoS One 7(10):e46896. https://doi.org/10.1371/journal.pone.0046896

Faria ML, Fernandes GW (2001) Vigour of a dioecious shrub and attack by a galling herbivore. Ecol Entomol 26:36–45

Fernandes GW, Carneiro MAA, Lara ACF, Allain LR, Julião GR, Reis TR, Andrade GI, Silva IM (1996) Galling insect on neotropical species of *Baccharis* (Asteraceae). Trop Zool 9:315–332

Fernandes GW, Silva JO, Espírito-Santo MM, Fagundes M, Oki Y, Carneiro MMA (2014) *Baccharis*: a Neotropical model system to study insect plant interactions. In: Fernandes GW, Santos JC (eds) Neotropical insect galls. Springer, Dordrecht, pp 193–220

Harris MS, Pannell JR (2008) Roots, shoots and reproduction: sexual dimorphism in size and costs of reproductive allocation in an annual herb. Proc R Soc B Biol Sci B 275:2595–2602. https://doi.org/10.1098/rspb.2008.0585

Heiden G, Antonelli A, Pirani JR (2019) A novel phylogenetic infrageneric classification of *Baccharis* (Asteraceae: Astereae), a highly diversified American genus. Taxon 68:1048–1081. https://doi.org/10.1002/tax.12128

Hemborg AM, Karlsson PS (1999) Sexual differences in biomass and nutrient allocation of first-year *Silene dioica* plants. Oecologia 118:453–460

Henry IM, Akagi T, Tao R, Comai L (2018) One hundred ways to invent the sexes: theoretical and observed paths to dioecy in plants. Annu Rev Plant Biol 69:553–575

Herms DA, Mattson WJ (1992) The dilemma of plants: to grow or to defend. Q Rev Biol 67:283–335

Imaji A, Seiwa K (2010) Carbon allocation to defense, storage and growth in seedlings of two temperate broad-leaved tree species. Oecologia 162:361–369

Jaenicke T, Morrow EH (2018) Operational sex ratio predicts the opportunity and direction of sexual selection across animals. Ecol Lett 23:384–391

Jing SW, Coley PD (1990) Dieocy and herbivory: the effect of growth rate on plant defense in *Acer negundo*. Oikos 58:369–377

Knauer AC, Schiesti FP (2017) The effect of pollinators and herbivores on selection for floral signals: a case study in *Brassica rapa*. Evol Ecol 31:285–304. https://doi.org/10.1007/s10682-016-9878-8

Krischik VA, Denno RF (1990) Patterns of growth, reproduction, defense, and herbivory in the dioecious shrub *Baccharis halimifolia* (Compositae). Oecologia 83:182–190

Leroy CJ, Hobbs JMR, Claeson SM, Moffett J, Garthwaite I, Criss N, Walker L (2020) Plant sex influences aquatic-terrestiral interactions. Ecosphere 11:1–10. https://doi.org/10.1002/ecs2.2994

Lovett-Doust J, O'Brien G, Doust LL (1987) Effect of density on secondary sex characteristics and sex ratio in *Silene alba* (Caryophyllaceae). Am J Bot 74:40–46

Loyd DG, Webb CJ (1977) Secondary sex characters in plants. Bot Rev 43:177–216

Madeira BG, Cornelissen TG, Faria ML, Fernandes GW (1997) Insect herbivore preference for plant sex and modules in *Baccharis concinna*, Barroso (Asteraceae). In: Raman A (ed) Ecology and evolution of plant-feeding insects in natural and man-made environments. International Scientific Publications, New Delhi, pp 135–143

Mani MS (1964) The ecology of plant galls. Junk, The Hague

Marques AR (1997) Sexo, distribuição espacial e alocação de recursos em *Baccharis concinna*, uma espécie dióica e endêmica do Espinhaço. MSc Dissertation Universidade Federal de Minas Gerais. Belo Horizonte-MG, Brazil

Marques ESA, Fernandes GW (2016) The gall inducing insect community on *Baccharis concinna* (Asteraceae): the role of shoot growth rates and seasonal variations. Lundiana 12:17–26

Marques AR, Fernandes GW, Reis IA, Assunção RM (2002) Distribution of adult male and female *Baccharis concinna* (Asteraceae) in the rupestrian fields of Serra do Cipó, Brasil. Plant Biol 4:94–103

Moreira X, Romero-Perez A, Luna-Chaparro E, Orona-Tamayo D, Quintana-Rodirguez E, Reyes-Chilpa R, Abdala-Roberts L, Cano-Santana Z, Hernandez-Cumplido J (2019) Effects of plant sex on insect abundance across three trophic levels in the perennial shrub *Buddleja cordata*. Entomol Exp Appl 167:950–956. https://doi.org/10.1111/eea.12845

Moritz KK, Bjorkman C, Parachnowitsch AL, Stenberg JA (2017) Plant sex effects on insect herbivores and biological control in a Short Rotation Coppice willow. Biol Control 115:30–36

Multikainen P, Delph LF (1996) Effects of herbivory on male reproductive success in plants. Oikos 75:353–358

Obeso JR (2002) The cost of reproduction in plants. New Phytol 155:321–348. https://doi.org/10.1046/j.1469-8137.2000.00571.x

Pearson JT, Sparrow AD, Lange RT (1990) Prolonged exposure to sheep grazing reduces palatability on Australian saltbush populations. Aust J Ecol 35:337–344

Pfeiffer T, Schmidt L, Roschanski AM, Schnittler M (2019) A battle of the sexes? Sex ratio and performance at small scales in dioecious *Mercurialis perennis*. Acta Oecol 100:103462. https://doi.org/10.1016/j.actao.2019.103462

Popp JW, Reinartz IA (1988) Sexual dimorphism in biomass allocation and clonal growth of *Xanthoxylum americanum*. Am J Bot 75:1732–1741

Ribeiro-Mendes HN, Marques ESA, Silva IM, Fernandes GW (2002) Influence of host-plant sex and habitat on survivorship of insect galls within the geographical range of the host plant. Trop Zool 15:5–15

Rivkin LR, Barrett SCH, Johnson MTJ (2018) Effects of plant sexual system and latitude on resistance to herbivores. Am J Bot 105:977–985. https://doi.org/10.1002/ajb2.1098

Rohfritsch O (1992) Patterns in gall development. In: Shorthouse JD, Rohfritsch O (eds) Biology of insect-induced galls. Oxford University Press, Oxford, pp 60–86

Sagers CL, Coley PD (1995) Benefits and costs of defense in a Neotropical shrub. Ecology 76:1835–1843

Silva JO, Espírito-Santo MM, Melo GA (2012) Herbivory on *Handroanthus ochraceus* (Bignoniaceae) along a successional gradient in a tropical dry forest. Arthropod-Plant Interact 6:45–57

Sokal RR, Rohlf FJ (2012) Biometry: the principles and practice of statistics in biological research. Freeman, New York

Stanton ML, Snow AA, Handel SN (1986) Floral evolution: attractiveness to pollinator increases male fitness. Science 232:1625–1627

Sutherland S, Delph LF (1984) On the importance of male fitness in plants: patterns of fruit set. Ecology 65:1093–1104

Vaughton G, Ramsey M (1998) Floral display, pollinator visitation and reproductive success in the dioecious perennial herb *Wurmbea dioica* (Liliaceae). Oecologia 115:93–101

Vega-Frutis R, Munguía-Rosas MA, Varga S, Kytöviita MM (2013) Sex-specific patterns of antagonistic and mutualistic biotic interactions in dioecious and gynodioecious plants. Perspect Plant Ecol Evol Syst 15:45–55

Wallace CS, Rundel PW (1979) Sexual dimorphism and resource allocation in male and female shrubs of *Simondsia chinensis*. Oecologia 44:34–39

Wallace BC, Lajeunesse MJ, Dietz G, Dahabreh IJ, Trikalinos TA, Schmid CH, Gurevitch J (2017) OpenMEE: intuitive, open-source software for meta-analysis in ecology and evolutionary biology. Methods Ecol Evol 8:941–947

Weis AE, Walton R, Greco CL (1988) Reactive plant tissue sites and the population biology of gall makers. Annu Rev Entomol 33:467–486

Wolfe LM (1997) Differential flower herbivory and gall formation on males and females of *Neea psychotrioides*, a dioecious tree. Biotropica 29:169–174

Chapter 5
Multitrophic and Indirect Interactions in the *Baccharis dracunculifolia* System

Milton Barbosa, Graziella França Monteiro, and G. Wilson Fernandes

Abstract The *Baccharis dracunculifolia* (Asteraceae) system is a reasonably self-contained community including the host plant and its rich fauna of arthropods. Over the last three decades, ecological researchers in Brazil have intensively used the system as a study model. Several interactions, across multiple trophic levels, among the plant species *B. dracunculifolia* and associated arthropods, have been empirically demonstrated. Half of these interactions are indirect and include effects propagated via changes in density, trait, habitat or interaction strength. Here we describe the components of this system and summarize their interactions in a network. We hope that this work will spark new ideas for experimental tests of ecological hypotheses on the role of indirect interactions at the level of the whole community in the field.

Keywords Experimental ecology · Insect herbivory · Interaction modification · Network · Non-trophic inter0actions

1 Multitrophic and Indirect Interactions

Species affect each other in countless ways. The most evident and easily visible influences are the direct trophic interactions, those involving feeding relationships between two species. On the other hand, indirect trophic interactions require the presence of a third species as an intermediary (Wootton 1994) and are propagated through feeding relationships. In a trophic cascade, a predator affects the density and/or behaviour of a prey and, therefore, indirectly affects the rate of consumption of a species in the next lower trophic level (Estes and Palmisano 1974). These interaction chains or indirect effects mediated by density (e.g. trophic cascades, keystone species, apparent competition) have been broadly demonstrated (e.g. Wootton 1994, 2002; Morris et al. 2004; van Veen et al. 2006). Indirect trophic interactions

M. Barbosa (✉) · G. F. Monteiro · G. W. Fernandes
Ecologia Evolutiva & Biodiversidade/DGEE, Universidade Federal de Minas Gerais, Belo Horizonte, Minas Gerais, Brazil

G. W. Fernandes et al. (eds.), *Baccharis*,
https://doi.org/10.1007/978-3-030-83511-8_5

play an important role in structuring communities and have generally not been taken into account in traditional studies (Bukovinszky et al. 2008).

Indirect interactions between species can also happen in non-trophic ways (e.g. via habitat, trait or behaviour modification: Wootton 1994). However, these interactions remain largely ignored (Ogushi 2008). Indirect effects can cause changes in species density (e.g., interference competition, facilitation: Prugh and Brashares 2012; Wetzel et al. 2016) or changes in species interactions (interaction modification: Wootton 1994, 2002; Prasad and Snyder 2010; Pagès et al. 2012; Hammill et al. 2015). Species can modify interactions by modifying a trait of one of the interacting species (e.g. fear effects associated with predator presence; plant-mediated interactions: Prasad and Snyder 2010; Hammill et al. 2015) or by changing the biotic or abiotic physical environment where the interaction takes place (Pagès et al. 2012; Barbosa et al. 2019).

Indirect effects have received increasing attention in ecological research, and their role in structuring ecological communities has been demonstrated experimentally (Tompkins et al. 2000; Morris et al. 2004; Sanders et al. 2015; Barbosa et al. 2019). For example, the positive indirect effects that consumer species have on each other, reducing competition among their respective resource species, can help maintain the diversity of consumer species (Sanders et al. 2015). In addition, indirect interactions can be the cause of unexpected results often found in experimental manipulations of direct interactions (Doak et al. 2008). Barbosa et al. (2017), for instance, found that the experimental reduction in abundance of a species affects other unrelated species in a food web. In addition, many of the effects that have been interpreted in the past as the result of competition can be caused by other indirect interactions (Holt 1977). For example, in the so-called apparent competition (Holt 1977; Morris et al. 2004), the negative effect of one herbivore on another, mediated by a common predator, may produce a result that resembles competition. Thus, it has become clear that the structure and dynamics of ecological communities cannot be fully understood without taking indirect interactions into account.

2 The *Baccharis dracunculifolia* System

The *Baccharis dracunculifolia* D.C. (Asteraceae) system is a self-contained assemblage of arthropods living on this host shrub species (Figs. 5.1 and 5.2). The plant species is a perennial, evergreen, dioecious shrub, 2–3 m high, widely distributed in the south-central portion of South America (Espírito-Santo et al. 2003). In some parts of Brazil, *B. dracunculifolia* produces flowers twice a year from March to June and from November to December (Collevatti and Sperber 1997). *B. dracunculifolia* frequently forms distinct patches that vary from 18 to 12,000 m^2 in area (Collevatti and Sperber 1997). This plant is fundamental to natural succession and regeneration (Fernandes et al. 2016) playing an important role in promoting biodiversity and ecosystem functioning (Perea et al. 2019).

Fig. 5.1 Species engaged in multitrophic and indirect interactions in the *Baccharis dracunculifolia* system: (**a**) *B. dracunculifolia*; (**b**) and (**c**) galls induced by *Baccharopelma dracunculifoliae*; (**d**) nymph of the galler *B. dracunculifoliae*; (**e**) parasitoid wasp; (**f**) aphids *Uroleucon tucumani* invading a gall; (**g**) ant tending aphids inside a post-emergence gall; (**h**) aphids inside a post-emergence gall; (**i**) ants removing a free-feeding herbivore. Photos by Milton Barbosa

The species *B. dracunculifolia* hosts many species of free-feeding herbivores, 17 gall-inducing species and also many species of predators, including the orders Araneae, Coleoptera, Mantodea and Hymenoptera (Fagundes et al. 2005; Fagundes and Fernandes 2011; Barbosa et al. 2017, 2019; Monteiro et al. 2020a). In Brazil,

Fig. 5.2 Species engaged in multitrophic and indirect interactions in the *Baccharis dracunculifolia* system: (**a**) spider using a post-emergence gall as shelter; (**b**) ladybird; (**c**) weevil; (**d–f**) mistletoe *Struthanthus flexicaulis* and its germinating seeds. (Photos a–c by Milton Barbosa and d-f by António Cruz)

over the past three decades *B. dracunculifolia* has been extensively used in community ecology research. The diverse associated fauna, and the wide distribution, usually at high frequency and through gradients (elevational, higrothermal, and habitat disturbance), make the species an attractive study model (e.g. Fernandes and Price 1988; Ribeiro-Mendes et al. 2002; Fagundes et al. 2005; Julião et al. 2005; Barbosa et al. 2017, 2019; Monteiro et al. 2020a, b). As a result, several multitrophic and indirect interactions have been revealed in the system, involving *B. dracunculifolia*, its associated arthropods and a mistletoe species (Collevatti and Sperber 1997; Fernandes et al. 1999; Espírito-Santo and Fernandes 2002; Fagundes et al. 2005; Neves et al. 2011; Bahia et al. 2015; Monteiro et al. 2020b). Below, we describe the *B. dracunculifolia* system and its interactions, summarizing them in an interaction network (Table 5.1, Fig. 5.3).

Table 5.1 Empirically demonstrated interactions involving arthropods and plants in the *Baccharis dracunculifolia* system, in Brazil. Interaction codes correspond to the links in the interaction network (Fig. 5.3)

Interaction code	Affecting	Affected	Interaction type	Outcome	Effect
1	Free-feeding herbivores	Host plant	Direct trophic	Negative	Herbivory (Neves et al. 2011)
2	Galler	Host plant	Direct trophic	Negative	Nutrient sinking (Espirito Santo and Fernandes 2002)
3	Host plant	Galler	Direct non-trophic	Negative	Plant resistance; gall dropping (Espirito-Santo and Fernandes 2002)
4	Parasitoid	Galler	Direct trophic	Negative	Parasitism (Tavares and Perioto 1993)
5	Parasitoid	Galler	Direct trophic	Positive	Unparasitized nymphs sharing a gall with parasitized nymphs are larger (Espírito-Santo et al. 2004)
6	Parasitoid	Host plant	Indirect trophic	Positive	Parasitism controls gall population (Espírito-Santo et al. 2004)
7	Parasitoid	Host plant	Indirect non-trophic	Negative	Increase gall size and nutrient sinking (Espírito-Santo et al. 2004)
8	Galler	Aphid	Indirect non-trophic	Positive	Gall provides shelter (Collevatti and Sperber 1997)
9	Aphid/galler	Galler/aphid	Indirect non-trophic	Negative	Competition for plant photoassimilates (Fagundes et al. 2005)
10	Aphid	Galler	Direct non-trophic	Negative	Inquilines kill galler nymphs (Espirito Santo and Fernandes 2002)
11	Aphid	Interaction 4	Interaction modification	Negative	Inquiline aphids preferably kill parasitized galling larvae (Barbosa et al. 2019)
12	Post-emergence gall	Interaction 8	Interaction modification	Positive/Negative	Providing extra habitat for aphids (Barbosa et al. 2019)

(continued)

Table 5.1 (continued)

Interaction code	Affecting	Affected	Interaction type	Outcome	Effect
13	Post-emergence gall	Interaction 11	Modification of an interaction modification	Positive/Negative	Interference with the negative effect of aphids on parasitism (Barbosa et al. 2019)
14	Ants/Aphids	Aphids/Ants	Direct non-trophic	Positive	Trophobiosis (Fagundes et al. 2005; Neves et al. 2011)
15	Ants	Galler	Direct non-trophic	Negative	Fewer nymphs per gall due to interference with oviposition (Neves et al. 2011)
16	Ants	Free-feeding herbivores	Direct trophic; Direct non-trophic	Negative	Predation or interference (Neves et al. 2011)
17	Aphid	Free-feeding herbivores	Indirect non-trophic	Negative	Ants attracted reduce number of herbivores (Neves et al. 2011)
18	Aphid	Host plant	Direct trophic	Negative	Sap sucking; nutrient sinking affects shoot growth (Neves et al. 2011)
19	Mistletoe	Host plant	Direct trophic	Negative	Parasitism; increased plant mortality (Bahia et al. 2015; Monteiro et al. 2020a, b)
20	Mistletoe	Interaction 2	Interaction modification	Positive	Increase abundance of galls of *B. dracunculifoliae* (Bahia et al. 2015)

Plant–Free-Feeding Herbivore Interaction

The most frequent orders of free-feeding herbivores found on *B. dracunculifolia* are Hemiptera, Coleoptera and Orthoptera (Interaction 1; see Table 5.1 and Figs. 5.1, 5.2 and 5.3, as for all interactions described hereafter). Among the chewing insects, the most common families are Chrysomelidae and Curculionidae. Chrysomelidae is one of the most numerous herbivorous insect families and has the greatest ecological and economic importance (e.g. monoculture pests: Riley et al. 2002, Chaboo 2007). The sucking insects most commonly found on *B. dracunculifolia* are from the families Cicadellidae, Membracidae and Psyllidae. These sucking insects often have a trophobiontic relationship with ants, which chase away other free-feeding herbivores. Ant effects depend on the context, the environment and the species of ants involved and influence the presence of other insects (Fagundes et al. 2005; Neves et al. 2011).

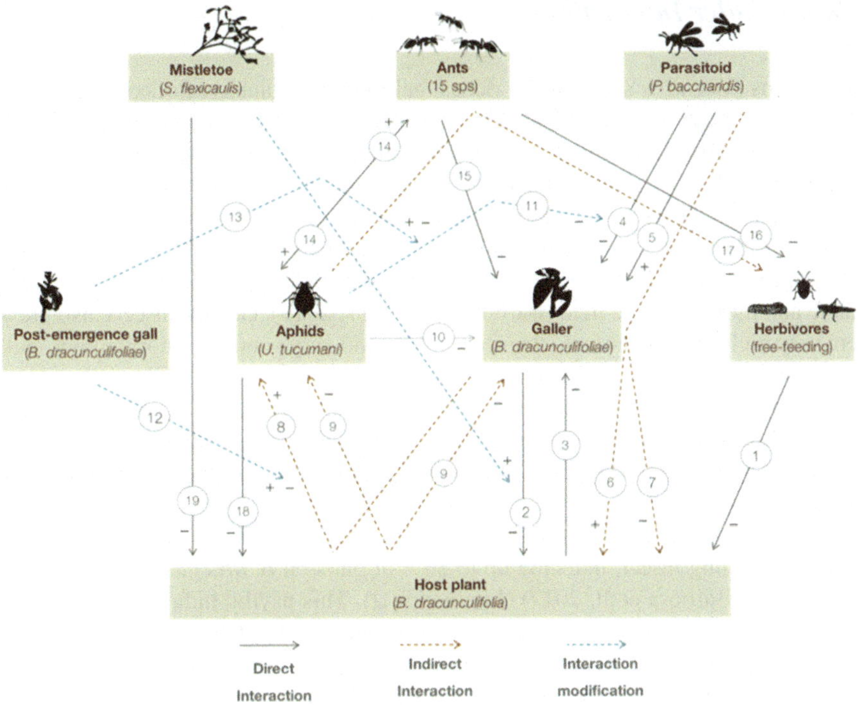

Fig. 5.3 Interaction network of arthropods and plants in the *Baccharis dracunculifolia* system, in Brazil. Links are categorized into direct interactions (solid arrows), indirect interactions (dashed orange arrows) and interaction modifications (dashed blue arrows). The link numbers refer to the interaction descriptions in Table 5.1

Interactions between herbivorous insects and their host plants can be affected by characteristics of the host plants or by changes in environmental conditions (Novotny et al. 2003; Fernandes 2016; Nunes et al. 2016a, b; Peters et al. 2016; Longino and Branstetter 2018). Monteiro et al. (2020a) investigated how the interaction between insect herbivores and *B. dracunculifolia* changed along an elevational gradient in the *campo rupestre* (rupestrian grassland) in Serra do Cipó, Brazil. In this study, the composition, diversity of interactions and specialization of insect herbivores changed with altitude. The richness and abundance of chewing insects increased with elevation, while the richness and abundance of sucking insects were higher at intermediate elevations. The change in diversity of insects associated with *B. dracunculifolia* along the elevation was due to abiotic and biotic factors. Abiotic factors such as increased wind speed and solar radiation and decreased temperature at higher elevations functioned as an environmental filter for many species. On the other hand, biotic factors such as the relationship between the richness of trophobiont ants and herbivores associated with *B. dracunculifolia* influenced the distribution of insect herbivore species.

Plant–Galler Interactions

The genus *Baccharis* supports the richest galling insect fauna recorded so far in the Neotropics, with 121 galling insect species found on 40 host species (Fernandes et al. 1996, 2014; Carneiro et al. 2009a, b). Among the several host species, *B. dracunculifolia* has the largest fauna of galling insects, with a total of 17 species recorded considering various locations (Fernandes et al. 1996, 2014; Barbosa et al. 2017). Galling insects represent appropriate model systems as they are usually straightforward to identify and count, which makes their populations easy to manipulate and monitor in the field (Barbosa et al. 2017, 2019). Galling insects also present very high specialization to their host plants and very persistent morphology, and, therefore, taxonomic identification can be indirectly made through the gall's external morphology and through the identity of their host (Rohfritsch and Shorthouse 1982; Carneiro et al. 2009b). In addition, the fact that parasitoids do not cause immediate damage to the galler allows the quantification of interaction frequency (van Veen et al. 2006).

Baccharopelma dracunculifoliae (Sternorrhyncha: Psyllidae) is by far the most common galling insect, inducing up to 83% of galls on *B. dracunculifolia* (Araújo et al. 1995; Barbosa et al. 2017) (Interaction 2). This psyllid induces a gall on the central rib of the leaf, which folds over itself until the borders are joined, forming an elliptical, green, glabrous, unicameral gall, which usually houses up to 4 nymphs, although up to 21 have already been recorded (Lara and Fernandes 1994; Arduin et al. 2005). The gall is not completely sealed because the edges of the leaves are joined but not fused. The galls of *B. dracunculifoliae* are colonized by several inquiline arthropods (species that occupy a living space produced by another species – e.g., a gall), such as Diptera, Hemiptera (Aphididae), Neuroptera (Chrysopidae), Thysanoptera and Acarina immature (Collevatti and Sperber 1997; Espírito Santo and Fernandes 2002; Barbosa et al. 2019). The post-emergence galls can persist on the plant and dry out but are still occupied by inquilines (Espírito-Santo and Fernandes 1998; Barbosa et al. 2019). These post-emergence galls can trigger indirect effects that feedback to the galler modifying its interactions with other species (discussed below in Sect. 2.4; Barbosa et al. 2019).

According to Espírito-Santo and Fernandes (2002), the host plant *B. dracunculifolia* has strong bottom-up control over the galling population. Plant-mediated mortality in the galling insect *B. dracunculifoliae*, due to gall drop and plant resistance, represented 40.7% of the initial cohort studied (Interaction 3).

Galler–Parasitoid Interaction

The performance of the galling psyllid *B. dracunculifoliae* is strongly controlled by top-down effects. Espírito-Santo et al. (2004) found that parasitoid attack was the main cause (45.2%) of mortality of the galls of *B. dracunculifoliae* that survived plant defences. At least ten wasp species have been known to parasitize galls of *B. dracunculifoliae* (Barbosa et al. 2017), the main ones are *Lyrcus* sp. (Hymenoptera: Pteromalidae), *Brasema* sp. (Hymenoptera: Eupelmidae), Platygastridae sp. (Hymenoptera: Platygastridae) and *Psyllaephagus baccharidis* (Hymenoptera: Encyrtidae) (Tavares and Perioto 1993; Sperber and Collevatti 1996; Barbosa et al. 2017).

Psyllaephagus baccharidis is the most common parasitoid of the galling psyllid. In the survey carried out by Espírito-Santo et al. (2004), this parasitoid was responsible for 93% of the parasitism recorded in galls of *B. dracunculifoliae* (Interaction 4). *P. baccharidis* is a solitary koinobiont endoparasitoid that consumes the nymph and pupates inside the host's cuticle, causing a so-called 'nymph mummification' (Espírito-Santo et al. 2004). The parasitoid attacks galls in the early stages of development, making them increase in size. Mortality by parasitism is clearly recognizable as parasitized nymphs have a distinct colour and shape (Espírito-Santo et al. 2004).

Apparently, the parasitoid *P. baccharidis* stimulates the nymphs to feed, increasing their size and also the size of the gall induced by *B. dracunculifoliae* (Espírito-Santo et al. 2004). This can decrease the performance of the plant, since larger galls represent larger nutrient sinks. In addition, unparasitized nymphs of *B. dracunculifoliae* sharing a gall with parasitized nymphs were larger than the ones found in unparasitized galls. This can result in adults of the gall inducer with greater survival and reproductive success (Espírito-Santo et al. 2004) (Interaction 5). Thus, the parasitoid *P. baccharidis* can have a positive indirect effect on the host plant, controlling the population of the gall inducer *B. dracunculifoliae* (Interaction 6), and a negative indirect effect, increasing the sinking of nutrients by the gall and the fitness of the galler (Espírito-Santo et al. 2004) (Interaction 7).

Barbosa et al. (2017) investigated the role of direct and indirect interactions in the structure and robustness (tolerance to species loss) of galler-parasitoid food webs on *B. dracunculifolia*. Theoretical models from previous works suggest that the loss or reduction in abundance of individual species can lead to secondary and cascading extinctions (Saavedra et al. 2008; Staniczenko et al. 2010). The authors experimentally manipulated quantitative host-parasitoid food webs to reduce the abundance of the main galler, *B. dracunculifoliae*. The perturbation resonated throughout the food web, affecting the food web structure and robustness (tolerance to species loss). Since there was no possibility for these effects to be propagated directly or indirectly via the documented trophic interactions, the effects must have spread non-trophically and/or through trophic links not included in the webs. The results emphasize that, even for a relatively simple ecological community, indirect interactions (trophic and non-trophic) are fundamental to their structure and dynamics (Fontaine et al. 2011; Kéfi et al. 2012).

Galler–Aphid Interaction

In addition to parasitoids, galls of *B. dracunculifoliae* can be occupied by many inquiline arthropods (Collevatti and Sperber 1997; Espírito-Santo and Fernandes 2002; Barbosa et al. 2019). These inquilines use the gall as shelter and sometimes as a food source, consuming silk made by the galler or other plant products. Some inquilines can enter the pre-emergence galls through the longitudinal opening, while others enter the gall through the characteristic exit hole left in the gall wall when parasitoids emerge (Espírito-Santo et al. 2004). Thus, the colonization of parasitized galls by inquilines can be facilitated by parasitoids (MB *unpublished data*). Post-emergence galls remain attached to the plant, sometimes for a few generations, and gradually become dry and woody (Lara and Fernandes 1994; Espírito-Santo and Fernandes 1998) and are still colonized by many species.

The aphid *Uroleucon tucumani* (Sternorrhyncha: Aphididae) is the main inquiline species found in galls of *B. dracunculifoliae* (Collevatti and Sperber 1997; Fagundes et al. 2005) (Interaction 8). This aphid species also feeds and reproduces on the apical meristems of the host plant, forming dense colonies that produce honeydew (sugary secretions). Fagundes et al. (2005) showed that aphids can have a negative indirect effect on gall development: in their absence, galls were heavier. Gall size may be related to the performance of the galling insect (Weis 1988). Therefore, this suggests exploitation competition between the galler *B. dracunculifoliae* and the aphid *U. tucumani* for sap assimilates and young leaves from terminal buds (Fagundes et al. 2005) (Interaction 9). Inquilinism by the aphid *U. tucumani* is also common and can be indirectly responsible for the death of the nymphs of the galler *B. dracunculifoliae* (Espírito-Santo and Fernandes 2002) (Interaction 10).

Theory suggests that indirect effects may be an important mechanism for community stability and persistence, but empirical data are scarce (Strauss 1991; Kéfi et al. 2012). Barbosa et al. (2019) examined the propagation of indirect effects in the *B. dracunculifolia* system. The authors investigated whether the indirect effects initiated by ecosystem engineering – physical changes in the biotic or abiotic environment caused by a species – can feedback to the engineer, changing the magnitude and direction of its interactions with other species. Gallers can be seen as ecosystem engineers since galls serve as habitat for other species, particularly aphids, which occupy hatches and unhatched galls of the most common galler on B. *dracunculifolia*. Barbosa et al. (2017) raised the hypothesis that the hatched galls could generate feedbacks on the galler – e.g. increasing the availability of shelter for aphids and, therefore, positively affecting the galler by decreasing the occupation of live galls by aphids. In a field experiment, they generated treatments with reduced or elevated ecosystem engineering, removing or adding hatched galls. Inquilism by aphids negatively affected parasitism rates (interaction modification) likely by killing parasitized galling larvae (Interaction 11) since they preferentially colonize parasitized galls (MB *unpublished data*). Post-emergence galls changed the interaction between aphid inquilines and the galler (positively or negatively, depending on gall density), probably by providing extra shelter for aphids (Interaction 12) and also altered (positively or negatively, depending on gall density) the negative effect that aphids had on parasitism (modification of an interaction modification) (Interaction 13). The results show that hatched galls of the dominant galler can trigger indirect interactions that feedback to the galler, modifying its interactions with parasitoids and inquiline aphids. In addition, the results suggest that these interaction modifications are dependent on the context, which changes with species densities.

Ant–Aphid Interaction

At least 15 species of ants have been recorded foraging on *B. dracunculifolia*, many of them tend and protect the aphid *U. tucumani* in a trophobiotic relationship (Fagundes et al. 2005; Neves et al. 2011; MB *unpublished data*) (Interaction 14). Fagundes et al. (2005) experimentally excluded ants, aphids or both from shoots of *B. dracunculifolia*. Ants had a direct negative effect on the performance of *B. dracunculifoliae*. When ants were present, a smaller number of nymphs were found in each gall of *B. dracunculifoliae*, most likely because ants interfered with the female galler during oviposition (Fagundes et al. 2005) (Interaction 15). Neves et al. (2011) observed that the presence of ants and *U. tucumani* aphids on *B. dracunculifolia* also decreased the abundance of other free-feeding herbivores and that the presence of aphids decreased the growth of plant shoots (Interactions 16, 17, and 18).

Monteiro et al. (2020a) studying the community of insect herbivores associated with *B. dracunculifolia* also found that ant abundance was positively related to the abundance of trophobionts and negatively associated with the richness of chewing herbivores. Ants protecting aphids can directly affect herbivores negatively, preying or interfering with them (Abe 1988; Fernandes et al. 1999). However, aphids on their own can reduce the abundance of fluid-sucking and chewing insects due to exploitation competition or by altering the nutritional quality of the host plant (Fay et al. 1996; Larson and Whitham 1997). Thus, aphids can potentially reduce herbivory and indirectly benefit the host plant. Also, as chewing insects can attack the gall and kill the galling nymph, aphids can also positively affect the galling species *B. dracunculifolia* (Espírito-Santo and Fernandes 2002).

Mistletoe–Plant Interaction

The hemiparasitic *Struthanthus flexicaulis* frequently colonizes *B. dracunculifolia* (Monteiro et al. 2020b; Fig. 5.2). *S. flexicaulis* represents an important stress factor for the host plant. Over time *S. flexicaulis* can cause significant reduction in growth and fitness or irreversible and sublethal damage to the host plants, increasing mortality (Press and Phoenix 2005; Cameron et al. 2008; Bahia et al. 2015; Mourão et al. 2016; Monteiro et al. 2020b) (Interaction 19).

Bahia et al. (2015) showed that *S. flexicaulis* can reduce the number of leaves of *B. dracunculifolia* and cause the death of occupied branches or even of the entire plant. The water imbalance caused by this hemiparasite may be a factor responsible for affecting growth and increased mortality of parasitized host plants (Bahia et al. 2015; Mourão et al. 2016; Monteiro et al. 2020b). In addition, when the plant faces water restriction, there is a tendency to decrease leaf size to prevent water loss through transpiration (Lincoln and Zeiger 2013). The effect of lower water availability on the vegetative development of host plants can also decrease the specific leaf area (SLA) of the host or even lead to increased senescence of the leaves (e.g. Figueirôa et al. 2004; Lincoln and Zeiger 2013). This water imbalance can affect the plant's ability to store water in the leaves and can also compromise leaf succulence (Cruz et al. 2018) due to the lower water supply for the hydration of the leaves (e.g. Scatena and Scremin-Dias 2003, Cruz et al. 2018).

A study by Monteiro et al. (*unpublished*) found that the rate of fluorescence in *B. dracunculifolia* is affected by the presence of *S. flexicaulis*. The photosynthetic rate of *B. dracunculifolia* was reduced in the most critical period of the day (12:00 noon to 3:00 pm); however, the host plant was unable to recover itself in the early evening. The percentage of chlorophyll was not affected by the presence of the hemiparasite on individuals of the host *B. dracunculifolia*. The percentage of nitrogen balance (NBI) was higher in non-parasitized individuals than in parasitized individuals. The percentage of flavonoids was higher in parasitized individuals than in non-parasitized individuals. Thus, parasitism affected the N/flavonoid balance of the host plant, representing a source of damage. The *S. flexicaulis* mistletoe has a much higher rate of transpiration than the host *B. dracunculifolia*. In addition, the mistletoes have the homeostatic control of their hosts and intensify the withdrawal of water throughout the day (Glatzel and Geils 2009). Plants parasitized by the mistletoe have a higher water potential than non-parasitized plants as a way to guarantee their hydration, since the mistletoe removes a large volume of water at noon.

The effects of parasitism on physiology also affect the architecture of the host *B. dracunculifolia*. Monteiro et al. (2020b) showed that individuals parasitized by the hemiparasite presented lower crown growth and height. These effects on the host are related to the fact that the hemiparasite obtains water and nutrients directly from the xylem, reducing the resources available for the host's own metabolism (Press and Phoenix 2005; Westwood et al. 2010; Guerra and Pizo 2014), which can affect their fitness and even lead to their death (Mallams and Mathiasen 2010; Mourão et al. 2016; Monteiro et al. 2020b). In fact, Monteiro et al. (2020b) found a high mortality of parasitized *B. dracunculifolia* individuals, 61%, after 2 years. The high mortality of abundant species such as the host *B. dracunculifolia* can lead to a restructuring of the community of woody plants in the environment in which they occur in high frequency, increasing plant diversity.

In addition, the presence of the hemiparasite *S. flexicaulis* can affect the host in indirect ways. The presence of the parasitic plant alters the interactions of *B. dracunculifolia* with its associated insect community. Bahia et al. (2015) found that branches occupied by mistletoe showed a greater abundance of galls induced by *B. dracunculifoliae* (Interaction 20). Although parasitic plants are known to cause deleterious effects on plant communities, a broader view of these relationship reveals several positive direct and indirect effects on the community (e.g. Watson 2009; Watson et al. 2011; Mellado et al. 2016; Ndagurwa et al. 2016; Mellado and Zamora 2017; Hódar et al. 2018). Mistletoes can be an important source of nectar and fleshy fruits, arthropods, nesting sites and perches for birds (Guerra and Pizo 2014; Watson et al. 2011). The abundance of ants and trophobionts can be increased since insects can benefit from the large flow of water and nutrients that come from parasites and, therefore, become more abundant (Guerra et al. 2011; Freitas and Rossi 2015). Mistletoes can also have positive impacts on soil microbial communities, increasing the diversity, quality and quantity of organic matter that enters the soil next to the host (Mellado et al. 2016). This mechanism also promotes an increase in diversity and in leaf cover of herbaceous vegetation next to the host tree (Watson 2009; Ndagurwa et al. 2016), which can lead to the presence of vertebrates in the area (Hódar et al. 2018).

3 Concluding Remarks

The *B. dracunculifolia* system is well studied and comprehended, as described here. The focal community represents a very appropriate model system for the study of indirect interactions under natural field conditions. The aggregated distribution of the host plant combined with a diverse and fairly specialized fauna of arthropods creates a discrete and highly self-contained multi-trophic community. These characteristics not only facilitate the manipulation and monitoring of species densities but also increase the chances of observing their effects, since they are more likely to be a result of local ecological processes, rather than being entangled with external processes (e.g. compensatory migration). In addition, contrary to experimental manipulations in laboratory conditions, where sets of interacting species are studied in isolation, in this system the propagation of systemic indirect effects is possible. We hope that this work inspires and facilitates the design of further empirical investigations on the role of indirect interactions in community structure and dynamics.

Acknowledgements M.B. and G.F.M. were supported by Coordenação de Aperfeiçoamento de Pessoal de Nível Superior (CAPES), and G.W.F. by CNPq and Fapemig. We also thank Reserva Vellozia and Planta Ltda.

References

Abe Y (1988) Trophobiosis between the gall wasp, *Andricus symbioticus*, and the gall-attending ant, *Lasius niger*. Appl Entomol Zool 23:41–44

Araújo AM, Fernandes GW, Bedê LC (1995) Influência do sexo e fenologia de *Baccharis dracuncufolia* D. C. (Asteraceae) sobre insetos herbívoros. Rev Bras Entomol 39:347–353

Arduin M, Fernandes GW, Kraus JE (2005) Morphogenesis of galls induced by *Baccharopelma dracunculifoliae* (Hemiptera: Psyllidae) on *Baccharis dracunculifolia* (Asteraceae) leaves. Braz J Biol 65:559–571. https://doi.org//S1519-69842005000400002

Bahia T d O, Zúñiga IG, Souza ML et al (2015) Hemiparasitism effect on *Baccharis dracunculifolia* DC. and consequences to its major galling herbivore. Acta Bot Bras 29:339–345. https://doi.org/10.1590/0102-33062014abb0008

Barbosa M, Fernandes GW, Lewis OT, Morris RJ (2017) Experimentally reducing species abundance indirectly affects food web structure and robustness. J Anim Ecol 86:327–336. https://doi.org/10.1111/1365-2656.12626

Barbosa M, Fernandes GW, Morris RJ (2019) Interaction engineering: non-trophic effects modify interactions in an insect galler community. J Anim Ecol 88:1168–1177. https://doi.org/10.1111/1365-2656.13025

Bukovinszky T, Van Veen FJF, Jongema Y, Dicke M (2008) Direct and indirect effects of resource quality on food web structure. Science 319:804–807. https://doi.org/10.1126/science.1148310

Cameron DD, Geniez JM, Seel WE et al (2008) Suppression of host photosynthesis by the parasitic plant *Rhinanthus minor*. Ann Bot 101:573–578. https://doi.org/10.1093/aob/mcm324

Carneiro MAA, Borges RAX, Araújo APA, Fernandes GW (2009a) Insetos indutores de galhas da porção sul da Cadeia do Espinhaço, Minas Gerais, Brasil. Rev Bras Entomol 53:570–592. https://doi.org/10.1590/s0085-56262009000400007

Carneiro MAA, Branco CSA, Braga CED et al (2009b) Are gall midge species (Diptera, Cecidomyiidae) host-plant specialists? Rev Bras Entomol 53:365–378. https://doi.org/10.1590/S0085-56262009000300010

Chaboo CS (2007) Biology and phylogeny of the Cassidinae Gyllenhal sensu lato (tortoise and leaf-mining beetles) (Coleoptera: Chrysomelidae). Bull Am Mus Nat Hist 350:189–233

Collevatti RG, Sperber CF (1997) The gall maker *Neopelma baccharidis* Burck. (Homoptera:Psyllidae) on *Baccharis dracunculifolia* DC. (Asteraceae): individual, local, and regional patterns. An Soc Entomol Bras 26:45–53. https://doi.org/10.1590/S0301-80591997000100006

Cruz AF, da Silva G, Silva EFF et al (2018) Stress index, water potentials and leaf succulence in cauliflower cultivated hydroponically with brackish water. Rev Bras Eng Agríc Ambient 22:622–627. https://doi.org/10.1590/1807-1929/agriambi.v22n9p622-627

Doak DF, Estes JA, Halpern BS et al (2008) Understanding and predicting ecological dynamics: are major surprises inevitable. Ecology 89:952–961. https://doi.org/10.1890/07-0965.1

Espirito-Santo MM, Fernandes GW (1998) Abundance of *Neopelma baccharidis* (Homoptera: Psyllidae) Galls on the Dioecious Shrub *Baccharis dracunculifolia* (Asteraceae). Environ Entomol 27:870–876

Espírito-Santo MM, Fernandes GW (2002) Host plant effects on the development and survivorship of the galling insect *Neopelma baccharidis* (Homoptera: Psyllidae). Austral Ecol 27:249–257. https://doi.org/10.1046/j.1442-9993.2002.01185.x

Espírito-Santo MM, Faria ML, Fernandes GW (2004) Parasitoid attack and its consequences to the development of the galling psyllid *Baccharopelma dracunculifoliae*. Basic Appl Ecol 5:475–484. https://doi.org/10.1016/j.baae.2004.04.010

Espírito-Santo MM, Madeira BG, Neves FS et al (2003) Sexual differences in reproductive phenology and their consequences for the demography of *Baccharis dracunculifolia* (Asteraceae), a dioecious tropical shrub. Ann Bot 91:13–19. https://doi.org/10.1093/aob/mcg001

Estes JA, Palmisano JF (1974) Sea otters: their role in structuring nearshore communities. Science 185:1058–1060. https://doi.org/10.1126/science.185.4156.1058

Fagundes M, Fernandes GW (2011) Insect herbivores associated with *Baccharis dracunculifolia* (Asteraceae): responses of gall-forming and free-feeding insects to latitudinal variation. Rev Biol Trop 59:1419–1432

Fagundes M, Neves FS, Fernandes GW (2005) Direct and indirect interactions involving ants, insect herbivores, parasitoids, and the host plant *Baccharis dracunculifolia* (Asteraceae). Ecol Entomol 30:28–35. https://doi.org/10.1111/j.0307-6946.2005.00668.x

Fay PA, Hartnett DC, Knapp AK (1996) Plant tolerance of gall-insect attack and gall-insect performance. Ecology 77:521–534. https://doi.org/10.2307/2265627
Fernandes GW (2016) Ecology and conservation of mountaintop grasslands in Brazil. Springer, Switzerland
Fernandes GW, Price PW (1988) Biogeographical gradients in galling species richness – tests of hypotheses. Oecologia 76:161–167. https://doi.org/10.1007/BF00379948
Fernandes GW, Carneiro MAA, Lara ACF et al (1996) Galling insects on neotropical species of *Baccharis* (Asteraceae). Trop Zool 9:315–332. https://doi.org/10.1080/03946975.1996.10539315
Fernandes GW, Fagundes M, Woodman RL, Price PW (1999) Ant effects on three-trophic level interactions: plant, galls, and parasitoids. Ecol Entomol 24:411–415. https://doi.org/10.1046/j.1365-2311.1999.00218.x
Fernandes GW, Silva JO, Espírito-Santo MM et al (2014) *Baccharis*: a Neotropical model system to study insect plant interactions. In: Fernandes WG, Santos CJ (eds) Neotropical insect galls. Springer Netherlands, Dordrecht, pp 193–219
Fernandes GW, Toma TSP, Angrisano P, Overbeck G (2016) Challenges in therestoration of quartzitic and ironstone rupestrian grasslands. In: Fernandes GW (Ed.) Ecology and conservation of mountaintop grasslands in Brazil. Springer, Switzerland, pp. 449–477
Figueirôa JM, Barbosa DCA, Simabukuro EA (2004) Crescimento de plantas jovens de (*Myracrodurum urundeuva* Allemão) Anacardiaceae sob diferentes regimes hídricos. Acta Bot Bras 18:573–580. https://doi.org/10.1590/S0102-33062004000300015
Fontaine C, Guimarães PR, Kéfi S et al (2011) The ecological and evolutionary implications of merging different types of networks. Ecol Lett 14:1170–1181. https://doi.org/10.1111/j.1461-0248.2011.01688.x
Freitas JD, Rossi MN (2015) Interaction between trophobiont insects and ants: the effect of mutualism on the associated arthropod community. J Insect Conserv 19:627–638. https://doi.org/10.1007/s10841-015-9785-2
Glatzel G, Geils BW (2009) Mistletoe ecophysiology: host-parasite interactions. Botany 87:10–15. https://doi.org/10.1139/B08-096
Guerra TJ, Pizo MA (2014) Asymmetrical dependence between a Neotropical mistletoe and its avian seed disperser. Biotropica 46:285–293. https://doi.org/10.1111/btp.12112
Guerra TJ, Camarota F, Castro FS et al (2011) Trophobiosis between ants and *Eurystethus microlobatus* Ruckes 1966 (Hemiptera: Heteroptera: Pentatomidae) a cryptic, gregarious and subsocial stinkbug. J Nat Hist 45:1101–1117. https://doi.org/10.1080/00222933.2011.552800
Hammill E, Kratina P, Vos M et al (2015) Food web persistence is enhanced by non-trophic interactions. Oecologia 178:549–556. https://doi.org/10.1007/s00442-015-3244-3
Hódar JA, Lázaro-González A, Zamora R (2018) Beneath the mistletoe: parasitized trees host a more diverse herbaceous vegetation and are more visited by rabbits. Ann For Sci 75:77. https://doi.org/10.1007/s13595-018-0761-3
Holt RD (1977) Predation, apparent competition, and the structure of prey communities. Theor Popul Biol 12:197–229. https://doi.org/10.1016/0040-5809(77)90042-9
Julião GR, Fernandes GW, Negreiros D et al (2005) Insetos galhadores associados a duas espécies de plantas invasoras de áreas urbanas e peri-urbanas. Rev Bras Entomol 49:97–106. https://doi.org/10.1590/S0085-56262005000100010
Kéfi S, Berlow EL, Wieters EA et al (2012) More than a meal… integrating non-feeding interactions into food webs. Ecol Lett 15:291–300. https://doi.org/10.1111/j.1461-0248.2011.01732.x
Lara CF, Fernandes GW (1994) Distribuição de galhas de *Neopelma baccharidis* (Homoptera: Psyllidae) em *Baccharis dracunculifolia* (Asteraceae). Rev Bras Biol 54:661–668
Larson KC, Whitham TG (1997) Competition between gall aphids and natural plant sinks: plant architecture affects resistance to galling. Oecologia 109:575–582. https://doi.org/10.1007/s004420050119
Lincoln T, Zeiger E (2013) Fisiologia Vegetal, 5th edn. Artmed, Porto Alegre

Longino JT, Branstetter MG (2018) The truncated bell: an enigmatic but pervasive elevational diversity pattern in Middle American ants. Ecography 41:1–12. https://doi.org/10.1111/ecog.03871

Mallams KM, Mathiasen RL (2010) Mistletoes on hardwoods in the United States. Forest Insect & Disease Leaflet, Portland, pp 1–12

Mellado A, Zamora R (2017) Parasites structuring ecological communities: the mistletoe footprint in Mediterranean pine forests. Funct Ecol 31:2167–2176. https://doi.org/10.1111/1365-2435.12907

Mellado A, Morillas L, Gallardo A et al (2016) Temporal dynamic of parasite-mediated linkages between the forest canopy and soil processes and the microbial community. New Phytol 211:1382–1392. https://doi.org/10.1111/nph.13984

Monteiro GF, Macedo-Reis LE, Dáttilo W et al (2020a) Ecological interactions among insect herbivores, ants and the host plant *Baccharis dracunculifolia* in a Brazilian mountain ecosystem. Austral Ecol 45:158–167. https://doi.org/10.1111/aec.12839

Monteiro GF, Novais S, Barbosa M, et al (2020b) The mistletoe Struthanthus flexicaulis reduces dominance and increases diversity of plants in campo rupestre. Flora 271:151690. https://doi.org/10.1016/j.flora.2020.151690

Morris RJ, Lewis OT, Godfray HCJ (2004) Experimental evidence for apparent competition in a tropical forest food web. Nature 428:310–313. https://doi.org/10.1038/nature02394

Mourão FA, Pinheiro RBP, Jacobi CM et al (2016) Host preference of the hemiparasite *Struthanthus flexicaulis* (Loranthaceae) in ironstone outcrop plant communities Southeast Brazil. Acta Bot Bras 30:41–46. https://doi.org/10.1590/0102-33062015abb0177

Ndagurwa HG, Ndarevani P, Muvengwi J et al (2016) Mistletoes via input of nutrient-rich litter increases nutrient supply and enhance plant species composition and growth in a semi-arid savanna, Southwest Zimbabwe. Plant Ecol 217:1095–1104. https://doi.org/10.1007/s11258-016-0635-4

Neves FS, Fagundes M, Sperber CF, Fernandes GW (2011) Tri-trophic level interactions affect host plant development and abundance of insect herbivores. Arthropod Plant Interact 5:351–357. https://doi.org/10.1007/s11829-011-9139-2

Novotny V, Basset Y, Kitching R (2003) Herbivore assemblage and their food resources. In: Basset Y, Novotny V, Miller S, Kitching R (eds) Arthropods of tropical forests: Spatio-temporal dynamics and resource use in the canopy. Cambridge University Press, Cambridge

Nunes CA, Quintino AV, Constantino R et al (2016a) Patterns of taxonomic and functional diversity of termites along a tropical elevational gradiente. Biotropica 49:186–194. https://doi.org/10.1111/btp.12365

Nunes CA, Braga RF, Figueira JEC et al (2016b) Dung beetles along a tropical altitudinal gradient: environmental filtering on taxonomic and functional diversity. PLoS One 11:e0157442

Ohgushi T (2008) Herbivore-induced indirect interaction webs on terrestrial plants: the importance of non-trophic, indirect, and facilitative interactions. Entomol Exp Appl 128:217–229. https://doi.org/10.1111/j.1570-7458.2008.00705.x

Pagès JF, Farina S, Gera A et al (2012) Indirect interactions in seagrasses: fish herbivores increase predation risk to sea urchins by modifying plant traits. Funct Ecol 26:1015–1023. https://doi.org/10.1111/j.1365-2435.2012.02038.x

Perea R, Cunha JS, Spadeto C et al (2019) Nurse shrubs to mitigate plant invasion along roads of montane Neotropics. Ecol Eng 136:193–196. https://doi.org/10.1016/j.ecoleng.2019.06.024

Peters MK, Hemp A, Appelhans T et al (2016) Predictors of elevational biodiversity gradients change from single taxa to the multi-taxa community level. Nat Commun 7:13736. https://doi.org/10.1038/ncomms13736

Prasad RP, Snyder WE (2010) A non-trophic interaction chain links predators in different spatial niches. Oecologia 162:747–753. https://doi.org/10.1007/s00442-009-1486-7

Press MC, Phoenix GK (2005) Impacts of parasitic plants on natural communities. New Phytol 166:737–751. https://doi.org/10.1111/j.1469-8137.2005.01358.x

Prugh LR, Brashares JS (2012) Partitioning the effects of an ecosystem engineer: kangaroo rats control community structure via multiple pathways. J Anim Ecol 81:667–678. https://doi.org/10.1111/j.1365-2656.2011.01930.x
Ribeiro-Mendes HN, Marques ESA, Silva IM, Fernandes GW (2002) Influence of host-plant sex and habitat on survivorship of insect galls within the geographical range of the host-plant. Trop Zool 15:5–15. https://doi.org/10.1080/03946975.2002.10531163
Riley EG, Clark SM, Flowers RW et al (2002) Family 124. Chrysomelidae Latreille 180. In: Arnett RH Jr, Thomas MC, Skelley PE et al (eds) American beetles: Polyphaga: Scarabaeoidea through Curculionoidea. CRC Press, Boca Raton
Rohfritsch O, Shorthouse JD (1982) Molecular biology of plant tumors. New York : Academic Press
Saavedra S, Reed-Tsochas F, Uzzi B (2008) Asymmetric disassembly and robustness in declining networks. Proc Natl Acad Sci U S A 105:16466–16471. https://doi.org/10.1073/pnas.0804740105
Sanders D, Kehoe R, van Veen FJF (2015) Experimental evidence for the population-dynamic mechanisms underlying extinction cascades of carnivores. Curr Biol 25:3106–3109. https://doi.org/10.1016/j.cub.2015.10.017
Scatena VL, Scremin-Dias E (2003) Parênquima, Colênquima e Esclerênquima. In: Appezzato-da-Glória B, Carmello-Guerreiro SM (eds) Anatomia Vegetal. UFV, Viçosa, pp 109–127
Sperber CF, Collevatti RG (1996) The gall maker *Neopelma baccharidis* (Homoptera) on *Baccharis dracunculifolia* (Asteraceae): success and parasitoidism density dependence. An Soc Entomol Bras 25:59–63
Staniczenko PPA, Lewis OT, Jones NS, Reed-Tsochas F (2010) Structural dynamics and robustness of food webs. Ecol Lett 13:891–899. https://doi.org/10.1111/j.1461-0248.2010.01485.x
Strauss SY (1991) Indirect effects in community ecology: their definition, study and importance. Trends Ecol Evol 6:206–210. https://doi.org/10.1016/0169-5347(91)90023-Q
Tavares MT, Perioto NW (1993) *Psyllaephagus baccharidis* n. sp. (Hymenoptera, Encyrtidae), a nymphal parasitoid of *Neopelma baccharidis* Burckhardt (Homoptera, Psyllidae). Rev Bras Entomol 37:317–320
Tompkins DM, Draycott RAH, Hudson PJ (2000) Field evidence for apparent competition mediated via the shared parasites of two gamebird species. Ecol Lett 3:10–14. https://doi.org/10.1046/j.1461-0248.2000.00117.x
Van Veen FJF, Morris RJ, Godfray HCJ (2006) Apparent competition, quantitative food webs, and the structure of phytophagous insect communities. Annu Rev Entomol 51:187–208
Watson DM (2009) Parasitic plants as facilitators: more dryad than Dracula? J Ecol 97:1151–1159. https://doi.org/10.1111/j.1365-2745.2009.01576.x
Watson DM, McGregor HW, Spooner PG (2011) Hemiparasitic shrubs increase resource availability and multi-trophic diversity of eucalypt forest birds. Funct Ecol 150:889–899
Weis A (1988) Reactive plant tissue sites and the population biology of gall makers. Annu Rev Entomol 33:467–486. https://doi.org/10.1146/annurev.ento.33.1.467
Westwood JH, Yoder JI, de Pamplhilis CD (2010) The evolution of parasitism in plants. Trends Plant Sci 15:227–235. https://doi.org/10.1016/j.tplants.2010.01.004
Wetzel WC et al (2016) Ecosystem engineering by a gall-forming wasp indirectly suppresses diversity and density of herbivores on oak trees. Ecology 97:1–5. https://doi.org/10.1007/s13398-014-0173-7.2
Wootton JT (1994) The nature and consequences of indirect effects in ecological communities. Annu Rev Ecol Syst 25:443–466. https://doi.org/10.1146/annurev.es.25.110194.002303
Wootton JT (2002) Indirect effects in complex ecosystems: recent progress and future challenges. J Sea Res 48:157–172. https://doi.org/10.1016/S1385-1101(02)00149-1

Chapter 6
Endophytic Fungi of *Baccharis*

Yumi Oki, Lucas Arantes-Garcia, Renata Maia, Michel Stórquio Belmiro, Isabela Maria Nascimento, Ary Correa Junior, and G. Wilson Fernandes

Abstract Endophytic fungi are important mediators in the structure and dynamics of terrestrial plant communities and their relationships with associated fauna. Although endophytic fungi are found in all living plants, only 1% of all *Baccharis* species (*Baccharis artemisioides*, *B. coridifolia*, *B. dracunculifolia*, *B. megapotamica*, and *B. trimera*) have had their endophytic mycota studied. To date, 28 genera of endophytic fungi have been identified in association with species of *Baccharis*. Analysis of the enzymes and metabolites produced by this mycota indicates that these endophytes have numerous properties that may be related to better performance and resistance of their *Baccharis* host to several stressors and natural enemies. Many of these endophytes have properties that can be exploited for the development of beneficial applications in the fields of agronomy, pharmacology, and conservation, making them a particularly important group for the development of biotechnological products.

Keywords Antimicrobial activity · Bioprospecting · Fungal endophyte diversity · Plant performance · Secondary metabolites

1 Introduction

Endophytic fungi are a group of fungi that live inside plant tissues without causing harm to the host (Faeth and Fagan 2002; Hyde and Soytong 2008). They produce a variety of enzymes and secondary compounds, which favor nutrient cycling (Sun et al. 2011; Behie and Bidochka 2014), improve plant performance and resistance to adverse conditions such as droughts and high temperatures (Rodriguez et al.

Y. Oki (✉) · L. Arantes-Garcia · R. Maia · M. S. Belmiro · I. M. Nascimento · G. W. Fernandes
Ecologia Evolutiva & Biodiversidade, Departamento de Genética, Ecologia & Evolução/ICB, CP 486, Universidade Federal de Minas Gerais, Belo Horizonte, MG, Brazil

A. C. Junior
Departamento de Microbiologia, ICB, Universidade Federal de Minas Gerais, Belo Horizonte, MG, Brazil

G. W. Fernandes et al. (eds.), *Baccharis*,
https://doi.org/10.1007/978-3-030-83511-8_6

2004; Hubbard et al. 2014; Bilal et al. 2020), and minimize damage caused by herbivores and pathogenic microorganisms (Breen 1992, 1993, 1994; Fernandes and Price 1992; Hammon and Faeth 1992; Faeth and Hammon 1997; Raps and Vidal 1998; Faeth 2002; Meister et al. 2006; Oki et al. 2008, 2021; Grunseich et al. 2020). These cryptic organisms have been ignored for a long time, and their importance in species interactions and to biodiversity has been rarely studied until recently (Oki et al. 2016). According to a survey of the Web of Science database, it wasn't until the 1990s that scientific articles started to regularly use the term "endophytic fungi" (Fig. 6.1). Although the number of studies on endophytic fungal communities has progressively increased since then, there were still only 557 publications in the first decade of the twenty-first century. Approximately 75% of all the articles about endophytes retrieved by the survey (3043 articles published from 1945 to 2019) were published in the last 10 years (2279 articles between 2010 and 2019). The discovery of the importance of endophytes in nature has captured the attention of researchers and brought a wide prospective for research and applied perspectives.

Although scientific knowledge about these microorganisms is relatively recent, investigations have highlighted their remarkable diversity and wide distribution. Hundreds of species of endophytic fungi may inhabit a given host plant, with their richness being influenced by the phylogeny, ontogeny, and organs, among other factors, of the host species (Arnold et al. 2000; Cannon and Simmons 2002; Arnold and Herre 2003; Arnold and Lutzoni 2007; Banerjee 2011; Oki et al. 2016; Griffin and Carson 2018). The high diversity of endophytes indicates a wide variety of

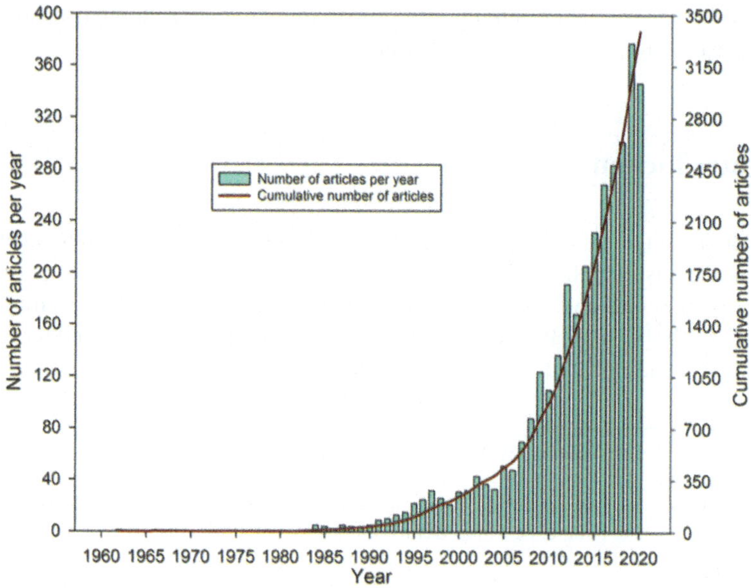

Fig. 6.1 Number of articles found in the Web of Science database (1945 to 2020) published with the term "endophytic fungi" per year and cumulatively

relationships with host plants and possibly a large number of ecological functions as well (Caruso et al. 2020). In this chapter we focus on the diversity and distribution of endophytic fungi found in association with species of *Baccharis*, a genus of shrubs and herbs that occupy many different habitats and ecosystems in their native range in the Americas, and highlight their ecological and economic potential.

2 Distribution and Diversity of Endophytic Fungi of Species of *Baccharis*

Despite there being 422 described species in the genus *Baccharis*, the associated endophytic fungal community is known for only 5: *Baccharis artemisioides*, *B. coridifolia*, *B. dracunculifolia*, *B. megapotamica*, and *B. trimera* (Table 6.1) (Heiden and Pirani 2016, see also Chap. 2 in this book). Nevertheless, 31 genera of endophytic fungi have already been reported from these 5 species: *Alternaria*, *Aureobasidium*, *Acremonium*, *Aspergillus*, *Biscogniauxia*, *Ceratopicnidium*, *Cercospora*, *Chaetomium*, *Cladosporium*, *Cochliobolus*, *Colletotrichum*, *Coriolopsis*, *Cylindrocladium*, *Diaporthe*, *Epicoccum*, *Eutypella*, *Fusarium*, *Glomerella*, *Myrothecium*, *Nigrospora*, *Penicillium*, *Phoma*, *Phomopsis*, *Podospora*, *Preussia*, *Rhizoctonia*, *Talaromyces*, *Trichoderma*, *Scopulariopsis*, *Sporormiella*, and *Xylaria* (Table 6.1). Even though none of these fungus genera were reported in all five *Baccharis* species, there are similarities among these plant species. *Baccharis dracunculifolia* and *B. trimera* were found to have the most similar endophytic fungi communities, sharing the following seven genera (Jaccard Index = 33%): *Chaetomium*, *Diaporthe*, *Nigrospora*, *Phoma*, *Phomopsis*, *Preussia*, and *Xylaria* (Fig. 6.2). Of the genera of endophytic fungi known for species of *Baccharis*, only 24 taxa have been identified to the species level: *Aureobasidium pullulans*, *Aureobasidium melanogenum*, *Aspergillus versicolor*, *Aspergillus spinulosporus*, *Ceratopicnidium baccharidicola*, *Cladosporium cladosporioides*, *Cladosporium halotolerans*, *Cladosporium endophytica*, *Cochliobolus lunatus*, *Coriolopsis rigida*, *Diaporthe phaseolorum*, *Epicoccum nigrum*, *Eutypella scoparia*, *Myrothecium verrucaria*, *Myrothecium roridum*, *Penicillium citrinum*, *Preussia africana*, *Preussia pseudominima*, *Talaromyces muroii*, *Trichoderma reesei*, *Xylaria adscendens*, *Xylaria apiculata*, and *Xylaria venosula*. Thus, the possibility exists for the description of unknown fungal species associated with *Baccharis* spp. and evidence of coevolution events.

Currently, the most studied species of *Baccharis* is *B. dracunculifolia*, with 24 endophytic fungus genera reported in 5 published studies: *Aureobasidium*, *Acremonium*, *Aspergillus*, *Biscogniauxia*, *Cercospora*, *Chaetomium*, *Cladosporium*, *Colletotrichum*, *Coriolopsis*, *Cylindrocladium*, *Diaporthe*, *Eutypella*, *Fusarium*, *Glomerella*, *Nigrospora*, *Penicillium*, *Phoma*, *Phomopsis*, *Preussia*, *Rhizoctonia*,

Table 6.1 Taxa of endophytic fungi found among species of *Baccharis*.

Endophytic genera	Endophytic taxa	Host plants	References
Alternaria	*Alternaria* sp.	*Baccharis trimera*	Vieira et al. (2014)
Aureobasidium	*Aureobasidium pullulans*	*Baccharis dracunculifolia*	Oki et al. (2009, 2020)
Acremonium	*Acremonium* sp.	*Baccharis dracunculifolia*	Cuzzi et al. (2012)
	Acremonium sp. strain D5-FB	*Baccharis dracunculifolia*	Onofre and Steilmann (2012)
Aspergillus	*Aspergillus* sp.	*Baccharis dracunculifolia*	Cuzzi et al. (2012)
	Aspergillus sp. strain D2-NC	*Baccharis dracunculifolia*	Onofre and Steilmann (2012)
Biscogniauxia	*Biscogniauxia* sp.	*Baccharis dracunculifolia*	Fernandes et al. (2018)
Ceratopicnidium	*Ceratopicnidium baccharidicola*	*Baccharis coridifolia*	Rizzo et al. (1997)
	Ceratopicnidium baccharidicola	*Baccharis artemisioides*	Rizzo et al. (1997)
Cercospora	*Cercospora* sp. strain D7-FB	*Baccharis dracunculifolia*	Onofre and Steilmann (2012)
Chaetomium	*Chaetomium* sp.	*Baccharis trimera*	Vieira et al. (2014)
	Chaetomium sp.	*Baccharis dracunculifolia*	Cuzzi et al. (2012)
Cladosporium	*Cladosporium cladosporioides*	*Baccharis dracunculifolia*	Oki et al. (2009)
	Cladosporium halotolerans	*Baccharis dracunculifolia*	Fernandes et al. (2018)
Cochliobolus	*Cochliobolus lunatus*	*Baccharis trimera*	Vieira et al. (2014)
Colletotrichum	*Colletotrichum* sp.	*Baccharis dracunculifolia*	Cuzzi et al. (2012)
	Colletotrichum sp. strain D4-FB	*Baccharis dracunculifolia*	Onofre and Steilmann (2012)
Coriolopsis	*Coriolopsis rigida*	*Baccharis dracunculifolia*	Fernandes et al. (2018)
Cylindrocladium	*Cylindrocladium* sp.	*Baccharis dracunculifolia*	Cuzzi et al. (2012)
	Cylindrocladium sp. strain D8-FB	*Baccharis dracunculifolia*	Onofre and Steilmann (2012)

(continued)

Table 6.1 (continued)

Endophytic genera	Endophytic taxa	Host plants	References
Diaporthe	*Diaporthe phaseolorum*	*Baccharis trimera*	Vieira et al. (2014)
	Diaporthe sp.	*Baccharis trimera*	Vieira et al. (2014)
Epicoccum	*Epicoccum nigrum*	*Baccharis trimera*	Vieira et al. (2014)
	Epicoccum sp.	*Baccharis trimera*	Vieira et al. (2014)
Fusarium	*Fusarium* sp.	*Baccharis dracunculifolia*	Cuzzi et al. (2012)
	Fusarium sp. strain D3-FB	*Baccharis dracunculifolia*	Onofre and Steilmann (2012)
Glomerella	*Glomerella* sp.	*Baccharis dracunculifolia*	Cuzzi et al. (2012)
Myrothecium	*Myrothecium verrucaria*	*Baccharis coridifolia*	Jarvis et al. (1987)
	Myrothecium roridum	*Baccharis coridifolia*	Jarvis et al. (1987)
	Myrothecium roridum	*Baccharis megapotamica*	Jarvis et al. (1987)
Nigrospora	*Nigrospora* sp.	*Baccharis trimera*	Vieira et al. (2014)
	Nigrospora sp.	*Baccharis dracunculifolia*	Oki et al. (2009)
	Nigrospora sp.	*Baccharis dracunculifolia*	Cuzzi et al. (2012)
Penicillium	*Penicillium* sp.	*Baccharis dracunculifolia*	Cuzzi et al. (2012)
	Penicillium sp.	*Baccharis dracunculifolia*	Oki et al. (2009)
Pestalotiopsis	*Pestalotiopsis* sp.	*Baccharis trimera*	Vieira et al. (2014)
Phoma	*Phoma* sp.	*Baccharis trimera*	Vieira et al. (2014)
Phomopsis	*Phomopsis* sp.	*Baccharis dracunculifolia*	Oki et al. (2009)
	Phomopsis sp.	*Baccharis trimera*	Vieira et al. (2014)
	Phomopsis sp.	*Baccharis dracunculifolia*	Cuzzi et al. (2012)
	Phomopsis sp. strain D10-NC	*Baccharis dracunculifolia*	Onofre and Steilmann (2012)
Podospora	*Podospora* sp.	*Baccharis trimera*	Vieira et al. (2014)

(continued)

Table 6.1 (continued)

Endophytic genera	Endophytic taxa	Host plants	References
Preussia	*Preussia africana*	*Baccharis dracunculifolia*	Oki et al. (2009); Fernandes et al. (2018), Oki et al. (2021)
	Preussia africana	*Baccharis trimera*	Vieira et al. (2014)
	Preussia pseudominima	*Baccharis trimera*	Vieira et al. (2014)
	Preussia sp.	*Baccharis trimera*	Vieira et al. (2014)
	Preussia sp.	*Baccharis dracunculifolia*	Fernandes et al. (2018)
Rhizoctonia	*Rhizoctonia* sp.	*Baccharis dracunculifolia*	Oki et al. (2009)
Trichoderma	*Trichoderma* sp.	*Baccharis dracunculifolia*	Cuzzi et al. (2012)
	Trichoderma reesei	*Baccharis dracunculifolia*	Onofre et al. (2014)
Scopulariopsis	*Scopulariopsis* sp.	*Baccharis dracunculifolia*	Cuzzi et al. (2012)
Sporormiella	*Sporormiella* sp.	*Baccharis trimera*	Vieira et al. (2014)
Xylaria	*Xylaria* sp.	*Baccharis dracunculifolia*	Oki et al. (2009)
	Xylaria sp.	*Baccharis dracunculifolia*	Fernandes et al. (2018)
	Xylaria sp.	*Baccharis trimera*	Vieira et al. (2014)
	Xylaria apiculate	*Baccharis dracunculifolia*	Fernandes et al. (2018)
	Xylaria venosula	*Baccharis dracunculifolia*	Oki et al. (2009), Fernandes et al. (2018), Oki et al. (2021)

Talaromyces, *Trichoderma*, *Scopulariopsis*, and *Xylaria*. The species of endophytic fungi known for *B. dracunculifolia* are *Aureobasidium pullulans*, *Aureobasidium melanogenum*, *Aspergillus versicolor*, *Aspergillus spinulosporus*, *Cladosporium cladosporioides*, *Cladosporium halotolerans*, *Cladosporium endophytica*, *Coriolopsis rigida*, *Diaporthe phaseolorum*, *Eutypella scoparia*, *Penicillium citrinum*, *Preussia africana*, *Talaromyces muroii*, *Trichoderma reesei*, *Xylaria adscendens*, *Xylaria apiculata*, and *Xylaria venosula* (Figs. 6.3 and 6.4). Many of these species (e.g., *A. pullulans*, *P. africana*, and *X. venosula*) play important ecological roles such as improving their hosts' defenses against herbivores and phytopathogens.

All together, these results indicate that the taxonomic diversity of endophytic fungi associated with species of *Baccharis* is enormous and likely underestimated. The array of functions associated with interactions between endophytic fungi and

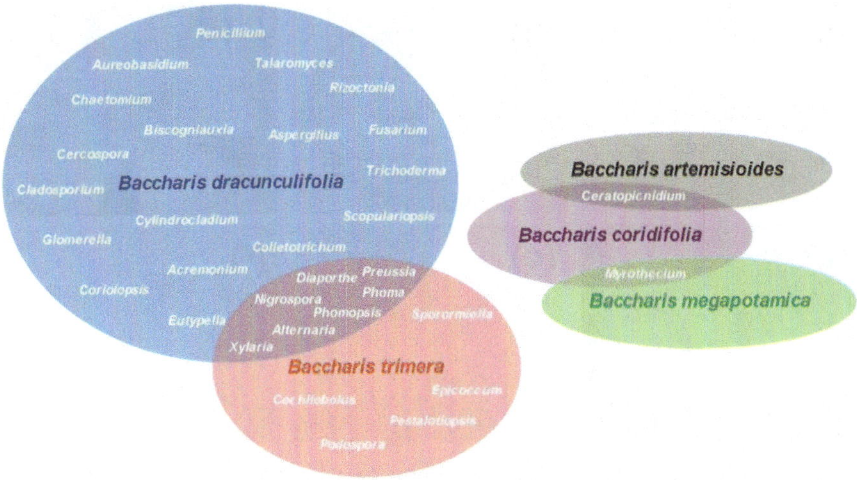

Fig. 6.2 Diagram showing fungal genera associated and shared among *Baccharis dracunculifolia, Baccharis trimera, Baccharis artemisioides, Baccharis coridifolia, and Baccharis megapotamica*

Baccharis hosts is also expected to be very large. Thus, there is a broad perspective for future studies to address, besides their biotechnological importance, the phylogenetic relationships of endophytes and hosts and the evolution and stability of associations across species and habitats.

3 Leaf Age, Plant Sex, and the Endophytic Fungus Community

The richness of endophytic fungi found among *Baccharis* spp. varies according to plant organ (Oki et al. 2009; Jia et al. 2016) and organ age (Arnold and Herre 2003; Fernandes et al. 2011; Sanchez-Azofeifa et al. 2012; Nascimento et al. 2015; Christian et al. 2019). For instance, endophytic fungus richness in mature leaves of *B. dracunculifolia* was seven times higher than that of leaves of intermediate age growing at high altitudes of Serra do Cipó, Brazil. No endophytic fungi were found in young leaves (Oki et al. 2008).

Another relevant factor that can affect the composition of endophytic mycota, particularly in the genus *Baccharis*, is host plant gender. The endophytic fungi found exclusively in female plants of *B. dracunculifolia* did not develop from

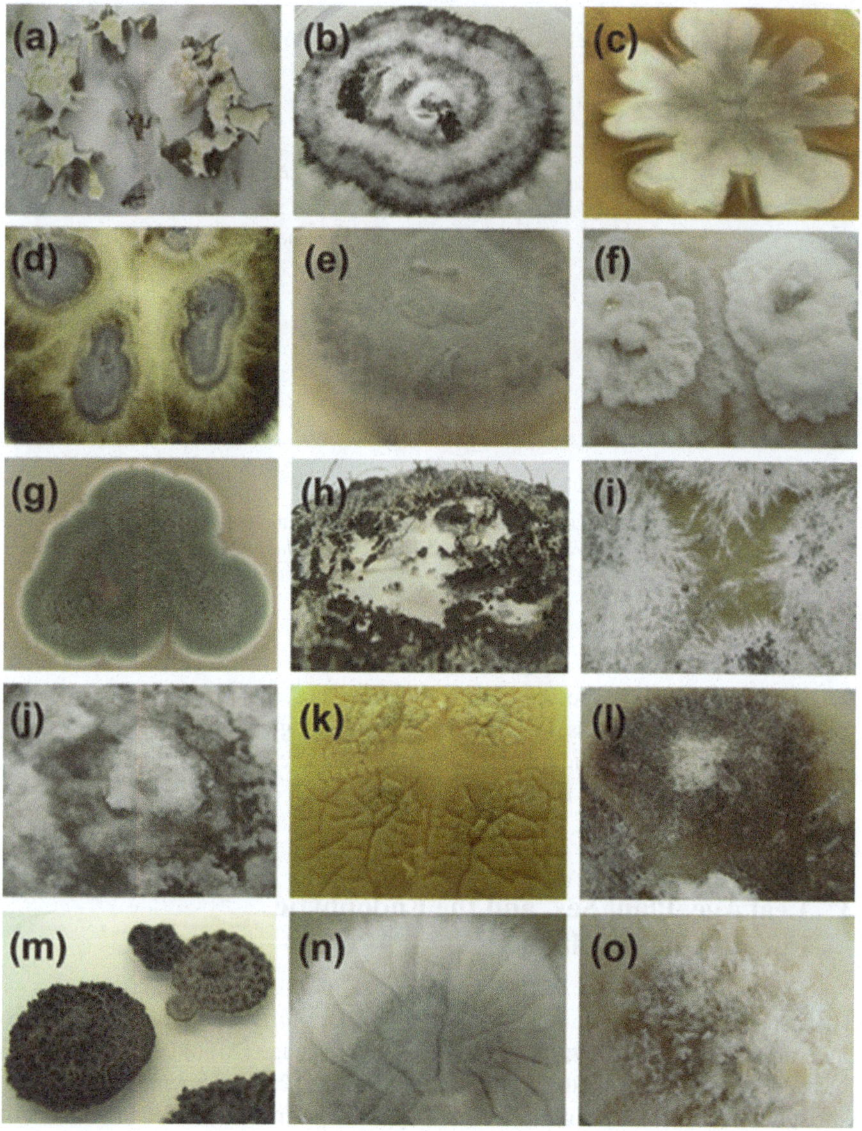

Fig. 6.3 Endophytic fungus genera found for *Baccharis dracunculifolia* (Asteraceae): (**a**) *Xylaria adscendens* (Xylariaceae); (**b**) *Biscogniauxia* sp. (Xylariaceae); (**c**) *Preussia africana* (Sporormiaceae); (**d**) *Aureobasidium pullulans* (Dothioraceae); (**e**) *Cladosporium endophytica* (Cladosporiaceae); (**f**) *Phomopsis* sp. (Diaporthaceae); (**g**) *Penicillium citrinum* (Trichocomaceae); (**h**) *Xylaria venosula* (Xylariaceae); (**i**) *Diaporthe phaseolorum* (Diaporthaceae); (**j**) *Nigrospora* sp. (Trichosphaeriaceae); (**k**) *Phoma* sp. (Didymellaceae); (**l**) *Fusarium* sp. (Nectriaceae); (**m**) *Aureobasidium melanogenum* (Dothioraceae); (**n**) *Aspergillus versicolor* (Trichocomaceae); (**o**) *Acremonium* sp. (Hypocreaceae)

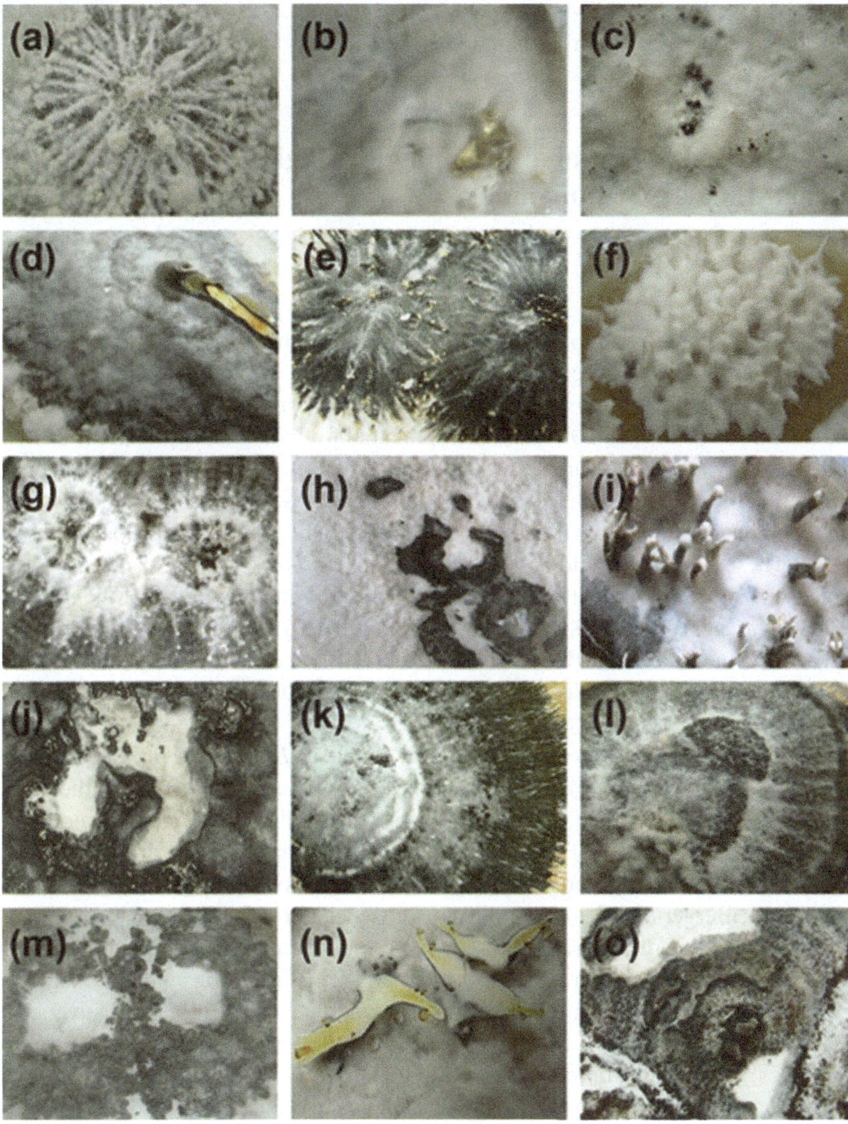

Fig. 6.4 Endophytic fungi of the family Xylariaceae found in leaves of *Baccharis dracunculifolia* (Asteraceae)

extracts from male individuals of the same species (Fernandes et al. 2018). It is likely that chemical differences between female and male plants (Darwin 1877; Wallace and Rundel 1979; Van Etten et al. 2008) determine the endophytic species that are capable of co-inhabiting these plants. Generally, female plants invest less in growth and more in resistance against herbivory than male plants (Wallace and

Rundel 1979; Herms and Mattson 1992; Van Etten et al. 2008) (but see Chap. 4). Nonetheless, these questions remain to be fully addressed in this system.

4 Untapped Potential of Endophytic Fungi: Enzymes and Bioactive Compounds

Enzymes

Despite the high diversity of endophytic fungi found among species of *Baccharis*, knowledge regarding their enzymatic makeup is scarce. This knowledge gap undermines a better understating of the physiological relationships between endophytes and their hosts (Sun et al. 2011) and their potential for industrial application (Côrrea et al. 2014). Endophytic fungi, such as species of the genus *Xylaria*, have a high production of enzymes such as cellulases and ligninases, which can degrade important components of plant cell walls (e.g., cellulose and lignin) (Carroll and Carroll 1978). These enzymes could assist in the decomposition of fallen leaves and promote nutrient cycling in nature. The accumulation of endophytic fungi themselves in leaves as age progresses (e.g., Fernandes et al. 2011; Sanchez-Azofeifa et al. 2012; Nascimento et al. 2015; Christian et al. 2019) seems to be important and in need of further evaluation, as well as the relevance of these fungi to key ecosystem functions such as decomposition.

The endophyte *Aureobasidium pullulans* found in *B. dracunculifolia* can produce lipases that belong to the class of serine hydrolases and do not need the presence of cofactors in order to act, unlike most extracellular enzymes of microbial origin (Nascimento 2010). The biological function of lipases is to hydrolyze triglycerides to form free fatty acids, mono- and diacylglycerols and glycerol (Kwon and Rhee 1986; Berger and Schnelder 1992; Bornscheuer 1995; Carvalho et al. 2003). Currently, lipases are of great interest to food (e.g., aroma improvement, food conservation, reduction of saturated fat), agricultural (e.g., herbicide synthesis), energy (e.g., biodiesel and hydrocarbon production), pharmaceutical (e.g., digestive aid), and cosmetics (e.g., active ingredient in the formulation and synthesis of specific cosmetics) industries, among others (Höfelmann et al. 1985; Cortez et al. 2017). Lipases can also play an important role in plant defense since lipids and lipid metabolites released in plant membranes function as signal molecules in the activation of plant defense responses (Shah 2005).

Recent studies have also indicated that some strains of endophytic fungi, such as *Fusarium* sp. and *Cercospora* sp. in *B. dracunculifolia*, can produce phenoloxidases that are capable of degrading phenolic compounds such as petroleum hydrocarbons and industrial effluents (Onofre and Steilmann 2012). These are important findings for bioremediation initiatives since these contaminants can be found at ca. 35 million metric tons per year in the oceans (Rosenberg and Ron 1996). These contaminants severely impact marine biodiversity over a time span of decades and even centuries (Goldberg and Bertine 2000; Pinheiro et al. 2019; Magris and Giarizzo 2020). Much of this contamination comes from oil in municipal and industrial waste and runoff, leaks in pipelines and storage tanks, and sewage and ballast water discharge (Telli Karakoç and Ediger 2020; Pokazeev et al. 2021).

Many endophytic fungi have proven to be excellent bioremediators of heavy metals that can harm human health, such as arsenic. Some of these endophytes are resistant to arsenic and arsenate and possess the ability to transform them into volatile arsenic gases (Páez-Espino et al. 2009). *Aspergillus, Cladosporium, Fusarium, Penicillium*, and *Trichoderma* are some of the endophytic fungus genera known for their bioremediation potential. They are frequently found in plants that have tolerance to, or the ability to bioaccumulate, heavy metals (Deng et al. 2014). Coincidentally, these fungus genera are also found among species of *Baccharis*, mainly *B. dracunculifolia*. *Baccharis dracunculifolia* has been shown to have great phytostabilization potential for areas contaminated with arsenic (Gilberti et al. 2014). This phytostabilization potential is likely associated with the endophytic fungi in this species; however, no studies have evaluated this relationship.

Endophytic Fungi as a Source of Bioactive Compounds

Endophytic fungi represent a promising source of natural bioactive products. Researchers around the world have been intrigued by the diversity of secondary compounds produced by endophytic fungi and by the similarity between the metabolites produced by endophytic fungi and their host plants (Kusari et al. 2013). This similarity in the production of secondary compounds may be due to several factors, including (1) host plant metabolism being induced by the fungus; (2) fungus metabolism being induced by the host plant; (3) fungus sharing specific biosynthesis pathways with the host plant; and (4) host plant being able to metabolize substances of fungal origin or vice versa (e.g., Ludwig-Müller 2015).

Among the best-known examples of bioactive compounds produced by endophytic fungi is palictaxel (commercialized as Taxol) produced by *Taxomyces andreanae* from the medicinal tree *Taxus brevifolia* (Stierle et al. 1993). Palictaxel is one of the most effective chemotherapeutic compounds used in the treatment of various types of cancer (i.e., ovary, breast, and lung) (Stierle et al. 1993, see Chap. 18). Before this discovery, the supply of this diterpenic substance was limited to the slow-growing barks of *T. brevifolia* that grow in moist soils close to lakes and rivers in some regions of the Pacific Northwest (Guchelaar et al. 1994). Thus, the possibility of extracting palictaxel from *T. andreanae* significantly reduced the production costs of this compound and increased its supply and availability. In addition, other species of endophytic fungi have also been reported to produce palictaxel in plant species of the genera *Taxus* and *Podocarpus* and even in *Ginkgo biloba* (Zhao et al. 2010; Naik 2019). One of these endophytic fungi is *Cladosporium cladosporioides* isolated from *Taxus media* (Zhang et al. 2009), which, coincidentally, is also found in *B. dracunculifolia*, although there is no information on whether it produces palictaxel. Future studies could attempt to propagate this fungus species from *B. dracunculifolia* and isolate its products to evaluate if the chemotherapeutic palictaxel is also produced by *C. cladosporioides* when it occurs in different host plant species. Such studies would generate a cascade of interesting and innovative investigations of major economic and medical relevance.

Among the endophytic fungi most frequently found among species of *Baccharis*, the genera *Xylaria* and *Preussia* deserve to be highlighted with regard to their metabolism. Species of *Xylaria* are known to produce several chemical constituents of the terpene class (Smith et al. 2002), xanthones (Healy et al. 2004), cyclopeptides (Huang et al. 2007), and xyloketals (Lin et al. 2001), among others. Species of this genus are found in other plant species of the family Asteraceae and are known for their inhibitory activity against phytopathogens such as *Penicillium expansum* (Bleicher and Bernardi 1985; Costa and Veiga 1996) and *Aspergillus niger* (Lock 1962; Santos et al. 2010). Furthermore, species of the genus *Preussia* are known to produce the metabolite preussomerin A, which is active against pathogens (Chen et al. 2009). A variety of substances produced by *Preussia* spp. have been identified in the last 10 years, including coumarins (Gonzalez-Menendez et al. 2017), anthraquinones (Gonzalez-Menendez et al. 2017), chronomes (Zhang et al. 2012; Gonzalez-Menendez et al. 2017), and preussochromones (Zhang et al. 2012). Some preussochromones have shown activity against lung cancer cell lines (Zhang et al. 2012).

Unfortunately, knowledge about the metabolites produced by endophytic fungi of species of *Baccharis* is still incipient and restricted to the endophytes found in *B. megapotamica*, *B. coridifolia*, and *B. dracunculifolia*. For instance, *B. megapotamica* and *B. coridifolia* produce certain macrocyclic trichothecenes, which are also produced by their endophytic fungi *Myrothecium verrucaria* and *Myrothecium roridum* (Jarvis et al. 1987, see Chaps. 14 and 15). These substances have been responsible for causing the death of cattle through necrosis of their ruminal epithelium and some lymphoid tissues (Varaschin et al. 1998, see also Chap. 15). On the other hand, some of these trichothecenes have been reported as effective in the treatment of lymphocytic leukemia (Kupchan et al. 1976; Jarvis et al. 1987, Carvalho et al. 2016, see also Chap. 14).

Among the endophytic fungi found in *B. dracunculifolia*, *A. pullulans* and *Xylaria venosula* stand out for the production of phenols and triterpenes, while *Preussia africana* for fatty acids (Oki et al. 2016, 2021). Chromatographic analysis showed that a group of triterpenes found in extracts from *A. pullulans* was similar to those found in *B. dracunculifolia*.

In addition, host plant metabolism can induce endophytic fungi of different genera and classes to produce similar secondary compounds (Ludwig-Müller 2015). A study with endophytic fungi of species of *Baccharis* revealed that a group of triterpenes present in extracts of *A. pullulans* was similar to those found in extracts of *X. venosula* (Oki et al. 2021). This similarity in the production of secondary compounds between different genera of endophytic fungi from the same plant species indicates a synergistic biochemical relationship between fungus species. Explanations for these findings include (1) endophytes sharing specific biosynthesis pathways with the host plant, and (2) endophytes producing the same secondary compounds as the host plant (see Ludwig-Müller 2015; Stierle and Stierle 2015). *Aureobasidium pullulans* is also used in the production of aureobasidin A, a cyclic peptide substance that has antifungal properties (Takesako et al. 1993). This fungus also releases volatile organic compounds that are capable of suppressing the growth of phytopathogens (Don et al. 2020).

Due to its arsenal of secondary metabolites, *A. pullulans* has been considered an effective biological control agent against several phytopathogenic fungi that affect numerous agricultural crops, including *Alternaria alternata* (Don et al. 2020), *Botrytis cinerea* (Di Francesco et al. 2015; Don et al. 2020; Oki et al. 2021), *Colletotrichum acutatum* (Di Francesco et al. 2015; Oki et al. 2021), *Neofusicoccum parvum* (Rusin et al. 2019), *Penicillium digitatum* (Di Francesco et al. 2015; Oki et al. 2021), *Penicillium expansum* (Di Francesco et al. 2015), *Penicillium italicum* (Di Francesco et al. 2015), and *Rhizoctonia solani* (Di Francesco et al. 2020). *Aureobasidium pullulans* is also highlighted for stimulating the growth of beans and soybean (Di Francesco et al. 2020), which may be attributed to improving host plant nitrogen fixation or a greater release of hormones such as auxins, gibberellins, and cytokinins by *A. pullulans* (Ali et al. 2019). Furthermore, this fungus can improve plant resistance in soil with heavy metals through the release of enzymes that assist in reducing metal absorption and enhance the plant's antioxidant system (Ali et al. 2019).

5 Endophytic Fungi-*Baccharis*-Herbivore Interaction

Some endophytic mycota (i.e., *A. pullulans*, *P. africana*, and *X. venosula*) from *B. dracunculifolia* can produce secondary compounds that reduce the survival of herbivores such as the aphid *Uroleucon erigeronensis* (Oki et al. 2021). On the other hand, a study with *B. dracunculifolia* leaf buds indicated that herbivory can be a gateway for endophytic fungi through horizontal transmission (Fernandes et al. 2018).

Although herbivores often favor the infection of the host plant with endophytic fungi, this was not found to be the case for galling insects on *Baccharis reticularia*. Individuals of *B. reticularia* with and without galls showed no difference in endophytic fungus richness (Formiga 2013). However, a higher richness of endophytic fungi was found in gall samples that had a higher content of nitrogen and potassium (Formiga 2013). Nevertheless, a relationship between endophytic fungus richness and nitrogen and phosphorus content was not observed. These results suggest that endophytes can increase the nutritional status of galled plants by improving plant vigor and reducing the effects of gall infestation (Formiga 2013). This mutualistic relationship has not been reported so far in the literature and certainly needs further studies.

All together, these findings suggest that although endophytic fungi are often imperceptible inside plants, they are important mediators in the relationships between herbivores and their host plants and apparently very important among species of the genus *Baccharis*.

6 Climate Change and Endophytic Fungus Diversity

Growing concerns about the impacts that climate change will have on endophytes have increased research interest in this group with searches for strategies that could mitigate these effects. Recent studies on the endophytes of *B. dracunculifolia* and *Baccharis platypoda* grown under increased CO_2 concentration did not find differences in mycota richness (Oki et al. 2020). On the other hand, endophytic community composition changed by 50% when compared to individual plants developed under conditions of ambient CO_2 (Oki et al. 2016). Thus, several species of endophytic fungi that play fundamental roles in plant performance and resistance may disappear or be replaced by other species of unknown functional roles. These changes in species composition are associated with structural and chemical changes (Sanchez-Azofeifa et al. 2012) that may occur in plants under increased CO_2 concentration, such as increased biomass, greater leaf thickness, and higher phenolic content, among other aspects (Oki et al. 2020). Despite these recent studies of endophytic fungi under conditions of predicted climate change, there is still no clear long-term notion of the impacts that increased atmospheric CO_2 concentration might exert on the symbiosis between endophytic fungi and their host plants.

7 Final Considerations and Ways Forward

Despite the limited number of studies on endophytic fungus communities of species of *Baccharis*, this chapter has shown how relevant these microorganisms are to ecological relationships, as well as their great potential in industrial research and developments involving their bioactive compounds. Further investigations into the diversity of endophytic fungi among species of *Baccharis* are needed to better understand the phylogenetic, ecological, and metabolic relationships among them, as well as their prevalence under certain environmental conditions. The relationship between endophyte and host attributes is also of major relevance, and the *Baccharis* system represents an interesting system to be evaluated in this regard. Some of these aspects could be intrinsic to the host plant (e.g., genetics, sex, age, resistant/susceptibility) or of the environment (e.g., seasonality, microhabitat conditions, climate change influence). The association between a host plant and its endophytes may be subject to certain genetic expressions that can vary seasonally and/or spatially (Faeth 2002, Mejía et al. 2014). Another relevant aspect yet to be studied in detail is the ability of endophytic fungi to signal one another triggering silent biosynthetic pathways (Scherlach and Hertweck 2009). Understanding these very interesting relationships and their intrinsic and extrinsic factors would certainly contribute to solidifying knowledge of endophytic fungi and contribute to better exploration of their potential uses for industrial/pharmaceutical purposes.

Acknowledgments We thank for CNPq, FAPEMIG, Planta and CAPES for grants and scholarships to the authors.

References

Ali A, Bilal S, Khan AL, Mabood F, Al-Harrasi A, Lee IJ (2019) Endophytic *Aureobasidium pullulans* BSS6 assisted developments in phytoremediation potentials of *Cucumis sativus* under Cd and Pb stress. J Plant Interact 14:303–313. https://doi.org/10.1080/17429145.2019.1633428

Arnold AE, Herre EA (2003) Canopy cover and leaf age affect colonization by tropical fungal endophytes: ecological pattern and process in *Theobroma cacao* (Malvaceae). Mycologia 95:388–398. https://doi.org/10.1080/15572536.2004.11833083

Arnold AE, Lutzoni F (2007) Diversity and host range of foliar fungal endophytes: are tropical leaves biodiversity hotspots? Ecology 88:541–549

Arnold AE, Maynard Z, Gilbert GS, Coley PD, Kursar TA (2000) Are tropical fungal endophytes hyperdiverse? Ecol Lett 3:267–274

Banerjee D (2011) Endophytic fungal diversity in tropical and subtropical plants. Res J Microbiol 6:54–62

Behie SW, Bidochka MJ (2014) Nutrient transfer in plant–fungal symbioses. Trends Plant Sci 19:734–740. https://doi.org/10.1016/j.tplants.2014.06.007

Berger M, Schnelder MP (1992) Enzymatic esterification of glycerol II. Lipase-catalyzed synthesis of regioisomerically pure 1 (3)-rac-monoacylglycerols. J Am Oil Chem Soc 69:961–965

Bilal S, Shahzad R, Imran M, Jan R, Kim KM, Lee IJ (2020) Synergistic association of endophytic fungi enhances *Glycine max* L. resilience to combined abiotic stresses: heavy metals, high temperature and drought stress. Ind Crop Prod 143:1–10. https://doi.org/10.1016/j.indcrop.2019.111931

Bleicher J, Bernardi J (1985) Podridões da maçã e seu controle na pós-colheita. Technical Bulletin. Empresa Catarinense de Pesquisa Agropecuária (EMPASC), Florianópolis, 24 p

Bornscheuer UT (1995) Lipase-catalyzed syntheses of monoacylglycerols. Enzym Microb Technol 17:578–586

Breen JP (1992) Temperature and seasonal effects on expression of *Acremonium* endophyte-enhanced resistance to *Schizaphis graminum* (Homoptera: Aphididae). Environ Entomol 21:68–74. https://doi.org/10.1093/ee/21.1.68

Breen (1993) Enhanced resistance to three species of aphids (Homoptera: Aphididae) in *Acremonium* endophyte-infected turf grasses. J Econ Entomol 86:1279–1286. https://doi.org/10.1093/jee/86.4.1279

Breen JP (1994) *Acremonium* endophyte interactions with enhanced plant resistance to insects. Annu Rev Entomol 86:401–423. https://doi.org/10.1146/annurev.en.39.010194.002153

Cannon PF, Simmons CM (2002) Diversity and host preference of leaf endophytic fungi in the Iwokrama Forest Reserve, Guyana. Mycologia 94:210–220

Carroll GC, Carroll FE (1978) Studies on the incidence of coniferous needle endophytes in the Pacific NorthWest. Can J Bot 56:3034–3043. https://doi.org/10.1139/b78-367

Caruso G, Abdelhamid MT, Kalisz A, Sekara A (2020) Linking endophytic fungi to medicinal plants therapeutic activity. A case study on Asteraceae. Agriculture 10:1–23. https://doi.org/10.3390/agriculture10070286

Carvalho PDO, Campos PRB, Noffs MDA, Oliveira JGD, Shimizu MT, Silva DMD (2003) Application of microbial lipases to concentrate polyunsaturated fatty acids. Quim Nova 26:75–80

Carvalho MP, Weich H, Abraham WR (2016) Macrocyclic trichothecenes as antifungal and anticancer compounds. Curr Med Chem 23:23–35. https://doi.org/10.2174/0929867323666151117121521

Chen X, Shi Q, Lin G, Guo S, Yang J (2009) Spirobisnaphthalene analogues from the endophytic fungus *Preussia* sp. J Nat Prod 72:1712–1725. https://doi.org/10.1021/np900302w

Christian N, Herre EA, Clay K (2019) Foliar endophytic fungi alter patterns of nitrogen uptake and distribution in *Theobroma cacao*. New Phytol 222:1573–1583. https://doi.org/10.1111/nph.15693

Corrêa RCG, Rhoden SA, Mota TR, Azevedo JL, Pamphile JA, Souza CGM, Polizeli MLTM, Bracht A, Peralta RM (2014) Endophytic fungi: expanding the arsenal of industrial enzyme producers. J Ind Microbiol Biotechnol 41:1467–1478. https://doi.org/10.1007/s10295-014-1496-2

Cortez DV, Castro HFD, Andrade GS (2017) Potencial catalítico de lipases ligadas ao micélio de fungos filamentosos em processos de biotransformação. Quim Nova 40:85–96

Costa IFD, Veiga P (1996) Inibição do desenvolvimento de *Penicillium expansum* (Link) Thom. por fungicidas, in vitro. Ciênc Rural 26:357–360. https://doi.org/10.1590/S0103-84781996000300002

Cuzzi C, Link S, Vilani A, Sartori C, Onofre SB (2012) Endophytic fungi of the "vassourinha" (*Baccharis dracunculifolia* D. C. – Asteraceae). Rev Bras Bioci 10:135–139

Darwin C (1877) The various contrivances by which orchids are fertilised by insects. John Murray

Deng Z, Zhang R, Shi Y, Hu L, Tan H, Cao L (2014) Characterization of Cd-, Pb-, Zn-resistant endophytic Lasiodiplodia sp. MXSF31 from metal accumulating *Portulaca oleracea* and its potential in promoting the growth of rape in metal-contaminated soils. Environ Sci Pollut Res 21:2346–2357

Di Francesco A, Ugolini L, Lazzeri L, Mari M (2015) Production of volatile organic compounds by *Aureobasidium pullulans* as a potential mechanism action against postharvest fruit pathogens. Biol Control 81:8–14. https://doi.org/10.1016/j.biocontrol.2014.10.004

Di Francesco A, Zajc J, Gunde-Cimerman N, Aprea E, Gasperi F, Placì N, Caruso F, Baraldi E (2020) Bioactivity of volatile organic compounds by *Aureobasidium* species against gray mold of tomato and table grape. World J Microb Biotechnol 36:1–11. https://doi.org/10.1007/s11274-020-02947-7

Don SY, Schmidtke LM, Gambetta JM, Steel CC (2020) *Aureobasidium pullulans* volatilome identified by a novel, quantitative approach employing SPME-GC-MS, suppressed *Botrytis cinerea* and *Alternaria alternata in vitro*. Sci Rep 10:4498. https://doi.org/10.1038/s41598-020-61471-8

Faeth SH (2002) Are endophytic fungi defensive plant mutualists? Oikos 98:25–36. https://doi.org/10.1034/j.1600-0706.2002.980103.x

Faeth SH, Fagan WF (2002) Fungal endophytes: common host plant symbionts but uncommon mutualists. Integr Comp Biol 42:360–368. https://doi.org/10.1093/icb/42.2.360

Faeth SH, Hammon KE (1997) Fungal endophytes in oak trees: long-term patterns of abundance and associations with leafminers. Ecology 78:810–819

Fernandes GW, Price PW (1992) The adaptive significance of insect gall distribution: survivorship of species in xeric and Mesic habitats. Oecologia 90:14–20

Fernandes GW, Oki Y, Sanchez-Azofeifa A, Faccion G, Amaro-Arruda HC (2011) Hail impact on leaves and endophytes of the endemic threatened *Coccoloba cereifera* (Polygonaceae). Plant Ecol 212:1687–1697. https://doi.org/10.1007/s11258-011-9941-z

Fernandes GW, Oki Y, Belmiro MS, Resende FM, Correa AC Jr, Azevedo JL (2018) Multitrophic interactions among fungal endophytes, bees, and *Baccharis dracunculifolia*: resin tapering for propolis production leads to endophyte infection. Arthropod Plant Interact 12:329–337. https://doi.org/10.1007/s11829-018-9597-x

Formiga A (2013) Distribuição temporal e respostas celulares de *Baccharis reticularia* DC. (Asteraceae) a interações bióticas. PhD Thesis, Universidade Federal de Minas Gerais, Belo Horizonte, Brazil

Gilberti L, Menezes A, Rodrigues AC, Fernandes GW, Berbara RLL, Marota HB (2014) Effects of arsenic on the growth, uptake and distribution of nutrients in the tropical species *Baccharis dracunculifolia* DC (Asteraceae). J Toxicol Sci 2014:1–18

Goldberg ED, Bertine KK (2000) Beyond the mussel watch—new directions for monitoring marine pollution. Sci Total Environ 247:165–174

Gonzalez-Menendez V, Martin J, Siles JA, Gonzalez-Tejero MR, Reyes F, Platas G, Tormo JR, Genillour O (2017) Biodiversity and chemotaxonomy of *Preussia* isolates from the Iberian Peninsula. Mycol Prog 16:713–728. https://doi.org/10.1007/s11557-017-1305-1

Griffin EA, Carson WP (2018) Tree endophytes: cryptic drivers of tropical forest diversity. In: Pirttilä A, Frank A (eds) Endophytes of Forest trees. Springer, pp 63–103. https://doi.org/10.1007/978-3-319-89833-9_4

Grunseich JM, Thompson MN, Aguirre NM, Helms AM (2020) The role of plant-associated microbes in mediating host-plant selection by insect herbivores. Plan Theory 9:1–23. https://doi.org/10.3390/plants9010006

Guchelaar HJ, Ten Napel CHH, De Vries EGE, Mulder NH (1994) Clinical, toxicological and pharmaceutical aspects of the antineoplastic drug taxol: a review. Clin Oncol 6:40–48

Hammon KE, Faeth SH (1992) Ecology of plant-herbivore communities: a fungal component? Nat Toxins 1:197–208. https://doi.org/10.1002/nt.2620010307

Healy PC, Hocking A, Tran-Dinh N, Pitt JI, Shivas RG, Mitchell JK, Kotiw M, Davis RA (2004) Xanthones de um microfungus do gênero *Xylaria*. Phytochemistry 65:2373–2378. https://doi.org/10.1016/j.phytochem.2004.07.019

Heiden G, Pirani JR (2016) Novelties towards a phylogenetic classification of *Baccharis* (Asteraceae, Astereae). Phytotaxa 289:285–290. https://doi.org/10.11646/phytotaxa.289.3.9

Herms DA, Mattson WJ (1992) The dilemma of plants: to grow or defend. Q Rev Biol 67:283–335

Höfelmann M, Hartmann J, Zink A, Schreier P (1985) Isolation, purification, and characterization of lipase isoenzymes from a technical *Aspergillus niger* enzyme. J Food Sci 50:1721–1725. https://doi.org/10.1111/j.1365-2621.1985.tb10574.x

Huang H, She Z, Lin Y, Vrijmoed LLP, Lin W (2007) Cyclic peptides from an endophytic fungus obtained from a mangrove leaf (*Kandelia candel*). J Nat Prod 70:1696–1699. https://doi.org/10.1021/np0605891

Hubbard M, Germida JJ, Vujanovic V (2014) Fungal endophytes enhance wheat heat and drought tolerance in terms of grain yield and second-generation seed viability. J Appl Microbiol 116:109–122. https://doi.org/10.1111/jam.12311

Hyde KD, Soytong K (2008) The fungal endophyte dilemma. Fungal Divers 33:163–173

Jarvis BB, Comezoglu SN, Amomon HL, Breedlove CK, Miller RW, Woode MK, Streelman DR, Sneden AT, Dailey RG, Kupchan SM (1987) New macrocyclic trichothecenes from *Baccharis megapotamica*. J Nat Prod 50:815–828. https://doi.org/10.1021/np50053a008

Jia M, Chen L, Xin HL, Zheng CJ, Rahman K, Han T, Qin LP (2016) A friendly relationship between endophytic fungi and medicinal plants: a systematic review. Front Microbiol 7:906. https://doi.org/10.3389/fmicb.2016.00906

Kupchan SM, Uchida I, Branfman AR, Dailey RG, Fei BY (1976) Antileukemic principles isolated from Euphorbiaceae plants. Science 191:571–572. https://doi.org/10.1126/science.1251193

Kusari S, Pandey SP, Spiteller M (2013) Untapped mutualistic paradigms linking host plant and endophytic fungal production of similar bioactive secondary metabolites. Phytochemistry 91:81–87. https://doi.org/10.1016/j.phytochem.2012.07.021

Kwon DY, Rhee JS (1986) A simple and rapid colorimetric method for determination of free fatty acids for lipase assay. J Am Oil Chem Soc 63:89–92

Lin Y, Wu X, Feng S, Jiang G, Luo J, Zhou S, Vrijmoed LLP, Jones EBG, Krohn K, Steingröver K, Zsila F (2001) Five unique compounds: xyloketals from mangrove fungus *Xylaria* sp. from the South China Sea coast. J Org Chem 66:6252–6256. https://doi.org/10.1021/jo015522r

Lock GW (1962) Sisal: twenty-five years sisal research. Longmans, London, 355p

Ludwig-Müller J (2015) Plants and endophytes: equal partners in secondary metabolite production? Biotechnol Lett 37:1325–1334

Magris RA, Giarrizzo T (2020) Mysterious oil spill in the Atlantic Ocean threatens marine biodiversity and local people in Brazil. Mar Pollut Bull 153:1–11

Meister B, Krauss J, Härri SA, Schneider MV, Muller CB (2006) Fungal endosymbionts affect aphid population size by reduction of adult life span and fecundity. Basic Appl Ecol 7:244–252. https://doi.org/10.1016/j.baae.2005.06.002

Mejía LC, Herre EA, Sparks JP, Winter K, García MN, Van Bael SA, Stitt J, Zi S, Shang Y, Guiltinan MJ, Maximova SN (2014) Pervasive effects of a dominant foliar endophytic fungus on host genetic and phenotypic expression in a tropical tree. Front Microbiol 5:1–17. https://doi.org/10.3389/fmicb.2014.00479

Naik BS (2019) Developments in taxol production through endophytic fungal biotechnology: a review. Orient Pharm Exp Med 19:1–13. https://doi.org/10.1007/s13596-018-0352-8

Nascimento IM (2010) Avaliação da atividade antipatogênica e enzimática de três fungos endofíticos de *Baccharis dracunculifolia* (Asteraceae). Monograph. Pontifícia Universidade Católica de Minas Gerais. 36p

Nascimento TL, Oki Y, Lima DMM, Almeida-Cortez JS, Fernandes GW, Souza-Motta CM (2015) Biodiversity of endophytic fungi in different leaf ages of *Calotropis procera* and their antimicrobial activity. Fungal Ecol 14:79–86. https://doi.org/10.1016/j.funeco.2014.10.004

Oki Y, Fernandes GW, Corrêa Júnior A (2008) Fungos: Amigos ou inimigos? Ciênc Hoje 42:64–66

Oki Y, Soares NR, Belmiro MS, Correa-Junior A, Fernandes GW (2009) The influence of the endophytic fungi on the herbivores from *Baccharis dracunculifolia* (Asteraceae). Neotrop Biol Conserv 4:83–88. https://doi.org/10.4013/5119

Oki Y, Goto BT, Jobim K, Rosa LH, Ferreira MC, Coutinho ES, de Xavier JHA, Carvalho F, de Souza MFM, Berbara RLL, Fernandes GW (2016) Arbuscular mycorrhiza and endophytic fungi in ruspestrian grasslands. In: Fernandes GW (ed) Ecology and conservation of mountaintop grasslands in Brazil. Springer, Switzerland, pp. 157–179

Oki Y, Arantes-Garcia L, Costa MB, Nunes BC, Silveira BR, Gélvez-Zúñiga I, Franco A, Fernandes GW (2020) CO_2 fertilizer effect on growth, polyphenols, and endophytes in two *Baccharis* species. Braz Arch Biol Technol 63:e20190302. https://doi.org/10.1590/1678-4324-2020190302

Oki Y, Nascimento IM, Costa NB, Maia RA, Takahashi JA, Ferraz V, Fernandes GW (2021) Effectiveness of endophytic fungi from *Baccharis dracunculifolia* against sucking insects and fungal pathogens. In: Rosa LH (ed) Neotropical endophytic fungi. Springer. pp.337–349

Onofre SB, Steilmann P (2012) Phenoloxidases produced by endophytic fungi isolated from *Baccharis dracunculifolia* D. C. (Asteraceae). Resour Environ 2:271–274. https://doi.org/10.5923/j.re.20120206.04

Onofre SB, Bonfante T, Santos ZMQ, Moura MC, Cardoso AF (2014) Cellulase production by endophytic strains of *Trichoderma reesei* from *Baccharis dracunculifolia* DC (Asteraceae). Adv Microbiol 4:275–283. https://doi.org/10.4236/aim.2014.45034

Páez-Espino D, Tamames J, Lorenzo V, Canovas D (2009) Microbial responses to environmental arsenic. Biometals 22:117–130. https://doi.org/10.1007/s10534-008-9195-y

Pinheiro HT, Teixeira JB, Francini-Filho RB, Soares-Gomes A, Ferreira CEL, Rocha LA (2019) Hope and doubt for the world's marine ecosystems. Perspect Ecol Conser 17:19–25

Pokazeev K, Sovga E, Chaplina T (2021) Main natural and anthropogenic sources of pollution of the Black Sea, its shelf zones and small water reservoirs. In: Pokazeev K, Sovga E, Chaplina T (eds) Pollution in the Black Sea. Springer Oceanography. Springer, Cham. https://doi.org/10.1007/978-3-030-61895-7_9

Raps A, Vidal S (1998) Indirect effects of an unspecialized endophytic fungus on specialized plant-herbivorous insect interactions. Oecologia 114:541–547. https://doi.org/10.1007/s004420050478

Rizzo I, Varsavky E, Haidukowski M, Frade H (1997) Macrocyclic trichothecenes in *Baccharis coridifolia* plants and endophytes and *Baccharis artemisioides* plants. Toxicon 35:753–757. https://doi.org/10.1016/S0041-0101(96)00149-3

Rodriguez RJ, Redman RS, Henson JM (2004) The role of fungal symbioses in the adaptation of plants to high stress environments. Mitig Adapt Strateg Glob Chang 9:261–272

Rosenberg E, Ron EZ (1996) Bioremediation of petroleum contamination. In: Crawford RL, Crawford DL (eds) Bioremediation: Principles and applications. Cambridge University Press, Cambridge, pp 100–124

Rusin C, Di Francesco A, Di Foggia M, D'Aquino S, Rombolà A, Tugnoli V, Bothelo RV, Baraldi E (2019) An emerging problem affecting apple production: *Neofusicoccum parvum*. *Aureobasidium pullulans* Le 1 and L8 strains as an alternative control strategy. Biol Control 134:157–162. https://doi.org/10.1016/j.biocontrol.2019.04.015

Sanchez-Azofeifa A, Oki Y, Fernandes GW, Ball RA, Gamon J (2012) Relationships between endophyte diversity and leaf optical properties. Trees 26:291–299. https://doi.org/10.1007/s11829-018-9597-x

Santos MB, Santos CY, Almeida MA, Santos CRS, Sant'Anna HLS, Santos OSN, Silva F, Martins GN (2010) Efeito inibitório in vitro de extrato vegetal de *Allium sativum* sobre *Aspergillus niger* Tiegh. Rev Bras Pl Med 12:13–17. https://doi.org/10.1590/1983-084X/12_150

Scherlach K, Hertweck C (2009) Triggering cryptic natural product biosynthesis in microorganisms. Org Biomol Chem 7:1753–1760. https://doi.org/10.1039/B821578B

Shah J (2005) Lipids, lipases and lipid modifying enzymes in plant disease resistance. Annu Rev Phytopathol 43:229–260. https://doi.org/10.1146/annurev.phyto.43.040204.135951

Smith CJ, Morin NR, Bills GF, Dombrowski AW, Salituro GM, Smith SK, Zhao A, MacNeil DJ (2002) Novel sesquiterpenoids from the fermentation of *Xylaria persicaria* are selective ligands for the NPY Y5 receptor. J Org Chem 67:5001–5004. https://doi.org/10.1021/jo011054+

Stierle AA, Stierle DB (2015) Bioactive secondary metabolites produced by the fungal endophytes of conifers. Nat Prod Commun 10:1671–1682. https://doi.org/10.1177/1934578X1501001012

Stierle A, Strobel G, Stierle D (1993) Taxol and taxane production by *Taxomyces andreanae*, an endophytic fungus of Pacific yew. Science 260:214–216. https://doi.org/10.1126/science.8097061

Sun X, Guo LD, Hyde KD (2011) Community composition of endophytic fungi in *Acer truncatum* and their role in decomposition. Fungal Divers 47:85–95. https://doi.org/10.1007/s13225-010-0086-5

Takesako K, Kuroda H, Inoue T, Haruna F, Yoshikawa Y, Kato I, Uchida K, Hiratani T, Yamaguchi H (1993) Biological properties of aureobasidin a, a cyclic depsipeptide antifungal antibiotic. J Antibiot 46:1414–1420. https://doi.org/10.7164/antibiotics.46.1414

Telli Karakoç F, Ediger D (2020) Oil pollution of the surrounding waters of Turkey. In: The handbook of environmental chemistry. Springer, Berlin/Heidelberg. pp. 1–26. https://doi.org/10.1007/698_2020_477

Van Etten ML, Prevost LB, Deen AC, Ortiz BV, Donovan LA, Chang SM (2008) Gender differences in reproductive and physiological traits in a gynodioecious species, *Geranium maculatum* (Geraniaceae). Int J Plant Sci 169:271–279. https://doi.org/10.1086/524109

Varaschin MS, Barros CS, Jarvis BB (1998) Intoxicação experimental por *Baccharis coridifolia* (Compositae) em bovinos. Pesq Vet Bras 18:65–68. https://doi.org/10.1590/S0100-736X1998000200006

Vieira ML, Johann S, Hughes FM, Rosa CA, Rosa LH (2014) The diversity and antimicrobial activity of endophytic fungi associated with medicinal plant *Baccharis trimera* (Asteraceae) from the Brazilian savannah. Can J Microbiol 60:847–856. https://doi.org/10.1139/cjm-2014-0449

Wallace CS, Rundel PW (1979) Sexual dimorphism and resource allocation in male and female shrubs of *Simmondsia chinensis*. Oecologia 44:34–39. https://doi.org/10.1007/BF00346394

Zhang P, Zhou PP, Yu LJ (2009) An endophytic taxol-producing fungus from Taxus media, *Cladosporium cladosporioides* MD2. Curr Microbiol 59:227–232

Zhang F, Li L, Niu S, Si Y, Guo L, Jiang X, Che Y (2012) A thiopyranchromenone and other chromone derivatives from an endolichenic fungus, *Preussia africana*. J Nat Prod 75:230–237. https://doi.org/10.1021/np2009362

Zhao J, Zhou L, Wang J, Shan T, Zhong L, Liu X, Gao X (2010) Endophytic fungi for producing bioactive compounds originally from their host plants. In: Méndez-Vilas A (ed) Current research, technology and education topics in applied microbiology and microbial biotechnology. Formatex Research Center, Badajoz, pp 567–576

Chapter 7
Baccharis as Nurse Plants

Ramón Perea, Marta Peláez, and G. Wilson Fernandes

Abstract The genus *Baccharis* includes important shrub species that facilitate the regeneration and performance of other species that grow underneath. Some *Baccharis* species represent crucial nurse shrubs that provide shelter from abiotic or biotic stress to other "beneficiary" plants. Here, we uncover the role of *Baccharis* species as facilitators, particularly in stressful and herbivore-dominated environments. We highlight that the net facilitative effect of *Baccharis* strongly depended on multiple factors (its size and architectural form, the type and intensity of stress, the ontogenetic stage of beneficiary plants, the presence of conspecifics, etc.). We particularly focus on *Baccharis* as tree recruitment microsites that enhance both recruitment quantity and quality along many systems of the Americas (from California to Chile) and boost vegetation dynamics towards late successional stages and functioning systems. In addition, we highlight the roles of some *Baccharis* species in facilitating native vs. exotic herbaceous species and their possible role in reducing the colonization and expansion of invasive plants. Thus, the genus *Baccharis* includes extraordinary interesting species from an ecological, conservation, and restoration point of view due to their ability to work as nurse plants that favor the regeneration of keystone species and reduce the proliferation of invasive plants.

Keywords Plant facilitation · Plant-plant interactions · Regeneration microsites · Restoration ecology

R. Perea (✉) · M. Peláez
Departamento de Sistemas y Recursos Naturales, Universidad Politécnica de Madrid, Madrid, Spain
e-mail: ramon.perea@upm.es

G. W. Fernandes
Ecologia Evolutiva and Biodiversidade, Departamento de Genética, Ecologia & Evolução/ICB, CP 486, Universidade Federal de Minas Gerais, Belo Horizonte, Minas Gerais, Brazil

G. W. Fernandes et al. (eds.), *Baccharis*,
https://doi.org/10.1007/978-3-030-83511-8_7

1 Plant Facilitation and the Nursing Phenomenon

Plant facilitation is a positive interaction where a nurse plant modifies the local abiotic and biotic conditions, improving the lifetime fitness of other, beneficiary plant species (Callaway 1995; Bertness and Leonard 1997; Bronstein 2009; Soliveres et al. 2011). Plant facilitation is, therefore, an important form of ecological nursing, where some plants (the "nurse" or "benefactors") facilitate seed arrival, germination, establishment, growth, or development of other plant species (the "beneficiary") by providing shelter from one or both sources of stress: (1) abiotic (e.g., drought, salinity, soil toxicity, frosts) or (2) biotic (e.g., herbivory, seed predation, pathogens; Franco and Nobel 1988; Valiente-Banuet and Ezcurra 1991; Brooker et al. 2008; Gómez-Aparicio et al. 2008; Perea and Gil 2014).

Plant facilitation has been mostly studied under abiotic stressful conditions, like those exerted in resource-limited or constraining environments (Stachowicz 2001; Tirado and Pugnaire 2005; but see Spadeto et al. 2017). As a result, some studies have shown strong evidence to support the stress gradient hypothesis (SGH), which predicts that the frequency of facilitation directly increases with abiotic stress (Bertness and Callaway 1994; Callaway 2007). However, other authors argue that the largest absolute facilitative effects will always occur under less abiotic stressful conditions (Holmgren and Scheffer 2010), leading to a strong debate on the predictions and extension of the SGH (Michalet et al. 2013; He et al. 2013; Soliveres and Maestre 2014; Soliveres et al. 2015).

A common consequence of plant facilitation is the formation of vegetation patches surrounded by open space (Prentice and Werger 1985; Castillo et al. 2010) or the proliferation of plant recruits (seedlings and saplings) under the cover of conspicuous nurse plants such as trees or shrubs (Callaway 1992; Gómez-Aparicio et al. 2008; Perea et al. 2016). Nurse shrubs are considered key elements on the facilitative process of plant-plant interactions as they are able to assemble ecological communities (Armas and Pugnaire 2005) and enhance the recruitment rates of many tree species that develop underneath (Castro et al. 2004; Smit et al. 2007; Perea et al. 2017; Spadeto et al. 2017). Among shrub species, the genus *Baccharis* L. is the largest genus in the family Asteraceae, with over 440 shrub species widely distributed throughout the Americas (Heiden et al. 2019). Their pioneer character (rapid growth, low longevity, high production of wind-dispersed seeds, and the capability to easily establish on disturbed sites; see Fernandes et al. 2014) stands for a strong potential to facilitate the progressive ecological shift towards later stages of plant succession where the nursing phenomenon plays an essential role.

This chapter synthesizes the nursing role of some shrub species in the genus *Baccharis* (Table 7.1) as they represent paradigmatic species within the plant facilitation process in the Americas (McBride 1974; Callaway and Davis 1998; Kitzberger et al. 2000; Duarte et al. 2006; Zavaleta and Kettley 2006; van Zonneveld et al. 2012; Perea et al. 2017, 2019; Macek et al. 2018; Peláez et al. 2019). We also highlight their ecological role as a key facilitator of native plant species over those invasive (Brennan et al. 2018; Perea et al. 2019), revealing a strong potential of *Baccharis*

Table 7.1 Summary of the different species of the genus *Baccharis* known to facilitate the regeneration and performance of different woody species along many systems of the Americas

Country	Habitat	Nurse plant *Baccharis* spp.	Beneficiary tree/ shrub species	References
United States	California grassland	*B. pilularis*	*Quercus agrifolia, Quercus douglasii, Q. lobata*	McBride (1974), Callaway and D'Antonio (1991), Zavaleta and Kettley (2006) and Perea et al. (2017)
United States	California coastal dunes	*B. pilularis*	*Lupinus arboreus*	Rudgers and Maron (2003)
Mexico	Fir forest clearings	*B. conferta*	*Abies religiosa*	Sánchez-Velásquez et al. (2011)
Mexico	Highlands of Chiapas	*B. vaccinioides*	*Quercus crassifolia Quercus rugosa*	Ramírez-Marcial et al. (1996)
Venezuela	Old fields in the tropical Andes	*B. prunifolia*	*Vallea stipularis Berberis discolor*	Bueno and Llambí (2015)
Brazil	*Pampa* grasslands	*B. uncinella*	*Araucaria angustifolia*	Duarte et al. (2006)
Brazil	*Pampa* grasslands	*B. mesoneura*	*Araucaria angustifolia*	Duarte et al. (2006)
Argentina	Patagonian xeric woodlands	*B. rhomboidalis*	*Austrocedrus chilensis*	Kitzberger et al. (2000)
Chile	Semiarid zone of the Chilean coast	*B. vernalis*	*Aextoxicon punctatum Myrceugenia correifolia*	Macek et al. (2018)
Chile	Temperate rain forest boundary	*B. vernalis*	*Myrceugenia correifolia Griselinia scandens*	van Zonneveld et al. (2012)

plants to prevent and control plant invasion. Finally, this chapter aims to provide ecologists and managers with new possible conservation practices based on the nursing ability of *Baccharis* shrubs to facilitate the recruitment of native species in harsh or degraded environments.

2 *Baccharis* as Tree Regeneration Microsite

The perpetuation of any natural ecosystem mainly depends on its regeneration ability. Those species unable to regenerate hamper their continuity in the ecosystem (Schemske et al. 1994). It is well known that the abundance, survival, and development of seedlings vary across microsites (Whittaker and Levin 1977; Collins and Good 1987; López-Sánchez et al. 2019). As a result, certain microsites increase the overall probability of plant recruitment by facilitating seed germination, seedling survival, or growth. These particularly favorable microsites for regeneration are known as "regeneration microsites" following the regeneration niche concept *sensu* Grubb (1977). Seedling establishment is indeed one of the most critical stages in the regeneration process of trees, with high mortality rates during this phase (Clark et al. 1999; Silvertown and Charlesworth 2001; Perea and Gil 2014). Multiple abiotic and biotic agents influence seedling survival and performance, among which are climatic conditions (Dreyer et al. 2001; Rodríguez-Calcerrada et al. 2008), soil resources (Baraloto et al. 2006), competition (Davis et al. 1998), and herbivory (Crawley 1983; Smit et al. 2006). However, all these factors might vary widely among microsites due to the great spatial and temporal heterogeneity of most ecosystems (González-Rodríguez et al. 2011; López-Sánchez et al. 2019).

Nevertheless, some particular microsites such as those underneath nurse shrubs have shown a significant advantage over other microsites (Callaway 1992; Pugnaire et al. 1996; Smit et al. 2007, 2008; Perea and Gil 2014). This is the paradigmatic case of *Baccharis* shrubs in open oak woodlands or savannas of North America (McBride 1974; Callaway and Davis 1998; Zavaleta and Kettley 2006; Perea et al. 2017) where there is a well-known facilitative relationship between the coyote brush (*Baccharis pilularis* DC) and the beneficiary coast live oak (*Quercus agrifolia* Neé), which strongly depends upon *Baccharis* cover to regenerate. Coyote brush is an evergreen, much-branched, and colonizer shrub that grows in shrub communities close to foothill woodlands of California (Steinberg 2002). These shrubs have shown a great ability to protect oak recruits from herbivory (Peláez et al. 2019) probably due to their low palatability and nutritious value (McBride and Heady 1968; Smither-Kopperl 2016; Fig. 7.1). In addition, shoot mortality attributed to water or temperature stress was 17% under shrubs and 63% in the open grassland (Callaway and D'Antonio 1991), suggesting a strong reduction of abiotic stress for those seedlings growing under shrub cover. Furthermore, abundances of all herbaceous species declined greatly after *Baccharis* formed a closed canopy at 2–3 years, and little seed of herbaceous species was either dispersed into shrub stands or stored in the soil (Hobbs and Mooney 1986).

Interestingly, Zavaleta and Kettley (2006) found that oaks established only under mature shrubs, suggesting that features of mature shrubs and their effects on site characteristics benefit establishing oaks. In fact, Peláez et al. (2019) found an interesting relationship between nurse plant size and the probability of facilitating oak recruitment, revealing that *Baccharis* started to be an efficient facilitator (probability of facilitation >0.5) when crown diameter was >2.5 m (age of 7 years) in heavily browsed areas and almost twice (> 5 m diameter; equivalent age of 15 years) at low herbivore stress levels (Fig. 7.2). This indicates that the herbivore pressure is a strong driver in determining plant facilitative effects, increasing overall plant facilitation at higher levels of biotic stress (herbivory).

Baccharis shrubs may not only contribute to increasing the number of recruits (regeneration abundance) but also to enhancing the recruit quality. Thus, the development of tree recruits may strongly depend on the protective effect of shrubs. In particular, *Baccharis* were proved to favor the growth and the adequate architecture

Fig. 7.1 *Baccharis pilularis* facilitating the establishment and growth of coast live oaks (*Quercus agrifolia*, left) and blue oaks (*Quercus douglasii*, right) in California oak woodlands. Oaks eventually surpass the shrub height and outcompete the *Baccharis* shrubs. At that stage, oak trees have grown sufficiently to avoid intense browsing damage aboveground and to reduce soil water deficit belowground. (Photos: Ramón Perea)

of oak saplings in herbivore-dominated environments by reducing the detrimental effect of browsers, which typically reduce the plant height/diameter ratio producing stunted individuals (Peláez et al. 2019; Fig. 7.3). The protective effect of *Baccharis* against browsers facilitates the advance of oak saplings to the next ontogenetic stages, ensuring the sexual reproduction of trees (production of flowers and fruits) and the plant community dynamics towards late successional stages (Fig. 7.3). The recruit quality was also affected by the *Baccharis* size, as sapling recruits under larger *Baccharis* shrubs had proportionally higher height/diameter ratios (Peláez et al. 2019). Another study in Mexico also showed that *Baccharis conferta* shrubs improved the growth of coniferous trees (*Abies religiosa* seedlings) growing underneath (Sánchez-Velásquez et al. 2011). Thus, *Baccharis conferta* shrubs were found to promote the seedling growth of commercial conifer plantations (Sánchez-Velásquez et al. 2011) as well as *Baccharis pilularis* has been shown useful for restoration purposes of the highly valuable oak woodlands (López-Sánchez et al. 2019). Similarly, other congeneric species such as *Baccharis uncinella* and *Baccharis mesoneura* have been argued to be facilitators of *Araucaria angustifolia* trees in the Brazilian Atlantic Forest (Duarte et al. 2006). In line with other studies, Duarte et al. (2006) found more forest species seedlings beneath the canopies of

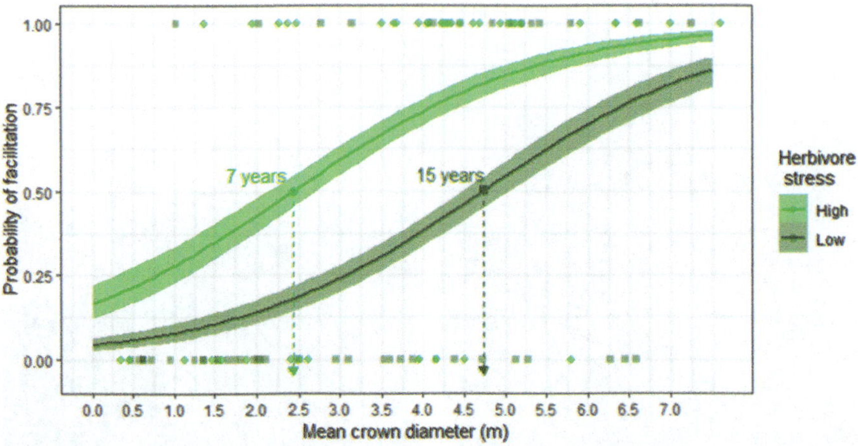

Fig. 7.2 Probability of plant facilitation in relation to nurse plant size (*Baccharis* mean crown diameter) and herbivory stress level (high or low deer densities). Arrows represent the threshold nurse plant size at which facilitation probability is 50%. Age of the nurse plant, in years, was annotated for both threshold values. (Adapted from Peláez et al. 2019)

Fig. 7.3 (**a**) Stunted individual of the coast live oak (*Quercus agrifolia*) growing in an open microsite, with low height/diameter ratio and modified architecture. (**b**) Oak juvenile of the same species growing under the protection of *Baccharis pilularis*, with adequate architecture and greater height/diameter ratio. (Photos: Ramón Perea)

nurse plants compared with open field grassland, creating new forest colonization sites in *Pampa* grasslands of Brazil. Other studies also point out shrub cover protection from direct sunlight as the main factor that enhanced the germination and survival of *Austrocedrus chilensis* in northern Patagonia (Argentina) where *Baccharis rhomboidalis* is a fairly common shrub that favorably influences tree regeneration (Kitzberger et al. 2000). Positive effects of *Baccharis prunifolia* shrubs were also found on the regeneration of old fields at the tropical Andean forest, where species richness, vegetation cover, and the density of dominant forest trees were higher under the shrub canopy than in the inter-shrub spaces (Bueno and Llambí 2015). Another study in semiarid Chile showed that tree species become established in a herbaceous matrix thanks to *Baccharis vernalis* patches, most likely due to a combination of fog-interception capacity, soil nutrient availability, and low competition (Macek et al. 2018).

Similarly, in the rain forest boundary of Chile, *Baccharis vernalis* was found to be the most successful nurse shrub along the arid scrubland-temperate rainforest, sheltering the most abundant and diverse seedling community (van Zonneveld et al. 2012). Interestingly, this study also highlights that dead shrubs also play an important role in facilitating tree seedling establishment although proportionally more seedlings were recruited under dead shrubs in the scrubland than in the forest borders, suggesting a stronger competition between living shrubs and establishing seedlings in drier (more stressed) environments (van Zonneveld et al. 2012). Previous studies demonstrated that dead nurse plants are able to ameliorate thermal stress without the negative cost of reducing soil water content (Anthelme et al. 2007). This occurs not only in dry and harsh environments but also after disturbances such as fires (Castro et al. 2011). Hence, dead shrubs may also play an important role in plant facilitation mostly through the abiotic mechanism of microsite amelioration and, thus, may represent key elements in the ecological restoration process after a fire or other disturbances.

Interestingly, *Baccharis* shrubs are also known to facilitate other shrubs. For instance, Rudgers and Maron (2003) recorded facilitative interactions between *Baccharis pilularis* and the coastal dune shrub *Lupinus arboreus*, an important nitrogen-fixer in California dunes. The relationship depended on the genotype of *B. pilularis*; only the prostrate architectural form of this species benefited seedling emergence, survival, and growth of *Lupinus arboreus* but had no effect on post-dispersal seed predation or adult establishment (Rudgers and Maron 2003). Importantly, by facilitating an important nitrogen-fixer, *Baccharis* shrubs also had effects that cascade to other members of the coastal plant community (Rudgers and Maron 2003), revealing a strong influence of *Baccharis* shrubs beyond pairwise interactions.

3 *Baccharis* as Facilitator of Native and Exotic Herbaceous Species

Many studies have documented the facilitative effect of *Baccharis* shrubs on native plant communities dominated by herbs, forbs, and grasses (van Zonneveld et al. 2012; Brennan et al. 2018; Perea et al. 2019). For instance, coyote brush (*Baccharis pilularis*) has been documented invading grasslands where non-native species were gradually replaced by not only coyote brush but also several other noteworthy native species (Brennan et al. 2018). This study finds that over the 37-year timeframe, exotic grasses gradually decline, while native plant cover increases in California grassy landscapes invaded by coyote brush (Brennan et al. 2018). Another recent study in two abandoned pasture areas of Brazil showed that plant diversity was significantly higher in the restored environment (after planting the nurse *Baccharis dracunculifolia* shrubs) compared to the degraded environment (Siqueira et al. in review). In addition, fewer ruderal and exotic species were recorded in the restored sites with *B. dracunculifolia*, concluding that restoration with *Baccharis* planting had a positive effect on the restructuring of the native plant community (Siqueira et al. in review, see also Fernandes et al. 2018).

Nevertheless, some studies have documented the opposite facilitative pattern, where neighbor grasses may act as facilitators of *Baccharis* shrubs by improving, for instance, water balance (Dechoum et al. 2018). Thus, facilitative interactions between herbaceous plants and *Baccharis* shrubs may be reciprocal and difficult to quantify when both groups coexist and, as a result, competition and facilitation are typically co-occurring. This has been documented in some field patterns where changes in resources (e.g., water) may shift from competitive to facilitative and vice versa (Holmgren et al. 1997). Although more rarely documented, few studies revealed that some invasive grasses may benefit from the *Baccharis* facilitative effect. For instance, the invasive South African grass, *Ehrharta calycina*, escaped herbivory by associating with *Baccharis pilularis*, showing greater performance (growth and aboveground biomass) than unassociated individuals of the California dune system (Cushman et al. 2011). Similarly, Perea et al. (2019) found that *Baccharis* facilitated the occurrence of the invasive *Hyparrhenia rufa* along the montane roads of Brazil although, overall, *Baccharis* favored the presence of native herbaceous plants over those exotic, with 61% greater probability of facilitation for native species than for exotic species (Fig. 7.4).

Some studies have shown that once *Baccharis* shrubs are established, they change the light and water availability under their canopy and provide cover for small mammals. These mammals, in turn, remove the grasses from under the shrub canopy and from its immediate ecotone (McBride and Heady 1968; Bartholomew 1970), thus preventing grasses from invading the shrubland. This mechanism may be behind the greater ability of *Baccharis* to facilitate native vs. exotic herbaceous plants although further studies are needed. Similarly, in Chile, the lower biomass of grasses under the canopy of *Baccharis linearis*, compared to where the shrubs are absent, also suggested inhibition of shrubs on the herbaceous stratum (Martínez and

Fig. 7.4 Adjacent areas without (left) and with (right) *Baccharis* shrubs along montane roads of rupestrian grasslands of Brazil. The establishment of transects allowed the evaluation of *Baccharis* as facilitator of invasive vs. native herbaceous plants (see Perea et al. 2019 for more details). Pioneer nurse shrubs such as a *Baccharis dracunculifolia* DC were found to alleviate the environmental shift generated by the construction and use of roads (e.g., disturbed soils with low nutrient content) (see Fernandes 2016) and represent an interesting alternative to mitigate exotic plant invasion along roadsides

Fuentes 1993). However, the interactive effects of invasive species and *Baccharis* plants may depend on the life-cycle stage at which the interactions occurred. For instance, in Chile, the early emergence of the invasive species *Centaurea solstitialis* L. enabled established plants to competitively displace the late-emerging *Baccharis linearis* and *B. paniculata* (Gómez-González et al. 2009). However, the presence of the invasive *C. solstitialis* (individuals or seeds) did not affect negatively the seed germination of the two abovementioned *Baccharis* species (Gómez-González et al. 2009). Interestingly, these authors also found that the biomass of both *Baccharis* species increased under conspecific competition compared to control (growing alone). Facilitation among conspecific plants of similar age or size is a kind of interaction that can be essential for seedling establishment in some arid and semiarid ecosystems (Goldberg et al. 2001; Franks 2003). Further studies should address whether seedling survival of *Baccharis* species is really being facilitated by conspecifics as they may better resist plant invasion at high densities (Gómez-González et al. 2009). This facilitative conspecific response may also contribute to explain the marked ecotones between *Baccharis* shrublands and annual grasslands in the California chaparral (McBride and Heady 1968; Hobbs and Mooney 1986) and central Chile (Martínez and Fuentes 1993).

Interestingly, *Baccharis* is a dioecious genus, i.e., with male and female individuals. Da Costa Fonseca et al. (2017) found that male and female *Baccharis platypoda* adults presented an aggregate pattern at smaller scales but random and uniform patterns for larger scales (>20 m). In addition, they found that male individuals preferred higher moisture soils probably due to distinct environmental preferences. Preliminary analysis (Perea et al. unpublished data) in the montane Neotropics showed that there were no significant differences in species richness between male and female *Baccharis* when facilitating herbaceous plants. However, there was

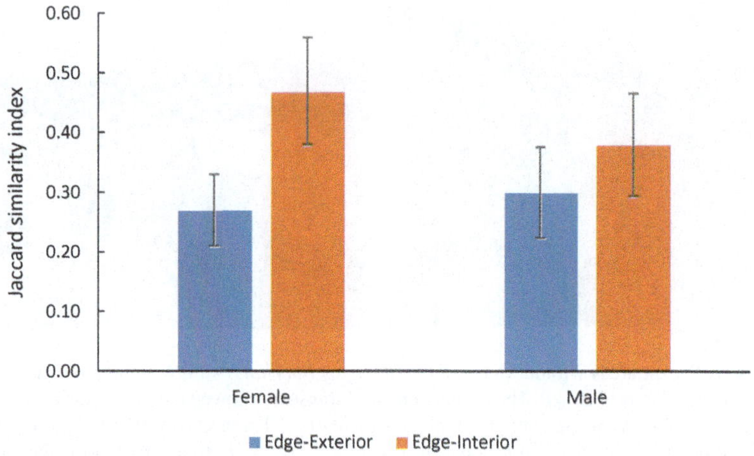

Fig. 7.5 Jaccard similarity index ± SE between communities located at the edge and the exterior and interior part of the nurse shrub for female and male *Baccharis* in Brazil. Notice how female individuals caused greater shifts in species turnover (Jaccard index) than male individuals. (Perea et al. unpublished results)

greater species turnover between exterior and edge position of the shrub than between interior and edge positions (Fig. 7.5), suggesting that female shrubs have a greater effect in species turnover with only 0.27 similarity between edge and exterior communities, 20% lower similarity than between interior and edge communities, whereas for males this turnover difference was only 8% (Fig. 7.5). These results reveal a possible differential pattern in species turnover for males and females which needs further analysis to corroborate a greater enhancement (facilitation) of plant community heterogeneity by female individuals as compared to male individuals (Perea et al. unpublished results). These sex-related differences in dioecious species tend to be associated with their distinct nutritional requirements (Marques et al. 2002), their sex-biased plant-animal interactions (Verdú and García-Fayos 2003), or their dissimilar physiology (Boecklen et al. 1990).

4 Conclusions

Overall, we summarize numerous facilitative interactions associated with the genus *Baccharis*, which are strongly dependent on multiple intrinsic and extrinsic factors (e.g., its size, sex and architectural form, the type and intensity of stress, the ontogenetic stage of beneficiary plants, the presence of conspecifics). Thus, the genus *Baccharis* includes extraordinary interesting species from an ecological, conservation, and restoration point of view due to their ability to work as nurse plants that favor the regeneration of keystone species and help reduce the proliferation of invasive plants in highly diverse systems. We encourage further research into the ecological and conservational role of *Baccharis* species as they are widespread and pioneer plants that are typically easy and inexpensive to propagate (Gomes and Fernandes 2002). As a result, plants in the genus *Baccharis* have an enormous potential to restore degraded and disturbed areas and to control plant invasion or recuperate areas already invaded by exotic plants. The facilitation process represents an incredible restoration mechanism that should be considered in future environmental and ecological projects where *Baccharis* species can play a prominent role in many areas of the Americas.

Acknowledgments We thank AEET (Spanish Society of Terrestrial Ecology) and the European Commission's Framework Programme 7 through the Marie Curie Actions (FP7-PEOPLE-2013-IOF-627450) for their financial support. MP acknowledges the financial support of the Spanish Ministry of Education (FPU13/00567 and EST16/00095). We thank the students from Stanford University (USA) and Universidade Federal de Minas Gerais (Brazil) for help with fieldwork. GWF thanks CNPq, Fapemig, Vale, and Anglo American for grant support.

References

Anthelme F, Michalet R, Saadou M (2007) Positive associations involving the tussock grass *Panicum turgidum* Forssk. in the Aïr-Ténéré Reserve, Niger. J Arid Environ 68:348–362

Armas C, Pugnaire FI (2005) Plant interactions govern population dynamics in a semi-arid plant community. J Ecol 93:978–989

Baraloto C, Bonal D, Goldberg DE (2006) Differential seedling growth response to soil resource availability among nine neotropical tree species. J Trop Ecol 22:487–497

Bartholomew B (1970) Bare zone between California shrub and grassland communities: the role of animals. Science 170:1210–1212

Bertness MD, Callaway R (1994) Positive interactions in communities. Trends Ecol Evol 9:191–193

Bertness MD, Leonard G (1997) The role of positive interactions in communities: lessons from the intertidal. Ecology 78:1978–1989

Boecklen WJ, Price PW, Mopper S (1990) Sex and drugs and herbivores: sex-biased herbivory in arroyo willow (*Salix lasiolepis*). Ecology 71:581–588

Brennan S, Laris PS, Rodrigue CM (2018) Coyote brush as facilitator of native California plant recovery in the Santa Monica Mountains. Madrono 65:47–60

Bronstein JL (2009) The evolution of facilitation and mutualism. J Ecol 97:1160–1170

Brooker RW, Maestre FT, Callaway RM, Lortie CL, Cavieres LA, Kunstler G, Liancourt P, Tielbörger K, Travis JMJ, Anthelme F et al (2008) Facilitation in plant communities: the past, the present, and the future. J Ecol 96:18–34

Bueno A, Llambí LD (2015) Facilitation and edge effects influence vegetation regeneration in old-fields at the tropical Andean forest line. Appl Veg Sci 18:613–623

Callaway RM (1992) Effect of shrubs on recruitment of *Quercus douglasii* and *Quercus lobata* in California. Ecology 73:2118–2128

Callaway RM (1995) Positive interactions among plants. Bot Rev 61:306–349

Callaway RM (2007) Positive interactions and interdependence in plant communities. Springer, Dordrecht

Callaway RM, D'Antonio CM (1991) Shrub facilitation of coast live oak establishment in central California. Madrono 38:158–169

Callaway RM, Davis FW (1998) Recruitment of *Quercus agrifolia* in central California: the importance of shrub-dominated patches. J Veg Sci 9:647–656

Castillo JP, Verdú M, Valiente-Banuet A (2010) Neighborhood phylodiversity affects plant performance. Ecology 91:3656–3663

Castro J, Zamora R, Hódar JA, Gómez JM, Gómez-Aparicio L (2004) Benefits of using shrubs as nurse plants for reforestation in Mediterranean mountains: a 4-year study. Restor Ecol 12:352–358

Castro J, Allen CD, Molina-Morales M, Marañón-Jiménez S, Sánchez-Miranda A, Zamora R (2011) Salvage logging versus the use of burnt wood as a nurse object to promote post-fire tree seedling establishment. Restor Ecol 19:537–544

Clark JS, Beckage B, Camill P, Cleveland B, HilleRis-Lambers J, Lichter J, McLachlan J, Mohan J, Wyckoff P (1999) Interpreting recruitment limitation in forests. Am J Bot 86:1–16

Collins SL, Good RE (1987) The seedling regeneration niche: habitat structure of tree seedlings in an oak-pine forest. Oikos 48:89–98

Crawley MJ (1983) Herbivory. The dynamics of plant–animal interactions. Blackwell Scientific Publications, Oxford

Cushman JH, Lortie CJ, Christian CE (2011) Native herbivores and plant facilitation mediate the performance and distribution of an invasive exotic grass. J Ecol 99:524–531

Da Costa Fonseca D, de Oliveira MLR, Pereira IM, Gonzaga APD, de Moura CC, Machado ELM (2017) Spatial pattern of *Baccharis platypoda* shrub as determined by sex and life stages. Acta Oecol 85:33–43

Davis MA, Wrage KJ, Reich PB (1998) Competition between tree seedlings and herbaceous vegetation: support for a theory of resource supply and demand. J Ecol 86:652–661

Dechoum MS, Peroni N, Pugnaire FI (2018) Factors controlling shrub encroachment in subtropical montane systems. Appl Veg Sci 21:190–197

Dreyer E, Le Roux X, Montpied P, Daudet FA, Masson F (2001) Temperature response of leaf photosynthetic capacity in seedlings from seven temperate tree species. Tree Physiol 21:223–232

Duarte LDS, Dos-Santos MM, Hartz SM, Pillar VD (2006) Role of nurse plants in Araucaria Forest expansion over grassland in south Brazil. Austral Ecol 31:520–528

Fernandes GW (2016) The shady future of the rupestrian grassland: major threats to conservation and challenges in the Anthropocene. In: Fernandes GW (ed) Ecology and conservation of mountain-top grasslands in Brazil, vol 1. Springer, Switzerland, pp 545–559

Fernandes GW, Silva JO, Espírito-Santo MM, Fagundes M, Oki Y, Carneiro MAA (2014) *Baccharis*: a Neotropical model system to study insect plant interactions. In: Fernandes GW, Santos JC (eds) Neotropical insect galls. Springer, Dordrecht, pp 193–219

Fernandes GW, Banhos A, Barbosa NPU, Barbosa M, Bergallo HG, Loureiro CG, Overbeck GE, Solar R, Strassburg BBN, Vale MM (2018) Restoring Brazil's road margins could help the country offset its CO_2 emissions and comply with the Bonn and Paris agreements. Perspect Ecol Consev 16:105–112

Franco AC, Nobel PS (1988) Interactions between seedlings of *Agave deserti* the nurse plant *Hilaria Rigida*. Ecology 69:1731–1740

Franks SJ (2003) Competitive and facilitative interactions within and between two species of coastal dune perennials. Can J Bot 81:330–337

Goldberg DE, Turkington R, Olsvig-Whittaker L, Dyer AR (2001) Density dependence in an annual plant community: variation among life history stages. Ecol Monogr 71:423–446

Gomes V, Fernandes GW (2002) Germinação de aquênios de *Baccharis dracunculifolia* D.C. (Asteraceae). Acta Bot Bras 16:421–427

Gómez-Aparicio L, Zamora R, Castro J, Hódar JA (2008) Facilitation of tree saplings by nurse plants: microhabitat amelioration or protection against herbivores? J Veg Sci 19:161–172

Gómez-González S, Cavieres LA, Torres P, Torres-Díaz C (2009) Competitive effects of the alien invasive *Centaurea solstitilis* L. on two Chilean *Baccharis* species at different life-cycle stages. Gayana Bot 66:71–83

González-Rodríguez V, Villar R, Casado R, Suárez-Bonnet E, Quero JL, Navarro-Cerrillo RM (2011) Spatio-temporal heterogeneity effects on seedling growth and establishment in four *Quercus* species. Ann For Sci 68:1217–1232

Grubb PJ (1977) The maintenance of species-richness in plant communities: the importance of the regeneration niche. Biol Rev 52:107–145

He Q, Bertness MD, Altieri AH (2013) Global shifts towards positive species interactions with increasing environmental stress. Ecol Lett 16:695–706

Heiden G, Antonelli A, Pirani JR (2019) A novel phylogenetic infrageneric lassification of *Baccharis* (Asteraceae: Astereae), a highly diversified American genus. Taxon. https://doi.org/10.1002/tax.12128

Hobbs RJ, Mooney HA (1986) Community changes following shrub invasion of grassland. Oecologia 70:508–513

Holmgren M, Scheffer M (2010) Strong facilitation in mild environments: the stress gradient hypothesis revisited. J Ecol 98:1269–1275

Holmgren M, Scheffer M, Huston MA (1997) The interplay of facilitation and competition in plant communities. Ecology 78:1966–1975

Kitzberger T, Steinaker DF, Veblen TT (2000) Effects of climatic variability on facilitation of tree establishment in northern Patagonia. Ecology 81:1914–1924

López-Sánchez A, Peláez M, Dirzo R, Fernandes GW, Seminatore M, Perea R (2019) Spatio-temporal variation of biotic and abiotic stress agents determines seedling survival in assisted oak regeneration. J Appl Ecol 56:2663–2674

Macek P, Schöb C, Núñez-Ávila M, Hernández Gentina IR, Pugnaire FI, Armesto JJ (2018) Shrub facilitation drives tree establishment in a semiarid fog-dependent ecosystem. Appl Veg Sci 21:113–120

Marques AR, Fernandes GW, Reis IA, Assunção RM (2002) Distribution of adult male and female *Baccharis concinna* (Asteraceae) in the rupestrian fields of Serra do Cipó, Brazil. Plant Biol 4:94–103

Martínez E, Fuentes E (1993) Can we extrapolate the California model of grassland-shrubland ecotone? Ecol Appl 3:417–423

McBride JR (1974) Plant succession in the Berkeley Hills, California. Madrono 22:317–380

McBride J, Heady HF (1968) Invasion of grassland by *Baccharis pilularis* DC. J Range Manag 21:106–108

Michalet R, Schob C, Lortie CJ, Brooker RW, Callaway RM (2013) Partitioning net interactions among plants along altitudinal gradients to study community responses to climate change. Funct Ecol 28:75–86

Peláez M, Dirzo R, Fernandes GW, Perea R (2019) Nurse plant size and biotic stress determine quantity and quality of plant facilitation in oak savannas. For Ecol Manag 437:435–442

Perea R, Gil L (2014) Shrubs facilitating seedling performance in ungulate-dominated systems: biotic versus abiotic mechanisms of plant facilitation. Eur J For Res 133:525–534

Perea R, López-Sánchez A, Roig S (2016) The use of shrub cover to preserve Mediterranean oak dehesas: a comparison between sheep, cattle and wild ungulate management. Appl Veg Sci 19:244–256

Perea R, López-Sánchez A, Dirzo R (2017) Differential tree recruitment in Californian oak savannas: are evergreen oaks replacing deciduous oaks? For Ecol Manag 399:1–8

Perea R, Cunha JS, Spadeto C, Gomes VM, Moura AL, Rúbia B, Fernandes GW (2019) Nurse shrubs to mitigate plant invasion along roads of montane Neotropics. Ecol Eng 136:193–196

Prentice IC, Werger MJA (1985) Clump spacing in a desert dwarf shrub community. Vegetatio 63:133–139

Pugnaire FI, Haase P, Puigdefábregas J (1996) Facilitation between higher plant species in a semi-arid environment. Ecology 77:1420–1426

Ramírez-Marcial N, González-Espinosa M, García-Moya E (1996) Establecimiento de *Pinus* spp y *Quercus* spp. en matorrales y pastizales de Los Altos de Chiapas. Agrociencia 30:249–257

Rodríguez-Calcerrada J, Pardos JA, Gil L, Beich PB, Aranda I (2008) Light response in seedlings of a temperate (*Quercus petraea*) and a sub-Mediterranean species (*Quercus pyrenaica*): contrasting ecological strategies as potential keys to regeneration performance in mixed marginal populations. Plant Ecol 195:273–285

Rudgers JA, Maron JL (2003) Facilitation between coastal dune shrubs: a non-nitrogen fixing shrub facilitates establishment of a nitrogen-fixer. Oikos 102:75–84

Sánchez-Velásquez LR, Domínguez-Hernández D, Pineda-López MDR, Lara-González R (2011) Does *Baccharis conferta* shrub act as a nurse plant to the *Abies religiosa* seedling? Open For Sci J 4:67–70

Schemske DW, Husband BC, Ruckelshaus MH, Goodwillie C, Parker IM, Bishop JG (1994) Evaluating approaches to the conservation of rare and endangered plants. Ecology 75:584–606

Silvertown J, Charlesworth D (2001) Introduction to plant population biology, 4th edn. Blackwell, Oxford

Smit C, den Ouden J, Muller-Scharer H (2006) Unpalatable plants facilitate tree sapling survival in wooded pastures. J Appl Ecol 43:305–312

Smit C, Vandenberghe C, Ouden den J, Müller-Schärer H (2007) Nurse plants, tree saplings and grazing pressure: changing facilitation along a biotic environmental gradient. Oecologia 152:265–273

Smit C, den Ouden J, Díaz M (2008) Facilitation of holm oak recruitment by shrubs in Mediterranean open woodlands. J Veg Sci 19:193–200

Smither-Kopperl M (2016) Plant guide for coyotebrush (*Baccharis pilularis*). USDA-Natural Resources Conservation Service, Lockeford Plant Materials Center, Lockeford

Soliveres S, Maestre FT (2014) Plant-plant interactions, environmental gradients and plant diversity: a global synthesis of community-level studies. Perspect Plant Ecol Evol Syst 16:154–163

Soliveres S, Eldridge DJ, Maestre FT, Bowker MA, Tighe M, Escudero A (2011) Microhabitat amelioration and reduced competition among understorey plants as drivers of facilitation across environmental gradients: towards a unifying framework. Perspect Plant Ecol Evol Syst 13:247–258

Soliveres S, Smit C, Maestre FT (2015) Moving forward on facilitation research: response to changing environments and effects on the diversity, functioning and evolution of plant communities. Biol Rev 90:297–313

Spadeto C, Fernandes GW, Negreiros D, Kunz SH (2017) Facilitative effects of tree species on natural regeneration in an endangered biodiversity hotspot. Braz J Bot 40:943–950

Stachowicz JJ (2001) Mutualism, facilitation, and the structure of ecological communities: positive interactions play a critical, but underappreciated, role in ecological communities by reducing physical or biotic stresses in existing habitats and by creating new habitats on which many species depend. Bioscience 51:235–246

Steinberg PD (2002) *Baccharis pilularis*. In: Fire Effects Information System, [Online]. U.S. Department of Agriculture, Forest Service, Rocky Mountain Research Station, Fire Sciences Laboratory (Producer)

Tirado R, Pugnaire FI (2005) Community structure and positive interactions in constraining environments. Oikos 111:437–444

Valiente-Banuet A, Ezcurra E (1991) Shade as a cause of the association between the cactus *Neobuxbaumia tetezo* and the nurse plant *Mimosa luisana* in the Tehuacan Valley, Mexico. J Ecol 79:961–971

van Zonneveld MJ, Gutiérrez JR, Holmgren M (2012) Shrub facilitation increases plant diversity along an arid scrubland–temperate rain forest boundary in South America. J Veg Sci 23:541–551

Verdú M, García-Fayos P (2003) Frugivorous birds mediate sex-biased facilitation in a dioecious nurse plant. J Veg Sci 14:35–42

Whittaker RH, Levin SA (1977) The role of mosaic phenomena in natural communities. Theor Popul Biol 12:117–139

Zavaleta ES, Kettley LS (2006) Ecosystem change along a woody invasion chronosequence in a California grassland. J Arid Environ 66:290–306

Chapter 8
Biological Invasion by *Baccharis*

Adrián Lázaro-Lobo, Gary N. Ervin, Lidia Caño, and F. Dane Panetta

Abstract In this chapter, we present and discuss information regarding biological invasions by species in the genus *Baccharis* L. around the world: in native, expansive, and introduced distributional ranges. *Baccharis halimifolia* L. is the invasive species par excellence of this genus. Therefore, we dedicate a great part of the chapter to describe (1) its distribution and introduction history; (2) abiotic and biotic factors that affect its invasion; (3) types of ecosystems invaded and environmental, economic, and social impacts; and (4) management of the species. Lastly, we collate all the available information in the literature regarding other species of this genus that are considered invasive or potentially invasive in both native and introduced areas. Those species are *Baccharis coridifolia* DC., *Baccharis dracunculifolia* DC., *Baccharis neglecta* Britton., *Baccharis pilularis* DC., *Baccharis pteronioides* DC., *Baccharis salicifolia* (Ruiz & Pav.) Pers., *Baccharis salicina* Torr. & A.Gray, *Baccharis sarothroides* A.Gray, *Baccharis spicata* (Lam.) Baill., and *Baccharis ulicina* Hook. & Arn.

Keywords Exotic species · Invasive character · Invasion history · Management Strategies · Plant distribution

A. Lázaro-Lobo (✉) · G. N. Ervin
Mississippi State University, Mississippi State, MS, USA

L. Caño
Universidad del País Vasco, San Sebastián, Spain

F. D. Panetta
The University of Melbourne, Melbourne, Australia

G. W. Fernandes et al. (eds.), *Baccharis*,
https://doi.org/10.1007/978-3-030-83511-8_8

1 Invasive Character of the Genus *Baccharis*

Multiple species of the genus *Baccharis* share life-history characteristics common to many invasive species, including effective dispersal mechanisms, adaptations to pioneer stages of succession, high competitive ability, and production of allelopathic compounds (Westman et al. 1975; Ibáñez and Zoppolo 2011; Caño et al. 2013a). Female shrubs produce numerous seeds that can travel long distances via several vectors (Panetta 1977; Weber 2003; USDA 2018) and rapidly colonize disturbances that occur over multiple scales. Seeds germinate under a wide range of environmental conditions and, as time passes, dense stands develop that prevent other species from establishing (Fried and Panetta 2016). While plants in this genus reproduce mainly by seeds, several species have the capacity to reproduce vegetatively by sprouting after cutting or burning (Westman et al. 1975; Hobbs and Mooney 1985; Grace et al. 2001). Thus, these species respond well to any form of disturbance, such as fire, animal activity (e.g., grazing and burrowing), or the biomass removal pursuant to targeted control.

Another characteristic that makes some species of this genus invasive is their generalist behavior, based upon morphological and physiological plasticity (Panetta 1977; Caño et al. 2016), which allows them to thrive in a wide variety of environmental conditions and endure rapid changes in the environment (Westman et al. 1975; Tucat 2015; Haque et al. 2008). However, the species of this large genus have evolved to invade different types of ecosystems. For instance, *B. halimifolia* sometimes forms dense monospecific stands in places with relatively high salinity levels (Caño et al. 2016), *B. salicifolia* is commonly found growing along waterways (Dimmit 2000), and *B. sarothroides* dominates desert regions partly because of its long root system which allows it to reach water and nutrients stored in deep parts of the soil (Dimmit 2000; Haque et al. 2008). Although the genus is generally evergreen, some species can also be deciduous as an adaptation to withstand less favorable environmental conditions. For example, *B. halimifolia* is deciduous in the cooler parts of its distributional range (Sims-Chilton and Panetta 2011), and *B. sarothroides* loses its leaves under drought conditions during the summer period (Virginia Tech 2018). Allelopathy is another mechanism by which invasive plants displace other species (Orr et al. 2005). Several studies show that some species of this genus, such as *B. dracunculifolia* and *B. ulicina*, produce secondary metabolites that negatively impact neighboring plants (Tucat 2015; Ibáñez and Zoppolo 2011). Lastly, this plant genus generally has a low palatability for herbivores, and some of its species are even toxic to them (Boldt 1989; Jarvis et al. 1991; USDA 2018), which increases their potential to outcompete other plants that are preferred by herbivores.

All the abovementioned characteristics make multiple species of *Baccharis* particularly important from an ecological and an economic point of view, not only in the invaded regions of the world but also within their native distributional ranges. The major invasive *Baccharis* species are shown in Fig. 8.1.

Fig. 8.1 Photos of the major invasive *Baccharis* species (with the exception of *B. pteronioides*): (**a**) *B. coridifolia*, (**b**) *B. dracunculifolia*, (**c**) *B. halimifolia*, (**d**) *B. neglecta*, (**e**) *B. pilularis*, (**f**) *B. salicifolia*, (**g**) *B. salicina*, (**h**) *B. sarothroides*, (**i**) *B. spicata*, and (**j**) *B. ulicina*. (Photos courtesy of G. Heiden)

Fig. 8.1 (continued)

2 *Baccharis halimifolia* L. (Eastern *Baccharis*, Groundsel Bush, Saltbush)

Distribution

Native Range

Baccharis halimifolia is native to the Atlantic and Gulf Coasts of North and Central America (Cronquist 1980; Sundberg and Bogler 2006; Fig. 8.2). It is widely distributed in areas of Nova Scotia (southeastern Canada), eastern and southern United States, eastern Mexico (especially northeastern, but it can also be found in areas near Veracruz, southeastern Mexico), the Bahamas, and Cuba (USDA 2018). This shrub occurs in areas from 0 to 100 m above sea level (Sundberg and Bogler 2006). During the last century, it has been expanding its native distributional range towards

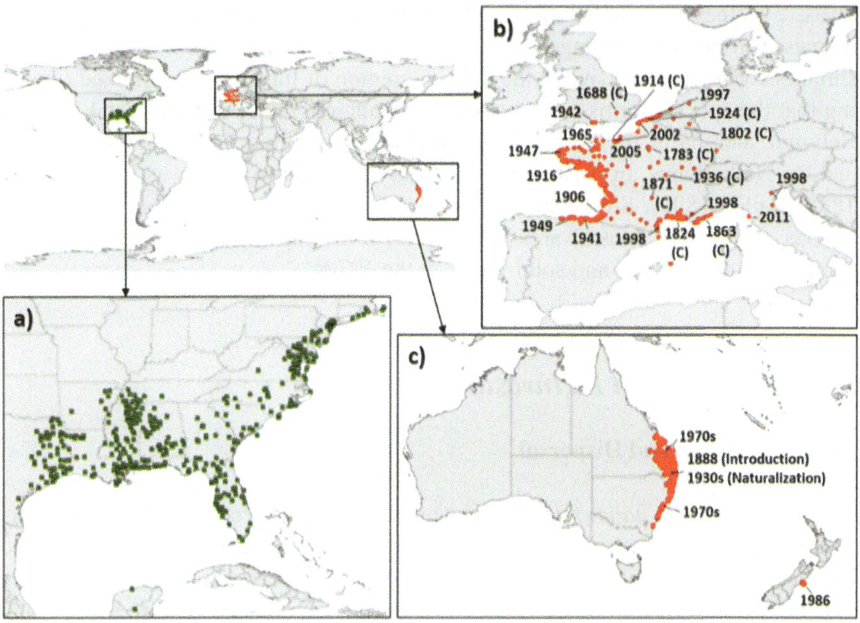

Fig. 8.2 Distributional ranges of *B. halimifolia* and earliest dates when this species was reported in the invasive regions. Regional maps correspond to (**a**) North America (occurrence points taken from Ervin (2009) and SERNEC Data Portal (2018)), (**b**) Europe (occurrence points taken from Fried et al. (2016) and dates from Caño et al. (2013a) and Fried et al. (2016)), and (**c**) Australia (occurrence points taken from Atlas of Living Australia website (2018) and dates from Bailey (1899), Sims-Chilton and Panetta (2011) , and Atlas of Living Australia website (2018)). In panel (**b**), records of *B. halimifolia* accompanied by "(C)" indicate regions where *B. halimifolia* is known only from cultivated populations

interior areas of the United States (Duncan 1954; Estes 2004), following human-induced disturbances such as reductions in tree canopy cover and increases in edge habitat (Ervin 2009).

Introduced Range

This species has invaded multiple regions across the world, including western Europe (France, northern Spain, northern Italy, southern England, Belgium, and the Netherlands), eastern Australia (Queensland and New South Wales), New Zealand, and the Republic of Georgia, where it has been introduced for ornamental use, soil stabilization, windbreaks, or aesthetic purposes (Caño et al. 2013a; Fried et al. 2016). Caño et al. (2013a) and Fried et al. (2016) conducted a comprehensive study regarding the history of *B. halimifolia* introduction and spread throughout western Europe. They pointed out that *B. halimifolia* became naturalized in northern Spain at the end of the nineteenth century. In France, this species was considered as locally invasive in the 1940s, reached the Mediterranean coast in the 1980s, and rapidly increased in numbers during the 1990s. Fried et al. (2016) also indicated that this shrub is considered invasive in the Tuscany region of Italy, it is naturalized (meaning that the species forms self-sustaining populations) in multiple coastal areas of Belgium, and that scattered individuals have been reported from southern England and the Netherlands. In Australia, it is believed that the species was introduced in 1888 (Bailey 1899). Sims-Chilton and Panetta (2011) specified that it became a serious problem in coastal areas of southeastern Queensland by the 1930s and spread both northwards and southwards in the 1970s.

Factors That Affect Its Invasion

Seed Production and Dispersal

Female shrubs of *B. halimifolia* produce numerous viable seeds annually, beginning as early as 3 years after germination (USDA 2018). Westman et al. (1975) pointed out that each plant can produce up to 1.5 million seeds per year. Boldt (1989) found that 4-year-old plants produced 31% more seeds than plants that were 9 years old, indicating that seed production decreases as the plants age. Seeds can travel long distances carried by the wind, water, animals, or vehicles (Panetta 1977; Weber 2003); however, Parsons and Cuthbertson (1992) indicated that with a steady breeze of 16 kph, most seeds disperse less than 6 meters from their mother shrub. Diatloff (1964) recorded that 2-meter-high plants can disperse their seeds to distances of 140 m. Seeds germinate without undergoing a period of dormancy and under a wide range of environmental conditions (Westman et al. 1975; USDA 2018), which will be explained in more detail in subsequent sections of the chapter.

Abiotic Factors

Temperature The most suitable areas for *B. halimifolia* in both the native and the invasive range occur in temperate to subtropical regions, including the Mediterranean areas in Europe and Australia (Sims-Chilton et al. 2010; Caño et al. 2013a; Fried et al. 2016). Modelling has shown a higher probability of presence when Maximum Temperature of Warmest Month is between 20 and 30 °C and Minimum Temperature of Coldest Month is above 0 °C (Fried et al. 2016). These data, along with the fact that the species is not present at high latitudes (up to 42° in North America and 51° in Europe), suggest that cold temperatures and frosts could limit its extension northwards in the Northern Hemisphere. However, the species has been also considered to be resistant to −15 °C (Huxley 1992). The cold resistance of the species has not been directly assessed in the field, neither has it been experimentally tested. Likewise, little is known about the physiological responses of *B. halimifolia* to temperature variations. Only at the seed stage Westman et al. (1975) and Panetta (1979a) demonstrated that optimal germination occurs between 15 and 20 °C. At the population level, Sims-Chilton et al. (2009) found that plant density increases with temperature, whereas plants' size seems to be unaffected.

Light Availability The fact that *B. halimifolia* colonizes open and disturbed habitats, often associated with other woody plants, as well as forests or pine plantations, demonstrates its high plasticity in response to light availability.

Experiments under controlled conditions have demonstrated that there is indeed a degree of shade tolerance during establishment (Panetta 1977). Plastic responses of leaf traits (e.g., increased specific leaf area) allow high shade tolerance (17% daylight) during the first stages of seedling growth, but older seedlings' shade tolerance decreases, probably because the morphological response is no longer enough to compensate for the decrease in net assimilation rate (Panetta 1977). The plasticity of leaf traits, such as specific leaf area and stomatal conductance, has been also found to contribute to the shade tolerance of *B. halimifolia* in estuarine communities (Pivovaroff et al. 2015). However, viable seed production is highly reduced by canopy closure (Westman et al. 1975; Panetta 1979b). Interestingly, Lázaro-Lobo et al. (2020) found shade-adaptive effects of parental environment on *B. halimifolia* offspring. In their study, progeny from low maternal light conditions performed better in the shade treatment than did those offspring from maternal plants grown under high light conditions, whereas the opposite pattern was found in high light conditions.

In the field, colonization patterns might not only reflect the ability to respond to light conditions but also could reflect the outcome of different competitive interactions and responses to disturbance or stressors (see sections below). For instance, Panetta (1979c) indicated that, since canopy closure did not affect population size structure in pine stands, other factors such as litter accumulation or the occurrence of disturbances at the soil level might determine the survival or productivity of this species.

Nutrient Availability The presence of *B. halimifolia* does not seem to be limited by nutrient availability since it is able to colonize different types of soils. In Australia, it is recorded from dry infertile forest soils to rich volcanic loams and low-lying clay soils with high moisture content (Winders 1937, cited in Sims-Chilton and Panetta 2011). In coastal Mississippi, USA, *B. halimifolia* occurrence was not affected by carbon-to-nitrogen ratio (Paudel and Battaglia 2015). Likewise, the species was found to colonize different sites in Queensland where soil nitrogen and phosphorus ranged, respectively, from 560 to 5500 ppm and from 4 to 73 ppm (Westman et al. 1975). In contrast, *B. halimifolia*'s productivity does respond positively to high levels of N availability under experimental conditions (Vick and Young 2013) and in the field (Connor and Wilson 1968). However, demand for P may increase under high levels of N concentration, moderating any potential increase in *B. halimifolia*'s growth (Westman et al. 1975; Vick and Young 2013). Also, recurrent flooding in coastal communities can reduce P assimilation and root growth (McKee et al. 2002). However, since, in its native range, the roots of

B. halimifolia are colonized by arbuscular mycorrhizal fungi (AMF, Paudel et al. 2014), it is possible that mutualistic associations with AMF would enhance nutrient uptake or contribute to salt or flooding stress tolerance (Neto et al. 2006).

Salinity *Baccharis halimifolia* is highly tolerant to salinity, but it is a facultative halophyte, i.e., its optimal growth occurs in the absence of salinity (Caño et al. 2016). For this reason, this species colonizes different types of soils where salinity values range from 0 to sea water salt concentrations (Westman et al. 1975; Young et al. 1994; Caño et al. 2013b, 2014; Frau et al. 2014), although the highest occurrence of *B. halimifolia* has been recorded under low to moderate levels of salinity (Young et al. 1994; Caño et al. 2014; Paudel and Battaglia 2015). On the other hand, threshold levels of tolerance to salt stress exposure can prevent colonization through massive mortality at the halophilous end of the gradient (Caño et al. 2013a). As in most non-obligate halophytic species, exposure of *B. halimifolia* to salt stress actually reduces root and shoot biomass and triggers a set of physiological responses affecting leaf traits, water relations, photosynthesis, and osmolyte accumulation, both in the field and under controlled conditions (Young et al. 1994; Tolliver et al. 1997; Zinnert et al. 2012; Fuertes-Mendizabal et al. 2014; Caño et al. 2013b, 2014, 2016).

Experiments under controlled conditions have demonstrated low mortality at high salinity levels (20 g NaCl/L) and an ability to germinate and reproduce under saline conditions (Tolliver et al. 1997; Paudel and Battaglia 2013; Caño et al. 2016). Paudel and Battaglia (2013) also reported that germination can be unaffected by salinity concentrations up to 20 g/L. However, others have shown that percent germination of *B. halimifolia* can be drastically reduced by salinity levels as low as 10 g/L (Young et al. 1994; Lázaro-Lobo et al. 2020). Environmental salinity of the maternal plants can highly affect progeny tolerance to salinity. For example, Lázaro-Lobo et al. (2020) found that seeds produced by maternal plants growing in saline areas had higher and faster germination in saline environments than seeds from plants growing in non-saline or subsaline areas. Moreover, seedlings are more likely to establish in saline environments if they are exposed to salinity since germination and, therefore, have more time to acclimate to this abiotic stressor (Lázaro-Lobo et al. 2020). Salinity levels at the upper end of the tolerance range of *B. halimifolia* are likely to delay or suppress flowering, but moderate salinity levels can trigger flowering, both in the greenhouse (Caño et al. 2016) and in the field (Caño et al. 2013b).

A set of physiological adaptations that characterize halophytic species and a degree of plasticity have been shown to underlie the high salt tolerance displayed by this species. *B. halimifolia* reduces specific leaf area and increases both leaf succulence and stomatal density under saline conditions, which may facilitate a better regulation of stomatal function and transpiration in order to avoid a greater NaCl concentration in tissues (Fuertes-Mendizabal et al. 2014). This shrub behaves as a salt-accumulating plant (Zinnert et al. 2012; Caño et al. 2016), and it has the ability to synthesize high levels of proline that likely act as an osmolyte or osmoprotectant

(Fuertes-Mendizabal et al. 2014). It can resist high salt concentrations in its tissues and simultaneously maintain low photosynthetic activity without necrosis, although it shows no capacity for salt exclusion (Zinnert et al. 2012; Fuertes-Mendizabal et al. 2014).

Water Availability (Floods and Droughts) Drought stress or anoxic stress by waterlogging might depend on precipitation and evapotranspiration in the upper zone of marshes and non-saline soils and on tidal influence in moderate-high salinity communities (Caño et al. 2013b; Pivovaroff et al. 2015). Pivovaroff et al. (2015) found that at the landward edge of *B. halimifolia*'s range, physiological adjustments to water stress were achieved through greater cavitation resistance. Concerning other hydrodynamic variables, Frau et al. (2014) showed that invasive estuarine populations in Spain occur in areas that are inundated <26% of the year, with water speed and water flow <0.1 m/s and <0.85 m^3/s, respectively.

Biotic Factors

Competition with Other Plants Despite the wide ecological amplitude of *Baccharis halimifolia* regarding light or nutrient availability and different stressors, establishment in the field is often determined by the competitive relations established with neighboring plants in different kinds of habitats. *B. halimifolia* can easily outcompete herbaceous species and take advantage of disturbance, but competition with woody species limits its establishment and reproduction (Ervin 2009; Caño et al. 2013a). In coastal communities in North America, interspecific differences in response to flooding and salinity underlie the zonation among different shrubs, and thus *B. halimifolia* typically inhabits the intermediate marsh zone, together with the salt-tolerant shrub *Iva frutescens* (Young et al. 1994; Tolliver et al. 1997). Although competition between these species can prevent the formation of monospecific stands at the subhalophilous zone of the marsh in North America, shrub and tree vegetation are absent in these communities in Europe. Here *B. halimifolia* outcompetes the dominant native herbaceous species (*Juncus maritimus*, *Elytrigia* spp., and eventually *Phragmites australis*), both in the Atlantic and Mediterranean coastal wetlands. It establishes in small gaps and spreads in these communities, even in almost undisturbed sites (Caño et al. 2013a; Fried et al. 2014; Fried and Panetta 2016). While this facultative halophyte could potentially perform better in upper marsh and non-saline sites, competition with highly productive estuarine alder forests (*Alnus glutinosa*, *Salix atrocinerea*) in such sites has been shown to prevent *B. halimifolia*'s colonization in coastal communities in Spain (Caño et al. 2013a).

Herbivory In its native range, various species of Lepidoptera, Hemiptera, and Coleoptera feed on *Baccharis halimifolia* as larvae or adults. Surveys conducted in North America have identified up to 133 phytophagous insects feeding on *B. halimifolia*, of which 14 were considered restricted to the genus *Baccharis* and 11 specific to *B. halimifolia* (Palmer 1987; Palmer and Bennett 1988). Some of these insects have been introduced in Australia as biological control agents (Sims-Chilton et al. 2010; see management section). In Europe, natural enemies identified are mostly mealybugs (*Ceroplastes sinensis* and *Saissetia oleae*), aphids (*Aphis fabae*, *A. spiraecola*), and sooty molds, but the presence of an undetermined Agromyzidae has been also reported (Dauphin and Matile-Ferrero 2003; Fried et al. 2013; Caño et al. 2013b).

Despite the abundant phytophagous species found on *B. halimifolia*, the level of impact and consumption in both the native (Westman et al. 1975) and the invasive ranges does not seem to reduce the species' performance (Westman et al. 1975; Fried et al. 2013; Caño et al. 2013b; Lovet 2015). However, Kraft and Denno (1982) reported that populations defoliated by the specialist beetle *Trirhabda bacharidis* failed to flower. Acetone soluble resins in the leaves of *B. halimifolia* have been found to act as deterrents for generalist herbivores, despite being tolerated by specialists such as the beetle *T. bacharidis* (Kraft and Denno 1982).

Finally, the trophic dynamics in *B. halimifolia* have been found to be affected in a qualitative and a quantitative way by plant sex (Krischik and Denno 1990b), plant and population characteristics, climatic variables (Sims-Chilton et al. 2009), environmental factors modifying leaf chemistry or quality (Younginger et al. 2009; Caño et al. 2013b), and mutualistic associations with mycorrhizal fungi (Moon et al. 2013).

Parasites, Fungi, Bacteria, and Viruses There are almost no serious diseases affecting *B. halimifolia* in its native range (Gilman 1999). A rust fungus specific to the *Baccharis* genus (*Puccinia evadens*, Groundsel bush rust) causes defoliation during summer and winter, and, in extreme cases, stems can die back over summer (Sims-Chilton and Panetta 2011; F.D. Panetta, pers. obs.). Also, different species of nematodes have been found associated with *B. halimifolia* in Florida and Australia, but high variability across years and experimental conditions obscures any clear differences between the native and the exotic range (Porazinska et al. 2014). Although *B. halimifolia* is a natural host of the bacterium *Xylella fastidiosa*, the latter does not cause any disease in this species (Hopkins and Adlerz 1988).

Type of Ecosystems Invaded and Impacts

Baccharis halimifolia colonizes a great variety of ecosystems in both the native and the introduced distributional ranges. In the southern United States, it typically invades coastal prairies and marshes (Penfound and Hathaway 1938; Harcomb 1989; Bruce et al. 1995), overgrazed rangelands (USDA 2018), disturbed grasslands (Penfound and Hathaway 1938; Allen 1950), desert areas (Boldt 1989), hedgerows and fallow fields (Krischik and Denno 1990a), former industrial sludge basins (Olson and Fletcher 2000), and roadsides, pine plantations, and forest edges (Ervin 2009). In Australia, this shrub invades not only human-disturbed and human-managed areas such as pastures (McFadyen 1973) and exotic pine plantations (Panetta 1979c) but also *Melaleuca* swamps and dry eucalypt forests that experience periodic natural disturbances such as fire and flooding (Westman et al. 1975). In Europe, *B. halimifolia* colonizes human-disturbed areas such as road and rail networks, irrigation channels, and wastelands (Le Moigne and Magnanon 2009), as well as subhalophilous coastal communities, which are part of the protected habitat "Atlantic Salt Meadows" included in the Habitats Directive 92/43/EEC (Caño et al. 2013a). In Spain, *B. halimifolia* occurs in up to 18 sites of community importance (Campos et al. 2014), and multiple endangered halophilous species restricted to these areas are threatened by *B. halimifolia* invasion (Caño et al. 2013a, 2014).

In the invaded regions, this species causes negative effects on biodiversity, ecosystem functioning, and human activities in multiple ways. It has a direct impact on the surrounding vegetation due to its ability to form dense monospecific thickets that prevent other species from establishing (Fried and Panetta 2016). *Baccharis halimifolia* not only reduces the herbaceous diversity of the coastal prairies and estuarine communities that it invades but also converts the native herbaceous vegetation into a landscape of monospecific woody stands (Harcomb 1989; Campos and Herrera 2009; Caño et al. 2014; Fried et al. 2014; Fried and Panetta 2016). Fried et al. (2016) indicated more details about what plant communities are more affected by *B. halimifolia* invasions in their work "Monographs on Invasive Plants in Europe: *Baccharis halimifolia* L."

Impacts that this species has on vegetation also have been shown to have consequences for some animal species. For example, Arizaga et al. (2013) showed that *B. halimifolia* causes perceptible changes on bird assemblages by promoting woodland species and potentially affects migrant species associated with ecosystems that have been invaded. Mallard (2008) pointed out that insect species richness and abundance were lower in stands of *B. halimifolia* than in native woody plant species. Furthermore, the impacts caused by this invasive species are aggravated by its potential ability to affect ecosystem processes such as sedimentation dynamics (Campos and Herrera 2009), fire regimen (Sinnassamy 2004), light interception, and succession (Campos et al. 2004). Thus, Campos and Herrera (2009) considered *B. halimifolia* as a "transformer" species (sensu Richardson et al. 2000) due to its potential ability to transform the structure and function of littoral ecosystems.

In addition to its impacts on natural ecosystems, this early successional species is also considered as a pest because it rapidly invades rangelands used for livestock grazing (Westman et al. 1975; Sims-Chilton and Panetta 2011). Hence, it reduces the productivity of the pasture and limits cattle movement (Palmer and Sims-Chilton 2012). Furthermore, *B. halimifolia* has a low palatability for herbivores and is even toxic to them (Boldt 1989; USDA 2018; see also Chap. 15 in this volume). Nevertheless, seedling establishment can be greatly affected by domestic cattle (Caño et al. 2013a; Fried et al. 2016). In Spain, the progressive reduction of cattle farming promotes the invasion of disturbed communities located in the salt marsh area. However, in grazed or managed meadows, *B. halimifolia* is totally absent due to periodic disturbance (Caño et al. 2013a). This shrub also causes problems in forestry plantations due to its capability to outcompete pine seedlings (Palmer and Sims-Chilton 2012) and in salt production areas by decreasing wind velocity and evaporation of water (Fried et al. 2016). Lastly, the pollen from this shrub can reach high concentrations in the air and potentially cause allergies in sensitive persons (Green et al. 2011).

Management

Mechanical Control

Mechanical control objectives in the management of *B. halimifolia* are basically twofold: (1) to kill individual plants and (2) to suppress flowering and thereby reduce seed production and spread. Young plants (less than 1 m in height) can be pulled up with little risk of sprouting, especially when the soil profile is moist. This approach may be feasible and cost-effective for small, incipient infestations. Larger plants often regrow from any parts of the root system that have not been removed. Such a strong vegetative regeneration capacity essentially limits the scope for mechanical control when used alone for this species, but methods that combine mechanical with chemical control are more effective (see below). Timely slashing of infestations can reduce seed production, and burning may also be an effective method of control, but rapid regrowth is common (Allain and Grace 2001). Flooding for several months during winter can eliminate adult plants; permanent flooding has been used effectively in Spanish estuarine environments (Campos et al. 2014; Fried et al. 2016).

Chemical Control

Given that *B. halimifolia* was first recognized as a serious weed in Australia (having been declared noxious in the 1950s), chemical control methods were first developed there and became a major component of its management. The plant was readily controlled by overall spray application of either 0.2% salts or esters of 2,4-D ((2,4-dichlorophenoxy) acetic acid) or 2,4,5-T ((2,4,5-trichlorophenoxy) acetic acid). Basal barking with esters in oil and cut-stumping using salts in water and esters in water or oil were also effective control methods (Sims-Chilton and Panetta 2011). Other work has reported effectiveness of dicamba plus MCPA, glyphosate, picloram plus 2,4-D, and triclopyr (Weber 2003). In hardwood forest plantations in southeastern Arkansas, Gann et al. (2012) found triclopyr to be more efficient in controlling *B. halimifolia* than imazamox, aminopyralid, and glyphosate.

During the 1950s in Australia, "brushing" was the most common method used, consisting of cutting plants and swabbing their stems with chemicals. Today cut-stumping, i.e., application of relatively concentrated herbicide solutions to the stumps of plants just after cutting, is a method commonly used in environmentally sensitive areas, such as nature reserves. When used properly this method presents less risk of off-target damage than foliar applications of more dilute solutions. Glyphosate is the most common herbicide used for cut-stumping in France and Spain (Fried et al. 2016). Follow-up treatment is essential because infestations of *B. halimifolia* develop persistent seed banks (Panetta 1979a). This will likely combine hand pulling of young plants with foliar spraying of regrowth from plants not killed by the previous treatment.

Biological Control

Surveys for potential biological control agents for *B. halimifolia* were initiated in the southern United States in the 1960s. These surveys, undertaken by researchers from the Queensland Government, continued for several decades. Overall, 35 agents were imported into Australia for testing, 14 were released, and 7 have established (Sims-Chilton et al. 2009). One of these was the pathogen *Puccinia evadens*, which has established over most of *B. halimifolia* distribution in Australia (Sims-Chilton et al. 2009). The other six were insects, including three species of Lepidoptera (*Aristotelia ivae* (Gelechiidae), *Bucculatrix ivella* (Bucculatricidae), and *Hellinsia balanotes* (Pterophoridae)); two species of Coleoptera (*Megacyllene mellyi* (Cerambycidae) and *Trirhabda bacharidis* (Chrysomelidae)); and a dipteran (*Rhopalomyia californica* (Cecidomyiidae)). All are native to North America, except for *M. mellyi* which is South American.

Biological control impacts on *B. halimifolia* vary in relation to several environmental factors (Sims-Chilton et al. 2009), but overall there has been a marked decline in the abundance of the weed throughout its Australian range. This decline is at least partially due to a long-term decrease in the climatic suitability of the invaded areas (Sims-Chilton et al. 2010). Regardless, the Australian experience reveals several potentially effective agents in the event that biological control is attempted elsewhere. In France, sheep have intentionally been used to control sprouting after application of physical methods on large areas (Fried et al. 2016).

3 Species of the Genus *Baccharis* That Are Invasive Within Their Native Distributional Range

Baccharis coridifolia *DC. (Mio-mio)*

Baccharis coridifolia is native to Paraguay, northern and central Argentina, Uruguay, and southern Brazil (Mongelli et al. 1997; Rizzo et al. 1997). This species can readily colonize abandoned fields (Sione et al. 2016) and thrives in pastures where livestock reduce the cover of palatable grasses (Berretta 2001). *Baccharis coridifolia* is primarily considered invasive due to its toxic effects in livestock, especially during the flowering season (April–May; Rizzo et al. 1997). In fact, it is one of the most important toxic plants within its native range (de Almeida et al. 2009), and several works have documented its negative effects in livestock (e.g., Tokarnia and Döbereiner 1975; Habermehl et al. 1985; Costa et al. 1995), as it is explained in another section of the book (Chap. 15).

Baccharis dracunculifolia *DC. (White Chilca)*

Baccharis dracunculifolia occurs in the southern region of South America, including Argentina, Paraguay, southern Brazil, Uruguay, and Bolivia (Lombardo 1964; Barroso 1976; cited in Ibáñez and Zoppolo 2011; Müller 2006). This shrub has high reproductive rates and dispersal capacity (Frenedozo 2004). Female plants produce large numbers of seeds that germinate readily (Gomes and Fernandes 2002). It has been recognized as an invasive and colonizing species on several occasions in some areas within its native range, partly because of its efficient establishment and growth in disturbed habitats (Dos Santos et al. 2008; Galindez et al. 2009). Frenedozo (2004) included *B. dracunculifolia* as one of the few colonizer species that appeared at limestone mining quarries. Ibáñez and Zoppolo (2011) examined the allelopathic properties of *B. dracunculifolia* and its phytotoxic effects on other species. They concluded that germination and growth of other plants were significantly inhibited by the essential oil of this shrub, which would explain the reduction of weeds near *B. dracunculifolia*. Even though *B. dracunculifolia* is an obligate seeder species and does not sprout after fire (Overbeck and Pfadenhauer 2007), its high seedling establishment after burning allows this shrub to compete with other vegetatively propagated species (Galindez et al. 2009). Furthermore, *B. dracunculifolia* is adapted to a wide range of soil conditions. It efficiently colonizes high fertility agriculture fields (Macedo et al. 2003) and degraded areas with low nutrient availability (Dos Santos et al. 2008). Negreiros et al. (2012) examined the survival and early growth of *B. dracunculifolia* seedlings grown across a gradient of nutrient availability. The results showed that seedlings growing on less fertile soils exhibited the highest survival rates. However, seedlings had a higher growth rate and accumulation of biomass on more fertile substrates. Finally, *B. dracunculifolia* has been proposed, among other species, as a helpful species to regenerate disturbed areas such as arsenic-contaminated areas and overburden piles produced by surface mining, as part of revegetation programs within its native range (Dos Santos et al. 2008; Gilberti et al. 2014).

Baccharis neglecta *Britton. (Roosevelt Weed)*

Baccharis neglecta is mainly found in open habitats from the southwestern and south-central United States to Coahuila, Chihuahua, and Durango, northern Mexico (Matuda 1957; Correll and Johnston 1979; cited in Boldt and Robbins 1987; Van

Auken and Bush 1990). The species has been reported as invasive within its native distributional range on several occasions (Hamilton et al. 2004, U.S. Department of Homeland Security 2004), primarily in overgrazed or disturbed productive rangelands, where it causes negative economic effects (Everitt et al. 1978). According to Mutz et al. (1979), livestock can occasionally graze upon young plants, but the species has little or no nutritional value.

Van Auken and Bush (1990) conducted an experiment to evaluate the light requirements of *B. neglecta* seedlings, obtaining higher values of number of leaves, stem length, basal diameter, and above- and belowground biomass from the plants submitted to the highest light treatment (photon flux densities (PPFD) of 611 $\mu M \cdot m^{-2} \cdot sec^{-1}$). Furthermore, seedling mortality was very high under the lower light treatments (PPFD <1 and PPFD = 53 $\mu M \cdot m^{-2} \cdot sec^{-1}$). All deaths occurred within 3 weeks of the start of the experiment, and all the plants died under the lowest light level. The above results show that *B. neglecta* is a sun-tolerant plant, or heliophyte, which is favored by removal of native vegetation by disturbances such as heavy grazing (Van Auken and Bush 1990; Hamilton et al. 2004).

It appears that *B. neglecta* can be controlled by maintaining a solid grass cover, decreasing the water available in the upper soil profile (Van Auken and Bush 1990). Other alternatives to prevent the establishment and spread of *B. neglecta* are mechanical removal, chemical treatments, and biological control. Individuals can be temporarily controlled by shredding or burning, but they will sprout in a short period of time (Boldt and Robbins 1987). Herbicides such as 2,4-D, picloram, and tebuthiuron are very effective against this species, but expensive (Scifres 1980). Boldt and Robbins (1987) evaluated the potential of host-specific insects from South America for the biological control of *B. neglecta*. They found that 17 insect species used this invasive shrub as an alternate feeding host, but only 3 species consistently caused damage to individual plants. Finally, *B. neglecta*, as with *B. sarothroides*, has been considered as a potential candidate for phytoremediation of mine tailings (Mendez and Maier 2008) due to its arsenic (As) tolerance and accumulating capacity (Flores-Tavizon et al. 2003).

Baccharis pilularis *DC. (Coyote Brush)*

Baccharis pilularis is probably the second most studied shrub of this genus. Its native distributional range corresponds to northwestern Mexico and the western United States (Hickman 1993; Ross 2004; USDA 2018). It is commonly found in chaparral, coastal scrub, and foothill woodland communities (Hobbs and Mooney 1985; Underwood et al. 2007), but *B. pilularis* can also grow in harsh serpentine soils (Hickman 1993).

McBride and Heady (1968) described the invasive character of this species for the first time in annual grasslands of northern California, and they studied the influence of grazing and burning on its expansion process. The results suggested that the spread of *B. pilularis* was favored by the reduction of wildfire frequency and the elimination of grazing livestock since the establishment of Regional Parks in 1934. McBride (1974) also reported the widespread distribution of this shrub in the region and emphasized the pivotal role that livestock play in preventing its invasion, even though mature shrubs have a low palatability. Multiple later studies have deepened our understanding of the invasions by this species in coastal ranges of California. Keeley (2005) reviewed the fire history of the grasslands mentioned above and concluded that changes in the fire regime were not as important as cessation of grazing to explain *B. pilularis* invasion. Further studies showed that the invasion of annual grasslands by this species often fails because seedling roots cannot reach deep parts of the soil profile before the summer drought occurs (Williams and Hobbs 1989). However, unusually favorable temperatures and adequate soil moisture conditions during this period (in years of abundant spring rainfall) allow for successful colonization (Williams et al. 1987; Williams and Hobbs 1989). Laris et al. (2016) argued that the use of mechanical treatments for the practice of grazing, including bulldozing or disking, caused long-lasting impacts on the region's vegetation dynamics, and, therefore, shrub advancement rates were lower in the least intensively disturbed sites, such as upper and steeper slopes.

Hobbs and Mooney (1986) studied the impacts of grassland colonization by *B. pilularis* and found that the abundances of all herbaceous species declined significantly after *Baccharis* stands formed a closed canopy at 2–3 years. They suggest that this result could be due to both the reduction of light penetrating the canopy and herbivory by small mammals, which are known to seek shelter in *Baccharis* stands. However, scattered *B. pilularis* individuals did not cause a great reduction of grassland species abundances. This shrub can also increase the fuel load of the invaded areas, thus altering their fire regimes (Russell and Tompkins 2005). Allelopathy was suggested to be another potential mechanism by which *B. pilularis* affects the surrounding vegetation (Hobbs and Mooney 1986).

In contrast to studies of managed grasslands, Cushman et al. (2011) showed, in a coastal foredune system in northern California (USA), that *B. pilularis* was the only native plant that dominated the site, along with the exotic species *Ammophila arenaria* and *Carpobrotus edulis*. However, herbaceous exotic species can greatly impact the growth of *Baccharis* seedlings, especially under dry conditions (Da Silva and Bartolome 1984).

Management of this species is very important to avoid changes from grassland to shrubland (Hobbs and Mooney 1985). Mechanical removal has proven to have low effectiveness at halting the spread of *B. pilularis*. The shrubs, especially 3–4-year-old plants, sprout readily from the base after cutting or burning, due to the development of an extensive root system for storage and water capture (Hobbs and Mooney 1985). However, seedlings and younger plants are susceptible to fire damage (Ross 2004). Biological methods could be used to control invasion of this species. For example, as described above, livestock grazing effectively decreases the establishment of new individuals (McBride and Heady 1968; McBride 1974). Seedling herbivory by slugs also has a considerable effect on seedling performance (Zavaleta 2006), and a fungal parasite called *Diaporthopsis* causes witches'-broom and dieback in mature shrubs (Bonar 1966). Chemical control is an effective but expensive control measure because it requires large quantities of herbicides (Elmore et al. 1968; Hyland and Holloran 2005; Ogden and Rejmánek 2005).

The effects of climate change on this shrub have also been studied by Zavaleta (2006), who suggested that seed germination could be increased by higher CO_2 concentrations and accelerated by water addition.

Baccharis pteronioides *DC. (Yerba de Pasmo)*

Baccharis pteronioides is a drought deciduous shrub that dominates some semidesert areas of its native distributional range (Bock and Bock 1992; Stegelmeier et al. 2009), corresponding to southern United States (Texas, New Mexico and Arizona) and northern Mexico (Kearney and Peebles 1969; Lamb 1975, cited in Kenney et al. 1986). It is considered troublesome in Arizona because it invades grasslands used for livestock grazing (Kenney et al. 1986). An experiment conducted by Kenney et al. (1986) demonstrated that the exclusion of domestic cattle increased the population density of *B. pteronioides*, even though livestock do not feed on this shrub when other plants are present. *Baccharis pteronioides*, as many other species from this genus, is fire resistant (Tellman 2002), which decreases the effectiveness of burning to prevent its spread. Furthermore, several livestock poisoning incidents caused by this species due to lack of alternative forage have been reported in the southwestern United States (Stegelmeier et al. 2009), which is explained in more detail in another section of the book (Chap. 15).

Baccharis salicifolia *(Ruiz & Pav.) Pers. (Mule-Fat)*

Baccharis salicifolia is widely distributed from the southwestern United States to Patagonia (Boldt and Robbins 1990; Müller 2006). However, it is believed that this species originated along the Andes Mountains in Argentina (Cuatrecasas 1968; Benson and Darrow 1981; Nesom 1988, cited in Boldt and Robbins 1990). It commonly grows along waterways, forming dense stands (Dimmit 2000). Several studies have pointed out the dominant role of this shrub in some areas within its native range. A study conducted in southern California by Boland (2014) showed that seedlings of *B. salicifolia* were widely distributed throughout the area, but adults comprised the dominant shrub species in the upper elevations of riparian woodlands. Seedling survivorship was very poor in the low-elevation riparian zone during the first winter, and adult survivorship was poor in the intermediate riparian zone in later years. DiPietro et al. (2002) indicated that *B. salicifolia* dominated southern California riparian southern willow scrub habitats. In the North Andean-Patagonian region of Argentina, Serra et al. (2013) specified that terrestrial vegetation was dominated by this shrub, among others.

Baccharis salicifolia is a phreatophyte, and its deep root system allows the plant to draw groundwater from deeper zones; the resulting additional use of water in semiarid basins can become problematic for neighboring species (Gatewood et al. 1950; Fletcher and Elmendorf 1955). McGuire (2005) concluded that transpiration through the leaves and stems of *B. salicifolia* was greater than precipitation during the growing season and that transpiration rate was comparable to the overstory cottonwood. Parker (1972) included this shrub in his book *An Illustrated Guide to Arizona Weeds*. Furthermore, *B. salicifolia* has a moderate tolerance to salinity (Glenn et al. 1998), which allows for occurrence in a wider range of environments. Humans have also influenced the spread of this species, planting it along waterways to control erosion due to its rapid growth rate and deep root system (Boldt and Robbins 1990; USDA 2018).

As with other *Baccharis* species, *B. salicifolia* has a low palatability to livestock or wildlife (Boldt and Robbins 1990; USDA 2018) and can sprout after fire, which increases its potential to behave as an invasive species. Boldt and Robbins (1990) evaluated the possible biological control of *B. salicifolia* by insects and found that this shrub was the host or alternate host for 106 species of phytophagous insects, which fed on the plant and caused moderate damage in localized areas.

Baccharis salicina *Torr. & A.Gray (Willow Baccharis)*

Baccharis salicina is native to the south-central United States (Colorado, Kansas, Oklahoma, Texas, New Mexico, and Arizona) and northern Mexico (Boldt and Robbins 1994; USDA 2018). However, it was identified by Polacik and Maricle (2013) as a non-native species in the Cedar Bluff Reservoir (Kansas, USA). *Baccharis salicina* requires light to germinate, is adapted to numerous soil types, and grows in moist disturbed areas and along saline waterways, forming narrow riparian strips (Ungar 1968; Boldt and Robbins 1994). Skousen et al. (1990) sampled the vegetation growing in unreclaimed surface mine sites in east-central Texas, finding that *B. salicina* established soon after mining and was the dominant woody plant species on 5–30-year-old sites.

Currently, *B. salicina* is invading riparian areas and lake basins throughout Texas, along with *Tamarix ramosissima*, due to both species' ability to outcompete other native vegetation in riparian areas (Muñoz et al. 2017). This species also is known to invade rangelands, where it has little or no value to livestock, making it undesirable in these systems. Managers of invaded rangelands have implemented different methods to control the rapid expansion of this shrub, including prescribed burning, herbicides, and mechanical treatments, with little success (Muñoz et al. 2017). There have also been attempts to promote the biological control of this species. Boldt and Robbins (1994) assessed the insects occurring and feeding on *B. salicina* in its native distributional range and found that the species was the host or alternate host for 61 species of phytophagous insects, of which only 19 occurred at densities greater than 1 per plant. Muñoz et al. (2017) studied the control of *B. salicina* and

T. ramosissima with goats and hypothesized that exposure to the plants at weaning would improve acceptance and consumption of the plant species by these animals. They found that goats consumed both invasive plant species but preferred *T. ramosissima* over *B. salicina.* This selective browsing could leave *B. salicina* without competitors in rangeland areas, allowing it to become the dominant shrub. Muñoz et al. (2017) also mentioned the possible toxic effects of *B. salicina* on goats and observed that the animals would consume the plant until the consumption reached toxic levels.

Baccharis sarothroides *A.Gray (Desert Broom)*

Baccharis sarothroides is native to northwestern Mexico and southwestern United States. It is considered invasive in some areas of its native range due to its ability to grow in harsh environments and highly disturbed areas (Mendez and Maier 2008). This species, like others in the genus, can reach water and nutrients stored in deep parts of the soil with its long root system, and its rapid growth allows it to withstand partial burial by sand (Haque et al. 2008). All the above characteristics make *B. sarothroides* one of the most dominant plants of sandy floodplains in areas of its native distributional range (Dimmit 2000). Another adaptation of this shrub to semiarid regions is that it loses its leaves under drought conditions, allowing it to survive during the summer dry period (Virginia Tech 2018). Haque et al. (2008) examined the phytoremediation potential of this species on mine tailings in Arizona and found that it was not affected by the excessive metal and metalloid concentration in the soil. This result suggests that *B. sarothroides* possesses certain metal adaptability and resistance, which could allow the colonization of other inhospitable environments. As with other *Baccharis* species, in some areas *B. sarothroides* is considered a bothersome plant due to its aggressive, invasive nature (Dimmit 2000).

Baccharis ulicina *Hook. and Arn. (Yerba de la Oveja)*

The distributional range of *B. ulicina* includes portions of Argentina and Bolivia. The major study regarding the biology and management of this species was documented by Tucat (2015) in his doctoral dissertation. All the information included in this section was obtained from that work. *Baccharis ulicina* is not palatable by cattle, which has facilitated its presence and dominance in agricultural systems. In fact, it is widespread in areas used for domestic livestock grazing, especially in pastures of the semiarid zone of Argentina. On average, one plant can produce from 900 to 1300 capitula per year. Each capitulum contains an average of 24 seeds, giving this species high reproductive potential. Moreover, seeds germinate in a short period of time after dispersal, and germination occurs under a wide range of environmental conditions. Experimental results showed that the germination rate was very high between temperatures of 10 and 28 °C and under any light level. This species also seems to possess some allelopathic activity. Specifically, *B. ulicina* negatively affected the establishment of other plants, such as the cultivated species *Avena sativa*, *Lolium perenne*, and *Raphanus sativus*, as well as the native species *Nassella clarazzi*. Finally, Tucat (2015) concluded that mechanical removal efforts and burning did not effectively control the spread of this species, due to its ability to sprout after the disturbance from stem buds near the ground. However, chemical control with glyphosate proved to be a good management tool.

Other Species of the Genus Baccharis *That Colonize Disturbed Areas and Regeneration Patches*

There are scattered studies that show the high capacity of other species from the genus *Baccharis* to colonize disturbed areas, as well as their important role in the regeneration of the vegetation within their native distributional ranges. These traits suggest that the introduction of these species to other areas could have detrimental consequences for the native vegetation. However, the competitive abilities of these species are not well understood, nor are their abiotic requirements for growth.

Holmgren et al. (2000) analyzed the recolonization of shrub species in patches that were either burned or cleared in a coastal area of central Chile. They found that *Baccharis linearis* (Ruiz & Pav.) Pers. was one of the two dominant species after the clearing of vegetation, but seed availability in burned patches was very low, probably because its seeds were unable to survive even low-intensity fires. They also stated that *B. linearis* is frequently found in abandoned agricultural fields.

Safford (2001) studied the postfire vegetation development in the surroundings of Rio de Janeiro (Brazil) and concluded that *Baccharis glaziovii* Baker, *Baccharis reticularia* DC., and especially *Baccharis platypoda* DC. successfully colonized burned areas, due to evolutionary adaptations to fire. However, the regeneration and postfire colonization rates were highly influenced by biotic and physical variables, such as altitude, aspect, and slope.

Limited research has been published on another handful of species in this genus. *Baccharis singularis* (Vell.) G.M.Barroso, for example, was found to occur commonly in fallow areas in southeastern Brazil (Salimon and Negrelle 2001). Slocum et al. (2004) showed that *Baccharis myrsinites* (Lam.) Pers. colonized areas previously occupied by fern thickets in the Dominican Republic, but only after the fern species were mechanically removed. *Baccharis punctulata* DC., *Baccharis notosergila* Griseb., and *Baccharis coridifolia* were found to be the dominant invasive species in abandoned fields of northeastern Argentina (Sione et al. 2016). *Baccharis punctulata* also was shown to invade natural forests and degraded grasslands, also in northeastern Argentina (Casermeiro and Spahn 1999; Marchesini 2003; Sione et al. 2016). Mechanical treatments such as shrub removal had to be applied multiple times to successfully control the spread of this species (Sabattini et al. 2012).

Finally, Lazarotto et al. (2017) considered *Baccharis psiadioides* (Less.) Joch.Müll. as a dominant species that forms dense stands within its native distributional range in southern Brazil and Uruguay (Deble et al. 2005). They further found that its essential oil affects seeds and seedlings of other plant species, such as *Arabidopsis thaliana*.

4 Sporadic Introductions

Baccharis spicata *(Lam.) Baill*

Baccharis spicata is native to southeastern South America (northeastern and central Argentina, Uruguay, Paraguay, and southern Brazil), where it occurs in grasslands, steppes, arable lands, river margins, disturbed coastal areas, abandoned paddy fields, and urbanized sites (Giuliano and Plos 2014; Verloove et al. 2018). Several localities in South America have recognized its potential invasive tendencies throughout its native distributional range and have organized campaigns to eradicate the species from susceptible areas. This species has recently invaded disturbed areas in Portugal but has not been reported in natural areas for the moment. However, considering its ecology in its native distributional range and its high capacity to disperse long distances by wind, it is likely that the species will spread to nearby natural areas (Verloove et al. 2018). Two naturalized populations were spotted in the surroundings of Porto (Matosinhos and Vila do Conde) in 2015. Verloove et al. (2018) provided details about those records of *B. spicata* and described the potential of this species to invade Europe. They stated that the species was especially abundant in Vila do Conde, where the stand probably consisted of 500–1000 individuals. This location was affected by an excavation in 2016, which favored the invasion of the area by *B. spicata*. They claimed that after the disturbance, the vegetation was dominated by this species due to its ability to sprout and germinate readily. Multiple hypotheses about this introduction are provided, but accidental introduction and subsequent naturalization seem to be the most likely. Moreover, it is likely that this species produces allelopathic compounds that affect negatively other plant species growing nearby (Damasceno et al. 2010).

Other Sporadic Introductions

Bartoli et al. (2008) reported the presence of *Baccharis pingraea* DC. and *Baccharis articulata* (Lam.) Pers. in southern Spain. The latter species was eradicated with herbicide treatments. The introduction of both species was associated with timber imported from South America (Verloove et al. 2018).

5 Use of *Baccharis* as Bioherbicide to Control Other Weeds

Extracts from a few species of the genus *Baccharis* have been used as an alternative bioherbicide to control the establishment and spread of weeds in agroecosystems. For example, the allelopathic compounds of *Baccharis trimera* (Less.) DC. and *Baccharis uncinella* DC. were used to control the invasive species *Eragrostis plana* and *Bidens pilosa*, respectively, in southern Brazil (Gonçalvez 2014; Dias et al. 2017).

6 Future Research on the Invasiveness of *Baccharis*

Further research is needed to deepen our understanding of the mechanisms of invasion by the genus *Baccharis*. There are multiple potential studies that would provide pivotal pieces of information about the invasiveness of *Baccharis*. For example, little is known about the importance of human activities for expansion of most of the species mentioned in this chapter. It is also uncertain whether there are differences in plant traits and demographic stages among the native, expansive, and invasive distributional ranges of *B. halimifolia*. In this regard, two of the authors of this chapter (AL-L and GNE) are investigating possible variations in early demographic stages, physiological tolerance, and plasticity among the abovementioned distributional ranges of *B. halimifolia*. Nothing is known about genetic diversity in *B. halimifolia*, either within or between its populations. Another potential field of study would be the evaluation of the effects of climate change on future distributions and invasiveness of *Baccharis* species. Furthermore, research could explore whether other congeners have characteristics like those of the known invasive *Baccharis* species in order to identify other high-risk plants. Lastly, although it is well known that the genus *Baccharis* produces essential oils composed mainly of monoterpenoids and sesquiterpenoids, which have multiple biological activities (e.g., antibacterial, antifungal, antiprotozoal, repellent, and cytotoxic properties; see Chap. 9 in this volume), there is little information on the ecological consequences in natural ecosystems of those compounds, such as chemical defense against natural enemies or allelopathic effects against plant competitors.

References

Allain L, Grace JB (2001) Changes in density and height of the shrub *Baccharis halimifolia* following burning in coastal tallgrass prairie. In: Bernstein NP, Ostrander LJ (eds) Proceedings of the 17th North American Prairie Conference. North Iowa Area Community College, Mason City, pp 66–72

Allen PF (1950) Ecological bases for land use planning in gulf coast marshlands. J Soil Water Conserv 5:57–85

Arizaga J, Unamuno E, Clarabuch O et al (2013) The impact of an invasive exotic bush on the stopover ecology of migrant passerines. Anim Biodivers Conserv 36:1–11

Atlas of Living Australia website (2018). http://www.ala.org.au. Accessed 8 Nov 2018

Bailey FM (1899) The Queensland flora, part 1. Diddams, Brisbane

Barroso G (1976) Compositae-Subtribo Baccharinidae Hoffmann. Estudo das espécies ocorrentes no Brasil. Rodriguésia 40:2–273 (original work not seen, see instead Ibáñez and Zoppolo 2011)

Bartoli A, Sánchez Gullón E, Weickert P et al (2008) Plantas americanas nuevas para la flora adventicia del sur de España [New American plants for the adventive flora of southern Spain]. Acta Bot Malacit 32:276–282

Benson L, Darrow R (1981) Trees and shrubs of the southwestern deserts. University of Arizona Press, Tucson (original work not seen, see instead Boldt and Robbins, 1990)

Berretta EJ (2001) Ecophysiology and management response of the subtropical grasslands of Southern South America. In: Gomide JA, Mattos WRS, Silva SC (eds) Proceedings of the XIX International Grassland Congress. Piracicaba, Brazil

Bock JH, Bock CE (1992) Vegetation responses to wildfire in native versus exotic Arizona grassland. J Veg Sci 3:439–446

Boland JM (2014) Factors determining the establishment of plant zonation in a southern Californian riparian woodland. Madrono 61:48–63

Boldt PE (1989) *Baccharis* (Asteraceae): a review of its taxonomy, phytochemistry, ecology, economic status, natural enemies and the potential for its biological control in the United States. Texas Agricultural Experiment Station, Texas A&M University, College Station

Boldt PE, Robbins TO (1987) Phytophagous and pollinating insect fauna of *Baccharis neglecta* (Compositae) in Texas. Environ Entomol 6:887–895

Boldt PE, Robbins TO (1990) Phytophagous and flower-visiting insect fauna of *Baccharis salicifolia* (Asteraceae) in the southwestern United States and Northern Mexico. Environ Entomol 19:515–523

Boldt PE, Robbins TO (1994) Phytophagous insect faunas of *Baccharis salicina*, *B. pteronioides*, and *B. bigelovii* (Asteraceae) in the south-western United States and northern Mexico. Environ Entomol 23:47–57

Bonar L (1966) A new *Diaporthopsis* causing brooming in *Baccharis*. Am J Bot 53:181–184

Bruce KA, Cameron GN, Harcombe PA (1995) Initiation of a new woodland type on the Texas Coastal Prairie by the Chinese tallow tree (*Sapzum sebiferum* (L.)) Roxb. Bull Torrey Bot Club 122:215–225

Campos J, Herrera M (2009) Invasive exotic flora assessment in the Basque Country (In Spanish). Dirección de Biodiversidad y Participación Ambiental. Departamento de Medio Ambiente y Ordenación del Territorio. Gobierno Vasco. Bilbao

Campos JA, Herrera M, Biurrun I et al (2004) The role of alien plants in the natural coastal vegetation in central-northern Spain. Biodivers Conserv 13:2275–2293

Campos JA, Caño L, Herrera M (2014) La invasión de *Baccharis halimifolia* en la costa cantábrica. Ambienta 109:78–91

Caño L, Campos JA, García-Magro D et al (2013a) Replacement of estuarine communities by an exotic shrub: distribution and invasion history of *Baccharis halimifolia* in Europe. Biol Invasions 15:1183–1188

Caño L, García-Magro D, Herrera M (2013b) Phenology of the dioecious shrub *Baccharis halimifolia* along a salinity gradient: consequences for the invasion of Atlantic subhalophilous communities. Plant Biosyst 147:1128–1138

Caño L, Campos JA, García-Magro D et al (2014) Invasiveness and impact of the non-native shrub *Baccharis halimifolia* in sea rush marshes: fine-scale stress heterogeneity matters. Biol Invasions 16:2063–2077

Caño L, Fuertes-Mendizabal T, García-Baquero G et al (2016) Plasticity to salinity and transgenerational effects in the nonnative shrub *Baccharis halimifolia*: insights into an estuarine invasion. Am J Bot 103:808–820

Casermeiro AL, Spahn E (1999) Caracterización de los recursos forrajeros nativos del norte entrerriano. In: Casermeiro J, Spahn E (eds) Sistemas agroforestales para pequeños productores de zonas húmedas. UNER-UNC-Gobierno de Entre Ríos-CERIDE, Paraná, pp 77–82

Connor DJ, Wilson GL (1968) Response of a coastal Queensland heath community to fertilizer application. Aust J Bot 16:117–123

Correll DS, Johnston MC (1979) Manual of the vascular plants of Texas. University of Texas, Dallas (original work not seen, see instead Boldt and Robbins 1987)

Costa ER, Costa JN, Armién AG, Barbosa JD, Peixoto PV (1995) Intoxicação experimental por *Baccharis coridifolia* (Compositae) em eqüinos. Pesqui Vet Bras 15:19–26

Cronquist A (1980) Vascular flora of the Southeastern United States. University of North Carolina Press, Chapel Hill

Cuatrecasas J (1968) Notas adicionales, taxonómicas y corológicas, sobre *Baccharis*. Rev Acad Colomb Cienc Exact Fis Nat 13:201–226 (original work not seen, see instead Boldt and Robbins, 1990)

Cushman JH, Lortie CJ, Christian CE (2011) Native herbivores and plant facilitation mediate the performance and distribution of an invasive exotic grass. J Ecol 99:524–531

Da Silva PG, Bartolome JW (1984) Interaction between a shrub, *Baccharis pilularis* subsp. *consanguinea* (Asteraceae) and an annual grass, *Bromus mollis* (Poaceae), in coastal California. Madrono 31:93–101

Damasceno FC, Nicolli KP, Caramão EB et al (2010) Changes in the volatile organic profile of *Schinus polygamus* (Anacardiaceae) and *Baccharis spicata* (Asteraceae) induced by galling psyllids. J Braz Chem Soc 21:556–563

Dauphin P, Matile-Ferrero D (2003) Présence de *Ceratoplastes sinensis* Del Guercio (Homoptera Coccidae) sur *Baccharis halimifolia* L. (Astéracées) en Gironde. Bull Soc Linn Bordx 31:261–263

de Almeida MB, Schild AL, Brasil ND, de Souza Quevedo P, Fiss L, Pfister JA, Riet-Correa F (2009) Conditioned aversion in sheep induced by *Baccharis coridifolia*. Appl Anim Behav Sci 117:197–200

Deble LP, Oliveira AS, Marchiori JNC (2005) O gênero *Baccharis* Lessing e táxones afins. Balduinia 1:1–20

Dias MP, Nozari RM, Santarem ER (2017) Herbicidal activity of natural compounds from *Baccharis* spp. on the germination and seedlings growth of *Lactuca sativa* and *Bidens pilosa*. Allelopath J 42:21–36

Diatloff G (1964) How far does groundsel seed travel? Queensland Agric J 90:354–356

Dimmit M (2000) Sunflower family. In: Steven JP, Patricia WC (eds) A natural history of the Sonoran Desert. Arizona-Sonora Desert Museum Press, Tucson, pp 174–175

DiPietro D, Ustin SL, Underwood EC (2002) Mapping the invasive plant *Arundo donax* and associated riparian vegetation using AVIRIS. In: Proceedings 11th Airborne visible/infrared image spectrometer (AVIRIS) Workshop: Jet Propulsion Laboratory, Pasadena

Dos Santos R, Citadini-Zanette V, Leal-Filho LS et al (2008) Spontaneous vegetation on overburden piles in the Coal Basin of Santa Catarina, Brazil. Restor Ecol 16:444–452

Duncan WH (1954) More and more weeds in Georgia. Bull GA Acad Sci 12:99–103

Elmore CL, Humphrey WA, Kretchum T (1968) Preemergence weed control in ground cover plantings. Calif Turfgrass Cult 18:9–10

Ervin GN (2009) Distribution, habitat characteristics, and new county-level records of *Baccharis halimifolia* L. on a portion of its present US range boundary. Southeast Nat 8:293–304

Estes D (2004) Noteworthy collections: middle Tennessee. Castanea 69:69–74

Everitt JH, Gerbermann AH, Akers DG (1978) Chemical control of dryland willow in the Lower Rio Grande Valley of Texas. J Rio Grande Val Hortic Soc 32:89–93

Fletcher H, Elmendorf H (1955) Phreatophytes – a serious problem in the West. In: Yearbook of agriculture. USDA, Washington, DC, pp 423–429

Flores-Tavizon E, Alarcon-Herrera MT, Gonzalez-Elizondo S et al (2003) Arsenic tolerating plants from mine sites and hot springs in the semi-arid region of Chihuahua, Mexico. Acta Biotechnol 23:113–119

Frau F, Ondiviela Eizaguirre B, Galván Arbeiza C et al (2014) The role of the hydrodynamic regime in the distribution of the invasive shrub *Baccharis halimifolia* (Compositae) in Oyambre Estuary (Cantabria, Spain). Limnetica 33:1–12

Frenedozo RC (2004) Plant reproductive phenology and dispersal patterns after natural regeneration in a limestone mining spoil banks. Braz Arch Biol Technol 47:261–271

Fried G, Panetta FD (2016) Comparing an exotic shrub's impact with that of a native life form analogue: *Baccharis halimifolia* L. versus *Tamarix gallica* L. in Mediterranean saltmarsh communities. J Veg Sci 27:812–823

Fried G, Balmès V, Germain JF (2013) A contribution to the inventory and study of the impacts of phytophagous insects found on *Baccharis halimifolia* in its introduced range in France. EPPO Bull 43:285–289

Fried G, Laitung B, Pierre C et al (2014) Impact of invasive plants in Mediterranean habitats: disentangling the effects of characteristics of invaders and recipient communities. Biol Invasions 16:1639–1658

Fried G, Caño L, Brunel S et al (2016) Monographs on invasive plants in Europe: *Baccharis halimifolia* L. Bot Lett 163:127–153

Fuertes-Mendizabal T, Caño L, González-Murua C et al (2014) Plasticity in the physiological response to salinity in the invasive species *Baccharis halimifolia*. Plant Biology Europe FESPB/EPSO Congress, Dublin

Galindez G, Biganzoli F, Ortega-Baes P et al (2009) Fire responses of three co-occurring Asteraceae shrubs in a temperate savanna in South America. Plant Ecol 202:149–158

Gann B, Thompson L, Schuler JL (2012) Control and management of eastern *Baccharis* in a recently established bottomland hardwood plantation. In: Butnor JR (ed) Proceedings of the 16th biennial southern silvicultural research conference e-general technical report SRS-156, United States Department of Agriculture Forest Service, Southern Research Station, Asheville

Gatewood J, Robinson T, Colby B et al (1950) Use of water by bottom-land vegetation in lower Safford Valley, Arizona. Geological survey water supply paper, vol 1103. U.S. Department of the Interior, Washington, DC

Gilberti LI, Menezes AL, Rodrigues AC, Fernandes GW, Berbara RL, Marota HB (2014) Effects of arsenic on the growth, uptake and distribution of nutrients in the tropical species *Baccharis dracunculifolia* DC (Asteraceae). J Toxicol Sci 2014:1–18

Gilman EF (1999) *Baccharis halimifolia* Salt Bush, Groundsel Bush. University of Florida, Gainesville

Giuliano DA, Plos A (2014) Género *Baccharis* [The genus *Baccharis*]. In: Zuloaga FO, Belgrano MJ, Anton AM (eds) Flora Argentina: flora vascular de la República Argentina, vol 7. Instituto de Botánica Darwinion, San Isidro, pp 43–123

Glenn E, Tanner R, Mendez S et al (1998) Growth rates, salt tolerance and water use characteristics of native and invasive riparian plants from the delta of the Colorado River, Mexico. J Arid Environ 40:281–294

Gomes V, Fernandes GW (2002) Germinação de aquênios de *Baccharis dracunculifolia* DC. (Asteraceae). Acta Bot Bras 16:421–427

Gonçalvez CEP (2014) Alelopatia de carqueja (*Baccharis trimara* Less) e ação de fungos em capim-annoni (*Eragrostis plana* Ness). Dissertação de Mestrado, Universidade Federal de Santa Maria

Grace JB, Smith MD, Grace SL et al (2001) Interactions between fire and invasive plants in temperate grasslands of North America. In: Galley KEM, Wilson TP (eds) Proceedings of the invasive species workshop: the role of fire in the control and spread of invasive species. Tall Timbers Research Station, Tallahassee, pp 40–65

Green B, Simpson R, Dettmann M (2011) Assessment of airborne Asteraceae pollen in Brisbane, Australia. Aerobiologia 28:295–301

Habermehl GG, Busam L, Heydel P, Mebs D, Tokarnia CH, Döbereiner J, Spraul M (1985) Macrocyclic trichothecenes: cause of livestock poisoning by the Brazilian plant *Baccharis coridifolia*. Toxicon 23:731–745

Hamilton WT, McGinty A, Ueckert DN et al (2004) Brush management: past, present, future. A & M University Press, College Station

Haque N, Peralta-Videa JR, Jones GL et al (2008) Screening the phytoremediation potential of desert broom (*Baccharis sarothroides* Gray) growing on mine tailings in Arizona, USA. Environ Pollut 153:362–368

Harcomb PA (1989) Reports progress of three prairie restoration/management projects in Houston area (Texas). Restor Manag Notes 7:35

Hickman JC (ed) (1993) The Jepson manual: higher plants of California. University of California Press, Berkeley

Hobbs RJ, Mooney HA (1985) Vegetative regrowth following cutting in the shrub *Baccharis pilularis* ssp. *consanguinea* (DC.) Wolf. Am J Bot 72:514–519

Hobbs RJ, Mooney HA (1986) Community changes following shrub invasion of grassland. Oecologia 70:508–513

Holmgren M, Segura AM, Fuentes ER (2000) Limiting mechanisms in the regeneration of the Chilean matorral. Plant Ecol 147:49–57

Hopkins DL, Adlerz WC (1988) Natural hosts of *Xylella fastidiosa* in Florida. Plant Dis 72:429–431

Huxley A (1992) The new RHS dictionary of gardening. MacMillan/Stockton Press, London

Hyland T, Holloran P (2005) Controlling European beachgrass (*Ammophila arenaria*) using prescribed burns and herbicide. In: California Invasive Plant Council Symposium 2005. Chico, CA

Ibáñez F, Zoppolo R (2011) Assessment of allelopathic properties of *Baccharis dracunculifolia* DC in laboratory and field conditions. Allelopath J 28:77–86

Jarvis BB, Mokhtari-Rejali N, Schenkel EP et al (1991) Trichothecene mycotoxins from Brazilian *Baccharis* species. Phytochemistry 30:789–797

Kearney TH, Peebles RH (1969) Arizona flora. University California Press, Berkeley (original work not seen, see instead Kenney et al. 1986)

Keeley JE (2005) Fire history of the San Francisco East Bay region and implications for landscape patterns. Int J Wildland Fire 14:285–296

Kenney WR, Bock JH, Bock CE (1986) Responses of the shrub, *Baccharis pteronioides*, to livestock exclosure in southeastern Arizona. Am Midl Nat 116:429–431

Kraft SK, Denno RF (1982) Feeding responses of adapted and non-adapted insects to the defensive properties of *Baccharis halimifolia* L (Compositae). Oecologia 52:156–163

Krischik VA, Denno RF (1990a) Differences in environmental response between the sexes of the dioecious shrub, *Baccharis halimifolia* (Compositae). Oecologia 83:176–181

Krischik VA, Denno RF (1990b) Patterns of growth, reproduction, defense, and herbivory in the dioecious shrub *Baccharis halimifolia* (Compositae). Oecologia 83:182–190

Lamb SH (1975) Woody plants of the Southwest. The Sunstone Press, Santa Fe (original work not seen, see instead Kenney et al. 1986)

Laris P, Brennan S, Engelberg K (2016) The coyote brush invasion of Southern California grasslands and the legacy of mechanical disturbance. Geogr Rev 107:640–660

Lázaro-Lobo A, Herrera M, Campos JA, Caño L, Goñi E, Ervin GN (2020) Influence of local adaptations, transgenerational effects and changes in offspring's saline environment on *Baccharis halimifolia* L. under different salinity and light levels. Environ Exp Bot 177:104134

Lazarotto DC, Da Silva ER, Pawlowski A et al (2017) Phytotoxic effects of *Baccharis psiadioides* (Asteraceae) volatiles on different phases of plant development. J Essent Oil Res 29:313–319

Le Moigne G, Magnanon S (2009) Le séneçon en arbre (*Baccharis halimifolia*). Conservatoire Botanique National de Brest, Brest

Lombardo A (1964) Flora Arbórea y Arborescente del Uruguay. Concejo Departamental de Montevideo, Montevideo (original work not seen, see instead Ibáñez and Zoppolo 2011)

Lovet J (2015) Lutte biologique contre un arbuste invasif en France: Données de biologie et impact de la cochenille *Ceroplastes sinensis* (Hem., Coccidae) sur *Baccharis halimifolia* (Asteraceae) en Camargue. MSc dissertation, Ecole des Métiers de l'Environnement (Rennes, France)

Macedo JF, Brandão M, Lara JFR (2003) Plantas daninhas na pós-colheita de milho nas várzeas do rio São Francisco, em Minas Gerais. [Weed survey on corn post-harvest under meadow holm conditions along São Francisco river, Minas Gerais, Brazil]. Planta Daninha 21:239–248

Mallard F (2008) Effet d'une espèce végétale introduite envahissante le Séneçon en arbre *Baccharis halimifolia* sur le peuplement d'arthropodes dans le Golfe du Morbihan. MSc dissertaion, Université de Rennes

Marchesini E (2003) Control de chilcas. Hoja Informativa Electrónica. Estación Experimental Agropecuaria INTA. Concepción del Uruguay, 86

Matuda E (1957) El género *Baccharis* en Mexico. Ann Inst Biol Mex 28:143–174. (original work not seen, see instead Boldt and Robbins 1987)

McBride JR (1974) Plant succession in the Berkeley Hills, California. Madrono 22:317–329

McBride JR, Heady HF (1968) Invasion of grassland by *Baccharis pilularis* DC. J Range Manag 21:106–108

McFadyen PJ (1973) Insects for groundsel bush control. Queensland Agric J 99:607–611

McGuire RD (2005) Quantifying consumptive water use by seep willow (*Baccharis salicifolia*) within the San Pedro Riparian National Conservation Area (SPRNCA). Master's theses, The University of Arizona

McKee KL, Feller IC, Popp M et al (2002) Mangrove isotopic (δ15N and δ13C) fractionation across a nitrogen vs. phosphorus limitation gradient. Ecology 83:1065–1075

Mendez MO, Maier RM (2008) Phytostabilization of mine tailings in arid and semiarid environments – an emerging remediation technology. Environ Health Perspect 116:278–283

Mongelli E, Desmarchelier C, Talou JR, Coussio J, Ciccia G (1997) In vitro antioxidant and cytotoxic activity of extracts of *Baccharis coridifolia* DC. J Ethnopharmacol 58:157–163

Moon DC, Barnouti J, Younginger B (2013) Context-dependent effects of mycorrhizae on herbivore density and parasitism in a tritrophic coastal study system. Ecol Entomol 38:31–39

Müller J (2006) Systematics of *Baccharis* (Compositae-Astereae) in Bolivia, including an overview of the genus. Syst Bot Monogr 76:1–341

Muñoz A, Garcia A, Scott C et al (2017) Consumption of salt cedar and willow *Baccharis* by Boercross goats. Rangel Ecol Manag 70:374–379

Mutz JL, Scifres CJ, Mohr WC et al (1979) Control of willow *Baccharis* and spiny aster with pelleted herbicides. Texas Agric Exp Stn Bull–1194

Negreiros D, Esteves D, Fernandes GW et al (2012) Growth-survival tradeoff in the widespread tropical shrub *Baccharis dracunculifolia* (Asteraceae) in response to a nutrient gradient. Trop Ecol 55:167–176

Nesom G (1988) *Baccharis monoica* (Compositae: Astereae), a monoecious species of the *B. salicifolia* complex from Mexico and Central America. Phytologia 65:160–164 (original work not seen, see instead Boldt and Robbins, 1990)

Neto D, Carvalho LM, Cruz C et al (2006) How do mycorrhizas affect C and N relationships in flooded *Aster tripolium* plants? Plant Soil 279:51–63

Ogden JAE, Rejmánek M (2005) Recovery of native plant communities after the control of a dominant invasive plant species, *Foeniculum vulgare*: implications for management. Biol Conserv 125:427–439

Olson PE, Fletcher JS (2000) Ecological recovery of vegetation at a former industrial sludge basin and its implications to phytoremediation. Environ Sci Pollut Res 7:195–204

Orr SP, Rudgers JA, Clay K (2005) Invasive plants can inhibit native tree seedlings: testing potential allelopathic mechanisms. Plant Ecol 181:153–165

Overbeck GE, Pfadenhauer J (2007) Adaptive strategies in burned subtropical grassland in southern Brazil. Flora 202:27–49

Palmer WA (1987) The phytophagous insect fauna associated with *Baccharis halimifolia* L. and *B. neglecta* Britton in Texas, Louisiana and northern Mexico. P Entomol Soc Wash 89:185–199

Palmer WA, Bennett FD (1988) The phytophagous insect fauna associated with *Baccharis halimifolia* L. in the eastern United States. P Entomol Soc Wash 90:216–228
Palmer WA, Sims-Chilton NM (2012) *Baccharis halimifolia* L. – groundsel bush. In: Biological control of weeds in Australia. CSIRO Publishing, Melbourne, pp 86–95
Panetta FD (1977) The effect of shade upon seedling growth in groundsel bush (*Baccharis halimifolia* L.). Aust J Agric Res 28:681–690
Panetta FD (1979a) Germination and seed survival in the woody weed, groundsel bush (*Baccharis halimifolia* L.). Aust J Agric Res 30:1067–1077
Panetta FD (1979b) The effects of vegetation development upon achene production in the woody weed, groundsel bush (*Baccharis halimifolia* L.). Aust J Agric Res 30:1053–1065
Panetta FD (1979c) Shade tolerance as reflected in population structures of the woody weed, groundsel bush (*Baccharis halimifolia* L.). Aust J Bot 27:609–615
Parker K (1972) An illustrated guide to Arizona weeds. University of Arizona Press, Tucson
Parsons WT, Cuthbertson EG (1992) Groundsel bush. In: Parsons WT, Cuthbertson EG (eds) Noxious weeds of Australia. Inkata Press, Melbourne, pp 256–259
Paudel S, Battaglia LL (2013) Germination responses of the invasive *Triadica sebifera* and two co-occurring native woody species to elevated salinity across a gulf coast transition ecosystem. Wetlands 33:527–535
Paudel S, Battaglia LL (2015) The role of light, soil and human factors on the probability of occurrence of an invasive and three native plant species in coastal transitions of coastal Mississippi, USA. J Plant Ecol 8:491–500
Paudel S, Baer SG, Battaglia LL (2014) Arbuscular mycorrhizal fungi (AMF) and success of *Triadica sebifera* invasion in coastal transition ecosystems along the northern Gulf of Mexico. Plant Soil 378:337–349
Penfound WT, Hathaway ES (1938) Plant communities in the marshlands of southeastern Lousiana. Ecol Monogr 8:3–56
Pivovaroff AL, Swift C, Battaglia LL et al (2015) Physiological profiles as indicators of response to hurricane disturbance for three coastal wetland species. J Coast Res 31:986–993
Polacik KA, Maricle BR (2013) Effects of flooding on photosynthesis and root respiration in saltcedar (*Tamarix ramosissima*), an invasive riparian shrub. Environ Exp Bot 89:19–27
Porazinska DL, Fujisaki I, Purcell MF et al (2014) Plant invasions from a belowground nematocentric perspective. Soil Biol Biochem 77:213–220
Richardson DM, Pysek P, Rejmanek M et al (2000) Naturalization and invasion of alien plants: concepts and definitions. Divers Distrib 6:93–107
Rizzo I, Varsavky E, Haidukowski M, Frade H (1997) Macrocyclic trichothecenes in *Baccharis coridifolia* plants and endophytes and *Baccharis artemisioides* plants. Toxicon 35:753–757
Ross C (2004) *Baccharis pilularis* DC. In: Francis JK (ed) Wildland shrubs of the United States and its territories: thamnic descriptions GTR IITF–GTR–26. USDA Forest Service International Institute of Tropical Forestry and USDA Forest Service Rocky Mountain Research Station, San Juan, PR and Fort Collins, CO, pp 103–104
Russell W, Tompkins R (2005) Estimating biomass in coastal *Baccharis pilularis* dominated plant communities. Fire Ecol 1:20–27
Sabattini RA, Ledesma SG, Sione SMJ (2012) Dinámica de la cobertura de *Melica macra* Nees y *Baccharis punctulata* D.C en un bosque nativo sometido a desarbustado mecánico. Rev Fac Cienc Agrar 12:13–19
Safford HD (2001) Brazilian páramos. III. Patterns and rates of postfire regeneration in the Campos de altitude. Biotropica 33:282–302
Salimon CI, Negrelle RRB (2001) Natural regeneration in a quaternary coastal plain in southern Brazilian Atlantic rain forest. Braz Arch Biol Technol 44:155–163
Scifres CJ (1980) Brush management. Texas A&M University, College Station
SERNEC Data Portal (2018). https://sernecportal.org/portal/index.php. Accessed 16 Nov 2018

Serra MN, Albarino R, Diaz Villanueva V (2013) Invasive *Salix fragilis* alters benthic invertebrate communities and litter decomposition in northern Patagonian stream. Hydrobiologia 701:173–188

Sims-Chilton NM, Panetta FD (2011) The biology of Australian weeds 58. *Baccharis halimifolia* L. Plant Prot Q 26:114–123

Sims-Chilton NM, Zalucki MP, Buckley YM (2009) Patchy herbivore and pathogen damage throughout the introduced Australian range of groundsel bush, *Baccharis halimifolia*, is influenced by rainfall, elevation, temperature, plant density and size. Biol Control 50:13–20

Sims-Chilton NM, Zalucki MP, Buckley YM (2010) Long term climate effects are confounded with the biological control programme against the invasive weed *Baccharis halimifolia* in Australia. Biol Invasions 12:3145–3155

Sinnassamy JM (2004) *Baccharis halimifolia*. In: Plantes invasives en France. Coordinated by Muller, S. Muséum national d'Histoire naturelle, Paris

Sione SMJ, Ledesma SG, Rosenberger LJ et al (2016) Banco de semillas del suelo en un área de bosques nativos sujeta a cambio en el uso de la tierra (Entre Ríos, Argentina). FAVE Sección Ciencias Agrarias 15:87–104

Skousen JG, Call CA, Knight RW (1990) Natural revegetation of an unreclaimed lignite surface mine in East-Central Texas. Southwest Nat 35:434–440

Slocum MG, Aide TM, Zimmerman JK et al (2004) Natural regeneration of subtropical montane forest after clearing fern thickets in the Dominican Republic. J Trop Ecol 20:483–486

Stegelmeier BL, Sani Y, Pfister JA (2009) *Baccharis pteronioides* toxicity in livestock and hamsters. J Vet Diagn Investig 21:208–213

Sundberg SD, Bogler DJ (2006) *Baccharis*. In: Flora of North America Editorial committee 1993+ (ed) Flora of North America North of Mexico. New York/Oxford, pp 23–28

Tellman B (ed) (2002) Invasive exotic species in the Sonoran region. University of Arizona Press, Arizona

Tokarnia CH, Döbereiner J (1975) Intoxicação experimental em bovinos por "mio-mio", *Baccharis coridifolia*. Pesqui Agropec Bras 10:79–97

Tolliver KS, Martin DW, Young DR (1997) Freshwater and saltwater flooding response for woody species common to barrier island swales. Wetlands 17:10–18

Tucat G (2015) Estudios sobre la biología de *Baccharis ulicina* Hook & Arn y su susceptibilidad a estrategias de manejo en el sur bonaerense. Doctoral dissertation, Universidad Nacional del Sur (Bahía Blanca, Argentina)

U.S. Department of Homeland Security (2004) Environmental impact statement for operation Rio Grande. CBP, Washington, DC

Underwood EC, Ustin SL, Ramirez CM (2007) A comparison of spatial and spectral image resolution for mapping invasive plants in coastal California. J Environ Manag 39:63–83

Ungar IA (1968) Species-soil relationships on the Great Salt Plains of northern Oklahoma. Am Midl Nat 80:392–406

USDA, NRCS (2018) The PLANTS database. http://plants.USDA.gov, 12 April 2018. National Plant Data Team, Greensboro, NC 27401–4901 USA

Van Auken OW, Bush JK (1990) Influence of light levels soil nutrients and competition on seedling growth of *Baccharis neglecta* Asteraceae. Bull Torrey Bot Club 117:438–444

Verloove F, Dana ED, Alves P (2018) *Baccharis spicata* (Asteraceae), a new potentially invasive species to Europe, Plant Biosyst 152:416–426

Vick JK, Young DR (2013) Comparative responses of a non-N-fixing shrub and an actinorhizal N-fixing shrub to N fertilization. Plant Soil 371:377–385

Virginia Tech (2018) Department of Forest Resources and Environmental Conservation. https://frec.vt.edu/

Weber E (2003) Invasive plant species of the world: a reference guide to environmental weeds. CABI Publishing, Oxon

Westman WE, Panetta FD, Stanley TD (1975) Ecological studies on reproduction and establishment of the woody weed groundsel bush (*Baccharis halimifolia* L.: Asteraceae). Aust J Agric Res 26:855–870

Williams K, Hobbs RJ (1989) Control of shrub establishment by springtime soil water availability in an annual grassland. Oecologia 81:62–66

Williams K, Hobbs RJ, Hamburg SP (1987) Invasion of an annual grassland in Northern California by *Baccharis pilularis* ssp. *consanguinea*. Oecologia 72:461–465

Winders CW (1937) Groundsel-bush in southeastern Queensland. Queensland Agric J 48:656-664. (original work not seen, see instead Sims-Chilton and Panetta 2011)

Young DR, Erickson DL, Semones SW (1994) Salinity and the small-scale distribution of three barrier island shrubs. Can J Bot 72:1365–1372

Younginger B, Barnouti J, Moon DC (2009) Interactive effects of mycorrhizal fungi, salt stress, and competition on the herbivores of *Baccharis halimifolia*. Ecol Entomol 34:580–587

Zavaleta E (2006) Shrub establishment under experimental global changes in a California grassland. Plant Ecol 184:53–63

Zinnert JC, Nelson JD, Hoffman AM (2012) Effects of salinity on physiological responses and the photochemical reflectance index in two co-occurring coastal shrubs. Plant Soil 354:45–55

Part II
Structure and Chemistry of *Baccharis*

Part II of the book invites you to learn the most cutting-edge discoveries on the diversity and chemical structure of *Baccharis*. This genus, which mainly comprises dioecious shrubs, has a wide range of morphology. These morphological traits (such as growth shape, branching pattern, shape, size and margins of leaves, type of indumentum, shape and arrangement of flower heads, and number of nerves in the cypselas/fruits) are essential for the identification of many species. The genus also exhibits a wide variety of chemical constituents (essential oils, phenolics, flavonoids, diterpenoids, triterpenoids, trichothecenes, coumarins, benzofurans) that provide unique signatures for each species. These chemicals have a great potential for the treatment of various inflammatory diseases of microbiological and parasitic origin, such as tumors, diabetes, arthritis, liver disorders, dental caries, stomach and prostate diseases, among others. The main constituents are artepillin C, (E) -nerolidol, spathulenol, and caffeic acid, which are also found in the green propolis produced from the resins of *Baccharis dracunculifolia*. Another highlight of *Baccharis*' chemistry involves the production of macrocyclic trichothecenes by *Baccharis megapotamica* and *Baccharis cordifolia*, which can lead to intoxication and lethality in various vertebrate organisms.

***Baccharis macrophylla*. Illustration by Patrícia Angrisano**

Chapter 9
Morpho-anatomical Characteristics of Species of *Baccharis*

Jane Manfron, Paulo Vitor Farago, Ikhlas Ahmed Khan, and Vijayasankar Raman

Abstract *Baccharis* L. is one of the largest and most diversified genera of Asteraceae. Two groups of species, namely, "vassouras" and "carquejas," are generally recognized based on their habit and branching pattern. "Vassouras" plants possess regular stems and leaves and are often broom-like in appearance, whereas "carquejas" have stems that are modified as cladodes and are devoid of leaves or have diminutive, scalelike leaves. Despite these major differences in their growth forms, stems, and leaves, the species within these groups show close resemblances in their morphologies. This chapter provides a comprehensive review of the morphology and anatomy of the genus and discusses the main morpho-anatomical features helpful in the identification of various *Baccharis* species. Further research involving comparative morpho-anatomical studies are required to better understand the species diversity as well as to develop a more accurate classification of *Baccharis* as well as of Asteraceae.

1 Introduction

Morpho-anatomy is the study of morphological and anatomical forms and structures with emphasis on features, which may be useful in distinguishing the species. A morphological examination is considered the first and preferred method of botanical authentication when the required features are available.

Plant anatomy is valuable for the evaluation of dried plant material, especially for fragmented or powdered material. Dried or fragmented plant parts do not result in the loss of most microscopic characteristics, which makes plant anatomy the

J. Manfron (✉) · P. V. Farago
Department of Pharmaceutical Sciences, State University of Ponta Grossa, Ponta Grossa, Paraná, Brazil

I. A. Khan · V. Raman
National Center for Natural Products Research, School of Pharmacy, The University of Mississippi, Oxford, MS, USA

G. W. Fernandes et al. (eds.), *Baccharis*,
https://doi.org/10.1007/978-3-030-83511-8_9

most stable technique for the evaluation of plant characteristics when it comes to authentication.

Morphological and anatomical analyses are inherent procedures in almost all pharmacopoeias and are the main identification tests required for the herbal industry. Even though the individual structural elements are relatively common within the same type of plant parts, how the elements are organized gives a plant species its characteristic fingerprint (Upton et al. 2011). Additionally, morphology and microscopy of botanical samples and microscopic descriptions of impurities are included in most pharmacopoeias demonstrating the importance of these characteristics in the quality control of medicinal plants.

Previous workers have stated that many *Baccharis* species have analogous morphologies making their morphological identification problematic (Freire et al. 2007; Budel et al. 2008). Species within the section *Caulopterae*, such as *B. crispa* Spreng. and *B. articulata* (Lam.) Pers., show a high degree of phenotypic plasticity leading to difficulties in species identification even in flowering stages (Simões-Pires et al. 2005).

In this context, microscopic characteristics can provide helpful information in solving taxonomic problems (Bobek et al. 2016; Budel et al. 2018a).

Another problem is that different species of *Baccharis* are called by the same folk names and used indiscriminately for the same therapeutic indications (Budel and Duarte 2010; Bobek et al. 2015a, 2016; Budel et al. 2018a). Considering traditional names, "carquejas" (plants with winged stems or cladodes) and "vassouras" (plants with broom-like and without winged stems) are the most common, creating confusion in the identification of *Baccharis* species. The confusion due to similar popular names is one of the main causes of misidentification of plant drugs (Upton et al. 2011).

2 Morphology

Baccharis L. is one of the largest genera in the subfamily Asteroideae Lindley within the family Asteraceae (Compositae) and comprises about 440 species (Heiden et al., 2019). All of the species are native to New World with about 90% of them occurring in South America (Nesom 1990). It is also introduced into several parts of the world including Europe, Transcaucasus, and Australia. *Baccharis* is commonly called "Groundsel tree" although the name "Groundsel" is also applied to different species of *Senecio* L. (Asteraceae). Species of *Baccharis* comprise mostly dioecious shrubs and show a wide range of morphological diversity. The most important morphological characters useful in *Baccharis* species identification include the growth form; branching pattern; the shape, size, and margins of leaves; type of indumentum; shape and arrangement of flower heads; and number of nerves on cypselae (fruits). Both staminate and pistillate specimens are important for species identification. Natural hybridization between different species occurs, resulting in the development of intermediate forms showing interspecific variations. For example, hybridization between *B. halimifolia* L. and *B. neglecta* Britton and between *B. halimifolia* and *B. angustifolia* Michx. has been reported. Intermediate forms between *B. thesioides* Kunth and *B. bigelovii* A. Gray have been observed in southern Arizona (eFloras 2008).

The genus is broadly characterized by the tufted indumentums of the leaves and stems, which consists of fused trichomes with a single adjoining basal cell. Dioecy is a common occurrence (Müller 2006).

The genus *Baccharis* has not been taxonomically revised, as a whole. Due to the high number of species, the revisionary works in the past have been carried out at the level of geographical regions. In South America, there are works in Colombia (Cuatrecasas 1969), central Argentina (Ariza-Espinar 1973), and Brazil (Barroso and Bueno 2002). Barroso (1976) and Oliveria et al. (2006) have studied the species for Brazil and reported 125 and 146 species, respectively. Nesom (1990) studied 43 species from the USA, Mexico, and Central America and grouped those under 6 sections.

Several authors have proposed different infrageneric classifications of *Baccharis* mainly based on morphological characters. De Candolle (1836) was the first author to classify *Baccharis* into eight sections based on leaf morphology. Baker (1882–1884) classified the Brazilian species into six series also using leaf characteristics. Cuatrecasas (1967) revised the Colombian species into sections. Giuliano (2001) subdivided 96 Argentinean species of *Baccharis* into 15 sections. Of these, the sect. *Caulopterae* DC. is characterized by the presence of longitudinal wings on the stems. This section was previously named *Alatae* Less. (Ariza-Espinar 1973) and *Trimera* group (Barroso 1976).

Heering (1904) provided the first subgeneric classification of *Baccharis* with five subgenera, namely, *Baccharis*, *Molina*, *Pteronioides*, *Stephananthus*, and *Tarchonanthoides*. In the most recent subgeneric classification of the genus, Müller (2006) accepted four of the five subgenera proposed by Heering (1904) and placed the subgenus *Stephananthus* as *incertae sedis* ("of uncertain placement"). Heiden and Pirani (2012) provided a taxonomic revision of the subgen. *Tarchonanthoides* with four sections, namely, *Canescentes*, *Coridifoliae*, *Curitybensis*, and *Tarchonanthoides*. This subgenus consists of 21 species and is characterized mainly by the corollas of the female florets with 5 papillose teeth and the male florets with a stigma nearly fully divided into lanceolate or ovate branches (Heiden and Pirani 2012). Recently, based on phylogenetic studies, Heiden et al. (2019) classified *Baccharis species into 7 subgenera and 47 sections.*

In some instances, different species of *Baccharis* exhibit similar morphological features leading to confusion in species identification. For example, *B. brevifolia* DC., *B. microdonta* DC., *B. pauciflosculosa* DC., and *B. trilobata* A.S. Oliveira & Marchiori are all commonly called "vassouras" (meaning "broom") in Brazil, due to their similar appearances (Bobek et al. 2016) as well as their use as brooms to clean the house. The name "carqueja" is applied to multiple species, including *B. crispa*, *B. articulata*, and *B. pentaptera* (Less.) DC. (Budel et al. 2005), due to their close resemblances in the morphological features. As a result, different species are used interchangeably (Gianello et al. 2000).

The number of striations (ribs) on the stems can be useful in recognizing species. Bobek et al. (2016) observed 5-6- ribbed stems in B. *brevifolia and B. paucifloscu-losa and 4-ribbed stems in B. microdonta* and *B. trilobata*, using anatomical evaluation. The shape of the stem in the cross section is determined by the number of ribs.

Morphological Description

Baccharis are perennial subshrubs, shrubs, or trees, growing 0.1–6 m tall, usually dioecious and rarely monoecious, usually woody at base, rarely rhizomatous [e.g., *B. acaulis* (Wedd. ex R.E.Fr.) Cabrera and *B. davidsonii* Cuatrec.]. In the "vassouras" group of species, including *B. sarothroides* A. Gray, *B. brevifolia* (Fig. 9.1a), *B. dracunculifolia* DC. (Fig. 9.1b), *B. microdonta* (Fig. 9.1c), *B. paucifloscuIosa* (Fig. 9.1d), and *B. trilobata*, the plants are often much-branched and broom-like (Fig. 9.1a–d).

Stems in *Baccharis* are erect, ascending, procumbent (*B. caespitosa* (Ruiz & Pav.) Pers., *B. alpina* Kunth, *B. humifusa* Kunth, and *B. pumila* Joch. Müll.) or rarely prostrate (*B. acaulis*), short or well developed, simple (*B. texana* (Torrey & A.Gray) A.Gray) or many-branched, usually striated, rarely terete and smooth, glabrous to hispidulous or villous (*B. plummerae* A.Gray subsp. *plummerae*), often resinous (*B. plummerae* A.Gray subsp. *glabrata*, *B. pteronioides* DC.); young stems are usually green.

In the "carqueja" group of species, the stems are modified into cladodes with 2–5 wings (*B. glaziovii*, *B. junciformis* DC. (syn. *B. usterii* Heering), *B. trimera*, *B. pentaptera* (Fig. 9.2a), and *B. sagittalis* (Less.) DC.), which run along the longitudinal axis of the stem. These stems are either leafless or with sparse leaves that are reduced to scales, such as in *B. glaziovii* (Fig. 9.2b).

Fig. 9.1 Habit of "vassouras" plants. *Baccharis brevifolia* (**a**), *B. dracunculifolia* (**b**), *B. microdonta* (**c**), and *B. paucifloscuIosa* (**d**). Scale bar: b = 1 cm; a, c, d = 5 cm. (**a**, **c**, **d** reproduced with permission from Bobek et al. 2016)

Fig. 9.2 Habit of "carquejas" showing winged stem cladodes. *Baccharis pentaptera* (**a**), *B. glaziovii* (**b**). Scale bar = 1.5 cm. (Reproduced with permission from Budel et al. 2015 (**a**) and Jasinski et al. 2014 (**b**))

Leaves cauline, sometimes deciduous and sparse or absent at flowering (*B. vanessae* R.M.Beauch.), sometimes reduced to scales distally (*B. sarothroides*), simple, alternate, rarely in rosettes, sessile (*B. sergiloides* A. Gray), or petiolate; lamina 1–3-nerved, linear, oblong, lanceolate, ovate, obovate or rhomboid, glabrous or rarely hispidulous or villous, often gland-dotted and resinous, usually acute or obtuse at apex, acute or tapering at base, entire or finely to coarsely serrate along margins (eFloras 2008).

Inflorescences are head-like, sessile or stalked, unisexual, discoid, solitary (*B. acaulis*, *B. alpina*, *B. davidsonii*, *B. pumila*, and *B. tola* Phil.), or in paniculiform, corymbiform, or racemiform (*B. pteronioides*) arrays; involucres cylindrical to campanulate or hemispheric, 3–9 mm in diameter; phyllaries 15–40 arranged in 2–5 series, mid usually green, sometimes red or purple, 1-nerved, ovate to lanceolate, oblique, scarious along margins, often erose, ciliate or keeled, midrib conspicuous or not, obtuse, acute, acuminate or keeled at apex, glabrous or hispid; receptacles flat, tholiform or conical, pitted or smooth, glabrous, tomentose or glandular, and usually epaleate. Staminate florets 10–50; corollas white to pale yellow, tubular; lobes 5, spreading-reflexed, triangular to ovate; pappi 20–40, equal, minutely barbellate or distally with plumose bristles. Pistillate florets 20–150; corollas whitish, narrowly tubular; lobes 5, spreading-reflexed, triangular to ovate; styles glabrate, flattened, and unappendaged (eFloras 2008).

Fruits are cypselae or achenes, obovoid to cylindrical, slightly compressed or 5-angled, 5–10-nerved, glabrous or hispid, light brown; pappi 25–50, whitish, tawny or brownish, minutely barbellate, apically attenuate bristles in 1–3 series, persistent or deciduous, accrescent in fruit (eFloras 2008).

3 Anatomy

Anatomical characteristics are important in the authentication of plants, especially when the plants are dried or fragmented. The drying and the fragmentation of the plants do not result in the loss of the most relevant microscopic characteristics, that is, the anatomical markers. Therefore, these are the reliable characters helpful in the identification and quality control of the botanical samples. Also, microscopy has the advantage of requiring small amounts of material. Microscopic analysis is helpful in the detection of inorganic materials adhered to parts of the plant, such as roots. Microscopy can also detect when different parts of the same plant are present in the sample.

Certain anatomical characteristics helpful in the discrimination of species of *Baccharis* include the epidermal features (contour of anticlinal epidermal cell walls, type of stomata, type of trichomes), the organization of mesophyll and vascular tissues, and the secretory ducts and morphotypes of crystals.

Epidermal Characters

The epidermis is a permanent and complex tissue, including different kinds of cells. Epidermal features can show many different anatomical markers, which may help in the diagnosis of the plant. Several authors have contributed to resolving taxonomic problems in *Baccharis* through the micromorphological analysis of leaves (Muller 2006; Freire et al. 2007; Bobek et al. 2016; Budel et al. 2018b; Almeida et al. 2021).

Epidermal characteristics, such as cuticle, epidermal cells, stomata, and trichomes, have been identified as important tools in the delimitation of some complex genera, such as *Passiflora* (Wosch et al. 2015) and *Piper* (Gogosz et al. 2012), and so is *Baccharis*. They also help in distinguishing medicinal species, since drugs of pharmaceutical use are made up of dried and fragmented parts in which the different macroscopic features of the species are not generally distinguishable (Jackson and Snowdon 1990; Jorge 2000).

Epidermal Cells

The epidermis is the outermost cell layer of a plant protecting against evaporation and attack by microorganisms and herbivores. In a surface view of the epidermis, the anticlinal cell walls may be straight, wavy, or sinuous. Cutin, in the form of the cuticle layer, is deposited on the outer surface of the epidermis to offer a waterproof surface enabling water retention. The cuticle may have a flat surface or display characteristic ornamentation.

In *Baccharis*, the leaves have straight (Fig. 9.3a) to wavy (Fig. 9.3b) anticlinal epidermal cell walls as seen in *B. illinita* DC. (Fig. 9.3a), *B. microdonta*, *B. punctulata* DC. (Fig. 9.3b), *B. sphenophylla* Dusén ex Malme (Budel et al. 2018a), *B. uncinella* DC. (Budel and Duarte 2008a), *B. junciformis* (Budel and Duarte 2010), *B. trilobata*, and *B. brevifolia* (Bobek et al. 2016). However, sinuous anticlinal epidermal walls (Fig. 9.3c) were reported for *B. anomala* DC. (Budel and Duarte 2008b), *B. decussata* (Klatt) Hieron., and *B. pentlandii* DC. (Freire et al. 2007). All of these species belong to *Baccharis* subgen. Molina (Pers.) Heering. This characteristic can be helpful for the identification of the species, as well as for the delimitation of *Baccharis* sections.

In *Baccharis*, the cuticle is typically striated (Fig. 9.3e–i), but it can also be smooth as observed in the leaves of *B. microdonta* (Fig. 9.3d), and winged stems of

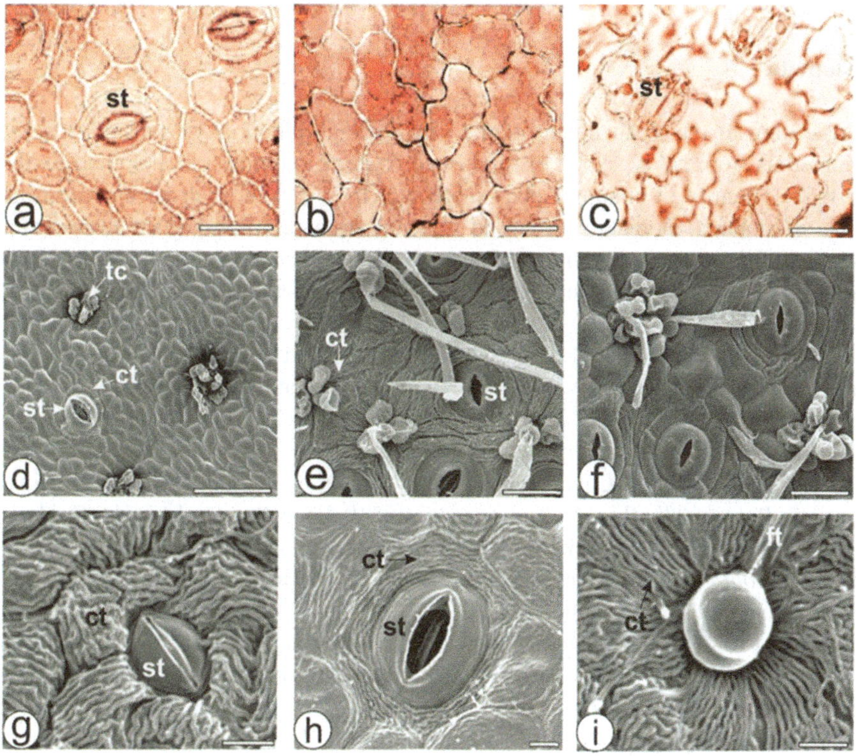

Fig. 9.3 Leaf epidermis in *Baccharis*. [(**a–c**) Light microscopy, stained in safranin; (**d–i**) scanning electron microscopy]. *B. anomala* (**c**), *B. illinita* (**a**), *B. microdonta* (**d**), *B. paucifloscuIosa* (**e**, **i**), *B. punctulata* (**b**, **g**), *B. sphenophylla* (**h**), and *B. trilobata* (**f**). Abaxial side (**c**, **g**, **h**) and adaxial side (**d**, **f**, **i**). Scale bars: b, f = 50 μm, a, c–e, g–i = 100 μm. (**a**, **b**, **d**, **e**, **g**, **h**, **i** reproduced with permission from Budel and Duarte 2008b; from Bobek et al. 2016; and from Budel et al. 2018a; **c**)

B. articulata (Cortadi et al. 1999) and *B. myriocephala* Baker (Sá and Neves 1996). The striate cuticle can occur on both surfaces or either on the abaxial or the adaxial sides (Bobek et al. 2016; Budel et al. 2018a; Almeida et al. 2021). Besides, species of this genus can show different cuticle orientation around the stomata (Fig. 9.3g, h) and the trichomes (Fig. 9.3i). For example, *B. punctulata* presents the cuticle striations arranged in a perpendicular orientation around stomata (Fig. 9.3g); cuticle striations can also occur perpendicular to trichome bases (Fig. 9.3i) or as concentric rings around the stomata (Fig. 9.3h), as reported in *B. pentaptera* (Budel et al. 2015). The cuticle is generally smooth in *B. microdonta* (Fig. 9.3d), but it is striated and radiated around stomata and trichome bases (Budel et al. 2018a).

In *Baccharis*, the leaf epidermis is usually unilayered and covered by a thin cuticle (Budel et al. 2018a). However, a thick cuticle was observed on both sides of *B. coridifolia* DC. (Budel and Duarte 2007), *B. ochracea* Spreng. (Barreto et al. 2015), and *B. spicata* (Lam.) Baill. (Oliveira et al. 2011) and on the adaxial leaf surface of *B. uncinella* (Budel and Duarte 2008a). Thick cuticles are also observed in the stems of some *Baccharis* species, for example, *B. spicata* (Oliveira et al. 2011).

Stomata

According to the arrangement of the surrounding epidermal or subsidiary cells, numerous stomatal types are distinguished. Types of stomata can help narrow the possible identity of unknown plant material. The presence or absence of stomata, size of stomata, stomatal index, and types of stomata are important features for characterizing and differentiating the species. In that sense, Rodriguez et al. (2010) reported that the density of stomata can help differentiate *B. articulata* out of *B. crispa*.

In *Baccharis*, anomocytic (Fig. 9.4a, e) and anisocytic (Fig. 9.4b) stomata types are the most common. However, other types of stomata have also been reported, such as cyclocytic type in *B. articulata*, *B. brevifolia*, *B. illinita*, *B. microdonta* (Fig. 9.4c), and *B. notosergila* Griseb.; actinocytic in *B. brevifolia* (Fig. 9.4d), *B. boliviensis* (Wedd.) Cabrera, *B. conferta* Kunth, and *B. pauciflosculosa*; hexacytic in *B. brevifolia* (Fig. 9.4h); tetracytic (Fig. 9.4f) in *B. reticularioides* Deble & A.S.Oliveira, *B. sphenophylla*, and *B. trilobata*; and staurocytic in *B. conferta*, *B. microdonta* (Fig. 9.4g), and *B. pauciflosculosa* (Freire et al. 2007; Pereira et al. 2014; Bobek et al. 2016; Budel et al. 2018a).

Amphistomatic leaves frequently occur in *Baccharis* (Freire et al. 2007; Molares et al. 2009; Budel et al. 2013; Bobek et al. 2015a, b; Barreto et al. 2015). However, some species, including *B. coridifolia* (Budel and Duarte 2007) and *B. punctulata* (Budel et al. 2018a), possess hypostomatic leaves. This characteristic is helpful in the diagnosis of *Baccharis* species, as demonstrated in the studies of Oliveira et al. (2011), Bobek et al. (2016), and Budel et al. (2018a).

Micro-measurements of stomata revealed that the majority of the species have stomata between 20 and 60 μm long (Freire et al. 2007; Rodriguez et al. 2013; Budel et al. 2018a). However, some species, such as *B. articulata* (60–75 μm) and

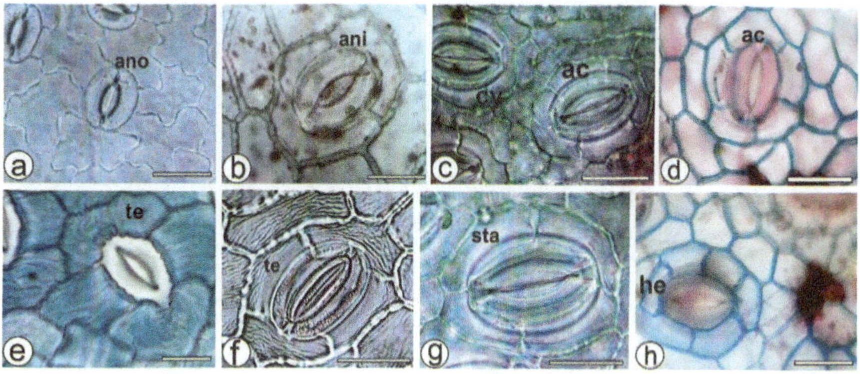

Fig. 9.4 Stomata in *Baccharis* species. *B. brevifolia* (**d, h**), *B. curitybensis* Heering ex Malme (**e**), *B. microdonta* (**c, g**), *B. junciformis* (**b**), *B. punctulata* (**a**), and *B. reticularioides* (**f**). [ano, anomocytic; ani, anisocytic; ac, actinocytic; cy, cyclocytic; he, hexacytic; te, tetracytic; sta, staurocytic]. Scale bars: a, g–i = 50 μm, b–f = 100 μm. (Reproduced with permission from Budel et al. 2018a (**a, c, f, g**); Oliveira et al. 2011 (**b, e**); and Bobek et al. 2016 (**d, h**))

B. illinita (75–105 μm), have larger stomata measuring more than 60 μm in length (Freire et al. 2007). The value of the stomatal index (percentage stomata out of the total number of epidermal cells plus stomata) is reported to range from 5 (*B. articulata*) to 16 (*B. punctulata*) (Rodriguez et al. 2013; Budel et al. 2018a).

Trichomes

Trichomes are small hairs or other outgrowths from the epidermis of plants. The two primary types of trichomes are glandular and non-glandular. Glandular trichomes possess the apical cells modified to secrete or store substances. Commonly, the secretions of these trichomes are responsible for biological activities. The recognition of the type and form of trichomes has been helpful in species identification. Non-glandular trichomes, or the covering trichomes, are generally characterized by having an acute apical cell. In *Baccharis*, trichomes are more common in leaves, but they are also sometimes present in stems and flowers.

Trichomes are considered to be the most important anatomical markers for the diagnosis of *Baccharis* species, followed by stomata type and epidermal cell walls (Freire et al. 2007). They usually appear isolated (Figs. 9.3i, and 9.5a, c–g) or in clusters (Figs. 9.3d, e, and 9.5h, i, k, l), arising from epidermal depressions (Fig. 9.5f, g, j). Simple non-glandular trichomes (Fig. 9.5a) formed by around six cells were present only in *B. anomala* (Budel and Duarte 2008b). Considering glandular trichomes, flagelliform (Figs. 9.3e, and 9.5b, c, e, f, g, i) and biseriate (Fig. 9.5f, h–l) types of trichomes have been commonly reported in *Baccharis* species (Budel et al. 2005; Freire et al. 2007; Bobek et al. 2016; Budel et al. 2018a; Almeida et al. 2021).

Biseriate glandular trichomes are formed by two pairs of basal cells and a head with up to four pairs of secretory cells. They appear singly as in *B. uncinella* (Budel

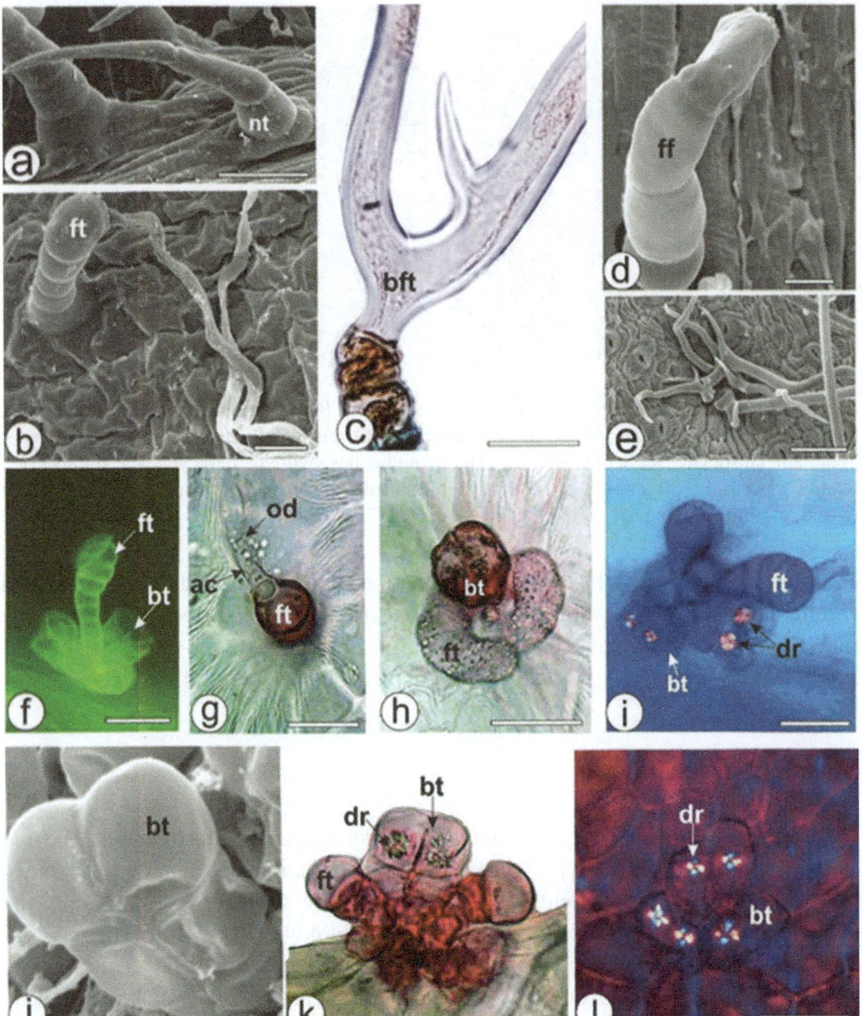

Fig. 9.5 Trichomes in *Baccharis* species [(**a**, **b**, **d**–**f**, **j**) SEM; (**i**, **k**, **l**) polarized light; (**f**) fluorescent light; all others: ordinary light microscopy]. *B. anomala* (**a**, **b**), *B. articulata* (**j**), *B. erioclada* (**e**), *B. microdonta* (**f**, **k**), *B. megapotamica* Spreng. (**d**), *B. pauciflosculosa* (**g**), *B. punctulata* (**h**), *B. sphenophylla* (**i**, **l**), *B. spicata* (**b**), and *B. uncinella* (**c**). ac, apical cell; bft, branched flagelliform trichome; bt, biseriate trichome; dr, druses; ff, filiform flagelliform trichome; ft., simple flagelliform trichome; nt, non-glandular trichome; od, oil droplets. Scale bars: c = 100 μm, a, e, f, g, h = 50 μm, b, i, k, l = 25 μm, d, j = 10 μm. (Reproduced with permission from Budel and Duarte 2008b (**a**, **b**); Bobek et al. 2015a (**e**); and Budel et al. 2018a (**f**, **g**, **k**))

and Duarte 2008a), in clusters of a single trichome type (Fig. 9.5l) as in *B. sphenophylla* (Budel et al. 2018a) or in clusters mixed with flagelliform trichomes (Fig. 9.5h, i, k) as seen in *B. illinita* (Budel et al. 2018a). Some species of *Baccharis* possess a pair of druses within each of the secretory head cells (Fig. 9.5i, k, l), as observed in *B. microdonta*, *B. punctulata*, and *B. sphenophylla* (Budel et al. 2018a).

Flagelliform glandular trichomes have some variations, including simple flagelliform with straight body (Fig. 9.5b, e, f, g, i), as seen in *B. microdonta* (Budel et al. 2018a), *B. pentaptera* (Budel et al. 2015), *B. ochracea* (Budel et al. 2012), *B. singularis* (Souza et al. 2011), *B. spicata* (Oliveira et al. 2011), *B. trilobata* (Bobek et al. 2016), *B. aracatubaensis*, and *B. organensis* (Zuccolotto et al. 2019); branched with straight body (Fig. 9.3c, e), as observed in *B. coridifolia* (Budel and 2007), *B. uncinella* (Fig. 9.5c), and *B. erioclada* (Fig. 9.5e); aseptate simple flagellate in *B. artemisioides* Hook. & Arn. (Freire et al. 2007) and *B. caprariifolia* DC. (Bobek et al. 2015a); filiform flagellate with pointed terminal cell in *B. multiflora* Kunth or pear-like and rounded at the apex (Fig. 9.5d) in *B. megapotamica* (Budel et al. 2012); and flagellate with C-shaped curved body (Fig. 9.5h), as observed in *B. punctulata* (Budel et al. 2018a). The body in these trichomes is secretory, voluminous, and made up of 3–9 cells. The apical cell is whip-like, tubular, and translucent, containing dense oil substances (Fig. 9.5g).

Mesophyll

In *Baccharis*, the organization of leaf mesophyll has correspondence to the chlorenchyma arrangement in the wings of cladodes. The majority of the species of *Baccharis* possesses isobilateral mesophyll (Fig. 9.4a). However, dorsiventral arrangement (Fig. 9.4b) was observed in *B. anomala* (Budel and Duarte 2008b), *B. singularis* (Vell.) (Souza et al. 2011), *B. ochracea* (Fig. 9.6b), and *B. punctulata* (Budel et al. 2018a).

Oil bodies in the leaf mesophyll (Fig. 9.6c) are present in some *Baccharis* species, e.g., *B. illinita*, *B. microdonta*, *B. punctulata*, *B. reticularioides*, and *B. sphenophylla*. They are specifically located in palisade parenchyma cells and some spongy parenchyma cells (Budel et al. 2018a). Minor collateral vascular bundles surrounded by an endodermis traverse the spongy parenchyma (Fig. 9.6d) in all *Baccharis* species studied (Bobek et al. 2016; Budel et al. 2018a).

In the "carquejas," the chlorenchyma in the wing of cladodes consists of palisade parenchyma, comprising approximately three layers of short cells beneath both sides of the epidermis, and spongy parenchyma in the central region (Fig. 9.6e). The isobilateral arrangement of the photosynthetic parenchyma is in correspondence to what is described for the wings of *B. articulata*, *B. myriocephala*, *B. crispa* (Rodriguez et al. 2008), and *B. sagittalis* (Less.) DC. (Petenatti et al. 2007).

At the wing edges, "carquejas" usually have 2–3 layers of angular collenchyma below the epidermis, a collateral vascular bundle with a perivascular fiber cap adjoining the phloem and secretory ducts (Fig. 9.6f), as observed in *B. junciformis*

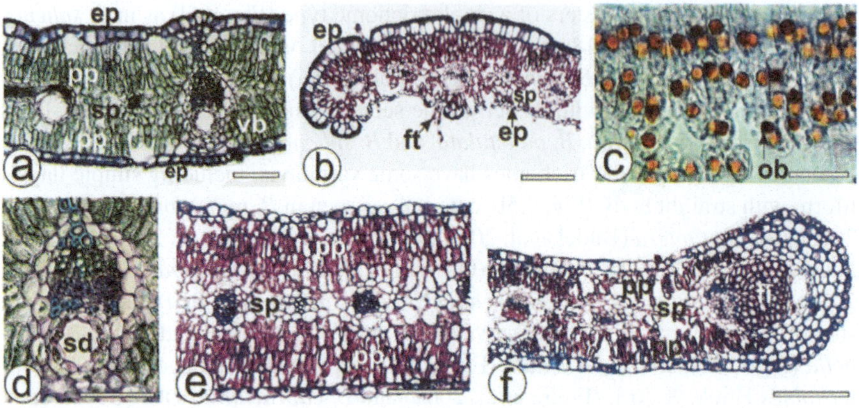

Fig. 9.6 Leaf anatomy of *Baccharis* spp.: leaf blade in cross section [(**c**) treated with Sudan III to stain lipophilic/oil content; (**a**, **d**) stained with toluidine blue; (**e**, **f**) stained with astra blue and basic fuchsin] . *B. ochracea* (**b**), *B. pauciflosculosa* (**a**, **d**), *B. reticularioides* (**c**), *B. microcephala* (**e**), *B. junciformis* (**f**). ep, epidermis; fi, fibers; ob, oil bodies; pp., palisade parenchyma; sd, secretory duct; sp., spongy parenchyma; vb, vascular bundle. Scale bars: a, b, d, f = 100 μm; c, e = 50 μm. (Reproduced with permission from Barreto et al. 2015 (**b**); Budel et al. 2018a (**d**); and Budel and Duarte 2010 (**e**, **f**))

(Budel and Duarte 2010). However, only sclerenchymatous tissue consisting of fibers can be found at the wing edges of some species, such as *B. myriocephala* (Sá and Neves 1996).

Midrib Shape in Cross Section

The shape of the midrib in a transverse section helps in the diagnosis of species as reported for different genera, including *Passiflora* (Wosch et al. 2015), *Mikania* (Almeida et al. 2017), and *Baccharis* (Bobek et al. 2016; Budel et al. 2018a) (Fig. 9.7a–i). Different shapes were reported for *Baccharis* species, such as plano-convex and prominently rounded on the abaxial side (Fig. 9.7a) in *B. anomala*; biconvex in *B. illinita* (Fig. 9.7c), *B. junciformis* (Fig. 9.7i), *B. cognata* (Budel et al. 2013), and *B. reticularioides* (Budel et al. 2018a); biconvex with a rounded projection on the adaxial side in *B. pauciflosculosa* (Budel et al. 2018a); slightly concave-convex in *B. rufescens* (Fig. 9.7f); concave-convex in *B. microdonta* (Fig. 9.7h); almost flat-convex in *B. caprariifolia* (Fig. 9.7b) and *B. megapotamica* (Fig. 9.7c); flat-convex, but truncate on the abaxial side, in *B. ochracea* (Fig. 9.7e); flat on both sides in *B. trilobata* (Bobek et al. 2016); and convex-flat in *B. cultrata* (Fig. 9.7g).

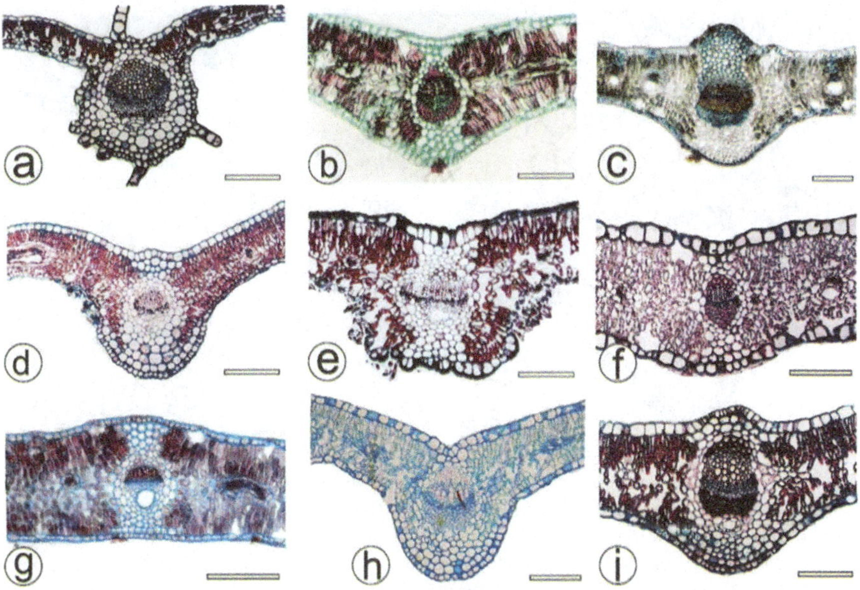

Fig. 9.7 Midrib anatomy in *Baccharis*. *B. anomala* (**a**), *B. caprariifolia* (**b**), *B. illinita* (**c**), *B. megapotamica* (**d**), *B. ochracea* (**e**), *B. rufescens* (**f**), *B. cultrata* (**g**), *B. microdonta* (**h**), and *B. junciformis* (**i**). Scale bars: a = 25 μm; b–h = 100 μm. (Reproduced with permission from Budel and Duarte 2008a (**a**); Bobek et al. 2015a (**b**); Barreto et al. 2015 (**e**); Budel et al. 2018a (**c**); and Bobek et al. 2016 (**g, h**))

Vascular System

In the leaves, collateral vascular bundles traverse the mesophyll and are commonly encircled by a parenchymatous sheath (Fig. 9.8a). The midrib vascular system is frequently represented by one collateral and circular bundle that is surrounded by ground parenchyma with lignified perivascular fibers abutting the xylem and phloem, as observed in *B. illinita* (Fig. 9.8b) and *B. microdonta* (Fig. 9.8c).

In the petiole, the vascular system is collateral and commonly similar to the midrib. Otherwise, they can show different organization, such as in *B. singularis* that presents one vascular bundle in open arc (Souza et al. 2011), whereas three or more collateral vascular bundles arranged in an open arc (Fig. 9.8d) are observed in *B. spicata* (Oliveira et al. 2011), *B. glaziovii* (Jasinski et al. 2014), and *B. microdonta* (Bobek et al. 2016).

In the regular stem (Fig. 9.8e) as well as the central axis of the cladodes (Fig. 9.8f), the vascular cylinder presents phloem outward and xylem inward. The xylem tracheary elements are arranged in radial rows and separated by parenchyma cells and fibers (Oliveira et al. 2011; Souza et al. 2011; Jasinski et al. 2014; Budel et al. 2015; Bobek et al. 2015a, b, 2016).

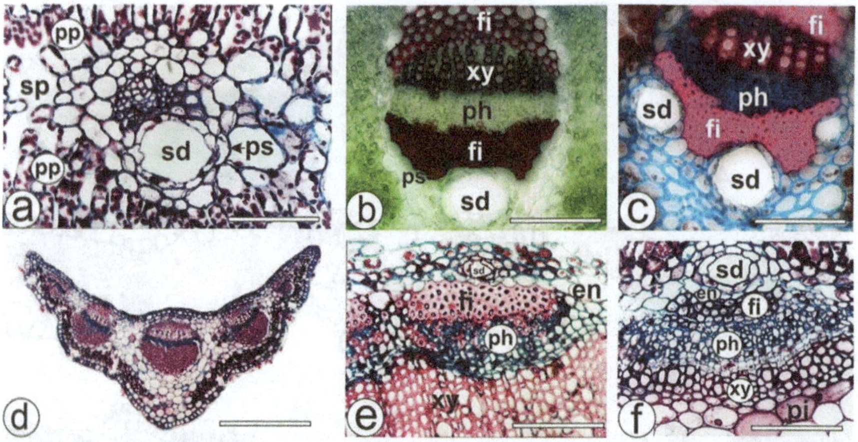

Fig. 9.8 Anatomy of *Baccharis.* (**a**) Leaf blade. (**b, c**) Midrib. (**d**) Petiole. (**e**) Stem. (**f**) Caulinar axis. (**a, c–f**) Stained in astra blue and basic fuchsin; (**b**) treated with phloroglucinol/HCl. *B. pentaptera* (**a, f**), *B. illinita* (**b**), *B. reticularioides* (**c**), *B. spicata* (**d**), *B. coridifolia* (**e**). [fi, fibers; ph, phloem; pi, pith; pp., palisade parenchyma; ps, parenchyma sheath; sd, secretory duct; sp., spongy parenchyma; xy, xylem]. Scale bars: a–f = 50 µm. (Reproduced with permission from Budel et al. 2015 (**a, f**); Oliveira et al. 2011 (**d**); and Budel and Duarte 2007 (**e**))

Secretory Ducts

Internal secretory structures may have various forms; some are isodiametric, such as the secretory cells and cavities, while others are elongated, such as the secretory ducts. In *Baccharis*, these ducts have uniseriate epithelium of 6–10 cells (Fig. 9.8a, b, c, e, f) containing dense cytoplasm and lipophilic contents. In the mesophyll, the secretory ducts are associated with the minor vascular bundles (Fig. 9.8a), adjacent to the phloem and the parenchyma sheath (Fig. 9.8a, b). In the midrib, usually, a single secretory duct occurs near the phloem (Fig. 9.8b), as observed in *B. coridifolia* (Budel and Duarte 2007), *B. ochracea* (Barreto et al. 2015), and *B. aracatubaensis* Malag. (Zuccolotto et al. 2019). However, *B. pauciflosculosa* (Budel et al. 2018a) has two secretory ducts, and *B. organensis* Baker has three secretory ducts in the midrib (Zuccolotto et al. 2019). Independent of whether regular stems or cladodes are examined, the secretory ducts are localized in the inner portion of the cortex next to the parenchyma sheathes (Fig. 9.8e, f) in all *Baccharis* species (Budel et al. 2003, 2004a, b, 2013; Budel and Duarte 2009; Oliveira et al. 2011; Souza et al. 2013; Pereira et al. 2014; Bobek et al. 2015a, b; Almeida et al. 2021).

Crystals

Several plants accumulate a variety of shapes and sizes of insoluble calcium salts, which, based on their morphology, can be crystalline sand, druses, raphides, styloids, and prismatic crystals. Typically, the morphotypes of crystals as well as their distribution in plant tissues are constant within a specific taxon (Franceschi and Horner 1980; Nakata 2003).

Crystals in plants have several functions, including the elimination of the excess of calcium ions, detoxification of heavy metals and aluminum, cellular ion balance, osmotic regulation, tissue mechanical support, and promotion of mechanical defense (Franceschi and Nakata 2005). Although calcium oxalate is more common in plants, calcium sulfate, magnesium oxalate, and other elements, such as potassium and silicon, can also be found in the crystals (He et al. 2012).

The identification of calcium oxalate crystals is done using a light microscope (LM), polarizing microscope, or scanning electron microscope (SEM). Energy-dispersive X-ray spectroscopy (EDS) coupled with SEM (Brito et al. 2021) is used to identify the elemental chemical composition of the crystals (as shown in Fig. 9.9).

Crystals of calcium oxalate are considered important for authentication purposes because the form, shape, and occurrence of these crystals in plants are species- and tissue-specific; hence, the presence or absence of a particular type of crystal can be used as a taxonomic character (Lersten and Horner 2011; Horner et al. 2012). They are also relevant in the systematic investigations and phylogenetic and ecophysiological characteristics of several plant families (Lersten and Horner 2011).

Different crystals morphotypes have been reported for several species of *Baccharis* and are often found in the stem pith (Budel and Duarte 2008b; Souza et al. 2011; Oliveira et al. 2011; Jasinski et al. 2014; Barreto et al. 2015; Bobek et al. 2015a, b; Budel et al. 2015; Almeida et al. 2021). The most common types are styloids (Fig. 9.10a, d, e, h, i, j, k), bipyramids (Fig. 9.10b, c, e, i), square bipyramids (Fig. 9.10f, l), and elongated square bipyramids (Fig. 9.10j); however, other types such as tile-shaped crystals (Fig. 9.10g) are also rarely observed.

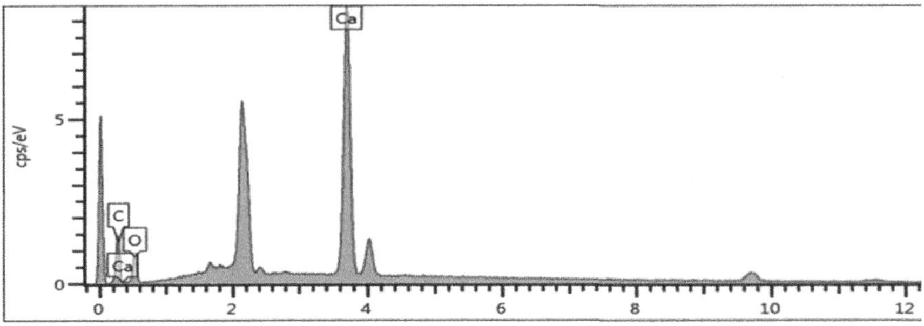

Fig. 9.9 EDS spectrum of a bipyramidal crystal in the stem of *B. pluricaptulata*. The unlabeled peaks in the spectra represent conductive metal used for coating the samples for SEM analysis

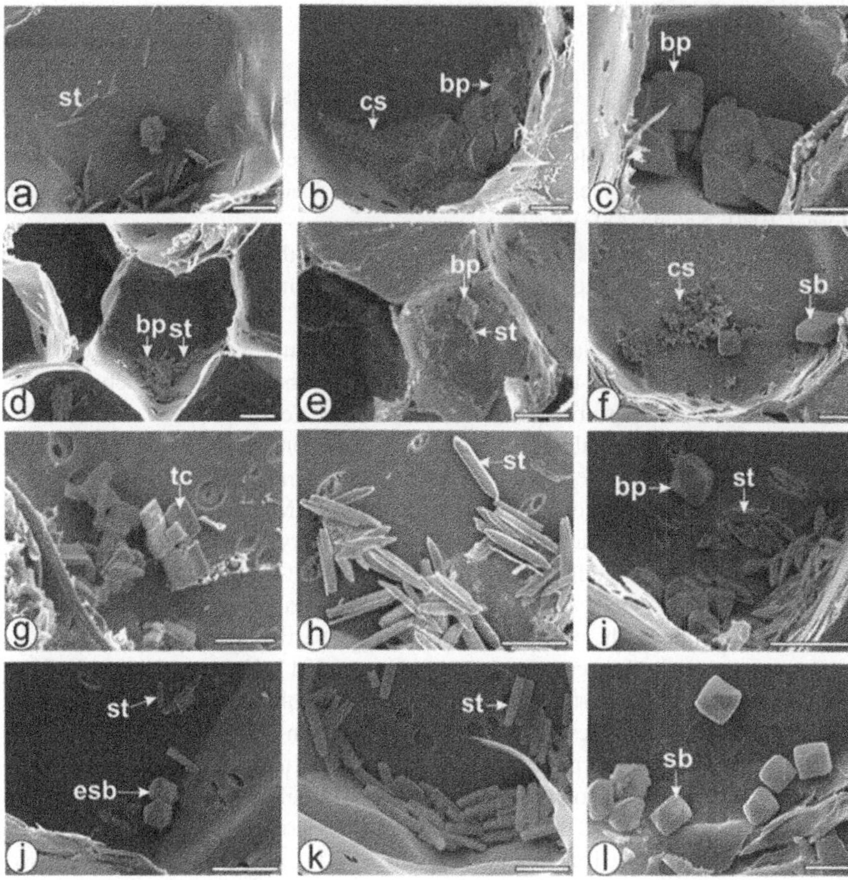

Fig. 9.10 Calcium oxalate crystals in *Baccharis* stem pith (SEM). *B. microdonta* (**a**), *B. pauciflosculosa* (**b**, **c**), *B. punctulata* (**d**, **e**), *B. reticularioides* (**f**, **g**), *B. sphenophylla* (**h**, **i**, **j**), *B. trilobata* (**k**, **l**). [bp, bipyramidal; cs, crystal sand; esb, elongated square bipyramidal; sb, square bipyramidal; st, styloids; tc, tile-shaped crystal]. Scale bar: a–f, h-j = 10 μm; g, k, l = 5 μm

Several species of *Baccharis* show different shapes of crystals within the same species. For example, crystal sand, styloids and square dipyramids in *B. brevifolia*; rare styloids in *B. microdonta*; crystal sand and square dipyramids in *B. pauciflosculosa*; and crystal sand, square dipyramids, elongated square dipyramids, styloids, and tabular crystals in piles that look like a tower in *B. trilobata* (Bobek et al. 2016) are observed. Additionally, a pair of druses is found within the secretory head of biseriate glandular trichomes (Fig. 9.5i, k, l) in some species of *Baccharis* (Budel et al. 2018a).

Morphotypes of calcium oxalate crystals have been described in the caulinar axis of *carquejas*, for instance, crystal sand and square bipyramids in *B. junciformis* (Budel and Duarte 2010); elongated square bipyramids in *B. articulata*; elongated square bipyramids, square bipyramids, cubes, and tetragonal prisms in *B. crispa*

(Cortadi et al. 1999); raphides and hexagonal and tetragonal prisms in *B. triangularis*; raphides and tetragonal prisms in *B. sagittalis* (Petenatti et al. 2007) and square bipyramids in *B. microcephala* (Budel and Duarte 2009); and raphides, styloids, and elongated square bipyramids in *B. glaziovii* (Jasinski et al. 2014).

Other Stem Characteristics

The cortex in *Baccharis*, not only in the stems but also in the caulinar axis, has collenchyma in alternation with chlorenchyma in several species, e.g., *B. anomala*, *B. microcephala*, *B. ochracea*, *B. spicata*, *B. stenocephala*, *B. uncinella*, and *B. junciformis*. This characteristic was also found in *B. sagittalis* and *B. triangularis* (Petenatti et al. 2007). However, a continuous stratum of collenchyma could be observed in some members of the genus, such as *B. caprariifolia* and *B. singularis* (Souza et al. 2011). In addition, one to five layers of angular collenchyma were found in *B. brevifolia*, *B. microdonta*, *B. pauciflosculosa*, and *B. trilobata*, particularly in the ribs (Bobek et al. 2016).

The endodermis is present in stems and roots and occurs as a continuous uniseriate layer separating the central cylinder from the cortical tissues. Ariza-Espinar (1973) reported that Casparian strips were absent in *Baccharis*. On the contrary, endodermis with Casparian strips was observed in the central axis of *B. myriocephala* (Sá and Neves 1996) and *B. crispa* (Cortadi et al. 1999).

Root Characteristics

Anatomical studies involving *Baccharis* roots are uncommon in literature, except for the root of *B. crispa* (syn. *B. trimera* (Less.) DC.), which has been previously studied. In a transverse section, the roots of this species exhibit early secondary growth with persistent unilayered rhizodermis. The cortex has 6–8 layers of assimilation parenchyma containing starch grains, followed by endodermis and pericycle layers. The vascular system is formed by a continuous ring of phloem with fiber caps. The xylem is porous and diffuse with solitary or rarely small groups of vessel elements and numerous fibers (Minteguiaga et al. 2018).

4 Final Considerations

Even though the two groups, carquejas and vassouras, are easily distinguishable by their stem and leaf morphological features, it is often difficult to distinguish the species within the same group due to their close resemblances. To address this problem, anatomical and micromorphological features of the whole plants and their parts can be used. The most important features that can help in the identification of different species of *Baccharis* include the habit of the plant, branching pattern, features of stems, presence or absence of cladodes and wings, leaf characters, and arrangement of the floral heads.

Anatomical and micromorphological features have played important roles in the identification of complex species. Numerous reports stress the usefulness of anatomical characteristics in the identification of *Baccharis* species. Noteworthy features that can help in the quality control and authentication of *Baccharis* species include the epidermis (contour of anticlinal epidermal cell walls and type of stomata and trichomes), mesophyll, vascular tissue, secretory ducts, and the type and macro pattern of calcium oxalate crystals.

Despite the importance of morpho-anatomical studies in solving taxonomic problems, a large portion of the genus remains unexplored. Therefore, future studies focusing on comparative anatomy and micromorphology of *Baccharis* species from different sections will not only help in the quality control and species identification but also aid in the development of a more accurate classification of this large and diversified genus.

References

Almeida VP, Hirt AA, Raeski PA et al (2017) Comparative morphoanatomical analysis of *Mikania* species. Rev Bras Farmacogn 27:9–19. https://doi.org/10.1016/j.bjp.2016.05.002

Almeida VP, Heiden G, Raman V et al (2021) Microscopy and histochemistry of leaves and stems of *Baccharis* subgenus *Coridifoliae* (Asteraceae) through LM and SEM-EDS. Microsc Microanal 27:1273–1289

Ariza-Espinar L (1973) Las especies de *Baccharis* (Compositae) de Argentina Central. Bol Acad Nac Cienc Córdoba 50:176–305

Barreto IF, Paula JP, Farago PV et al (2015) Pharmacobotanical study of leaves and stems of *Baccharis ochracea* Spreng for quality control. Lat Am J Pharm 34:1497–1502

Barroso GM (1976) Compositae – Subtribo Baccharidinae Hoffmann – Estudo das espécies ocorrentes no Brasil. Rodriguésia 28:1–273

Barroso GM, Bueno OL (2002) Compostas. Subtribo: Baccharidinae. Flora Ilustrada Catarinense. Herbário Barbosa Rodrigues, Santa Catarina, pp 1–302

Bobek VB, Almeida VP, Pereira CB et al (2015a) Comparative pharmacobotanical analysis of *Baccharis caprariifolia* DC. and *B. erioclada* DC. from Campos Gerais, Parana, Southern of Brazil. Lat Am J Pharm 34:1396–1402

Bobek VB, Budel JM, Nakashima T (2015b) Contribution to the pharmacobotanical study of *Baccharis*: *B. cultrata* Baker. Visão Acadêmica 16:18–29. https://doi.org/10.5380/acd.v16i1.40371

Bobek VB, Heiden G, Oliveira CF et al (2016) Comparative analytical micrographs of vassouras (*Baccharis*, Asteraceae). Rev Bras Farmacogn 26:665–672. https://doi.org/10.1016/j.bjp.2016.05.001

Brito PS, Sabedotti C, Flores TB, Raman V, Bussade J, Farago P, Manfron J (2021) Light and scanning electron microscopy, energy dispersive X-ray spectroscopy, and histochemistry of *Eucalyptus tereticornis*. Microsc Microanal 27:1295–1303

Budel JM, Duarte MR (2007) Caracteres morfoanatômicos de partes vegetativas aéreas de *Baccharis coridifolia* DC. (Asteraceae-Astereae). Lat Am J Pharm 26:723–731

Budel JM, Duarte MR (2008a) Estudo farmacobotânico de folha e caule de *Baccharis uncinella* DC., Asteraceae. Lat Am J Pharm 27:740–746

Budel JM, Duarte MR (2008b) Estudo farmacobotânico de partes vegetativas aéreas de *Baccharis anomala* DC., Asteraceae. Rev Bras Farmacogn 18:761–768. https://doi.org/10.1590/S0102-695X2008000500022

Budel JM, Duarte MR (2009) Análise morfoanatômica comparativa de duas espécies de carqueja: *Baccharis microcephala* DC. e *B. trimera* (Less.) DC., Asteraceae. Braz J Pharm Sci 45:75–85. https://doi.org/10.1590/S1984-82502009000100009

Budel JM, Duarte MR (2010) Macro and microscopic characters of the aerial vegetative organs of carqueja: *Baccharis usterii* Heering. Braz Arch Biol Technol 53:123–131. https://doi.org/10.1590/S1516-89132010000100016

Budel JM, Duarte MR, Santos CAM et al (2003) Macro and microscopical identification of four species of *Baccharis* from *Trimera* group. Rev Bras Farmacogn 13:42–43. https://doi.org/10.1590/S0102-695X2003000400014

Budel JM, Duarte MR, Santos CAM et al (2004a) Morfoanatomia foliar e caulinar de *Baccharis dracunculifolia* DC., Asteraceae. Acta Farm Bonaer 23:477–483

Budel JM, Duarte MR, Santos CAM (2004b) Stem morphoanatomy of *Baccharis cylindrica* (Less.) DC., Asteraceae. Rev Bras Cienc Farm 40:93–99. https://doi.org/10.1590/S1516-93322004000100014

Budel JM, Duarte MR, Santos CAM et al (2005) O progresso da pesquisa sobre o gênero *Baccharis*, Asteraceae: I – estudos botânicos. Rev Bras Farmacogn 15:268–271. https://doi.org/10.1590/S0102-695X2005000300018

Budel JM, Matzenbacher NI, Duarte MR (2008) Genus *Baccharis* (Asteraceae): a review of chemical and pharmacological studies. In: Singh VK, Govil JN (eds) Phytopharmacology and therapeutic values III. Studium Press LLC, Houston, pp 1–18

Budel JM, Duarte MR, Döll-Boscardin PM et al (2012) Composition of essential oils and secretory structures of *Baccharis anomala, B. megapotamica* and *B. ochracea*. J Essent Oil Res 24:19–24. https://doi.org/10.1080/10412905.2012.645634

Budel JM, Farago PV, Duarte MR (2013) Pharmacobotanical study of *Baccharis cognata* DC. (Asteraceae: Astereae). Lat Am J Pharm 32:540–544

Budel JM, Paula JP, Santos VLP et al (2015) Pharmacobotanical study of *Baccharis pentaptera*. Rev Bras Farmacogn 25:314–319. https://doi.org/10.1016/j.bjp.2015.07.007

Budel JM, Raman V, Monteiro LM et al (2018a) Foliar anatomy and microscopy of six Brazilian species of *Baccharis* (Asteraceae). Microsc Res Tech 88:832–842. https://doi.org/10.1002/jemt.23045

Budel JM, Wang M, Raman V et al (2018b) Essential oils of five *Baccharis* species: investigations on the chemical composition and biological activities. Molecules 23:1–19. https://doi.org/10.3390/molecules23102620

Cortadi A, Di Sapio O, McCargo J et al (1999) Anatomical studies of *Baccharis articulata, Baccharis crispa* and *Baccharis trimera*, "Carquejas" used in folk medicine. Pharm Biol 37:357–365. https://doi.org/10.1076/phbi.37.5.357.6054

Cuatrecasas J (1967) Revisión de las especies colombianas del género *Baccharis*. Rev Acad Colomb Cienc Exact Fis Nat 13:5–102

Cuatrecasas J (1969) Notas adicionales, taxonómicas y corológicas sobre *Baccharis*. Revista de la Academia Colombiana de Ciencias Exactas, Físicas y Naturales 13:201–226

De Candolle AP (1836) Compositae: *Baccharis*. In: Candolle D (ed) Prodromus systematis naturalis regni vegetabilis 5. Treuttel & Würtz, Paris, pp 398–429

eFloras (2008) Published on the Internet. http://www.efloras.org. Accessed 4 Mar 2019

Franceschi VR, Horner-Jr HT (1980) Calcium oxalate crystals in plants. Bot Rev 46:361–427

Franceschi VR, Nakata PA (2005) Calcium oxalate in plants: formation and function. Annu Rev Plant Biol 56:41–71. https://doi.org/10.1146/annurev.arplant.56.032604.144106

Freire SE, Urtubey E, Giuliano DA (2007) Epidermal characters of *Baccharis* (Asteraceae) species used in traditional medicine. Caldasia 29:23–38

Gianello JC, Ceñal JP, Giordano OS et al (2000) Medicamentos herbarios en el centro-oeste argentino II. "Carquejas": control de calidad de las drogas oficiales e sustituyentes. Acta Farm Bonaer 19:99–103

Giuliano DA (2001) Classificación infragenérica de las espécies Argentinas de *Baccharis* (Asteraceae, Astereae). Darwin 39:131–154

Gogosz AM, Boeger MRT, Negrelle RRB (2012) Anatomia foliar comparativa de nove espécies do gênero *Piper* (Piperaceae). Rodriguésia 63:405–417. https://doi.org/10.1590/S2175-78602012000200013

He H, Bleby TM, Veneklaas EJ et al (2012) Morphologies and elemental compositions of calcium crystals in phyllodes and branchlets of *Acacia robeorum* (Leguminosae: Mimosoideae). Ann Bot 109:887–896. https://doi.org/10.1093/aob/mcs004

Heering W (1904) Die Baccharis-Arten des Hamburger Herbars. Jahrbuch der Hamburgischen Wissenschaftlichen Anstalten 21:1–46

Heiden G, Pirani JR (2012) Novelties towards a phylogenetic infrageneric classification of *Baccharis* (Asteraceae, Astereae). Phytotaxa 289:285–290. https://doi.org/10.11646/phytotaxa.289.3.9

Heiden G, Antonelli A, Pirani JR (2019) A novel phylogenetic infrageneric classification of *Baccharis* (Asteraceae: Astereae), a highly diversified American genus. Taxon 68:1048–1081

Horner HT, Wanke S, Samain MS (2012) A comparison of leaf crystal macropatterns in the two sister genera *Piper* and *Peperomia* (Piperaceae). Am J Bot 99:983–997

Jackson BP, Sonwdon DW (1990) Atlas of microscopy of medicinal plants, culinary herbs and spices. CBS, New Delhi

Jasinski VCG, Silva RZ, Pontarolo R (2014) Morpho-anatomical characteristics of *Baccharis glaziovii* in support of its pharmacobotany. Rev Bras Farmacogn 24:609–616. https://doi.org/10.1016/j.bjp.2014.11.003

Jorge LIF (2000) Botânica aplicada ao controle de qualidade de alimentos e de medicamentos. Atheneu, São Paulo

Lersten NR, Horner HT (2011) Unique calcium oxalate "duplex" and "concretion" idioblasts in leaves of tribe *Naucleeae* (Rubiaceae). Am J Bot 98:1–11. https://doi.org/10.3732/ajb.1000247

Minteguiaga M, Mercado MI, Ponessa GI et al (2018) Morphoanatomy and essential oil analysis of *Baccharis trimera* (Less.) DC. (Asteraceae) from Uruguay. Ind Crop Prod 112:488–498. https://doi.org/10.1016/j.indcrop.2017.12.040

Molares S, Gonzalez SB, Ladio A et al (2009) Etnobotánica, anatomía y caracterización físico-química del aceite esencial de *Baccharis obovata* Hook. et Arn. (Asteraceae: Astereae). Acta Bot Bras 23:578–589. https://doi.org/10.1590/S0102-33062009000200030

Müller J (2006) Systematics of *Baccharis* (Compositae, Astereae) in Bolivia, including an overview of the genus. Syst Bot Monogr 76:1–341. https://doi.org/10.2307/25027945

Nakata PA (2003) Advances in our understanding of calcium oxalate crystal formation and function in plants. Plant Sci 164:901–909. https://doi.org/10.1016/S0168-9452(03)00120-1

Nesom GL (1990) Infrageneric taxonomy of North and Central American *Baccharis* (Asteraceae: Astereae). Phytologia 69:40–46. https://doi.org/10.5962/bhl.part.19647

Oliveira AMA, Santos VLP, Franco CRC et al (2011) Comparative morpho-anatomical study of *Baccharis curitybensis* Heering ex Malme and *Baccharis spicata* (Lam.). Baill. Lat Am J Pharm 30:1560–1566

Oliveira AS de, Deble LP, Shneider AA et al (2006) Checklist do gênero *Baccharis* L. para o Brasil (Asteraceae-Astereae). Balduinia 9:17–27

Pereira CB, Farago PV, Budel JM et al (2014) New contribution to the pharmacognostic study of carquejas: *Baccharis milleflora* DC., Asteraceae. Lat Am J Pharm 33:841–847

Petenatti EM, Petenatti ME, Cifuente DA et al (2007) Medicamentos herbarios en el centro-oeste Argentino. VI. Caracterización y control de calidad de dos espécies de "carquejas": *Baccharis sagittalis* y *B. triangularis* (Asteraceae). Lat Am J Pharm 26:201–208

Rodriguez MV, Gattuso S, Gattuso M (2008) *Baccharis crispa* y *Baccharis trimera* (Asteraceae): revisión y nuevos aportes para su normalización micrográfica. Lat Am J Pharm 27:387–397

Rodriguez MV, Martínez ML, Cortadi AA et al (2010) Characterization of three sect. *Caulopterae* species (*Baccharis*-Asteraceae) inferred from morphoanatomy, polypeptide profiles and spectrophotometry data. Plant Syst Evol 286:175–190. https://doi.org/10.1007/s00606-010-0297-y

Rodriguez MV, Gattuso SJ, Gattuso MA (2013) Micrographic standarization of *Baccharis* L. species (Asteraceae). Dominguezia 29:39–54

Sá MFA, Neves LJ (1996) Contribuição ao estudo das plantas medicinais: *Baccharis myriocephala* DC. Rev Bras Farm 77:88–96

Simões-Pires CA, Debenedetti S, Spegazzini E et al (2005) Investigation of the essential oil from eight species of *Baccharis* belonging to sect. *Caulopterae* (Asteraceae, Astereae): a taxonomic approach. Plant Syst Evol 253:23–32. https://doi.org/10.1007/s00606-005-0296-6

Souza CA, Farago PV, Duarte MR et al (2011) Pharmacobotanical study of *Baccharis singularis* (Vell.) G.M. Barroso, Asteraceae. Lat Am J Pharm 30:311–317

Souza JP, Santos VLP, Franco CRC et al (2013) *Baccharis rufescens* Spreng. var. *tenuifolia* (DC.) Baker: contribuição ao estudo farmacognóstico. Rev Bras Plantas Med 15:566–574. https://doi.org/10.1590/S1516-05722013000400013

Upton R, Graff A, Jolliffe G, Länger R, Williamson E (eds) (2011) American herbal pharmacopoeia botanical pharmacognosy – microscopic characterization of botanical medicines. CRC Press, Boca Raton

Wosch L, Imig DC, Cervi AC et al (2015) Comparative study of *Passiflora* taxa leaves. I. A morpho-anatomic profile. Rev Bras Farmacogn 25:328–343. https://doi.org/10.1016/j.bjp.2015.06.004

Zuccolotto T, Bressana J, Lourenço AVF et al (2019) Chemical, antioxidant, and antimicrobial evaluation of essential oils and an anatomical study of the aerial parts from *Baccharis* species (Asteraceae). Chem Biodiv 16:e1800547. https://doi.org/10.1002/cbdv.201800547

Chapter 10
Essential Oils of *Baccharis*: Chemical Composition and Biological Activities

Jane Manfron, Vijayasankar Raman, Ikhlas Ahmed Khan, and Paulo Vitor Farago

Abstract This chapter provides a review of the characteristics of essential oils of *Baccharis* species with special emphasis on their chemical composition and biological activities. Species of *Baccharis* show a great diversity in their morpho-anatomical and ecological features and also exhibit a wide range of chemical diversity in their essential oils. Several medicinally beneficial properties and biological activities, including anti-inflammatory, antimicrobial, antiulcerogenic, antimalarial, antioxidant, antitrypanosomal, cytotoxic, insecticidal, leishmanicidal, schistosomicidal, sedative, and adjuvant properties, have been reported for the essential oils of *Baccharis* species. Essential oils of *Baccharis* species and their major compounds have been reported to have medicinal properties and have shown significant activities against a range of insect pests, microorganisms, and parasites. Also, we illustrate the micromorphic features of various secretory structures that are responsible for biosynthesizing and storing the essential oils in *Baccharis* species, including glandular trichomes and secretory ducts.

Keywords Chemical composition · Medicinal properties · Insecticidal activity · Parasiticidal activities · Secretory structures

1 Introduction

Currently, it is estimated that more than 3000 essential oils (EOs) are known, and 300 of them have commercial value. EOs from a number of species are commonly used in pharmaceutical industries, aromatherapy and aromachology, household

---

J. Manfron (✉) · P. V. Farago
State University of Ponta Grossa, Department of Pharmaceutical Sciences, Ponta Grossa, PR, Brazil

V. Raman · I. A. Khan
National Center for Natural Products Research, School of Pharmacy, University, MS, USA

G. W. Fernandes et al. (eds.), *Baccharis*,
https://doi.org/10.1007/978-3-030-83511-8_10

cleaning products, as flavoring and antimicrobial agents in food products, and as flavoring agents in cigarettes, drinks, perfumeries, and cosmetics. EOs are also used in air fresheners and deodorizers as well as in balneology and massage therapy. Isolated compounds of the EOs are obtained by extraction from the plant or synthesized (Calo et al. 2015; Schmidt 2016).

The genus *Baccharis* has been widely studied mainly on the chemical composition of EOs since the early 1900s. The genus has provided valuable biomolecules in the discovery of new medicinal natural products (Abad and Bermejo 2007). The most important classes of organic compounds present in the EOs are phenolics and terpenoids. Considering the terpenoids, monoterpenoids, sesquiterpenoids, and diterpenoids especially *neo*-clerodane, labdane, and kaurene are found in the EOs of *Baccharis* (Campos et al. 2016).

Species of *Baccharis* produce EOs composed mainly of monoterpenoids and sesquiterpenoids. Sesquiterpenoids are generally more abundant in the majority of the species. However, EOs of some species contained more monoterpenoids than sesquiterpenoids (Budel et al. 2018b). A bibliographic review stated that about 60 compounds (concentrations above 10%) identified in the EOs of 16 species of *Baccharis* have shown important biological activities (Campos et al. 2016).

In *Baccharis*, there are two economically important groups of species occurring in South America, namely, carquejas and vassouras. Carquejas is the folk name referring to the plants possessing cladodes. The most important and well-studied carqueja is *B. crispa* Spreng. (syn. *B. trimera* (Less.) DC.), which has been included in the latest edition of the Brazilian Pharmacopeia. The major compound of its EO is carquejyl acetate (Minteguiaga et al. 2018a). In Argentina, the medicinal carquejas recognized by the 6th Argentinean Pharmacopeia are *B. crispa* and *B. articulata* (Lam.) Pers. *B. crispa* is in traditional medicine used as stomachic and diuretic (Budel et al. 2008). Vassouras is the common name given to species without cladodes, having leaves and normal stems. The main example is *B. dracunculifolia* DC., which is used in folk medicine to treat gastric disorders and largely employed in the fragrance industry due to the high content of *E*-nerolidol in its EO (Budel et al. 2008).

Several medicinal properties defined for *Baccharis* are attributed to their EOs. In that context, the EOs of species of *Baccharis* presented many beneficial properties, such as antibacterial, antifungal (Valarezo et al. 2015; Negreiros et al. 2016; Perera et al. 2017), antitrypanosomal (Budel et al. 2018b), antiviral, antioxidant (Sobrinho et al. 2016; Zuccolotto et al. 2019; Oliveira et al. 2019), anti-inflammatory (Florão et al. 2012), schistosomicidal (Oliveira et al. 2012), cytotoxic (Pereira et al. 2017), sedative (Ascari et al. 2012), larvicidal against *Aedes aegypti* (Botas et al. 2017), and insecticidal actions against bed bugs (Budel et al. 2018b).

2 Essential Oils

General Characteristics

Essential oils (EOs) are complex mixtures of mainly low-molecular-weight components biosynthesized and stored in specialized secretory structures of plants and are extracted by different methods from whole plants or plant parts. They can act as chemical signals in the plant kingdom and as chemical defense against the animal kingdom, presenting a biological function vital to the survival and adaptation of the plants to the environment. There is a great variation in the chemical composition of EOs from different taxa of plants.

EOs of aromatic plants present major volatile components biosynthesized through three different biosynthetic pathways, the methylerythritol pathway leading to mono- and diterpenoids, the mevalonate pathway leading to sesquiterpenoids, and the shikimic acid pathway leading to phenylpropanoids (Franz and Novak 2016).

EOs can be extracted by different processes, depending on the part of the plant, amount of the plant material, and the quality required. The usual methods used in EO extraction include steam distillation, hydrodistillation, cold mechanical processing, Soxhlet extraction, solvent extraction, microwave-assisted hydrodistillation, supercritical solvent, and headspace techniques (Elshafie and Camele 2017). They are usually analyzed by GC/MS and/or GC/FID, and the compounds are identified by comparison of their retention indexes (RIs) and mass spectra with literature data.

The yield and the chemical composition of the EOs can be influenced by the plant genotype, development stage, environmental conditions (day length, irradiance, temperature, and water supply), phenological factors, physiological variations inherent to the plant, genetic features of the cultivars, plant nutrition, stress during growth or maturity, application of fertilizers, drying conditions of the plant material, storage conditions, grinding method, and the EO extraction methods (Gobbo-Neto and Lopes 2007; Tischer et al. 2017). Considering all the factors that influence the chemical composition, chemotaxonomic reports and conclusions have to be based on comparable plant material, grown and harvested under analogous conditions (Budel et al. 2018b).

Also, phytochemical polymorphism is frequently the case between different plant organs. Even though several species store qualitatively similar compounds in each organ, some produce different components, resulting from the same biosynthetic pathway. At the same time, there are also species in which the chemical composition of different plant organs presents no close connection between their biosynthetic origins (Németh-Zámboriné 2016). It is important to highlight that EOs from different plant parts such as leaves, stems, flowers, roots, and fruits may exhibit different properties due to differences in their chemical compositions.

Polymorphisms can also be observed when comparing the chemical profiles of individual plants of the same species and are based on the genetic characteristics of the plants. Therefore, it is difficult to ascertain if the differences in the chemical compositions are related to specific chemotypes or due to the environmental conditions of the plant (Franz and Novak 2016).

Secretory Structures

EOs are produced by several species of plants. The ability to store EOs is not universal in plants, yet widely present in the plant kingdom, especially in some families, including Alliaceae, Apiaceae, Asteraceae, Lamiaceae, Myrtaceae, Piperaceae, Poaceae, Rutaceae, and Zingiberaceae. EOs are usually found in the leaves, flowers, and fruits and less frequently in the roots, rhizomes, corks, and seeds.

Independent of their chemical composition, EOs are usually stored in oil ducts, resin ducts, oil cells, glands, or glandular trichomes of the plants. In Asteraceae, EOs are biosynthesized and accumulated in different secretory structures, such as oil cavities, idioblast oil cells, secretory ducts, and glandular trichomes. In *Baccharis*, EOs can be found in the leaves, stems, flowers, and roots and are stored in glandular trichomes (Fig. 10.1a, b, c) and/or secretory ducts (Fig. 10.1d, e, f).

In *Baccharis*, there are different types of glandular trichomes (Freire et al. 2007). The most common are the biseriate glandular trichomes (Fig. 10.1a, b) and flagelliform glandular trichomes (Fig. 10.1c). Biseriate glandular trichome comprises two pairs of basal cells and a head with up to four pairs of secretory cells containing

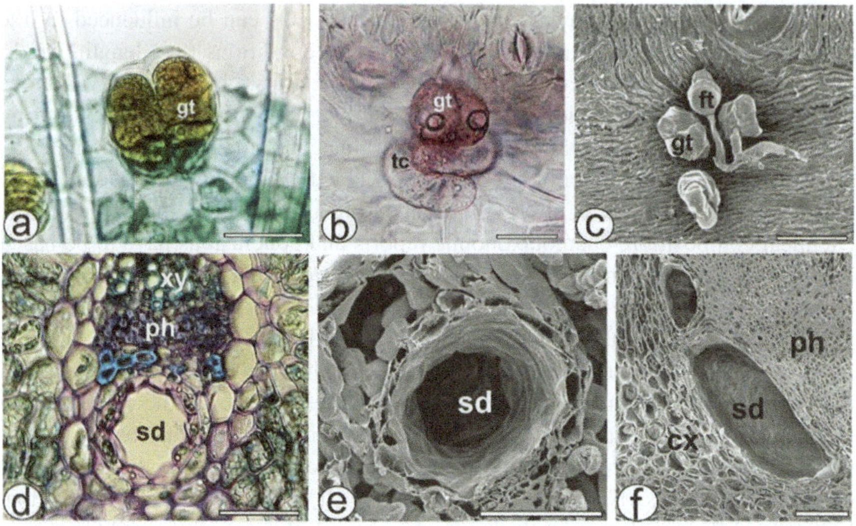

Fig. 10.1 Secretory structures of *Baccharis* species [(**a**, **b**, **d**) light microscopy; (**c**, **e**, **f**) scanning electron microscopy)]. *B. uncinella* (**a**), *B. punctulata* (**b**), *B. illinita* (**c**, **f**), *B. pauciflosculosa* (**d**, **e**). [gt, biseriate trichome; cx, cortex; tc, C-shaped flagelliform trichome; ft., straight flagelliform trichome; ph, phloem; sd, secretory duct; xy, xylem]. Scale bar: 50 μm. (Figure 10.1d reproduced from Budel et al. 2018a)

dense cytoplasm and covered by a cuticle layer. This trichome occurs in solitary, in groups with similar trichomes or groups with other flagelliform trichomes. The flagelliform glandular trichome consists of a voluminous body of about 10 secretory cells and an apical cell. The apical cell is elongated, whip-like, tubular, and translucent and contains dense oil substances (Budel et al. 2018a, b). The secretory body can be straight as seen in *B. illinita* DC. (Fig. 10.1c) or C-shaped as in *B. punctulata* DC. (Fig. 10.1b).

Secretory ducts are elongated and have large extracellular storage spaces containing secretion. In general, all *Baccharis* species possess secretory ducts. They are usually associated with the endodermis in the conducting system and occur next to the phloem. These ducts have uniseriate epithelium of 6–12 cells containing dense cytoplasm and lipophilic substances as observed in *B. pauciflosculosa* DC. (Fig. 10.1d, e) and *B. illinita* (Fig. 10.1f).

Essential Oils in Baccharis *Species*

In general, EOs of *Baccharis* species are liquid, clear or colored, volatile (with strong and characteristic aroma), and generally of lesser density than water. EO of *B. punctulata* is green, whereas it is yellowish in *B. microdonta* DC. (Budel et al. 2018b). EO extracted from aerial parts of *B. obtusifolia* Kunth had relative density of 0.8742 (Valarezo et al. 2015) and 0.8834 (Arze et al. 2004), whereas the relative density of the oils extracted from the leaves was 0.9151 for *B. dracunculifolia* and 0.9147 for *B. uncinella* DC. (Fabiane et al. 2008).

The olfactive profile of *B. anomala* DC. EO collected in Rio Grande do Sul, Brazil, was analyzed by gas chromatography-olfactometry (GC-O), and it was described to have a sweet, resinous, and woody odor. The major components of the EO were β-selinene (40.8%), caryophyllene oxide (9.9%), and spathulenol (6.8%) (Xavier et al. 2013).

The yield of EOs in *Baccharis* ranged from 0.08% to 2.82%. *B. obovata* Hook. & Arn. collected in Argentina presented the highest yield (Malizia et al. 2005), whereas the lowest content was obtained from *B. lateralis* Baker (syn. *B. schultzii* Baker) collected in Brazil (Lago et al. 2008).

Pretreatment processes such as drying and grinding are frequently applied to the plant material before the extraction of EOs for greater homogeneity. These processes must be carefully evaluated to provide reproducible results in EO investigation. In a recent study, Tischer et al. (2017) applied different grinding methods and compared their efficiencies as well as EO yields. They used cryogenics, knife (with and without cooling), and ball mills for grinding the materials of *B. articulata* and determined the yield and chemical compositions of the EOs. This study showed that cryogenic milling was found to be more efficient than the other methods in achieving lower particle size by disrupting the secretory structures (secretory ducts and glandular trichomes). However, this method of grinding resulted in the lower yields of EOs in comparison with other grinding methods (Tischer et al. 2017).

Another important factor affecting the EO yield is the plant part used in the extraction. EO extracted from the leaves of *B. microdonta* showed yields of 0.06–0.35% (Sayuri et al. 2010) and 0.08–0.21% (Lago et al. 2008), whereas Budel et al. (2018b) achieved 0.93% (*v/w*) of EO from the same species by extracting mixed parts of leaves and stems. Budel et al. 2018a observed large secretory ducts in the cortex of the stem. This anatomical feature contributed to the higher yield of EO.

Considering the dioecious nature of *Baccharis* species, there may be differences in the chemical compositions of EOs extracted from male and female plants and at different phenological stages. Some authors have reported slight differences in the EOs extracted at different growing stages (Zunino et al. 2004; Lago et al. 2008; Ascari et al. 2019). However, Besten et al. (2012) observed clear similarities between the EOs extracted from male and female specimens, during as well as outside the flowering periods, in five taxa of *Baccharis*, namely, *B. caprariifolia* DC., *B. dracunculifolia*, *B. coridifolia* DC., *B. semiserrata* var. *elaegnoides* (Steud. ex Baker) G.M.Barroso, and *B. pentaptera* (Less.) DC.

Differences in the EO chemical compositions have been reported for *B. punctulata* collected from different geographical locations. The EOs from aerial parts of the plants collected from Uruguay have shown β-phellandrene (5.2%), bornyl acetate (5.2%), α-cadinol (4.2%), δ-elemene (3.7%), and the ketone shyobunone (3.5%) as the major compounds (Minteguiaga et al. 2018b), whereas the EO from the leaves sourced from Guaíba, Brazil, have comprised bicyclogermacrene (9.73%), cis-cadin-4-en-7-ol (6.77%), and *(Z)*-ocimene (6.33%) (Schossler et al. 2009). Recent studies have also shown differences in the chemical composition of *B. punctulata* EO. α-Bisabolol was found in higher concentration (23.63%) in the aerial parts of plants collected in Paraná, Brazil (Budel et al. 2018b), whereas EO obtained from leaves of male plants showed δ-elemene (14.29%), germacrene D (11.29%), and bicyclogermacrene (10.90%), and in female plants bicyclogermacrene (42.44%), germacrene D (21.18%), and β-caryophyllene (14.06%) were found as major compounds (Ascari et al. 2019). Even though the chemical composition of EOs is often associated with environmental and phenological influences, it is necessary to investigate whether these variations in *B. punctulata* are also possibly linked to different chemotypes.

Chemically, EOs are generally composed of terpenoids and phenylpropanoids as the major compounds in addition to few aromatic and aliphatic constituents. Monoterpenes and sesquiterpenes, and their oxygenated derivatives, form the largest group of chemical entities in EOs. In *Baccharis* EOs, monoterpenes and sesquiterpenes are often found (Figs. 10.2 and 10.3). Sesquiterpenes appeared to be more plentiful in the majority of the species (Campos et al. 2016; Bogo et al. 2016; Zuccolotto et al. 2019; Oliveira et al. 2019; Tomazzoli et al. 2021), as observed in *B. anomala*, *B. ochracea* Spreng., *B. megapotamica* Spreng. (Budel et al. 2012), and *B. punctulata* (Ascari et al. 2019).

Sesquiterpenoid cyclic alcohols such as ledol, spathulenol, viridiflorol, and palustrol are not only important in the perfume industry due to their agreeable aromatic notes but also have taxonomic value (Minteguiaga et al. 2015).

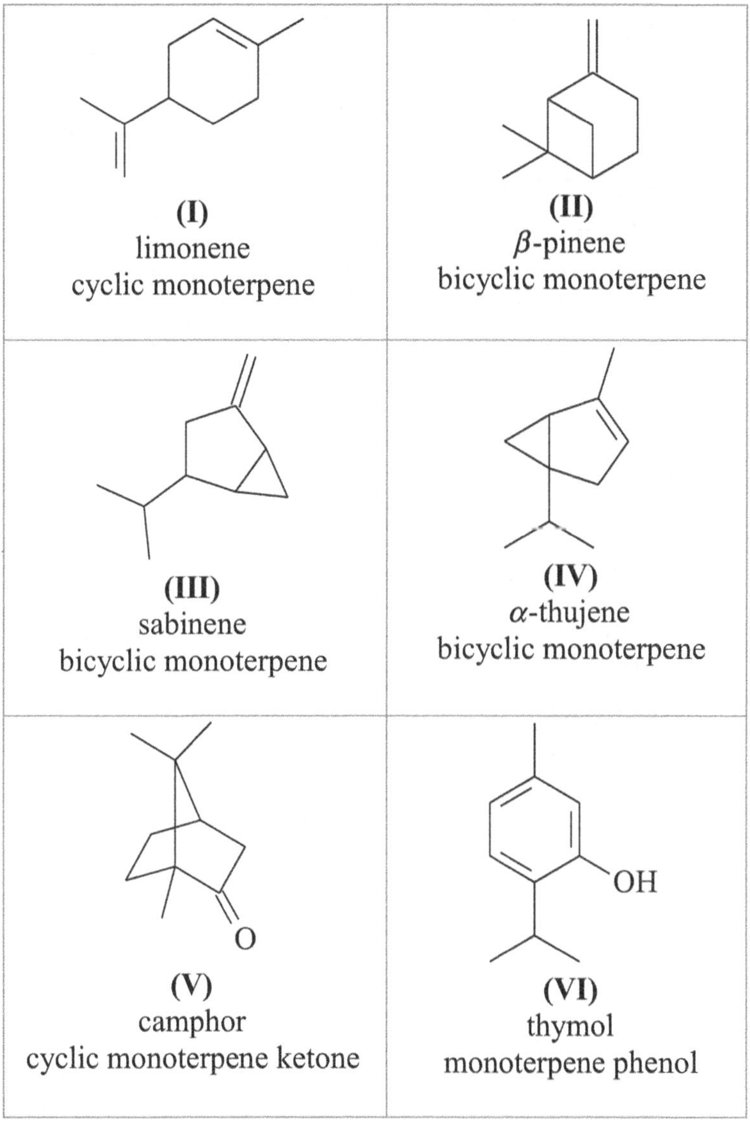

Fig. 10.2 Main monoterpenes with biological activities in *Baccharis*

Trombin-Souza et al. (2017) investigated EOs of 10 species of *Baccharis*, *B. anomala*, *B. articulata*, *B. axillaris* DC., *B. calvescens* DC., *B. mesoneura* DC., *B. milleflora* (Less.) DC., *B. myriocephala* DC., *B. oblongifolia* Pers., *B. crispa*, and *B. uncinella*. The major compound present in all species was limonene, whereas α-cadinol, α-thujene, α-pinene, β-pinene, *p*-cymene, *(E)*-β-ocimene, γ-terpinene, limonene, myrcene, sabinene, and spathulenol were identified in all samples. In addition, the chemical similarity was highest for *B. anomala*, *B. articulata*, *B.*

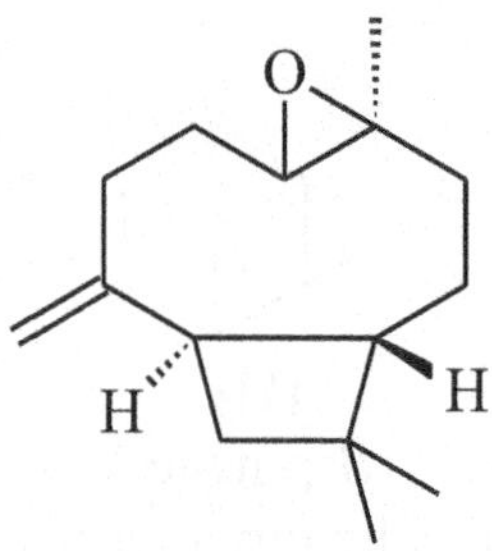

(VII)
caryophyllene oxide
bicyclic sesquiterpene

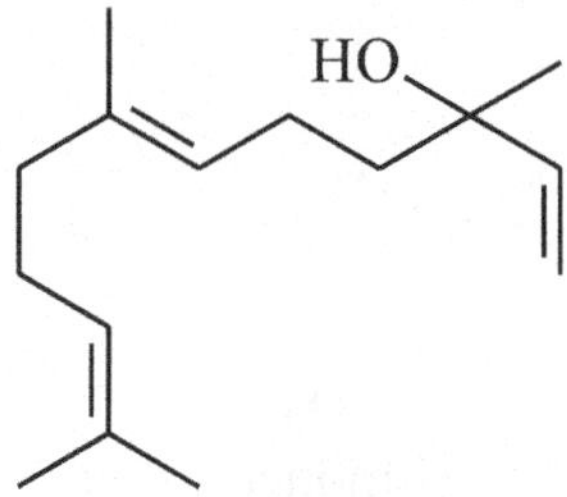

(VIII)
nerolidol
sesquiterpene alcohol

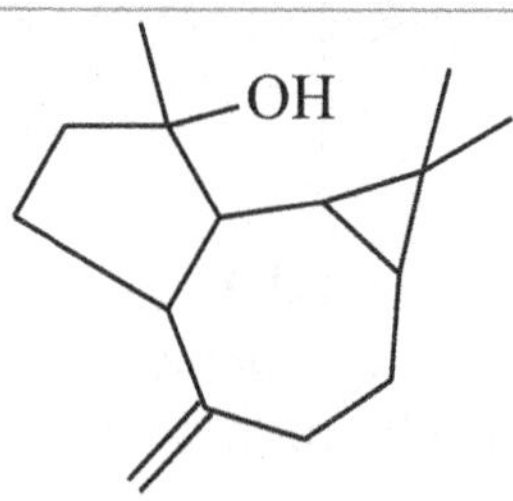

(IX)
spathulenol
tricyclic sesquiterpene
alcohol

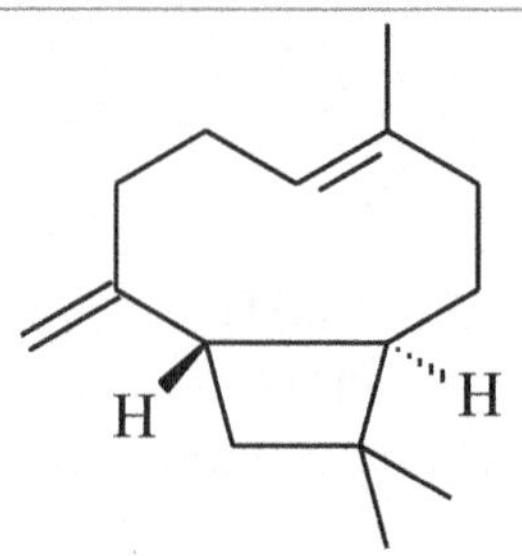

(X)
β-caryophyllene
bicyclic sesquiterpene
lactone

Fig. 10.3 Main sesquiterpenes with biological activities in *Baccharis*

calvescens, *B. milleflora*, *B. myriocephala*, *B. crispa*, *B. oblongifolia*, and *B. uncinella*.

The main components that were evidenced in a recent review that considered only EOs with biological activities were *α*-thujene (IV), *β*-caryophyllene (X), *β*-pinene (II), camphor (V), caryophyllene, caryophyllene oxide (VII), limonene (I), nerolidol (VIII), thymol (VI), thymol acetate, thymol methyl ether, sabinene (III), and spathulenol (IX) (Campos et al. 2016). According to Abad and Bermejo (2007), *E*-nerolidol, limonene, and spathulenol are the most abundant compounds in the EOs of *Baccharis*.

A recent investigation identified chemical markers for five *Baccharis* species with the major compounds spathulenol (22.74%) and kongol (22.22%) in *B. microdonta*; *β*-pinene (18.33%) and limonene (18.77%) in *B. pauciflosculosa*; *α*-bisabolol (23.63%) in *B. punctulata*; *α*-pinene (24.50%) in *B. reticularioides* Deble & A.S.Oliveira; and *α*-pinene (10.74%), *β*-pinene (15.24%), limonene (14.33%), and spathulenol (13.15%) in *B. sphenophylla* Dusén ex Malme. It is important to highlight that kongol was reported for *Baccharis* genus for the first time (Budel et al. 2018b).

Biological Activities of Baccharis *Essential Oils*

The chemical composition of the EOs generally shows a complex mixture of 20–60 compounds in different concentrations with 2 or 3 of the components in higher concentrations. In general, the major compounds present in the EOs are responsible for the biological activities of the oils (Guimarães et al. 2012). For example, nerolidol, a sesquiterpene found in many EOs, shows insecticidal and repellent activities, and the EOs containing this as a major compound exhibit these properties as well (Priestley et al. 2006).

Some plants possess major compounds characterizing the EOs. For example, *B. tricuneata* (L.f.) Pers. var. *ruiziana* presents more than 68% of *E*-nerolidol (Arze et al. 2004), making the correlation between the chemical composition and biological activities more evident. However, compounds present in lower concentration can act synergistically with other components, contributing to the action (Henriques et al. 2009).

Antioxidant Activity

The EO of *B. trinervis* Pers. showed significant antioxidant activity with IC_{50} values of 49.0 mg/mL and 28.87 mg/mL by the free radical DPPH$^{\bullet}$ scavenging assay and β-carotene/linoleic-acid oxidation model system, respectively. The chemical profile showed α-phellandrene (27.79%), *(Z)*-lachnophyllum (14.04%), sabinene (13.03%), *(Z)*-β-ocimene (8.13%), and α-thujene (6.65%) as the main components (Sobrinho et al. 2016).

EO of *B. milleflora* also evidenced antioxidant activity by DPPH$^{\bullet}$ methods with IC_{50} between 15.45 ± 0.52 and 21.06 ± 0.15 µg/mL, phosphomolybdenum with AAR 77.9 ± 0.90% – 79.81 ± 2.30%, TBARS with IA 12.60 ± 0.78% – 29.06 ± 1.47%, and ABTS$^{\bullet+}$ with IC_{50} between 3.85 ± 0.10 and 4.60 ± 0.07 µg/mL (Pereira et al. 2016b). The chemical profile of the EO collected in the four seasons was *trans*-caryophyllene (7.65–13.41%), germacrene-D (6.83–11.18%) and bicyclogermacrene (9.99–12.89%) (Pereira et al. 2016b).

EO extracted from leaves of *B. oreophila* Malme evidenced antioxidant capacity by three different methods, FRAP (4.09 µmol $FeSO_4$ E/mL), ABTS$^{\bullet+}$ (1.45 µmol TE/mL), and DPPH$^{\bullet}$ (1.04 µmol TE/mL).

Anti-inflammatory Effects

EOs of several species of *Baccharis* have been studied to assess their anti-inflammatory potentials. Florão et al. (2012) analyzed EOs extracted from the aerial parts of *B. articulata*, *B. crispa*, *B. dracunculifolia*, and *B. gaudichaudiana* DC. focusing on their immunomodulatory activities. All species except *B. articulata* inhibited expressively the proliferation of their phytohemagglutinin-stimulated counterparts. *B. dracunculifolia* showed the best anti-inflammatory effects, inhibiting significantly the casein-induced human granulocyte chemotaxis. The major compounds identified were spathulenol in *B. articulata*, *B. dracunculifolia*, and *B. gaudichaudiana*, as well as τ-gurjunene in *B. gaudichaudiana* and palustrol in *B. articulata* (Florão et al. 2012).

In a recent study, Ascari et al. (2019) subjected the leaf EOs of *B. punctulata* (male and female) to in vivo anti-inflammatory and in vitro antioxidant tests. Topical administration of both EOs was able to inhibit the formation of TPA-induced edema in the treated groups. Histological analysis evidenced that topical application of TPA promoted intense cellular infiltration. The results found in the ROS and DPPH$^{\bullet}$ tests suggest that both samples were able to reduce the inflammatory cells influx and had in vitro antioxidant properties, respectively. The major compounds found in the EO of *B. punctulata* were δ-elemene (14.29%), germacrene D (11.29%), and bicyclogermacrene (10.90%) in the male sample and bicyclogermacrene (42.44%), germacrene D (21.18%), and β-caryophyllene (14.06%) in the female sample (Ascari et al. 2019).

Antiulcerogenic Activity

EO of *B. dracunculifolia* was subjected to a test for antiulcerogenic action. The treatment in the doses of 50, 250, and 500 mg/kg of EO expressively reduced the lesion index, the total lesion area, and the percentage of lesions in comparison with both positive and negative control groups. The major compounds of EO were nerolidol (23.58%), germacrene D (21.54%), bicyclogermacrene (19.24%), *trans*-caryophyllene (7.12%), and spathulenol (6.03%) (Massignani et al. 2009).

Cytotoxic Effects

Cytotoxic activity of EO sourced from the cladodes of *B. milleflora* was investigated in relation to Jurkat, Raji, and HL-60 cells, as well as the cell mechanisms. All the tumor cells showed IC_{50} values lower than 50 μg/mL at 24, 48, and 72 h by MTT assay. The decrease in cell DNA content was demonstrated due to the inhibition of the proliferation of Jurkat, Raji, and HL-60 cells. Raji cells evidenced the greatest inhibition of cell proliferation. The EO acted via both necrotic and apoptotic mechanisms. The chemical profile showed bicyclogermacrene (12.16%), germacrene D

(11.18%), *(E)*-caryophyllene (9.28%), and α-humulene (8.05%) as the main components (Pereira et al. 2017).

Sedative Effects

B. uncinella is reported to be used by the Laklaño Indians in Santa Catarina State in Brazil for sedative purposes. Ascari et al. (2012) studied EO of *B. uncinella* collected from different locations in Paraná and Santa Catarina. Both of these samples significantly decreased locomotion and body temperature, as well as increased sleeping time. However, the hypnotic activity was sensitive to the differences in monoterpene composition. Sedative activity was observed better in *B. uncinella* that was collected in Santa Catarina because it presented a higher monoterpene/sesquiterpene ratio (0.31) in comparison to the other sample that showed a lower monoterpene/sesquiterpene ratio (0.004). The main compounds in the EO from Santa Catarina were caryophyllene (26.13%), spathulenol (13.39%), caryophyllene oxide (13.26%), limonene (7.21%), and α-pinene (6.42%), whereas the EO from Parana showed spathulenol (32.93%), caryophyllene oxide (27.78%), viridiflorol (5.29%), and α-cadinol (2.42%) as the major components (Ascari et al. 2012).

Antimicrobial Effects

EO of *B. dracunculifolia* was tested against *Candida* strains isolated from infants and their mothers during the lactation period to verify its enzymatic action and sensitivity. All strains were sensitive to EO of *B. dracunculifolia* with MIC between 0.2 and 6.25 mg/mL. The EO inhibited the growth of all strains, including the ones resistant to commercial antifungal agents (Pereira et al. 2011).

EO of *B. tridentata* Baker was observed in vitro to inhibit mycelial growth of the plant pathogens *Fusarium oxysporum*, *Colletotrichum gloeosporioides*, and *Rhizoctonia solani*. The chemical composition of the EO evidenced α-thujene (22.93%), β-pinene (20.33%), and β-felandrene (16.15%) as major compounds (Souza et al. 2011).

EO of *B. darwinii* Hook. & Arn. showed strong antifungal activity against yeast and dermatophytes of clinical relevance, including some fungi such as *Candida* spp. and *Trichophyton* spp. with MIC values between 62.5 and 125 μg/mL (Kurdelas et al. 2012).

EO sourced from the twigs of *B. semiserrata* DC. presented moderate antibacterial activity against *Staphylococcus aureus*, while EO from the leaves showed weak activity against *S. aureus* and *Bacillus cereus*. EOs from leaf and twig were active against *Microsporum gypseum*, *Candida albicans*, *Epidermophyton floccosum*, *Trichophyton mentagrophytes*, and *Cryptococcus neoformans*. The major compounds identified in the leaf EO included β-pinene (11.4%), spathulenol (9.8%), and *E*-nerolidol (9.6%), while the EO from twigs showed spathulenol (25.1%), limonene (9.1%), and caryophyllene oxide (8.0%) as major compounds. *Baccharis semiserrata* leaf EO was also active against *Trichophyton rubrum* (Vannini et al. 2012). This species and *T. mentagrophytes* also showed sensitivity when treated with EO from *B. latifolia* (Valarezo et al. 2013).

EO of *B. coridifolia* was tested for its antibiotic properties (ampicillin, cephalothin, chloramphenicol, gentamicin, and tetracycline) alone and in combination with the EO (4% v/v) through the disk diffusion susceptibility test. The occurrence of the synergistic or antagonistic effect was observed in both bacterial strains assessed, *S. aureus* and *E. coli* (Onofre et al. 2013).

The EO from the aerial parts of *B. obtusifolia* exhibited a moderate antibacterial effect against *Klebsiella pneumoniae* and *Enterococcus faecalis* and good antifungal activity against *Trichophyton rubrum* and *T. mentagrophytes*. The major compounds of the EO were limonene (28.3%), germacrene-D (9.8%), α-pinene (9.0%), β-pinene (8.2%), bicyclogermacrene (6.2%), and δ-cadinene (5.7%) (Valarezo et al. 2015).

Antimicrobial and antibiofilm activities of *B. psiadioides* (Less.) Joch.Müll. EO were tested against antibiotic-resistant *E. faecalis* strains. The oil inhibited the growth of the multidrug-resistant *E. faecalis* strains and was also effective when evaluated against biofilms. However, its activity was stronger when inhibiting the formation of a biofilm than when applied over established biofilms. The chemical profile showed β-pinene as the major component (33.13%), followed by D-3-carene (11.41%), limonene (5.97%), *(E)*-ocimene (3.82%), and α-pinene (2.58%) (Negreiros et al. 2016).

EO of *B. dracunculifolia* was studied to determine the minimal inhibitory concentration (MIC) against planktonic cultures of *S. mutans* and its antibacterial activity in biofilms formed in the discs of composite resin. The MIC of the *B. dracunculifolia* EO to planktonic growth of *S. mutans* was 6%. In biofilms of *S. mutans* clinical isolates, EO (6%) and chlorhexidine resulted in reductions of 53.3–91.1% and 79.1–96.6%, respectively. For the biofilm formed by the *S. mutans* reference strain, the reductions achieved with *B. dracunculifolia* EO and chlorhexidine were 39.3% and 88.1%, respectively (Pereira et al. 2016a).

EO from leaves of *B. oreophila* showed antimicrobial effects against *S. aureus* (10.33 ± 0.5 mm, MIC = 1250 μg/mL) and *C. albicans* (8.66 ± 0.5 mm, MIC >2500 μg/mL). The major compounds of the EO were khusimone (16.37%) and spathulenol (16.12%).

Although there are several EOs presenting antimicrobial activities, no activity was reported for the EO of *B. dracunculifolia* tested against methicillin-resistant *S. aureus* and *Mycobacterium intracellulare* (Parreira et al. 2010). Besides, EO of *B. uncinella* was inactive against all bacteria tested (Vannini et al. 2012).

EOs from *B. organensis* Baker, *B. burchellii* Baker, and *B. aracatubaensis* Malag. (male and female specimens) were tested against *S. aureus*, *E. coli*, *P. aeruginosa*, and *C. albicans*. They also did not exhibit any antimicrobial activity (Zuccolotto et al. 2019).

Insecticidal Activities

Allelochemical effects of EO of *B. salicifolia* (Ruiz & Pav.) Pers. and some isolated compounds were tested against adult red flour beetles, *Tribolium castaneum*. The EO showed toxicity and repellence activities. *β*-Pinene and pulegone were observed to be the most acutely toxic compounds after 3 days of treatment, whereas *α*-terpineol was identified as the most repellent compound (García et al. 2005).

EO of *B. spartioides* (Hook. & Arn. ex DC.) J. Rémy in Gay was evaluated for its repellency against *Aedes aegypti*. At concentrations of 12.5%, the EO showed the longest repellency. The major chemical compound of the EO was camphor (50.5%) (Gillij et al. 2008).

Baccharis salicifolia (Ruiz & Pav.) Pers. EO collected in Argentina showed *α*-pinene (21.7%) and spathulenol (14.4%) as the major compounds. EO was tested against *A. aegypti* and presented moderate repellency (Gleiser et al. 2011). This species was also collected in two other locations in San Luis, Argentina. In this study, the chemical composition of the EO was different; the major compounds were *(Z)-b*-ocimene, germacrene D, muuroladiene, and *β*-cubebene, with the addition of *α*-thujene and *α*-phellandrene, in location A and isoledene in location B. The EO from *B. salicifolia* from location A exhibited post-ingestive toxicity to *Spodoptera littoralis* larvae without antifeedant effects (Sosa et al. 2012).

EO of *B. darwinii* was tested in vitro to evaluate its insecticidal properties. It showed insecticidal activity against the Mediterranean fruit fly, *Ceratitis capitata*, with LD_{50} values of 19.9 and 31.0 μg/fly for males and females, respectively, at 72 h. The EO also displayed repellent activity against Chagas disease vector, *Triatoma infestans*, with an average repellence rate of 92%. The major components with recognized insecticidal and antimicrobial activities were limonene (47.1%), thymol (8.1%), and 4-terpineol (6.4%) (Kurdelas et al. 2012).

EO and its major compounds from aerial parts of *B. dracunculifolia* were tested against unengorged larvae and engorged females of the cattle tick *Rhipicephalus microplus*. In the larval packet test, the EO, as well as the pure compound nerolidol, exhibited high activity showing more than 90% mortality at concentrations from 15.0 and 10.0 mg/mL, respectively, whereas limonene did not show acaricidal activity. In the female immersion test, the EO and nerolidol also caused a reduction in the quantity and quality of eggs produced with a control rate of 96.3% and 90.3% at concentrations of 60.0 and 50.0 mg/mL, respectively (Lage et al. 2015).

EOs of *B. milleflora* collected during four different seasons showed repellency action IR 80.65 ± 6.69%, well above the IR 36.39 ± 21.00 found for the EO of *Cymbopogon nardus*, which was used as a standard. The fumigant activity showed a KT50 (min) of 10.63 ± 2.68 and 22.70 ± 3.40 for the EOs of *B. milleflora* and *Melaleuca alternifolia*, respectively (Pereira et al. 2016b).

Baccharis sphenophylla collected in South Brazil exhibited strong toxicity to *Cimex lectularius* (bed bug) in the fumigation bioassay, causing 66.67% mortality in "Bayonne" and 83.33% in "Ft.Dix" (Budel et al. 2018b).

Parasiticidal Activities

Antitrypanosomal Effects

EOs of five *Baccharis* species (*B. microdonta*, *B. pauciflosculosa*, *B. reticularioides*, *B. punctulata*, and *B. sphenophylla*) were submitted to test their antitrypanosomal activities against trypomastigotes cultures of *T. brucei*. All *Baccharis* species exhibited remarkable antitrypanosomal activities. *B. pauciflosculosa* demonstrated the highest effect, 0.31 μg/mL (IC_{50}) and 0.52 μg/mL (IC_{90}), followed by *B. reticularioides* at 0.96 μg/mL (IC_{50}) and 2.49 μg/mL (IC_{90}), and *B. sphenophylla* at 1.14 μg/mL (IC_{50}) and 2.38 μg/mL (IC_{90}) (Budel et al. 2018b).

Antimalarial Effects

EOs of five species of *Baccharis* were investigated against chloroquine-sensitive (D6) and chloroquine-resistant (W2) strains of *Plasmodium falciparum*. The results were, respectively, 10.90 μg/mL ± 0.98 and 14.20 μg/mL ± 1.08 for *B. pauciflosculosa*, 20.32 μg/mL ± 4.37 and 34.35 ± 10.15 for *B. reticularioides*, and 27.58 ± 1.64 and 32.53 ± 16.5 for *B. sphenophylla*, showing moderate antimalarial activities. *B. microdonta* and *B. punctulata* exhibited cytotoxicity to Vero cells (selectivity control) (Budel et al. 2018b).

EO of *B. dracunculifolia* was inactive in the antiplasmodial assay. Chloroquine and artemisinin were used as positive control. The IC_{50} values were 0.018 and 0.014 mg/mL, respectively, against *P. falciparum* (D6 clone) (Parreira et al. 2010).

Leishmanicidal Activity

EO of *B. dracunculifolia* showed activity against promastigote forms of *Leishmania donovani* with IC_{50} values of 42 μg/mL. The major compounds of the EO were *E*-nerolidol (33.51%) and spathulenol (16.24%). Pentamidine and amphotericin B were used as positive controls, which showed IC_{50} values of 1.9 mg/mL and 0.65 mg/mL, respectively (Parreira et al. 2010).

Schistosomicidal Activity

EO of *B. dracunculifolia* also displayed high schistosomicidal activity against the trematode parasite, *Schistosoma mansoni*, killing all pairs of adult worms after incubation with the EO (10, 50, and 100 μg/mL) (Parreira et al. 2010).

EO of *B. crispa* showed in vitro schistosomicidal activities. Its effects on the reproductive fitness and the tegumental alteration of adult worms were similar to those induced by praziquantel. The results evidenced a significant decline in the motility of the worms and a mortality rate of 100% 30 h after their exposure to the

EO in the concentration of 130 μg/mL. As for the morphological changes, the EO of *B. crispa* induced a peeling on the tegument surface, as well as the destruction of tubercles and spines, which resulted in smooth areas on the body surface. The EO also caused tegument destruction in female worms, in addition to the destruction of the oral and acetabular suckers (Oliveira et al. 2012).

Adjuvant Property

Leaf EO of *B. dracunculifolia* was assessed for its inhibitory properties on the coagulating and fibrinogenolysis activities induced by *Lachesis muta*, *Bothrops atrox*, and *Bothrops moojeni* snake venoms. The EO caused 100% inhibition on the fibrinogenolysis induced by *B. moojeni* and *L. muta* venoms, evidencing that the EO can be used as adjuvants for the treatment of snakebites (Miranda et al. 2016).

3 Final Considerations

Essential oils of *Baccharis* species have numerous medicinally beneficial properties. Several pharmacological properties and biological activities, including anti-inflammatory, antimicrobial, antiulcerogenic, antimalarial, antioxidant, antitrypanosomal, cytotoxic, insecticidal, leishmanicidal, schistosomicidal, sedative, and adjuvant properties, have been reported for the EOs of *Baccharis*. EOs and their major compounds of many *Baccharis* species have shown significant activities against many insect pests, microorganisms, and human parasites.

Baccharis EOs mainly contain monoterpenes and sesquiterpenes, while the latter compounds are more plentiful in the majority of the species. The most abundant compounds in the EOs of *Baccharis* are *E*-nerolidol, limonene, and spathulenol. EOs of many of the species have unique compounds, which can be used as chemical markers for species identification and quality control.

The genus *Baccharis* comprises about 406 species and almost all species produce EOs of varying chemical compositions and medicinal properties. However, this review of the literature indicates that only less than 10% of the *Baccharis* oils have been studied. Even among these studied species, only very few, such as *B. dracunculifolia*, are studied in detail. A large portion of this medicinally important genus is still unexplored. Future research focusing on the taxonomy and chemical and biological aspects of the EOs and major compounds of different species of *Baccharis* could bring out the hidden wealth and usefulness of this important genus.

References

Abad MJ, Bermejo P (2007) *Baccharis* (Compositae): a review update. ARKIVOC 7:76–96. https://doi.org/10.3998/ark.5550190.0008.709

Arze JBL, Garneau FX, Collin G et al (2004) Essential oils from Bolivia. I. Asteraceae: *Baccharis tricuneata* (L.f.) Pers. var. *ruiziana* Cuatrecassas. J Essent Oil Res 16:429–431. https://doi.org/10.1080/10412905.2004.9698763

Ascari J, Sens SL, Nunes DS et al (2012) Sedative effects of essential oils obtained from *Baccharis uncinella*. Pharm Biol 50:113–119. https://doi.org/10.3109/13880209.2011.634423

Ascari J, Oliveira MS, Nunes DS et al (2019) Chemical composition, antioxidant and anti-inflammatory activities of the essential oils from male and female specimens of *Baccharis punctulata* (Asteraceae). J Ethnopharmacol 24:1–7

Besten MA, Jasinski VCG, Costa AGLC et al (2012) Chemical composition similarity between the essential oils isolated from male and female specimens of each five *Baccharis* species. J Braz Chem Soc 23:1041–1047. https://doi.org/10.1590/S0103-50532012000600007

Bogo CA, Andrade MH, Paula JP et al (2016) Comparative analysis of essential oils of *Baccharis* L.: a review. Rev Strict Sensu 1:1–11. https://doi.org/10.24222/2525-3395.2016v1n2p001

Botas GS, Cruz RAS, Almeida FB et al (2017) *Baccharis reticularia* DC. and limonene nanoemulsions: promising larvicidal agents for *Aedes aegypti* (Diptera: Culicidae) control. Molecules 22:1–14. https://doi.org/10.3390/molecules22111990

Budel JM, Matzenbacher NI, Duarte MR (2008) Genus *Baccharis* (Asteraceae): a review of chemical and pharmacological studies. In: Singh VK, Govil JN (eds) Recent progress in medicinal plants – phytopharmacology and therapeutic value. Studium Press LLC, Houston, pp 1–18

Budel JM, Duarte MR, Döll-Boscardin PM et al (2012) Composition of essential oils and secretory structures of *Baccharis anomala, B. megapotamica* and *B. ochracea*. J Essent Oil Res 24:19–24. https://doi.org/10.1080/10412905.2012.645634

Budel JM, Raman V, Monteiro LM et al (2018a) Foliar anatomy and microscopy of six Brazilian species of *Baccharis* (Asteraceae). Microsc Res Tech 81:832–842. https://doi.org/10.1002/jemt.23045

Budel JM, Wang M, Raman V et al (2018b) Essential oils of five *Baccharis* species: investigations on the chemical composition and biological activities. Molecules 23:1–19. https://doi.org/10.3390/molecules23102620

Calo JR, Crandall PG, O'Bryan CA et al (2015) Essential oils as antimicrobials in food systems – a review. Food Control 54:111–119. https://doi.org/10.1016/j.foodcont.2014.12.040

Campos FR, Bressan J, Jasinski VCG et al (2016) *Baccharis* (Asteraceae): chemical constituents and biological activities. Chem Biodivers 13:1–17. https://doi.org/10.1002/cbdv.201400363

Elshafie HS, Camele I (2017) An overview of the biological effects of some Mediterranean essential oils on human health. Biomed Res Int 27:1–14. https://doi.org/10.1155/2017/9268468

Fabiane KC, Ferronatto R, Santos AC, Onofre SB (2008) Physicochemical characteristics of the essential oils of *Baccharis dracunculifolia* and *Baccharis uncinella* DC. (Asteraceae). Rev Bras Farmacogn 18:197–203

Florão A, Budel JM, Duarte MR et al (2012) Essential oils from *Baccharis* species (Asteraceae) have anti-inflammatory effects for human cells. J Essent Oil Res 24:561–570. https://doi.org/10.1080/10412905.2012.728081

Franz C, Novak J (2016) Sources of essential oils. In: Baser KHC, Buchbauer G (eds) Handbook of essential oils: science, technology, and applications, 2nd edn. CRC Press, Boca Raton, pp 43–86

Freire SE, Urtubey E, Giuliano DA (2007) Epidermal characters of *Baccharis* (Asteraceae) species used in traditional medicine. Caldasia 29:23–38

García M, Donadel OJ, Ardanaz CE et al (2005) Toxic and repellent effects of *Baccharis salicifolia* essential oil on *Tribolium castaneum*. Pest Manag Sci 61:612–618. https://doi.org/10.1002/ps.1028

Gillij YG, Gleiser RM, Zygadlo JA (2008) Mosquito repellent activity of essential oils of aromatic plants growing in Argentina. Bioresour Technol 99:2507–2515. https://doi.org/10.1016/j.biortech.2007.04.066

Gleiser RM, Bonino MA, Zygadlo JA (2011) Repellence of essential oils of aromatic plants growing in Argentina against *Aedes aegypti* (Diptera: Culicidae). Parasitol Res 108:69–78. https://doi.org/10.1007/s00436-010-2042-4

Gobbo-Neto L, Lopes NP (2007) Plantas medicinais: Fatores de influência no conteúdo de metabólitos secundários. Quim Nova 30:374–381

Guimarães AG, Oliveira AP, Ribeiro EAN et al (2012) Atividade farmacológica de monoterpenos. In: Souza GHB, Melo JCP, Lopes NP (eds) Farmacognosia: Coletânea Científica. UFOP, Minas Gerais, pp 219–250

Henriques AT, Simões-Pires CA, Apel MA (2009) Óleos essenciais: importância e perspectivas terapêuticas. In: Yunes RA, Cechinel-Filho V (eds) Química de produtos naturais, novos fármacos e a moderna farmacognosia, 2nd edn. UNIVALI, Santa Catarina, pp 219–256

Kurdelas RR, López S, Lima B et al (2012) Chemical composition, anti-insect and antimicrobial activity of *Baccharis darwinii* essential oil from Argentina, Patagonia. Ind Crop Prod 40:261–267. https://doi.org/10.1016/j.indcrop.2012.03.024

Lage TCA, Montanari RM, Fernandes AS et al (2015) Chemical composition and acaricidal activity of the essential oil of *Baccharis dracunculifolia* De Candole (1836) and its constituents nerolidol and limonene on larvae and engorged females of *Rhipicephalus microplus* (Acari: Ixodidae). Exp Parasitol 148:24–29. https://doi.org/10.1016/j.exppara.2014.10.011

Lago JHG, Romoff P, Fávero OA et al (2008) Chemical composition of male and female *Baccharis trimera* (Less.) DC. (Asteraceae) essential oils. Biochem Syst Ecol 36:737–740. https://doi.org/10.1016/j.bse.2008.05.009

Malizia RA, Cardell DA, Molli JS et al (2005) Volatile constituents of leaf oils from the genus *Baccharis*. Part II: *Baccharis obovata* hooker et Arnott and *B. salicifolia* (Ruiz et Pav.) Pers. species from Argentina. J Essent Oil Res 17:194–197. https://doi.org/10.1080/10412905.2005.9698873

Massignani JJ, Lemos M, Maistro EL et al (2009) Antiulcerogenic activity of the essential oil of *Baccharis dracunculifolia* on different experimental models in rats. Phytother Res 23:1355–1360. https://doi.org/10.1002/ptr.2624

Minteguiaga M, Umpiérrez N, Xavier V et al (2015) Recent findings in the chemistry of odorants from four *Baccharis* species and their impact as chemical markers. Chem Biodivers 12:1339–1348. https://doi.org/10.1002/cbdv.201400329

Minteguiaga M, Mercado MI, Ponessa GI et al (2018a) Morphoanatomy and essential oil analysis of *Baccharis trimera* (Less.) DC. (Asteraceae) from Uruguay. Ind Crop Prod 112:488–498. https://doi.org/10.1016/j.indcrop.2017.12.040

Minteguiaga M, González A, Cassel E et al (2018b) Volatile constituents from *Baccharis* spp. L. (Asteraceae): chemical support for the conservation of threatened species in Uruguay. Chem Biodivers 15:e1800017. https://doi.org/10.1002/cbdv.201800017

Miranda CASF, Cardoso MG, Marcussi S et al (2016) Clotting and fibrinogenolysis inhibition by essential oils from species of the Asteraceae Family. Braz Arch Biol Technol 59:e16150775. https://doi.org/10.1590/1678-4324-2016150775

Negreiros MO, Pawlowski A, Zinic CA et al (2016) Antimicrobial and antibiofilm activity of *Baccharis psiadioides* essential oil against antibiotic-resistant *Enterococcus faecalis* strains. Pharm Biol 54:3272–3279. https://doi.org/10.1080/13880209.2016.1223700

Németh-Zámboriné É (2016) Natural variability of essential oil components. In: Baser KHC, Buchbauer G (eds) Handbook of essential oils: science, technology, and applications, 2nd edn. CRC Press, Boca Raton, pp 87–126

Oliveira RN, Rehder VLG, Oliveira ASS et al (2012) *Schistosoma mansoni*: in vitro schistosomicidal activity of essential oil of *Baccharis trimera* (Less.) DC. Exp Parasitol 132:135–143. https://doi.org/10.1016/j.exppara.2012.06.005

Oliveira CT, Maia BHLNS, Ferriani AP et al (2019) Chemical characterization, antioxidant capacity and antimicrobial potential of essential oil from the leaves of *Baccharis oreophila* Malme. Chem Biodivers 16:e1800372. https://doi.org/10.1002/cbdv.201800372

Onofre SB, Canton M, Pires PA (2013) Action of essential oils obtained from *Baccharis coridifolia* DC. (Asteraceae-Astereae) on the activity of antibiotics. Adv Microbiol 3:166–170. https://doi.org/10.4236/aim.2013.32026

Parreira NA, Magalhães LG, Morais DR et al (2010) Antiprotozoal, schistosomicidal, and antimicrobial activities of the essential oil from the leaves of *Baccharis dracunculifolia*. Chem Biodivers 7:993–1001. https://doi.org/10.1002/cbdv.200900292

Pereira CA, Costa ACBP, Machado AKS et al (2011) Enzymatic activity, sensitivity to antifungal drugs and *Baccharis dracunculifolia* essential oil by *Candida* strains isolated from the oral cavities of breastfeeding infants and in their mothers' mouths and nipples. Mycopathologia 171:103–109. https://doi.org/10.1007/s11046-010-9353-y

Pereira CA, Costa ACBP, Liporoni PCS et al (2016a) Antibacterial activity of *Baccharis dracunculifolia* in planktonic cultures and biofilms of *Streptococcus mutans*. J Infect Public Health 9:324–330. https://doi.org/10.1016/j.jiph.2015.10.012

Pereira CB, Farago PV, Borsato DM et al (2016b) Chemical composition and biological activities of *Baccharis milleflora* essential oil. Lat Am J Pharm 35:2225–2233

Pereira CB, Kanunfre CC, Farago PV et al (2017) Cytotoxic mechanism of *Baccharis milleflora* (Less.) DC. essential oil. Toxicol in Vitro 42:214–221. https://doi.org/10.1016/j.tiv.2017.04.031

Perera WH, Bizzo HR, Gama PE et al (2017) Essential oil constituents from high altitude Brazilian species with antimicrobial activity: *Baccharis parvidentata* Malag., *Hyptis monticola* Mart. ex Benth. and *Lippia origanoides* Kunth. J Essent Oil Res 29:109–116. https://doi.org/10.1080/10412905.2016.1210039

Priestley CM, Burges IF, Williamson EM (2006) Lethality of essential oil constituents towards the human louse, *Pediculus humanus*, and its eggs. Fitoterapia 77:303–309. https://doi.org/10.1016/j.fitote.2006.04.005

Sayuri VA, Romoff P, Fávero OA et al (2010) Chemical composition, seasonal variation, and biosynthetic considerations of essential oils from *Baccharis microdonta* and *B. elaeagnoides* (Asteraceae). Chem Biodivers 7:2771–2782. https://doi.org/10.1002/cbdv.201000083

Schmidt E (2016) Production of essential oils. In: Baser KHC, Buchbauer G (eds) Handbook of essential oils: science, technology, and applications, 2nd edn. CRC Press, Boca Raton, pp 127–164

Schossler P, Schneider GL, Wunsch D et al (2009) Volatile compounds of *Baccharis punctulata*, *Baccharis dracunculifolia* and *Eupatorium laevigatum* obtained using solid phase microextraction and hydrodistillation. J Braz Chem Soc 20:277–287. https://doi.org/10.1590/S0103-50532009000200012

Sobrinho ACN, Souza EB, Rocha MFG et al (2016) Chemical composition, antioxidant, antifungal and hemolytic activities of essential oil from *Baccharis trinervis* (Lam.) Pers. (Asteraceae). Ind Crop Prod 84:108–115. https://doi.org/10.1016/j.indcrop.2016.01.051

Sosa ME, Lancelle HG, Tonn CE et al (2012) Insecticidal and nematicidal essential oils from Argentinean *Eupatorium* and *Baccharis* spp. Biochem Syst Ecol 43:132–138. https://doi.org/10.1016/j.bse.2012.03.007

Souza SP, Cardoso MG, Souza PE et al (2011) Óleo essencial de *Baccharis tridentata* Vahl: composição química, atividade antioxidante e fungitóxica, e caracterização morfológica das estruturas secretoras por microscopia eletrônica de varredura. Rev Bras Pl Med 13:456–466. https://doi.org/10.1590/S1516-05722011000400011

Tischer B, Vendruscolo RG, Wagner R et al (2017) Effect of grinding method on the analysis of essential oil from *Baccharis articulata* (Lam.) Pers. Chem Pap 71:753–761. https://doi.org/10.1007/s11696-016-0052-0

Tomazzoli MM, Amaral W, Cipriano RR et al (2021) Chemical composition and antioxidant activity of essential oils from populations of *Baccharis dracunculifolia* DC. in southern Brazil. Braz Arch Biol Technol 64: e21190253. https://doi.org/10.1590/1678-4324-2021190253

Trombin-Souza M, Trombin-Souza M, Amaral W et al (2017) Chemical composition of the essential oils of *Baccharis* species from southern Brazil: a comparative study using multivariate statistical analysis. J Essent Oil Res 29:400–406. https://doi.org/10.1080/10412905.2017.1322007

Valarezo E, Rosillo M, Cartuche L et al (2013) Chemical composition, antifungal and antibacterial activity of the essential oil from *Baccharis latifolia* (Ruiz & Pav.) Pers. (Asteraceae) from Loja, Ecuador. J Essent Oil Res 25:233–238. https://doi.org/10.1080/10412905.2013.775679

Valarezo E, Rosales J, Morocho V et al (2015) Chemical composition and biological activity of the essential oil of *Baccharis obtusifolia* Kunth from Loja, Ecuador. J Essent Oil Res 27:212–216. https://doi.org/10.1080/10412905.2015.1007217

Vannini AB, Santos TG, Fleming AC et al (2012) Chemical characterization and antimicrobial evaluation of the essential oils from *Baccharis uncinella* DC and *Baccharis semiserrata* DC. (Asteraceae). J Essent Oil Res 24:547–554. https://doi.org/10.1080/10412905.2012.728092

Xavier VB, Vargas RMF, Minteguiaga M et al (2013) Evaluation of the key odorants of *Baccharis anomala* DC. essential oil: new applications for known products. Ind Crop Prod 49:492–496. https://doi.org/10.1016/j.indcrop.2013.05.011

Zuccolotto T, Bressan J, Lourenço AVF et al (2019) Chemical, antioxidant, and antimicrobial evaluation of essential oils and an anatomical study of the aerial parts from *Baccharis* species (Asteraceae). Chem Biodivers 64:e1800547. https://doi.org/10.1002/cbdv.201800547

Zunino MP, Lopez ML, Zygadlo JA et al (2004) Essential oil composition of *Baccharis articulata* (Lam.) Pers. J Essent Oil Res 16:29–30. https://doi.org/10.1080/10412905.2004.9698643

Chapter 11
Flavonoids of *Baccharis*

Simone dos Santos Grecco, Deborah Pelosi Sessa, and João Henrique Ghilardi Lago

Abstract Despite more than 400 species of *Baccharis* occurring worldwide, only less than 20% of the species have chemically been studied. In the *Baccharis* genus, within terpenes and other phenolic compounds (see Chaps. 12 and 13), flavonoids are largely accumulated as aglycone, being apigenin, genkwanin, hispidulin, kaempferol (flavones), quercetin (flavonol), naringenin, sakuranetin (flavanones), and others widely distributed. Additionally, some flavonoid glycosides such as quercitrin, rutin, and others are also found, but in minor frequency. Flavonoids are compounds with a basic skeleton of 15 carbons (C_6-C_3-C_6) arranged in two aromatic rings linked through a three-carbon moiety. The oxidation degree of the C_3 moiety is directly related to the classification of flavonoids into flavanones, flavones, isoflavones, flavanonols, and flavonols. With respect to biological activity, flavonoids from *Baccharis* display a significant antioxidant potential, especially for the capacity of suppression of ROS formation, ROS scavenging, and upregulation of antioxidant defenses. In this chapter, the distribution of flavonoids in *Baccharis* is so justified through this antioxidant effect since those species are inserted in areas with direct incidence of sunrays, such as in montane savannas. In addition, flavonoids display hepatoprotective, antimicrobial, anti-inflammatory, antitumoral, antiviral, and other activities. In this chapter, the occurrence and distribution of flavonoids in *Baccharis* species are discussed, as well as their biosynthesis and biological aspects.

Keywords Biological activity · Biosynthesis · Flavonoid distribution · Modulation of redox balance · Structure and composition

1 Introduction

Although more than 400 *Baccharis* species are distributed worldwide, only a reduced number of plants belonging to this genus were chemically and/or biologically investigated. Among the chemical compounds present in these species, the

S. dos Santos Grecco · D. P. Sessa · J. H. G. Lago (✉)
Centro de Ciências Naturais e Humanas, Universidade Federal do ABC, São Paulo, Brazil
e-mail: joao.lago@ufabc.edu.br

G. W. Fernandes et al. (eds.), *Baccharis*,
https://doi.org/10.1007/978-3-030-83511-8_11

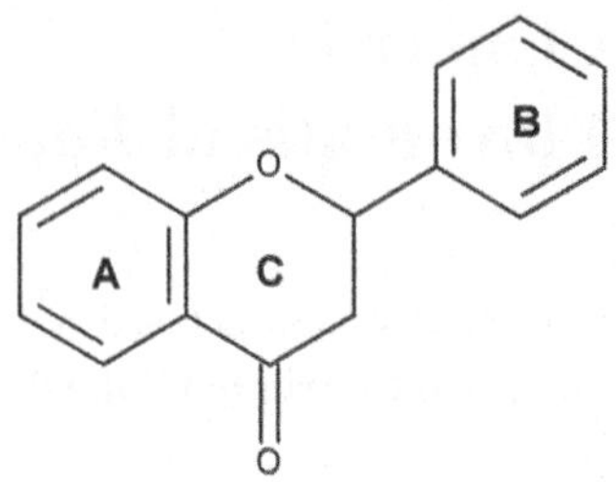

Fig. 11.1 Basic skeleton of a flavonoid

occurrence of flavonoids is reported in about 15% of *Baccharis* species (Campos et al. 2016). Flavonoids are a class of phenolic compounds, widely distributed within the kingdom Plantae, with a basic skeleton of 15 carbons (Fig. 11.1) arranged as two aromatic rings (A and B) connected by a three-carbon moiety (C_6-C_3-C_6).

The classification of flavonoids is directly related to the degree of oxidation of the C_3 portion. Flavonoids can be classified into flavanones, flavononols, flavones, isoflavones, flavonols, catechins, and anthocyanidins.

2 Biosynthesis

Secondary metabolites, also known as special metabolites, are substances from natural sources that do not participate in the essential functions of these organisms (reproduction, growth, or development). Most are substances produced in processes of interaction between plants and the environment in which they are inserted. It is important to note that the production of secondary metabolites is also associated with the various conditions to which plant species are subjected. Since these substances are produced to benefit these individuals, depending on the environment surrounding them, it acts as protection against the action of predators, coloring and volatiles agents, attracting pollinators, and also as competition agents among plants in the same habitat. Therefore, the production of these metabolites is fundamental in the maintenance of these species (Dewick 2009).

Besides being responsible for the production of substances that act with different functions in the interaction of these plant species and the environment, which are inserted, the secondary metabolism is also responsible for providing the majority of natural products with pharmacological activity. The secondary metabolism acts as substrates originating from the metabolic pathways that form the primary metabolism (photosynthesis, glycolysis, and Krebs/ citric acid cycle). These substrates are involved in different metabolic pathways, responsible for the synthesis of several classes of secondary metabolites.

Flavonoids are considered mixed pathway metabolites, being synthesized from precursors of two metabolic routes; shikimic acid (shikimate) and acetate-malonate. From the glycolysis process, the synthesis of phosphoenylpyruvate, together with erythrose-4-phosphate from the pentose-phosphate pathway, is responsible for the synthesis of shikimic acid. In glycolysis, phosphoenylpyruvate, through the enzyme pyruvate kinase, transfers the phosphate ion to a molecule of ADP, generating pyruvate and ATP. This pyruvate is responsible for the synthesis of acetyl-CoA (acetyl coenzyme A). From the shikimic acid path, *p*-coumaric acid (4-coumaric acid) is obtained, resulting from the ammonia elimination from L-phenylalanine side chain (precursors of the C_6-C_3 portions), and precursor of *p*-coumaric alcohol. The elimination of ammonia occurs in the presence of PAL (phenylalanine ammonia lyase) into cinnamic acid, which is converted into *p*-coumaric acid by direct hydroxylation, in the presence of cinnamate-4-hydroxylase (C4H), which is converted to 4-hydroxycinnamoyl-CoA by the 4-coumarate CoA ligase (4CL), by a process known as the general pathway to the formation of phenylpropanoids. The enzyme involved in the consensation process of malonyl-CoA units to form flavonoids is chalcone synthase, through the acetyl-CoA carboxylase (ACC) mediates carboxylation reaction. Therefore, the formation of flavonoids (Fig. 11.2) occurs from 4-hydroxycinnamoyl-CoA units, derived from the pathway of shikimic acid with the addition of malonyl-CoA units (for chain elongation), the route of acetate-malonate (Davies and Schwinn 2006; Dewick 2009).

After the addition of malonyl-CoA units, two pathways could be involved in the formation of different metabolites – flavonoids or stilbenes (e.g., resveratrol). After the formation of the polyketide, different enzymes are involved in aldol or Claisen type condensations to form aromatic ring A. For the formation of flavonoids, the enzyme involved in the condensation process of malonyl-CoA units is chalcone synthase. Chalcones are the precursors of flavonoids (I), followed by enolization (II) of the polyketide (Fig. 11.3).

For the synthesis of flavonoids, chalcones undergo a nucleophilic attack reaction of the OH group to the α,β-unsaturated ketone, forming a heterocyclic ring (C-ring) yielding flavanones. Acidic environments may favor the synthesis of flavanones,

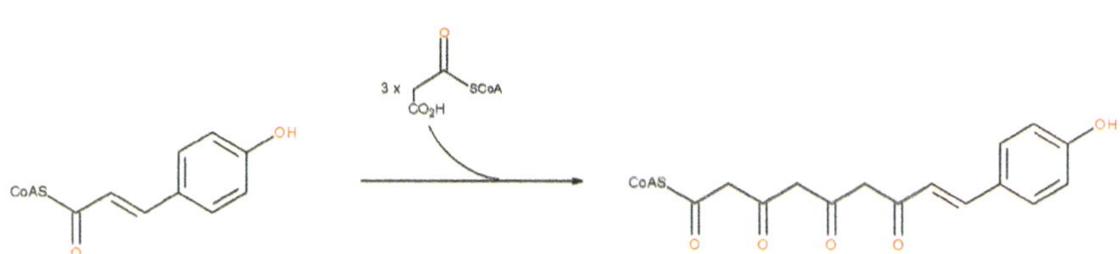

Fig. 11.2 Formation of the polyketide side chain, precursor of flavonoids, by acetate-malonate route

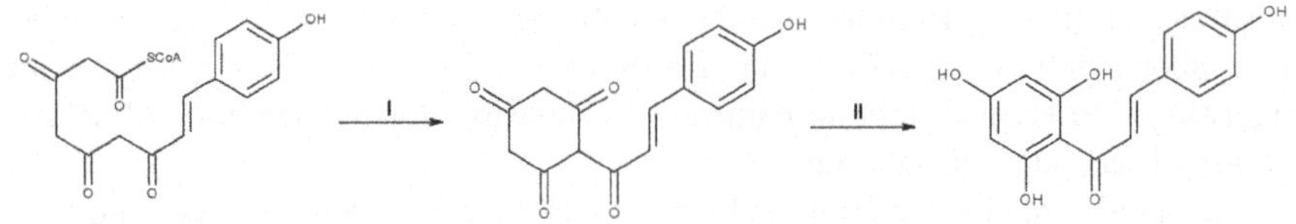

Fig. 11.3 Steps for chalcone biosynthesis from polyketide side chain

while alkaline environments favor the production of chalcones. However, it is worth emphasizing that in nature, these processes occur under very specific conditions, in the presence of stereospecific enzymes that prevent the formation of enantiomers. From the flavanones, a large variety of flavonoids can be synthesized, such as flavones, flavonols, and anthocyanidins (Davies and Schwinn 2006).

According to the basic skeleton of flavonoids, the variations between the different classes of flavonoids are related to the oxygenations and substituents of rings B and C. However, many flavonoids can also lose one or two hydroxyl groups in ring A, a process related to the action of chalcone reductase and chalcone synthase enzymes. The enzymatic complex involved in the biosynthesis of the different classes of flavonoids is broad and with high specificity, through this complex, the alterations of the oxygenation patterns in the aryl moiety are realized. In addition, methylation, glycosylation, and dimethylation processes are responsible for increasing the possibilities of formation of different compounds, increasing the diversity of flavonoids distributed in different plant species (Dewick 2009).

Flavanones, the first group of flavonoids synthesized from chalcones, are the precursors of the other groups of flavonoids. Flavones are synthesized from reactions in the presence of flavone synthase I, oxygen and 2-oxo-glutarate and flavone synthase IIe, oxygen and NADPH. In the presence of oxygen, 2-oxoglutarate, and flavanone 3-hydroxylase enzyme, dihydroflavonols are produced, which are the precursors of flavonols and flavandiols. Finally, flavodiols act as precursors of the catechins and anthocyanidins, through specific enzymatic processes that occur in the presence of oxygen and NADPH, and by water elimination (Davies and Schwinn 2006; Dewick 2009). Figure 11.4 briefly illustrates the biosynthesis of flavonoids.

3 Flavonoid Composition

As described above, the main composition of *Baccharis* are flavonoids, diterpenes, and other phenolic compounds. From the *Baccharis* genus, which is composed of 441 species, only approximately 20% have been investigated in chemical aspects. The occurrence of flavonoids was reported in 86 distinct species of *Baccharis*; these studies lead to the identification of 129 flavonoids separated into 16 flavanones (Table 11.1), 11 flavanonols (Table 11.2), 46 flavones (Table 11.3), and 55 flavonols (Table 11.4).

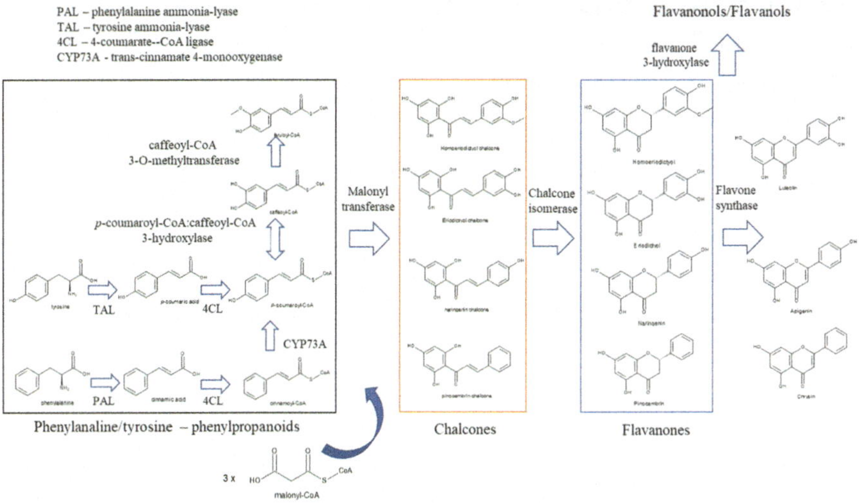

Fig. 11.4 Scheme of flavonoid biosynthesis

4 Flavanones

The main precursor of all flavonoids, naringenin (**1**), was found in aerial parts of *B. alaternoides* (Bohlmann et al. 1979), *B. conferta* (Weimann et al. 2002), *B. ligustrina* (Moreira et al. 2003a, b; Abad and Bermejo 2007), *B. polycephala* (Davila et al. 2013), *B. retusa* (Grecco et al. 2012a, b; Campos et al. 2016), *B. salzmannii* (Bohlmann et al. 1981a, b), and *B. varians* (Bohlmann et al. 1981a). Compound **1** was also found in the flowers of *B. illinita* (Verdi et al. 2004; Abad and Bermejo 2007; Campos et al. 2016), leaves of *B. dracunculifolia* (Fukuda et al. 2006; Campos et al. 2016) and *B. pseudotenuifolia* (Moreira et al. 2003a, b; Abad and Bermejo 2007). The methylation of position C-7 leads to sakuranetin (**2**) accumulated in the aerial parts of *B. concinna* (Wollenweber et al. 2006), *B. marginalis* (Faine et al. 1987), *B. retusa* (Herz et al. 1977; Grecco et al. 2012a, b, 2014a, b; Rodriguez et al. 2012; Taguchi et al. 2015a, b; Sakoda et al. 2016; Bittencourt-Mernak et al. 2017; Ueno et al. 2018), *B. salicifolia* (del Corral et al. 2012; Campos et al. 2016), *B. serrulata* (Bohlmann et al. 1981a), and *B. trinervis* var. *rhexioides* (Bohlmann et al. 1979). This compound was also found in the roots of *B. leptocephala* (Bohlmann et al. 1981a) and *B. intermixta* (Bohlmann et al. 1981a) and leaves of *B. teindalensis*. The methylation at position C-4′ rather than C-7 of naringenin (**1**) affords isosakuranetin (**3**) identified in aerial parts of *B. alaternoides* (Bohlmann et al. 1979), *B. conferta* (Weimann et al. 2002), *B. dracuncunfolia*, within its roots and vegetative gems (da Silva Filho et al. 2004, 2008; Park et al. 2004; de Alencar et al. 2005; Lemos et al. 2007; Missima et al. 2007; de Sousa et al. 2009; Guimaraes et al. 2012; Figueiredo-Rinhel et al. 2013; Campos et al. 2016) and in *B. polycephala* (Davila et al. 2013), in the roots of *B. leptocephala* (Bohlmann et al. 1981b), and in the

Table 11.1 Flavanones identified in parts of species from genus *Baccharis*

	Compounds	Species	Parts	References
1	Naringenin	*B. alaternoides*	Above ground	Bohlmann et al. (1979)
		B. conferta	Aerial parts	Weimann et al. (2002)
		B. ligustrina	Aerial parts	Moreira et al. (2003a, b) and Abad and Bermejo (2007)
		B. polycephala	Aerial parts	Davila et al. (2013)
		B. retusa	Aerial parts	Campos et al. (2016) and Grecco et al. (2012a)
		B. salzmannii	Aerial parts	Bohlmann et al. (1981a) and Campos et al. (2016)
		B. varians	Aerial parts	Bohlmann et al. (1981a)
		B. illinita	Flowers	Verdi et al. (2004), Abad and Bermejo (2007) and Campos et al. (2016)
		B. dracunculifolia	Leaves	Fukuda et al. (2006) and Campos et al. (2016)
		B. pseudotenuifolia	Leaves, shrub	Moreira et al. (2003a, b), Abad and Bermejo (2007) and Campos et al. (2016)
2	Sakuranetin	*B. concinna*	Aerial (leaves and stems)	Wollenweber et al. (2006)
		B. marginalis	Aerial parts	Faine et al. (1987)
		B. retusa	Aerial parts	Herz et al. (1977), Grecco et al. (2012b, 2014a, b), Toledo et al. (2013), Taguchi et al. (2015a, b), Sakoda et al. (2016), Bittencourt-Mernak et al. (2017) and Ueno et al. (2018)
		B. salicifolia	Aerial parts	Campos et al. (2016) and del Corral et al. (2012)
		B. serrulata	Aerial parts	Bohlmann et al. (1981a)
		B. trinervis var. *rhexioides*	Above ground	Bohlmann et al. (1979)
		B. leptocephala	Roots	Bohlmann et al. (1981b)
		B. intermixta	Roots	Bohlmann et al. (1981a)
		B. teindalensis	Leaves	Vidari et al. (2003)
3	Isosakuranetin	*B. alaternoides*	Above ground	Bohlmann et al. (1979)
		B. conferta	Aerial parts	Weimann et al. (2002)
		B. dracunculifolia	Aerial, leaves, roots, and vegetative gems	da Silva Filho et al. (2004, 2008), Park et al. (2004), de Alencar et al. (2005), Lemos et al. (2007), Missima et al. (2007), de Sousa et al. (2009), Guimaraes et al. (2012), Figueiredo-Rinhel et al. (2013) and Campos et al. (2016)

(continued)

Table 11.1 (continued)

	Compounds	Species	Parts	References
		B. polycephala	Aerial parts	Davila et al. (2013)
		B. leptocephala	Roots	Bohlmann et al. (1981b)
		B. leptophylla	Shrub	Almanza et al. (2000) and Mollinedo et al. (2001)
4	5,6,7-trihydroxy-4′-methoxyflavanone	*B. conferta*	Aerial parts	Weimann et al. (2002)
		B. retusa	Aerial parts	Grecco et al. (2010a, b)
		B. teindalensis	Leaves	Vidari et al. (2003)
		B. viminea		Wollenweber et al. (1997)
5	5-hydroxy-4′,7-dimethoxyflavanone	*B. conferta*	Aerial parts	Weimann et al. (2002)
		B. polycephala	Aerial parts	Davila et al. (2013)
6	4′-hydroxy-5,7-dimethoxyflavanone	*B. alaternoides*	Above ground	Bohlmann et al. (1979)
7	Pinocembrin	*B. concinna*	Roots	Bohlmann et al. (1981b)
		B. oxyodonta	Roots	Bohlmann et al. (1981b) and Abad Martinez et al. (2005)
		B. viminea		Wollenweber et al. (1997)
8	5,7-dihydroxy-3′-methoxyflavanone	*B. truncata*	Roots	Bohlmann et al. (1981b)
9	Dihydrooroxylin A	*B. uncinella*	Aerial parts	Campos et al. (2016)
10	Eriodictyol	*B. concinna*	Aerial parts	Wollenweber et al. (2006)
		B. confertifolia	Aerial parts	Wollenweber et al. (2006)
		B. marginalis	Aerial parts	Faine et al. (1987)
		B. retusa	Aerial parts	Campos et al. (2016) and Grecco et al. (2012a)
		B. pseudotenuifolia	Shrub, leaves	Moreira et al. (2003a, b), Abad and Bermejo (2007) and Campos et al. 2016)
11	Eriodictyol-7-methylether	*B. concinna*	Aerial parts	Wollenweber et al. (2006)
12	Homoeriodictyol	*B. calliprinos*	Aerial parts	Gianello et al. (1999)
13	Eriodictyol-3′,4′-dimethylether	*B. calliprinos*	Aerial parts	Gianello et al. (1999)
14	Eriodictyol 7,3′,4′-trimethyl ether	*B. confertifolia*	Aerial parts	Wollenweber et al. (2006)
15	Filifolin	*B. concinna*	Aerial parts	Wollenweber et al. (2006)
		B. boliviensis	Aerial parts	Campos et al. (2016)
16	8- methoxyeriodictyol	*B. concinna*	Aerial parts	Wollenweber et al. (2006)

Table 11.2 Flavanonols identified in parts of species from genus *Baccharis*

	Compounds	Species	Parts	References
17	Aromadendrin/ dihydrokaempferol	*B. dracunculifolia*	Leaves	Guimaraes et al. (2012)
		B. pseudotenuifolia	Shrub	Moreira et al. (2003a, b), Abad and Bermejo (2007) and Campos et al. (2016)
		B. illinita	Leaves and flowers	Verdi et al. (2004), Abad and Bermejo (2007) and Campos et al. (2016)
		B. retusa	Aerial parts	Campos et al. (2016) and Grecco et al. (2012a)
18	Aromadendrin-7-methyl ether	*B. dracunculifolia*	Aerial (leaves and stems)	Campos et al. (2016)
		B. illinita	Aerial (leaves and stems)	Campos et al. (2016) and Pizzolatti et al. (2006)
19	Dihydrokaempferide	*B. conferta*	Aerial parts	Weimann et al. (2002)
		B. dracunculifolia	Leaves, roots, flowers, buds, and stems	Figueiredo-Rinhel et al. (2013), da Silva Filho et al. (2004, 2008), Missima et al. (2007), Lemos et al. (2007); de Sousa et al. (2009, 2011), Resende et al. (2007), Rezende et al. (2014), Cestari et al. (2011), Midorikawa et al. (2003) and Kumazawa et al. (2003)
		B. leptophylla		Almanza et al. (2000) and Mollinedo et al. (2001)
20	3-acetoxy-4′,5,7-trihydroxyflavanone	*B. varians*	Aerial parts	(Bohlmann et al. (1981a) and Abad Martinez et al. (2005)
21	3- acetoxy-4′,5-dihydroxy-7-methoxyflavanone	*B. dracunculifolia*	Leaves	Campos et al. (2016) and Fukuda et al. (2006)
22	Pinobanksin	*B. oxyodonta*	Roots	Bohlmann et al. (1981b) and Abad Martinez et al. (2005)
		B. dracunculifolia	Leaves and vegetative gems	Park et al. (2004) and de Alencar et al. (2005)
23	Pinobanksin-3-acetate	*B. dracunculifolia*	Leaf bud	Park et al. (2005)
		B. trinervis		Jakupovic et al. (1986)
24	Taxifolin	*B. illinita*	Flowers	Verdi et al. (2004), Abad and Bermejo (2007) and Campos et al. (2016)
		B. retusa	Aerial parts	Campos et al. (2016) and Grecco et al. (2012a)
25	Taxifolin-4′-methyl ether	*B. tola*	Aerial parts	Simirgiotis et al. (2016)
26	8-prenyltaxifolin	*B. tola*	Aerial parts	Simirgiotis et al. (2016)
27	Taxifolin-3-acetate	*B. varians*	Aerial parts	Bohlmann et al. (1981a) and Abad Martinez et al. (2005)

Table 11.3 Flavones identified in parts of species from genus *Baccharis*

	Compounds	Species	Parts	References
28	Apigenin	*B. bigelovii*	Aerial parts	Arriaga-Giner et al. (1986)
		B. crispa	Aerial parts	Palacios et al. (1983)
		B. dentata	Aerial parts	Sartor et al. (2013) and Campos et al. (2016)
		B. gaudichaudiana	Aerial parts	Fullas et al. (1994), Guo et al. (2007), Visintini et al. (2013) and Campos et al. (2016)
		B. notosergila	Aerial parts	Palacios et al. (1983)
		B. pedicellata	Aerial parts	Faine et al. (1987)
		B. retusa	Aerial parts	Campos et al. (2016) and Grecco et al. (2012a)
		B. salicifolia	Aerial parts	Bohlmann et al. (1981b), Wollenweber et al. (1986) and Campos et al. (2016)
		B. salzmannii	Aerial parts	Campos et al. (2016)
		B. trimera	Aerial parts and epigeous parts	Soicke and Leng-Eschlow (1987), Nakasugi and Komai (1998) and Nakasugi (1990)
		B. illinita	Flowers	Verdi et al. (2004), Abad and Bermejo (2007) and Campos et al. (2016)
		B. pseudotenuifolia	Leaves and shrub	Moreira et al. (2003a, b), Abad and Bermejo (2007) andCampos et al. (2016)
		B. teindalensis	Leaves	Vidari et al. (2003)
		B. dracunculifolia	Vegetative gems	de Alencar et al. (2005)
		B. genistelloides		Kuroyanagi et al. (1985), Abad Martinez et al. (2005) and Hennig et al. (2011)
		B. heterophylla		Wollenweber et al. (1986)
		B. pteronioides		Wollenweber et al. (1986)
		B. ramosissima		(Bohlmann et al. 1981b)
		B. tola		San Martin et al. (1982, 1983) and Abad Martinez et al. (2005)

(continued)

Table 11.3 (continued)

	Compounds	Species	Parts	References
		B. trinervis		Arriaga et al. (1982)
		B. vaccinioides		Wollenweber et al. (1986)
		B. viminea		Wollenweber et al. (1997)
29	Isoscutellarein	*B. pilularis*		Wollenweber et al. (1997)
30	Acacetin	*B. articulata*	Aerial parts	Gianello and Giordano (1984) and Cariddi et al. (2012)
		B. conferta	Aerial parts	Weimann et al. (2002)
		B. polycephala	Aerial parts	Davila et al. (2013)
		B. dracunculifolia	Leaves	Park et al. (2004) and da Silva Filho et al. (2009)
		B. rhomboidalis	Leaves	Silva et al. (1971)
		B. grandicapitulata		Bohlmann et al. (1985)
		B. patagonica		Zdero et al. (1986) and Rivera et al. (1988)
		B. salicifolia		Wollenweber et al. (1986)
		B. trinervis		Arriaga et al. (1982)
		B. vaccinioides		Wollenweber et al. (1986)
		B. viminea		Wollenweber et al. (1997)
31	Genkwanin	*B. crispa*	Aerial parts	Palacios et al. (1983)
		B. notosergila	Aerial parts	Palacios et al. (1983)
		B. pedicellata	Aerial parts	Faine et al. (1987)
		B. trimera	Aerial parts	Soicke and Leng-Eschlow (1987) and Nakasugi and Komai (1998)
		B. trinervis	Leaves	Herrera et al. (1996)
		B. genistelloides		Abad Martinez et al. (2005) and Hennig et al. (2011)
		B. pilularis var. *consanguinea*		Wollenweber et al. (1997)
32	Thevetiaflavone	*B. gaudichaudiana*	Aerial parts	Guo et al. (2007) and Campos et al. (2016)
33	7-hydroxy-5,4′-dimethoxyflavone	*B. articulata*		de Oliveira et al. (2014)
		B. usterii	Aerial parts	de Oliveira et al. (2014)

(continued)

Table 11.3 (continued)

	Compounds	Species	Parts	References
34	5-hydroxy-4′,7-dimethoxyflavone	*B. conferta*	Aerial parts	Weimann et al. (2002)
		B. crispa	Aerial parts	Bandoni et al. (1978)
		B. illinita	Aerial parts	Arriaga-Giner et al. (1986) and Campos et al. (2016)
		B. rhomboidalis	Leaves	Silva et al. (1971)
		B. teindalensis	Leaves	Vidari et al. (2003)
		B. tola	Leaves	San Martin et al. (1982)
		B. trinervis	Leaves	Herrera et al. (1996)
		B. latifolia		Salcedo et al. (2001)
35	Chrysin	*B. dracunculifolia*	Leaves	Paula et al. (2017) and Park et al. (2004)
		B. viminea		Wollenweber et al. (1997)
36	7-methylchrysin	*B. viminea*		Wollenweber et al. (1997)
37	2′-methoxychrysin	*B. illinita*	Flowers	Verdi et al. (2004), Abad and Bermejo (2007) and Campos et al. (2016)
38	Hispidulin	*B. flabellata*	Aerial parts	Saad et al. (1988)
		B. gaudichaudiana	Aerial parts	Campos et al. (2016), Fullas et al. (1994) and Akaike et al. (2003)
		B. genistelloides	Aerial parts	Daily et al. (1984) and San Martin et al. (2012)
		B. grisebachii	Aerial parts	Abad and Bermejo (2007) and Tapia et al. (2004)
		B. halimifolia	Aerial parts	Joshi et al. (1997) and Jakupovic et al. (1990)
		B. ligustrina	Aerial parts	Abad and Bermejo (2007), Moreira et al. (2003a, b) and Nogueira et al. (2016)
		B. magellanica	Aerial parts	Cordano et al. (1982)
		B. rhomboidalis	Aerial parts	Cordano et al. (1982)
		B. trimera	Aerial and epigeous	Nakasugi and Komai (1998), Soicke and Leng-Eschlow (1987), Nakasugi (1990) and Padua et al. 2014)
		B. uncinella	Aerial parts	Campos et al. (2016) and Grecco et al. (2010a, b, 2014a, b)

(continued)

Table 11.3 (continued)

	Compounds	Species	Parts	References
		B. pseudotenuifolia	Leaves and shrub	Moreira et al. (2003a, b), Abad and Bermejo (2007) and Campos et al. (2016)
		B. vaccinioides		Wollenweber et al. (1986)
39	Pectolinaringenin	*B. concava*	Aerial parts	Zamorano et al. (1987)
		B. conferta	Aerial parts	Weimann et al. (2002)
		B. decussata	Aerial parts	Morales Mendez et al. (1984) and Rojas and Morales (2000)
		B. grisebachii	Aerial parts	Gianello and Giordano (1987), Abad Martinez et al. (2005) and Feresin et al. (2003)
		B. uncinella	Aerial parts	Grecco et al. (2010a, b, 2014a, b), Passero et al. (2011), Zalewski et al. (2011) and Campos et al. (2016)
		B. pedunculata	Leaf	Rahalison et al. (1995)
		B. trinervis	Branches	Sharp et al. (2000) and Abad Martinez et al. (2005)
		B. macraei	Fresh	Faini et al. (1991)
		B. pilularis var. consanguinea		Wollenweber et al. (1997)
		B. vaccinioides		Wollenweber et al. (1986)
40	Cirsimaritin	*B. concava*	Aerial parts	Zamorano et al. (1987)
		B. concinna	Aerial parts	Wollenweber et al. (2006)
		B. conferta	Aerial parts	Weimann et al. (2002)
		B. genistelloides	Aerial parts	Suttisri et al. (1994), Abad Martinez et al. (2005) and Hennig et al. (2011)
		B. rhomboidalis	Aerial parts	Labbe et al. (1986)
		B. rufescens var. *rufescens.*	Aerial parts	Simirgiotis et al. (2003), Abad and Bermejo (2007) and Campos et al. (2016)
		B. trimera	Aerial parts	Nakasugi (1990) and Nakasugi and Komai (1998)

(continued)

Table 11.3 (continued)

	Compounds	Species	Parts	References
		B. elaeagnoides		Mesquita et al. (1985)
		B. halimifolia		Joshi et al. (1997)
		B. intermedia		Faini et al. (1991)
		B. macraei		Faini et al. (1991)
		B. pilularis var. *consanguinea*		Wollenweber et al. (1997)
		B. tricuneata		Wagner et al. (1978)
41	Scutellarein-7,4′-dimethyl ether	*B. tucumanensis*	Aerial parts	Tonn et al. (1982)
42	Salvigenin	*B. concava*	Aerial parts	Zamorano et al. (1987)
		B. conferta	Aerial parts	Weimann et al. (2002)
		B. rhomboidalis	Aerial parts	Labbe et al. (1986)
		B. scandens	Aerial parts	Cabrera et al. (2016)
		B. trinervis	Branches	Sharp et al. (2000) and Abad Martinez et al. (2005)
		B. macraei		Faini et al. (1991)
		B. pilularis		Wollenweber et al. (1997)
43	Desmethoxysudachitin	*B. grisebachii*	Aerial parts	Gianello and Giordano (1987), Tapia et al. (2004), Abad Martinez et al. (2005) and Campos et al. (2016)
		B. solierii	Aerial parts	Labbe et al. (1986)
44	Nevadensin	*B. decussata*	Aerial parts	Rojas and Morales (2000)
		B. grisebachii	Aerial parts	Gianello and Giordano (1987), Feresin et al. (2003), Tapia et al. (2004) and Abad Martinez et al. (2005)
		B. nitida	Aerial parts	Chidiak et al. (2007) and Campos et al. (2016)
45	Xantomicrol	*B. boliviensis*	Aerial parts	Calle et al. (2012) and Campos et al. (2016)
		B. nitida	Aerial parts	Chidiak et al. (2007) and Campos et al. (2016)
		B. patens	Aerial parts	Silva et al. (1985)
		B. scandens	Aerial parts	Cabrera et al. (2016)
		B. tucumanensis	Aerial parts	Tonn et al. (1982)
		B. pentlandii	Leaves	Tarqui et al. (2012)
		B. quitensis	Roots	Bohlmann et al. (1981b)
46	Gardenin B	*B. grisebachii*	Aerial parts	Feresin et al. (2003), Tapia et al. (2004) and Campos et al. (2016)

(continued)

Table 11.3 (continued)

	Compounds	Species	Parts	References
		B. scandens	Aerial parts	Cabrera et al. (2016)
47	Tangeretin	*B. illinita*	Flowers	Verdi et al. (2004), Abad and Bermejo (2007) and Campos et al. (2016)
48	Luteolin	*B. articulata*	Aerial parts	Cariddi et al. (2012)
		B. bigelovii	Aerial parts	Arriaga-Giner et al. (1986) and Wollenweber et al. (1986)
		B. concinna	Aerial parts	Wollenweber et al. (2006)
		B. confertifolia	Aerial parts	Wollenweber et al. (2006)
		B. gaudichaudiana	Aerial parts	Guo et al. (2007)
		B. genistelloides	Aerial parts	San Martin et al. (2012)
		B. incarum	Aerial parts	Zampini et al. (2009) and Campos et al. (2016)
		B. linearis	Aerial parts	Wollenweber et al. (2006)
		B. lycioides	Aerial parts	Wollenweber et al. (2006)
		B. reticularia	Aerial parts	Bohlmann et al. (1981a)
		B. trimera	Aerial parts	Soicke and Leng-Eschlow (1987), da Silva et al. (2016) and Menezes et al. (2016)
		B. trinervis	Aerial parts	Jaramillo-Garcia et al. (2018)
		B. varians	Aerial parts	Bohlmann et al. (1981a)
		B. illinita	Flowers	Verdi et al. (2004), Abad and Bermejo (2007) and Campos et al. (2016)
		B. halimifolia		Wollenweber et al. (1997)
		B. microcephala		Bohlmann et al. (1985)
		B. nitida		Bohlmann et al. (1985)
		B. pteronioides		Wollenweber et al. (1986)
49	Chrysoeriol	*B. salicifolia*	Aerial parts	Warning et al. (1986)
		B. pseudotenuifolia	Leaves and shrub	Moreira et al. (2003a, b), Abad and Bermejo (2007) and Campos et al. (2016)

(continued)

Table 11.3 (continued)

	Compounds	Species	Parts	References
		B. illinita	Flowers	Verdi et al. (2004), Abad and Bermejo (2007) and Campos et al. (2016)
50	Luteolin-4′,7-dimethylether	*B. trimera*	Aerial parts	Padua et al. (2014) and de Araujo et al. (2016)
		B. trinervis	Leaves	Herrera et al. (1996)
51	Luteolin-3′,7-dimethylether	*B. calliprinos*	Aerial parts	Gianello et al. (1999)
		B. rhetinodes	Aerial parts	Gianello et al. (1999)
		B. salicifolia	Aerial parts	Warning et al. (1986)
52	5-hydroxy-7,3′,4′-trimethoxyflavone, luteolin-3′,4′,7-trimethylether	*B. crispa*	Aerial parts	Bandoni et al. (1978)
		B. latifolia	Aerial parts	Salcedo et al. (2003) and Campos et al. (2016)
		B. trinervis	Leaves	Rivera et al. (1988)
53	6-hydroxyluteolin	*B. boliviensis*	Aerial parts	Abad and Bermejo (2007)
54	Nepetin; eupafolin	*B. concinna*	Aerial parts	Verdi et al. (2004)
		B. confertifolia	Aerial parts	Verdi et al. (2004)
		B. flabellata	Aerial parts	Bandoni et al. (1978)
		B. gaudichaudiana	Aerial parts	Salcedo Ortiz et al. (2001) and Abad and Bermejo (2007)
		B. genistelloides	Aerial parts	San Martin et al. (2012)
		B. linearis	Aerial parts	He (1995), He et al. (1996) and Wollenweber et al. (2006)
		B. lycioides	Aerial parts	Wollenweber et al. (2006)
		B. trimera	Aerial parts	Soicke and Leng-Eschlow (1987) and Simões-Pires et al. (2005)
55	Jaceosidin	*B. concinna*	Aerial parts	Wollenweber et al. (2006)
		B. flabellata	Aerial parts	Saad et al. (1988)
		B. gaudichaudiana	Aerial parts	Akaike et al. (2003) and Campos et al. (2016)
		B. grisebachii	Aerial parts	Tapia et al. (2004)
56	Desmethoxycentaureidin	*B. gaudichaudiana*	Aerial parts	Akaike et al. (2003) and Campos et al. 2016)
		B. solierii	Stems and leaves	Labbe et al. (1986)
		B. petiolata		Labbé et al. (1990)
		B. salicina		Parodi and Fischer (1988) and Quijano et al. (1998)

(continued)

Table 11.3 (continued)

	Compounds	Species	Parts	References
57	Cirsiliol	*B. concinna*	Leaves and stems	Wollenweber et al. (2006)
		B. confertifolia	Leaves and stems	Wollenweber et al. (2006)
		B. flabellata	Aerial parts	Saad et al. (1988)
		B. linearis	Leaves and stems	Wollenweber et al. (2006)
		B. rufescens var. *rufescens* .	Aerial parts	Simirgiotis et al. (2003), Abad and Bermejo (2007) and Campos et al. (2016)
		B. genistelloides		Kuroyanagi et al. (1985), Abad Martinez et al. (2005) and Hennig et al. (2011)
		B. tricuneata		Wagner et al. (1978)
58	Cirsilineol	*B. concinna*	Aerial parts	Wollenweber et al. (2006)
		B. salicifolia	Aerial parts	Warning et al. (1986)
		B. genistelloides		Abad Martinez et al. (2005) and Hennig et al. (2011)
59	5,6-Dihydroxy-3′,4′,7-trimethoxyflavone	*B. trimera*	All plants	Borella et al. (2006)
60	Eupatilin	*B. conferta*	Aerial parts	Weimann et al. (2002)
		B. gaudichaudiana	Aerial parts	Akaike et al. (2003) and Campos et al. 2016)
61	Eupatorin	*B. genistelloides*	Aerial parts	Suttisri et al. (1994) and Hennig et al. (2011)
		B. trimera	Leaves, stems, flowers, and fruits	Herz et al. (1977), de Mello and Petrovick (2000), Torres et al. (2000), da Silva et al. (2004), Silva et al. (2006), Padua et al. (2014) and de Araujo et al. (2016)
62	5-hydroxy-3′,4′,6,7-tetramethoxyflavone	*B. genistelloides*	Aerial parts	Suttisri et al. (1994)
		B. trimera	Leaves	Rendon and Vila (1995) and Silva et al. (2006)
63	Sideritiflavone	*B. patens*	Aerial parts	Silva et al. (1985)
		B. thymifolia	Aerial parts	Saad et al. (1987)
		B. pentlandii	Leaves	Tarqui et al. (2012) and Campos et al. (2016)

(continued)

Table 11.3 (continued)

	Compounds	Species	Parts	References
64	4′,5-dihydroxy-3′,6,7,8-tetramethoxyflavone, 7-methylsudachitin, 3′-methoxyxanthomicrol	*B. incarum*	Aerial parts	Faini et al. (1982a, b)
		B. salicifolia	Aerial parts	Warning et al. (1986)
		B. thymifolia	Aerial parts	Saad et al. (1987)
		B. pentlandii	Leaves	Tarqui et al. (2012) and Campos et al. (2016)
		B. oxyodonta	Roots	Bohlmann et al. (1981b)
		B. quitensis	Roots	Bohlmann et al. (1981b)
65	Gardenin D	*B. patens*	Aerial parts	Silva et al. (1985)
66	5-Hydroxy-3′,4′,6,7,8-pentamethoxytlavone	*B. thymifolia*	Aerial parts	Saad et al. (1987)
67	Nobiletin	*B. illinita*	Flowers	Abad and Bermejo (2007) and Campos et al. (2016)
68	Isoschaftoside	*B. gaudichaudiana*	Aerial parts	Akaike et al. (2003 and Campos et al. (2016)
69	Vicenin II	*B. trimera*	Aerial parts	Rabelo et al. (2017)
		B. trimera	Aerial parts	Rabelo et al. (2017)
70	6(8)-C-furanosyl-8(6)-C-hexosyl flavone	*B. trimera*	Aerial parts	de Araujo et al. (2016)
71	6(8)-C- hexosyl-8(6)-C-furanosyl flavone	*B. trimera*	Aerial parts	de Araujo et al. (2016) and Rabelo et al. (2017)
72	5,3′-dihydroxy-4′-methoxy-7-O-pyranosyl-furanosyl flavone	*B. trimera*	Aerial parts	Padua et al. (2014)

shrub *B. leptophylla*. The C-6 hydroxylation of isosakuranetin (**3**) leads to the formation of 5,6,7-trihydroxy-4′-methoxyflavanone (**4**) accumulated in the leaves and aerial parts of *B. conferta* (Weimann et al. 2002), *B. retusa* (Grecco et al. 2010a, b), *B. teindalensis* (Vidari et al. 2003), and *B. viminea* (Wollenweber et al. 1997). The dimethoxy derivative of naringenin (**1**), 5-hydroxy-4′,7-dimethoxyflavanone (**5**), was identified in the aerial parts of *B. conferta* (Weimann et al. 2002) and *B. polycephala* (Davila et al. 2013) and 4′-hydroxy-5,7-dimethoxyflavanone (**6**) in ground parts of *B. alaternoides* (Bohlmann et al. 1979). The 4′-dehydroxylated derivative, pinocembrin (**7**), was identified in *B. viminea* (Wollenweber et al. 1997), in the roots of *B. concinna* (Bohlmann et al. 1981b), and in *B. oxyodonta* (Bohlmann et al. 1981b; Abad Martinez et al. 2005). Less frequently isolated 3'- and 6-monomethoxylated derivatives of pinocembrin (**7**): 5,7-dihydroxy-3′-methoxyflavanone (**8**) and 5,7-dihydroxy-6-methoxyflavanone, known as dihydrooroxylin A (**9**), were identified in the roots of *B. truncata* [27] and aerial parts of *B. uncinella*, respectively (Grecco et al. 2010a, b; Campos et al. 2016).

Table 11.4 Flavanols identified in parts of species from genus *Baccharis*

	Compounds	Species	Parts	References
73	Kaempferol	*B. conferta*	Aerial parts	Weimann et al. (2002)
		B. dentata	Aerial parts	Sartor et al. (2013) and Campos et al. (2016)
		B. gaudichaudiana	Aerial parts	Campos et al. (2016)
		B. polycephala	Aerial parts	Davila et al. (2013)
		B. retusa	Aerial parts	Grecco et al. (2012a) and Campos et al. (2016)
		B. illinita	Flowers	Verdi et al. (2004), Abad and Bermejo (2007) and Campos et al. (2016)
		B. maritima	Flowering tops	Moreira et al. (1975)
		B. pseudotenuifolia	Shrub, leaves	Moreira et al. (2003a, b), Abad and Bermejo (2007) and Campos et al. (2016)
		B. trimera	Leaves	da Silva et al. (2016)
		B. dracunculifolia	Leaves, buds, stems and vegetative gems	Midorikawa et al. (2003), Park et al. (2004, 2005) and de Alencar et al. (2005)
		B. genistelloides		Daily et al. (1984)
		B. pilularis var. consanguinea		Wollenweber et al. (1997)
74	Isokaempferide	*B. linearis*	Aerial parts	Wollenweber et al. (2006) and Faini et al. (1999)
		B. lycioides	Aerial parts	Wollenweber et al. (2006)
		B. pedicellata	Aerial parts	Faine et al. (1987)

(continued)

Table 11.4 (continued)

	Compounds	Species	Parts	References
		B. papillosa	Leaves	Campos et al. (2016) and Escobar et al. (2009)
		B. intermedia		Faini et al. (1991)
		B. linearis		Faini et al. (1991)
		B. macraei		Faini et al. (1991)
		B. pilularis var. *consanguinea*		Wollenweber et al. (1997)
75	Kaempferide	*B. conferta*	Aerial parts	Weimann et al. (2002)
		B. polycephala	Aerial parts	Davila et al. (2013)
		B. dracunculifolia	Flowers, leaves, buds, stems and vegetative gems	Midorikawa et al. (2003), Park et al. (2004, 2005), de Alencar et al. (2005), Piantino et al. (2008), Campos et al. (2016) and Paula et al. (2017)
		B. leptophylla		Almanza et al. (2000) and Mollinedo et al. (2001)
		B. pilularis var. consanguinea		Wollenweber et al. (1997)
		B. viminea		Wollenweber et al. (1997)
76	Kaempferol-7-methyleter	*B. pilularis var. consanguinea*		Wollenweber et al. (1997)
77	Ermanin	*B. papillosa*	Leaves	Campos et al. (2016)
		B. dracunculifolia	Buds; leaves; stems	Midorikawa et al. (2003) and da Silva Filho et al. (2009)
		B. pilularis var. consanguinea		Wollenweber et al. (1997)
		B. viminea		Wollenweber et al. (1997)
78	Kaempferol-7,4′-dimethyl ether	*B. crispa*	Aerial parts	Bandoni et al. (1978)

(continued)

Table 11.4 (continued)

	Compounds	Species	Parts	References
		B. Illinita	Leaves and stems	Campos et al. (2016) and Pizzolatti et al. (2006)
		B. teindalensis	Leaves	Vidari et al. (2003)
79	Kaempferol-3,7-dimethylether	*B. pedicellata*	Aerial parts	Faine et al. (1987)
		B. santelicis	Aerial parts	Zdero et al. (1991)
		B. pilularis var. consanguinea		Wollenweber et al. (1997)
80	Kaempferol-3,4′,7-trimethylether	*B. illinita*	Aerial parts	Campos et al. (2016) and Pizzolatti et al. (2006)
		B. santelicis	Aerial parts	Zdero et al. (1991)
81	Galangin	*B. dracunculifolia*	Leaf bud	Park et al. (2005)
		B. viminea		Wollenweber et al. (1997)
82	Galangin-7-methylether	*B. viminea*		Wollenweber et al. (1997)
83	6-hydroxykaempferol	*B. pilularis var. consanguinea*		Wollenweber et al. (1997)
84	6-methoxykaempferol	*B. dracunculifolia*	Leaves	Kumazawa et al. (2003)
85	4′,6-dimethoxykaempferol	*B. conferta*	Aerial parts	Weimann et al. (2002)
86	6,7-dimethoxykaempferol	*B. pilularis var. consanguinea*		Wollenweber et al. (1997)
87	Penduletin	*B. pedunculata*	Leaves	Rahalison et al. (1995)
		B. trinervis	Branches	Sharp et al. (2000 and Abad Martinez et al. (2005)
		B. salicifolia		Wollenweber et al. (1986)
		B. sarothroides		Wollenweber et al. (1986)
		B. vaccinioides		Wollenweber et al. (1986)
88	Herbacetin-3-methyl ether	*B. pilularis var. consanguinea*		Wollenweber et al. (1997)
89	Quercetin	*B. articulata*	Aerial parts	Campos et al. (2016)

(continued)

Table 11.4 (continued)

	Compounds	Species	Parts	References
		B. bigelovii	Aerial parts	Wollenweber et al. (1986) and Grecco et al. (2014a, b)
		B. concinna	Aerial parts	Wollenweber et al. (2006)
		B. confertifolia	Aerial parts	Wollenweber et al. (2006)
		B. dentata	Aerial parts	Sartor et al. (2013) and Campos et al. (2016)
		B. genistelloides	Aerial parts	Montes et al. (1971) and Morales Mendez et al. (1984)
		B. grisebachii	Aerial parts	Moreira et al. (1975)
		B. linearis	Aerial parts	Wollenweber et al. (2006)
		B. lycioides	Aerial parts	Wollenweber et al. (2006)
		B. pteronioides	Aerial parts	Wollenweber et al. (1986) and Jakupovic et al. (1990)
		B. retusa	Aerial parts	Grecco et al. (2012a), Ferreira et al. (2015) and Campos et al. (2016)
		B. salicifolia	Aerial parts	Wollenweber et al. (1986)
		B. scandens	Aerial parts	Cabrera et al. (2016)
		B. thesioides	Aerial parts	Liu et al. (1993)
		B. trimera	Aerial parts	Soicke and Leng-Eschlow (1987), Simões-Pires et al. (2005), Padua et al. (2014), da Silva et al. (2016), de Araujo et al. (2016), Menezes et al. (2016) and Sabir et al. (2017)

(continued)

Table 11.4 (continued)

	Compounds	Species	Parts	References
		B. trinervis	Aerial parts	Jaramillo-Garcia et al. (2018)
		B. dracunculifolia	Leaf bud and flowers	Midorikawa et al. (2003) and Park et al. (2005)
		B. pseudotenuifolia	Leaves and shrubs	
		B. spicata	Leaves, stems, and flowers	Moreira et al. (2003a, b), Abad and Bermejo (2007) and Campos et al. (2016)
		B. Illinita	Flowers	Verdi et al. (2004), Abad and Bermejo (2007) and Campos et al. (2016)
		B. maritima	Flowering tops	Moreira et al. (1975)
		B. viminea		Wollenweber et al. (1997)
90	Rhamnetin	*B. confertifolia*	Leaves and stems	Wollenweber et al. (2006)
		B. pilularis var. consanguinea		Wollenweber et al. (1997)
91	Isorhamnetin	*B. tola*	Aerial parts	Simirgiotis et al. (2016)
		B. linearis	Leaves and stems	Wollenweber et al. (2006) and He (1995)
		B. lycioides	Leaves and stems	Wollenweber et al. (2006)
		B. pseudotenuifolia	Leaves and shrub	
		B. viminea		Wollenweber et al. (1997)
92	3-methylquercetin	*B. tola*	Aerial parts	Simirgiotis et al. (2016)
		B. trimera	Aerial parts	de Mello and Petrovick (2000)
		B. linearis	Leaves	Faini et al. (1999)
		B. papillosa	Leaves	Campos et al. (2016)

(continued)

Table 11.4 (continued)

	Compounds	Species	Parts	References
		B. illinita	Flowers	Verdi et al. (2004), Abad and Bermejo (2007) and Campos et al. (2016)
		B. halimifolia		Wollenweber et al. (1997)
		B. intermedia		Faini et al. (1991)
		B. linearis		Faini et al. (1991)
93	Rhamnazin	*B. tola*	Aerial parts	Simirgiotis et al. (2016)
		B. confertifolia	Leaves and stems	Wollenweber et al. (2006)
		B. latifolia		Salcedo Ortiz et al. (2001)
94	Isorhamnetin 3-methyl ether	*B. tola*	Aerial parts	Simirgiotis et al. (2016)
		B. linearis	Leaves and stems	Wollenweber et al. (2006) and Faini et al. (1999)
		B. lycioides	Leaves and stems	Wollenweber et al. (2006)
		B. intermedia		Faini et al. (1991)
		B. linearis		Faini et al. (1991)
95	3,4′-dimethoxy-3′,5,7-trihydroxyflavone	*B. sarothroides*		Abad Martinez et al. (2005)
96	3,7-dimethoxyquercetin	*B. triangularis*	Aerial parts	Gianello and Giordano (1989)
		B. pilularis var. consanguinea		Wollenweber et al. (1997)
97	3,7-dimethyl-isorhamnetin	*B. tola*	Aerial parts	Simirgiotis et al. (2016)
98	3,7,4′-trimethylquercetin	*B. illinita*	Leaves and stems	Campos et al. (2016, 46)
		B. grisebachii		Abad and Bermejo (2007)
99	3,5-dihydroxy-7,3′,4′-trimethoxyflavone	*B. latifolia*		Salcedo Ortiz et al. (2001)
100	Retusin	*B. illinita*	Leaves and stems	Campos et al. (2016, 46)

(continued)

Table 11.4 (continued)

	Compounds	Species	Parts	References
101	3-hydroxy-5,7,3′,4’tetramethoxyflavone	*B. latifolia*	Aerial parts	Campos et al. (2016, 121)
		B. conferta	Aerial parts	Weimann et al. (2002)
102	Patuletin	*B. concinna*	Leaves and stems	Wollenweber et al. (2006)
		B. confertifolia	Leaves and stems	Wollenweber et al. (2006)
		B. halimifolia		Wollenweber et al. (1997)
		B. pilularis var. consanguinea		Wollenweber et al. (1997)
103	3′-methylpatuletin	*B. halimifolia*		Wollenweber et al. (1997)
104	Axillarin	*B. incarum*	Leaves and stems	Campos et al. (2016)
		B. linearis	Leaves and stems	Wollenweber et al. (2006) and Labbe et al. (1986)
		B. solierii	Leaves and stems	Labbe et al. (1986)
		B. halimifolia		Wollenweber et al. (1997)
105	Eupatolitin	*B. confertifolia*	Leaves and stems	Wollenweber et al. (2006)
		B. gaudichaudiana	Aerial parts	Akaike et al. (2003)
106	Centaureidin	*B. sarothroides*	Leaves and twigs	Abad Martinezet al. (2005) and Montes et al. (1971)
		B. solierii	Leaves and stems	Labbe et al. (1986)
		B. salicina		Parodi and Fischer (1988) and Quijano et al. (1998)
107	5,7-dihydroxy-3,6,3′,4′-tetramethoxyflavone	*B. salicina*		Quijano et al. (1998)
108	3,3′-dimethylgossypetin	*B. linearis*	Leaves and stems	Wollenweber et al. (2006)
109	3,8-dimethylgossypetin	*B. linearis*	Leaves and stems	Wollenweber et al. (2006)
110	3,8,3′-trimethylgossypetin	*B. linearis*	Leaves and stems	Wollenweber et al. (2006)

(continued)

Table 11.4 (continued)

	Compounds	Species	Parts	References
111	3′,4′,5,7-tetrahydroxy-3,6,8-trimethoxyflavone	*B. incarum*	Leaves and stems	Campos et al. (2016)
112	4′,5,7-trihydroxy-3,3′,6,8-tetramethoxyflavone	*B. incarum*	Leaves and stems	Campos et al. (2016)
		B. solierii	Leaves and stems	Labbe et al. (1986)
113	3′,5-dihydroxy-3,4′,6,7,8-pentamethoxyflavone	*B. boliviensis*	Aerial parts	Campos et al. (2016)
114	4′,5-dihydroxy3′,3,6,7,8-pentamethoxyflavone	*B. incarum*	Leaves and stems	Faini et al. (1982a, b), Nuño et al. (2012) and Campos et al. (2016)
115	3-O-acetylmyricetin	*B. tola*	Aerial parts	Simirgiotis et al. (2016)
116	Myricetin 7,3′-dimethylether	*B. tola*	Aerial parts	Simirgiotis et al. (2016)
117	Myricetin 7,3′,5′-trimethylether	*B. tola*	Aerial parts	Simirgiotis et al. (2016)
118	Myricetin 3′,5′,7,8-tetramethylether	*B. tola*	Aerial parts	Simirgiotis et al. (2016)
119	6-hydroxy-7,3′,5′-trimethylmyricetin	*B. tola*	Aerial parts	Simirgiotis et al. (2016)
120	6-hydroxy-3,7,3′,5′-tetramethylmyricetin	*B. tola*	Aerial parts	Simirgiotis et al. (2016)
121	Apigenin 3-O-β-D-glucopyranoside/ astragalin	*B. Dracunculifolia*	Aerial parts	Nagatani et al. (2001)
		B. angustifolia		Wagner et al. (1972)
122	Kaempferol-3-O-rutinoside/ nicotiflorin	*B. antioquensis*	Leaves	Mejia-Giraldo et al. (2016)
123	Quercetin 3-O-[β-D-apiofuranosyl (1 → 2)α-Lrhamnopyranosyl (1 → 6)]-β-D-glucopyranoside	*B. thesioides*	Aerial parts	Liu et al. (1993)
124	Isoquercetin	*B. dracunculifolia*	Aerial parts	Nagatani et al. (2001)
		B. thesioides	Aerial parts	Kakehashu et al. (2016)
		B. trimera	Aerial parts	Simões-Pires et al. (2005)
		B. pseudotenuifolia	Leaves and shrub	Moreira et al. (2003a, b), Abad and Bermejo (2007) and Campos et al. (2016)

(continued)

Table 11.4 (continued)

	Compounds	Species	Parts	References
		B. angustifolia		Wagner et al. (1972)
		B. ochracea		Schenkel et al. (1997)
125	Hyperoside	*B. dracunculifolia*	Aerial parts	Nagatani et al. (2001)
		B. thesioides	Aerial parts	Nagatani et al. (2001)
126	Quercitrin	*B. gaudichaudiana*	Aerial parts	Campos et al. (2016) and Akaike et al. (2003)
		B. microdonta	Leaves	Toyama et al. (2014)
		B. pseudotenuifolia	Leaves and shrubs	Moreira et al. (2003a, b), Abad and Bermejo (2007) and Campos et al. (2016)
127	Rutin	*B. gaudichaudiana*	Aerial parts	Campos et al. (2016) and Akaike et al. (2003)
		B. thesioides	Aerial parts	Tarqui et al. (2012)
		B. trinervis	Aerial parts	Jaramillo-Garcia et al. (2018)
		B. trimera	Aerial parts and leaves	Gene et al. (1996), Menezes et al. (2016) and Sabir et al. (2017)
		B. antioquensis	Leaves	Mejia-Giraldo et al. (2016)
		B. dentata	Leaves and stems	Campos et al. (2016) and Sartor et al. (2013)
		B. spicata	Leaves, stems, and flowers	Agudelo et al. (2016)
		B. genistelloides		Hennig et al. (2011)
128	Quercetin -3-O-(4″-O-caffeoyl)-rhamnopyranosyl-(1 → 6)-galactopyranoside – 174	*B. antioquensis*	Leaves	Mejia-Giraldo et al. (2016)

Eriodictyol (**10**) was found in aerial parts of *B. concinna* (Bohlmann et al. 1981b), *B. confertifolia* (Wollenweber et al. 2006), *B. marginalis* (Faine et al. 1987), *B. retusa* (Campos et al. 2016; Grecco et al. 2012a, b), and *B. pseudotenuifolia* (Moreira et al. 2003a, b; Abad and Bermejo 2007; Campos et al. 2016). Six eriodictyol derivatives were found in *Baccharis* species: eriodictyol-7-methyl ether (**11**) at *B. concinna* (Wollenweber et al. 2006), homoeriodictyol (**12**) and eriodictyol-3′,4′-dimethyl ether (**13**) at *B. calliprinos* (Gianello et al. 1999), eriodictyol 7,3′,4′-trimethyl ether (**14**) at *B. confertifolia* (Wollenweber et al. 2006), filifolin (**15**) at *B. concinna* (Wollenweber et al. 2006) and *B. boliviensis* (Campos et al. 2016), and 8-methoxyeriodictyol (**16**) at *B. concinna* (Wollenweber et al. 2006). Flavanones identified at *Baccharis* species are summarized in Table 11.1, followed by their respective structures in Fig. 11.5.

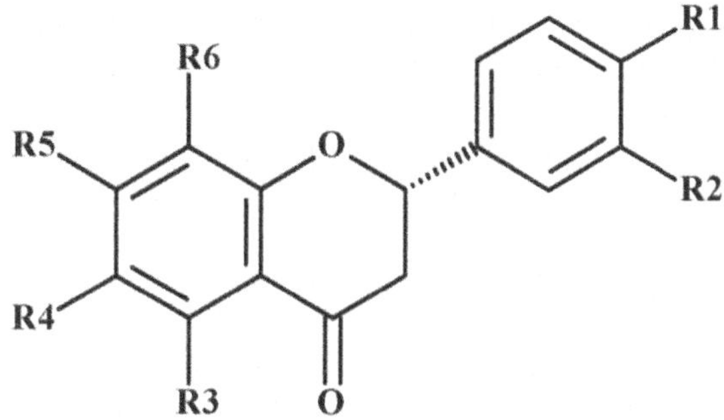

Fig. 11.5 Structures of flavanones **1–16** identified in *Baccharis* species

1 – R1 = R3 = R5 = OH; R2 = R4 = R6 = H
2 - R1 = R3 = OH; R5 = OMe; R2 = R4 = R6 = H
3 – R1 = OMe; R3 = R5 = OH; R2 = R4 = R6 = H
4 – R1 = OMe; R3 = R4 = R5 = OH; R2 = R6 = H;
5 – R1 = R5 = OMe; R3 = OH; R2 = R4 = R6 = H
6 – R1 = OH; R3 = R5 = OMe; R2 = R4 = R6 = H
7 – R3 = R5 = OH; R1 = R2 = R4 = R6 = H
8 – R2 = OMe; R3 = R5 = OH; R1 = R4 = R6 = H
9 – R3 = R5 = OH; R4 = OMe; R1 = R2 = R6 = H
10 – R1 = R2 = R3 = R5 = OH; R4 = R6 = H
11 – R1 = R2 = R3 = OH; R5 = OMe; R4 = R6 = H
12 – R1 = R3 = R5 = OH; R2 = OMe; R4 = R6 = H
13 – R1 = R2 = OMe; R3 = R5 = OH; R4 = R6 = H
14 – R1 = R2 = R5 = OMe; R3 = OH; R4 = R6 = H
15 – R1 = R2 = R3 = R5 = OH; R4 = OMe; R6 = H
16 – R1 = R2 = R3 = R5 = OH; R6 = OMe; R4 = H

5 Flavanonols

The 3-hydroxylation of flavanones through 3-hydroxylase enzymes leads to the biosynthesis of flavanonol derivatives. Naringenin-3-hydroxylase affords aromadendrin, also known as dihydrokaempferol (**17**), identified in leaves of *B. dracunculifolia* (Guimaraes et al. 2012), shrubs of *B. pseudotenuifolia* (Moreira et al. 2003a, b; Abad and Bermejo 2007; Campos et al. 2016), leaves and flowers of *B. illinita* (Verdi et al. 2004; Abad and Bermejo 2007; Campos et al. 2016), and in aerial parts of *B. retusa* (Campos et al. 2016; Grecco et al. 2012a). Aromadendrin-7-methyl ether (**18**) was isolated from aerial parts of *B. dracunculifolia* (Campos et al. 2016) and *B. illinita* (Pizzolatti et al. 2006). Dihydrokaempferide (**19**) was identified in the aerial parts of *B. conferta* (Weimann et al. 2002), in leaves, roots, flowers, buds, and stems of *B. dracunculifolia* (Figueiredo-Rinhel et al. 2013; da Silva Filho et al. 2004, 2008; Missima et al. 2007; Lemos et al. 2007; de Sousa et al. 2009, 2011; Resende et al. 2007; Rezende et al. 2014; Cestari et al. 2011Midorikawa et al. 2003; Kumazawa et al. 2003), and from *B. leptophylla* (Mollinedo et al. 2001; Almanza et al. 2000). 3-Acetylated derivative such as 3-acetoxy-4′ 5 7-trihydroxyflavanone (**20**) was isolated from aerial parts of *B. varians* (Bohlmann et al. 1981a; Abad Martinez et al. 2005), while 3-acetoxy-4′,5-dihydroxy-7-methoxyflavanone (**21**) was obtained from leaves of *B. dracunculifolia*, (Fukuda et al. 2006; Campos et al. 2016). A 4′-dehydroxylated derivative of aromadendrin, pinobanksin (**22**), was isolated from roots of *B. oxyodonta* (Bohlmann et al. 1981b; Abad Martinez et al. 2005) as well as from vegetative gems and leaves of *B. dracunculifolia* (Park et al. 2004; de Alencar et al. 2005). Additionally, its acetylated derivative, pinobanksin-3-acetate (**23**), was identified in *B. dracunculifolia* leaf bud (Park et al. 2005) and in *B. trinervis* (Jakupovic et al. 1986). Taxifolin (**24**) was identified in *B. illinita* (Abad and Bermejo 2007), especially in flowers (Verdi et al. 2004; Campos et al. 2016), as well as in aerial parts of *B. retusa* (Grecco et al. 2012a; Campos et al. 2016). 3-hydroxyhesperetin or taxifolin-4′-methyl ether (**25**), together with 8-prenyltaxifolin (**26**), were identified in aerial parts of *B. tola* (Simirgiotis et al. 2016). Furthermore, an acetylated derivative, taxifolin-3-acetate (**27**), was identified in aerial parts of *B. varians* (Bohlmann et al. 1981a; Abad Martinez et al. 2005). Flavanonols identified in *Baccharis* species are summarized in Table 11.2, followed by their respective structures in Fig. 11.6, Flavanonols identified in *Baccharis* species are summarized in Table 11.3, followed by their respective structures in Fig. 11.6.

6 Flavones

The enzymatic dehydration of flavanone naringenin (**1**) through flavone synthase affords apigenin (**28**) identified in the aerial parts of *B. bigelovii* (Arriaga-Giner et al. 1986), *B. crispa* (Palacios et al. 1983), *B. dentata* (Campos et al. 2016), *B. gaudichaudiana* (Fullas et al. 1994; Guo et al. 2007; Visintini et al. 2013; Campos

Fig. 11.6 Structures of flavanonols **17–27** identified in *Baccharis* species

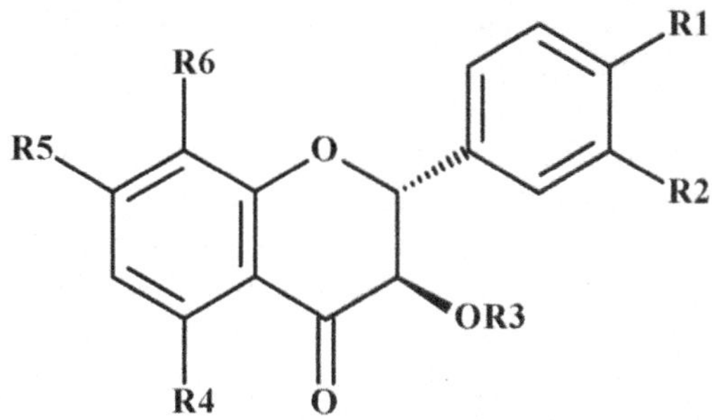

17 – R1 = R4 = R5 = OH; R2 = R3 = R6 = H
18 - R1 = R4 = OH; R5 = OMe; R2 = R3 = R6 = H
19 – R1 = OMe; R4 = R5 = OH; R2 = R3 = R6 = H
20 – R1 = R4 = R5 = OH; R3 = Acetyl; R2 = R6 = H
21 - R1 = R4 = OH; R3 = Acetyl; R5 = OMe; R2 = R6 = H
22 - R4 = R5 = OH; R1 = R2 = R3 = R6 = H
23 - R3 = Acetyl; R4 = R5 = OH; R1 = R2 = R6 = H
24 - R1 = R2 = R4 = R5 = OH; R3 = R6 = H
25 - R1 = OMe; R2 = R4 = R5 = OH; R3 = R6 = H
26 - R1 = R2 = R4 = R5 = OH; R6 = Prenyl; R3 = H
27 - R1 = R2 = R4 = R5 = OH; R3 = Acetyl; R6 = H

et al. 2016), *B. notosergila* (Palacios et al. 1983), *B. pedicellata* (Faine et al. 1987), *B. retusa* (Grecco et al. 2012a; Campos et al. 2016), *B. salicifolia* (Campos et al. 2016; Bohlmann et al. 1981b), *B. salzmannii* (Campos et al. 2016), as well as in the aerial and epigeous parts of *B. trimera* (Nakasugi 1990, 1998; Fullas et al. 1994). This compound was also identified in the flowers of *B. illinita* (Verdi et al. 2004; Abad and Bermejo 2007; Campos et al. 2016), leaves, and shrubs of *B. pseudotenuifolia* (Moreira et al. 2003a, b; Abad and Bermejo 2007; Campos et al. 2016) and *B. teindalensis* (Vidari et al. 2003), and vegetative gems of *B. dracunculifolia* (de Alencar et al. 2005). The flavone **28** was also found in *B. genistelloides* (Kuroyanagi et al. 1985; Abad Martinez et al. 2005; Hennig et al. 2011), *B. heterophylla* (Wollenweber et al. 1986), *B. pteronioides* (Wollenweber et al. 1986), *B. ramosissima* (Bohlmann et al. 1981b), *B. tola* (San Martin et al. 1982, 1983; Abad Martinez et al. 2005), *B. trinervis* (Arriaga et al. 1982), *B. vaccinioides* (Wollenweber et al. 1986), and *B. viminea* (Wollenweber et al. 1997). Its 8-hydroxylated derivative, isoscutellarein (**29**), was identified only in *B. pilularis* var. *consanguinea* (Wollenweber et al. 1997), while the 4′-methylated derivative, acacetin (**30**), was found in aerial parts of *B. articulata* (Gianello and Giordano 1984; Cariddi et al. 2012), *B. conferta* (Weimann et al. 2002), and *B. polycephala* (Davila et al. 2013), in the leaf parts of *B. dracunculifolia* (Park et al. 2004; da Silva Filho et al. 2009) and *B. rhomboidalis* (Silva et al. 1971) and from *B. grandicapitulata* (Bohlmann et al. 1985), *B. patagonica* (Zdero et al. 1986; Rivera et al. 1988); *B. salicifolia* (Wollenweber et al. 1986), *B. trinervis* (Arriaga et al. 1982), *B. vaccinioides* (Wollenweber et al. 1986), and *B. viminea* (Wollenweber et al. 1997). Also other two monomethylated derivatives were found: genkwanin (**31**) in aerial parts of *B. crispa* (Palacios et al. 1983), *B. notosergila* (Palacios et al. 1983), *B. pedicellata* (Faine et al. 1987), *B. trimera* (Nakasugi 1990; Nakasugi and Komai 1998), in leaves of *B. trinervis* (Herrera et al. 1996) as well as from *B. genistelloides* (Abad Martinez et al. 2005; Hennig et al. 2011) and *B. pilularis* (Wollenweber et al. 1997). Thevetiaflavone (**32**) was identified in aerial parts of *B. gaudichaudiana* (Guo et al. 2007; Campos et al. 2016). 4′,5- and 4′,7- dimethylated derivatives (**33** and **34** derivatives, respectively) were identified in some *Baccharis* species, 7-Hydroxy-5,4′-dimethoxyflavone (**33**) was isolated from aerial parts of *B. articulata* (de Oliveira et al. 2014) and *B. usterii* (Salcedo Ortiz et al. 2001), while 5-Hydroxy-4′,7-dimethoxyflavone (**34**), from aerial parts of *B. conferta* (Weimann et al. 2002), *B. crispa* (Bandoni et al. 1978), *B. illinita* (Pizzolatti et al. 2006; Campos et al. 2016), from leaves of *B. rhomboidalis* (Silva et al. 1971), *B. teindalensis* (Vidari et al. 2003), *B. tola* (San Martin et al. 1982), *B. trinervis* (Herrera et al. 1996), and from *B. latifolia* (Salcedo Ortiz et al. 2001). Flavones biosynthesized from phenylalanine-cinnamic acid-pinocembrin were also identified: chrysin (**35**) and 7-methylchrysin (**36**) from leaves of *B. dracunculifolia* (Park et al. 2004; Paula et al. 2017) and *B. viminea* (Wollenweber et al. 1997). Its 2′-methoxylated derivative, 2′-methoxychrysin (**37**), was also identified from flowers of *B. illinita* (Verdi et al. 2004; Abad and Bermejo 2007; Campos et al. 2016). Hispidulin (**38**) was identified in aerial parts of *B. flabellata* (Saad et al. 1988), *B. gaudichaudiana* (Fullas et al. 1994; Akaike et al. 2003; Campos et al. 2016), *B. genistelloides* (Daily et al. 1984; San Martin et al. 2012), *B. grisebachii* (Tapia et al. 2004; Abad and Bermejo 2007), *B. halimifolia* (Joshi et al. 1997; Jakupovic et al. 1990), *B. ligustrina* (Moreira et al. 2003a, b; Abad and Bermejo 2007; Nogueira Sobrinho et al. 2016), *B. magellanica* (Cordano et al. 1982), *B. rhomboidalis* (Labbe et al. 1986), *B. trimera* (Soicke and Leng-Eschlow 1987; Nakasugi 1990; Nakasugi and Komai 1998; Padua et al. 2014), and *B. uncinella* (Grecco et al. 2010a, b, 2014a, b; Campos et al. 2016). Flavonoid **38** was also identified in leaves and shrub of *B. pseudotenuifolia* (Moreira et al. 2003a, b; Abad and Bermejo 2007; Campos et al. 2016) and nonspecified parts of *B. vaccinioides* (Wollenweber et al. 1986). Pectolinaringenin (**39**) was identified in aerial parts of *B. concava* (Zamorano et al. 1987), *B. conferta* (Weimann et al. 2002), *B. decussata* (Morales Mendez et al. 1984; Rojas and Morales 2000), *B. grisebachii* (Gianello and Giordano 1987; Feresin et al. 2003; Abad Martinez et al. 2005), and *B. uncinella* (Grecco et al. 2010a, b, 2014a, b; Passero et al. 2011; Zalewski et al. 2011; Campos et al. 2016). Flavonoid (**39**) was also detected in the leaves of *B. pedunculata* (Rahalison et al. 1995), branches of *B. trinervis* (Sharp et al. 2000; Abad Martinez et al. 2005) [44, 103], and nonspecified parts of fresh *B. macraei* (Faini et al. 1991) [104], *B. pilularis* var. *consanguinea* (Wollenweber et al. 1997) [41] and *B. vaccinioides* (Wollenweber et al. 1986). Cirsimaritin (**40**) was identified in the aerial parts of *B. concava* (Zamorano et al. 1987), *B. concinna*

(Wollenweber et al. 2006), *B. conferta* (Weimann et al. 2002), *B. genistelloides* (Suttisri et al. 1994; Abad Martinez et al. 2005; Hennig et al. 2011), *B. rhomboidalis* (Labbe et al. 1986), *B. rufescens* var. *rufescens* (Simirgiotis et al. 2003; Abad and Bermejo 2007; Campos et al. 2016), *B. trimera* (Nakasugi 1990; Nakasugi and Komai 1998), *B. elaeagnoides* (Mesquita et al. 1985), *B. halimifolia* (Joshi et al. 1997), *B. intermedia* and *B. macraei* (Faini et al. 1991), *B. pilularis* var. *consanguinea* (Wollenweber et al. 1997), and *B. tricuneata* (Wagner et al. 1978). Scutellarein-7,4′-dimethyl ether (**41**) was identified from aerial parts of *B. tucumanensis* (Tonn et al. 1982), while salvigenin (**42**) was found in the aerial parts of *B. concava* (Zamorano et al. 1987), *B. conferta* (Weimann et al. 2002), *B. rhomboidalis* (Labbe et al. 1986), and *B. scandens* (Cabrera et al. 2016). Flavonoid **42** was identified in the branches of *B. trinervis* (Sharp et al. 2000; Abad Martinez et al. 2005), and in nonspecified parts of *B. macraei* (Faini et al. 1991) and *B. pilularis* var. *consanguinea* (Wollenweber et al. 1997). Desmethoxysudachitin (**43**) was identified in the aerial parts of *B. grisebachii* (Gianello and Giordano 1987; Tapia et al. 2004; Grecco et al. 2010a, b; Campos et al. 2016) and *B. solierii* (Labbe et al. 1986). Its 4′-methylated derivative, nevadensin (**44**), was found in the aerial parts of *B. decussata* (Rojas and Morales 2000), *B. grisebachii* (Gianello and Giordano 1987; Feresin et al. 2003; Tapia et al. 2004; Grecco et al. 2010a, b), and *B. nitida* (Chidiak et al. 2007; Campos et al. 2016). A methylated derivative of **43** in position C-7 afforded xantomicrol (**45**) found in aerial parts of *B. boliviensis* (Calle et al. 2012; Campos et al. 2016), *B. nitida* (Chidiak et al. 2007; Campos et al. 2016), *B. patens* (Silva et al. 1985), *B. scandens* (Cabrera et al. 2016), and *B. tucumanensis* (Tonn et al. 1982). Compound **45** was also found in the leaves of *B. pentlandii* (Tarqui et al. 2012) and roots of *B. quitensis* (Bohlmann et al. 1981b). Gardenin B (**46**) was identified in the aerial parts of *B. grisebachii* (Feresin et al. 2003; Tapia et al. 2004; Campos et al. 2016) and *B. scandens* (Cabrera et al. 2016). A pentamethoxylated derivative, tangeretin (**47**), was found in the flowers of *B. illinita* (Verdi et al. 2004; Abad and Bermejo 2007; Campos et al. 2016). The flavone luteolin (**48**), derived from dehydration of flavanone eriodictyol, was identified in aerial parts of *B. articulata* (Cariddi et al. 2012), *B. bigelovii* (Arriaga-Giner et al. 1986; Wollenweber et al. 1986), *B. concinna* (Wollenweber et al. 2006), *B. confertifolia* (Wollenweber et al. 2006), *B. gaudichaudiana* (Guo et al. 2007), *B. genistelloides* (San Martin et al. 2012), *B. incarum* (Zampini et al. 2009; Campos et al. 2016), *B. linearis* (Wollenweber et al. 2006), *B. lycioides* (Wollenweber et al. 2006), *B. reticularia* (Bohlmann et al. 1981a), *B. trimera* (Soicke and Leng-Eschlow 1987; da Silva et al. 2016; Menezes et al. 2016), *B. trinervis* (Jaramillo-Garcia et al. 2018), and *B. varians* (Bohlmann et al. 1981a). Furthermore, flavonoid **48** was isolated from flowers of *B. illinita* (Verdi et al. 2004; Abad and Bermejo 2007; Campos et al. 2016), and nonspecified parts of *B. halimifolia* (Wollenweber et al. 1997), *B. microcephala* (Bohlmann et al. 1985), *B. nitida* (Bohlmann et al. 1985), and *B. pteronioides* (Wollenweber et al. 1986). Methylated derivatives of luteolin, biosynthesized through luteolin-methyltransferase, were found in several studied *Baccharis* species. A monomethylated (C-3′) derivative, chrysoeriol (**49**), was identified in the flowers of *B. illinita* (Verdi et al. 2004; Abad and Bermejo 2007; Campos et al.

2016), shrub of *B. pseudotenuifolia* (Moreira et al. 2003a, b; Abad and Bermejo 2007; Campos et al. 2016), and aerial parts of *B. salicifolia* (Warning et al. 1986). 4′,7- and 3′,7- Dimethylated derivatives of Luteolin were also found in *Baccharis* genus, luteolin-4′,7-dimethylether (**50**) in aerial parts of *B. trimera* (Padua et al. 2014; de Araujo et al. 2016), and leaves of *B. trinervis* (Herrera et al. 1996) and luteolin-3′,7-dimethylether (**51**) in aerial parts of *B. calliprinos* (Gianello et al. 1999), *B. rhetinodes* (Gianello et al. 1999), and *B. salicifolia* (Warning et al. 1986). The trimethylated derivative, 5-hydroxy-7,3′,4′-trimethoxyflavone/luteolin-3′,4′,7-trimethylether (**52**), was identified in the aerial parts of *B. crispa* (Bandoni et al. 1978) [80], *B. latifolia* (Salcedo et al. 2003; Campos et al. 2016), and in the leaves of *B. trinervis* (Herrera et al. 1996). 6-Hydroxyluteolin (**53**) was found only in aerial parts of *B. boliviensis* (Campos et al. 2016), while its methylated derivatives were identified in several *Baccharis* species; nepetin, also known as eupafolin (**54**), was identified in aerial parts of *B. concinna* (Wollenweber et al. 2006), *B. confertifolia* (Wollenweber et al. 2006), *B. flabellata* (Saad et al. 1988), *B. gaudichaudiana* (Akaike et al. 2003; Campos et al. 2016), *B. genistelloides* (San Martin et al. 2012), *B. linearis* (He 1995; He et al. 1996; Wollenweber et al. 2006), *B. lycioides* (Wollenweber et al. 2006), and *B. trimera* (Soicke and Leng-Eschlow 1987; Simões-Pires et al. 2005). Jaceosidin (**55**) was isolated from aerial parts of *B. concinna* (Wollenweber et al. 2006), *B. flabellata* (Saad et al. 1988), *B. gaudichaudiana* (Akaike et al. 2003; Campos et al. 2016), and *B. grisebachii* (Tapia et al. 2004). Desmethoxycentaureidin (**56**) was obtained from aerial parts of *B. gaudichaudiana* (Akaike et al. 2003; Campos et al. 2016), *B. solierii* (Labbe et al. 1986), and non-specified parts of *B. petiolata* (Labbé et al. 1990) and *B. salicina* (Parodi and Fischer 1988; Quijano et al. 1998). Cirsiliol (**57**) was identified in the aerial parts of *B. concinna* (Wollenweber et al. 2006), *B. confertifolia* (Wollenweber et al. 2006), *B. flabellata* (Saad et al. 1988), *B. linearis* (Wollenweber et al. 2006), *B. rufescens* var. *rufescens* (Simirgiotis et al. 2003; Abad and Bermejo 2007; Campos et al. 2016), and in nonspecified parts of *B. genistelloides* (Kuroyanagi et al. 1985; Abad Martinez et al. 2005; Hennig et al. 2011) and *B. tricuneata*. Cirsilineol (**58**) was detected in aerial parts of *B. concinna* (Wollenweber et al. 2006), *B. salicifolia* (Warning et al. 1986), and nonspecified parts of *B. genistelloides* (Abad Martinez et al. 2005; Hennig et al. 2011), while 5,6-dihydroxy-3′,4′,7-trimethoxyflavone (**59**) from young parts of *B. trimera* (Borella et al. 2006). The aerial parts of *B. conferta* (Weimann et al. 2002) and *B. gaudichaudiana* (Akaike et al. 2003; Campos et al. 2016) afforded eupatilin (**60**), while aerial parts of *B. genistelloides* (Suttisri et al. 1994; Hennig et al. 2011) and leaves, stems, flowers, and fruits of *B. trimera* (Herz et al. 1977; de Mello and Petrovick 2000; Torres et al. 2000; da Silva et al. 2004; Silva et al. 2006; Padua et al. 2014; de Araujo et al. 2016) afforded eupatorin (**61**). 5-Hydroxy-3′,4′,6,7-tetramethoxyflavone (**62**) was identified from aerial parts of *B. genistelloides* (Suttisri et al. 1994) and leaves of *B. trimera* (Rendon and Vila 1995; Silva et al. 2006). Sideritiflavone (**63**) was detected in aerial parts of *B. patens* (Silva et al. 1985) and *B. thymifolia* (Saad et al. 1987) as well as from leaves of *B. pentlandii* (Tarqui et al. 2012; Campos et al. 2016;). 4′,5-Dihydroxy-3′,6,7,8-tetramethoxyflavone (**64**), known as 7-methylsudachitin or 3′-methoxyxanthomicrol, was identified from aerial parts of *B. incarum* (Faini et al. 1982a, b), *B. salicifolia*

(Warning et al. 1986) and *B. thymifolia* (Saad et al. 1987), from leaves of *B. pentlandii* (Tarqui et al. 2012; Campos et al. 2016), and roots of *B. oxyodonta* and *B. quitensis* (Bohlmann et al. 1981b). Gardenin D (**65**) was identified in aerial parts of *B. patens* (Silva et al. 1985). A pentamethoxylated derivative, 5-hydroxy-3′,4′,6,7,8-pentamethoxytlavone (**66**), was detected only in aerial parts of *B. thymifolia* (Saad et al. 1987). Nobiletin was identified only in the flowers of *B. illinita* (**67**). Although in less distribution, flavone glycosides were found in *Baccharis* species: isoschaftoside (**68**) was identified in the aerial parts of *B. gaudichaudiana* (Akaike et al. 2003; Campos et al. 2016), while vicenin II (**69**), 6(8)-C-furanosyl-8(6)-C-hexosyl flavone (**70**), 6(8)-C-hexosyl-8(6)-C-furanosyl flavone (**71**), and 5,3′-dihydroxy-4′-methoxy-7-O-pyranosyl-furanosyl flavone (**72**) were detected in the aerial parts of *B. trimera* (Padua et al. 2014; de Araujo et al. 2016; Rabelo et al. 2017). Flavones identified in *Baccharis* species are summarized in Table 11.3, followed by their respective structures in Fig. 11.7.

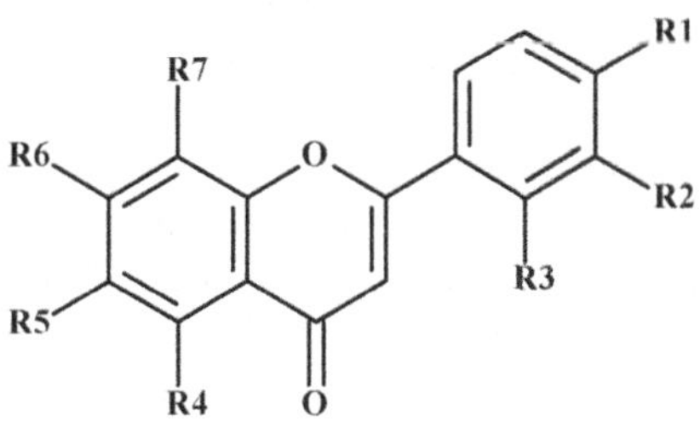

28 - R1 = R4 = R6 = OH; R2 = R3 = R5 = R7 = H	**29** - R1 = R4 = R6 = R7 = OH; R2 = R3 = R5 = H
30 - R1 = OMe; R4 = R6 = OH; R2 = R3 = R5 = R7 = H	**31** - R1 = R4 = OH; R6 = OMe; R2 = R3 = R5 = R7 = H
32 - R1 = R6 = OH; R4 = OMe; R2 = R3 = R5 = R7 = H	**33** - R1 = R4 = OMe; R6 = OH; R2 = R3 = R5 = R7 = H
34 - R1 = R6 = OMe; R4 = OH; R2 = R3 = R5 = R7 = H	**35** - R4 = R6 = OH; R1 = R2 = R3 = R5 = R7 = H
36 - R4 = OH; R6 = OMe; R1 = R2 = R3 = R5 = R7 = H	**37** - R3 = OMe; R4 = R6 = OH; R1 = R2 = R5 = R7 = H
38 - R1 = R4 = R6 = OH; R5 = OMe; R2 = R3 = R7 = H	**39** - R1 = R5 = OMe; R4 = R6 = OH; R2 = R3 = R7 = H
40 - R1 = R4 = OH; R5 = R6 = OMe; R2 = R3 = R7 = H	**41** - R1 = R6 = OMe; R4 = R5 = OH; R2 = R3 = R7 = H
42 - R1 = R5 = R6 = OMe; R4 = OH; R2 = R3 = R7 = H	**43** - R1 = R4 = R6 = OH; R5 = R7 = OMe; R2 = R3 = H
44 - R1 = R5 = R7 = OMe; R4 = R6 = OH; R2 = R3 = H	**45** - R1 = R4 = OH; R5 = R6 = R7 = OMe; R2 = R3 = H
46 - R1 = R5 = R6 = R7 = OMe; R4 = OH; R2 = R3 = H	**47** - R1 = R4 = R5 = R6 = R7 = OMe; R2 = R3 = H
48 - R1 = R2 = R4 = R6 = OH; R3 = R5 = R7 = H	**49** - R1 = R4 = R6 = OH; R2 = OMe; R3 = R5 = R7 = H
50 - R1 = R6 = OMe; R2 = R4 = OH; R3 = R5 = R7 = H	**51** - R1 = R4 = OH; R2 = R6 = OMe; R3 = R5 = R7 = H
52 - R1 = R2 = R6 = OMe; R4 = OH; R3 = R5 = R7 = H	**53** - R1 = R2 = R4 = R5 = R6 = OH; R3 = R7 = H
54 - R1 = R2 = R4 = R6 = OH; R5 = OMe; R3 = R7 = H	**55** - R1 = R4 = R6 = OH; R2 = R5 = OMe; R3 = R7 = H
56 - R1 = R5 = OMe; R2 = R4 = R6 = OH; R3 = R7 = H	**57** - R1 = R2 = R4 = OH; R5 = R6 = OMe; R3 = R7 = H
58 - R1 = R6 = OMe; R2 = R4 = R5 = OH; R3 = R7 = H	**59** - R1 = R2 = R6 = OMe; R4 = R5 = OH; R3 = R7 = H
60 - R1 = R2 = R5 = OMe; R4 = R6 = OH; R3 = R7 = H	**61** - R1 = R5 = R6 = OMe; R2 = R4 = OH; R3 = R7 = H
62 - R1 = R2 = R5 = R6 = OMe; R4 = OH; R3 = R7 = H	**63** - R1 = R2 = R4 = OH; R5 = R6 = R7 = OMe; R3 = H
64 - R1 = R4 = OH; R2 = R5 = R6 = R7 = OMe; R3 = H	**65** - R1 = R5 = R6 = R7 = OMe; R2 = R4 = OH; R3 = H
66 - R1 = R2 = R5 = R6 = R7 = OMe; R4 = OH; R3 = H	**67** - R1 = R2 = R4 = R5 = R6 = R7 = OMe; R3 = H

68 - R1 = R4 = R6 = OH; R5 = Ara; R7 = Glu; R2 = R3 = H

69 - R1 = R4 = R6 = OH; R5 = R7 = Glu; R2 = R3 = H

70 - R5 = Furanosyl; R7 = Hexosyl; R1 = R2 = R3 = R4 = R6 = H

71 - R5 = Hexosyl; R7 = Furanosyl; R1 = R2 = R3 = R4 = R6 = H

72 - R1 = OMe; R2 = R4 = OH; R6 = O-Pyranosyl-Furanosyl; R3 = R5 = R7 = H

Fig. 11.7 Structures of flavones **28**–**72** identified in *Baccharis* species

7 Flavanols

The biosynthesis of flavanols, the main group of flavonoids of genus *Baccharis*, occurs after the formation of flavanonols, through flavanol synthase enzyme (FLS). Thus, naringenin (**1**) is converted to aromadendrin/dihydrokaempferol (**17**) to finally afford the flavanol kaempferol (**73**). Compound **73** was identified in aerial parts of *B. conferta* (Weimann et al. 2002), *B. dentata* (Sartor et al. 2013; Campos et al. 2016), *B. gaudichaudiana* (Campos et al. 2016), *B. polycephala* (Davila et al. 2013), *B. trimera* (da Silva et al. 2016), and *B. retusa* (Campos et al. 2016; Grecco et al. 2012a). This compound was also identified in the flowers of *B. illinita* (Verdi et al. 2004; Abad and Bermejo 2007; Campos et al. 2016), flowering tops of *B. maritima* (Moreira et al. 1975), leaves of *B. pseudotenuifolia* (Moreira et al. 2003a, b; Campos et al. 2016), and buds, stems, and vegetative gems of *B. dracunculifolia* (Palacios et al. 1983; Park et al. 2004, 2005; de Alencar et al. 2005) and nonspecified parts of *B. genistelloides* (Daily et al. 1984), and *B. pilularis* var. *consanguinea* (Wollenweber et al. 1997). The 3-methylation of kaempferol afforded isokaempferide (**74**) identified in the aerial parts of *B. linearis* (Faini et al. 1999; Wollenweber et al. 2006), *B. lycioides* (Wollenweber et al. 2006), and *B. pedicellata* (Faine et al. 1987). Furthermore, this compound was also identfiied in the leaves of *B. papillosa* (Escobar et al. 2009; Campos et al. 2016), and nonspecified parts of fresh *B. intermedia* (Faini et al. 1991), *B. linearis* (Faini et al. 1991), *B. macraei* (Faini et al. 1991), and *B. pilularis* var. *consanguinea* (Wollenweber et al. 1997). The 4′-methylation affords kaempferide (**75**), found in aerial parts of *B. conferta* (Weimann et al. 2002) and *B. polycephala* (Davila et al. 2013) as well as in the flowers, leaves, buds, stems, and vegetative gems of *B. dracunculifolia* (Midorikawa et al. 2003; Park et al. 2004, 2005; de Alencar et al. 2005; Piantino et al. 2008; Campos et al. 2016; Paula et al. 2017), in no-nspecified parts of *B. leptophylla* (Almanza et al. 2000; Mollinedo et al. 2001) and in *B. pilularis* var. *consanguinea* and *B. viminea* (Wollenweber et al. 1997). Kaempferol-7-methyleter (**76**) was identified only in *B. pilularis* var. *consanguinea* (Wollenweber et al. 1997). The leaves of *B. papillosa* (Campos et al. 2016) afforded emanin (**77**), also identified in buds, leaves, and stems of *B. dracunculifolia* (Midorikawa et al. 2003; da Silva Filho et al. 2009) and from nonspecified parts of *B. pilularis* var. *consanguinea* and from *B. viminea* (Wollenweber et al. 1997). Kaempferol-7,4′-dimethyl ether (**78**) was detected in the aerial parts of *B. crispa* (Bandoni et al. 1978) and *B. illinita* (Pizzolatti et al. 2006; Campos et al. 2016) as well as from leaves of *B. teindalensis* (Vidari et al. 2003). Kaempferol-3,7-dimethyl ether (**79**) was identified in the aerial parts of *B. pedicellata* (Faine et al. 1987) and *B. santelicis* (Zdero et al. 1991) and nonspecified parts of *B. pilularis* var. *consanguinea* (Wollenweber et al. 1997). The trimethylated derivative, kaempferol-3,4′,7-trimethylether (**80**), was identified in the aerial parts of *B. illinita* (Pizzolatti et al. 2006; Campos et al. 2016) and *B. santelicis* (Zdero et al. 1991). Galangin (**81**) was detected in the leaf bud of *B. dracunculifolia* (Park et al. 2005) and nonspecified parts of *B. viminea* (Wollenweber et al. 1997), that also afforded galangin-7-methyleter (**82**). 6-Hydroxylated kaempferol and its derivatives were identified in few *Baccharis* species, 6-hydroxykaempferol (**83**) in *B. pilularis* var. *consanguinea* (Wollenweber et al. 1997); 6-methoxykaempferol (**84**) in leaves of *B. dracunculifolia* (Kumazawa et al. 2003); 4′,6-dimethoxykaempferol (**85**) in aerial parts of *B. conferta* (Weimann et al. 2002); and 6,7-dimethoxykaempferol (**86**) in *B. pilularis* var. *consanguinea* (Wollenweber et al. 1997). Penduletin (**87**) was identified in leaves of *B. pedunculata* (Rahalison et al. 1995), branches of *B. trinervis* (Sharp et al. 2000; Abad Martinez et al. 2005), and nonspecified parts of *B. salicifolia, B. sarothroides* and *B. vaccinioides* (Wollenweber et al. 1986), while herbacetin-3-methylether (**88**) was isolated from *B. pilularis* var. *consanguinea* (Wollenweber et al. 1997). Quercetin (**89**) was identified in the aerial parts of *B. articulata* (Campos et al. 2016), *B. bigelovii* (Arriaga-Giner et al. 1986; Wollenweber et al. 1986), *B. concinna* and *B. confertifolia* (Wollenweber et al. 2006), *B. dentata* (Sartor et al. 2013; Campos et al. 2016), *B. genistelloides* (Daily et al. 1984; San Martin et al. 2012), *B. grisebachii* (Tapia et al. 2004), *B. linearis* (Wollenweber et al. 2006, *B. lycioides* (Wollenweber et al. 2006), *B. pteronioides, B. retusa, B. salicifolia* (Wollenweber et al. 1986), *B. scandens* (Cabrera et al. 2016), *B. thesioides* (Liu et al. 1993), *B. trimera* (Soicke and Leng-Eschlow 1987; Simões-Pires et al. 2005; Padua et al. 2014; da Silva et al. 2016; de Araujo et al. 2016; Menezes et al. 2016; Sabir et al. 2017), *B. trinervis* (Jaramillo-Garcia et al. 2018), in the leaf bud of *B. dracunculifolia* (Park et al. 2005), leaves and shrubs of *B. pseudotenuifolia* (Moreira et al. 2003a, b; Abad and Bermejo 2007; Campos et al. 2016), leaves, stems, and flowers of *B. spicata* (Agudelo et al. 2016) flowers of *B. illinita* (Verdi et al. 2004; Abad and Bermejo 2007; Campos et al. 2016) and *B. maritima* (Moreira et al. 1975) as well as in the nonspecified parts of *B. viminea* (Wollenweber et al. 1997). Rhamnetin (**90**) was detected in leaves and stems of *B. confertifolia* (Wollenweber et al. 2006) and nonspecified parts of *B. pilularis var. consanguinea* (Wollenweber et al. 1997), while isorhamnetin (**91**) was detected in the aerial parts of *B. tola*, leaves and stems of *B. linearis* (He 1995; Wollenweber et al. 2006) and *B. lycioides* (Wollenweber et al. 2006), leaves and shrubs of *B. pseudotenuifolia* (Moreira et al. 2003a, b; Abad and Bermejo 2007; Campos et al. 2016), and in nonspecified parts of *B. viminea* (Wollenweber et al. 1997). The 3-methylated derivative of quercetin, 3-methylquercetin (**92**), was identified in the aerial parts of *B. tola* (Simirgiotis et al. 2016) and *B. trimera* (de Mello and Petrovick 2000), in the leaves of *B. linearis* (Faini et al. 1999), flowers of *B. illinita* (Verdi et al. 2004; Abad and Bermejo 2007; Campos et al. 2016), fresh parts of *B. intermedia* and *B. linearis* (Faini et al. 1991) as well as in the nonspecified parts of *B. halimifolia* (Wollenweber et al. 1997). Rhamnazin (**93**) was found in the aerial parts of *B. tola* (Simirgiotis et al. 2016), leaves and stems of *B. confertifolia* (Wollenweber et al. 2006), and nonspecified parts of *B. latifolia* (Salcedo et al. 2001). Isorhamnetin 3-methylether (**94**) was identified in the aerial parts of *B. tola* (Simirgiotis et al. 2016), leaves and stems of *B. linearis* (Faini et al. 1999; Wollenweber et al. 2006) and *B. lycioides* (Wollenweber et al. 2006), and in the nonspecified parts of fresh *B. intermedia* and *B. linearis* (Faini et al. 1991). *B. sarothroides* afforded 3,4′-dimethoxy-3′,5,7-trihydroxyflavone (**95**) (Abad Martinez et al. 2005), while 3,7-dimethoxyquercetin (**96**) was found in aerial parts of *B. triangularis* (Gianello and Giordano 1989) and nonspecified parts of *B. pilularis* var. *consanguinea* (Wollenweber et al. 1997). 3,7-Dimethylisorhamnetin (**97**) was isolated from aerial parts of *B. tola* (Simirgiotis et al. 2016), while 3,7,4′-trimethylquercetin (**98**) was obtained from leaves and stems of *B. illinita* (Campos et al. 2016) and nonspecified parts of *B. grisebachii* (Abad and Bermejo 2007). 3,5-dihydroxy-7,3′,4′-trimethoxyflavone (**99**) was detected in *B. latifolia*. Retusin (**100**) was isolated from leaves and stems of *B. illinita* (Pizzolatti et al. 2006; Campos et al. 2016), while 3-hydroxy-5,7,3′,4'tetramethoxyflavone (**101**) was obtained from aerial parts of *B. latifolia* (Salcedo et al. 2003; Campos et al. 2016) and *B. conferta* (Weimann et al. 2002). Patuletin (**102**) was detected in leaves and stems of *B. concinna* and *B. confertifolia* (Wollenweber et al. 2006) as well as in nonspecified parts of *B. halimifolia* and *B. pilularis* var. *consanguinea* (Wollenweber et al. 1997). Flavonoid 3′-methylpatuletin (**103**) was detected in *B. halimifolia* (Wollenweber et al. 1997), while axillarin (**104**) was identified in the leaves and stems of *B. incarum* (Campos et al. 2016), *B. linearis* (Labbe et al. 1986; Wollenweber et al. 2006), and *B. solierii* (Labbe et al. 1986) as well as in the nonspecified parts of *B. halimifolia* (Wollenweber et al. 1997). Eupatolitin (**105**) was isolated from leaves and stems of *B. confertifolia* (Wollenweber et al. 2006) and aerial parts of *B. gaudichaudiana* (Akaike et al. 2003). Centaureidin (**106**) was identified in the leaves and twigs of *B. sarothroides* (Montes et al. 1971; Abad Martinez et al. 2005), leaves and stems of *B. solierii* (Labbe et al. 1986), and nonspecified parts of *B. salicina* (Parodi and Fischer 1988; Quijano et al. 1998). 5,7-Dihydroxy-3,6,3′,4′-tetramethoxyflavone (**107**) was detected in *B. salicina* (Quijano et al. 1998), while three gossypetin derivatives (**108–110**) were identified in leaves and stems of *B. linearis* (Wollenweber et al. 2006): 3,3′-dimethyl- (**108**), 3,8-dimethyl- (**109**) and 3,8,3′-trimethyl-(**110**) derivatives. 3′,4′,5,7-Tetrahydroxy-3,6,8-trimethoxyflavone (**111**) and 4′,5,7-trihydroxy-3,3′,6,8-tetramethoxyflavone (**112**) were identified in the leaves and stems of *B. incarum* (Campos et al. 2016); compound **112** was also identified in the leaves and stems of *B. solierii* (Labbe et al. 1986). 3′,5-Dihydroxy-3,4′,6,7,8-pentamethoxyflavone (**113**) was detected in the aerial parts of *B. boliviensis* (Campos et al. 2016) and 4′,5-dihydroxy3′,,3,6,7,8-pentamethoxyflavone (**114**) in the leaves and stems of *B. incarum* (Faini et al. 1982a, b; Nuño et al. 2012; Campos et al. 2016). Six myricetin derivatives (**115–120**) were detected in leaves and stems of *B. tola* (Simirgiotis et al. 2016): 3-O-acetyl- (**115**), 7,3′-dimethyl- (**116**), 7,3′,5′-trimethyl- (**117**), 3′,5′,7,8-tetramethyl- (**118**), 6-hydroxy-7,3′,5′-trimethyl – (**119**) and 6-hydroxy-3,7,3′,5′-tetramethyl (**120**) derivatives. Additionally, nine flavonol glycosides (**121–129**) were identified in several species of *Baccharis*: apigenin 3-O-β-D-glucopyranoside, also known as astragalin (**121**), from aerial parts of *B. dracunculifolia* (Nagatani et al. 2001) and from *B. angustifolia* (Wagner et al. 1972), kaempferol-3-O-rutinoside, also known as nicotiflorin (**122**), from *B. antioquensis* (Mejia-Giraldo et al. 2016), quercetin 3-O-[β-D-apiofuranosyl-(1 → 2) α-Lrhamnopyranosyl-(1 → 6)]-β-D-glucopyranoside (**123**) from *B. thesioides* (Liu et al. 1993), isoquercetin (**124**) from aerial parts of *B. dracunculifolia* (Nagatani et al. 2001), *B. thesioides* (Kakehashu et al. 2016), *B. trimera* (Simões-Pires et al. 2005), *B. pseudotenuifolia* (Moreira et al. 2003a, b; Abad and Bermejo 2007; Campos et al. 2016), *B. angustifolia* (Wagner et al. 1972), and *B. ochracea* (Schenkel et al. 1997). Hyperoside (**125**) was detected in the aerial parts of *B. dracunculifolia* (Nagatani et al. 2001) and *B. thesioides* (Liu et al. 1993) [144], while quercitrin (**126**) was identified in *B. gaudichaudiana* (Akaike et al. 2003; Campos et al. 2016),

B. microdonta (Toyama et al. 2014), and *B. pseudotenuifolia* (Moreira et al. 2003a, b; Abad and Bermejo 2007; Campos et al. 2016). Rutin (**127**) was identified in the aerial parts of *B. gaudichaudiana* (Akaike et al. 2003; Campos et al. 2016), *B. thesioides* (Liu et al. 1993), *B. trinervis* (Jaramillo-Garcia et al. 2018), and *B. trimera* (Gene et al. 1996; Menezes et al. 2016; Sabir et al. 2017). Furthermore, flavonoid **127** was also isolated from leaves of *B. antioquensis* (Mejia-Giraldo et al. 2016), leaves and stems of *B. dentata* (Sartor et al. 2013; Campos et al. 2016), leaves, stems, and flowers *B. spicata* (Agudelo et al. 2016), and nonspecified parts of *B. genistelloides* (Hennig et al. 2011). Finally, quercetin-3-O-(4‴-O-caffeoyl)-rhamnopyranosyl-(1 → 6)-galactopyranoside (**128**) was identified in the leaves of *B. antioquensis* (Mejia-Giraldo et al. 2016). Flavanols identified in *Baccharis* species are summarized in Table 11.4, followed by their respective structures in Fig. 11.8.

Flavonoids display a significant role in biological functions, especially those that depend on the modulation of redox balance. Several studies describes the antioxidant and radical scavenger activities of extracts and isolated flavonoids of *Baccharis*, such as the antioxidant activities of isosakuranetin (**3**), dihydrokaempferide (**19**), kaempferol (**75**) (Mollinedo et al. 2001), isokaempferide (**74**), ermanin (**77**), 3-methylquercetin (**92**) (Escobar et al. 2009), luteolin (**48**), axillarin (**104**), 3′,4′,5,7-tetrahydroxy-3,6,8-trimethoxyflavone (**111**), 4′,5,7-trihydroxy-3,3′,6,8-tetramethoxyflavone (**112**), and 4′,5-dihydroxy-3′,3,6,7,8-pentamethoxyflavone (**114**). Antioxidant activities were also attributed to extracts of *B. spicata*, *B. tola*,

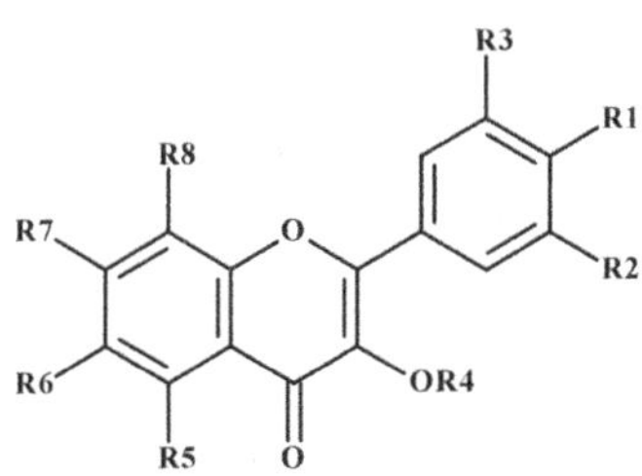

73 - R1= R5= R7= OH; R2= R3= R4= R6= R8= H
74 - R1= R5= R7= OH; R4= OMe; R2= R3= R6= R8 = H
75 - R1 = OMe; R5 = R7 = OH; R2 = R3 = R4 = R6 = R8 = H
76 - R1= R5= OH; R7= OMe; R2= R3= R4= R6= R8= H
77 - R1= OMe; R4= Me; R5= R7= OH; R2= R3= R6= R8= H
78 - R1= R7= OMe; R5= OH; R2= R3= R4=R6= R8= H
79 - R4= Me; R7= OMe; R1= R5= OH; R2= R3= R6= R8= H
80 - R1= R7= OMe; R4= Me; R5= OH; R2= R3= R6= R8= H
81 - R5= R7= OH; R1= R2= R3= R4= R6= R8= H
82 - R5= OH; R7= OMe; R1= R2= R3= R4= R6= R8= H
83 - R1= R5= R6= R7= OH; R2= R3= R4= R8= H
84 - R1= R5= R7= OH; R6= OMe; R2= R3= R4= R8= H
85 - R1= R6= OMe; R5= R7= OH; R2= R3= R4= R8= H
86 - R1= R5= OH; R6= R7= OMe; R2= R3= R4= R8= H
87 - R1= R5= OH; R4= Me; R6= R7= OMe; R2= R3= R4= R8= H
88 - R1= R5= R7= R8= OH; R4= Me; R2= R3= R6= H
89 - R1= R2= R5= R7= OH; R3= R4= R6= R8= H
90 - R1= R2= R5= OH; R7= OMe; R3= R4= R6= R8= H
91 - R1= R5= R7= OH; R2= OMe; R3= R4= R6= R8= H
92 - R1= R2= R5= R7= OH; R4= Me; R3= R6= R8= H
93 - R1= R5= OH; R2= R7= OMe; R3= R4= R6= R8= H
94 - R1= R5= R7= OH; R2= OMe; R4= Me; R3= R6= R8= H
95 - R1= OMe; R2= R5= R7= OH; R4= Me; R3= R6= R8= H
96 - R1= R2= R5= OH; R4= Me; R7= OMe; R3= R6= R8= H
97 - R1= R5= OH; R2= R7= OMe; R4= Me; R3= R6= R8= H
98 - R1= R7= OMe; R2= R5= OH; R4=Me; R3= R6= R8= H
99 - R1= R2= R7= OMe; R5= OH; R3= R4= R6= R8= H
100 - R1= R2= R7= OMe; R4= Me; R5= OH; R3= R6= R8= H
101 - R1= R2= R7= R5= OMe; R3= R4= R6= R8= H
102 - R1= R2= R5= R7= OH; R6= OMe; R3= R4= R8= H
103 - R1= R5= R7= OH; R2= R6= OMe; R3= R4= R8= H
104 - R1= R2= R5= R7= OH; R4= Me; R6= OMe; R3= R8= H
105 - R1= R2= R5= OH; R6= R7= OMe; R3= R4= R8= H
106 - R1= R6= OMe; R2= R5= R7= OH; R4= Me; R3= R8= H
107 - R1= R2= R6= OMe; R5= R7= OH; R4= Me; R3= R8= H
108 - R1= R5= R7= R8= OH; R2= OMe; R4= Me; R3= R6= H
109 - R1= R2= R5= R7= OH; R4= Me; R8= OMe; R3= R6= H
110 - R1= R5= R7= OH; R2= R8= OMe; R4= Me; R3= R6= H
111 - R1= R2= R5=R7= OH; R4= Me; R6= R8= OMe; R3= H
112 - R1= R5=R7= OH; R2= R6= R8= OMe; R4= Me; R3= H
113 - R1= R6= R7= R8= OMe; R2= R5= OH; R4= Me; R3= H
114 - R1= R5= OH; R2= R6= R7= R8= OMe; R4= Me; R3= H
115 - R1= R2= R3= R5= R7= OH; R4= Acyl; R6= R8= H
116 - R1= R3= R5= OH; R2= R7= OMe; R4= R6= R8= H
117 - R1= R5= OH; R2= R3= R7= OMe; R4= R6= R8= H
118 - R1= R5= OH; R2= R3= R7= OMe; R8= Me; R4= R6= H
119 - R1= R5= R6= OH; R2= R3= R7= OMe; R4= R8= H
120 - R1= R5= R6= OH; R2= R3= R7= OMe; R4= Me; R8= H
121 - R1= R5= R7= OH; R4= Glu;R2= R3= R6= R8= H
122 - R1= R5= R7= OH; R4= Rha-Glu;R2= R3= R6= R8= H
123 - R1= R2= R5= R7= OH; R4= (Apio)-Rha-Glu; R3= R6= R8= H
124 - R1= R2= R5= R7= OH; R4= Glu; R3= R6= R8= H
125 - R1= R2= R5= R7= OH; R4= Gal; R3= R6= R8= H
126 - R1= R2= R5= R7= OH; R4= Rha; R3= R6= R8= H
127 - R1= R2= R5= R7= OH; R4= Rha-Glu; R3= R6= R8= H
128 – R1= R2= R5= R7= OH; R4= Rha-Glu-caffeoyl; R3= R6= R8= H

Fig. 11.8 Structures of flavanols **73–128** identified in *Baccharis* species Biological activities and *B. trimera* (Agudelo et al. 2016; Simirgiotis et al. 2016; Sabir et al. 2017), while radical scavenger activity (DPPH, ROS) was attributed to *B. dentata* and *B. trimera* extracts (Sartor et al. 2013; Agudelo et al. 2016; Sabir et al. 2017). The last extract also inhibits ROS production through PKC and downregulates p47phox phosphorylation of NADPH oxidase in SK Hep-1 cells. Hispidulin (**38**), desmethoxysudachitin (**43**), jaceosidin (**55**), and quercetin (**89**) suppress/scavenge erythrocyte lipoperoxidation, **38** and **89** superoxide anions, and **89** DPPH (Tapia et al. 2004). Eupafolin (**54**), quercitrin (**127**), and rutin (**128**) moderately scavenge DPPH radical (Akaike et al. 2003). Compounds **48, 104, 111, 112,** and **114** also displayed antimicrobial activity against methicillin-resistant *Staphylococcus aureus* and *Enterococcus faecalis* (Zampini et al. 2009; Nuño et al. 2012). Antimicrobial activities were also attributed to some flavonoids: penduletin (**87**) presented antifungal activity against some human pathogenic and phytopathogenic fungi (Rahalison et al. 1995). Isosakuranetin (**3**) showed antifungal activities against *Neurospora crassa* (Almanza et al. 2000) and *Cryptococcus neoformans* (da Silva Filho et al. 2008); besides the antioxidant activity, previously described, compound **3** also demonstrated strong trypanocidal (da Silva Filho et al. 2004) and antiinflammatory activities (Figueiredo-Rinhel et al. 2013); both activities were also attributed to dihydrokaempferide (**19**). *B. dentata* extracts presented antibacterial activity against *Staphylococcus aureus* (Sartor et al. 2013), while *B. crispa* and *B. notosergila* ethanolic extracts presented antimicrobial activities against *Bacillus subtilis*, *Micrococcus luteus*, and *Staphylococcus aureus*, due to the presence of genkwanin (**31**) and apigenin (**28**). Sakuranetin (**2**) presented antifungal activity against pathogenic yeast belonging to the genus *Candida* (six species – *C. dubliniensis, C. tropicalis, C. glabrata, C. parapsilosis, C. krusei, and C. albicans*), *Cryptococcus* (two species/four serotypes – *C. neoformans* – serotypes A and D, *C. gattii* – serotypes B and *C. neoformans*), and *Saccharomyces cerevisiae* (Grecco et al. 2014a, b). Compound **2** demonstrated promising activities to prevent and treat several respiratory disorders, such as acute lung injury (Bittencourt-Mernak et al. 2017), chronic allergic pulmonary inflammation (Sakoda et al. 2016), emphysema, through regulation of NF-κB, oxidative stress, and metalloproteinases (Taguchi et al. 2015a, b),and reverses airway inflammation and remodelling in an asthma murine model (Toledo et al. 2013). Sakuranetin (**2**) displayed activity against four *Leishmania* species (*L. amazonensis, L. braziliensis, L. major,* and *L. chagasi*) and *Trypanosoma cruzi* (Davila et al. 2013) and also presented phytotoxic activities against *Panicum miliaceum* and *Raphanus sativus*, inhibiting its growth and germination (del Corral et al. 2012). Pectolinaringenin (**39**) and cirsimaritin (**40**) present spasmolytic activities (Weimann et al. 2002) and antiparasitic potential (antileishmanial and trypanocidal). This was attributed to 5,6,7-trihydroxy-4′-methoxyflavanone (**4**), hispidulin (**38**), spectolinaringenin (**39**), acacetin (**30**), and ermanin (**77**) (da Silva Filho et al. 2009; Passero et al. 2011; Grecco et al. 2010a, b, 2014a, b). Luteolin-3′,7-dimethylether (**51**) displayed anti-inflammatory activity (Gianello et al. 1999), which was also attributed to rutin (**128**), within its analgesic effect (Gene et al. 1996) and pectolinaringenin (**39**) (Zalewski et al. 2011). *B. gaudichaudiana* and *B. spicata* presented antiviral activities against poliovirus type 2 (PV2) and vesicular stomatitis virus; PV2 attributed to the presence of apigenin (**28**) (Visintini et al. 2013) This compound also enhances the action of nerve growth factor to stimulate neurite outgrowth from PC12D cells, that could be useful in the treatment of neurological disorders (Guo et al. 2007). *B. articulata* extract induced the death of human peripheral blood mononuclear cells (PBMCs) through apoptosis and exerted low mutagenic effects on mice, those could be related to the presence of luteolin (**48**) and acacetin (**30**) and other nonflavonoidic compounds (Cariddi et al. 2012). Antitumoral activities were attributed to 3,4′-dimethoxy-3′,5,7-trihydroxyflavone (**95**) and centaureidin (**106**) (Montes et al. 1971), while gardenin B (**46**) demonstrated cytotoxic activities against HL60 and U937 – human leukemia cell lines, leading to activation of extrinsic and the intrinsic apoptotic pathways (Cabrera et al. 2016). Kaempferol and quercetin glycosides (**123**, **128** and **129**) from *B. antioquensis* displayed photoprotective potential, while quercitrin (**127**) displayed antivenom effect (*Crotalus durissus terrificus* snake) through inhibition of sPLA2 enzyme, preventing myotoxicity and edematogenic effect. Genkwanin (**31**), cirsimaritin (**40**), hispidulin (**38**), and apigenin (**28**) showed antimutagenic activity (Nakasugi and Komai 1998), while quercetin (**89**), luteolin (**48**), nepetin (**54**), apigenin (**28**), and hispidulin (**38**) were responsible for antihepatotoxic properties of *B. trimera* (Soicke and Leng-Eschlow 1987), leading to a protective effect against acute hepatic injury induced by acetaminophen (Padua et al. 2014), an effect also detected in *B. dracunculifolia* leaves extract (Rezende et al. 2014). *B. teindalensis* displayed antiulcer and antidiarrhoeic effects (Vidari et al. 2003), also observed for *B. dracunculifolia* extracts, with inhibition of doxorubicin-induced mutagenicity (Resende et al. 2007). Finally, stems of *B. illinita* presented anticoagulant activity, leading to a significant effect on platelet aggregation (Pizzolatti et al. 2006) and *B. trinervis* displayed cytotoxic, genotoxic, and mutagenic activities that could be related to the presence of flavonoids (Jaramillo-Garcia et al. 2018).

References

Abad MJ, Bermejo P (2007) *Baccharis* (Compositae): a review update. ARKIVOC 7:76–96. https://doi.org/10.3998/ark.5550190.0008.709

Abad Martinez MJ, Bessa AL, Benito PB (2005) Biologically active substances from the genus *Baccharis* L. (Compositae). Stud Nat Prod Chem 30:703–759. https://doi.org/10.1016/s1572-5995(05)80045-6

Agudelo IJ, Isolabella AS, Filip R et al (2016) *Baccharis spicata* (lam) baill: polyphenol screening, determination of their antioxidant activity and their main polyphenolic metabolites. J Pharmacogn Phytochem 5:278–285

Akaike S, Sumino M, Sekine T et al (2003) A new Ent-clerodane diterpene from the aerial parts of *Baccharis gaudichaudiana*. Chem Pharm Bull 51:197–199. https://doi.org/10.1248/cpb.51.197

Almanza GR, Arduz C, Balderrama L et al (2000) Phytochemical study of *Baccharis leptophylla* bioguided against *Neurospora crassa*. Boliv J Chem 17:1–8

Arriaga FJG, Borges JC, Manresa MTF et al (1982) Compounds from salvadoran plants. III. Components of *Baccharis trinervis* Persoon. Rev Latinoam Quím 13:47–49

Arriaga-Giner FJ, Wollenweber E, Hradetzky D (1986) New flavonoids from the exudate of *Baccharis bigelovii* (Asteraceae). Z Naturforsch C J Biosci 41:946–948

Bandoni AL, Medina JE, Rondina RVD, Coussio JD (1978) Genus *Baccharis* L. I: phytochemical analysis of a non polar fraction from *B. crispa* Sprengel. Planta Med 34:328–331

Bittencourt-Mernak MI, Santana FPR, Guerreiro MP et al (2017) Prophylactic and therapeutic treatment with the flavonone sakuranetin ameliorates LPS-induced acute lung injury. Am J Phys Lung Cell Mol Phys 312:L217–L230. https://doi.org/10.1152/ajplung.00444.2015

Bohlmann F, Knauf W, King RM, Robinson H (1979) Naturally occurring terpene derivatives. 196 new diterpene and other compounds from *Baccharis* species. Phytochemistry 18:1011–1014

Bohlmann F, Zdero C, Grenz M et al (1981a) Naturally occurring terpene derivatives. 307 five diterpenes and other constituents from nive *Baccharis* species. Phytochemistry 20:281–286. https://doi.org/10.1016/0031-9422(81)85107-2

Bohlmann F, Kramp W, Grenz M et al (1981b) Naturally occurring terpene derivatives. Part 359. diterpenes from *Baccaris* species. Phytochemistry 20:1907–1913. https://doi.org/10.1016/0031-9422(81)84033-2

Bohlmann F, Banerjee S, Jakupovic J et al (1985) Clerodane and labdane diterpenoids from *Baccharis* species. Phytochemistry 24:511–515. https://doi.org/10.1016/S0031-9422(00)80758-X

Borella JC, Duarte DP, Novaretti AAG et al (2006) Seasonal variability in the content of saponins from *Baccharis trimera* (Less.) DC (carqueja and Isolation of Flavone). Rev Bras Farmacogn 16:557–561

Cabrera J, Saavedra E, del Rosario H et al (2016) Gardenin B-induced cell death in human leukemia cell involves multiple caspases but is independent of the generation of reactive oxygen species. Chem Biol Interact 256:220–227. https://doi.org/10.1016/j.cbi.2016.07.016

Calle A, Yupanqui J, Flores Y, Almanza GR (2012) Flavonoids of *Baccharis boliviensis*. Boliv J Chem 29:158–163

Campos FR, Bressan J, Jasinski VCG et al (2016) *Baccharis* (Asteraceae): chemical constituents and biological activities. Chem Biodivers 13:1–17. https://doi.org/10.1002/cbdv.201400363

Cariddi L, Escobar F, Sabini C et al (2012) Apoptosis and mutagenicity induction by a characterized aqueous extract of *Baccharis articulata* (Lam.) Pers. (Asteraceae) on normal cells. Food Chem Toxicol 50:155–161. https://doi.org/10.1016/j.fct2011.11.002

Cestari SH, Bastos JK, Di Stasi LC (2011) Intestinal anti-inflammatory activity of *Baccharis dracunculifolia* in the trinitrobenzenesulphonic acid model of rat colitis. J Evid Based Complementary Altern Med 50:1091–1097. https://doi.org/10.1093/ecam/nep081

Chidiak S, Guada M, Buitrago D, Morales A (2007) Flavonoids isolated from *Baccharis nitida* (Ruiz et Pavon Pers). Cienc 15:361–365

Cordano GF, Munoz de la Pena A, Medina JJ (1982) Scopoletin and hispidulin in *Baccharis magellanica*. J Nat Prod 45:653. https://doi.org/10.1021/np50023a029

da Silva Filho AA, de Sousa JPB, Soares S et al (2008) Antimicrobial activity of the extract and isolated compounds from *Baccharis dracunculifolia* D.C. (Asteraceae). Z Naturforsch C 63:40–46

da Silva Filho AA, Resende DO, Fukui MJ et al (2009) In vitro antileishmanial, antiplasmodial and cytotoxic activities of phenolics and triterpenoids from *Baccharis dracunculifolia* D. C. (Asteraceae). Fitoterapia 80:478–482. https://doi.org/10.1016/j.fitote.2009.06.007

da Silva DPL, da Silva Filho AA, Polizello ACM et al (2004) Comparative evaluation of in-vitro effects of Brazilian green Propolis and *Baccharis dracunculifolia* extracts on cariogenic factors of *Streptococcus mutans*. Biol Pharm Bull 27:1834–1839

da Silva ARH, Reginato FZ, Guez CG et al (2016) Acute and sub-chonic (28 days) oral toxicity evaluation of tincture *Baccharis trimera* (Less) baker in male and female rodent animals. Regul Toxicol Pharmacol 74:170–177. https://doi.org/10.1016/j.yrtph.2015.10.024

Daily A, Wagner H, Seligmann O (1984) Hispidulin and stigmasta-7,22-dien-3-ol from *Baccharis genistelloides*. Fitoterapia 55:236–238

Davies KM, Schwinn KE (2006) Molecular biology and biotechnology of flavonoid biosynthesis. In: Andersen OM, Markham KR (eds) Flavonoids, chemistry, biochemistry and applications. CRC Taylor & Francis Group, Boca Raton, pp 143–218

Davila M, Sterner O, Hinojosa N (2013) Flavonoids from *Baccharis polycephala* Weddell. Boliv J Chem 30:137–141

de Alencar SM, de Aguiar CL, Paredes-Guzmán J, Park YK (2005) Chemical composition of *Baccharis dracunculifolia*, the botanical source of propolis from the states of São Paulo and Minas Gerais, Brazil. Cienc Rural 35:909–915

de Araujo GR, Rabelo ACS Meira JS et al (2016) *Baccharis trimera* inhibits reactive oxygem species production through PKC and down-regulation p47phox phosphorylation of NADPH oxidase in SK Hep-1 Cells. Exp Biol Med 242:333–343. https://doi.org/10.1177/1535370216672749

de Mello JCP, Petrovick PR (2000) Quality control of *Baccharis trimera* (Less.) DC. (Asteraceae) hydroalcoholic extracts. Acta Farm Bonaer 19:211–215

de Oliveira SQ, Kappel VD, Pires VS et al (2014) Antioxidant properties of phenolic compounds from *Baccharis articulata* and *B. usterii*. Nat Prod Commun 9:941–942

de Sousa JPB, da Silva Filho AA, Bueno PCP et al (2009) A validated reverse-phase HPLC analytical method for the quantification of phenolic compounds in *Baccharis dracunculifolia*. Phytochem Anal 20:24–32

de Sousa JPB, Leite MF, Jorge R et al (2011) Seasonality role on the phenolics from cultivated *Baccharis dracunculifolia*. J Evid Based Complementary Altern Med 2011:1–9. https://doi.org/10.1093/ecam/nep077

del Corral S, Cuffini SL, Cardoso SG et al (2012) Phytotoxic halimanes isolated from *Baccharis salicifolia* (Ruiz & Pad.) Pers. Phytochem Lett 5:280–283. https://doi.org/10.1016/j.phytol.2012.02.001

Dewick PM (2009) Medicinal natural products: a biosynthetic approach, 3rd edn. Wiley, New York

Escobar Z, Flores Y, Tejeda L et al (2009) Phenolic compounds from *Baccharis papillosa* subsp. *papillosa*. Boliv J Chem 26:111–117

Faine F, Rivera P, Mahu M, Castilho M (1987) Neo-clerodane diterpenoids and other constituents from *Baccharis* species. Phytochemistry 26:3281–3283. https://doi.org/10.1016/S0031-9422(00)82487-5

Faini FA, Castillo M, Torres RM (1982a) Flavonoids of *Baccharis incarum*. J Nat Prod 45:501–502. https://doi.org/10.1021/np50022a027

Faini FA, Castillo M, Torres RM (1982b) Two unusual flavonoids in *Baccharis incarum*. Bol Sol Chil Quim 27:276–279

Faini F, Hellwig F, Labbe C, Castillo M (1991) Hybridization in the genus *Baccharis*: *Baccharis linearis* x B. *macraei*. Biochem Syst Ecol 19:53–57. https://doi.org/10.1016/0305-1978(91)90113-E

Faini F, Labbe C, Coll J (1999) Seasonal changes in chemical composition of epicuticular waxes from the leaves of *Baccharis linearis*. Biochem Syst Ecol 27:673–679. https://doi.org/10.1016/S0305-1978(98)00131-8

Feresin GE, Tapia A, Gimenez A et al (2003) Constituents of the Argentinian medicinal plant *Baccharis grisebachii* and their antimicrobial activity. J Ethnopharmacol 89:73–80. https://doi.org/10.1016/S0378-8741(03)00259-9

Ferreira JVN, Grecco SS, Lago JHG et al (2015) Ultrathin films of lipids to investigate the action of a flavonoid with cell membrane models. Mater Sci Eng 48:112–117. https://doi.org/10.1016/j.msec.2014.11.061

Figueiredo-Rinhel ASG, Kabeya LM, Bueno PCP et al (2013) Inhibition of the human neutrophil oxidative metabolism by *Baccharis dracunculifolia* DC (Asteraceae) is influenced by seasonality and the ratio of caffeic acid to other phenolic compounds. J Ethnopharmacol 150:655–664. https://doi.org/10.1016/j.jep.2013.09.019

Fukuda M, Ohkoshi E, Makino M, Fujimoto Y (2006) Studies on the constituents of the leaves of *Baccharis dracunculifolia* (Asteraceae) and their cytotoxic activity. Chem Pharm Bull 54:1465–1468

Fullas F, Hussain RA, Chai HB et al (1994) Cytotoxic constituents of *Baccharis gaudichaudiana*. J Nat Prod 57:801–807. https://doi.org/10.1021/np50108a017

Gene RM, Cartana C, Adzet T et al (1996) Anti-inflammatory and analgesic activity of *Baccharis trimera*. Identification of its active constituents. Planta Med 62:232–235. https://doi.org/10.1055/s-2006-9578666

Gianello JC, Giordano OS (1984) Chemical examination of six species of the *Baccharis* genus. Rev Latinoam Quim 15:84–86

Gianello JC, Giordano OS (1987) Constituents from *Baccharis grisebachii*. An Asoc Quim Argent 75:1–3

Gianello JC, Giordano OS (1989) Chemical components from *Baccharis triangularis*. An Asoc Quim Argent 77:353–357

Gianello JC, Cifuente DA, Giordano OS, Tonn CE (1999) Bioactive flavones and terpenes from *Baccharis calliprinos* and *B. rhetinodes* (Asteraceae). Acta Farm Bonaer 18:99–102

Grecco SS, Gimenes L, Ferreira MJP et al (2010a) Triterpenoids and phenolic derivatives from *Baccharis uncinella* C.DC. (Asteraceae). Biochem Syst Ecol 38:1234–1237. https://doi.org/10.1016/j.bse.2010.12.005

Grecco SS, Reimao JQ, Tempone AG et al (2010b) Isolation of an antileishmanial flavanone from the leaves of *Baccharis retusa* DC (Asteraceae). Parasitol Res 106:1245–1248

Grecco SS, Ferreira MJP, Romoff P et al (2012a) Phenolic derivatives from *Baccharis retusa* DC. (Asteraceae). Biochem Syst Ecol 42:21–24. https://doi.org/10.1016/j.bse.2011.12.014

Grecco SS, Reimao JQ, Tempone AG et al (2012b) In vitro antileishmanial and antitrypanosomal activities of flavanones from *Baccharis retusa* DC. (Asteraceae). Exp Parasitol 130:141–145. https://doi.org/10.1016/j.exppara.2011.11.002

Grecco SS, Dorigueto AC, Landre IM et al (2014a) Structural crystalline characterization of sakuranetin – an antimicrobial flavanone from twigs of *Baccharis retusa* (Asteraceae). Molecules 19:7528–7542. https://doi.org/10.3390/molecules19067528

Grecco SS, Felix MJP, Lago JHG et al (2014b) Anti-trypanosomal phenolic derivatives from *Baccharis uncinella*. Nat Prod Commun 9:171–173

Guimaraes NSS, Melo JC, Paiva JS et al (2012) *Baccharis dracunculifolia*, the main source of green propolis, exhibits potent antioxidant activity and prevents oxidative mitochondrial damage. Food Chem Toxicol 50:1091–1097. https://doi.org/10.1016/j.fct.2011.11.014

Guo Y, Li Y, Xu J et al (2007) Clerodane diterpenoids and flavonoids with NGF-potentiating activity from the aerial parts of *Baccharis gaudichaudiana*. Chem Pharm Bull 55:1532–1534

He K (1995) Phytochemical investigations of medicinal plants from Chile and the Fiji islands (*Baccharis linearis*, *Aristotelia chilensis*, *Dysoxylum lenticellare*). The University of Arizona, 182 pp

He K, Montenegro G, Hoffmann JJ, Timmermann BN (1996) Diterpenoids from *Baccharis linearis*. Phytochemistry 41:1123–1127. https://doi.org/10.1016/0031-9422(95)00780-6

Hennig L, Garcia GM, Giannis A, Bussmann RW (2011) New constituents of *Baccharis genistelloides* (Lam.) Pers. ARKIVOC 6:74–81. https://doi.org/10.3998/ark.5550190.0012.607

Herrera JC, Romero AJR, Crescente OE et al (1996) Analysis of 5-hydroxy-7-methoxyflavones by normal-phase high-performance liquid chromatography. J Chromatogr A 740:201–206. https://doi.org/10.1016/0021-9673(96)00108-2

Herz W, PIlotti AM, Soderholm AC et al (1977) New ent-clerodane-type diterpenoid from *Baccharis trimera*. J Organomet Chem 42:3913–3917. https://doi.org/10.1021/jo00444a026

Jakupovic J, Eid F, King RM (1986) Further ingredients of *Baccharis trinervis*. Pharmazie 41:157–158

Jakupovic J, Schuster A, Ganzer U et al (1990) Sesqui- and Diterpenes from *Baccharis* species. Phytochemistry 29(7):2217–2222. https://doi.org/10.1016/0031-9422(90)83041-X

Jaramillo-Garcia V, Trindade C, Lima E et al (2018) Chemical characterization and cytotoxic, genotoxic, and mutagenic properties of *Baccharis trinervis* (lam, Persoon) from Colombia and Brazil. J Ethnopharmacol 213:210–220. https://doi.org/10.1016/j.jep.2017.10.027

Joshi SB, Haider IS, Pelletier WS (1997) Flavonoids from *Baccharis halimifolia*, *Monarda didyma*, and *Gnaphalium dioicum*. J Indian Chem Soc 74:874–876

Kakehashu A, Ishii N, Fujioka M et al (2016) Ethanol-extracted Brazilian propolis exerts protective effects on tumorigenesis in wistar hannover rats. PLoS One 11:e0158654/1–e0158654/15. https://doi.org/10.1371/journal.pone.0158654

Kumazawa S, Yoneda M, Shibata I et al (2003) Direct evidence for the plant origin of Brazilian propolis by the observation of honeybee behavior and phytochemical analysis. Chem Pharm Bull 51:740–742. https://doi.org/10.1248/cpb.51.740

Kuroyanagi M, Fujita K, Kazaoka M et al (1985) Studies on the constituents of *Baccharis genistelloides*. Chem Pharm Bull 33:5075–5078. https://doi.org/10.1248/cpd.33.5075

Labbe C, Rovirosa J, Faini F et al (1986) Secondary metabolites from Chilean *Baccharis* species. J Nat Prod 49:517–518. https://doi.org/10.1021/np50045a025

Labbé C, Faini F, Castillo M (1990) Diterpenoids from Chilean *Baccharis* species. Phytochemistry 29:324–325

Lemos M, de Barros MP, Sousa JBP (2007) *Baccharis dracunculifolia*, the main botanical source of Brazilian green propolis, displays antiulcer activity. J Pharm Pharmacol 59: 603–608. https://doi.org/10.1211/jpp.59.4.0017

Liu YL, Timmermann BN, Hoffmann JJ et al (1993) A flavonol triglycoside from *Baccharis thesioides*. Phytochemistry 33:1549–1551. https://doi.org/10.1016/0031-9422(93)85134-D

Mejia-Giraldo JC, Sanchez-Zapata AM, Puertas-Mejia MA et al (2016) Photoprotective potential of *Baccharis antioquensis* (Asteraceae) as natural sunscreen. Photochem Photobiol 92:742–752

Menezes APS, da Silva J, Fisher C et al (2016) Chemical and toxicological effects of medicinal *Baccharis trimera* extract from coal burning area. Chemosphere 146:396–404. https://doi.org/10.1016/j.Chemosphere.2015.12.02

Mesquita AAL, Correa DB, de Padua AP et al (1985) Flavonoids from four Compositae species. Phytochemistry 25:1255–1256. https://doi.org/10.1016/S0031-9422(00)81599-X

Midorikawa K, Banskota AH, Tezuka Y et al (2003) Buds of *Baccharis dracunculifolia*: potente source of biologically active e caffeoyquinic acids and labdane-type diterpenes of Brazilian propolis. J Pharm Soc Jpn 20:187–194

Missima F, da Silva Filho AA, Nunes GA et al (2007) Effect of *Baccharis dracunculifolia* D.C. (Asteraceae) extracts and its isolated compounds on macrophage activation. J Pharm Pharmacol 59:463–468

Mollinedo P, Salcedo OL, Vila JL et al (2001) Evaluation of the antioxidant capacity of three Bolivian plant species. Boliv J Chem 18:59–64

Montes GM, Wikomirski FT, Valenzuela RL, Neira MR (1971) Alkaloids of *Baccharis linearis*. Rev R Acad de Cienc Exactas, Fis Nat Madr 65:499–511

Morales Mendez A, Nunez O, Rosquete C, Montilla A (1984) Flavones from *Baccharis decussata* (Klatt) Hieron. An Quim, Serie C: Quimica Organica y Bioquimica 80:98–99

Moreira EA, Keller CC, Krambeck R, Miguel OG (1975) Phytochemical analysis of *Baccharis maritima*. Preliminary chemical investigation, and identification of polyphenolic substances. Phytochem Anal 43:3–17

Moreira FPM, Branco A, Pizzolatti MG et al (2003a) Acid triterpenes and flavonoids from *Baccharis ligustrina* (Asteraceae). Biochem Syst Ecol 31:319–321. https://doi.org/10.1016/S0305-1978(02)00149-7

Moreira FPM, Voutinho V, Pimentel Montanher AB et al (2003b) Flavonoids and triterpenes from *Baccharis pseudotenuifolia* – bioactivity on *Artemia salina*. Quim Nova 26:309–311. https://doi.org/10.1590/S0100-40422003000300004

Nagatani Y, Warashina T, Noro T (2001) Studies on the constituents from the aerial part of *Baccharis dracunculifolia* DC. Chem Pharm Bull 49:1388–1394. https://doi.org/10.1248/cpd.49.1388

Nakasugi T (1990) Anti-mutagen agents. Japan. Kokai Tokkyo Koho JP 11180867 a 19990706

Nakasugi T, Komai K (1998) Antimutagens in the Brazilian folk medicinal plant carqueja (*Baccharis trimera Less.*). J Agric Food Chem 46:2560–2564. https://doi.org/10.1021/JF9711045

Nogueira Sobrinho AC, de Souza EB, Rocha MFG et al (2016) Chemical composition, antioxidant, antifungal and hemolytic activies of essential oil from *Baccharis trivervis* (Lam.) Pers. (Asteraceae). Ind Crop Prod 84:108–115. https://doi.org/10.1016/j.indcrop.2016.01.051

Nuño G, Zampini IC, Ordonez RM et al (2012) Antioxidant/antibacterial activities of a topical phytopharmaceutical formulation containing a standardized extract of *Baccharis incarum*, an extremophile plant species from Argentine Puna. Phytother Res 26:1759–1767. https://doi.org/10.1002/ptr.4633

Padua BC, Rossoni JVJ, Magalhaes CLB et al (2014) Protective effect of *Baccharis trimera* extract on acute hepatic injury in a modelo of inflammation induced by acetaminophen. Mediat Inflamm 196598:1–15. https://doi.org/10.1155/2014/196598

Palacios P, Gutkind G, Rondina RV et al (1983) Genus *Baccharis*. II. Antimicrobial activity of *B. crispa* and *B. notosergila*. Planta Med 49:128.

Park YK, Paredes-Guzman JF, Aguiar CL et al (2004) Chemical constituents in *Baccharis dracunculifolia* as the main botanical origin of southeastern Brazilian propolis. J Agric Food Chem 52:1100–1103

Park YK, Fukuda I, Ashida H et al (2005) Suppressive effects of ethanolic extracts from propolis and its main botanical origin on dioxin toxicity. J Agric Food Chem 53:10306–10309. https://doi.org/10.1021/jf058111a

Parodi FJ, Fischer NH (1988) Guaianolides from *Baccharis salicina*. Phytochemistry 27:2987–2988. https://doi.org/10.1016/0031-9422(88)80705-2

Passero LFD, Bonfim-Melo A, Corbett CEP et al (2011) Anti-leishmanial effects of purified compounds from aerial parts of *Baccharis uncinella* C. DC (Asteraceae). Parasitol Res 108:529–536

Paula JT, Sousa IMO, Foglio MA, Cabral FA (2017) Selective fractionation of supercritical extracts from leaves of *Baccharis dracunculifolia*. J Supercrit Fluids 127:62–70. https://doi.org/10.1016/j.supflu.2017.03.032

Piantino CR, Aquino FWB, Follegatti-Romero LA, Cabral FA (2008) Supercritical CO_2 extraction of phenolic compounds from *Baccharis dracunculifolia*. J Supercrit Fluids 47:209–214. https://doi.org/10.1016/j.supflu.2008.07.012

Pizzolatti MG, Verdi LG, Brighente IMC et al (2006) Anticoagulant effect and constituents of *Baccharis illinita*. Nat Prod Commun 1:37–42

Quijano L, Rios T, Fronczek FR, Fischer NH (1998) The molecular structure of maniladiol from *Baccharis salicina*. Phytochemistry 49:2065–2068. https://doi.org/10.1016/S0031-9422(98)00363-X

Rabelo ACS, de Araujo GR, Lucio KP et al (2017) Aqueous extract of *Baccharis trimera* improves redox status and decreases the severity of alcoholic hepatotoxicity. Rev Bras Farmacogn 27:729–738. https://doi.org/10.1016/j.bjp.2017.09.003

Rahalison L, Benathan M, Monod M et al (1995) Antifungal principles of *Baccharis pedunculata*. Planta Med 61:360–362. https://doi.org/10.1055/s-2006-958101

Rendon W, Vila JL (1995) Phytochemical study of *Baccharis trimera*. Boliv J Chem 12:13–15

Resende FA, Alves JM, Munari CC et al (2007) Inhibition of doxorubicin-induced mutagenicity by *Baccharis dracunculifolia*. Mutat Res 634:112–118

Rezende TP, Correa JOA, Aarestrup BJV et al (2014) Protective effects of *Baccharis dracunculifolia* leaves extract against carbon tetrachloride- and acetaminophen-induced hepatotoxicity in experimental animals. Molecules 19:9257–9272. https://doi.org/10.3390/molecules19079257

Rivera FA, Faini F, Castilho M (1988) 15α-hydroxy-β-amyrin and patogonic acid from *Baccharis magellanica* and *Baccharis paragonica*. J Nat Prod 51:155–157. https://doi.org/10.1021/np50055a026

Rodriguez MV, Martinez ML, Cortadi A et al (2012) Quality control of herbal medicines with spectrophotometry and chemometric techniques – application to *Baccharis* L. species belonging to sect – Caulopterae DC. (Asteraceae). In: Macro to nano spectroscopy. IntechOpen, Jamal Uddin, pp 399–420

Rojas J, Morales A (2000) Study of the chemical components of *Baccharis decussata* (K) Hieron. Ciencia 8:251–256

Saad JR, Pestchanker MJ, Giordano OS (1987) Furanditerpenes from *Baccharis thymifolia*. Phytochemistry 26:3033–3035. https://doi.org/10.1016/S0031-9422(00)84587-2

Saad JR, Davicino JG, Giordano OS (1988) A Diterpene and flavonoid of *Baccharis flabellata*. Phytochemistry 27:1884–1887. https://doi.org/10.1016/0031-9422(88)80471-0

Sabir SM, Athayde ML, Boligon AA, Rocha JBT (2017) Antioxidant activities and phenolic profile of *Baccharis* a commonly used medicinal plant from Brazil. S Afr J Bot 113:318–323. https://doi.org/10.1016/j.saj.2017.09.010

Sakoda CPP, Prado CM, de Toledo AC et al (2016) Sakuranetin reverses vascular peribronchial and lung parenchyma remodeling in a murine model of chronic allergic pulmonary inflammation. Acta Histochem 118:615–624. https://doi.org/10.1016/j/acthis.2016.07.001
Salcedo Ortiz L, Sterner O, Almanza Vega GR (2001) Phytochemical study of *Baccharis latifolia*. Boliv J Chem 18:43–46
Salcedo L, Pillco A, Rodrigo G et al (2003) Isolation of flavonoids and study of the toxic and antibacterial activity of *Baccharis latifolia* extracts. Boliv J Chem 20:43–48
San Martin A, Rovirosa J, Castillo M (1982) Diterpenes from *Baccharis tola*. Part II. Bol Soc Chil Quim 27:252–254
San Martin A, Rovirosa J, Castillo M (1983) Diterpenoids from *Baccharis tola*. Phytochemistry 22:1461–1463. https://doi.org/10.1016/S0031-9422(00)84036-4
San Martin AJ, Villanueva E, Cruz AT et al (2012) Phytochemical and preliminary sprectroscopic study of five medicinal plants of Carmen Pampa (Coroico) Bolivia. Boliv J Chem 29:121–130
Sartor T, Xavier VB, Falcao MA et al (2013) Seasonal changes in phenolic compounds and in the biological activities of *Baccharis dentata* (Vell.) G.M. Barroso. Ind Crop Prod 51:355–359. https://doi.org/10.1016/j.indcrop.2013.09.018
Schenkel EP, Sobottka AM, Lankhorst PP et al (1997) An unusual diterpene and other constituents from *Baccharis ochracea*. Phytochemistry 44:755–757. https://doi.org/10.1016/S0031-9422(96)00565-1
Sharp H, Bartholomew B, Bright C et al (2000) 6-Oxygenated flavones from *Baccharis trinervis* (Asteraceae). Biochem Syst Ecol 29:105–107. https://doi.org/10.1016/S0305-1978(00)00019-3
Silva Filho AA, Pires PCB, Gregorio LE et al (2004) In vitro trypanocidal activity evaluation of crude extract and isolated compounds from *Baccharis dracunculifolia* D.C. (Asteraceae). J Pharm Pharmacol 56:1195–1199. https://doi.org/10.1211/0022357044067
Silva M, Mundaca JM, Sammes PG (1971) Flavonoid and triterpene constituents of *Baccharis rhomboialis*. Phytochemistry 10:1942–1943. https://doi.org/10.1016/S0031-9422(00)86469-9
Silva GAB, Henriques A, Alice CB et al (1985) Methoxyflavones from *Baccharis* patens (Asteraceae). J Nat Prod 48:861. https://doi.org/10.1021/np50041a041
Silva FG, Januario AH, Pinto JEBP et al (2006) Flavonoid content in nature and cultivated populations of "carqueja" [*Baccharis trimera* (Less.) DC.] Harvested in dry and humid seasons. Rev Bras Plant Med 8:19–25
Simirgiotis MJ, Garcia M, Sosa ME et al (2003) Na ent-kaurene derivative from aerial parts of *Baccharis rufescens*. J Argent Chem Soc 91:109–116
Simirgiotis MJ, Quispe C, Borquez J et al (2016) High resolution metabolite fingerprinting of the resin of *Baccharis tola* Phill. from the Atacama desert and its antioxidant capacities. Ind Crop Prod 94:368–375. https://doi.org/10.1016/j.indcrop.2016.08.037
Simões-Pires CA, Queiroz EF, Henriques AT, Hostettmann K (2005) Isolation and on-line identification of antioxidant compounds from three *Baccharis* species by HPLC-UV-MS/MS with post-column derivatisation. Phytochem Anal 16:307–314
Soicke H, Leng-Eschlow E (1987) Characterization of flavonoids from *Baccharis trimera* and their antihepatotoxic properties. Planta Med 53:37–39
Suttisri R, Kighorn AD, Wright AD, Sticher O (1994) Neo-clerodane diterpenoids and other constituents from *Baccharis genistelloides*. Phytochemistry 35:443–446. https://doi.org/10.1016/S0031-9422(00)94778-2
Taguchi L, Pinheiro NM, Olivo CR et al (2015a) A flavanone from *Baccharis retusa* (Asteraceae) prevents elastase-induced emphysema in mice by regulating nf-κb, oxidative stress and metalloproteinases. Respir Res 16:1–15. https://doi.org/10.1186/s12931-015-0233-3
Taguchi L, Pinheiro NM, Olivo CR et al (2015b) Erratum: a flavanone from *Baccharis retusa* (Asteraceae) prevents elastase-induced emphysema in mice by regulating NF-κB, oxidative stress and metalloproteinases. Respir Res 16:113. https://doi.org/10.1186/s12931-015-0258-7
Tapia A, Rodriguez J, Theoduloz C et al (2004) Free radical scavengers and antioxidants from *Baccharis grisebachii*. J Ethnopharmacol 95:155–161
Tarqui ST, Segura YF, Veja GRA (2012) Polyoxygenated flavonoids from *Baccharis pentlandii*. Boliv J Chem 29:10–14

Toledo AC, Sakoda CPP, Perini A et al (2013) Flavonone treatment reverses airway inflammation and remodelling in an asthma murine model. Br J Pharmacol 168:1736–1749. https://doi.org/10.1111/bph.12062

Tonn CE, Rossomando PC, Giordano OS (1982) Batudioic acid and flavonoid from *Baccharis tucumanensis*. Phytochemistry 21:2599–2600. https://doi.org/10.1016/0031-9422(82)85268-0

Torres LMB, Gamberini MT, Roque NF et al (2000) Diterpene from *Baccharis trimera* with a relaxant effect on rat vascular smooth muscle. Phytochemistry 55:617–619. https://doi.org/10.1016/S0031-9422(00)00223-5

Toyama DO, Gaeta HH, Terashima de Pinho MV et al (2014) An evaluation of 3-rhamnosylquercetin, a glycosylated form of quercetin, against the myotoxic and edematogenic effects of sPLA2 from *Crotalus durissus terrificus*. Biomed Res Int 2014:1–11. https://doi.org/10.1155/2014/341270

Ueno AK, Barcello AF, Grecco SS et al (2018) Sesquiterpenes, diterpenes, alkenyl p-coumarates, and flavonoid from the aerial parts of *Baccharis retusa* (Asteraceae). Biochem Syst Ecol 78:39–42. https://doi.org/10.1016/jbase.2018.03.013

Verdi LG, Brighente IMC, Schripsema J et al (2004) Kaurene diterpenes and flavonoids from *Baccharis illinita* flowers. Biochem Syst Ecol 32:837–840. https://doi.org/10.1016/j.bse.2002.11.008

Vidari G, Vita FP, Zarzuelo A et al (2003) Antiulcer and antidiarrhoeic effect of *Baccharis teindalensis*. Pharm Biol 41:405–411. https://doi.org/10.1076/phbi.41.6.405.17832

Visintini JMF, Redko F, Muschietti L et al (2013) In vitro antiviral activity of plant extracts from Asteraceae medicinal plants. Virol J 10:1–10. https://doi.org/10.1186/1743-422x-10-245

Wagner H, Iyengar MA, Herz W (1972) Flavonol 3-glucosides in *Baccharis angustifolia* and *Berlandiera pumila*. Phytochemistry 11:444. https://doi.org/10.1016/S0031-9422(00)90045-1

Wagner H, Seitz R, Lotter H, Herz W (1978) New Furanoid Ent-clerodanes from *Baccharis tricuneata*. J Organomet Chem 43:3339–3345. https://doi.org/10.1021/jo00411a017

Warning U, Bohlmann F, Sanchez VH et al (1986) New constituents of *Baccharis salicifolia*. Rev Latinoam Quim 17:199–200

Weimann C, Goransson U, Pongprayoon-Claeson U et al (2002) Spasmolytic effects of *Baccharis conferta* and some of its constituents. J Pharm Pharmacol 54:99–104

Wollenweber E, Schober I, Dostal P et al (1986) Flavonoids and terpenoids from the exudates of some *Baccharis* species. J Biosci 41:87–93

Wollenweber E, Dorr M, Fritz H et al (1997) Exudate flavonoids in Asteraceae from Arizona, California, and Mexico. Z Naturforsch C J Biosc 52:301–307

Wollenweber E, Valant-Vetschera KM, Fernandes GW (2006) Chemodiversity of exudate flavonoids in *Baccharis concinna* and three further South-American *Baccharis* species. Nat Prod Commun 1:627–632

Zalewski CA, Passero LFD, Melo ASRB et al (2011) Evaluation of anti-inflammatory activity of derivatives from aerial parts of *Baccharis uncinella*. Pharm Biol 49:602–607. https://doi.org/10.3109/13880209.2010.537828

Zamorano R, Aguirre ME, Munoz de la Pena A et al (1987) Flavonoids from *Baccharis concava* Pers. Bol Soc Chil Quim 32:101–103

Zampini IC, Isla MI, Schmeda-Hirschmann G (2009) Antimicrobial and antioxidant compounds from the infusion and methanolic extract of *Baccharis incarum* (Wedd.) Perkins. J Chil Chem Soc 54:477–481. https://doi.org/10.4067/S0717-97072009000400035

Zdero C, Bohlmann F, King RM, Robinson H (1986) Diterpene glycosides and other constituents from Argentinian *Baccharis* species. Phytochemistry 25:2841–2855. https://doi.org/10.1016/S0031-9422(00)83754-1

Zdero C, Bohlmann F, Niemeyer HM (1991) An unusual dimeric sesquiterpene and other constituents from Chilean *Baccharis* species. Phytochemistry 30:1597–1601. https://doi.org/10.1016/0031/9422(91)86215-E

Chapter 12
Chemistry and Biological Activities of Phenolic Compounds from *Baccharis* Genus

Jairo Kenupp Bastos and Caroline Arruda

Abstract Plants belonging to *Baccharis* genus (Asteraceae) have been used in folk medicine since ancient times. Usually, different *Baccharis* species are used in folk medicine as infusion or tea for gastrointestinal diseases, inflammation, ulcers, as an analgesic, spasmolytic, and antimicrobial, among others. Examples of medicinal plants from *Baccharis* are *B. dracunculifolia* D.C., *B. illinita* D.C., and *B. trimera* (Less.) DC., among many others. Over the years, these plants have been more studied: both chemical composition and biological activities. There are approximately 500 *Baccharis* species spread across the American continent, especially in South and Central America, which are important sources of bioactive compounds. The chemistry of these plants is characterized mainly by the presence of monoterpenes and sesquiterpenes in their essential oils. The nonvolatile fraction is characterized by diterpenes, triterpenes, and phenolic compounds, among others. Phenolic compounds are represented by phenylpropanoids, prenylated phenylpropanoids, flavonoids, flavonoid glycosides, coumarins, and simple phenolic compounds. In *Baccharis* spp. luteolin, chlorogenic acid, apigenin, acacetin, quercetin, kaempferol, *p*-coumaric acid derivatives, and coumarins have also been found. Many *Baccharis* spp. crude extracts and some of their isolated compounds were correlated with several biological activities. One example is the antioxidant effect of Brazilian Green Propolis, which is composed mainly of *B. dracunculifolia* compounds, such as flavonoid aglycones and *p*-coumaric acid derivatives, like artepillin C, baccharin, and drupanin. *Baccharis* spp. extracts display trypanocidal, antimicrobial, and anti-inflammatory activities, corroborating many folk medicinal uses. Therefore, in this chapter, an overview of the chemical composition is presented, highlighting the phenolic compounds of *Baccharis* spp., as well as its ethnopharmacological uses, in the light of many published scientific studies, focusing on the corroboration of folk uses. Furthermore, the toxicity of *Baccharis* species is discussed, which is a very important issue that is not well discussed in folk medicine: *B. coridifolia*, for example, is a poisonous plant responsible for necrosis of gastrointestinal tissue of rabbits

J. K. Bastos (✉) · C. Arruda
School of Pharmaceutical Sciences of Ribeirão Preto, University of São Paulo, São Paulo, Brazil
e-mail: jkbastos@fcfrp.usp.br

G. W. Fernandes et al. (eds.), *Baccharis*,
https://doi.org/10.1007/978-3-030-83511-8_12

and horses. Also, the chromatographic analyses of these plant extracts are addressed due to the importance of their chemical composition, the content of active compounds, and certainty of the correct botanical identification. In the final part, we conclude and discuss future perspectives for *Baccharis* extracts and their isolated compounds in the development of efficacious and safe medicines.

Keywords Artepillin C · Baccharin · Drupanin · Pharmacological properties · Plant chemistry

1 Ethnopharmacology of *Baccharis* spp.

Baccharis is one of the largest genus belonging to the Asteraceae family (formerly known as Compositae), which is formed mainly by shrubs (Verdi et al. 2005). According to The Plant List database (theplantlist.org), there are more than 440 accepted *Baccharis* species names and several of them have synonyms. Furthermore, there are more than 100 *Baccharis* species with unresolved names. Plants from this genus can be found especially in South American countries, such as Argentina, Bolivia, Brazil, and Colombia. Plant heights vary from 0.5 to 4.0 meters high. Most of these species, around 120, are found in Brazil, especially in the South and Southeast regions, in the states of Santa Catarina, Paraná, São Paulo, and Rio Grande do Sul. In Brazil, they are known by the population under the names "carqueja," "vassoura," "vassourinha," "alecrim," and "vassourinha do campo" (Verdi et al. 2005; Abad and Bermejo 2006).

From the numerous *Baccharis* species already identified, many of them have been reported in folk medicine for their medicinal properties, such as *Baccharis anomala*, which is used in the treatment of general infections and as a wound-washing agent (De Souza et al. 2004). These plants are widely used in folk medicine to treat gastrointestinal and liver disorders, anemia, inflammation, infections, and diabetes. Most of the time, they are consumed as an infusion or decoction of the plant (Abad and Bermejo 2006). Some selected *Baccharis* species and their ethnopharmacological uses are summarized in Table 12.1.

On the one hand, these plants are popularly used for treating several types of pathologies and many of these biological effects have been proved and reported in the scientific literature. They can also be used as ornamental plants, as hedges and to avoid erosion. Some bees are attracted to *Baccharis* spp. and from collecting their resins and nectar, the bees produce high-quality honey and propolis, which are important to the apicultural economy. The essential oil of *Baccharis* spp. is used in the cosmetic industry. On the other hand, some species are invasive and others biosynthesize toxic metabolites, like trichothecenes, which can cause dizziness, tremors, diarrhea, and even death to cattle, for example. The *Baccharis* species that produce these metabolites are generally toxic to human beings too. (Martinez et al.

Table 12.1 Traditional uses of *Baccharis* species (Martinez et al. 2005; Abad and Bermejo 2006)

Baccharis species	Use in folk medicine	Preparation method	Part of the plant used
Baccharis articulata (lam.) Pers	Diuretic, digestive, antidiabetic	Decoction, infusion,	Aerial parts
Baccharis conferta Kunth	Stomachaches, laxative, urination stimulating, weight loss assistant	Infusion	Aerial parts
Baccharis coridifolia D.C.	Anti-inflammatory, horse's distemper, horse's external parasites	Decoction	Whole plant
Baccharis crispa Spreng.	Digestive and antiseptic	Decoction	Aerial parts
Baccharis dracunculifolia D.C.:	Health improving	Beverages with Brazilian Green Propolis (made by honeybees from *B. dracunculifolia* resinous material)	
Baccharis floribunda Kunth	Skin infections, diabetes, and rheumatism.	Decoction, infusion	Leaves, stems
Baccharis gaudichaudiana D.C.	Diabetes, tonic, gastrointestinal disorders	[a]	[a]
Baccharis trimera (Less.) DC.	Liver diseases, rheumatism, diabetes, digestive, kidney disorders, aphrodisiac	Decoction	Whole plant
Baccharis glutinosa Pers.	Gynecological problems, digestive disorders, skin diseases	Infusion	Leaves
Baccharis grisebachii Hieron	Gastric ulcers, as a digestive, local antiseptic, healing	Infusion	Aerial parts
Baccharis heterophylla Kunth	Gastrointestinal diseases	Infusion, decoction	Aerial parts
Baccharis illinita D.C.	Anti-inflammatory, skin and wound healing, antiulcer, anti-infectious	Infusion, dried leaves powder	Leaves and stems
Baccharis incarum (Wedd.) Perkins	Wound healing	Decoction, poultices, infusion	Leaves, stems
Baccharis latifolia (Ruiz & Pav.) Pers.	Rheumatism, liver disorders, wounds, ulcers	Decoction	Leaves, stems
Baccharis multiflora Kunth	Catarrhs, urinary disorders	Infusion,	Leaves
Baccharis obtusifolia Kunth	Rheumatism, liver disorders, wounds, ulcers	Decoction	Aerial parts
Baccharis pentlandii DC.	Anti-inflammatory, rheumatism	[a]	[a]

(continued)

Table 12.1 (continued)

Baccharis species	Use in folk medicine	Preparation method	Part of the plant used
Baccharis pseudovaccinioides Teodoro Luis	Gastrointestinal disorders	Infusion, decoction	Whole plant
Baccharis rubricaulis Rusby	Mucous	Infusion, decoction	Leaves, stems
Baccharis salicifolia (Ruiz & Pav.) Pers	Anti-inflammatory, women hygiene agent	Infusion, decoction	Leaves, branches, stems
Baccharis sarothroides A.	Cold and muscles injury treatments	Boiling	Twigs
Baccharis subalata Wedd.	Rheumatism, liver disorders, wounds, ulcers	Decoction	Aerial parts
Baccharis teindalensis Kunth	Anti-inflammatory, analgesic, antimicrobial	Infusion, decoction	Aerial parts
Baccharis tricuneata (L. f.) Pers.	Skin infection, diabetes	Decoction, infusion	Leaves, stems
Baccharis trinervis Pers.	Fever, edema, wounds, muscle injuries	Infusion	Aerial parts

[a]Not described

2005; Verdi et al. 2005). Therefore, the use of these plants in folk medicine should be done carefully, considering that according to their chemical composition, it may display a pharmacological desired effect, a toxic effect, or be innocuous.

2 Chemical Composition

Phenolic Compounds

In *Baccharis* spp. different classes of secondary metabolites can be identified, and among them, phenolic compounds are present in high amounts (Abad and Bermejo 2006). Compounds belonging to this class can also be found in several plant species and as a matter of fact, it is widespread throughout the plant kingdom. These phytochemicals are biosynthesized in plants by the shikimate pathway and are undoubtedly important to plant development, contributing toward the defense against herbivore insects and pathogens, as well by giving color and scent (Balasundram et al. 2006). Considering that phenolic compounds display important biological effects, like gastroprotective, anti-inflammatory, antioxidant, antimicrobial, and antiparasitic effects (Arruda et al. 2017; Berretta et al. 2017; Costa et al. 2018; Ribeiro et al. 2018), in the past few years, several studies were performed aiming to discover promising compounds for treating many diseases that the available medicines are not satisfactory. Examples of these diseases are the ones known as "neglected diseases," such as leishmaniasis, Chagas, malaria, and schistosomiasis. These diseases are usually caused by parasites or infectious agents, and most pharmaceutical companies do not support the development of novel medicines for their treatment. Therefore, the discovery of new compounds more effective and less toxic to treat these parasitic diseases is very important (Abad and Bermejo 2006; Grecco et al. 2010b; de Oliveira et al. 2012a, b, 2014). Due to their antioxidant activities, phenolic compounds have been correlated to the health effects of eating fruits and vegetables, as well as functional foods and beverages, whose consumption has increased in the last few years (Berretta et al. 2017).

Chemically, these compounds bear hydroxyl group(s) attached to a benzene ring, as a simple phenolic compound or phenolic polymers. Other phenols isolated from plants are the ones bearing one or more sugar moieties, along with esters and methyl ester derivatives. Some classes of phenolic compounds naturally occurring in plants are simple phenolics, quinones, hydroxybenzoic acids, phenylpropanoids, xanthones, stilbenes, lignans, lignins, tannins and flavonoids: flavonoids, tannins, and simple phenolic compounds are the most abundant in plants. Their chemical differences are in how many phenyl groups are present and the carbon side chain; for example, phenylpropanoids have an aromatic ring with a three-carbon side chain. The hydroxybenzoic acid derivatives have the aromatic ring attached to one-carbon side chain. Another class of phenolics are flavonoids, which have a mixed biosynthetic pathway and are formed by a C6-C3-C6 unit. Flavonoids can be found in most plant species and are subdivided into subclasses, depending on the ring C substitution: flavonols, flavones, flavanols, flavanones, isoflavones, flavanonols, and anthocyanidins (Balasundram et al. 2006).

Regarding *Baccharis* spp. phenolic compounds, several flavonoids, prenylated phenolics, simple phenolics, and phenolic compounds attached to sugar moieties can be found: *B. dracunculifolia* is probably the *Baccharis* species most studied from chemical and pharmacological point of view, and some of its major phenolic compounds are the prenylated phenolics baccharin, artepillin C, and drupanin (Lemos et al. 2007; De Sousa et al. 2011; Costa et al. 2018). Kupchan et al. (1976) described baccharin as a trichothecene triepoxide isolated from *B. megapotamica*, but more recently papers had named baccharin one of the prenylated phenolic compounds biosynthesized from *p*-coumaric acid (da Silva Filho et al. 2008; Cestari et al. 2011; De Sousa et al. 2011; Oliveira et al. 2011; Costa et al. 2018). Besides, *p*-coumaric, ferulic, *trans*-cinnamic, chlorogenic, and caffeic acids had also been identified in this plant species. The flavonoids, kaempferol, kaempferide, isosakuranetin, pinobanskin, chrysin, aromadendrin-4′-O-methyl ether, 11-hydroxy-10,11-dihydro-euparin, acacetin, and ermanin had also been detected in *B. dracunculifolia* extracts. Caffeoylquinic acids like 3,4-di-*O*-caffeoylquinic acid and 3,5-di-*O*-caffeoylquinic acid were reported, as well as the other phenolics, such as 6-hydroxy-tremetone, dihydrocumaric acid, 2,2-dimethyl-6-carboxyethenyl-2H-1-benzopyran acid, viscidone, protocatechuic acid, sinapic acid and (*E*)-3-(*E*)-3-hydroxy-3-methyl-1-butenyl-4-(2,3-dihydrocinnamoyloxy)-cinnamic acid (Abad and Bermejo 2006; Lemos et al. 2007; Nakajima et al. 2007; Barros et al. 2008; Chang et al. 2008; De Sousa et al. 2011; Costa et al. 2018).

Other well-studied *Baccharis* species is *B. trimera*, which also contains caffeoylquinic acids such as 5′-*O*-caffeoylquinic acid, 4-*O*-(*E*)- caffeoyl-1-methylquinic acid, 1′-5′-*O*-dicaffeoylquinic acid, 1,3-di-(*E*)-caffeoylquinic acid, 5-*O*-(*E*)-caffeoylquinic acid, 3,4-*O*-(*E*)-dicaffeoylquinic acid, 3,5-*O*-(*E*)-dicaffeoylquinic acid, 4,5-*O*-(*E*)-dicaffeoylquinic acid, and tricaffeoylquinic acid, along with the flavonoids eupafolin, hispidulin, quercetin, luteolin, and apigenin in its chemical composition (Aboy et al. 2012; Lívero et al. 2016a, b; de Araújo et al. 2017). Luteolin, acacetin, and quercetin, along with chlorogenic acid and 4′-*O*-β-D-gluco- pyranosyl-3′,5′-dimethoxybenzyl-caffeate, were found in *B. articulata* (Cariddi et al. 2012).

The phytochemical analyses of *B. chilco*, *B. darwinii*, and *B. dentata* revealed the presence of 5-*O*-[(*E*)-caffeoyl-quinic acid, 3,5-di-*O*-[(*E*)- caffeoyl]quinic acid and rosmarinic acid (Argoti et al. 2013), anisocoumarin, 7-geranyloxycoumarin and diversinin (Kurdelas et al. 2010), as well as caffeic acid, rutin, quercetin, apigenin, and kaempferol (Sartor et al. 2013), respectively. In *B. genistelloides* and *B. illinita*, several flavonoids have also been found, such as luteolin, identified in both species and eupatrin, cirsimaritin, cirsiliol, apigenin, genkwanim, eridictyol, hispidulin, quercetin, nepetin, rutin and eupatorin in *B. genistelloides*, and nobiletin, tangeretin, kaempferol, and naringenin in *B. illinita*. In the same way, only chlorogenic acid was reported in *B. oxyodonta* (Toyama et al. 2014).

B. pentladii, *B. retusa*, and *B. spicata* have revealed several flavonoids in their chemical composition: *B. pentladii*: 5,4′- dihydroxy-6,7,8,3′-tetramethoxyflavone, 8-Methoxycirsilineol, 5,4′-dihydroxy-6,7,8-trimethoxyflavone, xanthomicrol, 5,3′,4′-trihydroxy-6,7,8-trimethoxyflavone and sideritoflavone (Tarqui et al. 2012); *B. retusa*: sakuranetin, 5,6,7- trihydroxy-4′-methoxyflavanone and naringenin (Grecco et al. 2010b, 2012); and rutin in *B. spicata* (Agudelo et al. 2016). In *B. retusa* it was also found (7E, 18′Z)-hexacos- 18′-enyl coumarate and (7Z,18′Z)-hexacos-18′-enyl coumarate; and in *B. spicata* 3,5-dichlorogenic acid, 3,4-dichlorogenic acid, and 4,5-dichlorogenic acid (Agudelo et al. 2016; Ueno et al. 2018).

Regarding *B. trinervis*, five flavonoids were detected: rutin, luteolin, 5,7-Dihydroxy-6,4′-dimethoxyflavone, 5-Hydroxy-6,7,4′- trimethoxyflavone and 5,4′-dihydroxy-3,6,7-trimethoxyflavone, as well as caffeic, ellagic, and rosmarinic acids (Sharp et al. 2001; Jaramillo-García et al. 2018). Several compounds were identified in *B. uncinella* as well, including caffeic acid, ferulic acid, pectolinaringen, hispidulin, and dihydrooroxylin (Grecco et al. 2010a; Bocco et al. 2016). In *B. incarum*, chlorogenic acid, 3′,4′,5,7-tetrahydroxyflavone, dicaffeoyl quinic acid and 3′,4′,5,7-tetrahydroxy-3,6-dimethoxyflavone, 3′,4′,5,7-tetrahydroxy-3,6,8-trimethoxyflavone, 4′,5,7- trihydroxy-3′,3,6,8-tetramethoxyflavone, 4′,5-dihydroxy-3′,3,6,7,8-pentamethoxyflavone, chlorogenic acid, dicaffeoylquinic acid, and quercetin diglycoside were reported, as well (Zampini et al. 2009). The chemical compounds found in several *Baccharis species* and the biological activities displayed by these plant extracts are shown in Table 12.2.

By comparing the phenolic compounds present in different *Baccharis* species, it is possible to observe that some compounds occur in several species, like

Table 12.2 Chemical composition of *Baccharis* species and biological effects displayed by their extracts

Baccharis species	Phenolic compounds	Biological activity	References
Baccharis articulata	Chlorogenic acid, luteolin, acacetin, quercetin	Antidiabetic	Cariddi et al. (2012), Borgo et al. (2010) and Kappel et al. (2012)
Baccharis chilco	5-*O*-[(*E*)-caffeoyl]quinic acid, 3,5-di-*O*-[(*E*)- caffeoyl-quinic acid, rosmarinic acid	Antioxidant	Argoti et al. (2013)
Baccharis darwinii	Anisocoumarin, 7-geranyloxycoumarin, diversinin	Antifungal	Kurdelas et al. (2010)
Baccharis dentata	Caffeic acid, rutin, quercetin, apigenin, kaempferol	Antioxidant Antibacterial	Sartor et al. (2013)
Baccharis dracunculifolia	Baccharin, artepillin C, drupanin, *p*-coumaric acid, ferulic acid, *trans*-cinnamic acid, chlorogenic acid, caffeic acid, kaempferol, kaempferide, isosakuranetin, pinobanskin, chrysin, aromadendrin-4′*O*-methyl ether, 11-hydroxy-10,11-dihydro-euparin, 6- hydroxy-tremetone, dihydrocumaric acid, 2,2-dimethyl-6-carboxyethenyl-2H-1-benzopyran acid, acacetin, ermanin, viscidone, protocatechuic acid, sinapic acid, (*E*)-3-(*E*)-3-hydroxy-3-methyl-1--butenyl-4-(2,3-dihydrocinnamoyloxy)-cinnamic acid, 3,4-di-*O*-caffeoylquinic acid, 3,5-di-*O*-caffeoylquinic acid	Gastroprotective Antifungal Antibacterial Anti-inflammatory Antinociceptive Antiparasitic Antiobesity Insecticidal Antioxidant Cytotoxic Hepatoprotective	Li et al. (2007), Lemos et al. (2007), Missima et al. (2007), da Silva Filho et al. (2008, 2009), dos Santos et al. (2010), De Sousa et al. (2011), Hocayen et al. (2016), Da Silva et al. (2017), Paula et al. (2017) Guimarães et al. (2012), Szliszka et al. (2012), Rezende et al. (2014) and Abad and Bermejo (2006)
Baccharis genistelloides	Eupatrin, cirsimaritin, cirsiliol, apigenin, genkwanin, eriodictyol, Hispidulin, quercetin, luteolin, nepetin, rutin, eupatorin	Anti-arthritic	Coelho et al. (2004), Prasad et al. (2009), Abad and Bermejo (2006) and Hennig et al. (2011)
Baccharis illinita	Luteolin, nobiletin, tangeretin, kaempferol, naringenin	Gastroprotective Antinociceptive Anti-inflammatory	Baggio et al. (2003), Freitas et al. (2008, 2009) and Boller et al. (2010)

(continued)

Table 12.2 (continued)

Baccharis species	Phenolic compounds	Biological activity	References
Baccharis incarum	Chlorogenic acid, 3′,4′,5,7-tetrahydroxyflavone, dicaffeoyl quinic acid, 3′,4′,5,7-tetrahydroxy-3,6-dimethoxyflavone, 3′,4′,5,7-tetrahydroxy-3,6,8-trimethoxy flavone, 4′,5,7-trihydroxy-3′,3,6,8-tetramethoxyflavone, 4′,5-dihydroxy-3′,3,6,7,8-pentamethoxyflavone dicaffeoylquinic acid, quercetin diglycoside	Antioxidant, antimicrobial	Zampini et al. (2009) and Nuño et al. (2012)
Baccharis oxyodonta	Chlorogenic acid	Snake poisoning treatment inflammation-induced by secretory PLA2	Toyama et al. (2014)
Baccharis pentladii	5,4′- Dihydroxy-6,7,8,3′-tetramethoxyflavone 8-Methoxycirsilineol, 5,4′-dihydroxy-6,7,8-trimethoxyflavone Xanthomicrol, 5,3′,4′-trihydroxy-6,7,8-trimethoxyflavone Sideritoflavone	Anti-inflammatory	Tarqui et al. (2012) and Abad et al. (2006)
Baccharis retusa	Sakuranetin, (7E, 18′Z)-hexacos- 18′-enyl coumarate, (7Z,18′Z)-hexacos-18′-enyl coumarate, 5,6,7- trihydroxy-4′-methoxyflavanone, naringenin	Anti-emphysema Antiparasitic	Taguchi et al. (2015), Ueno et al. (2018) and Grecco et al. (2010b, 2012)
Baccharis spicata	Rutin, chlorogenic acid, and 3, 5 dichlorogenic acid, 3, 4 dichlorogenic acids, 4, 5 dichlorogenic acid	Antioxidant	De Oliveira et al. (2004) and Agudelo et al. (2016)
Baccharis trimera	5'-*O*-caffeoylquinic acid, 4-*O*-[*E*]-caffeoyl-1-methyl-quinic acid, 1′-5′-*O*-dicaffeoylquinic acid, 1,3-di-[*E*]-caffeoylquinic acid, Eupafolin, hispidulin 5-*O*-[*E*]-caffeoylquinic acid, 3,4-*O*-[*E*]-dicaffeoylquinic acid, 3,5-*O*-[*E*]-dicaffeoylquinic acid, 4,5-*O*-[*E*]-dicaffeoylquinic acid, tricaffeoylquinic acid, quercetin, luteolin, apigenin	Antacid Antiulcer Antioxidant Anti-inflammatory Antiparasitic Antiobesity Gastroprotective Antihepatotoxic Anti-alcoholic fatty liver disease	Biondo et al. (2011), de Araújo et al. (2017), De Oliveira et al. (2012a, b, 2014), Do Nascimento et al. (2017), dos Reis Lívero et al. (2016a, b), Herrerias et al. (2010), Aboy et al. (2012) and Soicke and Leng-Peschlow (1987)

(continued)

Table 12.2 (continued)

Baccharis species	Phenolic compounds	Biological activity	References
Baccharis trinervis	5,7-Dihydroxy-6,4′-dimethoxyflavone, 5-Hydroxy-6,7,4′-trimethoxyflavone, 5,4′-Dihydroxy-3,6,7-trimethoxyflavone,caffeic acid, ellagic acid, rosmarinic acid, rutin, luteolin	Antivirus	Sharp et al. (2001), Jaramillo-García et al. (2018) and Sanchez Palomino et al. (2002)
Baccharis uncinella	Caffeic acid, ferulic acid, pectolinaringen, hispidulin, dihydrooroxylin	Metabolic syndrome treatment Antiparasitic Anti-inflammatory	Bocco et al. (2016), Passero et al. (2011), Grecco et al. (2010a) and Zalewski et al. (2011)

chlorogenic acid, which is found in *B. spicata*, *B. dracunculifolia*, *B. articulata*, *B. oxyodonta*, and *B. incarum*. In the same way, luteolin and quercetin were identified in *B. trimera*, *B. articulata*, *B. dentata*, and *B. genistelloides*. Hispidulin, apigenin, and kaempferol are also described in different *Baccharis* species: hispidulin in *B. uncinella*, *B. genistelloides*, *B. trimera* and *B. uncinella*; apigenin in *B. trimera*, *B. dentata*, and *B. genistelloides*; and kaempferol in *B. illinita*, *B. dracunculifolia*, and *B. articulata*. Caffeic acid is reported in more than three *Baccharis* species, as well. These compounds, besides occurring in several *Baccharis* species, are widespread through the plant kingdom and are reported in high amounts, especially in edible plants. For example, chlorogenic acid, a hydroxycinnamic acid, is found in many sources, including coffee, apples, pears, berries, aubergines, etc. (Olthof et al. 2001). The flavonoids quercetin, kaempferol, luteolin, and apigenin are found in vegetables and fruits as well and, in most cases, are detected as a glycoside, although kaempferol, luteolin, and apigenin may be found as aglycones too (Miean and Mohamed 2001). Flavonoids and phenolic compounds usually display different biological effects, mainly antioxidants (De Oliveira et al. 2012a, b). *Baccharis dracunculifolia* also contains prenylated compounds, such as artepillin C, baccharin, and drupanin that are usually found in good amounts only in this plant species. Some of these compounds are important, because they are related to several biological effects, such as gastroprotective, anti-inflammatory, and cytotoxic against cancer cell lines (Lemos et al. 2007; Paulino et al. 2008; De Sousa et al. 2011; Szliszka et al. 2012; Costa et al. 2018). The chemical structures of these compounds, along with some flavonoids and other phenolics widespread in the *Baccharis* genus, are displayed in Fig. 12.1.

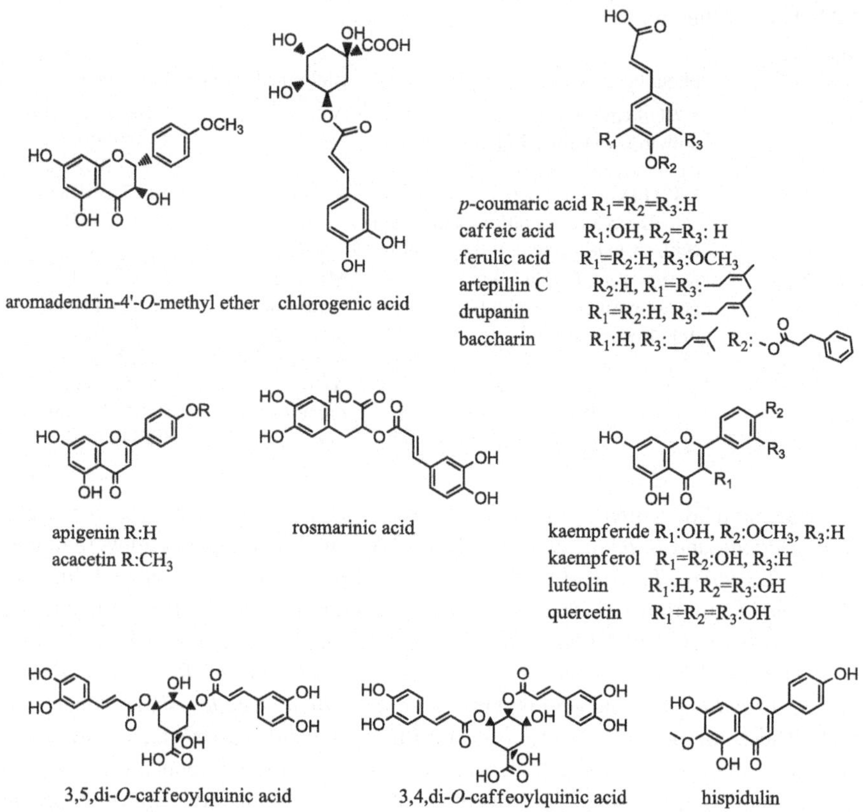

Fig. 12.1 Some phenolic compounds found in *Baccharis* spp.

3 Biological Activities

In the past few decades, numerous scientific publications have reported the biological activities displayed by *Baccharis* spp., and several of their isolated phenolic compounds for the treatment of miscellaneous diseases. Examples are the anti-angiogenesis effect displayed by artepillin C (Ahn et al. 2007), the insecticidal by *B. dracunculifolia* ethanolic extract (Da Silva et al. 2017), hepatoprotective by *B. dracunculifolia* leaves extract (Rezende et al. 2014), *B. trimera* and some isolated phenolic compounds (Soicke and Leng-Peschlow 1987; Lívero et al. 2016a, b), the anti-emphysema by sakuranetin, a flavonoid isolated from *B. retusa* (Taguchi et al. 2015), and antiarthritic by *B. genistelloides* aqueous extract (Coelho et al. 2004). Another activity reported is the antivirus of *B. trinervis* aqueous extract (Palomino et al. 2002). Taking that into account, in this chapter, some of the important biological effects displayed by *Baccharis* spp. extracts and its isolated phenolic compounds are discussed. The biological effects and chemical composition of several *Baccharis* species are displayed in Table 12.2.

Gastroprotective

Several *Baccharis* species have been described in folk medicine regarding their effects in the treatment of gastrointestinal disorders, including ulcer. Therefore, many researchers have evaluated the pharmacological potential of plants belonging to this genus: Baggio et al. (2003) found that the stems and leaves ethanolic extracts and the roots and flowers aqueous extracts of *B. illinita* by the oral route were able to furnish gastroprotection for lesions caused by ethanol from 0.3 to 1.0 g/kg of body-weight. These authors also found that the hydroalcoholic extracts did not show toxic effects up to 6 g/kg of body weight, which indicates the pharmacological potential of this plant. Freitas et al. (2008), considering the previous report of the gastroprotective effect of *B. illinita* extracts and the importance of determining its mechanism of action, evaluated the possible pathways that *B. illinita* flowers chloroform extract could be acting. Their studies indicated that this effect occurs by decreasing the gastric secretion through inhibition of the H+/K+ ATPase, and consequently reducing the acid secretion. Besides this, the flavonoid luteolin, present in this extract, was able to act on H+/K+ ATPase, as well.

The hydroalcoholic extract of *B. dracunculifolia*, the botanical source of Brazilian green propolis, displays a significant gastroprotective activity too: from 50 to 500 mg/kg of body-weight, the extract decreased in a significant manner the gastric lesion in comparison with the negative control, by reducing the gastric juice volume and increasing the stomach pH. The major compounds found in this extract were caffeic acid, ferulic acid, *p*-coumaric acid, aromadendrin-4′-*O*-methyl-ether, isosakuranetin, baccharin, and artepillin C (Lemos et al. 2007). Therefore, this effect is probably due to the presence of these phenolic compounds in the extract. To prove this hypothesis, Barros et al. (2008) assessed the potential of caffeic, ferulic, *p*-coumaric, and cinnamic acids for ulcer treatment and in furnishing gastroprotection. They found that these compounds from 50 to 250 mg/kg were able to decrease the gastric lesion caused by NSAID, ethanol, and stress by diminishing the gastric juice and increasing the stomach pH, despite being less effective than the positive control omeprazole. On the other hand, these authors have demonstrated that up to 2000 mg/kg, these compounds were not toxic. Artepillin C, drupanin and the flavonoids aromadendrin-4′-*O*-methyl ether and kaempferide also display gastroprotective activity: at doses from 0.3 to 3 mg/kg, these phytochemicals displayed an antiulcer effect and were capable of preventing ulcer induced by ethanol/HCl or indomethacin. They promote this effect by different mechanisms of actions (Costa et al. 2018). These phenolic compounds showed a potent gastroprotective effect considering that at lower doses (0.3–3 mg/kg), they displayed the pharmacological effect, whereas the positive control (omeprazole and carbenoloxone) doses were considerably higher: 20 and 200 mg/kg, respectively. It shows that these compounds indeed contribute to the gastroprotective effect of *B. dracunculifolia* extracts. Therefore, these compounds and, *B. dracunculifolia* extracts and Brazilian green propolis present pharmacological potential for the possible development of novel phytomedicines and phytotherapeutic agents.

B. trimera has potential for the treatment of gastrointestinal disorders as well: its aqueous extract displays antacid and antiulcer effects at 1 and 2 g/kg by reducing the gastric volume and acid secretion, as well as it shows protection against lesion caused by restraint at 4 °C. It was found that the extract is composed of chlorogenic acid, flavonoids, and compounds from other secondary metabolite classes, such as ent-clerodane diterpene and a dilactonic neo-clerodane diterpene. The main mechanism of action of *B. trimera* extract is by modifying the cholinergic pathway (Biondo et al. 2011). The hydroalcoholic extract of this plant was also evaluated: when orally administered, it decreased the stomach lesion and oxidative stress, and yet stimulated the healing effect on chronic ulcer. It acts by inhibiting free radicals production and, therefore, lipid peroxidation too. This extract has flavonoids and caffeoylquinic acid derivatives in its chemical composition, which may be responsible for at least part of the curative and protective effect of ulcer lesions (Lívero et al. 2016a, b).

Cytotoxic

Considering the importance of cancer, several researchers have been assessing natural products against cancer cells aiming to discover novel drugs or plants that have cytotoxic potential without significant side effects: *B. dracunculifolia* hydroalcoholic extract displayed a GI_{50} value (50% of cells growth inhibition) of 5.5 μg/mL in human prostate's primary malignant tumor cell lines. It also promoted a decrease in the cell viability of prostate's metastasis cells, although higher doses were necessary to achieve optimum efficacy. This extract displays this effect probably by affecting the cell's S phase arrest, and by regulating the expression of cyclins D1, CDK4, and B1 (Li et al. 2007). Artepillin C, one of the major compounds from *B. dracunculifolia*, possibly contributes to the cytotoxicity of this plant against cancer prostate cells and is capable of sensitizing the tumor necrosis factor–related apoptosis-inducing ligand, an important pathway of apoptosis in cancer cells (Szliszka et al. 2012).

Therefore, considering that promising extracts are the ones that show IC_{50} values (or ED_{50}/GI_{50}) lower than 20 μg/mL (Vijayarathna and Sasidharan 2012), *B. dracunculifolia* extract should be further studied for the development of new phytotherapeutic medicines to treat cancer. Furthermore, *Baccharis*' phenolic compounds, such as artepillin C, display anticancer potential as well.

Antiparasitic

In an attempt to overcome the problem of low investments of the pharmaceutical industry in the development of innovative drugs to treat neglected diseases, like the ones caused by parasites, several researchers, especially from universities, have evaluated the potential of plants against many parasites. Da Silva Filho et al. (2009) found that *B. dracunculifolia* dichloromethanic extract displays IC_{50} values of 45 μg/mL against *Leishmania donovani* and approximately 20 μg/mL against *Plasmodium falciparum*. *B. trimera* showed significant in vivo and in vitro effects against the juvenile and adult *Schistosoma mansoni* worms: dichloromethane and aqueous fraction at 130 μg/mL inhibited 100% of the female's oviposition and induced the death of *S. mansoni* worms by many morphological changes. In vivo, the samples at 40 mg/kg decreased by 75% (aqueous fraction) and 68% (dichloromethane fraction) the juvenile female worms, and by almost 100% the eggs in the feces. Studies like this are relevant due to the resistance, the numerous side effects, and low efficacy against the juvenile *Schistosoma* spp. of the drugs currently in the market, such as praziquantel. *B. trimera*. It is also a promising plant to be further studied regarding its antiparasitic effect (De Oliveira et al. 2014). Likewise, the aqueous extract of this species is effective against *Rhipicephalus microplus*, an ectoparasite that causes anemia and is responsible for the transmission of lethal diseases in cattle. Besides, the resistance to synthetic acaricides has been increasing, and *B. trimera* leaves aqueous extract at 150 and 200 mg/mL was able to reduce 100% of *R. microplus* egg hatching. Thus, it could be a new approach to the discovery of new acaricidal agents (Lázaro et al. 2013).

Regarding the antileishmanial and trypanocidal potential of *Baccharis* spp., the flavonoid 5,6,7- trihydroxy-4′-methoxyflavanone from *B. retusa* leaves methanolic extract was isolated, which displays a significant antiparasitic effect: the IC_{50} value against *T. cruzi* trypomastigotes found was 20.39 μg/mL, while benznidazole's was 47.54 μg/mL. Therefore, this flavonoid is more potent than benznidazole, a standard drug currently used to treat Chagas disease (Grecco et al. 2010b). Passero et al. (2011) evaluated the effect of caffeic acid and the flavonoid pectolineragenin against *L. amazonensis* and *L. braziliensis*, which displayed IC_{50} values of 190 ng/μL for caffeic acid against *L. amazonensis* promastigotes and 110 μg/μL for pectolineragenin against *L. braziliensis* promastigotes. Although, since amphotericin B has IC_{50} of 0.30 and 0.07 ng/μL against *L. amazonensis* and *L. braziliensis*, respectively, the concentrations of *Baccharis* isolated compounds able to inhibit the growth of 50% of the parasites are considerably high in comparison with the positive control.

Anti-inflammatory and Antinociceptive

Several reports have described the anti-inflammatory and antinociceptive potential of *Baccharis* spp. and their isolated phenolic compounds: the hexane, hydroalcoholic and aqueous fractions of *B. illinita* aerial parts decreased the nociceptive response in vivo at doses of 30–1000 mg/kg, presenting a dose-related response (Freitas et al. 2009). Moreover, *B. illinita* leaves crude extract is a topical anti-inflammatory agent capable of inhibiting ear edema caused by 12-*O*-tetradecanoilforbol acetate and arachidonic acid. It decreased the polymorphonuclear cells migration as well, showing a similar effect in comparison with dexamethasone (Boller et al. 2010).

B. dracunculifolia, which contains caffeic acid, *p*-coumaric acid, aromadendrin-4′-O-methyl ether, drupanin, artepillin C, and 2,2-dimethyl-6-carboxyethenyl-2H-1-benzopyran as major compounds, display in vivo anti-inflammatory and antinociceptive effects at doses ranging from 50 to 400 mg/kg of its leaves hydroalcoholic extract. It reduced the number of abdominal constrictions caused by acetic acid, glutamate of complete Freund adjuvant, and decreased the nociceptive induced by formalin. Moreover, it was effective as anti-hypernociceptive in the acute inflammation pain caused by carrageenan and inhibited the enzyme COX-2 (dos Santos et al. 2010). Therefore, considering that many phenolic compounds are found in *B. dracunculifolia* hydroalcoholic extract, they probably contribute in a significant way to the anti-inflammatory and antinociceptive effects displayed by this extract.

B. dracunculifolia ethyl acetate extract, which shows a similar chemical profile (major compounds: baccharin, artepillin C, drupanin, caffeic acid, *p*-coumaric acid and aromadendrin-4-O-methyl ether), displays intestinal anti-inflammatory activity: from 5 to 50 mg/kg, it reduced significantly the ulcerative colitis caused by trinitrobenzenesulfonic acid by avoiding glutathione depletion, inhibiting lipid peroxidation, and decreasing myeloperoxidase effect (Cestari et al. 2011). Paulino et al. (2008) reported the in vivo anti-inflammatory effect of artepillin C, which inhibited the paw edema by 38% at 10 mg/kg. At 1 mg/kg, it displayed a similar effect in comparison with the positive control, indomethacin, at 1 mg/kg. Its activity comes from reducing the neutrophils numbers and prostaglandin E2. In vitro, this compound reduces nitric oxide generation and NF-kβ. Besides this, these authors found that artepillin C is orally absorbed in vivo, which is important to biological activity. Therefore, artepillin C may be considered a promising anti-inflammatory natural compound and, since artepillin C is one of the phenolic compounds present in high amounts in *B. dracunculifolia*, it corroborates with the hypothesis that the phenolic compounds found in *Baccharis* spp. have a relevant contribution to this biological effect. To increase the potency of *B. dracunculifolia* leaves hydroalcoholic extract, low diameter and biocompatible liposomes with the sample were developed: the free extract had reduced the swelling, the leucocytes and neutrophil migration; and the levels of TNF-α and interleukins 6 and 1β. The liposomes containing the extract increased the anti-inflammatory activity in vivo by reducing the effective dose by almost six times. Caffeic acid liposomes also had their anti-inflammatory effect improved (de Figueiredo-Rinhel et al. 2018).

A phenolic fraction of *B. trimera* aerial parts ethanolic extract at 15 mg/kg reduced the acute inflammation caused by carrageenan in comparison with the negative control (De Oliveira et al. 2012a, b). *B. trimera* aqueous extract also displayed a significant anti-inflammatory effect on carrageenan-induced edema at intraperitoneal doses of 400 and 800 mg/kg, respectively. This extract promoted a decrease in inflammatory parameters, such as cell migration, edema, polymorphonuclear leukocytes, and proteins (Paul et al. 2009). Although the aqueous extract was able to act as an anti-inflammatory agent, its doses are considerably higher than the phenolic fraction of the ethanolic extract. It probably demonstrates that the phenolic fraction has more active and/or potent compounds than the aqueous extract, which may contain several compounds with no anti-inflammatory effect, such as sugars, for example. Likewise, most of *B. pentladii, B. obtusifolia, B. latifolia, and B. subulata* extracts (hexanic, dichloromethanic, ethanolic and aqueous) in concentrations ranging from 12.5 to 200 μg/mL were capable of reducing the inflammatory parameters as well, such as COX-2, nitric oxide and TNF-α production (Abad et al. 2006).

Antidiabetic and Antiobesity

Diabetes is a disorder that affects millions of people worldwide and is caused by low production or resistance to insulin, consequently increasing blood glucose levels. Because of that, the body organs and tissues like the liver and muscle cannot use glucose or store it in glycogen form. Once it becomes a chronic disease, several other consequences take place, such as damage to the eyes, blood vessels, and many other body parts. *B. articulata* butanolic fraction of the crude extract, which has mainly flavonoids, when administered orally in vivo, was able to improve the insulin production, with an effect similar to glipizide, a standard drug; and the liver and muscle glycogen levels were increased (Kappel et al. 2012). Artepillin C, one of the major prenylated phenolic compounds from *B. dracunculifolia* and Brazilian green propolis, shows high affinity to the nuclear receptor peroxisome proliferator-activated receptor, known as PPAR, and by activating this receptor, genes like aP2, adiponectin, and glucose transporter are expressed, increasing the body response to insulin in type 2 diabetes. Besides, it stimulated adipocytes differentiation, increasing the glucose uptake by the mature adipocytes (Choi et al. 2011). A clinical trial using 16 healthy people aged around 20-year-old was performed with *B. dracunculifolia* extracts at 20 mg/kg. The sample intake led to a 25% decrease in glucose blood content and no significant alterations in cardiovascular parameters, such as blood pressure and heart rate. Since the extracts have phenolic compounds as the ones in higher concentrations, they can probably be associated with the antidiabetic effect of this plant (Oliveira et al. 2014). The activity on glucose homeostasis and on insulin regulation reflects the antidiabetic potential of *Baccharis* spp. and the phenolic compounds isolated from them. Therefore, they can possibly be sources for new antidiabetic drugs or for the development of new phytotherapeutic medicines.

Considering that obesity and type 2 diabetes are related, the discovery of new agents anti-obesity is important, and *B. dracunculifolia* extract, after oral administration to rats, induced the secretion of serum insulin at 30% in obese rats (Hocayen et al. 2016). *B. trimera* aqueous extract showed potential in treating obesity by reducing the lipids and adipogenic transcriptional factors by 90% (Do Nascimento et al. 2017). The methanol extract of this plant also inhibited the enzyme pancreatic lipase by 78%, which is responsible for hydrolyzing triacylglycerols. Therefore, its inhibition contributes to the antiobesity effect of *B. trimera* (de Souza et al. 2011). Caffeic and ferulic acids, found in many *Baccharis* species, including in *B. uncinella* aerial parts, in a mixture containing the two of them, showed to be effective in decreasing biochemical parameters associated with obesity, like hyperglycemia, high cholesterol, and triglycerides levels and avoided the gain of body weight (Bocco et al. 2016).

These reports show that *Baccharis* spp. and many of their secondary metabolites can modulate biochemical parameters related to obesity and/or diabetes and, therefore, represent a possible plant material source of bioactive compounds. These plants could be used for the development of functional foods and/or herbal medicines too.

Antimicrobial and Antifungal

Due to the increasing resistance of bacteria and fungi strains to the antibiotics currently in the market, compounds isolated from plants and the plant extracts have been evaluated for their antimicrobial and antifungal effects: diversinin, a coumarin isolated from *B. darwinii* at 15.6 μg/mL, inhibited *Microsporum gypseum*, *Trichophyton rubrum*, and *Trichophyton mentagrophytes* strains. Although the MIC value of diversinin is not high, the isolated compounds considered to be promising are the ones with MIC below 10 μg/mL, and in the case of extracts, the ones with MIC below 100 μg/mL (Kurdelas et al. 2010). Da Silva Filho et al. (2008) described the antifungal and antimicrobial effect of *B. dracunculifolia* leaves extract in comparison with Brazilian green propolis, which is made by bees from *B. dracunculifolia*: the leaf extract showed IC_{50} values of 65 μg/mL for *C. krusei* and 40 μg/mL for *C. neoformans*, while propolis extract displayed better effects, with IC_{50} of 9 μg/mL for *C. krusei*. It shows that although the bioactive compounds are present in the *B. dracunculifolia* extract, they are probably more concentrated in the propolis, increasing its potency.

Taking into account the antibiotic resistance, Nuño et al. (2012) tested *B. incarum* extracts at ethanol 60% and 80% against clinic isolated methicillin-resistant *S. aureus* and *E. faecalis*, and found that their MIC values were promising, ranging from 40 to 80 μg GAE/mL. Therefore, using the 60% tincture, a topical formulation was developed, which also displayed an antimicrobial effect. The in vitro drug-releasing experiments revealed that the phenolic compounds chlorogenic acid and 4′,5-dihydroxy-3′,3,6,7,8-pentamethoxyflavone were the major ones found in the receptor solution, showing that probably, these compounds were responsible for the biological effect. Therefore, this formulation, after additional studies, may be useful for the development of new anti-acne agents or for topical treatment of tissues infected by *Propionibacterium acnes*, as well as by *S. aureus* and/or *E. faecalis* methicillin resistant.

4 Toxicology

In folk medicine, usually many people think that all natural products are safe for consumption, although several studies show that it is not true. For example, *B. pteronioides* contains the toxic compound trichothecenes (Stegelmeier et al. 2009). Therefore, it is important to assess the toxicity of plants before stating their safety. Da Silva et al. (2016) evaluated the acute toxicity of *B. trimera* tinctures in wistar rats by administration of a single dose of 2000 mg/kg and found that it was not able to induce significant hematological or biochemical changes (lipid peroxidation, δ-aminolevulinate dehydratase, and catalase); neither showed other signs of toxicity nor increased the animals mortality. The subchronic toxicity was assessed by oral administration of the sample at 100, 200, and 400 mg/kg for 28 consecutive days. *B. trimera* tincture, besides not showing any toxicity, decreased the liver enzymes alanine and aspartate aminotransferases, which are related to hepatic cells damage. Therefore, in the subchronic treatment, it additionally promoted a hepatoprotective effect. Considering that *B. trimera* contains many phenolic compounds in its chemical composition, such as gallic acid, ellagic acid, rutin, quercitrin, and quercetin, they are probably not toxic and responsible for the hepatoprotective activity. Therefore, *B. trimera* can probably be considered safe at these doses.

The aqueous extract of *B. genistelloides* at doses of 4.2 and 42 mg/kg, after 37 days of oral treatment, did not show genotoxicity to liver and kidney. Moreover, it did not induce alterations in the aspects of kidneys, liver, and lungs, like color and weight. However, it reduced the body and thymus weights and glucose and triglyceride levels. Therefore, this extract is not toxic at these doses and displays hypoglycemic and hypotriglyceridemic effects (Coelho et al. 2004). On the other hand, *B. dracunculifolia* aqueous extract at 500, 1000, and 2000 mg/kg for 3 days induced genotoxic and mutagenic effects by increasing blood and liver DNA damage and the frequency of micronucleus in bone marrow (Rodrigues et al. 2009). On the other hand, one of the major compounds found in *B. dracunculifolia*, artepillin C at 0.4, 0.8, and 1.6 mg/kg, was not genotoxic and presented a protective effect against liver cells DNA damage induced by methyl methanesulfonate (Monteiro Neto et al. 2011). In a similar way, baccharin at 0.12, 0.24, and 0.48 mg/kg was also able to reduce DNA damage in liver cells and the frequency of micronucleated polychromatic erythrocytes in mice (Oliveira et al. 2011). Caffeic, cinnamic, and ferulic acids, three major compounds from *B. dracunculifolia*, increased the frequency of micronucleated cells, demonstrating a clastogenic effect of these compounds, despite presenting a not genotoxic effect to rat hepatoma tissue cells (Maistro et al. 2011). Therefore, these compounds, probably more concentrated in the aqueous extract, may have contributed to the mutagenic effect in vivo found by Rodrigues et al. (2009). On the other hand, artepillin C and baccharin were not genotoxic in these tested doses and are more likely to be found in the ethanolic extract.

5 Chromatographic Analyses

The pharmacological potential of plant extracts is usually related to their chemical composition, since the presence of bioactive compounds and their concentrations are the most important parameters for displaying the biological effect. To assure plant extracts quality, many analytical methods were developed for the chromatographic analyses of *Baccharis* extracts: by using thin-layer chromatography; De Oliveira et al. (2006) reported a simple method to differentiate *B. articulata, B. cylindrica, B. spicata, B. trimera, and B. usterii*, by using the aqueous extract of the leaves and its butyl alcohol fraction, obtained by the partition of the aqueous extract. The samples are applied on silica gel plates of 20 × 20, and the chromatographic elution is performed using chloroform:ethanol:acetic acid in a proportion of 60:40:6 v/v as mobile phase. The detection is undertaken by using two colorimetric reagents: anisaldehyde: H_2SO_4 plus heating to 100 °C and diphenylboryloxyethylamine 1% methanol, PEG 400 (5% w/v). After spraying the colorimetric reagent on the plates, they were observed under long-wave UV and visible lights. According to the chromatographic profile, considering the retention factor, number and color of the spots on silica plates, these *Baccharis* species can be differentiated. Lonni et al. (2003), by using HPLC coupled to a photodiode array detector and chemometric, were able to differentiate between different *Baccharis* species as well. By using as stationary phase a C18 column, methanol as mobile phase, and detection at 254 nm, the obtained chromatographic profile of the ethanolic extracts allowed distinguishing among *B. genistelloides var. trimera*, *B. milleflora*, and *B. articulata*.

A validated RP-HPLC method using a C-18 reversed-phase column and a gradient consisting of acidified water and acetonitrile was developed to perform analyses of 5-O-[E]-caffeoylquinic acid, 3,4-O-[E]-dicaffeoylquinic acid, 3,5-O-[E]-dicaffeoylquinic acid, 4,5-O-[E]-dicaffeoylquinic acid, and a tricaffeoylquinic acid in *B. trimera* hydroalcoholic extracts. The quantification of these compounds is important, because they are related to the digestive effect of *B. trimera*. The limits of quantification for the compounds were below 12.5 μg/mL and the method presented suitable selectivity, linearity, robustness, precision, and recovery according to ICH validation guidelines (Aboy et al. 2012).

Regarding *B. dracunculifolia*, aiming at proving the botanical origin of Brazilian green propolis, Kumazawa et al. (2003), by using liquid chromatography coupled to a mass spectrometer, compared the chemical constituents identified in both samples and concluded that there was no significant difference in the chemical composition of *B. dracunculifolia* and Brazilian green propolis. Although qualitative methods are important to determine each compound present in the samples, the quantitative analysis is also relevant to perform, because the biological effect of the plant extracts usually relies on both the presence and amount of bioactive compounds. Therefore, for the development of phytotherapeutic agents, analytical methods able to perform both qualitative and quantitative analyses are necessary. Taking that into account, de Sousa et al. (2009) developed and validated a reversed-phase HPLC method to analyze 10 phenolic compounds in *B. dracunculifolia*: caffeic acid, coumaric acid, ferulic acid, cinnamic acid, aromadendrin-4-*O*-methyl ether, isosakuranetin, drupanin, artepillin C, baccharin, and 2,2-dimethyl-6- carboxyethenyl-2H-1-benzopyran acid. The method parameters adjusted were: stationary phase a C18 column, mobile phase: nonlinear gradient of acetonitrile and water with mobile phase modifiers, and since the standards are all phenolic compounds, the wavelength of detection was set at 280 nm. Considering that it is a validated method, it presented selectivity, linearity, accuracy, precision, and robustness.

Some analytical methods to perform *Baccharis* spp. phenolic compounds quality control had been reported in the literature, although they cover mainly qualitative analysis. Only for *B. dracunculifolia*, the most studied *Baccharis* species, was developed a RP-HPLC method able to furnish both qualitative and quantitative results. It is really important from the pharmacological potential point of view to have reliable validated analytical methods to quantify the active compounds in medicinal plants, such as the ones belonging to *Baccharis* genus.

6 Conclusion

Baccharis spp. are used in folk medicine for the treatment of several diseases, such as gastrointestinal disorders, fever, inflammation, type-2 diabetes, parasitoses, and arthritis and many researchers have been undertaking scientific studies to corroborate many of the plants folk uses. Some examples of diseases that *Baccharis* spp. have proven effects against are diabetes, obesity, gastric ulcer, parasites, bacteria, fungi, arthritis, and inflammation, among others. The mechanisms of action of many extracts and/or their isolated phenolic compounds have been already reported. Analytical methods to perform chromatographic analyses of the plant material have been developed too due to the importance of the presence and amount of bioactive compounds in the samples. Only one validated quantitative method for analyzing phenolic compounds in *Baccharis* species was found, which was developed for *B. dracunculifolia*, the botanical source of Brazilian green propolis. It shows that *Baccharis* quality control field is still lacking analytical methods for the other *Baccharis* species. Even though these plants display important biological effects, some *Baccharis* species have compounds that present some toxicity. Therefore, *Baccharis* spp. may be potential plants for the development of novel phytotherapeutic medicines and/or be sources of bioactive compounds. However, more studies should be performed to validate the pharmacological activities and to better assess their toxicity.

References

Abad MJ, Bermejo P (2006) *Baccharis* (Compositae): a review update. ARKIVOC 2007:76–96

Abad MJ, Bessa AL, Ballarin B et al (2006) Anti-inflammatory activity of four bolivian *Baccharis* species (Compositae). J Ethnopharmacol 103:338–344

Aboy AL, Apel MA, Debenedetti S et al (2012) Assay of caffeoylquinic acids in *Baccharis trimera* by reversed-phase liquid chromatography. J Chromatogr A 1219:147–153

Agudelo IJ, Isolabella SA, Filip R et al (2016) *Baccharis spicata* (Lam) Baill: polyphenol screening, determination of their antioxidant activity and their main polyphenolic metabolites. J Pharmacogn Phytochem 5:278–285

Ahn MR, Kunimasa K, Ohta T et al (2007) Suppression of tumor-induced angiogenesis by Brazilian propolis: major component artepillin C inhibits *in vitro* tube formation and endothelial cell proliferation. Cancer Lett 252:235–243

Argoti JC, Linares-Palomino PJ, Salido S et al (2013) On-line activity screening for radical scavengers from *Baccharis chilco*. Chem Biodivers 10:189–197

Arruda C, Eugênio D de S, Moreira MR et al (2017) Biotransformation of (-)-cubebin by *Aspergillus* spp. into (-)-hinokinin and (-)-parabenzlactone, and their evaluation against oral pathogenic bacteria. Nat Prod Res 6419:1–14

Baggio CH, Freitas CS, Rieck L, Marques MCA (2003) Gastroprotective effects of a crude extract of *Baccharis illinita* DC in rats. Pharmacol Res 47:93–98

Balasundram N, Sundram K, Samman S (2006) Phenolic compounds in plants and agri-industrial by-products: antioxidant activity, occurrence, and potential uses. Food Chem 99:191–203

Barros MP, Lemos M, Maistro EL et al (2008) Evaluation of antiulcer activity of the main phenolic acids found in Brazilian Green Propolis. J Ethnopharmacol 120:372–377

Berretta AA, Arruda C, Miguel FG et al (2017) Functional properties of Brazilian propolis: from chemical composition until the market technology & medicine. In: Waisundara V, Shiomi N (eds) Superfood and functional food – an overview of their processing and utilization. Intech, London, pp 56–98

Biondo TMA, Tanae MM, Coletta ED et al (2011) Antisecretory actions of *Baccharis trimera* (Less.) DC aqueous extract and isolated compounds: analysis of underlying mechanisms. J Ethnopharmacol 136:368–373

Bocco BM, Fernandes GW, Lorena FB et al (2016) Combined treatment with caffeic and ferulic acid from *Baccharis uncinella* C. DC. (Asteraceae) protects against metabolic syndrome in mice. Braz J Med Biol Res 49:3–9

Boller S, Soldi C, Marques MCA et al (2010) Anti-inflammatory effect of crude extract and isolated compounds from *Baccharis illinita* DC in acute skin inflammation. J Ethnopharmacol 130:262–266

Borgo J, Xavier CAG, Moura DJ et al (2010) Influência dos processos de secagem sobre o teor de flavonoides e na atividade antioxidante dos extratos de *Baccharis articulata* (Lam.) pers., Asteraceae. Braz J Pharmacogn 20:12–17

Cariddi L, Escobar F, Sabini C et al (2012) Apoptosis and mutagenicity induction by a characterized aqueous extract of *Baccharis articulata* (Lam.) Pers. (Asteraceae) on normal cells. Food Chem Toxicol 50:155–161

Cestari SH, Bastos JK, Di Stasi LC (2011) Intestinal anti-inflammatory activity of *Baccharis dracunculifolia* in the trinitrobenzenesulphonic acid model of rat colitis. Evidence-Based Complement Altern Med 524349:1–9

Chang R, Piló-Veloso D, Morais SAL, Nascimento EA (2008) Analysis of a brazilian green propolis from *Baccharis dracunculifolia* by HPLC-APCI-MS and GC-MS. Braz J Pharmacogn 18:549–556

Choi SS, Cha BY, Iida K et al (2011) Artepillin C, as a PPARγ ligand, enhances adipocyte differentiation and glucose uptake in 3T3-L1 cells. Biochem Pharmacol 81:925–933

Coelho MGP, Reis PA, Gava VB et al (2004) Anti-arthritic effect and subacute toxicological evaluation of *Baccharis genistelloides* aqueous extract. Toxicol Lett 154:69–80

Costa P, Almeida MO, Lemos M et al (2018) Artepillin C, drupanin, aromadendrin-4′-O-methyl-ether and kaempferide from Brazilian green propolis promote gastroprotective action by diversified mode of action. J Ethnopharmacol 226:82–89

Da Silva Filho AA, de Sousa JPB, Soares S et al (2008) Antimicrobial activity of the extract and isolated compounds from *Baccharis dracunculifolia* DC. (Asteraceae). Z Naturforsch C 63:40–46

Da Silva Filho AA, Resende DO, Fukui MJ et al (2009) *In vitro* antileishmanial, antiplasmodial and cytotoxic activities of phenolics and triterpenoids from *Baccharis dracunculifolia* D. C. (Asteraceae). Fitoterapia 80:478–482

da Silva ARH, Reginato FZ, Guex CG et al (2016) Acute and sub-chronic (28 days) oral toxicity evaluation of tincture *Baccharis trimera* (Less) Backer in male and female rodent animals. Regul Toxicol Pharmacol 74:170–177

Da Silva EM, Roel ARRR, Porto KRA et al (2017) Insecticidal effect of the ethanol extract of *Baccharis dracunculifolia* (Asterales: Asteraceae). Rev Biol Trop 65:517–523

De Araújo GR, Rabelo ACS, Meira JS et al (2017) *Baccharis trimera* inhibits reactive oxygen species production through PKC and down-regulation p47phoxphosphorylation of NADPH oxidase in SK Hep-1 cells. Exp Biol Med 242:333–343

De Figueiredo-Rinhel ASG, de Andrade MF, Landi-Librandi AP et al (2018) Incorporation of *Baccharis dracunculifolia* DC (Asteraceae) leaf extract into phosphatidylcholine-cholesterol liposomes improves its anti-inflammatory effect *in vivo*. Nat Prod Res 6419:1–5

De Oliveira SQ, Dal-Pizzol F, Moreira JCF et al (2004) Antioxidant activity of *Baccharis spicata*, *Baccharis trimera* and *Baccharis usterii*. Acta Farm Bonaer 23:365–368

De Oliveira SQ, Barbon G, Gosmann G, Bordignon S (2006) Differentiation of south Brazilian *Baccharis species* by TLC. J Liq Chromatogr Relat Technol 29:2603–2609

De Oliveira CB, Comunello LN, Lunardelli A et al (2012a) Phenolic enriched extract of *Baccharis trimera* presents anti-inflammatory and antioxidant activities. Molecules 17:1113–1123

De Oliveira RN, Rehder VLG, Santos Oliveira AS et al (2012b) *Schistosoma mansoni*: *in vitro* schistosomicidal activity of essential oil of *Baccharis trimera* (less) DC. Exp Parasitol 132:135–143

De Oliveira RN, Rehder VLG, Oliveira ASS et al (2014) Anthelmintic activity *in vitro* and *in vivo* of *Baccharis trimera* (Less) DC against immature and adult worms of *Schistosoma mansoni*. Exp Parasitol 139:63–72

De Sousa JPB, Leite MF, Jorge RF et al (2011) Seasonality role on the phenolics from cultivated *Baccharis dracunculifolia*. Evidence-Based Complement Altern Med 464289:1–8

De Souza GC, Haas APS, Von Poser GL et al (2004) Ethnopharmacological studies of antimicrobial remedies in the south of Brazil. J Ethnopharmacol 90:135–143

De Souza SP, Pereira LLS, Souza AA, dos Santos CD (2011) Inhibition of pancreatic lipase by extracts of *Baccharis trimera*: evaluation of antinutrients and effect on glycosidases. Braz J Pharmacogn 21:450–455

Do Nascimento DSM, Oliveira RM, Camara RBG et al (2017) *Baccharis trimera* (Less.) DC exhibits an anti-adipogenic effect by inhibiting the expression of proteins involved in adipocyte differentiation. Molecules 22:1–16

Dos Santos DA, Fukui MJ, Nanayakkara DNP et al (2010) Anti-inflammatory and antinociceptive effects of *Baccharis dracunculifolia* DC (Asteraceae) in different experimental models. J Ethnopharmacol 127:543–550

Freitas CS, Baggio CH, Finau J et al (2008) Inhibition of H^+/K^+ ATPase in the gastroprotective effect of *Baccharis illinita* DC. J Pharm Pharmacol 60:1105–1110

Freitas CS, Baggio CH, Dos Santos AC et al (2009) Antinociceptive properties of the hydroalcoholic extract, fractions and compounds obtained from the aerial parts of *Baccharis illinita* DC in Mice. Basic Clin Pharmacol Toxicol 104:285–292

Grecco SS, Gimenes L, Ferreira MJP et al (2010a) Triterpenoids and phenolic derivatives from *Baccharis uncinella* C.DC. (Asteraceae). Biochem Syst Ecol 38:1234–1237

Grecco SS, Reimão JQ, Tempone AG et al (2010b) Isolation of an antileishmanial and antitrypanosomal flavanone from the leaves of *Baccharis retusa* DC. (Asteraceae). Parasitol Res 106:1245–1248

Grecco SS, Reimão JQ, Tempone AG et al (2012) *In vitro* antileishmanial and antitrypanosomal activities of flavanones from *Baccharis retusa* DC. (Asteraceae). Exp Parasitol 130:141–145

Guimarães NSS, Mello JC et al (2012) *Baccharis dracunculifolia*, the main source of green propolis, exhibits potent antioxidant activity and prevents oxidative mitochondrial damage. Food Chem Toxicol 50:1091–1097

Hennig L, Garcia GM, Giannis A, Bussmann RW (2011) New constituents of *Baccharis genistelloides* (Lam.) Pers. ARKIVOC 2011:74–81

Herrerias T, Oliveira AAA, Belem ML et al (2010) Effects of natural flavones on membrane properties and citotoxicity of HeLa cells. Rev Bras Farmacogn 20:403–408

Hocayen PAS, Grassiolli S, Leite NC et al (2016) *Baccharis dracunculifolia* methanol extract enhances glucose-stimulated insulin secretion in pancreatic islets of monosodium glutamate induced-obesity model rats. Pharm Biol 54:1263–1271

Jaramillo-García V, Trindade C, Lima E et al (2018) Chemical characterization and cytotoxic, genotoxic, and mutagenic properties of *Baccharis trinervis* (Lam, Persoon) from Colombia and Brazil. J Ethnopharmacol 213:210–220

Kappel VD, Pereira DF, Cazarolli LH et al (2012) Short and long-term effects of *Baccharis articulata* on glucose homeostasis. Molecules 17:6754–6768

Kumazawa S, Yoned M, Shibata I et al (2003) Direct evidence for the plant origin of Brazilian Propolis by the observation of honeybee behavior and phytochemical analysis. Chem Pharm Bull 51:740–742

Kupchan SM, Jarvis BB, Dailey RG et al (1976) Baccharin, a novel potent antileukemic trichothecene triepoxide from *Baccharis megapotamica*. J Am Chem Soc 98:7092–7093

Kurdelas RR, Lima B, Tapia A et al (2010) Antifungal activity of extracts and prenylated coumarins isolated from *Baccharis darwinii* Hook & Arn. (Asteraceae). Molecules 15:4898–4907

Lázaro SF, Fonseca LD, Martins ER et al (2013) Effect of aqueous extracts of *Baccharis trimera* on development and hatching of *Rhipicephalus microplus* (Acaridae) eggs. Vet Parasitol 194:79–82

Lemos M, De Barros MP, Sousa JPB et al (2007) *Baccharis dracunculifolia*, the main botanical source of Brazilian green propolis, displays antiulcer activity. J Pharm Pharmacol 59:603–608

Li H, Kapur A, Yang JX et al (2007) Antiproliferation of human prostate cancer cells by ethanolic extracts of Brazilian propolis and its botanical origin. Int J Oncol 31:601–606

Lívero FAR, da Silva LM, Ferreira DM et al (2016a) Hydroethanolic extract of *Baccharis trimera* promotes gastroprotection and healing of acute and chronic gastric ulcers induced by ethanol and acetic acid. Arch Pharmacol 389:985–998

Lívero FAR, Martins GG, Telles JEQ et al (2016b) Hydroethanolic extract of *Baccharis trimera* ameliorates alcoholic fatty liver disease in mice. Chem Biol Interact 260:22–32

Lonni ASG, Scarminio IS, Silva LMC, Ferreira DT (2003) Differentiation of species of the *Baccharis* genus by HPLC and chemometric methods. Anal Sci 19:1013–1017

Maistro EL, Angeli JP, Andrade SF, Mantovani MS (2011) *In vitro* genotoxicity assessment of caffeic, cinnamic and ferulic acids. Genet Mol Res 10:1130–1140

Martinez MJA, Bessa AL, Benito PB (2005) Biologically active substances from the genus *Baccharis* L. (Compositae). Stud Nat Prod Chem 30:703–759

Miean KH, Mohamed S (2001) Apigenin content of edible tropical plants. J Agric Food Chem 49:3106–3112

Missima F, Da Silva Filho AA, Nunes GA et al (2007) Effect of *Baccharis dracunculifolia* D.C (Asteraceae) extracts and its isolated compounds on macrophage activation. J Pharm Pharmacol 59:463–468

Monteiro Neto MAB, de Souza Lima IM, Furtado RA et al (2011) Antigenotoxicity of artepillin C *in vivo* evaluated by the micronucleus and comet assays. J Appl Toxicol 31:714–719

Nakajima Y, Shimazawa M, Mishima S, Hara H (2007) Water extract of propolis and its main constituents, caffeoylquinic acid derivatives, exert neuroprotective effects via antioxidant actions. Life Sci 80:370–377

Nuño G, Zampini IC, Ordoñez RM et al (2012) Antioxidant/antibacterial activities of a topical phytopharmaceutical formulation containing a standardized extract of *Baccharis incarum*, an extremophile plant species from argentine Puna. Phyther Res 26:1759–1767

Oliveira PF, Monteiro Neto MAB, Leandro LF et al (2011) *In vivo* antigenotoxicity of baccharin, an important constituent of *Baccharis dracunculifolia* DC (Asteraceae). Basic Clin Pharmacol Toxicol 109:35–41

Oliveira TA, Michel RG, Snak AL et al (2014) *Baccharis dracunculifolia* with high levels of phenol compounds reduces blood glucose in healthy human. Afr J Pharm Pharmacol 8:670–673

Olthof MR, Hollman PCH, Katan MB (2001) Human nutrition and metabolism chlorogenic acid and caffeic acid are absorbed in humans. J Nutr 131:66–71

Palomino SP, Abad MJ, Bedoya LM et al (2002) Screening of south american plants against human immunodeficiency virus: preliminary fractionation of aqueous extract from *Baccharis trinervis*. Biol Pharm Bull 25:1147–1150

Passero LFD, Bonfim-Melo A, Corbett CEP et al (2011) Anti-leishmanial effects of purified compounds from aerial parts of *Baccharis uncinella* C. DC. (Asteraceae). Parasitol Res 108:529–536

Paul EL, Lunardelli A, Caberlon E et al (2009) Anti-inflammatory and immunomodulatory effects of *Baccharis trimera* aqueous extract on induced pleurisy in rats and lymphoproliferation *in vitro*. Inflammation 32:419–425

Paula JT, Sousa IMO, Foglio MA, Cabral FA (2017) Selective fractionation of supercritical extracts from leaves of *Baccharis dracunculifolia*. J Supercrit Fluids 127:62–70

Paulino N, Abreu SRL, Uto Y et al (2008) Anti-inflammatory effects of a bioavailable compound, Artepillin C, in Brazilian propolis. Eur J Pharmacol 587:296–301

Prasad NR, Jeyanthimala K, Ramachandran S (2009) Caffeic acid modulates ultraviolet radiation-B induced oxidative damage in human blood lymphocytes. J Photochem Photobiol B Biol 95:196–203

Rezende TP, Corrêa JODA, Aarestrup BJV et al (2014) Protective effects of *Baccharis dracunculifolia* leaves extract against carbon tetrachloride and acetaminophen-induced hepatotoxicity in experimental animals. Molecules 19:9257–9272

Ribeiro VP, Arruda C, Abd El-Salam M, Bastos JK (2018) Brazilian medicinal plants with corroborated anti-inflammatory activities: a review. Pharm Biol 56:253–268

Rodrigues CRF, Dias JH, Semedo JG et al (2009) Mutagenic and genotoxic effects of *Baccharis dracunculifolia* (D.C.). J Ethnopharmacol 124:321–324

Sartor T, Xavier VB, Falcão MA et al (2013) Seasonal changes in phenolic compounds and in the biological activities of *Baccharis dentata* (Vell.) G.M. Barroso. Ind Crop Prod 51:355–359

Sharp H, Bartholomew B, Bright C et al (2001) 6-Oxygenated flavones from *Baccharis trinervis* (Asteraceae). Biochem Syst Ecol 29:105–107

Soicke H, Leng-Peschlow E (1987) Characterisation of flavonoids from *Baccharis trimera* and their antihepatotoxic properties. Planta Med 53:37–39

Sousa JPB, Da Silva Filho AA, Bueno PCP et al (2009) A validated reverse-phase HPLC analytical method for the quantification of phenolic compounds in *Baccharis dracunculifolia*. Phytochem Anal 20:24–32

Stegelmeier BL, Sani Y, Pfister JA (2009) *Bacchavis pteronioides* toxicity in livestock and hamsters. J Vet Diagn Investig 21:208–213

Szliszka E, Zydowicz G, Mizgala E, Krol W (2012) Artepillin C (3,5-diprenyl-4-hydroxycinnamic acid) sensitizes LNCaP prostate cancer cells to TRAIL-induced apoptosis. Int J Oncol 41:818–828

Taguchi L, Pinheiro NM, Olivo CR et al (2015) A flavanone from *Baccharis retusa* (Asteraceae) prevents elastase-induced emphysema in mice by regulating NF-κB, oxidative stress and metalloproteinases. Respir Res 16:1–15

Tarqui ST, Segura YF, Vega GRA (2012) Polyoxygenated flavonoids from *Baccharis Pentlandii*. Rev Boliv Quim 29:10–14

Toyama DO, Ferreira MJP, Romoff P et al (2014) Effect of chlorogenic acid (5-Caffeoylquinic Acid) isolated from *Baccharis oxyodonta* on the structure and pharmacological activities of secretory phospholipase A2 from *Crotalus durissus terrificus*. Biomed Res Int 726585:1–10

Ueno AK, Barcellos AF, Grecco SS et al (2018) Sesquiterpenes, diterpenes, alkenyl p -coumarates, and flavonoid from the aerial parts of *Baccharis retusa* (Asteraceae). Biochem Syst Ecol 78:39–42

Verdi LG, Brighente IMC, Pizzolatti MG (2005) Gênero *Baccharis* (Asteraceae): Aspectos químicos, econômicos e biológicos. Quim Nova 28:85–94

Vijayarathna S, Sasidharan S (2012) Cytotoxicity of methanol extracts of *Elaeis guineensis* on MCF-7 and Vero cell lines. Asian Pac J Trop Biomed 2:826–829

Zalewski CA, Passero LFD, Melo ASRB et al (2011) Evaluation of anti-inflammatory activity of derivatives from aerial parts of *Baccharis uncinella*. Pharm Biol 49:602–607

Zampini IC, Isla MI, Schmeda-Hirschmann G (2009) Antimicrobial and antioxidant compounds from the infusion and methanolic extract of *Baccharis incarum* (WEDD.) Perkins. J Chil Chem Soc 54:477–481

Chapter 13
Baccharis Terpenoid Compounds

Sérgio Ricardo Ambrósio, Mário Ferreira Conceição Santos, Larissa Costa Oliveira, Ana Lúcia Bassi, Jairo Kenupp Bastos, and Rodrigo Cassio Sola Veneziani

Abstract *Baccharis* is an important genus of the Asteraceae family comprising more than 440 species, which are used in folk medicine for displaying important biological activities, such as analgesic, anti-inflammatory, antimicrobial, insect antifeedant, and antiparasitic, among others. There are several classes of metabolites produced by *Baccharis*, from which terpenoid stands out. The main volatile terpenes found in forty *Baccharis* species are reported in this chapter, pointing out *B. dracunculifolia*, the botanical source of green propolis, that contains (E)-nerolidol and spathulenol as major compounds, among others, giving the characteristic smell of green propolis. The combination of gas chromatography coupled with mass spectrometry is a powerful tool for the analyses of essential oils combined with the use of the Kovats index to determine the retention indexes using a homologous series of aliphatic hydrocarbons. Regarding diterpenes in *Baccharis* species, three main carbon skeleton types, kaurane, labdane, and neo-clerodane, have been reported. Many of these diterpenes display antimicrobial, antiparasitic, anti-inflammatory, analgesic, and cytotoxic activities, but the feeding-deterrent potential against insects displayed by neo-clerodane type-diterpenes stands out. In the final part, it is mentioned the triterpene and steroids, which are also found in this genus, and play important role in the reported biological activities of *Baccharis* species, as well as the occurrence of baccharinoids, a particular type of macrocyclic trichothecenes, which are associated with cattle poisoning in South America fed with *B. megapotamica*, but also display antiviral, anticancer, antimalarial, and antifungal activities.

Keywords Chemical structure · Nerolidol · Plant chemistry · Terpenes · Spathulenol

---

S. R. Ambrósio (✉) · M. F. C. Santos · L. C. Oliveira · A. L. Bassi · R. C. S. Veneziani
Programa de Pós-graduação em Ciências, Universidade de Franca, Franca, São Paulo, Brazil
e-mail: sergio.ambrosio@unifran.edu.br

J. K. Bastos
Departamento de Ciências Farmacêuticas, Faculdade de Ciências Farmacêuticas de Ribeirão Preto, Universidade de São Paulo, Ribeirão Preto, São Paulo, Brazil
e-mail: jkbastos@fcfrp.usp.br

G. W. Fernandes et al. (eds.), *Baccharis*,
https://doi.org/10.1007/978-3-030-83511-8_13

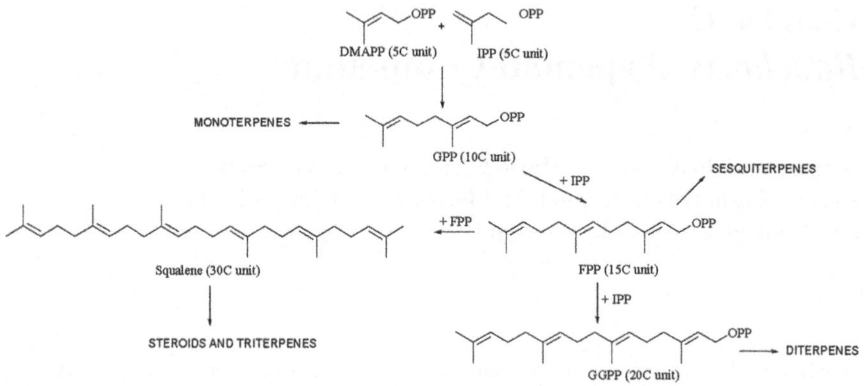

Fig. 13.1 Precursor units of different terpenoids, formed according to the number of isoprene units

1 General Aspects of Terpenes

Terpenoids are a large class of metabolites that can be subdivided according to the number of connected isoprenes (5C) units. Biosynthetically, such units are mostly bonded in a "head-to-tail" sequence in order to furnish the different types of terpenes. The dimethylallyl diphosphate (DMAPP) binds with an isopentenyl diphosphate (IPP) unit to form the 10C unit precursor of monoterpenes (GPP; geranyl diphosphate) [1] (Fig. 13.1). A GPP unit can be subsequently condensed with another IPP group to give farnesyl diphosphate (FPP), the precursor of sesquiterpenes. The addition of another IPP to FPP can furnish the 20C precursor of diterpenes – a geranylgeranyl diphosphate (GGPP) molecule. All these reactions are catalyzed by their respective *trans*-prenyl transferases. Alternatively, the junction of two FPP and two GGPP groups in a "head-to-head" binding can furnish a 30C squalene unit, and a 40C tetraterpene unit, which are the precursors of steroids/triterpenes and carotenoids, respectively (Veneziani et al. 2017; Furtado et al. 2017).

In this chapter, the main types of terpenoids found in *Baccharis* species are reported. Additionally, these compounds are discussed regarding their biological perspectives, focusing on their ecological, pharmacological, and toxicological reported data. Details on the chemical structures of some terpenes and their features are presented, as well.

2 *Baccharis* Essential Oils

Essential oils (EOs; or volatile oils) can be defined as mixtures of volatile and lipophilic, often odoriferous, substances produced by aromatic plants (Furtado et al. 2017). They are associated with a series of important ecological purposes, like molecular signaling among plants and pollinizers, as chemical defensives against predators (insects and other animals), as growth inhibitors of plants, and also as an antimicrobial against fungi or bacteria (Campos et al. 2016; Furtado et al. 2017). Also, these oils are obtained through steam distillation and are utilized in the manufacture of various products such as cosmetics, household cleaning products, air fresheners, and hygiene products, as well as in aromatherapy and in some paramedicinal practices due to their characteristic odor and flavor (Xavier et al. 2013; Campos et al. 2016).

Chemically, EOs are characterized by the presence of phenylpropanoids and relatively low molecular weight terpenoids like monoterpenes and sesquiterpenes. Monoterpenes (10C) present a great variety of structures, such as acyclic chains, monocyclic or aromatic derivatives, and even rigid bicyclic rings (Sarah et al. 2014; Furtado et al. 2017). Sesquiterpenes (15C) are less volatile than monoterpenes, and they can be found in the composition of many essential oils, including commercial ones (Campos et al. 2016; Veneziani et al. 2017) (Fig. 13.2).

The combination of gas chromatography coupled with mass spectrometry (GC-MS) is a powerful tool to analyze EOs in a fast, reliable, and simple way. The use of long capillary columns (up to 60 m) allows the separation of very complex mixtures of volatile compounds (Furtado et al. 2017). Also, this technique presents another important advantage when compared with other methods like thin layer and liquid chromatography, since the use of MS detection allows the identification of the components without the need of analytical standards. This is possible due to (Veneziani et al. 2017) the use of Kovats index as a more accurate form to determine the retention indexes using a homologous series of aliphatic hydrocarbons and (Furtado et al. 2017) the comparison between the mass spectra obtained for each compound and with those present in computational spectral databases (Shibamoto 1987; Furtado et al. 2017; Adams 2017).

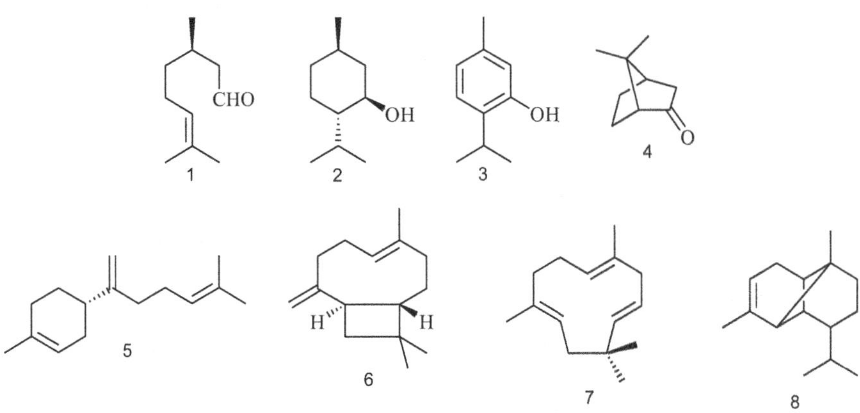

Fig. 13.2 Examples of monoterpenes (**1**-citronellal from *Cymbopogon citratus* "lemongrass", **2**-menthol from *Mentha piperita* "peppermint," **3**- thymol from *Thymus vulgaris* "Thyme," and **4**-camphor from *Rosmarinus officinalis* "Rosemary") and sesquiterpenes from *Copaifera sp.* "Copaíba" oilresins (**5**- β-bisabolene, **6**- *trans*-β-caryophyllene, **7**- α-humulene, and **8**- α-copaene)

Due to the above-mentioned features, GC-MS is broadly used to determine the qualitative and quantitative composition of essential oils of many species of aromatic plants, including the *Bachharis* genus. In this sense, it is possible to find in the literature a great number of studies that present the composition of many EOs of *Baccharis* species. However, for the purposes of this chapter, the focus will be kept on the terpenoids content of such products (Zunino et al. 1998; Albuquerque et al. 2004; Xavier et al. 2013). *Baccharis* species that had their EOs main monoterpenes and sesquiterpenes analyzed using GC-MS and their respective references are displayed in Table 13.1, as well as their chemical structures in Figs. 13.3a and 13.3b.

Table 13.1 *Baccharis* essential oils and their main terpenoids (mono and sesquiterpenes)

Specie	Compounds	References
B. anomala	Alpha-acorenol, spathulenol, caryophyllene oxide, limonene, alpha-thujene, alpha-pinene, sabinene, beta-pinene, myrcene, *p*-cymene, (E)-beta-ocimene, gamma-terpinene, alpha-cadinol, and beta-selinene	Budel et al. (2012), Xavier et al. (2013), and Trombin-Souza et al. (2017)
B. articulata	Spathulenol, palustrol, beta-pinene, beta-caryophyllene, caryophyllene oxide, (E)-nerolidol, and bicyclogermacrene	Zunino et al. (1998), Zunino et al. (2004), Florao et al. (2012), and Tischer et al. (2017)
B. axillaris	Limonene, alpha-thujene, alpha-pinene, sabinene, beta-pinene, myrcene, *p*-cymene, (E)-beta-ocimene, gamma-terpinene, spathulenol, and alpha-cadinol	Trombin-Souza et al. (2017)
B. calvescens	Alpha-thujene, alpha-pinene, sabinene, beta-pinene, myrcene, *p*-cymene, limonene, (E)-beta-ocimene, gamma-terpinene, spathulenol, and alpha-cadinol	Trombin-Souza et al. (2017)
B. caprariaefolia	Beta-caryophyllene, germacrene D, alpha-selinene, and spathulenol	Besten et al. (2012)
B. cordobensis	*Trans*-nerolidol, tau-cadinol, and cubenol	Zunino et al. (2000)
B. coridifolia	Isocaryophyllene, beta-caryophyllene, caryophyllene oxide, beta-selinene, spathulenol, viridiflorol, carotol, and alpha-bisabolol	Bailac et al. (2001), and Besten et al. (2012)
B. darwinii	Limonene, thymol, and 4-terpineol	Kurdelas et al. (2012)
B. dracunculifolia	Beta-caryophyllene, beta-humulene, germacrene D, beta-guaiene, delta-cadinene, (E)-nerolidol, spathulenol, limonene, bicyclogermacrene, beta-elemene, and mustakone	Fabiane et al. (2008), Lago et al. (2008), Parreira et al. (2010), Besten et al. (2012), Florao et al. (2012), Lage et al. (2015), and Salazar et al. (2018)
B. elaeagnoides	Viridiflorol, spathulenol, beta-caryophyllene, and germacrene D	Sayuri et al. (2010)

(continued)

Table 13.1 (continued)

Specie	Compounds	References
B. elaeoides	Gamma-cadinene, limonene, tau-cadinol, alpha-cadinol, beta-pinene, beta-patchoulene, alpha-muurolene, alpha-pinene, alpha-calacorene, gamma-muurolene, alpha-thujene, alpha-cubebene, and terpinen-4-ol	Simonsen et al. (2009)
B. gaudichaudiana	Spathulenol	Florao et al. (2012)
B. Genistelloides	Palustrol and spathulenol	Florao et al. (2012)
B. grisebachii	Thymol, thymol methyl ether, thymyl acetate, alpha-pinene, alpha-humulene, and globulol	Hadad et al. (2007)
B. latifolia	Limonene, beta-phellandrene, sabinene, beta-pinene, and alpha-pinene	Valarezo et al. (2013)
B. magellanica	4-hydroxyacetophenone, massoia lactone, alpha-cadinol, gamma-eudesmol, elemol, beta-eudesmol, and caryophyllene oxide	Simonsen et al. (2009)
B. megapotamica	Spathulenol, and caryophyllene oxide	Budel et al. (2012)
B. Mesoneura	Limonene, alpha-thujene, alpha-pinene, sabinene, beta-pinene, myrcene, *p*-cymene, (E)-beta-ocimene, gamma-terpinene, spathulenol, and alpha-cadinol	Trombin-Souza et al. (2017)
B. microdonta	Caryophyllene oxide, elemol, spathulenol, beta-caryophyllene, and germacrene D	Lago et al. (2008) and Sayuri et al. (2010)
B. Milleflora	Viridiflorol, beta-caryophyllene, germacrene-D, Bicyclogermacrene, alpha-humulene, limonene, (E)-beta-ocimene, gamma-terpinene, spathulenol, and alpha-cadinol	Besten et al. (2014), Pereira et al. (2016), Pereira et al. (2017), and Trombin-Souza et al. (2017)
B. Myriocephala	Alpha-thujene, alpha-pinene, sabinene, beta-pinene, myrcene, *p*-cymene, limonene, (E)-beta-ocimene, gamma-terpinene, spathulenol, and alpha-cadinol	Trombin-Souza et al. (2017)
B. myrtilloides	Germacrene D	Zunino et al. (1998)
B. Notosergila	Alpha-pinene, limonene, beta -caryophyllene, and spathulenol	Cobos et al. (2001)
B. oblongifolia	Alpha-thujene, alpha-pinene, sabinene, beta-pinene, myrcene, *p*-cymene, limonene, (E)-beta-ocimene, gamma-terpinene, spathulenol, and alpha-cadinol	Trombin-Souza et al. (2017)
B. obtusifolia	Limonene, germacrene-D, alpha-pinene, beta-pinene, bicyclogermacrene, and delta-cadinene	Valarezo et al. (2015)
B. ochracea	Spathulenol, and caryophyllene oxide	Budel et al. (2012)
B. patens	Beta-caryophyllene, aromadendrene, bicyclogermacrene, spathulenol, caryophyllene oxide, linalool and beta-pinene	da Silva et al. (2018)

(continued)

Table 13.1 (continued)

Specie	Compounds	References
B. pentaptera	Sabinene, himachalol, beta-pinene, and delta-3-carene	Perera et al. (2017)
B. Psiadioides	Beta-pinene, delta-3-carene, limonene, and ocimene	Negreiros et al. (2016)
B. Regnelli	Bicyclogermacrene, delta-cadinene, and delta-car-3-ene	Lago et al. (2008)
B. rufescens	Limonene, and *trans*-nerolidol	Zunino et al. (1998)
B. salicifolia	Alpha-pinene, camphene, beta-pinene, alpha-phellandrene, alpha-cubebene, beta-Cariophyllene, 6,9-guaiadiene, germacrone, (Z)-beta-ocimene, germacrene D, beta-cubebene, alpha-thujene	Flores et al. (2009), and Sosa et al. (2012)
B. schultzii	Spathulenol, and limonene	Lago et al. (2008)
B. semiserrata	Spathulenol, caryophyllene oxide, viridiflorol, carotol, alpha-cadinol, alpha-pinene, beta-pinene, limonene, beta-caryophyllene, gamma-muurolene), bicyclogermacrene, and (E)-nerolidol	Besten et al. (2012) and Vannini et al. (2012)
B. spartioides	Alpha-phellandrene, sabinene, alpha-pinene, camphor, limonene, citronellal, carvone, spathulenol, and 6R-7R-bisabolone	van Baren et al. (2002), Oliva et al. (2007), and Barud et al. (2014)
B. tenella	Spathulenol	Biurrun et al. (2005)
B. tricuneata	(E)-nerolidol	Arze et al. (2004)
B. trimera	Alpha-humulene, limonene, alpha-thujene, alpha-pinene, sabinene, beta-pinene, myrcene, *p*-cymene, (E)-beta-ocimene, gamma-terpinene, spathulenol, and alpha-cadinol	Lago et al. (2008)
B. trinervis	Alpha-thujene, alpha-pinene, sabinene, beta-pinene, beta-phellandrene, (E)-Lachnophyllum acid methylester, (Z)-lachnophyllum acid methyl ester, caryophyllene oxide, viridiflorol, germacrene D, germacrene B, spathulenol, delta-3-carene, globulol, *cis*-muurola-4(14), 5-diene, bicyclogermacrene, ar-curcumene, sabinene	Albuquerque et al. (2004), Sobrinho et al. (2016), and Chaverri and Ciccio (2017)
B. uncinella	Alpha-pinene, limonene, beta-caryophyllene, spathulenol, caryophyllene oxide, viridiflorol, alpha-cadinol, (E)-nerolidol, globulol, (E)-nerolidol, bicyclogermacrene, and terpinen-4-ol	Frizzo et al. (2001), Fabiane et al. (2008), and Ascari et al. (2012)

The essential oil of *B. dracunculifolia* is the most studied among all others EOs obtained from this genus (Lage et al. 2015; Lago et al. 2008; Fabiane et al. 2008; Parreira et al. 2010; Florao et al. 2012; Besten et al. 2014; Salazar et al. 2018). This fact must be due to the close relationship between this species and propolis, a resinous material produced by bees (*Apis mellifera*) that is used as a sanitizer and hive-repairing component (Parreira et al. 2010; Lage et al. 2015; Campos et al. 2016). To produce propolis, bees collect plant buds and mix them with wax: *B.*

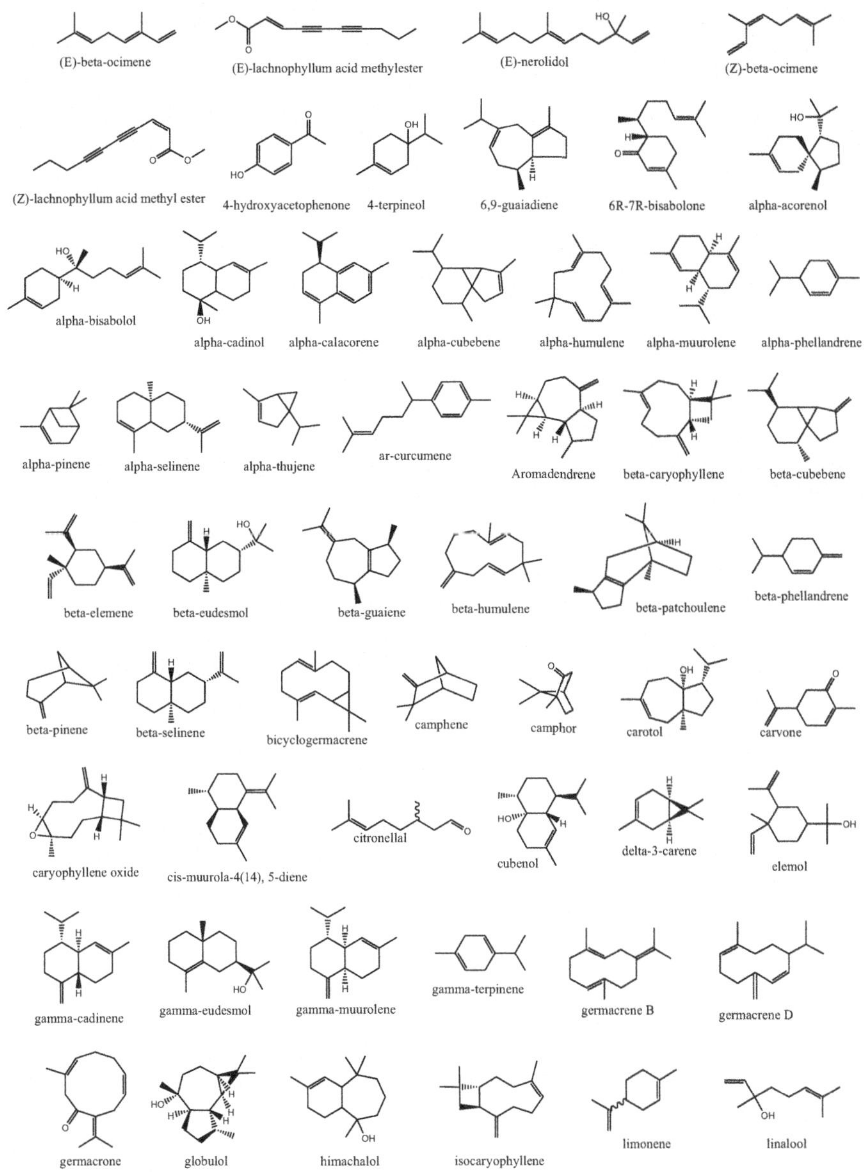

Fig. 13.3a Chemical structures of mono- and sesquiterpenes found in *Baccharis* essential oils. Source: NIST (National Institute of Standards and Technology) databank (Linstrom and Mallard 2018)

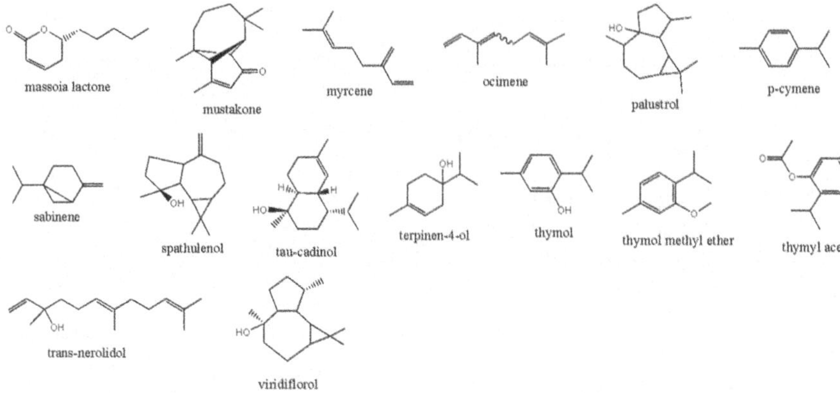

Fig. 13.3b Chemical structures of mono- and sesquiterpenes found in *Baccharis* essential oils. Source: NIST (National Institute of Standards and Technology) databank (Linstrom and Mallard 2018)

dracunculifolia is the main botanical source of Brazilian green propolis (BGP). In Brazil, BGP is the most studied and economically relevant tropical propolis, which is typically found in the South Eastern Region, where *B. dracunculifolia* popularly called "alecrim do campo" occurs (Parreira et al. 2010; Figueiredo-Rinhel et al. 2013; Lage et al. 2015; Campos et al. 2016). In fact, sesquiterpenes that are found in *B. dracunculifolia* like (E)-nerolidol, beta-caryophyllene, spathulenol, and γ-cadinene are also present in BGP and are responsible for the similar and peculiar aroma of both "alecrim do campo" EO and BGP. Moreover, such compounds are associated with several biological effects, including antimicrobial activity, among others. Especially in Asian markets, BGP prices can reach more than U$ 100 per kg and BGP extracts are incorporated in several "natural medicines," cosmetics, toiletries, food, and beverages, as well as in food supplements (Marostica et al. 2008; Figueiredo-Rinhel et al. 2013).

Although other *Baccharis* EOs are not associated with economically relevant products like propolis, studies of these EOs are significant from the ecological and academic points of view, since they possess a different biological potential and present geographical and seasonal variations (see references in Table 13.1). Spathulenol, limonene, beta-pinene, alpha-pinene, sabinene, beta-caryophyllene, alpha-cadinol, caryophyllene oxide, germacrene D, bicyclogermacrene, and alpha-thujene are, among others, the most frequently found terpenoids in these others *Baccharis* Eos (Table 13.1).

3 *Baccharis* Diterpenes

Diterpenes are a broad and diversified class of secondary metabolites, widely distributed among different organisms, mainly in plants and fungi, originated from the head-tail coupling of four C5 isoprene units (Figueiredo-Rinhel et al. 2013). This class of natural products presents a great structural variety, since the GGPP units (Fig. 13.1) can rearrange in many different manners. Several skeleton types and distinct chemical functionalization have been reported for diterpenoids, which are classified according to the number and cyclization patterns of their chemical skeletons (Garcia et al. 2007; de Sousa et al. 2018). Acyclic-, bicyclic-, tricyclic-, tetracyclic-, and macrocyclic-type diterpenes (Fig. 13.4) have been isolated and identified from a diverse number of natural sources (De Sousa et al. 2018).

Phytochemical studies have been performed with aerial parts, leaves, stems, flowers, and exudates from *Baccharis* species during the two last decades. Phenolic and terpenoid compounds are described as the main metabolites found in these

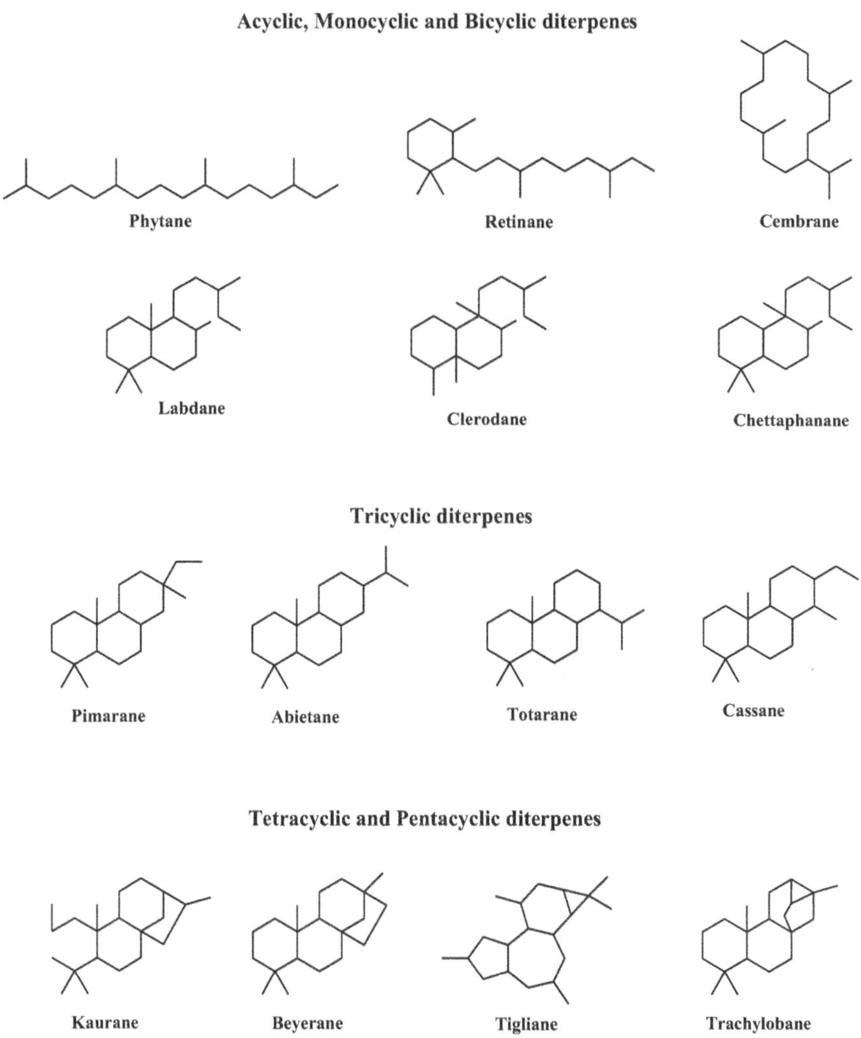

Fig. 13.4 Some examples of skeleton types of diterpenes related to their number of rings

botanical sources, in which the presence of flavonoids and diterpenes is noteworthy (Campos et al. 2016). Regarding diterpenes, the scientific literature pointed out the biosynthesis of three main carbon skeletons in *Baccharis* species – kaurane, labdane, and neo-clerodane type-diterpenes. Many diterpenes were isolated or identified from *Baccharis* species until 2018 (Table 13.2).

Table 13.2 Examples of diterpenes obtained through chemical studies from *Baccharis* species

Species, references, and structures
B. boliviensis (Campos et al. 2016)
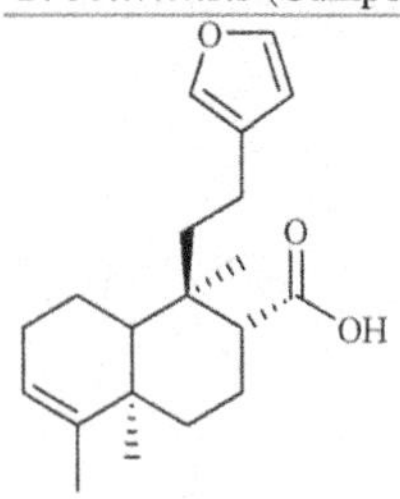 **Bbol1** 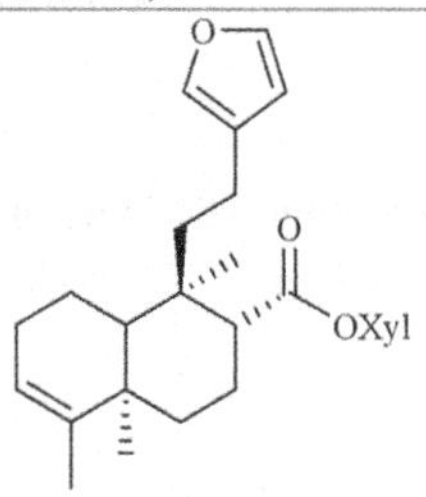 **Bbol2** 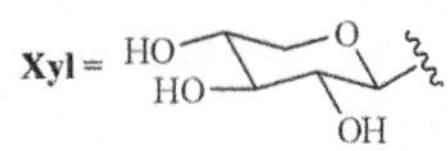
B. crispa (Cenal et al. 1997)
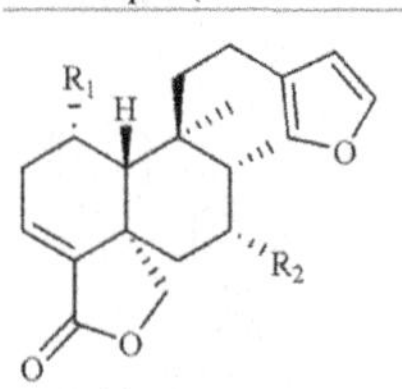 **Bcri1** - R1 = R2 = OH; **Bcri2** - R1 = H; R2 = OH; **Bcri3** - R1 = R2 = H 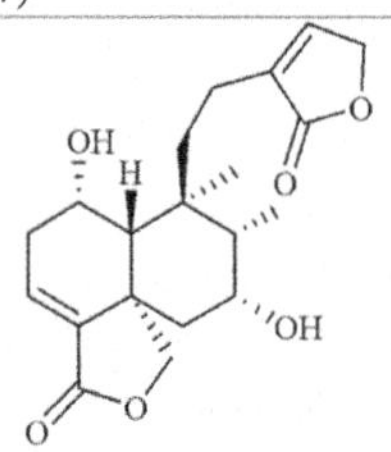 **Bcri4** 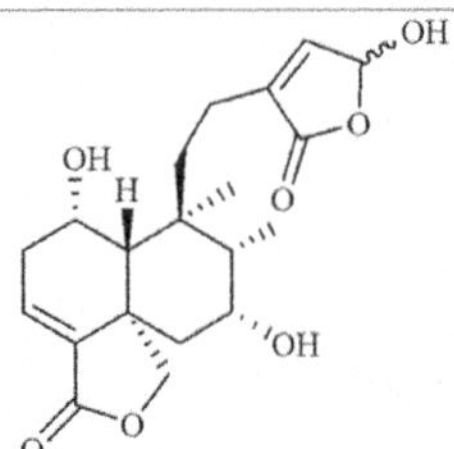 **Bcri5**
B. flabellata (Hikawczuk et al. 2006; Funes et al. (2018a, b)
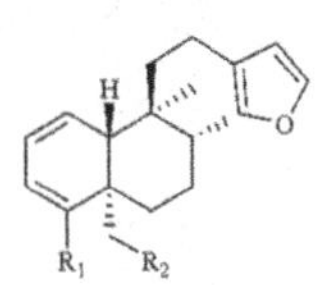 **Bfla1** - R1 = COOH; R2 = OCOCH3 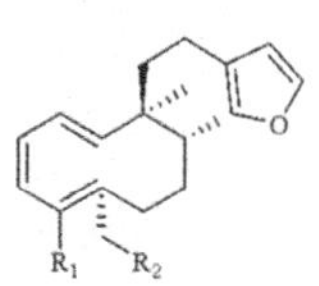 **Bfla2** - R1 = COOH; R2 = OCOCH3 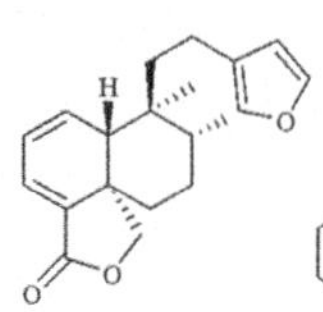 **Bfla4** 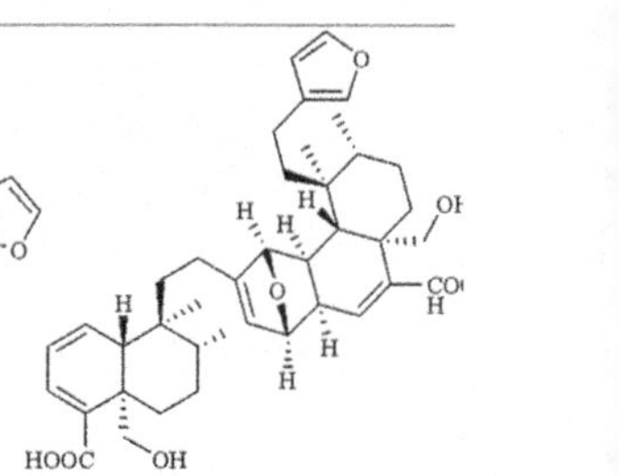 **Bfla5**
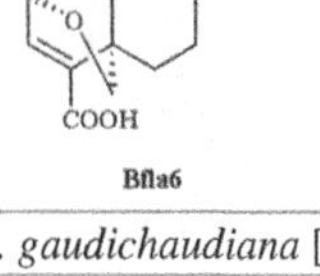 **Bfla6** 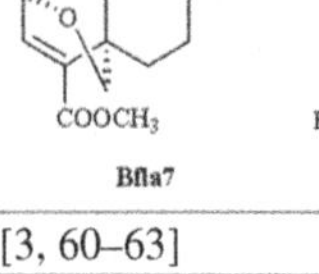 **Bfla7** 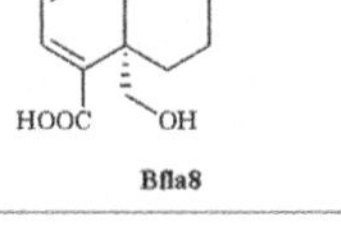 **Bfla8** 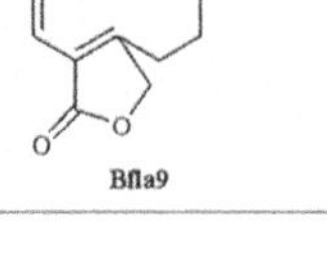 **Bfla9** 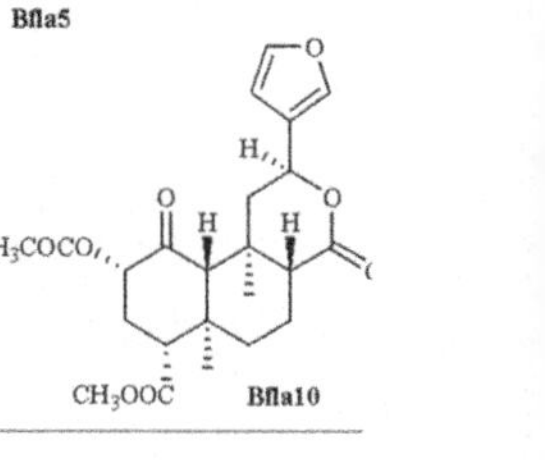 **Bfla10**
B. gaudichaudiana [3, 60–63]

(continued)

Table 13.2 (continued)

Species, references, and structures
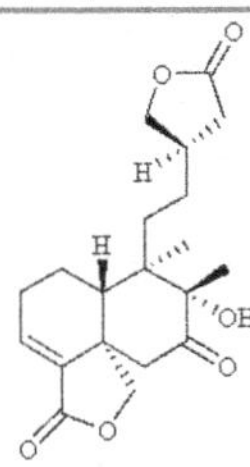 Bgau1 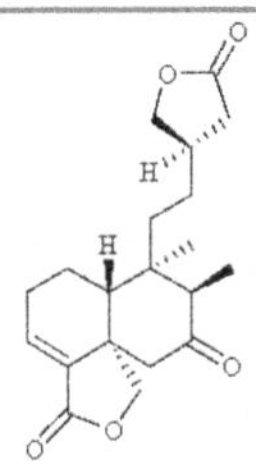 Bgau2 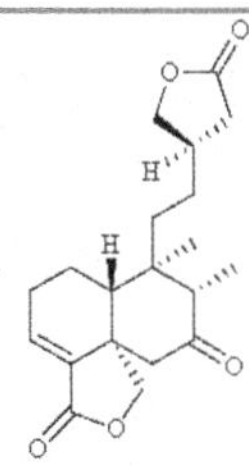 Bgau3 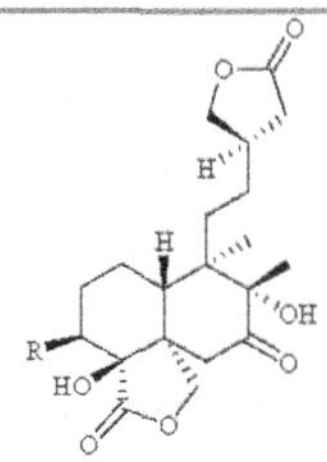 Bgau4 - R = *p*-bromobenzoyl 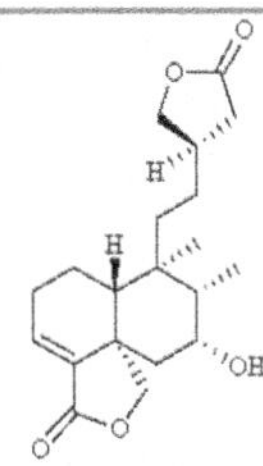 Bgau5
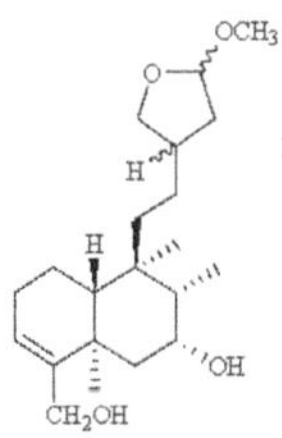 Bgau6 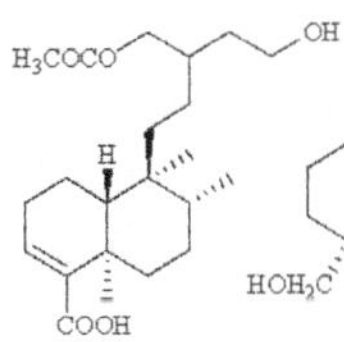 Bgau7 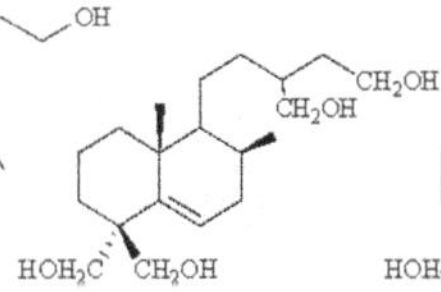 Bgau8 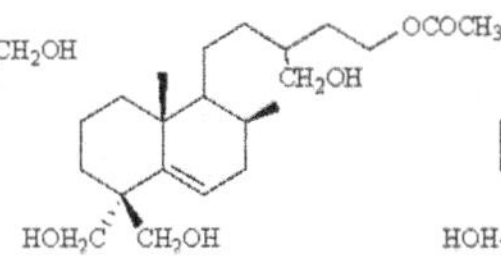 Bgau9 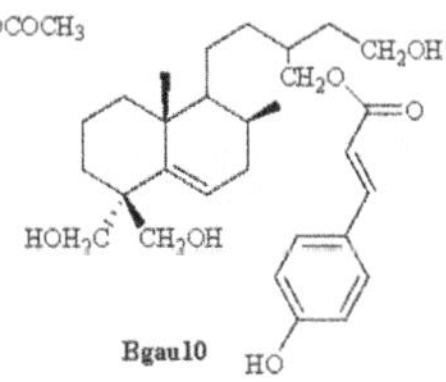 Bgau10
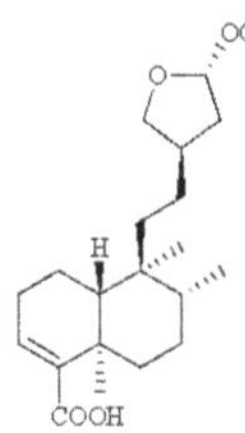 Bgau11 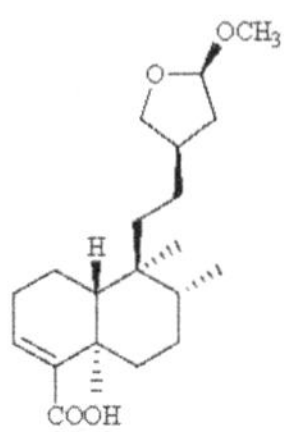 Bgau12 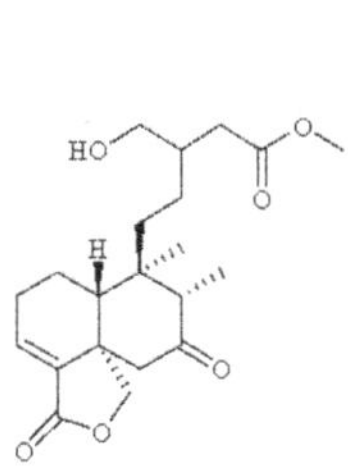 Bgau13 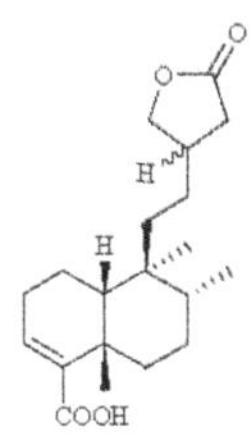 Bgau14 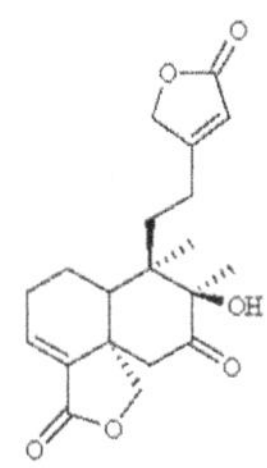 Bgau15
B. grisebachii (Feresin et al. 2003)
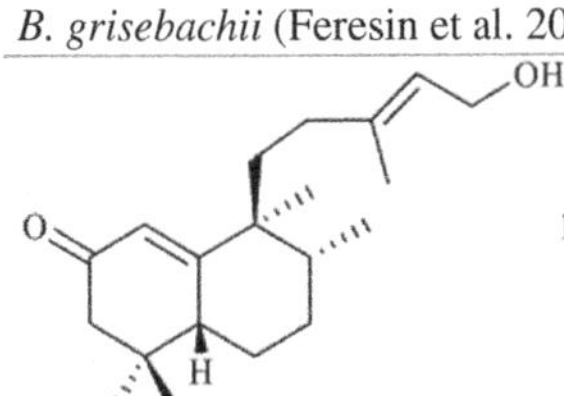 Bgri1 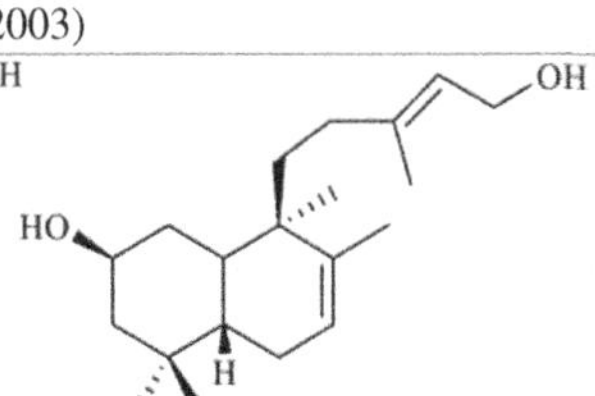 Bgri2
B. halimifolia (Jakupovic et al. 1990)
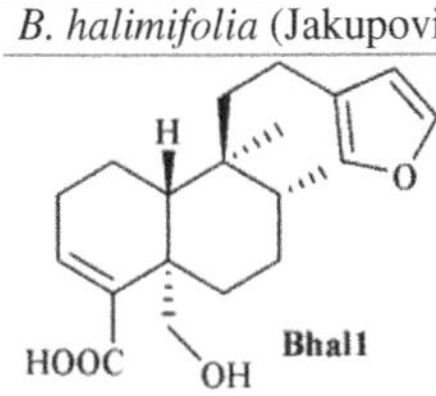 Bhal1 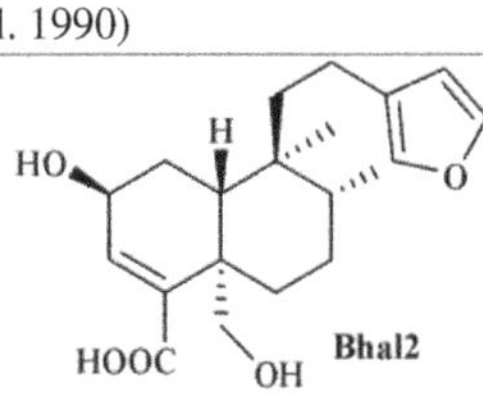 Bhal2 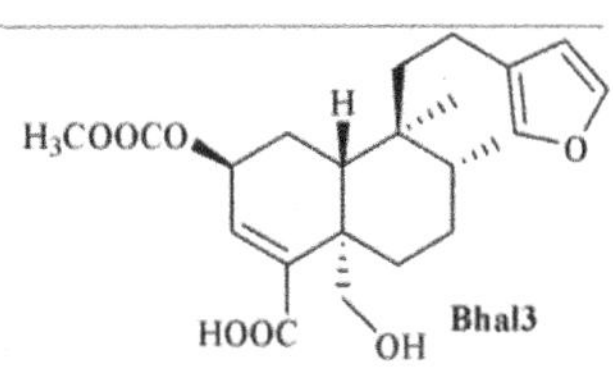 Bhal3
B. illinita (Campos et al. 2016; Verdi et al. 2004)

(continued)

Table 13.2 (continued)

Species, references, and structures

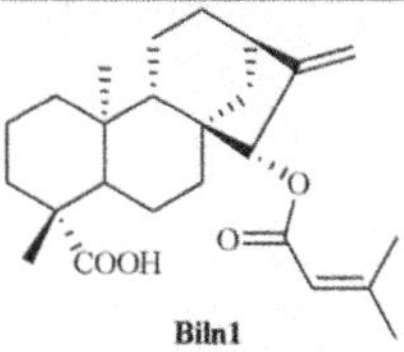

Biln1

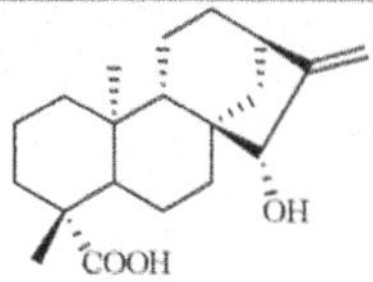

Biln2

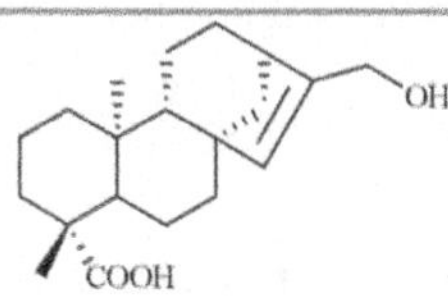

Biln3

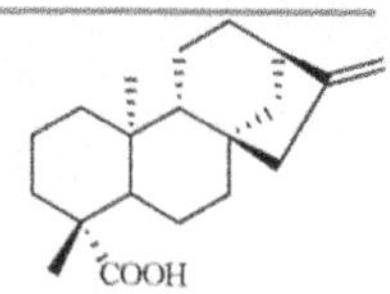

Biln4

B. marginalis (Campos et al. 2016; San-Martin et al. 2010)

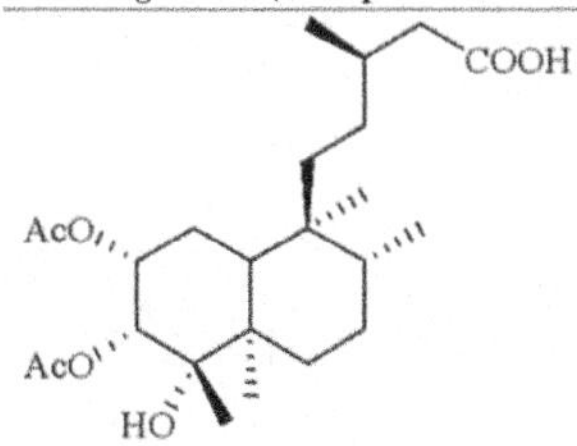

Bmar1

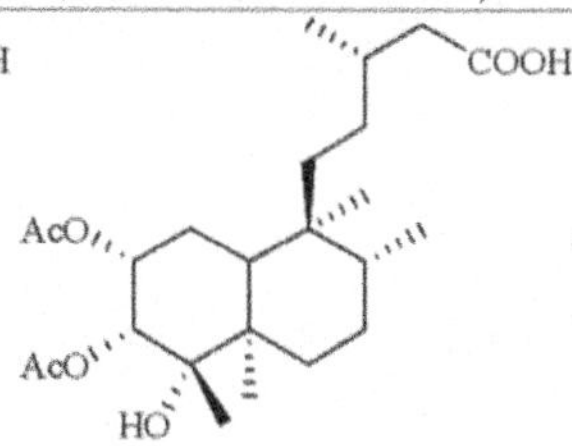

Bmar2

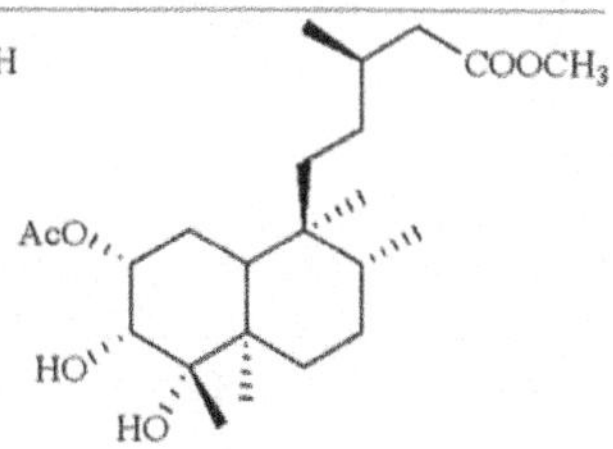

Bmar3

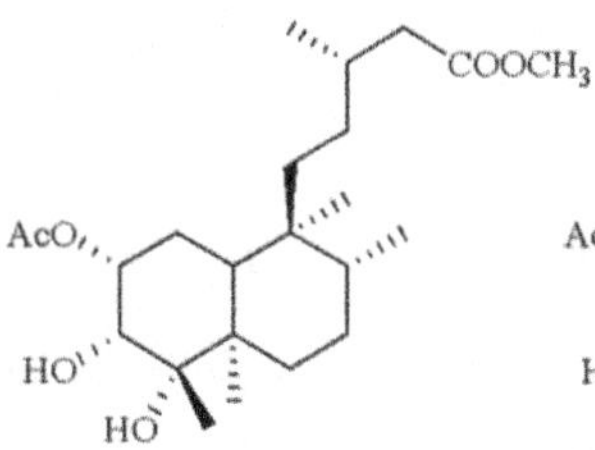

Bmar4

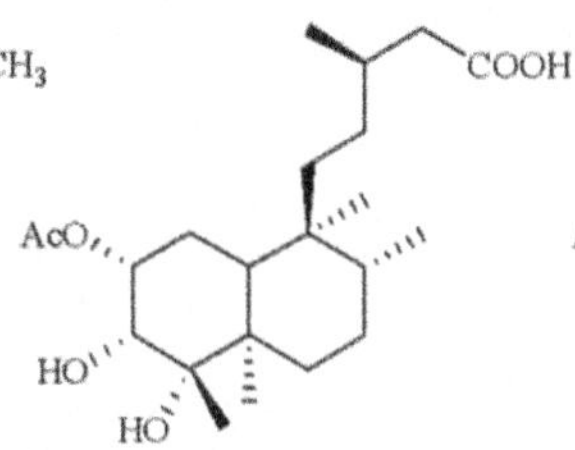

Bmar5

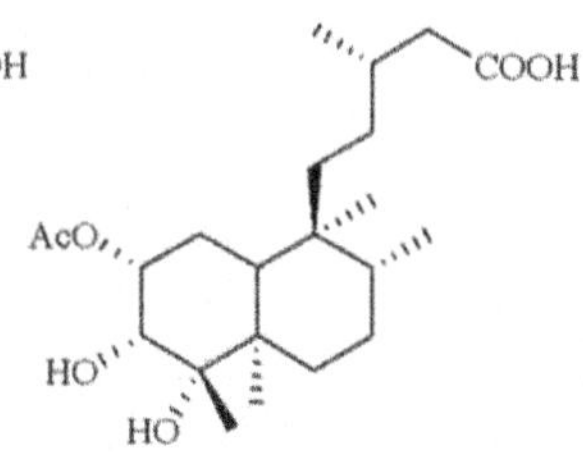

Bmar6

B. minutiflora (Bohlmann et al. 1982)

(continued)

Table 13.2 (continued)

Species, references, and structures

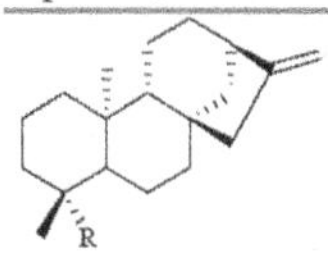

Bmin1 - R = CH_3
Bmin2 - R = CHO
Bmin3 - R = COOH
Bmin4 - R = CH_2OH

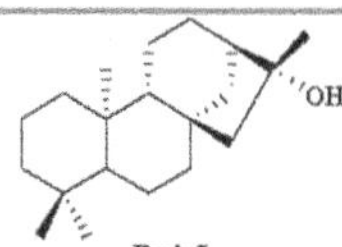

Bmin5

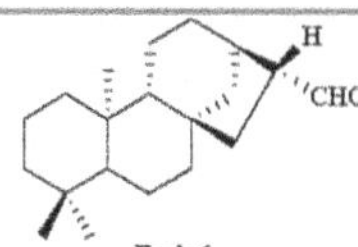

Bmin6

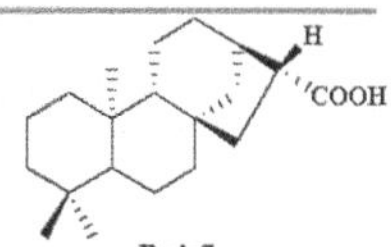

Bmin7

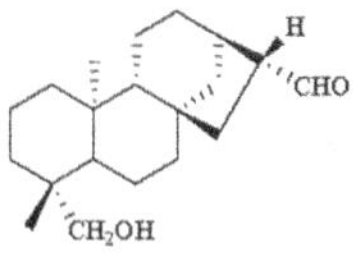

Bmin8

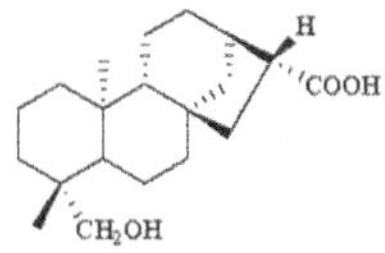

Bmin9

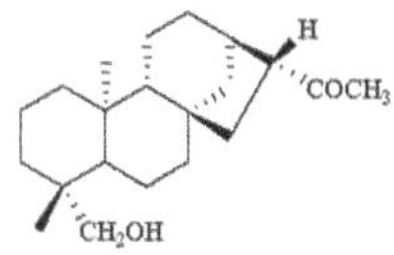

Bmin10

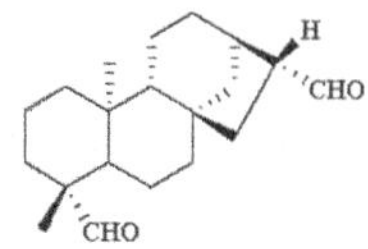

Bmin11

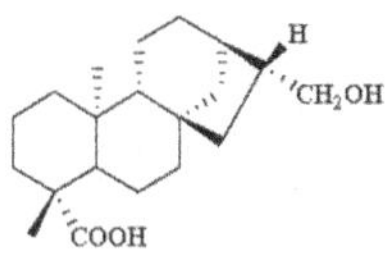

Bmin12

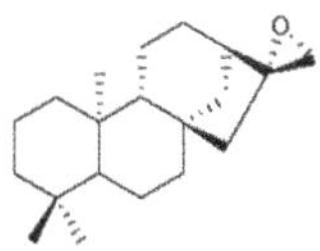

Bmin13

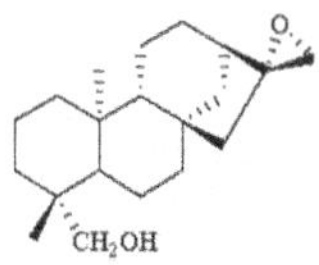

Bmin14

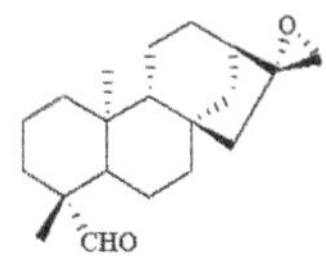

Bmin15

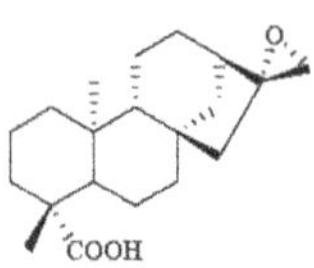

Bmin16

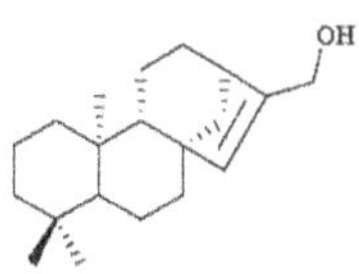

Bmin17

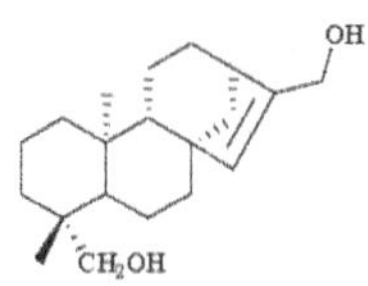

Bmin18

B. neglecta (Jakupovic et al. 1990)

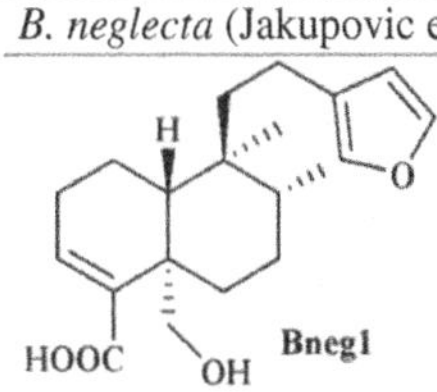

Bneg1

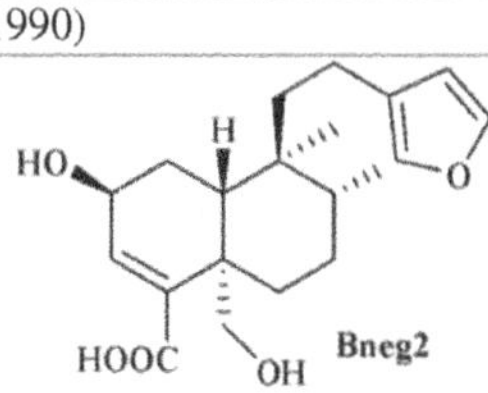

Bneg2

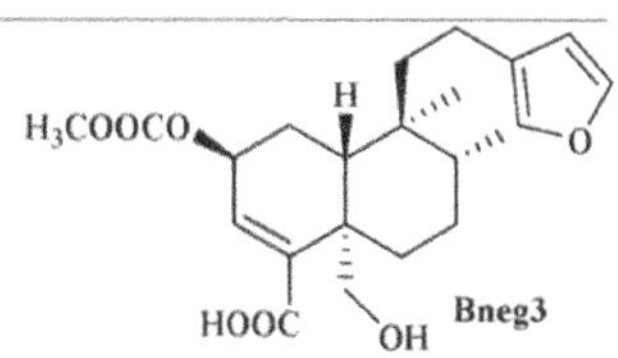

Bneg3

B. patens (Campos et al. 2016)

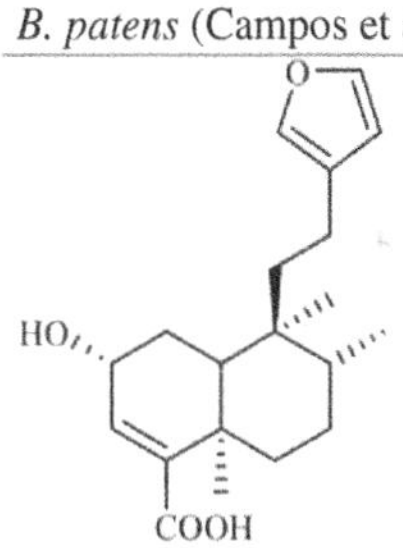

Bpat1

B. pingraea (Wachter et al. 1999)

(continued)

Table 13.2 (continued)

Species, references, and structures

B. pingraea (Wachter et al. 1999)

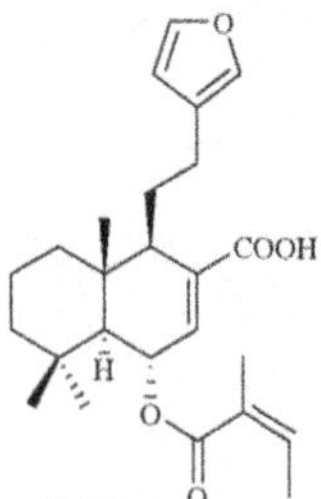

Bpin1

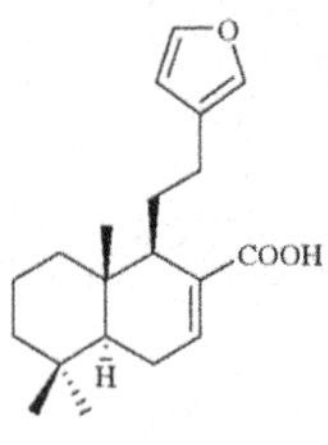

Bpin2

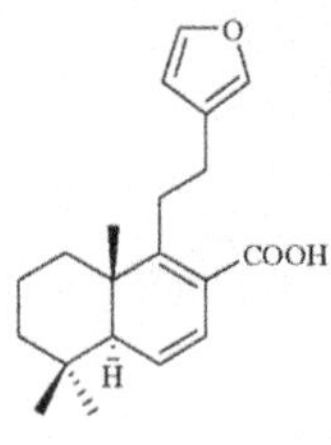

Bpin3

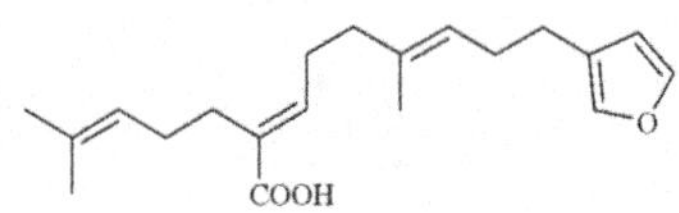

Bpin4

B. platypoda (Campos et al. 2016)

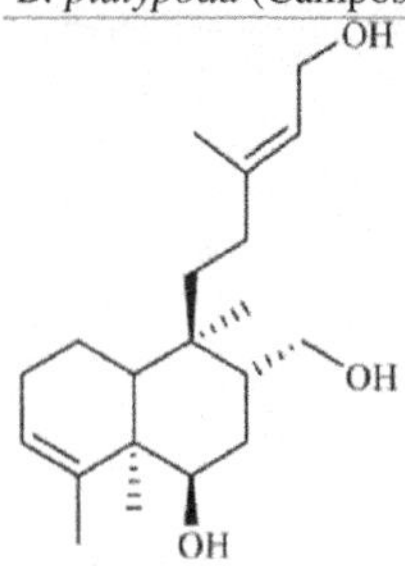

Bpla1

B. potosina (Jakupovic et al. 1990)

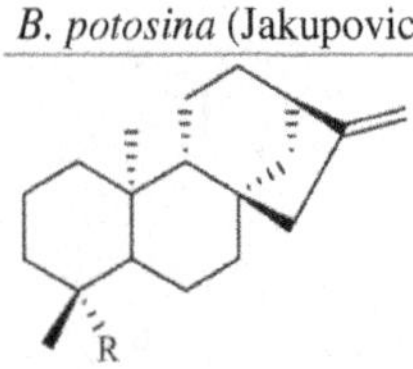

Bpot1 - R = CHO

Bpot2 - R = COOH

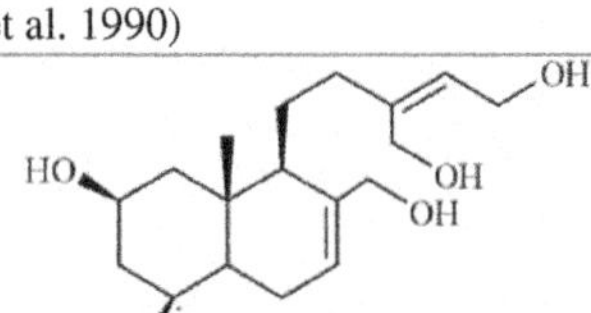

Bpot3

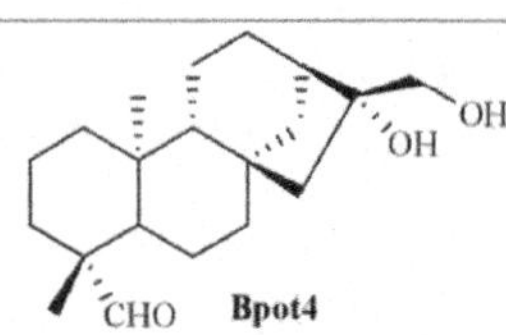

Bpot4

B. retusa (Ueno et al. 2018a, b)

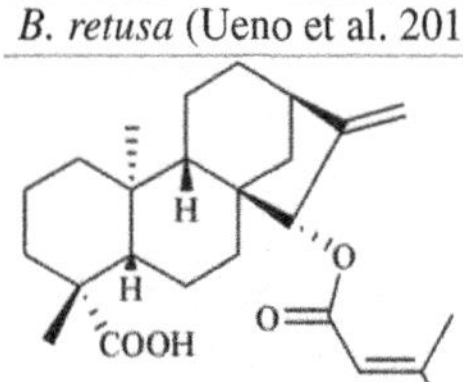

Bret1

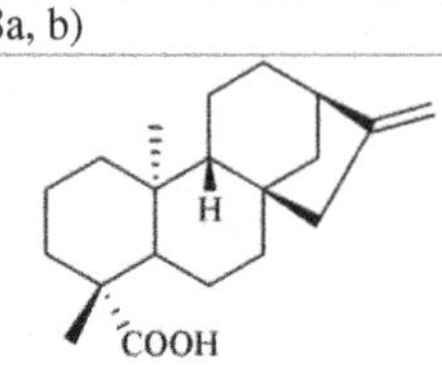

Bret2

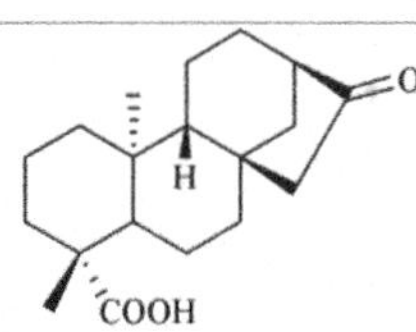

Bret3

B. rufescens (Campos et al. 2016)

(continued)

Table 13.2 (continued)

Species, references, and structures

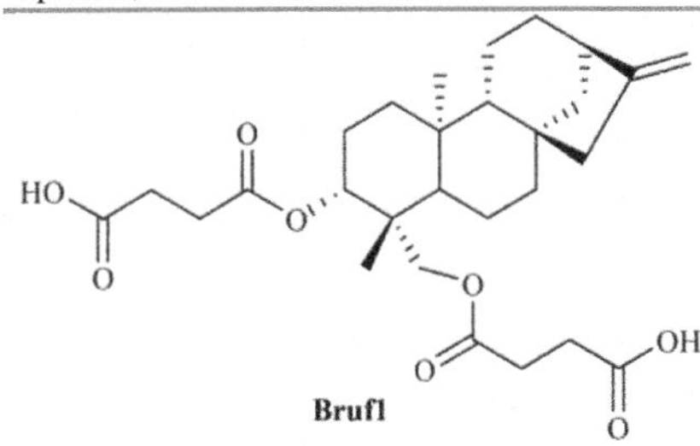

Bruf1

B. sagittalis (Campos et al. 2016)

Bsag1/Bsag2 - R =

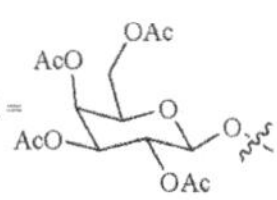

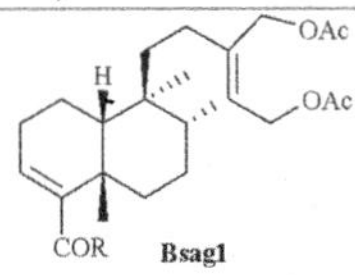

Bsag1

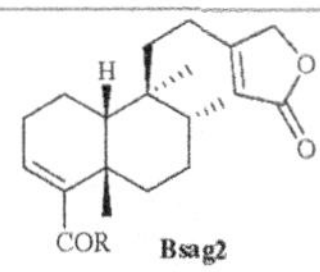

Bsag2

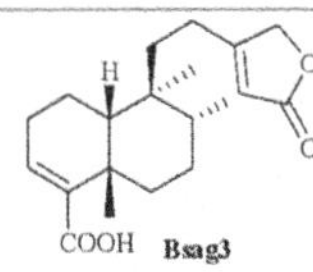

Bsag3

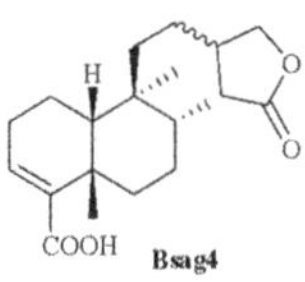

Bsag4

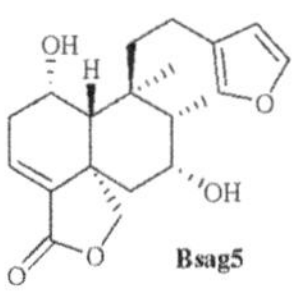

Bsag5

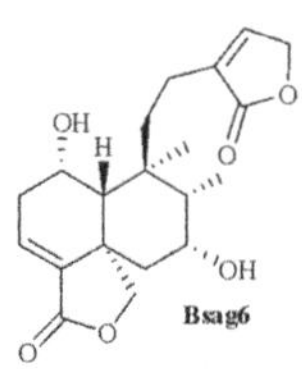

Bsag6

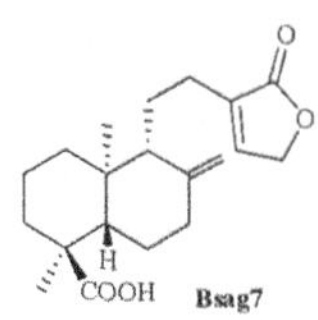

Bsag7

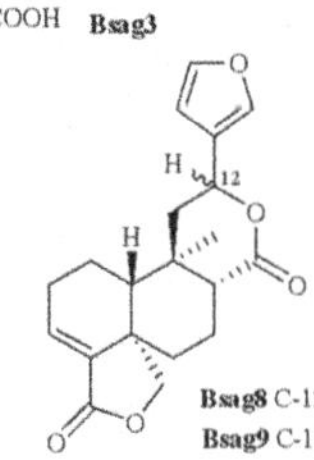

Bsag8 C-12 (*R*)
Bsag9 C-12 (*S*)

B. salicifolia (Campos et al. 2016; Jakupovic et al. 1990)

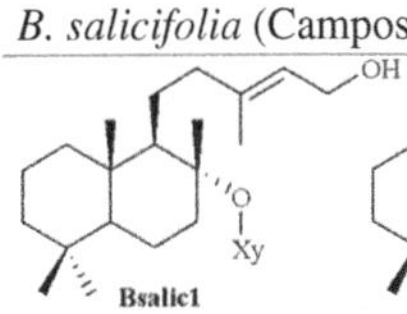

Bsalic1

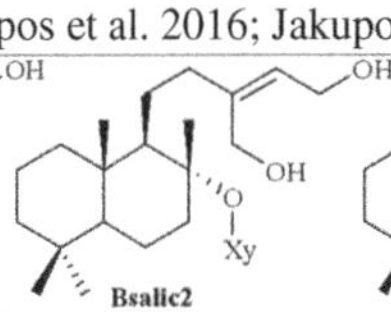

Bsalic2

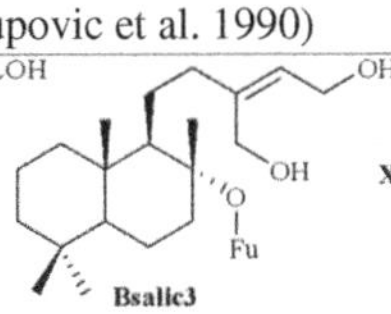

Bsalic3

Xy =

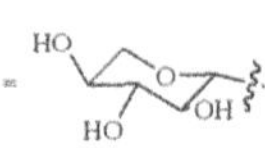

Fu =

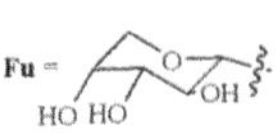

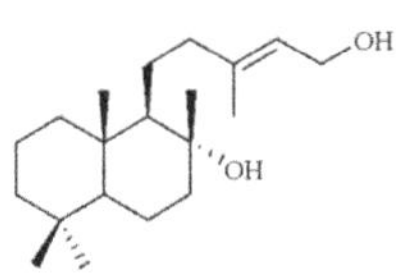

Bsalic4

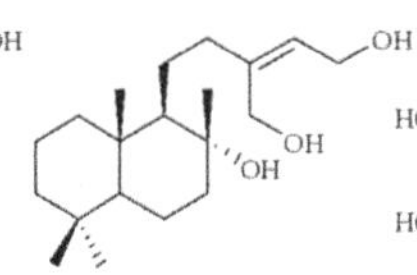

Bsalic5

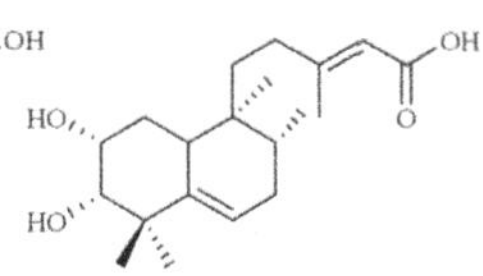

Bsalic6

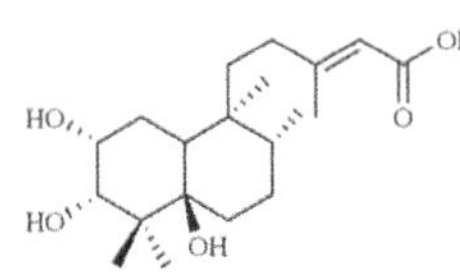

Bsalic7

B. salicina (Jakupovic et al. 1990)

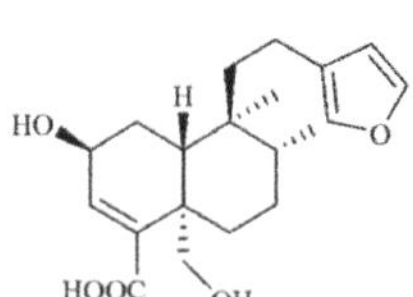

Bsal1

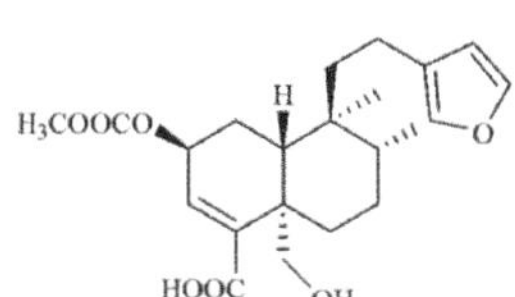

Bsal2

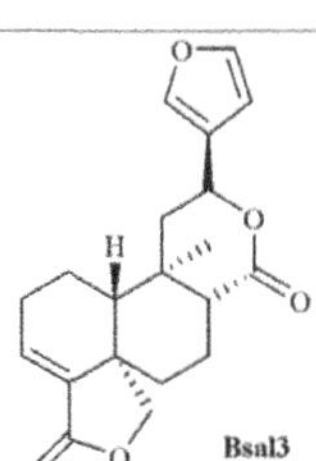

Bsal3

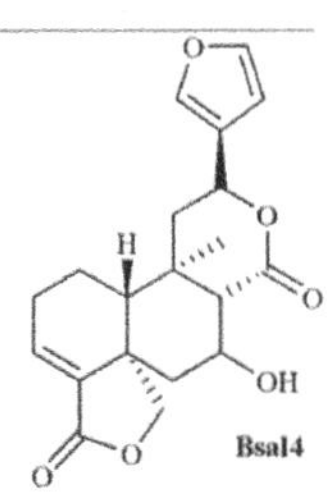

Bsal4

(continued)

Table 13.2 (continued)

Species, references, and structures
B. salzmannii (Campos et al. 2016)
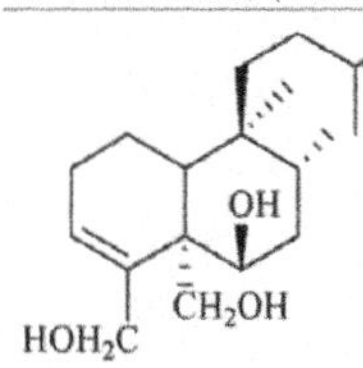 **Bsalz1** 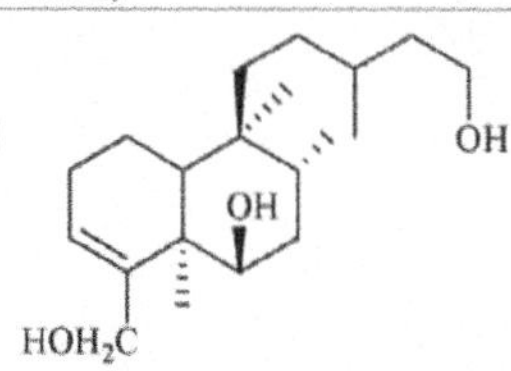 **Bsalz2** 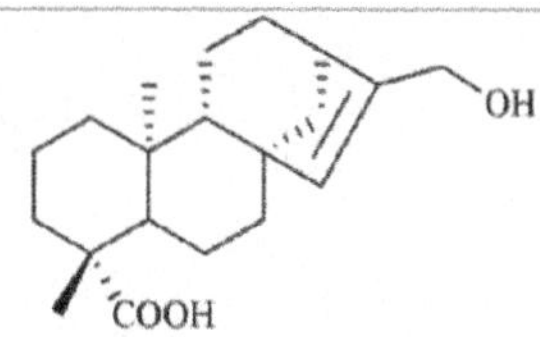 **Bsalz3**
B. sarathroides (Jakupovic et al. 1990)
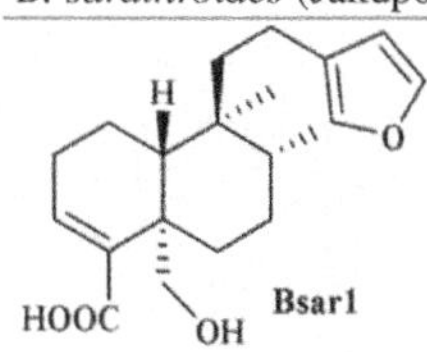 **Bsar1** 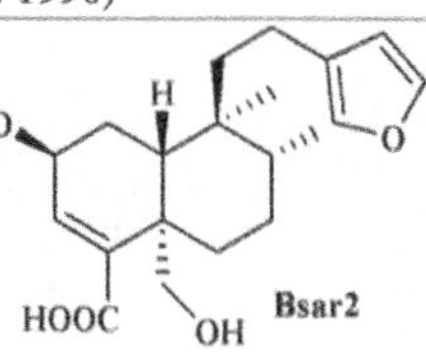 **Bsar2** 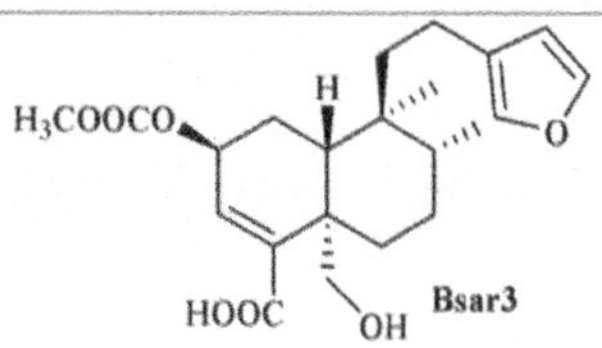 **Bsar3**
B. thymifolia (Hikawczuk et al. 2008)
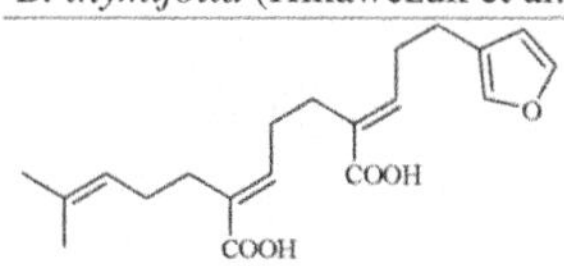 **Bthy1** 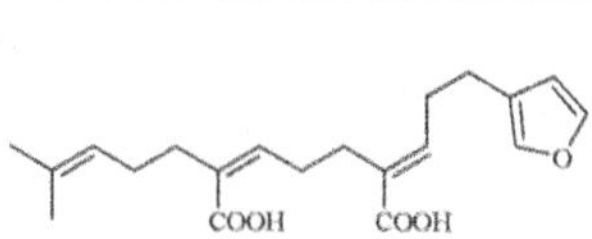 **Bthy2** 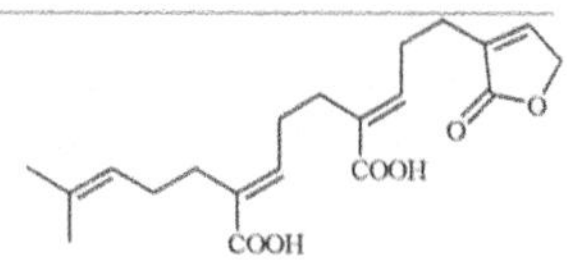 **Bthy3**
B. tola (Simirgiotis et al. 2016)
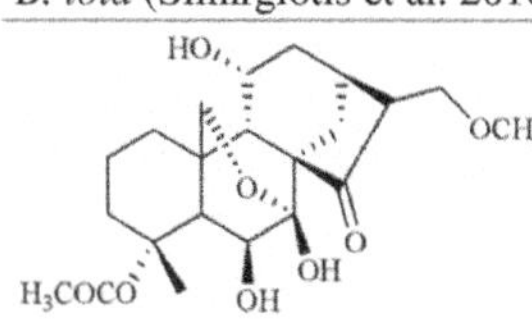 **Btog1** 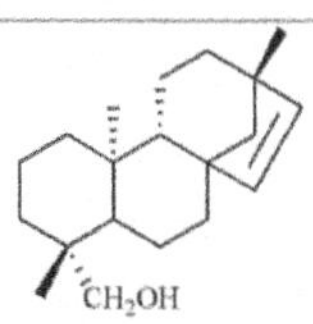 **Btog2** 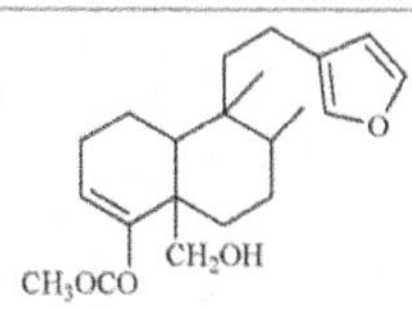 **Btog3** 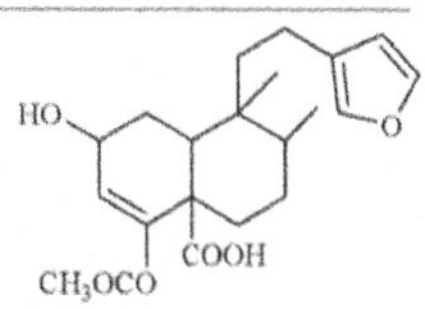 **Btog4**
B. trimera (Januario et al. 2004; Garcia et al. 2014; Campos et al. 2016)
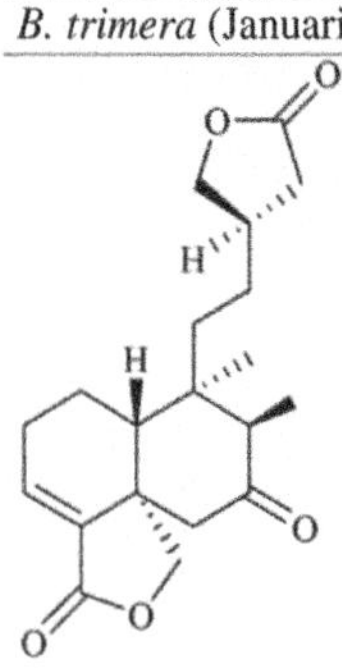 **Btri1** 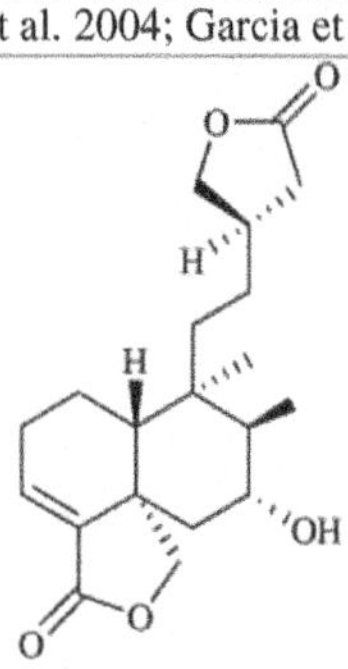 **Btri2** 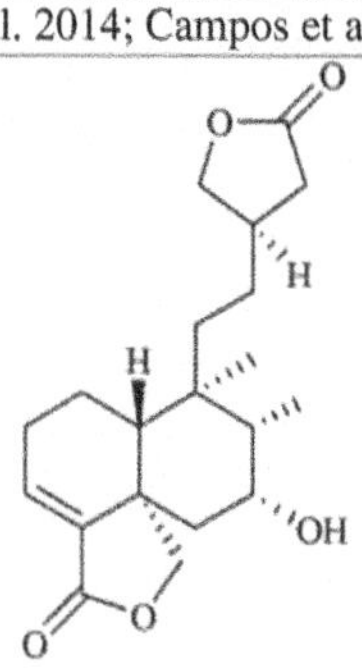 **Btri3**

Several diterpenes have been considered as the active compounds of various medicinal plants, due to their wide spectrum of pharmacological effects (de Sousa et al. 2018; Liu et al. 2018). In fact, scientific reports point out that this class of natural products possesses remarkable biological properties such as antiparasitic (Ambrosio et al. 2008), antihypertensive (Ambrosio et al. 2004; Tirapelli et al. 2010), anti-inflammatory and analgesic (Mizokami et al. 2012; Possebon et al. 2014), cytotoxicity against tumor cell lines (Batista et al. 2013; da Costa et al. 2018), and antimicrobial (Leandro et al. 2014; Severiano et al. 2010), among others (Kato et al. 2012; Xu et al. 2018; Liu et al. 2018).

Regarding the *Baccharis* diterpenes, several authors have investigated and reported their biological activities (Table 13.3).

From all biological properties that have been reported, the feeding-deterrent potential against insects displayed by neo-clerodane type-diterpenes from *Baccharis* species is highlighted in the literature (Hikawczuk et al. 2006; Cifuente et al. 2002; Hikawczuk et al. 2008; Sosa et al. 1994), and some efforts to understand the structure–activity relationships related with this class of compounds have been carried out (Hikawczuk et al. 2006; Cifuente et al. 2002).

Cifuente et al. (2002) evaluated the insect antifeedant activity of several clerodane-type diterpenes isolated from three *Baccharis* species (*B. sagittalis*, *B. crispa*, and *B. spicata*) against *Tenebrio molitor* larvae (Coleoptera: Tenebrionidae). The results obtained from this study allowed the authors to suggest that the presence of a β-substituted furan ring (A; Fig. 13.6) or a β-substituted butenolide group (B; Fig. 13.6) on the C-9 side chain plays an important role in the

Table 13.3 Relevant biological activities displayed by *Baccharis* diterpenes, reported in the last two decades

Biological activity	Active diterpenes	References
Insect antifeedant	Bsag3; Bsag5; Bsag8; Bthy1; Bthy3	Cifuente et al. (2002) and Hikawczuk et al. (2008)
Antimicrobial	Bgri2	Feresin et al. (2003)
Cytotoxicity against cancer cell lines	Bgau10	(Fullas et al. 1994)
ROS and RNS scavenging abilities	Bfla5; Bfla8	Funes et al. (2018a, b
Antinociceptive	Bfla5; Bfla8	Funes et al. (2018a, b)
Influx and mobilization of intracellular calcium	Btri1; Btri2	Garcia et al. (2014)
NGF (Nerve Growth Factor) potentiation	Bgau6; Bgau11; Bgau12; Bgau14	Guo et al. (2006) and Guo et al. (2007)
Antiproteolytic and antihemorrhagic	Btri3	Januario et al. (2004)
Relaxant effect on rat vascular smooth muscle	Btri3	Torres et al. (2000)
Antitrypanosomal	Bret1; Bret3	Ueno et al. (2018a, b)

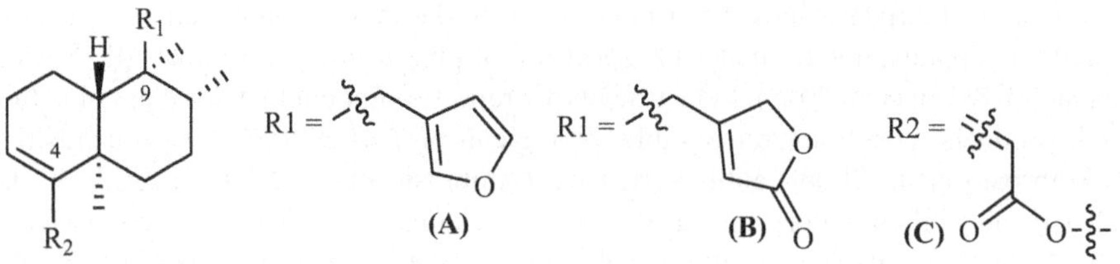

Fig. 13.5 Chemical groups presented in clerodane-type diterpenes related to their antifeedant activity (Cifuente et al. 2002; Enriz et al. 1994)

antifeedant bioactivity displayed by these diterpenes. Moreover, the literature also pointed out the presence of an α,β-unsaturated carbonyl system insert in the *trans*-decaline system at C-4 (C; Fig. 13.5) of clerodane diterpenes as a prerequisite to the antifeedant bioactivity (Enriz et al. 1994).

More recently, Hikawczuk et al. (2006) investigated the antifeedant activity of *neo*-clerodanes diterpenes from *Baccharis flabellata* against the grain stored insect *Tribolium castaneum*, and some considerations on the structure–activity relationship were also established by the authors. Among the nine diterpenes investigated in this study, compounds **6 and 8** (Fig. 13.6) were shown to be more effective against *T. castaneum*.

As previously described in the literature, the presence of a furan or a butenolide group at C-9 side chain and an α,β-unsaturated carbonyl system inserted in the *trans*-decaline at C-4 (or a C-4 *spiro*-epoxide) are basic structural requirements related to the antifeedant activity of the clerodane-type diterpenes (Cifuente et al. 2002; Enriz et al. 1994). However, conformational and electronic factors were also revealed to be important in the antifeedant activity displayed by this class of diterpenes (Hikawczuk et al. 2006).

From a molecular model investigation performed by Hikawczuk et al. (2006), it was possible to establish a correlation of the antifeedant activity and the distance between the heteroatom located at the C-9 side chain and the olefinic carbon at C-3. According to the authors, the optimum interatomic distance between these two moieties ranged from 8.117 to 9.694 Å (Hikawczuk et al. 2006; Li et al. 2016). This study was carried out taking into account the potential of the double bond at C-3 to act as a nucleophilic acceptor of proteins in a Michael-type addition reaction, which can be related to the ability of diterpenes to inhibit the feeding of *T. castaneum* (Hikawczuk et al. 2006). Finally, the authors observed positive values of the electrostatic charge in the olefinic carbons at C-3 for the active diterpenes (Fig. 13.6, compounds **6** and **8**), whereas negative values were observed for all inactive compounds (Hikawczuk et al. 2006).

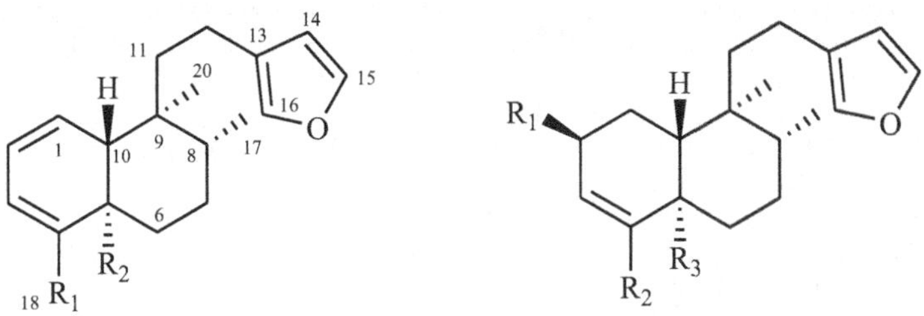

1 - R1 = COOH; R2 = CH_2OAc

2 - R1 = $COOCH_3$; R2 = CH_2OAc

3 - R1 = COOH; R2 = CH_2OH

4 - R1 = $COOCH_3$; R2 = CH_2OH

5 - R1 = CH_2OH; R2 = CH_2OH

6 - R1 = H; R2 = COOH; R3 = CH_2OH

7 - R1 = OAc; R2 = COOH; R3 = CH_2OH

8 - R1 = H; R2-R3 = $COOCH_2$

9 - R1 = OAc; R2-R3 = $COOCH_2$

Fig. 13.6 Chemical structures of clerodane-type diterpenes from *B. flabellata* investigated as antifeedant against *T. castaneum* (Hikawczuk et al. 2006)

4 Final Considerations

This chapter discussed in detail some chemical features and biological activities of the most representative terpenes found in *Baccharis* species (volatile terpenes and diterpenes). However, squalene-derived compounds, like steroids and triterpenes, are also found in this genus and can be associated with some biological activities, mainly antiparasitic (da Silva et al. 2009a, b; Passero et al. 2011), antinociceptive (Freitas et al. 2009), and anti-inflammatory (Boller et al. 2010).

Finally, it is important to mention the occurrence of a particular type of *Baccharis* trichothecenes, named baccharinoids. These compounds are associated with cattle poisoning in South America fed with *B. megapotamica* and also with several biological activities such as antiviral, anticancer, antimalarial, and antifungal (de Carvalho et al. 2016).

Acknowledgments Authors are thankful to São Paulo Research Foundation – FAPESP (process number 2017/04138-8), CNPq (processes numbers 306441/2017-9 and 306432/2017-0), and CAPES for funds and grants.

References

Adams RP (2017) Identification of essential oil components by gas chromatography / mass spectrometry. Texensis Publishing, Gruver

Albuquerque M, Souza EB, Lins M et al (2004) Composition and antimicrobial activity of the essential oil from aerial parts of *Baccharis trinervis* (Lam.) Pers. ARKIVOC 6:59–65

Ambrosio SR, Tirapelli CR, Coutinho ST et al (2004) Role of the carboxylic group in the antispasmodic and vasorelaxant action displayed by kaurenoic acid. J Pharm Pharmacol 56:1407–1413

Ambrosio SR, Arakawa NS, Esperandim VR et al (2008) Trypanocidal activity of pimarane diterpenes from *Viguiera arenaria* (Asteraceae). Phytother Res 22:1413–1415

Arze JBL, Garneau FX, Collin G et al (2004) Essential oils from Bolivia. I. Asteraceae: *Baccharis tricuneata* (L.f.) pers. var. *ruiziana* Cuatrecassas. J Essent Oil Res 16:429–431

Ascari J, Sens SL, Nunes DS et al (2012) Sedative effects of essential oils obtained from *Baccharis uncinella*. Pharm Biol 50:113–119

Bailac PN, Dellacasa AD, Bernasconi HO et al (2001) Essential oil of female plants of *Baccharis coridifolia* De Candole. J Essent Oil Res 13:23–24

Barud FJ, Lopez S, Tapia A et al (2014) Attractant, sexual competitiveness enhancing and toxic activities of the essential oils from *Baccharis spartioides* and *Schinus polygama* on *Ceratitis capitata* Wiedemann. Ind Crop Prod 62:299–304

Batista R, Garcia PA, Castro MA et al (2013) Synthesis, cytotoxicity and antiplasmodial activity of novel ent-kaurane derivatives. Eur J Med Chem 62:168–176

Besten MA, Jasinski VCG, Costa A et al (2012) Chemical composition similarity between the essential oils isolated from male and female specimens of each five *Baccharis* species. J Braz Chem Soc 23:1041–1047

Besten, Aparecida M, Nunes DS et al (2014) Chemical composition of essential oils from cladodes and inflorescences from male and female specimens of *Baccharis milleflora*. J Essent Oil-Bear Plants 17:899–905

Biurrun F, Juliani RH, Lopez ML et al (2005) Essential oil composition of *Baccharis tenelia* Hook. et Arn. J Essent Oil Res 17:122–123

Bohlmann F, Kramp W, Jakupovic J et al (1982) Naturally-occurring terpene derivatives. Diterpenes from *Baccharis* Species. Phytochemistry 21:399–403

Boller S, Soldi C, Marques MCA et al (2010) Anti-inflammatory effect of crude extract and isolated compounds from *Baccharis illinita* DC in acute skin inflammation. J Ethnopharmacol 130:262–266

Budel JM, Duarte MR, Doll-Boscardin PM et al (2012) Composition of essential oils and secretory structures of *Baccharis anomala, B. megapotamica* and *B. ochracea*. J Essent Oil Res 24:19–24

Campos FR, Bressan J, Jasinski VCG, Zuccolotto T, da Silva LE, Cerqueira LB (2016) *Baccharis* (Asteraceae): chemical constituents and biological activities. Chem Biodivers 13:1–17

Cenal JP, Giordano OS, Rossomando PC et al (1997) Neoclerodane diterpenes from *Baccharis crispa*. J Nat Prod 60:490–492

Chaverri C, Ciccio JF (2017) Essential oils of *Baccharis trinervis* (Asteraceae) from Costa Rica. Rev Biol Trop 65:1307–1321

Cifuente DA, Borkowski EJ, Sosa ME et al (2002) Clerodane diterpenes from *Baccharis sagittalis*: insect antifeedant activity. Phytochemistry 61:899–905

Cobos MI, Rodriguez JL, Oliva MD et al (2001) Composition and antimicrobial activity of the essential oil of *Baccharis notosergila*. Planta Med 67:84–86

da Costa RM, Bastos JK, Costa MCA et al (2018) In vitro cytotoxicity and structure-activity relationship approaches of ent-kaurenoic acid derivatives against human breast carcinoma cell line. Phytochemistry 156:214–223

da Silva AA, Resende DO, Fukui MJ et al (2009a) Constituents of *Baccharis dracunculifolia* DC (Asteraceae) with in vitro antileishmanial, antiplasmodial and cytotoxic activities. Planta Med 75:949–949

da Silva AA, Resende DO, Fukui MJ et al (2009b) In vitro antileishmanial, antiplasmodial and cytotoxic activities of phenolics and triterpenoids from *Baccharis dracunculifolia* D. C. (Asteraceae). Fitoterapia 80:478–482

da Silva ER, Lazarotto DC, Pawlowski A et al (2018) Antioxidant evaluation of *Baccharis* patens and *Baccharis psiadioides* essential oils. J Essent Oil-Bear Plants 21:485–492

de Carvalho MP, Weich H, Abraham WR (2016) Macrocyclic Trichothecenes as antifungal and anticancer compounds. Curr Med Chem 23:23–35

de Sousa IP, Teixeira MVS, Furtado NAJC (2018) An overview of biotransformation and toxicity of diterpenes. Molecules 23:1–32. https://doi.org/10.3390/molecules23061387
Enriz RD, Baldoni HA, Jauregui EA et al (1994) Structure-activity relationship of clerodane diterpenoids acting as antifeedant agents. J Agric Food Chem 42:2958–2963
Fabiane KC, Ferronatto R, dos Santos AC et al (2008) Physicochemical characteristics of the essential oils of *Baccharis dracunculifolia* and *Baccharis uncinella* DC (Asteraceae). Rev Bras Farmacogn 18:197–203
Feresin GE, Tapia A, Gimenez A et al (2003) Constituents of the Argentinian medicinal plant *Baccharis grisebachii* and their antimicrobial activity. J Ethnopharmacol 89:73–80
Figueiredo-Rinhel AS, Kabeya LM, Bueno PC et al (2013) Inhibition of the human neutrophil oxidative metabolism by *Baccharis dracunculifolia* DC (Asteraceae) is influenced by seasonality and the ratio of caffeic acid to other phenolic compounds. J Ethnopharmacol 150:655–664
Florao A, Budel JM, Duarte MD et al (2012) Essential oils from *Baccharis* species (Asteraceae) have anti-inflammatory effects for human cells. J Essent Oil Res 24:561–570
Flores RC, Ponzi M, Ardanaz C et al (2009) Chemical composition of essential oil of *Baccharis salicifolia* (Ruiz & Pavon) pers. and antibacterial activity. J Chil Chem Soc 54:475–476
Freitas CS, Baggio CH, dos Santos AC et al (2009) Antinociceptive properties of the hydroalcoholic extract, fractions and compounds obtained from the aerial parts of *Baccharis illinita* DC in mice. Basic Clin Pharmacol 104:285–292
Frizzo CD, Serafini LA, Dellacassa E et al (2001) Essential oil of *Baccharis uncinella* DC. from Southern Brazil. Flavour Fragr J 16:286–288
Fullas F, Hussain RA, Chai HB et al (1994) Cytotoxic constituents of *Baccharis gaudichaudiana.* J Nat Prod 57:801–807
Funes M, Garro MF, Tosso RD et al (2018a) Antinociceptive effect of neo-clerodane diterpenes obtained from *Baccharis flabellata.* Fitoterapia 130:94–99
Funes M, Tonn CE, Kurina-Sanz M (2018b) In vivo photoinduced [4+2] dimerization of a neo-clerodane diterpene in *Baccharis flabellata.* ROS and RNS scavenging abilities. J Photoch Photobio B 186:137–143
Furtado NAJC, Veneziani RCS, Ambrósio SR (2017) Farmacognosia. Atheneu, São Paulo
Garcia PA, de Oliveira AB, Batista R (2007) Occurrence, biological activities and synthesis of kaurane diterpenes and their glycosides. Molecules 12:455–483
Garcia FAD, Tanae MM, Torres LMB et al (2014) A comparative study of two clerodane diterpenes from *Baccharis trimera* (Less.) DC. on the influx and mobilization of intracellular calcium in rat cardiomyocytes. Phytomedicine 21:1021–1025
Guo YQ, Li YS, Xu J et al (2006) Bioactive ent-clerodane diterpenoids from the aerial parts of *Baccharis gaudichaudiana.* J Nat Prod 69:274–276
Guo Y, Li Y, Xu J et al (2007) Clerodane diterpenoids and flavonoids with NGF-potentiating activity from the aerial parts of *Baccharis gaudichaudiana.* Chem Pharm Bull 55:1532–1534
Hadad M, Zygadlo JA, Lima B et al (2007) Chemical composition and antimicrobial activity of essential oil from *Baccharis grisebachii* Hieron (Asteraceae). J Chil Chem Soc 52:1186–1189
Hikawczuk VEJ, Verrilli MAL, Borkowski EJ et al (2006) Antifeedant activity of neo-clerodane diterpenes from *Baccharis flabellata* Hook & Arn var. *flabellata* toward *Tribolium castaneum* Herbst: structure-activity relationships. Nat Prod Res 20:813–819
Hikawczuk VEJ, Saad JR, Giordano OS et al (2008) Insect growth regulatory effects of linear diterpenoids and derivatives from *Bacchatis thymifolia.* J Nat Prod 71:190–194
Jakupovic J, Schuster A, Ganzer U et al (1990) Sesquiterpenes and diterpenes from *Baccharis* species. Phytochemistry 29:2217–2222
Januario AH, Santos SL, Marcussi S et al (2004) Neo-clerodane diterpenoid, a new metalloprotease snake venom inhibitor from *Baccharis trimera* (Asteraceae): anti-proteolytic and anti-hemorrhagic properties. Chem Biol Interact 150:243–251
Kato FH, Viana NI, Santini CB et al (2012) Assessment of the in vitro and in vivo genotoxic and antigenotoxic effects of pimaradienoic acid in mammalian cells. Mutat Res Genet Toxicol Environ Mutagen 749:87–92

Kurdelas RR, Lopez S, Lima B et al (2012) Chemical composition, anti-insect and antimicrobial activity of *Baccharis darwinii* essential oil from Argentina, Patagonia. Ind Crop Prod 40:261–267

Lage TC, Montanari RM, Fernandes SA et al (2015) Chemical composition and acaricidal activity of the essential oil of *Baccharis dracunculifolia* De Candole (1836) and its constituents nerolidol and limonene on larvae and engorged females of *Rhipicephalus microplus* (Acari: Ixodidae). Exp Parasitol 148:24–29

Lago JHG, Romoff P, Favero OA et al (2008) Composition of essential oils from the leaves of six species of the B*accharis* genus from "Campos de altitude" of the Atlantic Forest of Sao Paulo. Quim Nova 31:727–730

Leandro LF, Oliveira Cardoso MJ, Caetano Silva SD et al (2014) Antibacterial activity of *Pinus elliottii* and its major compound, dehydroabietic acid, against multidrug-resistant strains. J Med Microbiol 63:1649–1653

Li RT, Morris-Natschke SL, Lee KH (2016) Clerodane diterpenes: sources, structures, and biological activities. Nat Prod Rep 33:1166–1226

Linstrom PJ, Mallard WG (eds) (2018) NIST Chemistry WebBook, NIST Standard Reference Database Number 69. National Institute of Standards and Technology, National Institute of Standards and Technology. https://doi.org/10.18434/T4D303. Accessed 11 Jun 2018.

Liu N, Wu C, Wang P et al (2018) Diterpenoids from liverworts and their biological activities. Curr Org Chem 22:1847–1860

Marostica MR, Daugsch A, Moraes CS et al (2008) Comparison of volatile and polyphenolic compounds in Brazilian green propolis and its botanical origin *Baccharis dracunculifolia*. Cienc Tecnol Aliment 28:178–181

Mizokami SS, Arakawa NS, Ambrosio SR et al (2012) Kaurenoic acid from *Sphagneticola trilobata* inhibits inflammatory pain: effect on cytokine production and activation of the NO-cyclic GMP-protein kinase G-ATP-sensitive potassium channel signaling pathway. J Nat Prod 75:896–904

Negreiros MD, Pawlowski A, Zini CA et al (2016) Antimicrobial and antibiofilm activity of *Baccharis psiadioides* essential oil against antibiotic-resistant *Enterococcus faecalis* strains. Pharm Biol 54:3272–3279

Oliva MM, Zunino MP, Lopez ML et al (2007) Variation in the essential oil composition and antimicrobial activity of *Baccharis spartioides* (H. et A.) J. Rimy from three regions of Argentina. J Essent Oil Res 19:509–513

Parreira NA, Magalhaes LC, Morais DR et al (2010) Antiprotozoal, schistosomicidal, and antimicrobial activities of the essential oil from the leaves of *Baccharis dracunculifolia*. Chem Biodivers 7:993–1001

Passero LFD, Bonfim-Melo A, Corbett CEP et al (2011) Anti-leishmanial effects of purified compounds from aerial parts of *Baccharis uncinella* C. DC. (Asteraceae). Parasitol Res 108:529–536

Pereira CB, Farago PV, Borsato DM et al (2016) Chemical composition and biological activities of *Baccharis milleflora* essential oil. Lat Am J Pharm 35:2225–2233

Pereira CB, Kanunfre CC, Farago PV et al (2017) Cytotoxic mechanism of *Baccharis milleflora* (less.) DC. Essential oil. Toxicol In Vitro 42:214–221

Perera WH, Bizzo HR, Gama PE et al (2017) Essential oil constituents from high altitude Brazilian species with antimicrobial activity: *Baccharis parvidentata* Malag., *Hyptis monticola* Mart. ex Benth. and *Lippia origanoides* Kunth. J Essent Oil Res 29:109–116

Possebon MI, Mizokami SS, Carvalho TT et al (2014) Pimaradienoic acid inhibits inflammatory pain: inhibition of NF-kappa B activation and cytokine production and activation of the NO-cyclic GMP-protein kinase G-ATP-sensitive potassium channel signaling pathway. J Nat Prod 77:2488–2496

Salazar GJT, de Sousa JP, Lima CNF et al (2018) Phytochemical characterization of the *Baccharis dracunculifolia* DC (Asteraceae) essential oil and antibacterial activity evaluation. Ind Crop Prod 122:591–595

San-Martin A, Astudillo L, Gutierrez M et al (2010) 13-epi-neoclerodanes from *Baccharis marginalis*. J Chil Chem Soc 55:118–120

Sarah KJ-C, Andrew TK, David TF (2014) Synthetic routes to methylerythritol phosphate pathway intermediates and downstream isoprenoids. Curr Org Chem 18:1050–1072

Sayuri VA, Romoff P, Favero OA et al (2010) Chemical composition, seasonal variation, and biosynthetic considerations of essential oils from *Baccharis microdonta* and *B. elaeagnoides* (Asteraceae). Chem Biodivers 7:2771–2782

Severiano ME, Simão MR, Porto TS et al (2010) Anticariogenic properties of ent-pimarane diterpenes obtained by microbial transformation. Molecules 15:8553–8566

Shibamoto T (1987) Retention indices in essential oil analysis, capillary gas chromatography in essential oil analysis. Huethig Verlag, New York, pp 259–274

Simirgiotis MJ, Quispe C, Borquez J et al (2016) High resolution metabolite fingerprinting of the resin of *Baccharis tola* Phil. from the Atacama Desert and its antioxidant capacities. Ind Crop Prod 94:368–375

Simonsen HT, Riedel C, Gade LB et al (2009) Chemical composition and antibacterial activity of the leaf essential oil of *Baccharis magellanica* (Lam.) Pers. and *Baccharis elaeoides* Remy from Chile. J Essent Oil Res 21:377–380

Sobrinho ACN, de Souza EB, Rocha MFG et al (2016) Chemical composition, antioxidant, antifungal and hemolytic activities of essential oil from *Baccharis trinervis* (Lam.) Pers. (Asteraceae). Ind Crop Prod 84:108–115

Sosa ME, Tonn CE, Giordano OS (1994) Insect antifeedant activity of clerodane diterpenoids. J Nat Prod 57:1262–1265

Sosa ME, Lancelle HG, Tonn CE et al (2012) Insecticidal and nematicidal essential oils from Argentinean *Eupatorium* and *Baccharis* spp. Biochem Syst Ecol 43:132–138

Tirapelli CR, Ambrosio SR, de Oliveira AM et al (2010) Hypotensive action of naturally occurring diterpenes: a therapeutic promise for the treatment of hypertension. Fitoterapia 81:690–702

Tischer B, Vendruscolo RG, Wagner R et al (2017) Effect of grinding method on the analysis of essential oil from *Baccharis articulata* (Lam.) Pers. Chem Pap 71:753–761

Torres LMB, Gamberini MT, Roque NF et al (2000) Diterpene from *Baccharis trimera* with a relaxant effect on rat vascular smooth muscle. Phytochemistry 55:617–619

Trombin-Souza M, Trombin-Souza M, Amaral W et al (2017) Chemical composition of the essential oils of *Baccharis* species from southern Brazil: a comparative study using multivariate statistical analysis. J Essent Oil Res 29:400–406

Ueno AK, Barcellos AF, Costa-Silva TA et al (2018a) Antitrypanosomal activity and evaluation of the mechanism of action of diterpenes from aerial parts of *Baccharis retusa* (Asteraceae). Fitoterapia 125:55–58

Ueno AK, Barcellos AF, Grecco SD et al (2018b) Sesquiterpenes, diterpenes, alkenyl p-coumarates, and flavonoid from the aerial parts of *Baccharis retusa* (Asteraceae). Biochem Syst Ecol 78:39–42

Valarezo E, Rosillo M, Cartuche L et al (2013) Chemical composition, antifungal and antibacterial activity of the essential oil from *Baccharis latifolia* (Ruiz & Pav.) Pers. (Asteraceae) from Loja, Ecuador. J Essent Oil Res 25:233–238

Valarezo E, Rosales J, Morocho V et al (2015) Chemical composition and biological activity of the essential oil of *Baccharis obtusifolia* Kunth from Loja, Ecuador. J Essent Oil Res 27:212–216

van Baren CM, Lira PD, Bandoni AL et al (2002) Composition of the essential oil of pichana *Baccharis spartioides* (Hook. et Arn.) Remy (Compositae) from different populations of the Patagonia, Argentina. J Essent Oil Res 14:183–186

Vannini AB, Santos TG, Fleming AC et al (2012) Chemical characterization and antimicrobial evaluation of the essential oils from *Baccharis uncinella* DC and *Baccharis semiserrata* DC (Asteraceae). J Essent Oil Res 24:547–554

Veneziani RCS, Ambrósio SR, Martins CHG, Lemes DC, Oliveira LC (2017) Antibacterial potential of diterpenoids. In: Atta-ur-Rahman BT-S (ed) Studies in natural products chemistry. Elsevier, Dordrecht, pp 109–139. https://doi.org/10.1016/B978-0-444-63929-5.00004-8

Verdi LG, Brighente IMC, Schripsema J et al (2004) Kaurene diterpenes and flavonoids from *Baccharis illinita* flowers. Biochem Syst Ecol 32:837–840

Wachter GA, Montenegro G, Timmermann BN (1999) Diterpenoids from *Baccharis pingraea*. J Nat Prod 62:307–308

Xavier VB, Vargas RMF, Minteguiaga M, Umpiérrez N, Dellacassa E, Cassel E (2013) Evaluation of the key odorants of *Baccharis anomala* DC essential oil: new applications for known products. Ind Crop Prod 49:492–496

Xu WH, Liu WY, Liang Q (2018) Chemical constituents from *Croton* species and their biological activities. Molecules 23:1–38. https://doi.org/10.3390/molecules23092333

Yamamoto ES, Campos BLS, Laurenti MD et al (2014) Treatment with triterpenic fraction purified from *Baccharis uncinella* leaves inhibits *Leishmania* (*Leishmania*) *amazonensis* mazonensis spreading and improves Th1 immune response in infected mice. Parasitol Res 113:333–339

Zunino MP, Newton MN, Maestri DM et al (1998) Essential oils of three *Baccharis* species. Planta Med 64:86–87

Zunino MP, Lopez ML, Faillaci SM et al (2000) Essential oil of *Baccharis cordobensis* Heering. Flavour Fragr J 15:151–152

Zunino MP, Lopez ML, Zygadlo JA et al (2004) Essential oil composition of *Baccharis articulata* (Lam.) Pers. J Essent Oil Res 16:29–30

Chapter 14
Macrocyclic Trichothecenes of *Baccharis*

Alessandra Caroline Montes Frade, Maria Beatriz de Oliveira Rabelo, Priscilla Rodrigues Valadares Campana, Rodrigo Maia de Pádua, and Fernão Castro Braga

Abstract Trichothecenes constitute a large group of sesquiterpene metabolites. They are classified into four different groups (types A, B, C, and D) according to their characteristic functional groups. These structural differences result in distinct biological activities. About 40 macrocyclic trichothecenes have been isolated so far from three species of *Baccharis* (*Baccharis megapotamica*, *Baccharis coridifolia*, and *Baccharis artemisioides*) and their chemical structures have been defined by spectroscopic analyses, especially by using nuclear magnetic resonance (NMR) and mass spectrometry (MS) data. It was initially supposed that trichothecenes found in *Baccharis* species originated from fungi present in the soil, such as *Myrothecium* species. Endophytic fungi synthesize exclusively simple trichothecenes, while some macrocyclic trichothecenes have been isolated from *B. coridifolia* and *B. megapotamica*, indicating that they are produced by the plants. The biosynthetic pathway for the production of macrocyclic trichothecenes is not fully elucidated and to date, only a few intermediates and final products have been isolated and characterized. Macrocyclic trichothecenes have been reported to possess antifungal, antimalarial, antitumor, and antiviral activities. The effects of trichothecenes on animal and plant cells include inhibition of DNA and RNA syntheses, inhibition of mitochondrial function, membrane destabilization, changes in cell division, and apoptosis. Additional phytochemical and biological studies with *Baccharis* species are demanded to better understand the chemical and medicinal properties of trichothecenes, aiming at future exploitation.

A. C. M. Frade · M. B. de Oliveira Rabelo · P. R. V. Campana
R. M. de Pádua · F. C. Braga (✉)
Department of Pharmaceutical Products, Faculty of Pharmacy, Universidade Federal de Minas Gerais, Belo Horizonte, Brazil
e-mail: fernaobraga@farmacia.ufmg.br

G. W. Fernandes et al. (eds.), *Baccharis*,
https://doi.org/10.1007/978-3-030-83511-8_14

Keywords Biological activities · Biosynthetic pathway · Chemical structure · Plant chemistry · Toxic effects

1 General Aspects of Trichothecenes

Trichothecenes comprise an important group of mycotoxins, usually responsible for human and animal intoxications (for more information, see Chap. 11 "Livestock intoxication by *Baccharis*"), which also plays a significant role in the pathogenicity of some plants (Jarvis et al. 1991,1996).

The accumulation of trichothecenes in agricultural products, such as rice, oats, rye, barley, and wheat, leads to great economic losses due to the destruction of crops by plant intoxication (McCormick et al. 2011). Their ingestion by humans and animals causes clinical symptoms of acute intoxication, such as subcutaneous hemorrhaging, alimentary toxic aleukia (ALA), deterioration of throat tissue, exhaustion of bone marrow, abortion of fetuses, and even death (Kumari et al. 2016).

Trichothecin (6) was the first isolated trichothecene, resulting from the screening of antifungal metabolites in *Trichothecium roseum* culture. This compound is also produced by several species of fungi from the genera *Fusarium, Myrothecium, Trichoderma*, and *Cephalosporium*. Subsequently, other trichothecenes, designate verrucarins and roridins, were isolated from *Myrothecium verrucaria* and *Myrothecium roridum*, respectively (Ueno 1980).

These mycotoxins are divided into two groups: simple trichothecenes, such as T-2 toxin (1), diacetoxiscirpenol (2), and deoxynivalenol (3), and macrocyclic trichothecenes, like baccarinoids, roridins, and verrucarins. These compounds have several biological activities, including antimicrobial, cytotoxic, and insecticidal properties, besides being phytotoxic (Jarvis et al. 1991).

Before discovering the presence of macrocyclic trichothecenes in *Baccharis megapotamica* Spreng, these compounds had been isolated from soil fungus species. Thereafter, other trichothecenes isolated from *Baccharis coridifolia,* popularly known as "mio-mio" or "romerillo," have been reported (Jarvis et al. 1987c). Later, Rizzo and collaborators (Rizzo et al. 1997) found roridins and verrucarin in *Baccharis artemisioides.*

The genus *Baccharis* (Asteraceae) encompasses about 500 species, mainly distributed in Brazil, Argentina, Paraguay, and Uruguay (Verdi et al. 2005). In those countries, *B. coridifolia* is considered to be one of the principal poisonous plants to herbivorous mammals, causing severe losses among cattle (Habermehl et al. 1985).

Throughout this chapter, all macrocyclic trichothecenes isolated so far from *Baccharis* species (*B. megapotamica*, *B. coridifolia*, and *B. artemisioides*) will be reviewed. Also, the biosynthesis of these compounds will be discussed, as well as their already investigated biological activities.

2 Classification and Structure of Trichothecenes

Trichothecenes constitute a large group of sesquiterpene metabolites. They generally have a common nucleus, represented by a rigid tetracyclic ring, consisting of a cyclohexene ring (ring A with a C9-C10 double bond) attached to a tetrahydropyran ring (ring B) and a cyclopentyl ring (ring C), in addition to an epoxide at the C12-C13 position (Shank et al. 2011). The rigidity of the system results in a distinct stereochemistry, in which ring A adopts a half-chair conformation and ring B is found in a chair conformation (Jarvis et al. 1990). These chemical structures are represented in Fig. 14.1. The trichothecenes differ otherwise in the patterns of side-chain oxygenation and esterification at carbons 3,4,7,8,15 and in the presence of a keto group at C8 (represented by R in Fig. 14.1a) (Desjardins et al. 2007).

The trichothecenes are classified into different groups according to their characteristic functional groups. Type A compounds have the simplest structures, the C8 position is esterified, hydroxylated, or unsubstituted, while type B trichothecenes have a carbonyl at the C8 position. Type C derivatives have an additional epoxide goup at C7-C8 or C8-C9 positions, while type D trichothecenes have a macrocyclic ring between the C4 and C15 positions (Wu et al. 2013; Terciolo et al. 2018). These structural differences result in distinct biological activities. Examples of chemical structures representative of each group of trichothecenes can be seen in Fig. 14.2.

Typically, trichothecenes are resistant to degradation by environmental factors, like air and light, but may be affected by the presence of bacteria or fungi. They are nonvolatile compounds with a low molecular weight (between 250-550 Da), naturally occurring as colorless, crystalline, and optically active solids. They are highly soluble in acetone, ethyl acetate, chloroform, dimethyl sulfoxide (DMSO), ethanol, methanol, and propylene glycol. Trichothecenes are not inactivated by autoclaving but can be effectively deactivated under strong acid or alkaline conditions (Wu et al. 2013; Sudakin 2003).

Several methods can be used for the analysis and identification of these substances. When trichothecin (6) was first isolated in 1948, its structure could not be

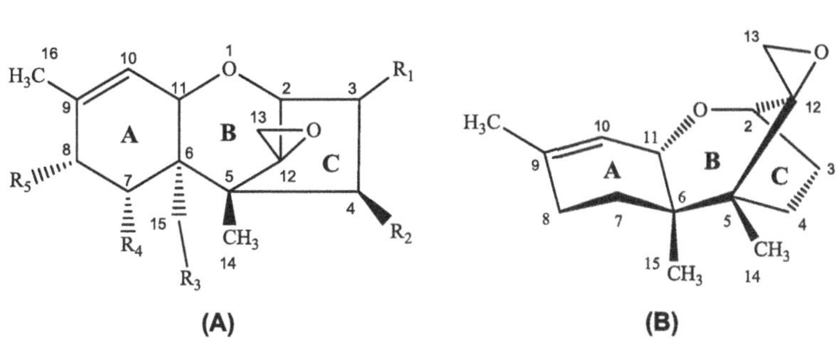

Fig. 14.1 (**a**) Chemical structure of the trichothecene core. (**b**) Three-dimensional stereochemistry of the trichothecene core. (Source: Shank, adapted (Shank et al. 2011))

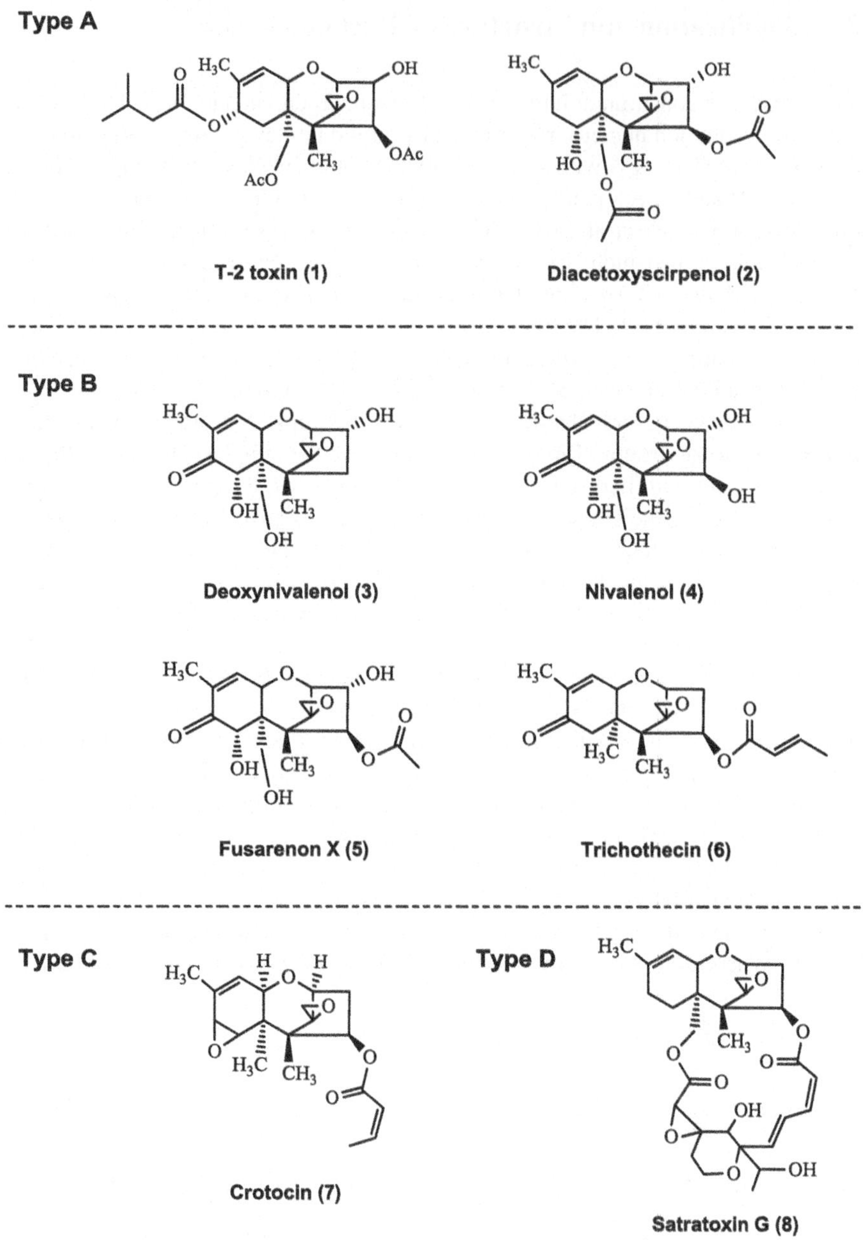

Fig. 14.2 Chemical structures of types A-D trichothecenes. (Source: Terciolo, adapted (Terciolo et al. 2018))

determined by the available methods (Shank et al. 2011). The chemical structure of trichothecin was reported only in 1959 by using chemical modifications, which allowed the determination of the toxin nucleus and substituent groups (Freeman et al. 1959).

The number of isolated trichothecenes with identified chemical structures started to increase after new structure elucidation techniques became available. A fundamental technique for structural elucidation and stereochemical assignment of trichothecenes was nuclear magnetic resonance (NMR) (Shank et al. 2011). Trichothecenes have some relevant structural features, easily identified by NMR data. For example, the ^{1}H methylene coupling of the epoxide ring is found between 2.7 to 3.4 ppm, with a scalar coupling constant of approximately 4 Hz. The methyl hydrogens at C-14 appear as a sharp singlet, while the methyl hydrogens at C-16 appear as a broad singlet, because of the long-range coupling with H-10. In addition, the rigidity of the ring system can explain the long-range coupling observed between H-7 and H-11 (Savard et al. 1987).

In the ^{1}H NMR spectra of macrocyclic trichothecenes, H-2′ resonates at 5.2 ppm in 2′-enes and at 3.4 ppm in 2′,3′-epoxides. In verrucarins, H-7′ and H-10′ appear as doublets between 5.8 and 6.2 ppm, H-9′ is found as a triplet between 6.5 and 6.6 ppm, and H-8′ is seen as a double doublet at 8.1 ppm. In roridins, the configuration at C-6′ determines the value of $J_{6',7'}$, which is 6 Hz in compounds where C-6′ has *S* configuration and 2 Hz if C-6′ has *R* configuration. Regarding ^{13}C NMR data, the chemical shift values reported for carbons of the trichothecene nucleus are very similar, whereas changes occur in the macrocyclic ring. For example, in (6′-*R*)-roridins, C-13′ resonates at higher frequencies (1-2 ppm) in 13′-*R* compounds in comparison to diastereoisomers with 13′-*S* configuration, as found in baccharinoids B13 (20) and B14 (21) (Grove 1993).

The ^{1}H and ^{13}C NMR spectral data reported for baccharin (9), a macrocyclic trichothecene, are depicted in Table 14.1.

Table 14.1 ^{1}H and ^{13}C NMR spectral data reported for baccharin (δ in ppm)[a]

Position	H	C	Position	H	C
2		78.1	1′		167.4
3	2.48	34.0	2′	3.37	56.0
4	5.80	73.8	3′		64.4
5		48.5	4′		75.3
6		42.3	5′		72.1
7		16.7	6′		86.5
8		25.7	7′	5.98	138.2
9		57.7	8′	7.48	125.1
10	3.11	56.9	9′	6.60	142.6
11		66.6	10′	5.82	117.4
12		65.2	11′		166.3
13	2.75 / 3.16	47.1	12′	1.65	11.6
14	0.75	6.5	13′		71.0
15	4.24 / 4.42	63.1	14′	1.20	17.7
16	1.83	21.6			

[a]Spectra measured in $CDCl_3$. Source: Kupchan (Kupchan et al. 1976; Kupchan et al. 1977)

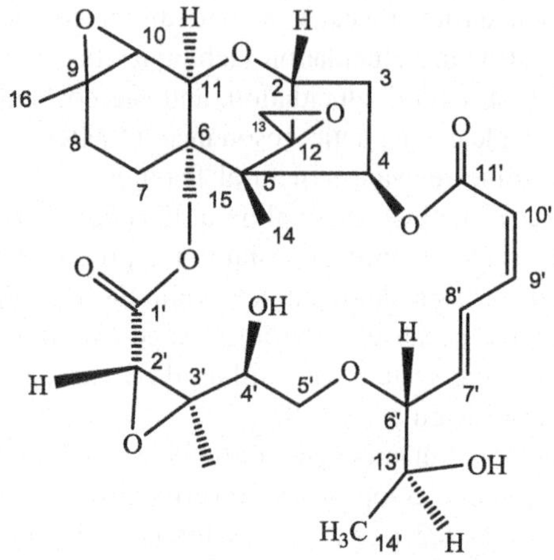

Baccharin (9)

Rosen et al. (1986) converted some macrocyclic trichothecenes into their trimethylsilyl derivatives for their analysis by gas chromatography coupled to mass spectrometry (GC/MS), since the presence of polar, labile ester bridges in the molecules difficult their direct analysis by this technique. Other methods of detection and quantification were posteriorly introduced, with improved sensitivity, selectivity, and accuracy, by using direct chemical ionization tandem mass spectrometric techniques with minimal amounts of samples (Krishnamurthy and Sarver 1989).

Roridins and baccharinoids were identified in Brazilian *Baccharis* species by a direct chemical ionization tandem mass spectrometry (MS/MS) method. A mixture of these compounds was subjected to chemical ionization in the presence of ammonia followed by sequential collisionally activated dissociation of the specific adducts using argon. The identification occurs by detection of specific daughter ions and their associated parent ions (Krishnamurthy and Sarver 1989). Table 14.2 lists some chemical ionization MS data reported for macrocyclic trichothecenes.

Later, high-performance liquid chromatography coupled to mass spectrometry (HPLC-MS), employing a reversed-phase column and thermospray ionization, allowed the unambiguous identification of isomeric baccharinoids (Krishnamurthy and Sarver 1989). The ions observed in the mass spectra revealed that structurally similar compounds undergo common fragmentation forms and neutral losses (Krishnamurthy and Sarver 1989). Thus, Krishnamurthyt and Sarver (1989) identified several daughter ions of macrocyclic trichothecenes. Table 14.3 lists the mass spectral data reported for the most significant daughter ions formed by thermospray ionization, whereas their chemical structures are depicted in Fig. 14.3.

The identification of daughter ions structures and neutral losses made it possible to propose fragmentation pathways for these compounds. Thus, it was observed that most of the fragments are formed by the bond break between the ester group and ether bonds in the exocyclic ring (Krishnamurthy and Sarver 1989).

Table 14.2 Chemical ionization mass spectra data (M^- ions) reported for some macrocyclic trichothecenes of *Baccharis*

Compound	MW	*m/z* (% relative abundance)
Baccharinol (11)	562	562 (3.8); 501 (10.1); 417 (100.0); 375 (2.4); 365 (29.9); 161 (1.4); 153 (19.9); 143 (12.8); 141 (1.2); 135 (12.9); 125 (12.2)
Roridin A (30)	532	402 (35.8); 401 (100.0); 153 (19.6); 145 (69.8)
Roridin D (31)	530	530 (1.7); 417 (1.3); 401 (100.0); 359 (2.3); 349 (1.0); 193 (2.5); 175 (1.1); 153 (7.1); 151 (1.2); 149 (2.5); 145 (3.1); 135 (13.1); 127 (1.2)
Roridin E (32)	514	514 (14.9); 484 (7.0); 470 (2.1); 403 (71.4); 386 (3.4); 385 (17.2); 359 (100.0); 154 (6.2); 153 (20.3); 136 (10.7); 134 (3.8); 111 (8.1)

Source: Krishnamurthy, adapted. (Krishnamurthy and Sarver 1989)

Table 14.3 Mass spectral data reported for the most significant daughter ions of some macrocyclic trichothecenes

Compound	MW	m/z
Baccharinol (11)	562	137, 247, 391
Roridin A (30)	532	249, 333, 403
Roridin D (31)	530	231, 249, 403
Roridin E (32)	514	137, 231, 361

Source: Krishnamurthy, adapted (Krishnamurthy and Sarver 1989)

Figure 14.4 shows the fragmentation pathway proposed for roridin A. Cleavage of C4-O and C6'-O bonds originates the neutral fragment of 154 Da, whereas cleavages between C3'-C4' and C6'-O result in neutral losses of 46 Da (ethanol). In its turn, cleavages at C1'-O and C6'-O bonds give neutral losses of 130 Da. The daughter ion at *m/z* 333 can be formed by neutral losses of 200 or 154 Da, followed by a loss of 46 Da. The other neutral losses observed are due to cleavages of bonds connected to C1'-O, C4-O, or C11'-O atoms (Krishnamurthy and Sarver 1989).

3 Biosynthesis of Trichothecenes

Trichothecenes are secondary metabolites mainly produced by filamentous fungi of the genera *Fusarium, Trichothecium, Stachybotrys, Mycothecium, Cephalosporium, Verticimonosporium*, and *Trichoderma* (Rocha et al. 2005; Arunachalam and Doohan 2013), being *Fusarium* spp. the principal producer (Kimura et al. 2007; Kumari et al. 2016).

Fusarium species are very diversified and have a wide geographic distribution (Smith 2007), commonly found as aerial parts colonizers of several vegetal species. They can be also found in association with roots of plants or in the soil, being part of the normal microflora or acting as pathogens (Nelson et al. 1994; Guarro 2013). Although fungi are the main producers of trichothecenes, some compounds

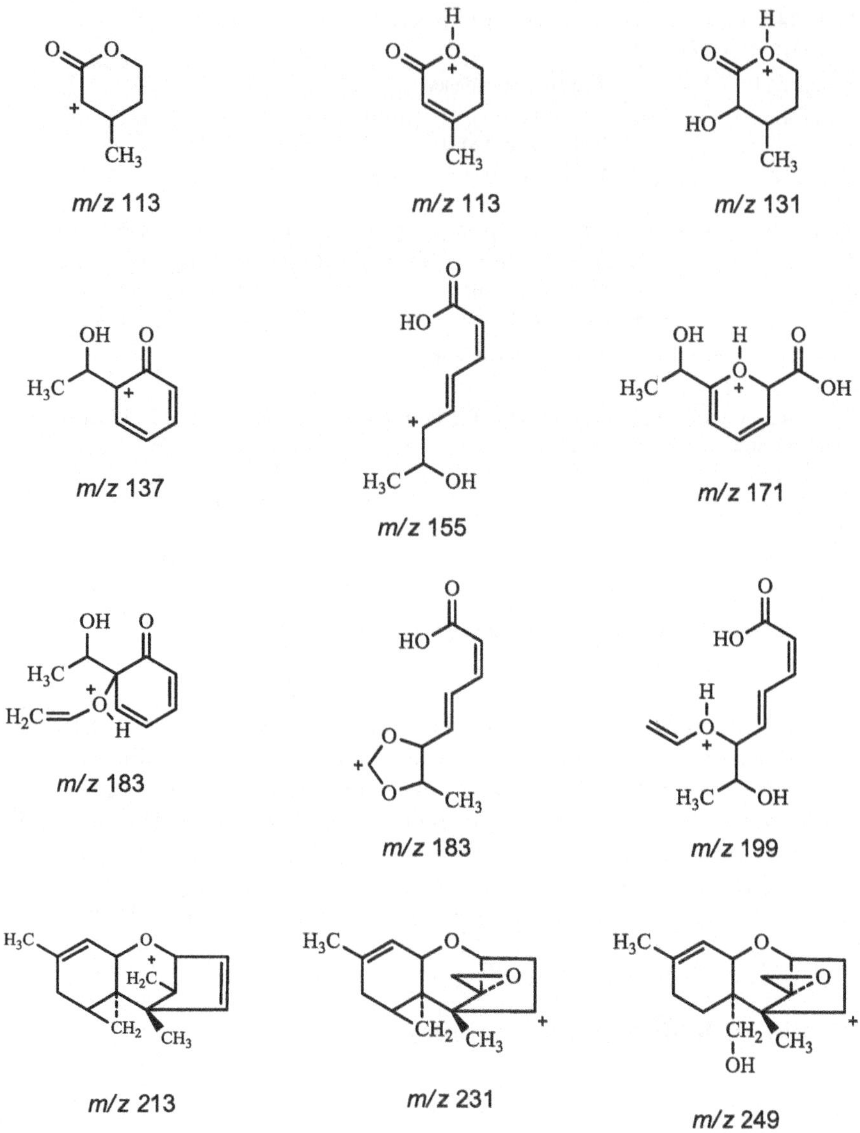

Fig. 14.3 Chemical structures of the most common daughter ions resultant from macrocyclic trichothecenes fragmentation by thermospray ionization. (Source: Krishnamurthy, adapted (Krishnamurthy and Sarver 1989))

of this class have already been isolated from plants of the genus *Baccharis*. It is important to mention that fungi exclusively produce simple trichothecenes, while some macrocyclic trichothecenes, such as baccharinoids, verrucarins A (36) and J (38), roridins A (30), D (31), and E (32), were isolated from *B. coridifolia* and *B. megapotamica* (Jarvis et al. 1991; Grove 1993).

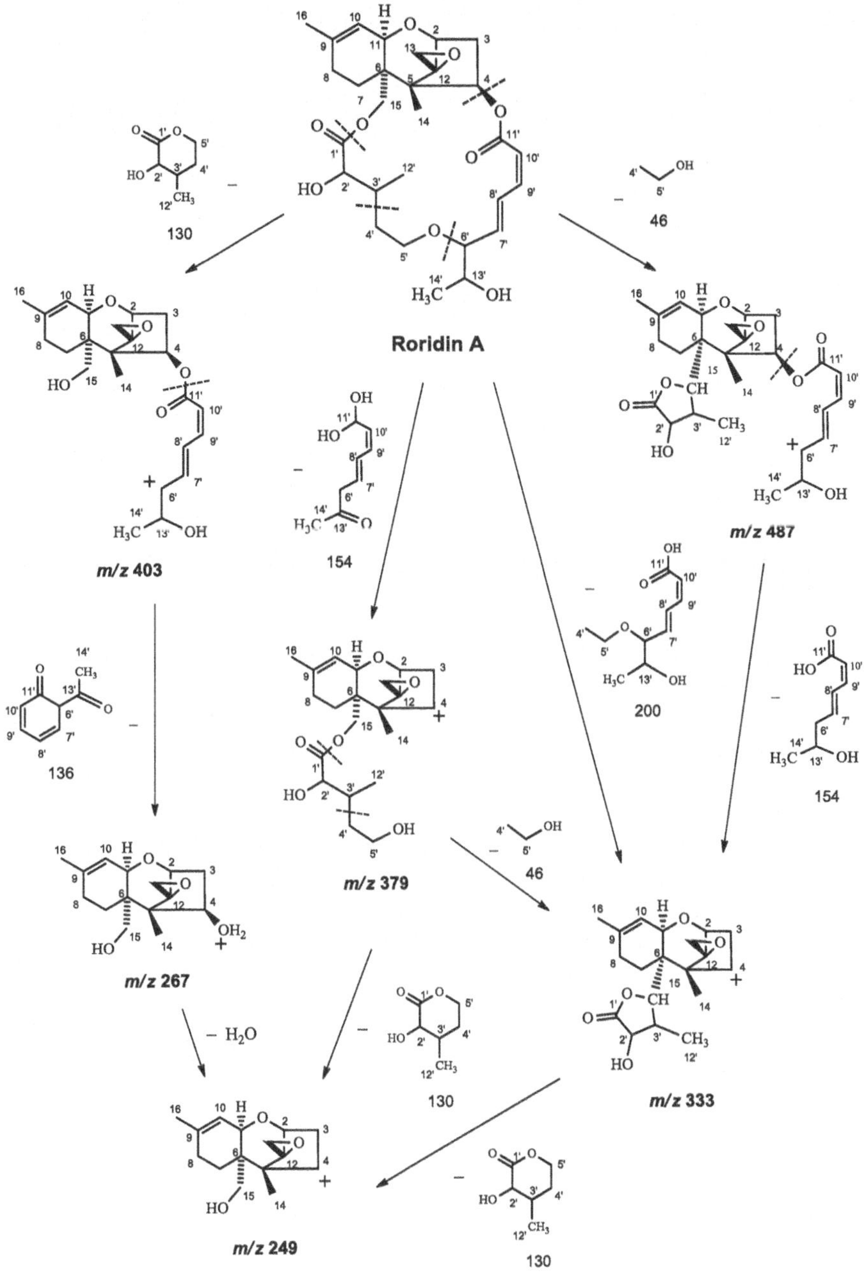

Fig. 14.4 Fragmentation pathway for roridin A. (Source: Krishnamurthy, adapted (Krishnamurthy and Sarver 1989))

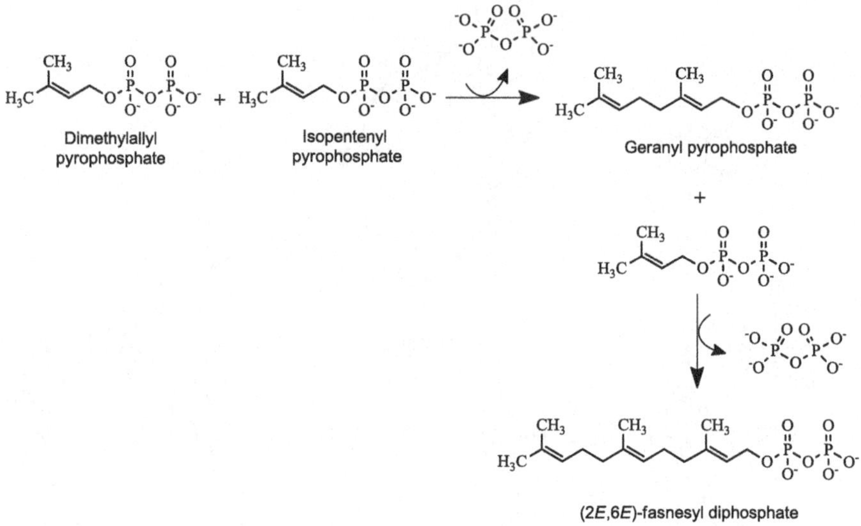

Fig. 14.5 Synthesis of farnesyl pyrophosphate. (Source: Dhar, adapted (Dhar et al. 2013))

The biosynthesis of trichothecenes in *Fusarium* and *Trichothecium* species was extensively studied. Trichothecenes have a carbon skeleton derived from farnesyl pyrophosphate (FPP), which is synthesized by the condensation reaction of dimethylallyl pyrophosphate with two units of isopentenyl pyrophosphate, having geranyl pyrophosphate as intermediate compounds (Kimura et al. 2007; Dhar et al. 2013), as shown in Fig. 14.5.

Figure 14.6 presents the biosynthetic pathway of trichothecenes proposed for *Fusarium* species. In the biosynthetic sequence, the cyclization reaction of an acyclic FPP molecule produces the bicyclic trichodiene (TDN), a precursor of trichothecenes (Kimura et al. 2007; Villafana et al. 2019). In its turn, TND is submitted to sequential and nonrandom oxidation reactions catalyzed by specific enzymes, at carbons 2, 12, and 13 (epoxidation reaction), along with C-11 and C-3, forming isotrichodiol. Subsequently, trichodiol formation occurs in a nonenzymatic reaction. The trichodiol molecule undergoes a second cyclization, giving rise to isotrichodermol, the first trichothecene of the biosynthetic pathway. Changes in the carbon skeleton, catalyzed by enzymes, lead to the formation of calonectrin, an important biosynthetic intermediate of different trichothecenes, such as deoxynivalenol (3), T-2 toxin (1), and nivalenol (4) (Villafana et al. 2019).

The biosynthetic pathway for the production of macrocyclic trichothecenes (Fig. 14.7) is not fully elucidated and to date, only a few intermediates and final products have been isolated and characterized from fungal cultures of *Myrothecium* species (Trapp et al. 1998). This pathway has isotrichodiol as a starting compound that undergoes nonenzymatic isomerization and cyclization to form 12,13-epoxytrichothec-9-ene. Next, hydroxylation occurs at C-15 and esterification takes place at C-4, with the introduction of a polyketide chain in this position. In the

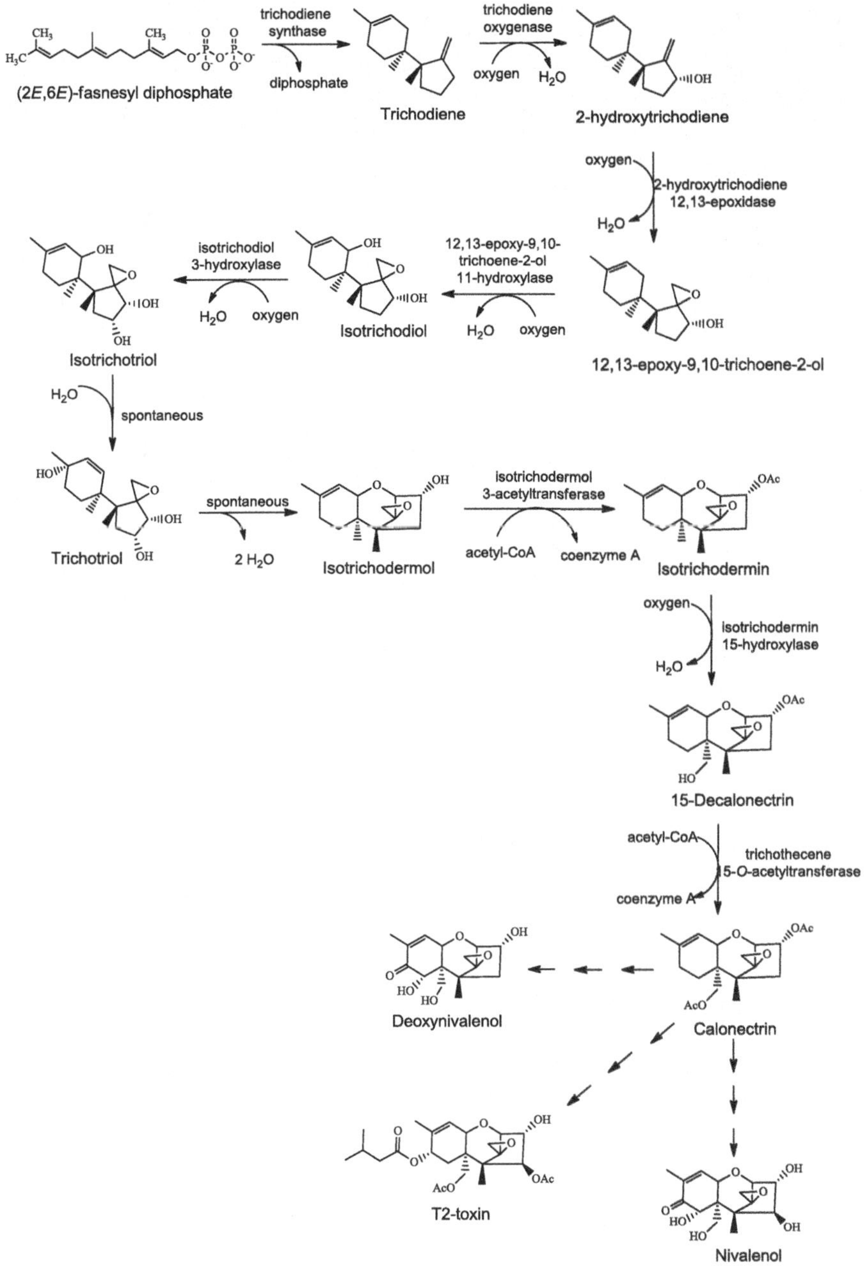

Fig. 14.6 Biosynthesis of trichothecenes in *Fusarium* species. (Source: Villafana, adapted (Villafana et al. 2019))

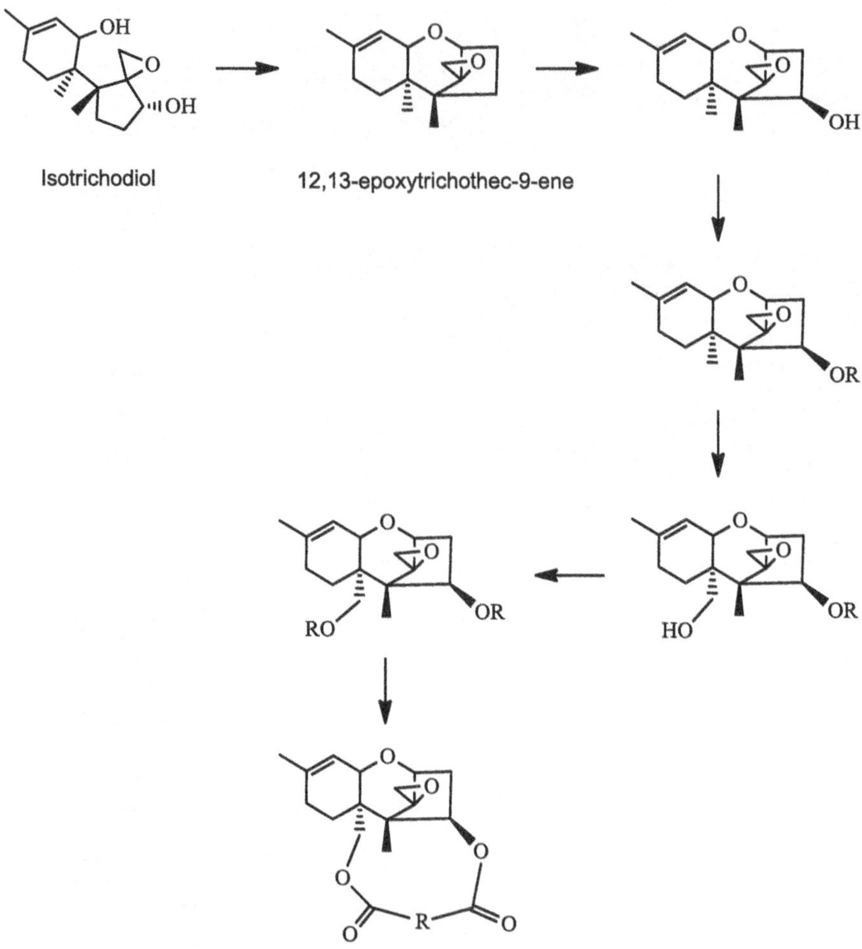

Fig. 14.7 Biosynthesis proposal for macrocyclic trichothecenes formation in *Myrothecium* species. (Source: McCormick, adapted (McCormick et al. 2011))

sequence, the hydroxyl group is esterified forming a polyketide ester at C-15. The condensation of the polyketide chains at C-4 and C-15 results in the formation of a macrocyclic trichothecene (McCormick et al. 2011).

It was initially supposed that trichothecenes found in *Baccharis* species originated from fungi present in the soil, such as *Myrothecium* species, which produce roridins. Thus, it was first accepted that in *B. megapotamica*, the roridins were absorbed, translocated, and metabolized in baccharinoids, whereas in *B. coridifolia*, they were only stored (Jarvis et al. 1981; Jarvis et al. 1987b, c).

However, in 1988, Jarvis et al. (1988b) refuted the original hypothesis of trichothecenes production by fungi in *Baccharis* species and claimed that these metabolites are synthesized by the plants themselves. The authors came to this conclusion after analyzing species of *B. coridifolia*, for which neither *Myrothecium verrucaria* nor *Myrothecium roridum* could be detected. In addition, they could not isolate endophytic fungi from the species. The authors also found high concentrations of roridins A (30) and E (32) in *B. coridifolia* seeds.

In a later study of 1990, Kuti et al. (1990) sought to demonstrate the role of phytotoxins in seed maturation and germination of *Baccharis* species. The authors observed that *Baccharis* species from Brazil (*B. megapotamica* and *B. coridifolia*), which produce trichothecenes, had their germination completely inhibited when the seed husks were removed in comparison to American *Baccharis* (*B. halimifolia* and *B. glutinosa*), which presented better germination rates. Based on these results, the authors suggested that macrocyclic trichothecenes play a regulatory role in the reproduction and germination of *B. megapotamica* and *B. coridifolia*.

In relation to the distribution of these toxins between male and female individuals, it was observed that female *B. coridifolia* plants have a higher concentration of trichothecenes than the male ones, whereas for *B. megapotamica*, no difference was found. It is important to note that male individuals of *B. coridifolia* have miotoxin C (41) as a major component, whereas in female individuals, there is a higher concentration of roridins A (30) and E (32) (Jarvis et al. 1991). According to Jarvis et al. (1996), female plants of *B. coridifolia* are more toxic than male plants, since they contain five to ten times more trichothecenes.

4 Macrocyclic Trichothecenes in *Baccharis* Species

In 1976, Kupchan et al. (1976) isolated the first macrocyclic trichothecene from a plant species. In a screening of antitumor compounds from natural sources, the researchers observed the inhibitory activity of *B. megapotamica* ethanolic extract against KB cells (nasopharyngeal carcinoma) in vitro and in a model of P-388 murine leukemia in vivo. The phytochemical investigation resulted in the isolation of baccharin (9), isobaccharin (10), baccharinol (11), and isobaccharinol (12), considered responsible for the antileukemic activity (Kupchan et al. 1977). After that, other baccharinoids were isolated from the same species, including the baccharinoids B1 (13), B2 (14), B3 (15), B7 (16), B9 (17), B10 (18), B12 (19), B13 (20), B14 (21), B16 (22), B17 (23), B20 (24), B21 (25), B23 (27), B24 (28), B25(26), and B27 (29) (Jarvis et al. 1987a, b).

Due to the toxic effects caused by *Baccharis coridifolia*, associated with poisoning in animals, Habermehl et al. (1985) identified nine macrocyclic trichothecenes, considered responsible for the toxic effect: roridins A (30) and E (32), miotoxins A (39), B (40), C (41), D (47), and iso-D (48), along with miophytocens A (50) and B (51). It is noteworthy that roridins had already been isolated as secondary metabolites from fungal cultures, while miotoxins and miophytocens were unknown compounds. Later, Jarvis et al. (1996) isolated and characterized new trichothecene glycosides from the same species, including roridin A β-glucoside (33), roridin D β-glucoside (34), roridin E β-glucoside (35), verrucarin A β-glucoside (37), miotoxin A β-glucoside (45), and miotoxin F β-glucoside (46), in addition to the new miotoxins E (42), F (43) and G (44). According to Rizzo et al. (1997), specimens of *B. coridifolia* collected in Argentina presented verrucarins A (36) and J (38), in addition to the roridins.

The first report on macrocyclic trichothecenes in *Baccharis artemisioides* occurred in 1977, when Rizzo et al. (1997) described the presence of roridins A (30), D (31) and E (32), along with verrucarins A (36) and J (38) in an Argentina specimen. Furthermore, the authors reported higher concentrations of toxins in *B. artemisioides* than in *B. coridifolia* (Rizzo et al. 1997).

The chemical structures of the macrocyclic trichothecenes here described are depicted in Fig. 14.8.

5 Biological Activities of Macrocyclic Trichothecenes

Antifungal, antimalarial, antitumor, and antiviral activities have been reported for macrocyclic trichothecenes. The toxic effects on eukaryotic cells are complex and varied; the toxicity and selectivity are dependent on the compound and cell type tested (Arunachalam and Doohan 2013).

The effect of macrocyclic trichothecenes on the growth of wheat (*Triticum aestivum* L.), bean (*Phaseolus vulgaris* L.), corn (*Zea mays* L.), and tobacco (*Nicotiana tabacum* L.) was investigated. The authors observed that roridin A (30), verrucarins A (36) and J (38) not only inhibited the growth but were also toxic to these species (Cutler and Jarvist 1985). In addition, roridins E (32) and H (49), along with verrucarins A (36) and J (38), presented phytotoxicity against cultures of duckweed (*Lemna pausicostata* L.) and kudzu (*Pueraria lobata* L.) (Abbas et al. 2002). In view of this phytotoxicity, the effects of roridins A (30) and E (32) on cell cultures of Brazilian *Baccharis* (*B. megapotamica* and *B. coridifolia*) and North American *Baccharis* (*B. halimifolia* and *B. glutinosa*) were compared, the last ones being more sensitive to the cytotoxic effects (Jarvis et al. 1988a).

Regarding the antifungal effect, roridins A (30) and D (31) were active against *Saccharomyces cerevisiae*, *Magnaporthe grisea*, and *Sclerotinia sclerotiorum* strains, with minimum inhibitory concentrations lower than or equal to fluconazole, used as positive control (Xie et al. 2008). Moreover, roridin A (30) and verrucarin A (36) inhibited *Candida albicans*, *Aspergillus niger*, and *Trichophyton rubrum* growth, at lower or equal concentrations of the positive control ketoconazole (Liu et al. 2006).

Isaka et al. (1999) demonstrated the antimalarial activity of roridins A (30) and E (32), and verrucarins A (36) and J (38); they all showed lower EC_{50} than the reference drug artemisinin. In the same study, the authors reported high cytotoxic activity of these compounds against human oral epidermoid carcinoma KB cells, human breast cancer BC1 cells, and Vero cells. Zhang et al. (2002) described the inhibition growth of *Plasmodium falciparum* strains induced by roridin E (32), with IC_{50} values lower than 1 ng/mL.

Roridins A (30) and E (32), along with verrucarins A (36) and J (38), were assayed against the Junin virus (JUNV), the etiologic agent of the Argentine hemorrhagic fever (AHF). All assayed trichothecenes inhibited virus replication at non-toxic concentrations, being verrucarin J (38) the most active compound, with a better selectivity index (García et al. 2002).

B. megapotamica (Kupchan et al. 1976, 1977; Jarvis et al. 1987a, b)

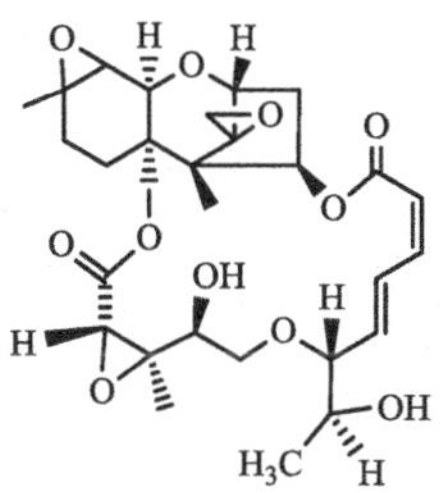

Baccharin (9)

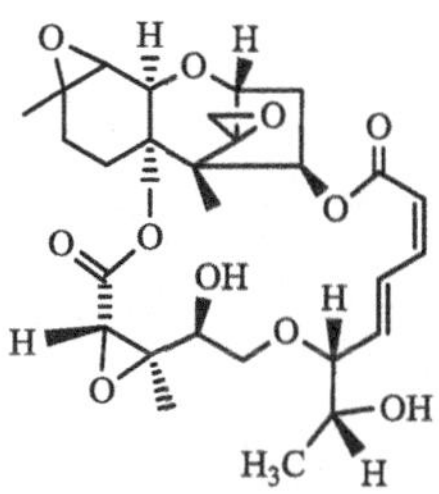

Isobaccharin (10)

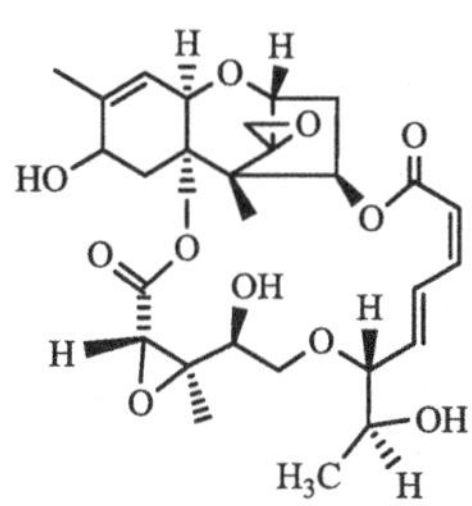

Baccharinol (11)

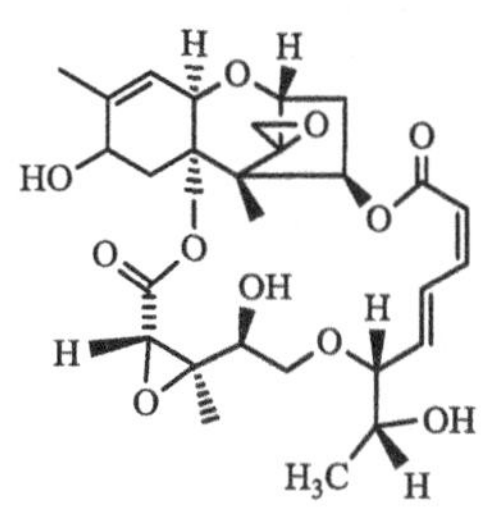

Isobaccharinol (12)

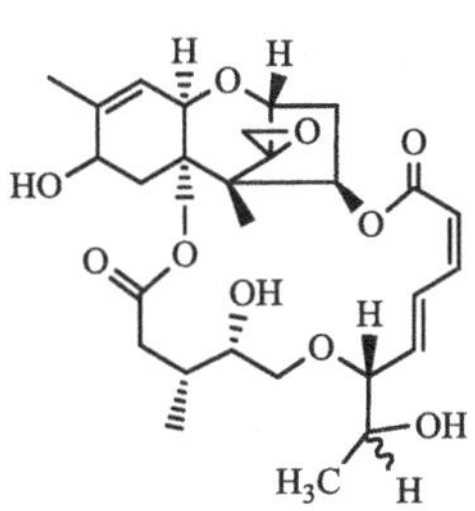

Baccharinoid B1 C13' *R* (13)
Baccharinoid B2 C13' *S* (14)

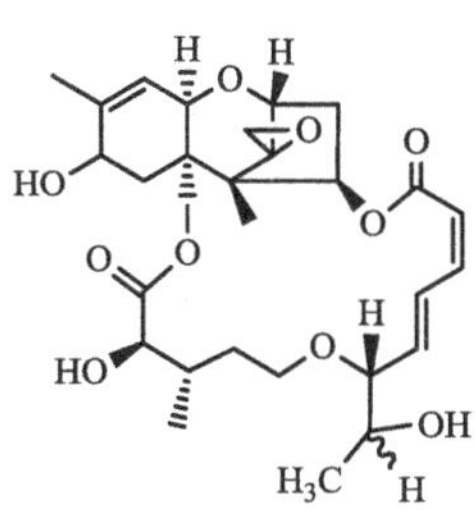

Baccharinoid B3 C13' *S* (15)
Baccharinoid B7 C13' *R* (16)

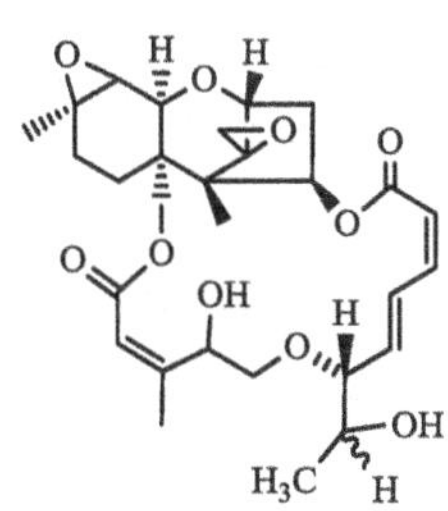

Baccharinoid B9 C13' *R* (17)
Baccharinoid B10 C13' *S* (18)

Fig. 14.8 Chemical structures of macrocyclic trichothecenes isolated from *Baccharis* species

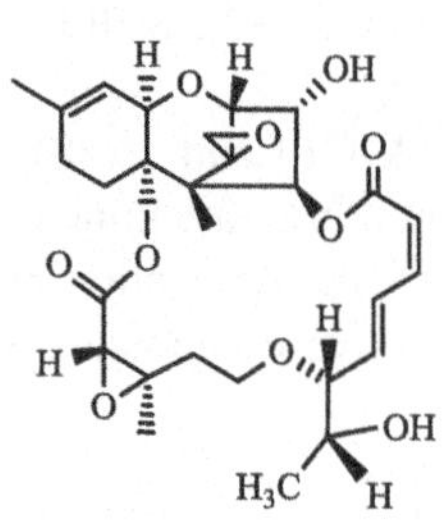

Baccharinoid B12 (19)

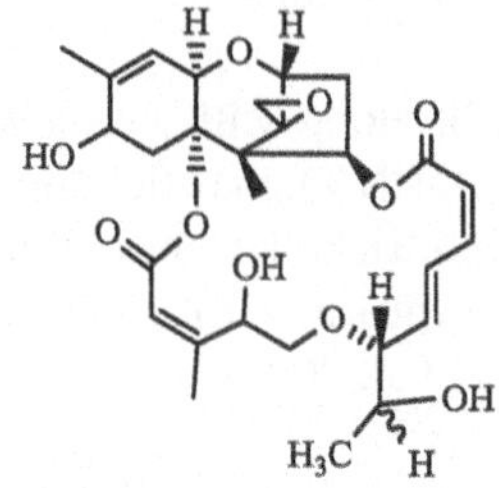

Baccharinoid B13 C13′ *R* (20)
Baccharinoid B14 C13′ *S* (21)

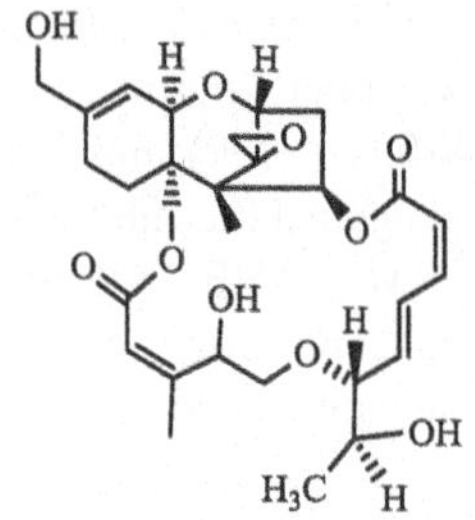

Baccharinoid B16 (22)

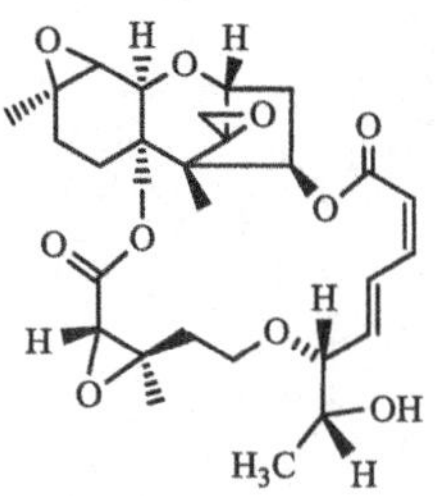

Baccharinoid B17 (23)

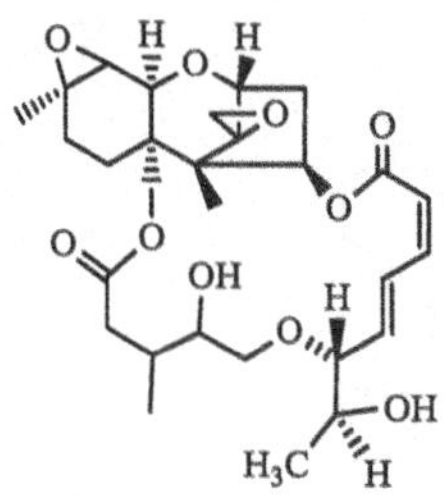

Baccharinoid B20 (24)

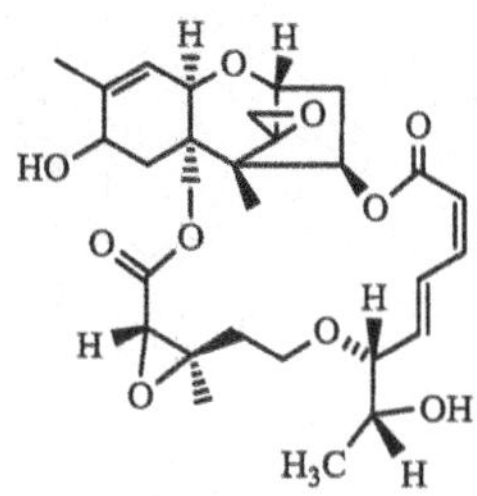

Baccharinoid B21 (25)

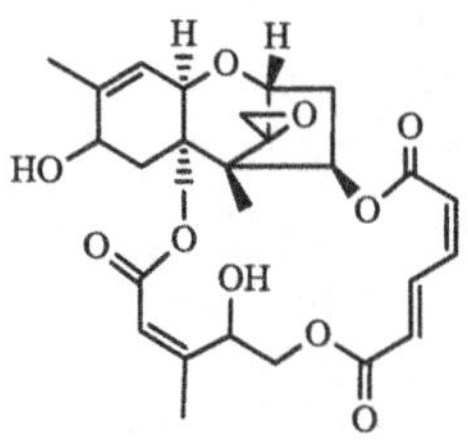

Baccharinoid B25 (26)

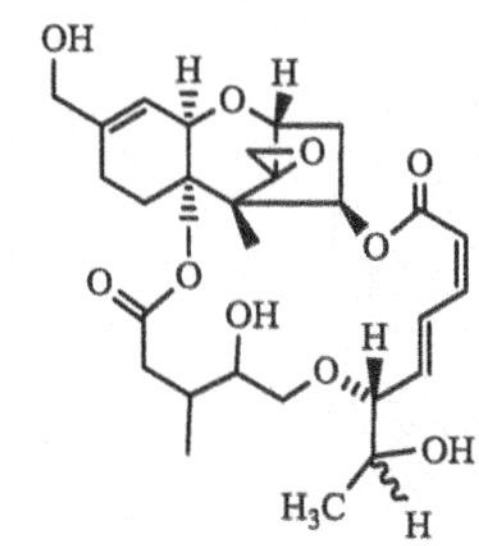

Baccharinoid B23 C13′ *R* (27)
Baccharinoid B24 C13′ *S* (28)

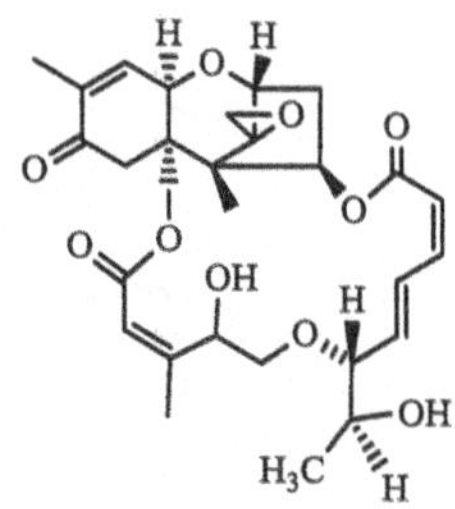

Baccharinoid B27 (29)

B. coridifolia (Habermehl et al. 1985; Jarvis et al. 1996; Rizzo et al. 1997)

Fig. 14.8 (continued)

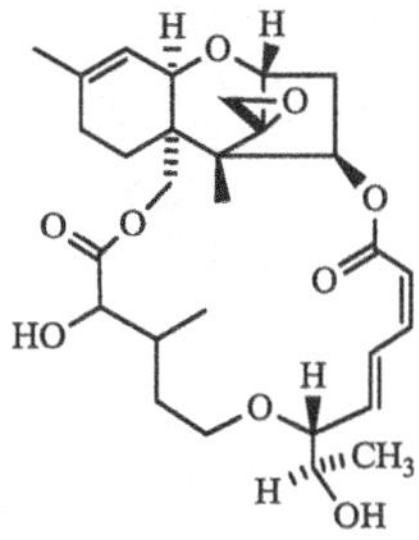

Roridin A (30)

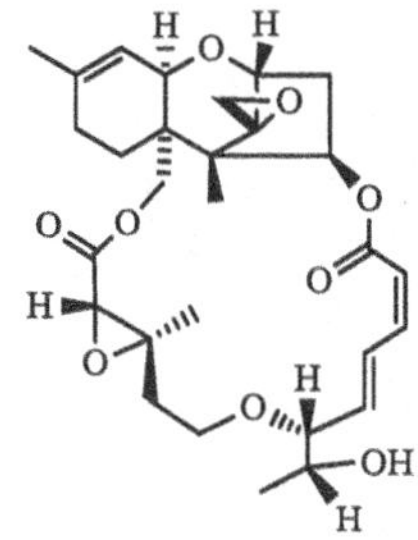

Roridin D (31)

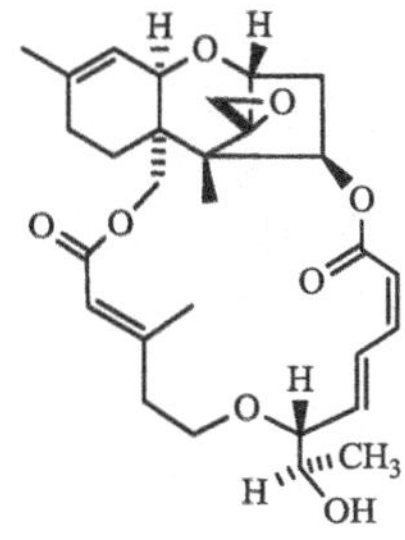

Roridin E (32)

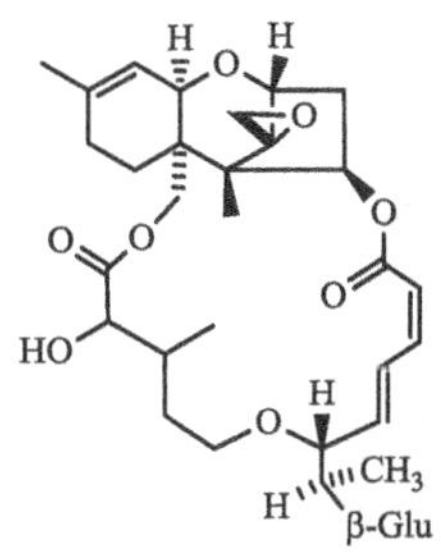

Roridin A b-Glucoside (33)

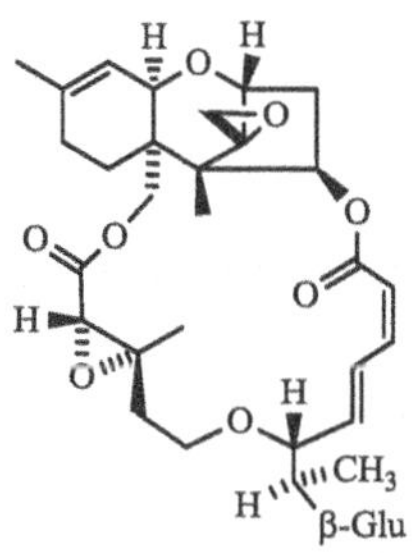

Roridin D b-Glucoside (34)

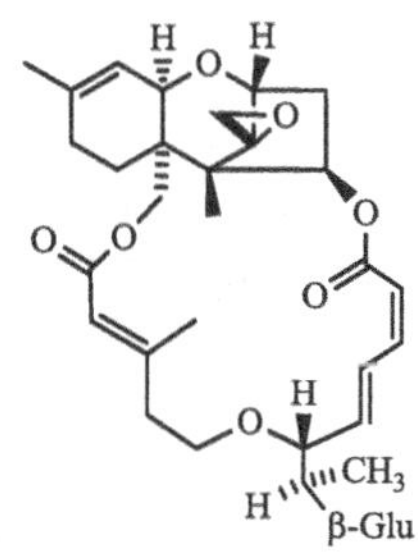

Roridin E b-Glucoside (35)

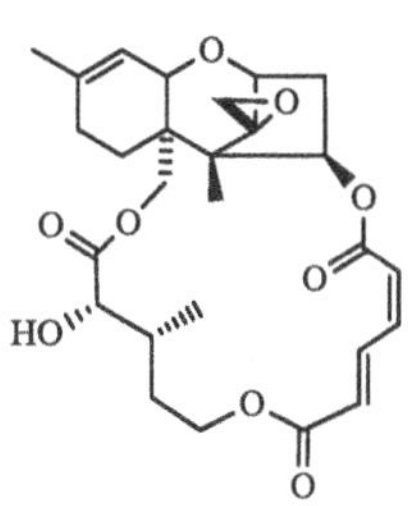

Verrucarin A (36)

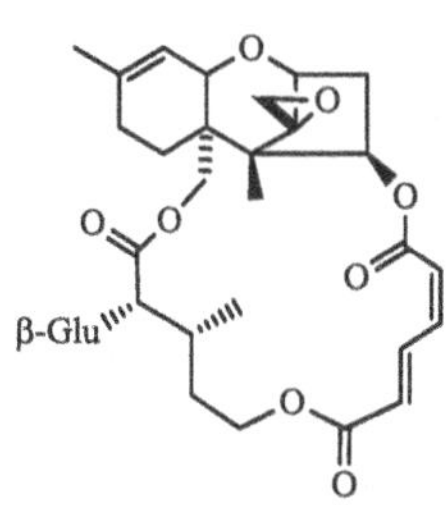

Verrucarin A b-Glucoside (37)

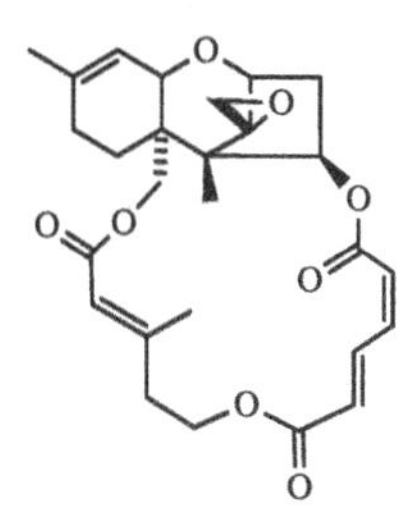

Verrucarin J (38)

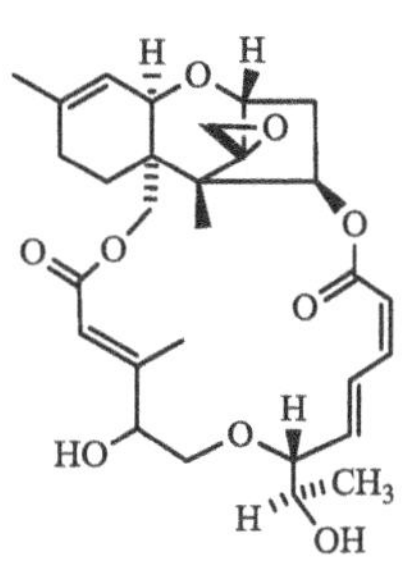

Miotoxin A (39)

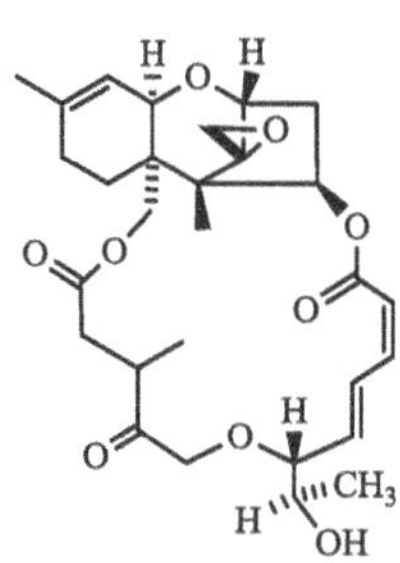

Miotoxin B (40)

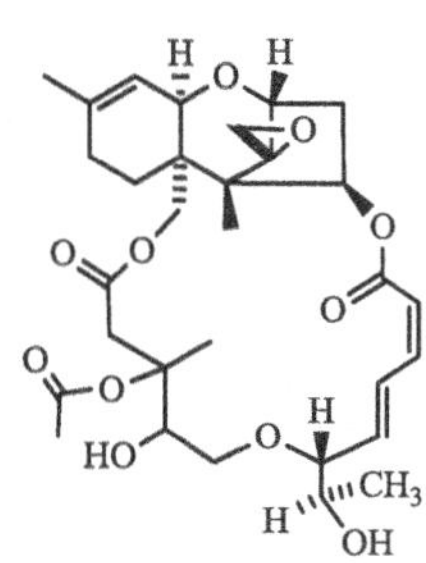

Miotoxin C (41)

Fig. 14.8 (continued)

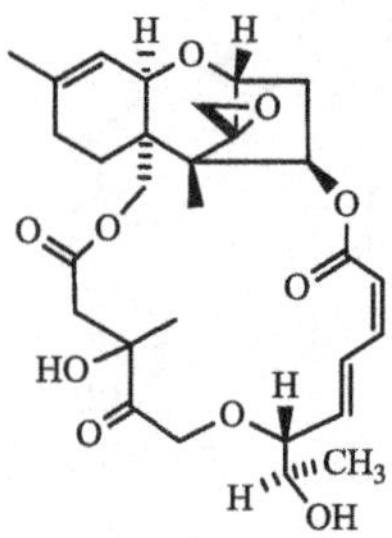

Miotoxin E (42)

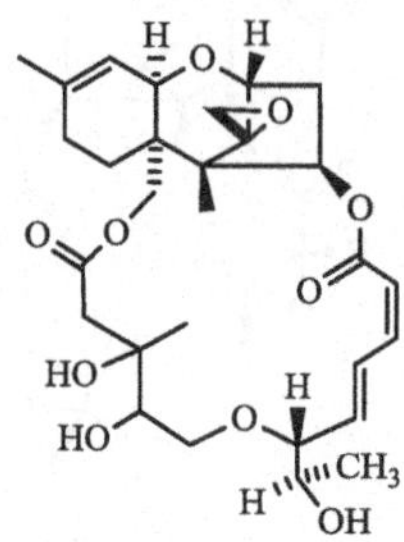

Miotoxin F (43)

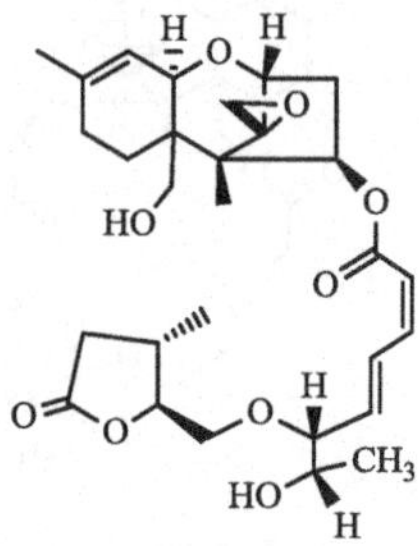

Miotoxin G (44)

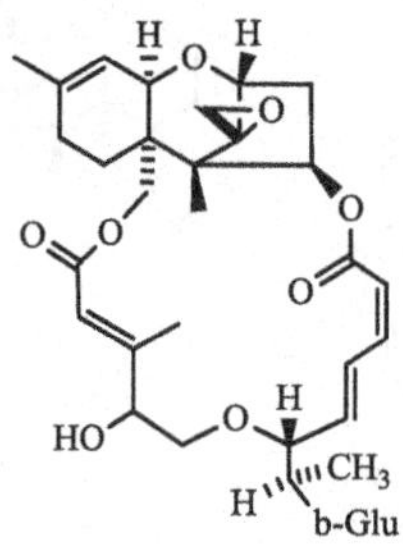

Miotoxin A b-Glucoside (45)

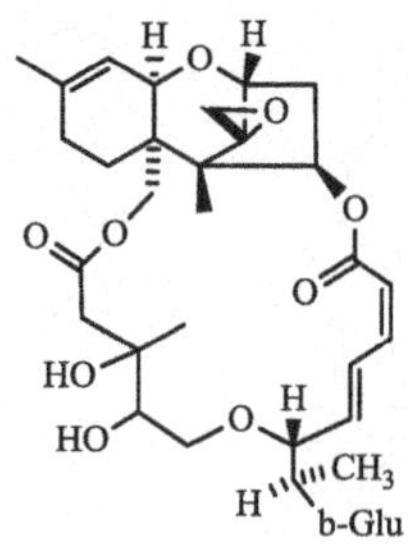

Miotoxin F b-Glucoside (46)

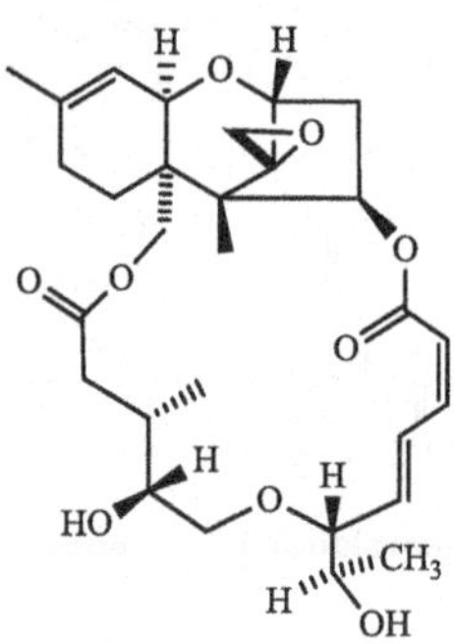

Miotoxin D (47)

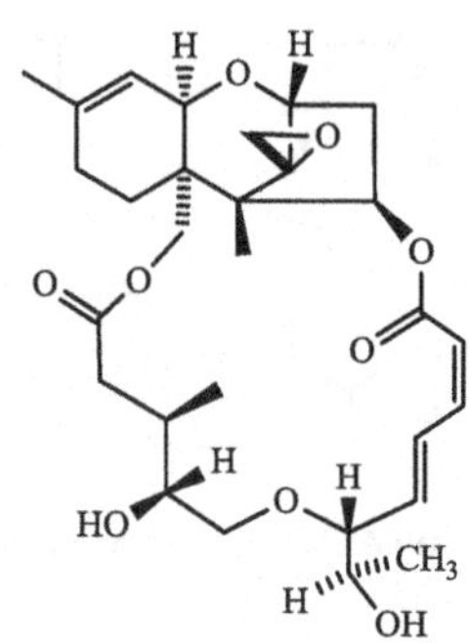

Iso - Miotoxin D (48)

Fig. 14.8 (continued)

The cytotoxic activity of several mycotoxins against NIH/3 T3 cells (Swiss mouse embryonic fibroblast) and BE 12-6 cells (bovine embryonic cells) was evaluated and disclosed verrucarin A (36) and roridin A (30) as the most cytotoxic compounds (Terse et al. 1993). The cytotoxicity of roridins E (32) and H (49), along with verrucarins A (36) and J (38), has been demonstrated in NIH3T3 cells (Swiss mouse embryonic fibroblast), KA31T cells (oncogenically transformed 3 T3 mouse fibroblasts), H4TG cells (rat hepatoma), and MDCK cells (Madin-Darby canine

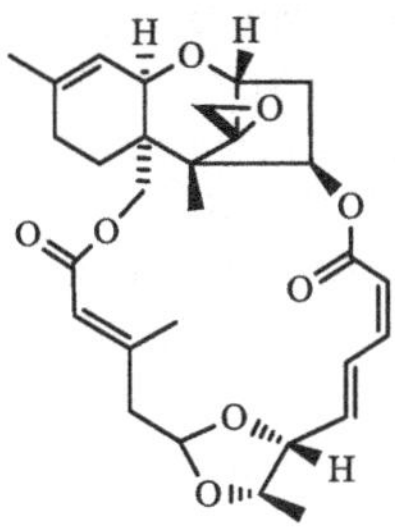

Roridin H (49)

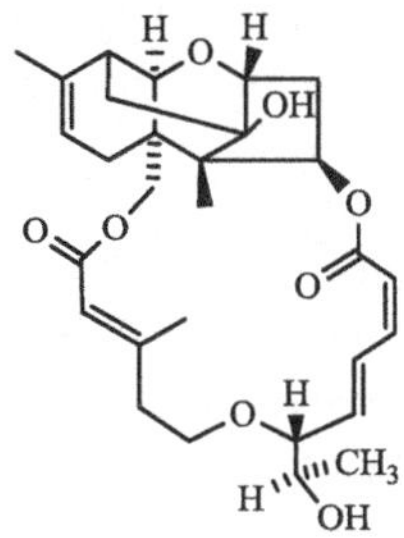

Miophytocen A (50)

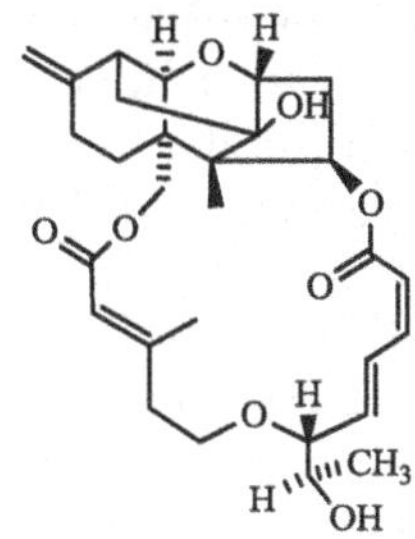

Miophytocen B (51)

B. artemisioides (Rizzo et al. 1997)

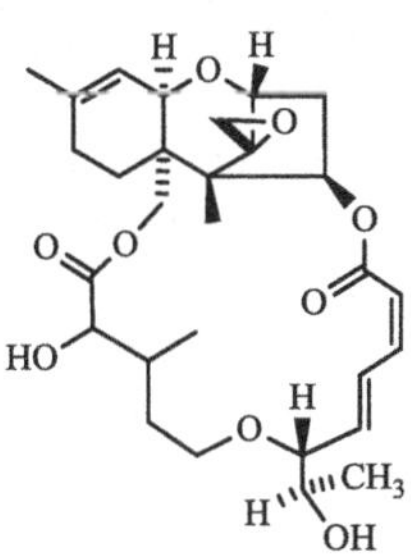

Roridin A (30)

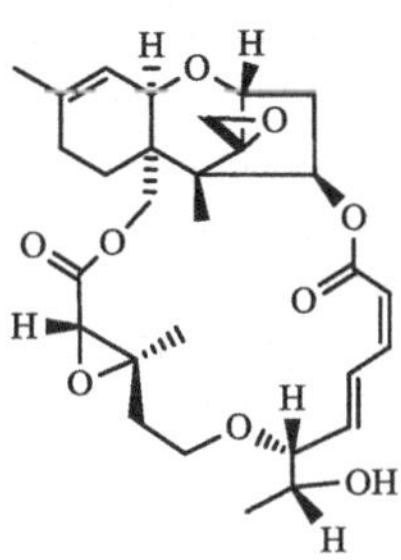

Roridin D (31)

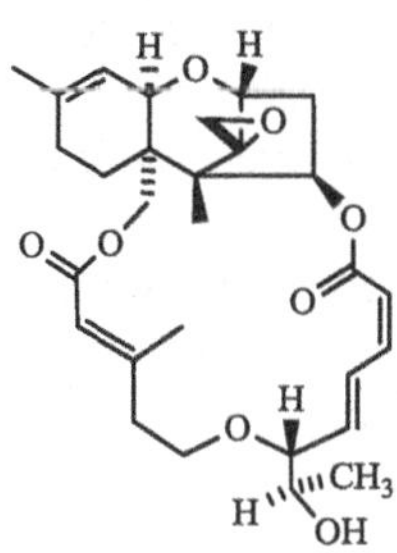

Roridin E (32)

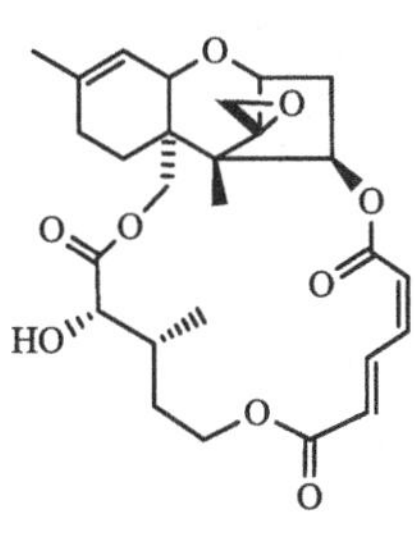

Verrucarin A (36)

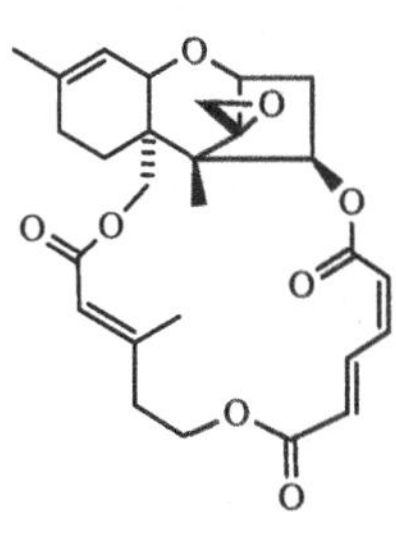

Verrucarin J (38)

Fig. 14.8 (continued)

kidney) (Abbas et al. 2002). In addition, baccharin (9), baccharinol (11), roridins A (30) and H (49), and verrucarins A (36) and J (38) were shown to be active in vivo against P388 mouse leukemia (Jarvis et al. 1984).

Hughes et al. (1990) evaluated the effects of roridins A (30), D (31), and E (32); baccharinoid B12 (19); and verrucarins A (36) and J (38) on the viability and mitogenesis of splenic lymphocytes of CD-1 mice. The trichothecenes showed cytotoxic effects at the concentration range from 10^{-6} to 10^{-4} M. Regarding the cell proliferation, roridin A (30) proved to be a good inhibitor, with concentrations lower than the maximum lethal effect.

Roridin E (32) was administered alone and in combination with linoleic acid to Sprague-Dawley rats. The combination produced more significant toxic effects, such as decreased blood glucose and glutathione levels, reduction of superoxide dismutase and glucose-6-phosphatase in the kidney, in addition to increased lipid peroxidation in the liver (El-Din et al. 1997).

6 Structure–Activity Relationship of Macrocyclic Trichothecenes

Most of the literature reporting structure-activity relationships is directed to simple trichothecenes, found in fungi species. As for macrocyclic trichothecenes, most studies were performed using two or more related compounds (de Carvalho et al. 2016). An example is the cytotoxic activity of roridin E (32) and its 12'-OH derivative against different leukemia lines (THP-1 cells – human monocytic leukemia and HL-60 cells – human promyelocytic leukemia) and against V79 cells (Chinese hamster lung fibroblast). Roridin E (32) showed higher cytotoxicity than the derivative, suggesting that a hydroxyl group at C-12′ position is required for the activity (Oda et al. 2010).

Amagata et al. (2003) studied the effect of some macrocyclic trichothecenes in tumor cell lines and in healthy cell lines, in order to analyze their cytotoxicity and selectivity. According to the authors, miophytocen C was the least potent compound and its lower potency would be related to the absence of the C12-C13 epoxide ring. On the other hand, trichothecenes like verrucarin A (36) and roridin A (30), which have an epoxide ring, were selectively active against the tumor cells. It can be therefore concluded that an epoxide ring at C12-C13 positions is required for selective cytotoxic effect. In addition, the authors concluded that the stereochemistry at C-6'and C-13'positions showed a small effect on cell selectivity.

Nineteen mycotoxins, all possessing the C12-C13 epoxide ring, were tested for their ability to inhibit protein synthesis in Vero cells and rat spleen lymphocytes. Trichothecenes with a C4-C15 hydrocarbon chain ring, namely, roridin A (30) and verrucarin A (36), showed higher potency than trichothecenes bearing open hydrocarbon chains at those positions, suggesting that the macrocyclic part is relevant for the cytotoxic effects (Thompson and Wannemacher 1986). Furthermore, 3D QSAR studies (Quantitative Structure–Activity Relationship) revealed that the toxicity is highly dependent on the conformation of the macrocyclic ring and that an additional epoxide ring at C9-C10 positions, as found in baccharin (9), increases the antileukemic activity (Steinmetz et al. 2009).

Verrucarins A (36) and J (38), along with roridin E (32), showed potent cytotoxic activity against *Artemia salina*. In this study, the verrucarins showed higher toxicity than roridin, suggesting that the presence of an ester at the C6'position is essential for the activity. In addition, since verrucarin A (36) was six times more active than verrucarin J (38), it seems that the C2'-OH group is capable of improving toxicity (Zhao et al. 2011).

The cytotoxicity of roridin H (49) and its 8α-OH and 8α-OAc derivatives was assayed against L929 cells (mouse fibroblast), MCF-7 cells (breast cancer), Hela KB3.1 cells, and A431 cells (skin cancer). The obtained results indicated that hydroxylation increased the cytotoxic activity against the tumor cells, whereas acetylation had the opposite effect. In addition, a hydroxyl group at C-8 position seems to be required for selectivity (de Carvalho et al. 2016).

7 Mode of Action of Macrocyclic Trichothecenes

Trichothecenes have several effects on animal and plant cells, including inhibition of DNA and RNA syntheses, inhibition of mitochondrial function, membrane destabilization, changes in cell division, and induction of apoptosis (Rocha et al. 2005). They also present effects on the inhibition of protein synthesis through interaction with ribosomes (Schindler et al. 1974; Cundliffe and Davies 1977).

In order to analyze the mode of action of some specific inhibitors of protein synthesis, trichothecenes were tested in H-HeLa cells. Roridin A (30) and verrucarin A (36) caused the accumulation of ribosomal structures, that were found to consist of 80S monosomes and 40S initiation complexes attached to the same mRNA strand, not normally present in control lysates. For this reason, it was suggested that these compounds can prevent the formation of 80S initiation complexes, in addition to inhibiting the function of ribosomes already attached to mRNA (Cundliffe and Davies 1977).

Loubresse et al. (2014) analyzed the interaction of verrucarin A (36) with the 80S ribosome of *Saccharomyces cerevisiae* using X-ray crystallography. According to the authors, the compound interacts with the A-site in the peptidyl transferase center. Also, the large macrocycle extends further toward the macrolide-binding site, thus confirming the relevance of stereochemistry on the biological effects of macrocyclic trichothecenes.

Trichothecenes are capable of inhibiting protein synthesis by binding to the ribosomal peptidyl transferase site. This inhibition results in cellular stress reactions, leading to the activation of mitogen-activated protein kinases (MAPKs), which are part of the apoptotic pathway. Gel electrophoresis revealed DNA fragmentation after the treatment of cells with some of these compounds, suggesting a role in apoptosis (Shifrin and Anderson 1999). This was confirmed by treatment of HL-60 cells (human promyelotic leukemia) with roridin A (30), which induced DNA fragmentation, with a minimal effective concentration of 0.001 μg/mL (Ueno et al. 1995).

Yang et al. (2000) investigated the relationship between the cytotoxic and apoptotic capacities of trichothecenes and the activation of MAPK. For this purpose, two myeloid models were employed: RAW 264.7 (murine macrophage) and U937 (human leukemic cells) cells. Upon evaluating the cytotoxic activity by the 3-(4,5-dimethyl-thiazol-2-yl) 2,5-diphenyl tetrazolium bromide (MTT) assay, roridin A (30) and verrucarin A (36) have been shown to be highly toxic and, by using DNA fragmentation and fluorescence microscopy assays, the results suggest that cytotoxicity was mediated through an apoptotic process. In addition, the assessment of MAPK activation using Western blot analysis revealed that these trichothecenes activated some proteins of this group.

There are some studies evaluating the role of macrocyclic trichothecenes in immunological functions. An example is the inhibitory effect of roridin A (30) and verrucarin A (36) on human lymphocyte transformation, analyzed by the mitogen-induced blastogenesis assay. Both compounds were able to inhibit B and T cell subsets stimulation, after induction with leukoagglutinin, concanavalin A, and pokeweed mitogen (Pestka and Forsell 1988).

In order to study the capacity of the mycotoxins to alter immune functions, the effects of roridin A (30) and verrucarin A (36) on interleukin 2 (IL-2) production and viability were evaluated in a murine T-cell model. It was observed that IL-2 levels were significantly increased in cultures incubated with low concentrations of these compounds. However, in the presence of higher concentrations, IL-2 levels were depressed. Besides, cell viability determined by the MTT method was significantly decreased by 0.5 ng/mL of roridin A (30) and verrucarin A (36). The biphasic behavior of these compounds in the deregulation of cytokine production, where low concentrations superinduce IL-2 production and higher concentrations suppress, complicates the interpretation of different effects in in vivo experiments (Lee et al. 1999).

Kankkunen et al. (2009) demonstrated that roridin A (30) and verrucarin A (36) can activate the inflammasome-associated caspase-1, required for the formation of IL-1β and IL-18 in LPS-primed cells. Later, the same researchers showed that these compounds trigger activation of NLRP3 inflammasome through P2X7R and Src tyrosine kinase signaling–dependent pathway, using human primary macrophages (Kankkunen et al. 2014). Moreover, verrucarin A (36) inhibited IL-8 and NF-κB activation in HL-60 cells (human promyelocytic leukemia), at noncytotoxic concentrations (Oda et al. 2005).

Table 14.5 lists the biological activities reported for macrocyclic trichothecenes, together with the IC_{50}, EC_{50}, MIC, and 50%PSI values, in different models and cell lines.

Table 14.5 Biological activities of macrocyclic trichothecenes from *Baccharis* species

Macrocyclic trichothecenes	Biological activities	Model / cell line / microorganism	IC_{50} / EC_{50} / MIC / 50%PSI	Ref.
Baccharin	Cytotoxic	P-388 leukemia in mice	5.0 mg/kg	Jarvis et al. (1984)
Baccharinol	Cytotoxic	P-388 leukemia in mice	2.5 mg/kg	Jarvis et al. (1984)
Roridin A	Antiviral	Junin virus (JUNV)	3.1 ng/mL	García et al. (2002)
	Antifungal	*Saccharomyces cerevisiae, Magnaporthe grisea, Sclerotinia sclerotiorum*	31.25 μg/mL 125 μg/mL 31.25 μg/mL	Xie et al. (2008)
		Candida albicans, Aspergillus niger, Trichophyton rubrum	125 μg/mL 31.25 μg/mL 62.5 μg/mL	Liu et al. (2006)
	Antimalarial	*Plasmodium falciparum*	0.31 ng/mL	Isaka et al. (1999)
	Cytotoxic	KB cells BC1 cells Vero cells	1.0 ng/mL 5.0 ng/mL 1.7 ng/mL	Isaka et al. (1999)
		NIH/3 T3 BE 12-6	0.00028 μg/mL 0.00096 μg/mL	Terse et al. (1993)
		P-388 leukemia in mice	0.06 mg/kg	Jarvis et al. (1984)
		HL-60	0.0005 μg/mL	Ueno et al. (1995)
	Protein synthesis inhibition	Vero cells Rat spleen lymphocytes	12 nM 5 nM	Thompson and Wannemacher (1986)
Roridin E	Antiviral	Junin virus (JUNV)	3.1 ng/mL	Matsumoto et al. (1977)
	Antimalarial	*Plasmodim falciparum*	0.15 ng/mL	(Isaka et al. (1999)
		P. falciparum clone W2 *P. falciparum* clone D6	0.6 ng/mL 0.2 ng/mL	Zhang et al. (2002)
	Cytotoxic	THP-1 HL-60 V79	9.1 nM 7.9 nM 0.74 nM	Oda et al. (2010)
		Artemia salina	0.221 μg/mL	Zhao et al. (2011)
		KB cells BC1 cells Vero cells	0.5 ng/mL 0.7 ng/mL 0.4 ng/mL	Isaka et al. (1999)
		NIH/3 T3 KA31T H4TG MDCK	2.30 nM 7.68 nM 3.45 nM 1.74 nM	Abbas et al. (2002)

(continued)

Table 14.5 (continued)

Macrocyclic trichothecenes	Biological activities	Model / cell line / microorganism	IC_{50} / EC_{50} / MIC / 50%PSI	Ref.
Roridin D	Antifungal	*Saccharomyces cerevisiae, Magnaporthe grisea, Sclerotinia sclerotiorum*	62.5 μg/mL 250 μg/mL 31.25 μg/mL	Xie et al. (2008)
Roridin H	Cytotoxic	NIH/3 T3 KA31T H4TG MDCK	9.84 nM 8.64 nM 10.8 nM 3.98 nM	Abbas et al. (2002)
		P-388 leukemia in mice	12.5 mg/kg	Jarvis et al. (1984)
Verrucarin A	Antiviral	Junin virus (JUNV)	4.9 ng/mL	García et al. (2002)
	Antifungal	*Candida albicans, Aspergillus niger, Trichophyton rubrum*	31.25 μg/mL 250 μg/mL 125 μg/mL	Liu et al. (2006)
	Antimalarial	*Plasmodim falciparum*	0.90 ng/mL	Isaka et al. (1999)
	Cytotoxic	*Artemia salina*	0.001 μg/mL	Zhao et al. (2011)
		KB cells BC1 cells Vero cells	1.2 ng/mL 4.2 ng/mL 1.1 ng/mL	Isaka et al. (1999)
		NIH/3 T3 BE 12-6	0.00033 μg/mL 0.00092 μg/mL	Terse et al. (1993)
		NIH/3 T3 KA31T H4TG MDCK	3.03 nM 1.94 nM 2.50 nM 0.96 nM	Abbas et al. (2002)
		P-388 leukemia in mice	2 mg/kg	Jarvis et al. (1984)
	Protein synthesis inhibition	Vero cells Rat spleen lymphocytes	12 nM 3 nM	Thompson and Wannemacher (1986)

(continued)

Table 14.5 (continued)

Macrocyclic trichothecenes	Biological activities	Model / cell line / microorganism	IC_{50} / EC_{50} / MIC / 50%PSI	Ref.
Verrucarin J	Antiviral	Junin virus (JUNV)	1.2 ng/mL	García et al. (2002)
	Antimalarial	*Plasmodim falciparum*	0.20 ng/mL	Isaka et al. (1999)
	Cytotoxic	*Artemia salina*	0.006 μg/mL	Zhao et al. (2011)
		KB cells *BC1 cells* *Vero cells*	3.9 ng/mL 14 ng/mL 1.2 ng/mL	Isaka et al. (1999)
		NIH/3 T3 KA31T H4TG MDCK	4.55 nM 3.14 nM 4.56 nM 2.02 nM	Abbas et al. (2002)
		P-388 leukemia in mice	0.8 mg/kg	Jarvis et al. (1984)

8 Final Considerations

Phytochemical studies with *Baccharis* species were of great relevance to demonstrate that plants, in addition to endophytic fungi, can produce trichothecenes. Therefore, *Baccharis* spp. represent new sources of trichothecenes and suitable models for studying the biosynthesis of this class of metabolites. Additionally, they allow investigating evolutionary aspects and ecological relations between fungi and plants (Rizzo et al. 1997).

The toxicity of trichothecenes is a matter of concern, being the main cause of animal deaths in southern Brazil. The occurrence of intoxication with *B. coridifolia* was reported in cattle, sheep, and horses, while *B. megapotamica* intoxication was recorded in cattle and sheep (Rissi et al. 2005; Pedroso et al. 2010).

It is important to note that the number of studies performed with *Baccharis* trichothecenes is still limited. Additional phytochemical and biological studies are thus demanded to better understand their chemical and medicinal properties, aiming at future exploitation.

References

Abbas HK, Johnson BB, Shier WT, Tak H, Jarvis BB, Boyette CD (2002) Phytotoxicity and mammalian cytotoxicity of macrocyclic trichothecene mycotoxins from *Myrothecium verrucaria*. Phytochemistry 59:309–313. https://doi.org/10.1016/S0031-9422(01)00464-2

Amagata T, Rath C, Rigot JF, Tarlov N, Tenney K, Valeriote FA, Crews P (2003) Structures and cytotoxic properties of trichoverroids and their macrolide analogues produced by saltwater culture of *Myrothecium verrucaria*. J Med Chem 46:4342–4350. https://doi.org/10.1021/jm030090t

Arunachalam C, Doohan FM (2013) Trichothecene toxicity in eukaryotes: cellular and molecular mechanisms in plants and animals. Toxicol Lett 217:149–158. https://doi.org/10.1016/j.toxlet.2012.12.003

Cundliffe E, Davies JE (1977) Inhibition of initiation, elongation, and termination of eukaryotic protein synthesis by trichothecene fungal toxins. Antimicrob Agents Chemother 11:491–499. https://doi.org/10.1128/AAC.11.3.491

Cutler HG, Jarvist BB (1985) Preliminary observations on the effects of macrocyclic trichothecenes on plant growth. Environ Exp Bot 25:115–128

de Carvalho M, Weich H, Abraham WR (2016) Macrocyclic trichothecenes as antifungal and anticancer compounds. Curr Med Chem 23:23–35. https://doi.org/10.2174/0929867323666151117121521

Desjardins AE, McCormick SP, Appell M (2007) Structure-activity relationships of trichothecene toxins in an *Arabidopsis thaliana* leaf assay. J Agric Food Chem 55:6487–6492. https://doi.org/10.1021/jf0709193

Dhar MK, Koul A, Kaul S (2013) Farnesyl pyrophosphate synthase: a key enzyme in isoprenoid biosynthetic pathway and potential molecular target for drug development. New Biotechnol 30:114–123. https://doi.org/10.1016/j.nbt.2012.07.001

El-Din MO, El-Sawi NM, Meki ARMA (1997) Acute toxicity of the mycotoxin roridin E on liver and kidney of rats. J Appl Anim Res 12:145–152. https://doi.org/10.1080/09712119.1997.9706200

Freeman GG, Gill JE, Waring WS (1959) The structure of trichothecin and its hydrolysis products. J Chem Soc:1105–1132

García CC, Rosso ML, Bertoni MD, Maier MS, Damonte EB (2002) Evaluation of the antiviral activity against junin virus of macrocyclic trichothecenes produced by the hypocrealean epibiont of *Baccharis coridifolia*. Planta Med 68:209–212. https://doi.org/10.1055/s-2002-23134

Grove JF (1993) Macrocyclic Trichothecenes. Nat Prod Rep 10:429–448

Guarro J (2013) Fusariosis, a complex infection caused by a high diversity of fungal species refractory to treatment. Eur J Clin Microbiol Infect Dis 32:1491–1500. https://doi.org/10.1007/s10096-013-1924-7

Habermehl GG, Busam L, Heydel P, Mebs D, Tokarnia CH, Döbereiner J, Spraul M (1985) Macrocyclic trichothecenes: cause of livestock poisoning by the Brazilian plant *Baccharis coridifolia*. Toxicon 23:731–743. https://doi.org/10.1016/0041-0101(85)90003-0

Hughes BJ, Jarvis BB, Sharma RP (1990) Effects of macrocyclic trichothecene congeners on the viability and mitogenesis of murine splenic lymphocytes. Toxicol Lett 50:57–67. https://doi.org/10.1016/0378-4274(90)90252-H

Isaka M, Punya J, Lertwerawat Y, Tanticharoen M, Thebtaranonth Y (1999) Antimalarial activity of macrocyclic trichothecenes isolated from the fungus *Myrothecium verrucaria*. J Nat Prod 62:329–331. https://doi.org/10.1021/np980323x

Jarvis BB, Midiwo JO, Tuthill D (1981) Interaction between the antibiotic trichothecenes and the higher plant *Baccharis megapotamica*. Science 214:460–462

Jarvis BB, Midiwo J, Mazzolat EP (1984) Antileukemic compounds derived by chemical modification of macrocyclic. J Med Chem 27:239–244

Jarvis BB, Cómezoglu SN, Ammon HL, Breedlove CK, Miller RW, Woode MK, Streelman DR, Sneden AT, Dailey RG, Kupchan SM (1987a) New macrocyclic trichothecenes from *Baccharis megapotamica*. J Nat Prod 50:815–828. https://doi.org/10.1021/np50053a008

Jarvis BB, Comezoglu SN, Rao MM, Pena NB, Boettner FE, Williams M, Forsyth G, Epling B (1987b) Isolation of macrocyclic trichothecenes from a large-scale extract of *Baccharis megapotamica*. J Org Chem 52:45–56. https://doi.org/10.1021/jo00377a007

Jarvis BB, Wells KM, Lee YW, Bean GA, Kommedahl T, Barros CS, Barros SS (1987c) Macrocyclic trichothecene mycotoxins in Brazil species of *Baccharis*. Phytopathology 77:980–984

Jarvis BB, Kuti J, Bean GA, Park C (1988a) Phytotoxicity of macrocyclic trichothecenes toward *Baccharis* cell lines. JSM Mycotoxins:199–203

Jarvis BB, Midiwo JO, Bean GA, Aboul-Nasr MB, Barros CS (1988b) The mystery of trichothecene antibiotics in *Baccharis* species. J Nat Prod 51:736–744. https://doi.org/10.1021/np50058a012

Jarvis BB, Mazzocchi DB, Ammon HL, Mazzola EP, Flippen-Anderson JL, Gilardi RD, George CF (1990) Conformational effects in trichothecenes: structures of 15-hydroxy C4 and C8 ketones. J Org Chem 55:660–3662

Jarvis B, Mokhtari-Rejali N, Schenkel EP, Barros CS, Matzencachers NI (1991) Trichothecene mycotoxins from Brazilian *Baccharis* species. Phytochemistry 30:789–797

Jarvis BB, Wang S, Cox C, Rao MM, Philip V, Varaschin MS, Barros CS (1996) Brazilian *Baccharis* toxins: livestock poisoning and the isolation of macrocyclic trichothecene glucosides. Nat Toxins 4:58–71. https://doi.org/10.1002/19960402NT2

Kankkunen P, Rintahaka J, Aalto A, Leino M, Majuri M-L, Alenius H, Wolff H, Matikainen S (2009) Trichothecene mycotoxins activate inflammatory response in human macrophages. J Immunol 182:6418–6425. https://doi.org/10.4049/jimmunol.0803309

Kankkunen P, Välimäki E, Rintahaka J, Palomäki J, Nyman T, Alenius H, Wolff H, Matikainen S (2014) Trichothecene mycotoxins activate NLRP3 inflammasome through a P2X7 receptor and Src tyrosine kinase dependent pathway. Hum Immunol 75:134–140. https://doi.org/10.1016/j.humimm.2013.11.010

Kimura M, Tokai T, Takahashi-Ando N, Ohsato S, Fujimura M (2007) Molecular and genetic studies of fusarium trichothecene biosynthesis: pathways, genes, and evolution. Biosci Biotechnol Biochem 71:2105–2123. https://doi.org/10.1271/bbb.70183

Krishnamurthy T, Sarver EW (1989) Mass spectral investigations on trichothecene mycotoxins -VIII - Thermospray ionization and collisionally activated dissociation of macrocyclic trichothecenes. Biomed Environ Mass Spectrom 18:287–300. https://doi.org/10.1016/s0021-9673(01)97321-2

Kumari I, Ahmed M, Akhter Y (2016) Multifaceted impact of trichothecene metabolites on plant-microbe interactions and human health. Appl Microbiol Biotechnol 100:5759–5771. https://doi.org/10.1007/s00253-016-7599-0

Kupchan SM, Jarvis BB, Dailey RG, Bright W, Bryan RF, Shizuri Y (1976) Baccharin, a novel potent antileukemic trichothecene triepoxide from *Baccharis megapotamica*. J Am Chem Soc 98:7092–7093. https://doi.org/10.1021/ja00438a066

Kupchan SM, Streelman DR, Jarvis BB, Dailey RG, Sneden AT (1977) Isolation of potent new antileukemic trichothecenes from *Baccharis megapotamica*. J Org Chem 42:4221–4225

Kuti JO, Jarvis BB, Mokhtari-Rejali N, Bean GA (1990) Allelochemical regulation of reproduction and seed germination of two Brazilian *Baccharis* species by phytotoxic trichothecenes. J Chem Ecol 16:3441–3453. https://doi.org/10.1007/BF00982109

Lee MG, Li S, Jarvis BB, Pestka JJ (1999) Effects of satratoxins and other macrocyclic trichothecenes on il-2 production and viability of el-4 thymoma cells. J Toxicol Environ Health A 57:459–474. https://doi.org/10.1080/009841099157548

Liu JY, Huang LL, Ye YH, Zou WX, Guo ZJ, Tan RX (2006) Antifungal and new metabolites of *Myrothecium* sp. Z16, a fungus associated with white croaker *Argyrosomus argentatus*. J Appl Microbiol 100:195–202. https://doi.org/10.1111/j.1365-2672.2005.02760.x

Loubresse GN, Prokhorova I, Holtkamp W, Rodnina MV, Yusupova G, Yusupov M (2014) Structural basis for the inhibition of the eukaryotic ribosome. Nature 513:517–522. https://doi.org/10.1038/nature13737

Matsumoto M, Minato H, Uotani N, Matsumoto K, Kondo E (1977) New antibiotics from *Cylindrocarpon* sp. J Antibiot 30:681–682

McCormick SP, Stanley AM, Stover NA, Alexander NJ (2011) Trichothecenes: from simple to complex mycotoxins. Toxins 3:802–814. https://doi.org/10.3390/toxins3070802

Nelson P, Dignani M, Anaissie EJ (1994) Taxonomy, biology, and clinical aspects of *Fusarium* species. Clin Microbiol Rev 7:479–504. https://doi.org/10.1128/CMR.7.4.479

Oda T, Namikoshi M, Akano K, Kobayashi H, Honma Y, Kasahara T (2005) Verrucarin A inhibition of MAP kinase activation in a PMA-stimulated promyelocytic leukemia cell line. Mar Drugs 3:64–73

Oda T, Xu J, Ukai K, Nakazawa T, Namikoshi M (2010) 12′-Hydroxyl group remarkably reduces roridin E cytotoxicity. Mycoscience 51:317–320. https://doi.org/10.1007/s10267-010-0035-x

Pedroso PMO, Bandarra PM, Feltrin C, Gomes DC, Watanabe TTN, Ferreira HH, Driemeier D (2010) Intoxicação por *Baccharis megapotamica* var. *weirii* em ovinos. Pesqui Vet Bras 30:403–405. https://doi.org/10.1590/S0100-736X2010000500005

Pestka JJ, Forsell JH (1988) Inhibition of human lymphocyte transformation by the macrocyclic trichothecenes roridin A and verrucarin A. Toxicol Lett 41:215–222. https://doi.org/10.1016/0378-4274(88)90057-4

Rissi DR, Rech RR, Fighera RA, Cagnini DQ, Kommers GD (2005) Intoxicação espontânea por *Baccharis coridifolia* em bovinos. Pesq Vet Bras 25:111–114

Rizzo I, Varsavky E, Haidukowski M, Frade H (1997) Macrocyclic trichothecenes in *Baccharis coridifolia* plants and endophytes and *Baccharis artemisioides* plants. Toxicon 35:753–757. https://doi.org/10.1016/S0041-0101(96)00149-3

Rocha O, Ansari K, Doohan FM (2005) Effects of trichothecene mycotoxins on eukaryotic cells: a review. Food Addit Contam 22:369–378. https://doi.org/10.1080/02652030500058403

Rosen JD, Rosen RT, Hartman TG (1986) Capillary gas chromatography - mass spectrometry of several macrocyclic trichothecenes. J Chromatogr A 355:241–251

Savard ME, Blackwell BA, Greenhalgh R (1987) An ^{1}H nuclear magnetic resonance study of derivatives of 3-hydroxy-12 ,13-epoxytrichothec-9-enes. Can J Chem 65:2254–2262

Schindler D, Grant P, Davies J (1974) Trichodermin resistance: mutation affecting eukaryotic ribosomes. Nature 248:535–536. https://doi.org/10.1038/248535a0

Shank RA, Foroud NA, Hazendonk P, Eudes F, Blackwell BA (2011) Current and future experimental strategies for structural analysis of trichothecene mycotoxins-A prospectus. Toxins 3:1518–1553

Shifrin VI, Anderson P (1999) Trichothecene mycotoxins trigger a ribotoxic stress response that activates c-Jun N-terminal kinase and p38 mitogen-activated protein kinase and induces apoptosis. J Biol Chem 274:13985–13992. https://doi.org/10.1074/jbc.274.20.13985

Smith SN (2007) An overview of ecological and habitat aspects in the genus *Fusarium* with special emphasis on the soil borne pathogenic forms. Plant Pathol Bull 16:97–120

Steinmetz WE, Rodarte CB, Lin A (2009) 3D QSAR study of the toxicity of trichothecene mycotoxins. Eur J Med Chem 44:4485–4489. https://doi.org/10.1016/j.ejmech.2009.06.012

Sudakin DL (2003) Trichothecenes in the environment: relevance to human health. Toxicol Lett 143:97–107. https://doi.org/10.1016/S0378-4274(03)00116-4

Terciolo C, Maresca M, Pinton P, Oswald IP (2018) Review article: Role of satiety hormones in anorexia induction by trichothecene mycotoxins. Food Chem Toxicol 121:701–714. https://doi.org/10.1016/j.fct.2018.09.034

Terse PS, Madhyastha MS, Zurovac O, Stringfellow D, Marquardt RR, Kemppainen BW (1993) Comparison of in vitro and in vivo biological activity of mycotoxins. Toxicon 31:913–919. https://doi.org/10.1016/0041-0101(93)90228-B

Thompson WL, Wannemacher RW (1986) Structure function relationship of trichothecenes in cell culture and in vivo. Toxicon 24:985–994

Trapp SC, Hohn TM, McCormick S, Jarvis BB (1998) Characterization of the gene cluster for biosynthesis of macrocyclic trichothecenes in *Myrothecium roridum*. Mol Gen Genet 257:421–432. https://doi.org/10.1007/s004380050666

Ueno Y (1980) Trichothecene mycotoxins mycology, chemistry, and toxicology. In: Draper HH (ed) Advances in nutritional research. Springer, Boston, pp 301–353. https://doi.org/10.1007/978-1-4757-4448-4_10

Ueno Y, Umemori K, Niimi E-C, Tanuma S-I, Nagata S, Sugamata M, Ihara T, Sekijima M, Kawai K-I, Ueno I, Tashiro F (1995) Induction of apoptosis by T-2 toxin and other natural toxins in HL-60 human promyelotic leukemia cells. Nat Toxins 3:129–137. https://doi.org/10.1002/nt.2620030303

Verdi LG, Maria I, Brighente C, Pizzolatti G (2005) Gênero *Baccharis* (Asteraceae): aspectos químicos, econômicos e biológicos. Quim Nova 28:85–94

Villafana RT, Ramdass AC, Rampersad SN (2019) Selection of *Fusarium trichothecene* toxin genes for molecular detection depends on TRI gene cluster organization and gene function. Toxins 11:1–25. https://doi.org/10.3390/toxins11010036

Wu Q, Dohnal V, Ku K, Yuan Z (2013) Trichothecenes : structure-toxic activity relationships. Curr Drug Metab 14:641–660

Xie LW, Jiang SM, Zhu HH, Sun W, Ouyang YC, Dai SK, Li X (2008) Potential inhibitors against *Sclerotinia sclerotiorum*, produced by the fungus *Myrothecium* sp. associated with the marine sponge *Axinella* sp. Eur J Plant Pathol 122:571–578. https://doi.org/10.1007/s10658-008-9326-x

Yang GH, Jarvis BB, Chung YJ, Pestka JJ (2000) Apoptosis induction by the satratoxins and other trichothecene mycotoxins: relationship to ERK, p38 MAPK, and SAPK/JNK activation. Toxicol Appl Pharmacol 164:149–160. https://doi.org/10.1006/taap.1999.8888

Zhang HJ, Tamez PA, Aydogmus Z, Tan GT, Saikawa Y, Hashimoto K, Nakata M, Van Hung N, Xuan LT, Cuong NM, Soejarto DD, Pezzuto JM, Fong HHS (2002) Antimalarial agents from plants. III. Trichothecenes from *Ficus fistulosa* and *Rhaphidophora decursiva*. Planta Med 68:1088–1091. https://doi.org/10.1055/s-2002-36350

Zhao L, Liu L, Wang N, Wang S-J, Hu J-C, Gao J-M (2011) Potent toxic macrocyclic trichothecenes from the marine-derived fungus *Myrothecium verrucaria* Hmp-F73. Nat Prod Commun 6:1915–1916. https://doi.org/10.1177/1934578x1100601231

Chapter 15
Livestock Intoxication by *Baccharis*

Claudio S. L. Barros and Bruce B. Jarvis

Abstract Relatively few *Baccharis* species have been associated with livestock poisoning. *Baccharis coridifolia, B. megapotamica*, and *B. weirii* are the most important causes of livestock loss in southern Brazil, Uruguay, Argentina, and Paraguay. *B. artemisioides* causes disease in cattle in a restricted zone of Argentina and the poison by *B. vulneraria* was reported once in dairy cattle from southern Brazil. The toxicosis has acute onset of a usually fatal clinical course consisting of muscular tremors, tachypnea, tachycardia, recumbency, and death. Lesions include reddening, edema, and necrosis of gastrointestinal mucosa and widespread lymphoid necrosis affecting lymphocytes from B-cell zones. *B. halimifolia, B. glomerulifolia*, and *B. pteronioides* are suspected of inducing toxicosis livestock in North America. *B. halimifolia* was also introduced from the USA into Australia and reportedly causes cattle poisoning there. The toxicosis caused by these three species was experimentally induced in chicks and laboratory rodents by feeding the plant, but the characterization of the spontaneous disease and toxic principles are less clearly established. The chemicals responsible for the phytotoxicities of *B. coridifolia* and *B. megapotamica* belong to the trichothecene class of well-studied fungal toxins (mycotoxins), specifically the macrocyclic trichothecenes. The levels of these toxins vary according to the time of year and the sex of these dioecious plants. The highest concentration (thousands of ppm) occurs in female plants at the time of flowering. Although there is good reason to think that these toxins are produced by fungal endophytes, the behavior of *B. coridifolia* and *B. megapotamica* with respect to these compounds is quite surprising. In addition, a survey of over 20 other Brazilian *Baccharis* plants found no further examples of plants with these toxins. In addition, feeding studies with *B. halimifolia* showed that this species is exquisitely sensitive to the toxic effects of these mycotoxins, a characteristic shared by virtually all other plants.

C. S. L. Barros (✉)
Laboratório de Patologia Animal, Universidade Federal de Mato Grosso do Sul, Campo Grande, Mato Grosso do Sul, Brazil

B. B. Jarvis
Department of Chemistry and Biochemistry, University of Maryland, College Park, MD, USA

G. W. Fernandes et al. (eds.), *Baccharis*,
https://doi.org/10.1007/978-3-030-83511-8_15

Keywords Cattle poisoning · Chemical structure · Macrocyclic trichothecenes · Poisoning · Toxicosis

1 Introduction

Relatively few *Baccharis* species have been associated with livestock poisoning. Among them, *Baccharis coridifolia* DC. is the most important (Barros 1993, 1998). Less frequently *B. megapotamica* Spreng. and *B. weirii* Baker, and *B. vulneraria* are involved in livestock deaths in southern Brazil (Driemeier et al. 2000; Pedroso et al. 2010; Oliveira Filho et al. 2011; Panziera et al. 2015), and *B. artemisioides* Hook. & Arn. (aka white romerillo) causes disease in cattle in a restricted zone of Argentina (Rizzo et al. 1997). *B. halimifolia* L. (Everist 1974), *B. glomerulifolia* L. (Duncan et al. 1957), and *B. pteronioides* Gray (Marsh et al. 1920; Stegelmeier et al. 2009) are mentioned as toxic to livestock in the USA and Australia.

2 *Baccharis coridifolia*

Baccharis coridifolia (Fig. 15.1) is one of the most recognized toxic plants in southern Brazil, where it grows in the states of Rio Grande do Sul, Santa Catarina, Paraná. The plant is also found in the Brazilian southeastern state of São Paulo, in large areas of Uruguay, and in northern Argentina and Paraguay (Barros 1993). The habitat of *B. coridifolia* is native grasslands in nonhumid areas (Tokarnia et al. 2012).

In Brazil, the plant is commonly known as "mio-mio" (Tokarnia et al. 2012) and, in the Spanish-speaking countries, as "romerillo" (Schang 1929). It is a dioecious plant that sprouts in the spring (October through November), and blooms in late summer and early fall (end of February to April). Although the plant is more toxic when flowering, the toxicosis can occur at any time of the year, and heavy cattle losses have occurred in the spring (Barros 1993).

The spontaneous poisoning occurs mainly in cattle (Barros et al. 1986; Barros 1993, 1998; Rissi et al. 2005), and less frequently in sheep (Rozza et al. 2006; Hammerschmitt et al. 2018), goats (Panziera et al. 2015), buffalo (Lértora and Negrette 2015), and horses (Alda et al. 2009). The poisoning has been experimentally reproduced in various animal species including cattle (Tokarnia and Döbereiner 1975; Varaschin et al. 1998), sheep (Tokarnia and Döbereiner 1976), horses (Costa et al. 1995), rabbits (Döbereiner et al. 1976; Rodrigues and Tokarnia 1995), and pigeons (Flores and Houssay 1917).

The toxic principles of *B. coridifolia* and *B. megapotamica* belong to the trichothecene class of well-studied fungal toxins (mycotoxins), specifically the macrocyclic trichothecenes (*see* ahead). The toxin levels in the plant vary according to the

Fig. 15.1 *Baccharis coridifolia*. Blooming female specimen. Photography taken in March in Rio Grande do Sul. (Courtesy Dr. Ana Lucia Schild, LRD, Pelotas, RS, Brazil)

time of year and the sex of the dioecious plants. The highest concentration (thousands of ppm) occurs in female specimens at blooming (Jarvis et al. 1991; Varaschin et al. 1998). Accordingly doses as low as 0.25–0.5 g/kg/bodyweight of the green flowering plant may cause deaths in cattle. In the sprouting period, 2 g/kg/body weight is required for the same effect (Tokarnia and Döbereiner 1975). Sheep are comparatively resistant, and the toxic dose for this species is about twice as great as that for cattle (Tokarnia and Döbereiner 1976). Horses are particularly susceptible. The administration of a single dose of 0.06 g/kg causes severe disease, and doses in the range 0.125–0.5 g/kg are lethal (Costa et al. 1995). When dried, *B. coridifolia* retains about 50% of its potency for at least 17 months (Tokarnia and Döbereiner 1975).

3 The Chemical Biology of *Baccharis*

In the fall of 1975, a 5 kg shipment of *B. megapotamica* arrived in the laboratory of Prof. Morris Kupchan at the University of Virginia. Kupchan had a contract with the US National Institutes of Health (NIH) to evaluate plants collected from around the world for the presence of anticancer compounds and had evaluated several *Baccharis* species in the past, none of which showed any promise. However, the extract of *B. megapotamica* proved to be quite interesting; it exhibited very high in vitro cytotoxicity, and even better had extraordinarily high in vivo activity against mouse leukemia, a property rarely seen with crude extracts (Kupchan et al. 1976).

Subsequently, extraction of a 54 kg sample of *B. megapotamica* yielded compounds that belonged to a class of natural products called trichothecenes,

Fig. 15.2 Chemical structure of baccharinoid B5

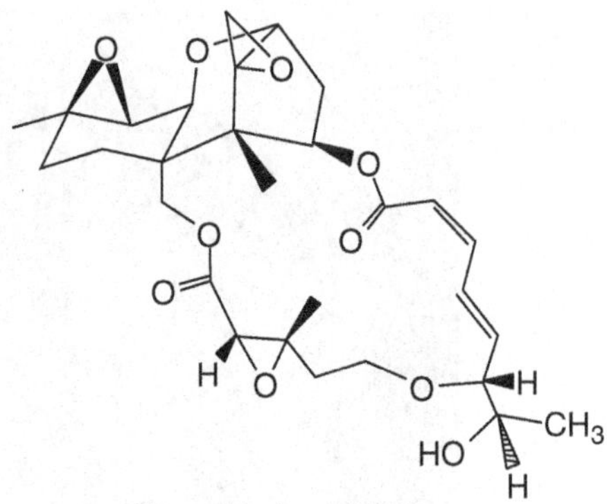

specifically macrocyclic trichothecenes, which were responsible for the anticancer properties (Kupchan et al. 1977). The NIH was keen to obtain larger quantities of the original member of this group, baccharinoid B5 (Fig. 15.2), and so an 1800 kg collection of the plant was extracted and processed to yield a large number (>30) of related macrocyclic trichothecenes (Jarvis et al. 1987a). Morris Kupchan passed away in the fall of 1976, and the work on *Baccharis* was transferred to the laboratory of BBJ.

With gram quantities of B5 in hand, the NIH carried out a number of cancer preclinical studies on B5. However, an earlier trichothecene, anguidine, which was in human cancer clinical trials, proved too toxic for further development. In light of the negative results with this trichothecene, the B5 trichothecene was dropped from further consideration as an anticancer drug. Although disappointing, from this latter 1800 kg extraction, many interesting compounds were isolated and their chemistry studied (Jarvis 1992).

Later, in the 1980s, Habermehl and coworkers reported that *B. coridifolia*, a plant held responsible for livestock poisoning in Brazil, also contained a series of macrocyclic trichothecenes, including roridins A and E (Habermehl et al. 1985, Fig. 15.3).

Trichothecenes are a very broad group (>200 compounds) of terpenes produced by fungi. They are well-recognized mycotoxins known to be associated with both human and livestock health issues; several of them are highly toxic, especially the macrocyclic compounds (McCormick et al. 2011). Relevant to *Baccharis* intoxications in South American livestock are similar intoxications of Eastern European livestock caused by their eating moldy hay and straw. The fungus responsible is *Stachybotrys chartarum*, which produces a series of macrocyclic trichothecenes closely related in structure to compounds found in some *Baccharis* species (Jarvis 2003). This fungus has also been found in mold-damaged buildings and held responsible, in part, for the health problems of the building occupants (Jarvis and Miller 2005).

Up until the report of their presence in *B. megapotamica*, trichothecenes had never been detected in plants except in plants with obvious fungal contamination.

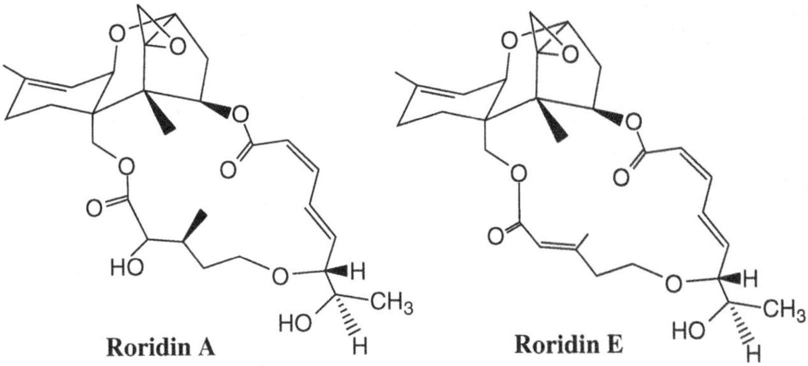

Fig. 15.3 Chemical structures of roridins A and E

Neither *B. megapotamica* nor *B. coridifolia* had any obvious fungal infections even though these plants were shown to contain from several hundred to thousands of ppm of macrocyclic trichothecenes (Jarvis 1991). Also later, several African plants were shown to contain trichothecenes, and these plants also showed no obvious signs of fungal activity (Loukaci et al. 2000; Zhang et al. 2002). Furthermore, trichothecenes are highly phytotoxic. When tested in plant cell cultures, macrocyclic trichothecenes were found to be among the most phytotoxic compounds known (Cutler and Jarvis 1985; Jarvis et al. 1988). All this raises the question of the sources of the plant-derived trichothecenes. Of immediate relevance is the question of whether trichothecenes are present in other species of *Baccharis*.

No further examples of macrocyclic trichothecene-containing *Baccharis* species have been found by chemical analysis of the following species: *B. anomala*, *B. articulata*, *B. cultrata*, *B. cylindrica* (a synonym of *B. trimera*), *B. dracunculifolia*, *B. microptera*, *B. myriocephala*, *B. ochracea*, *B. pseudotenuifolia* (a synonym of *B. linearifolia*), *B. spicata*, *B. tridentata*, *B. trimera*, *B. usterii* (a synonym of *B. junciformis*), (all from Brazil), *B. articulata* var. *gaudichaudiana* (a synonym of *B. articulata*) (Paraguay), *B. magellanica*, *B. patagonica* (both from Chile), *B. glutinosa*, *B. halimifolia*, *B. neglecta*, and *B. sarothroides* (from the United States) (Jarvis et al. 1991). This same paper reported that herbarium samples of *B. coridifolia* and *B. megapotamica* that dated back to the nineteenth century contained significant levels of the macrocyclic trichothecenes.

When *B. megapotamica* was grown in the greenhouse of the University of Maryland, no macrocyclic trichothecenes could be detected in the plant material. However, when fed roridin A, the plant absorbed roridin A through its roots, and the roridin A was readily transferred to the leaves where it was transformed into baccharinoid B7 (Fig. 15.4). This conversion occurs in two steps: first a rapid introduction of the OH group into 8-position followed by a slow epimerization of the 2'-OH group at the 2′-position. The plant was uninjured by this treatment; however, when the North American species, *B. halimifolia*, was subjected to the same feeding, it died within a day (Jarvis et al. 1981).

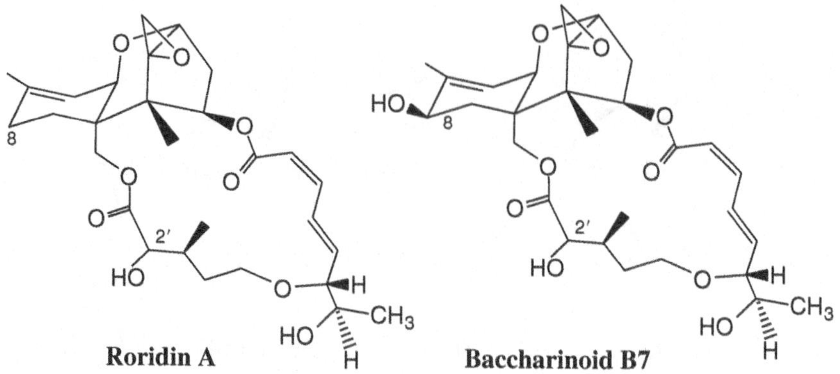

Fig. 15.4 Chemical structures of roridin A and baccharinoid B7

Trichothecenes bind to most eukaryotic ribosomes where they shut down protein biosynthesis (Dejardins et al. 1993). However, unpublished results from our laboratory showed that roridin A did not bind to the ribosomes of *B. megapotamica* (as is also the case with trichothecene-producing fungi) but did so with the ribosomes of *B. halimifolia*. In fungi, this resistance has been found to be due to methylation of ribosomes (Iglesias and Ballesta 1994), but nothing is yet known about the mechanism of resistance in the Brazilian plants.

The genus *Baccharis* is dioecious, and with respect to trichothecene content, there is no difference between the male and female plants of *B. megapotamica*; however, this is not the case with *B. coridifolia*. Until these plants go into flower, the levels of trichothecenes (<1000 ppm, mainly oridin A) are the same in the male and female plants of *B. coridifolia*, but as they go into flower, the levels of roridin A in the female plants increase dramatically to greater than 4000 ppm (Kuti et al. 1990). Furthermore, roridin A facilitates the germination of *B. coridifolia* seeds. When the seed coats of this plant are removed, the resulting decoated embryos do not germinate easily. Normally, removing the seed coats actually helps with germination since the seed coats contain germination-inhibiting compounds. When the decoated seeds of *B. coridifolia* are treated with a solution of 10^{-6} M roridin A, the seeds readily germinate. Decoated seeds of *B. halimifolia* and *B. glutinosa* (that readily germinate when their seed coats are removed) are killed by this solution of roridin A. This rather bizarre behavior seems to be related to the pollination of *B. coridifolia*. The seeds of *B. coridifolia* whose flowers were hand-pollinated with pollen from the male plant developed viable seeds with a high level (1300 ppm) of trichothecenes. Surprisingly, female flowers prevented from being pollinated produced seeds that would not germinate. When female *B. coridifolia* flowers were cross-pollinated with pollen from *B. halimifolia* and *B. megapotamica*, the resulting seeds exhibited significantly lower viability and decreased germination. Most interestingly, no trichothecenes were detected in the seeds from both these cross-pollinated experiments (Kuti et al. 1990).

There have been two quite different suggestions as to the source of the trichothecenes in the *Baccharis* plants. Because of the unusually high levels of these toxins in the plants as well as their bizarre effects (e.g., on germination and dependence of levels on the sex of *B. coridifolia* and its pollination of this plant), it has been suggested that the *Baccharis* plants themselves are biosynthesizing the trichothecenes (Kuti et al. 1990). However, this would require that the plants somehow acquire the trichothecene biosynthetic genes, presumably through horizontal gene transfer (HGT). Although HGT is common among microorganisms (Andam and Gogarten 2011), it is much less so between eukaryotic species such as plants and fungi (Richardson and Palmer 2007). Furthermore, the trichothecene biosynthetic pathway is complex and involves numerous genes (Desjardins et al. 1993; Trapp et al. 1998); it is difficult to see how the entire set of trichothecene biosynthetic genes could be transferred from a fungus (e.g., *Myrothecium*) to *Baccharis* plants, presumably on more than one occasion. Furthermore, these plants are found in two evolutionarily distinct lineages; *B. coridifolia* is placed in the clade of *Baccharis* subgen. *Coridifoliae*, while *B. megapotamica* and *B. weirii* are sister species in the clade of *Baccharis* subgen. *Baccharis* (Heiden et al. 2019).

The other possibility is that *B. megapotamica, B. weirii,* and *B. coridifolia* have an associated endophyte(s) that is actually the source of the plant-derived trichothecenes. In fact, such an endophyte was isolated from *B. coridifolia* and *B. artemisioides* and studied by Argentinean biologists (Rizzo et al. 1997). *B. artemisioides* joins the other three *Baccharis* species as a plant source for trichothecenes. These workers showed that the fungus they isolated from plant tissue produced macrocyclic trichothecenes in culture. However, in our hands, we found their fungus grew poorly in culture (not uncommon for endophytes), and we could detect no trichothecenes (unpublished results).

There are reports of two African plants found to contain appreciable levels of, in one case simple trichothecenes (Loukaci et al. 2000) and in the other case, macrocyclic trichothecenes (Zhang et al. 2002). Neither found any appreciable fungal activity to account for the presence of trichothecenes.

On the other hand, we (Jarvis et al. 1987b) and others (Habermehl et al. 1985; Habermehl 1989) have found *M. verrucaria* in the rhizosphere of the roots of *B. coridifolia*. In addition, roots of *B. coridifolia*, which have been surfaced sterilized and placed on agar plates, produce colonies of *M. verrucaria*, and only *M. verrucaria*, after a two-week period (Aboul-Nasr 1989).

Most if not all vascular plants have associated endophytes (Strobel and Daisy 2003) that play important roles in the lives of these plants. These endophytes produce a multitude of chemicals that have been isolated from plant extracts and were initially attributed to the biosynthetic machinery of the plants. However, a number of biologically active natural products isolated from plant extracts in the past are now recognized as being derived from endophytes (Newman and Cragg 2015). The trichothecene-containing *Baccharis* plants appear to have established a relationship with an endophyte that results in their accumulating the highly toxic macrocyclic trichothecenes. How this was done is an interesting question that awaits further investigation.

4 *Baccharis* spp. Poisoning

Typically, the intoxication occurs when livestock that is raised in zones free of the plant is transported and released into pastures infested with *B. coridifolia*. The risk of intoxication increases considerably if, during transport, animals are still subjected to long marches, stress, hunger, and thirst (Schang 1929; Barros et al. 1986; Barros 1993).

Although the situation described above is the one found in the vast majority of cases of *B. coridifolia* poisoning, there are reports of other less frequent conditions in which the disease may occur. The intoxication may, exceptionally, occur in lactating animals (mainly sheep) when they begin to graze (Tokarnia and Döberener 1975; Hammerschmitt et al. 2018). Others maintain that, during periods of drought and overcrowding, hunger forces some animals to ingest the plant (Flores and Houssay 1917), especially if they are placed in pastures where the plant is sprouting after burning (Tokarnia and Döberener 1975). Most of these observations are reported by farmers who are by no means unanimous. Many believe that, even with extreme hunger, livestock raised in pastures where *B. coridifolia* already exists will not ingest the plant (Barros 1993).

The magnitude of economic losses from the death of livestock poisoned by *B. coridifolia* is difficult to assess. Many cases are probably not communicated to diagnostic laboratories probably due mainly to two factors; most farmers and field veterinary practitioners recognize the disease without the aid of the laboratory, or the incidence of the poisoning has decreased in face of awareness about the toxicosis and the prophylactic methods used. Data on morbidity and mortality ratios are scarce in the literature; in general, both are high. Mortality rates of 20% and 50% in a period of respectively 72 and 30 h after cattle were introduced into a pasture highly infested with *B. coridifolia* have been reported (Barros et al. 1986; Barros 1993).

The toxicosis runs an acute and usually fatal clinical course. In fatal cases in cattle and sheep, the clinical signs begin, respectively, 5–29 h and 3–23 h after the ingestion of the fresh green plant, and death occurs respectively between 3–34 h and 2–42 h after the onset of the signs (Tokarnia and Döbereiner 1975, 1976; Barros 1993). In cattle, there is loss of appetite, mild-to-moderate bloat, instability of the hind limbs, muscle tremors, dry muzzle, ocular secretion, lack of rumination, dry small amounts of feces, mild sialorrhea, thirst, rapid and laborious breathing, groans, tachycardia, and restlessness (the animal lies down and lifts repeatedly). Finally, it remains most of the time in sternal decubitus, assuming lateral decubitus 30 min to 1 h before death. In nonfatal cases, the clinical signs are similar; there is anorexia, the absence of ruminal movements, catarrhal discharge through the nasal and conjunctival fossa with mild sialorrhea, and constipation followed by diarrhea (Tokarnia and Döbereiner 1975). Some authors maintain that diarrhea appears to be a sign of a good prognosis, indicating a recovery within a few hours to up to 2 weeks (Tokarnia and Döbereiner 1975; Rissi et al. 2005). In our experiments with *B.*

coridifolia in cattle, however, death was preceded by severe diarrhea in 4 cases with acute clinical courses (Varaschin et al. 1998).

In sheep, the clinical signs are similar. The animal stops grazing, moves away from the lot and assumes sternal decubitus for intermittent and progressively more extended periods, and shows apathy, muscle tremors, and wheezing. Just a few hours before death, there is lateral decubitus with paddling movements (Tokarnia and Döbereiner 1976).

The predominant sign in horses is colic associated with increased heart rate, anorexia, small intestinal hypermotility and distension of the colon, and cecum by gas and, in some cases, diarrhea (Alda et al. 2009).

Necropsy findings are restricted to the digestive system, mainly to the gastrointestinal tract, and are similar in cattle, sheep, and buffalo (Tokarnia and Döbereiner 1975, 1976; Rissi et al. 2005; Rozza et al. 2006; Lértora and Negrette 2015). The lesions are not very specific. The rumen is distended by liquid. There are varying degrees of reddening, edema, and erosions in the mucous membranes of the forestomachs (Fig. 15.5), redness, and petechiae in the mucosa of the abomasum and small intestine, which has a liquid content. The liver may be yellowish, lighter than usual, or can present a slight accentuation of the lobular pattern. Hemorrhages occur in the epi- and endocardium; although these are nonspecific lesions, they occur in the vast majority of cases. Additionally, in our cases of experimental intoxication by *B. coridifolia* in cattle, we occasionally observed marked edema and reddening of the mucosa large intestine and hemorrhagic content in the cecum and proximal colon. In another case, macroscopic lesions were not observed (Varaschin et al. 1998). Brain edema was reported in buffalo (Lértora and Negrette 2015).

In the spontaneous poisoning in horses, gross lesions included hemorrhages, and edema in the glandular parts of the stomach and marked edema and hemorrhages in the mucosa ileum, cecum (Fig. 15.6), and large colon (Alda et al. 2009).

The main histological lesions consist of degeneration, necrosis, and detachment of the epithelium lining the forestomachs (Tokarnia and Döbereiner 1975, 1976; Rissi et al. 2005; Rozza et al. 2006; Lértora and Negrette 2015; Hammerschmitt

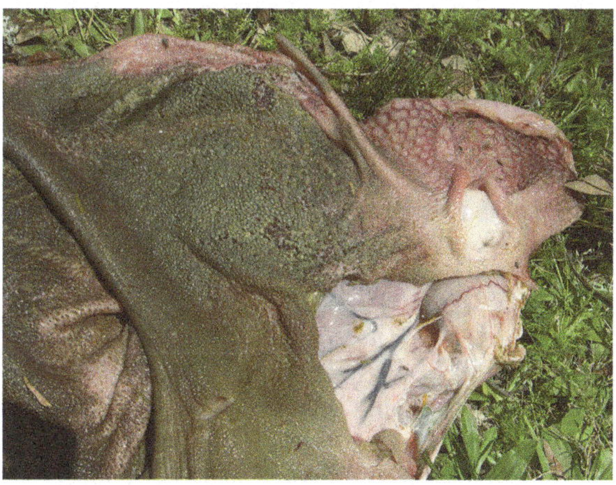

Fig. 15.5 Poisoning by *Baccharis coridifolia*. Forestomachs, lamb. The ruminal mucosa is ulcerated, and the mucosa of the reticulum is markedly reddish. (Courtesy Drs. Welden Panziera and Márcia Hammerschmitt, Setor de Patologia Veterinária, Faculdade de Veterinária, Universidade Federal do Rio Grande do Sul)

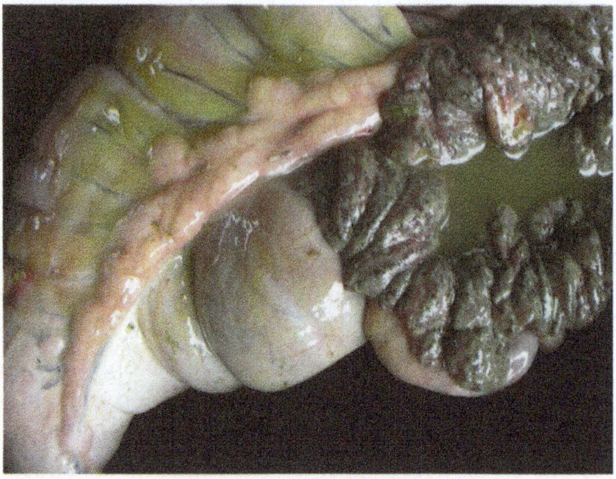

Fig. 15.6 Poisoning by *Baccharis coridifolia*, cecum, horse. There is marked transmural edema and ulceration of the mucosa. (Reproduced with permission from Alda et al. (2009) Pesqu Vet Bras 29:409–414)

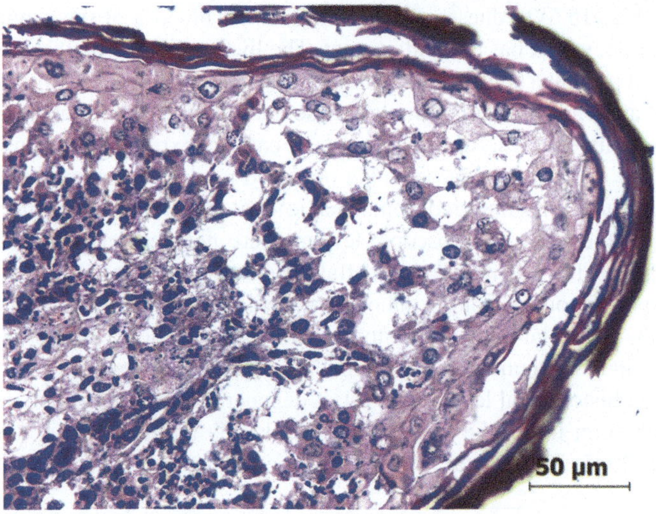

Fig. 15.7 Poisoning by *Baccharis coridifolia*, histopathology, ruminal mucosa, lamb. Marked necrosis of the stratified lining epithelium of the rumen. (Courtesy Drs. Welden Panziera and Márcia Hammerschmitt, Setor de Patologia Veterinária, Faculdade de Veterinária, Universidade Federal do Rio Grande do Sul)

et al. 2018). This lesion (Fig. 15.7) begins with acute cellular swelling, progressing to ballooning degeneration, the formation of vesicles within the epithelial lining, necrosis, and transient leukocyte infiltration. The degenerative-necrotic lesions frequently cause separation between the epithelium and the lamina propria. In experiments with cattle (Varaschin et al. 1998), and laboratory mice (Varaschin and Alessi 2003), necrotic lesions of secondary lymphoid organs (in zones of B-lymphocytes) have been consistently observed (Fig. 15.8). Occasionally, neutrophilic infiltrates and perivascular hemorrhages are observed respectively in the intestinal mucosa and in the brain (Tokarnia and Döbereiner 1975). Although perivascular

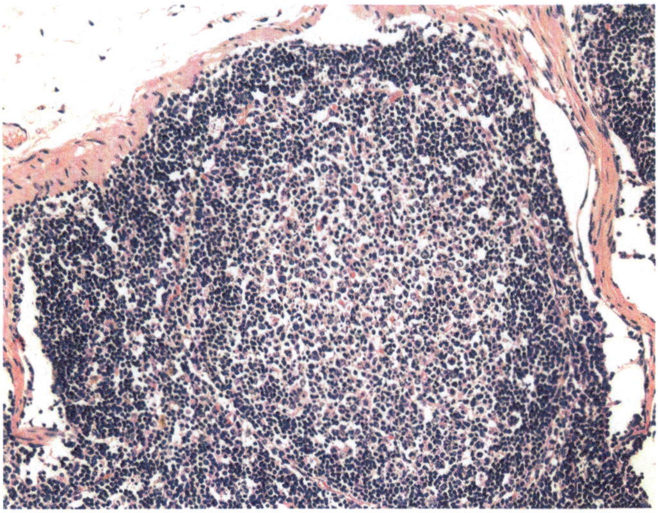

Fig. 15.8 Poisoning by *Baccharis coridifolia*, histopathology, lymph node, lamb. Necrosis at the germinal centers of secondary follicles. (Courtesy Drs. Welden Panziera and Márcia Hammerschmitt, Setor de Patologia Veterinária, Faculdade de Veterinária, Universidade Federal do Rio Grande do Sul)

hemorrhages in the brain can be potentially produced by the direct action of the toxicants, we have observed similar findings in brains of cattle and sheep that have died from different causes, and we believe that, at least in some situations, they may be mechanical artifacts produced in the brain during the opening of the calvarium with the ax or due to agonic ischemia.

Additional histological lesions have been described in *B. coridifolia* poisoning in buffalo (Lértora and Negrette 2015). The vascular component of the gastrointestinal mucosa had degenerative changes associated with thrombosis. The liver showed diffuse hepatocellular swelling, and there were brain edema and multifocal cerebrocortical necrosis (Lértora and Negrette 2015).

In the spontaneous poisoning in horses, histopathological observations were restricted to the gastrointestinal tract and included gastritis and enteritis with necrosis of the lining epithelium of the stomach, ileum, cecum, and colon, and marked inflammatory infiltrate, edema, and dilatation of submucosal lymphatics (Alda et al. 2009).

5 Other Species of *Baccharis*

B. megapotamica, B. weirii, and *B. artemisioides* produce the same toxic effects in livestock as *B. coridifolia*, and they share the same class of toxic principles (Oliveira Filho et al. 2012; Rizzo et al. 1997). Cattle from a relatively small area in Argentina, northwest of Buenos Aires and southeast of Cordoba, are spontaneously poisoned by the ingestion of *B. artemisioides* (Rizzo et al. 1997).

Poisoning by *B. megapotamica* and its sister species *B. weirii* is observed in livestock in southern Brazil, states of Rio Grande do Sul, and Santa Catarina. Unlike *B. coridifolia*, it inhabits swampy areas and is therefore known as marsh mio-mio. These two species are both toxic to livestock. Spontaneous outbreaks of intoxication have been described only with *Baccharis weirii* in cattle (Driemeier et al. 2000), sheep (Pedroso et al. 2010), buffalo (Oliveira-Filho et al. 2011), and goats (Panziera et al. 2015). Intoxication by *B. megapotamica* and *B. weirii* has been experimentally produced in cattle (Tokarnia et al. 1992) sheep (Armién et al. 1993), goats (Barbosa et al. 1994), and buffalo (Oliveira-Filho et al. 2012).

An outbreak of poisoning by *Baccharis vulneraria* (formerly *Baccharidastrum triplinervium*) in dairy cows was reported from Paraná, Brazil. Out of 58 cows at risk, 15 got sick, and 6 died after a clinical course of 12–60 h. The disease was clinically and pathologically similar to those caused by other species of *Baccharis* and was reproduced by force-feeding three calves with the aerial fresh parts of the plant (20–30 g/kg/bw), but no macrocyclic trichothecenes were detected in the chemical analysis of the plant (Langohr et al. 2005).

Baccharis pteronioides is a North American species of *Baccharis* that grows in clusters in gravelly soil in Texas, New Mexico, Arizona, and northern Mexico (Kingsbury 1964). It has been associated with cattle poisoning, for almost a century (Marsh et al. 1920; Manley et al. 1982). However, there are no convincing descriptions of the gross or histologic lesions of the affected cattle to support a definitive association with the ingestion of the plant. In 2004, in New, Mexico, USA, 80 from a herd of 100 cattle became sick and more than 40 died from what was presumptively diagnosed as *B. pteronioides* poisoning (Stegelmeier et al. 2009). Postmortem examination was delayed, and the findings described seem to indicate autolysis and putrefaction. An attempt to reproduce the intoxication in hamsters by feeding large doses of *B. pteronioides* resulted in a disease characterized histologically by severe necrotizing vasculitis with vascular thrombosis of hepatic and renal vessels (Stegelmeier et al. 2009). The authors concluded that, at high doses, *B. pteronioides* is toxic to hamsters and produces lesions that are very similar to bacterial endotoxin-produced vasculitis and infarction. However, there is yet no definite demonstration that *B. pterinoides* was the causative agent in spontaneous cattle outbreaks in New Mexico. No information is available concerning the toxin content in *B. pteronioides*.

Other North American species such as *B. halimifolia* and *B. glomeruliflora* are suspect of being toxic to livestock, but the association of these plants with livestock loss is unclear at best. *B. halimifolia* was introduced from the USA into Australia and Europe and reportedly causes cattle poisoning there (Everist 1974), and it was experimentally toxic to chicks (Duncan et al. 1957). There was no determination of the toxic principle either in *B. halimifolia* or *B. glomeruliflora*.

6 Treatment and Prophylaxis

Although some recommend the oral administration of activated charcoal to affected cattle in order to protect the gastrointestinal mucosa, there is no good or practical treatment available.

Livestock born and raised in pastures where the *B. coridifolia* exists will not ingest the plant (Barros 1993, 1998). Based on this common-sense knowledge, empirical methods to prevent *B. coridifolia* poisoning in ruminants were developed.

The Spaniards that colonized South America already used methods that consisted of being careful when introducing animals from *B. coridifolia*-free pastures to regions where the plant exists (Cobo 1653). These methods included burning dried mio-mio (to produce smoke) next to the nostrils of the animal, as well as rubbing fresh mio-mio on the muzzle and gums of the animal several times, or even satiated the animals and giving them small amounts of the plant before releasing them in the pasture infested by mio-mio.

When there is a large number of cattle to set free in a mio-mio-infested pasture, an alternative method is used by farmers, whereby the animals are gradually introduced into the pasture under a person's supervision, allowing the animals to remain gradually more time in the field every day, always taking care in not allowing them to graze too long, and keeping them in this management until ingestion of the plant is no longer noted, which will take 5–10 days (Tokarnia et al. 2012).

Recent studies concluded that an efficient method to induce aversion to the plant is the force-feeding of sublethal doses to naïve cattle before setting them loose in a *B. coridifolia*-infested-pasture (Almeida et al. 2011, 2013). For better results, cattle should remain there at least 24-h after dosing.

The same authors concluded that these methods may be impractical when a large number of bovines must be moved from a *B. coridifolia*-free to a *B. coridifolia*-infested pastures due to excessive management needed. In these instances, cattle should first be held in pastures with a low density of mio-mio and remain there for approximately 7–30 days before being transferred to the dangerous pasture.

It is apparent that future studies on mio-mio toxicosis should touch two points. (1) The pathogenesis of the disease, that is, the mechanisms that lead the animals to death so quickly and (2) on the methods to control the disease.

References

Aboul-Nasr MB (1989) Studies on *Baccharis coridifolia-Myrothecium* species interaction. PhD thesis, University of Maryland, College Park

Alda JL, Sallis ESV, Nogueira CEW et al (2009) Intoxicação espontânea por *Baccharis coridifolia* (Compositae) em equinos no Rio Grande do Sul. Pesqui Vet Bras 29:409–414

Almeida MB, Assis-Brasil ND, Schild AL et al (2011) Conditioned aversion induced by *Baccharis coridifolia* in sheep and cattle. In: Riet-Correa F, Pfister JA, Schild AL, Wierenga

T (eds) Poisoning by plants, mycotoxins and related toxins. CABI International, Wallingford, pp 613–616
Almeida MB, Schild AL, Pfister JA et al (2013) Methods of inducing conditioned food aversion to *Baccharis coridifolia* (mio-mio) in cattle. Ciên Rural 43:1866–1871
Andam, CP, Gogarten, JP (2011) Biased gene transfer in microbial evolution. Nat Rev Microbiol 9:543–555
Armién AG, Peixoto PV, Tokarnia CH (1993) Intoxicação experimental por *Baccharis megapotamica* var. *megapotamica* e var. *weirii* (Compositae) em ovinos. Pesqui Vet Bras 13:5–20
Barbosa JD, Armién AG, Tokarnia CH (1994) Intoxicação experimental por *Baccharis megapotamica* var. *weirii* (Compositae) em caprinos. Pesqui Vet Bras 13:5–20
Barros CSL (1993) Intoxicação por *Baccharis coridifolia*. In: Riet-Correa F, Méndez MC, Schild AL (eds) Intoxicações por Plantas e Micotoxicoses em Animais Domésticos. Editorial Agropecuaria del Hemisferio Sur, Uruguay, pp 159–169
Barros CSL (1998) Livestock poisoning by *Baccharis coridifolia*. In: Garland T, Barr AC (eds) Toxic plants and other natural toxicants. CAB International, New York, pp 569–572
Barros CSL, Weiblen R, Santurio JM et al (1986) Intoxicação por "mio-mio" (*Baccharis coridifolia*) em bovinos. In: Relatório do Centro de Diagnóstico Veterinário da Universidade Federal de Santa Maria. Imprensa Universitária, pp 20–21
Cobo B (1653) Historia del nuevo mundo Biblioteca de Autores Españoles, [re-edited in 1964]. Madrid 92:226–227
Costa ER, Costa JN, Armién AG et al (1995) Intoxicação experimental por *Baccharis coridifolia* (Compositae) em eqüinos. Pesqui Vet Bras 15:19–26
Cutler HC, Jarvis BB (1985) Preliminary observations on the effects of macrocyclic trichothecenes on plant growth. Exp Environ Bot 25:115–128
Desjardins AE, Hohn TM, McCormick SP (1993) Trichothecene biosynthesis in *Fusarium* species: chemistry, genetics, and significance. Microbiol Rev 57:595–604
Driemeier D, Cruz CEF, Loretti AP (2000) *Baccharis megapotamica* var *weirii* poisoning in Brazilian cattle. Vet Hum Toxicol 42:220–221
Döbereiner ML, Rezende AML , Tokarnia CH (1976) Intoxicação experimental por *Baccharis coridifolia* em coelhos. Pesqui Agrop Bras, Ser Vet 11:27–35
Duncan WH, Piercy PL, Feurt SD, Starling R (1957) Toxicological aspects of southeastern plants. II. Compositae. Econ Bot 11:75–85
Everist SL (1974) Poisonous plants of Australia. Angus & Robertson, Sidney
Flores C, Houssay BA (1917) Estudios sobre mio-mio o romerillo *Baccharis coridifolia* D.C. Revta Hist Bacteriol Dep Nac Hygiene 1:59–100
Habermehl G (1989) Isolation, structure of new toxins from plants. Pure Appl Chem 61:377–380
Habermehl GG, Busam L, Hydel P et al (1985) Macrocyclic trichothecenes: cause of livestock poisoning by the Brazilian plant *Baccharis coridifolia*. Toxicon 23:731–745
Hammerschmitt ME, Panziera W, Vielmo A et al (2018) Spontaneous *Baccharis coridifolia* poisoning in suckling lambs. Acta Sci Vet 46(Suppl 1):1–4
Heiden G, Antonelli A, Rubens JP (2019) A novel phylogenetic infrageneric classification of *Baccharis* (Asteraceae: Astereae), a highly diversified American genus. Taxon 68:1048–1081
Iglesias M, Ballesta JF (1994) Mechanism of resistance to the antibiotic trichothecin in the producing fungi. Eur J Biochem 223:447–453
Jarvis BB (1991) Macrocyclic Trichothecenes. In: Sharma RP, Salunkhe DK (eds) Mycotoxins and Phytoalexins. CRC Press, Boca Raton, pp 361–421
Jarvis BB (1992) Macrocyclic trichothecenes from Brazilian *Baccharis* species: from microanalysis to large scale isolation. Phytochem Anal 3:241–249
Jarvis BB (2003) *Stachybotrys chartarum*: a fungus for our time. Phytochemistry 64:53–60
Jarvis BB, Miller JD (2005) Mycotoxins as harmful indoor contaminants. Appl Microbiol Biotechnol 66:367–372
Jarvis BB, Midiwo JO, Tuthill D et al (1981) Interaction between the antibiotic trichothecenes and the higher plant *Baccharis megapotamica*. Science 214:460–462

Jarvis BB, Çömezoglu SN, Rao MM et al (1987a) Isolation of macrocyclic trichothecenes from a large scale extract of *Baccharis megapotamica*. J Organomet Chem 52:45–56
Jarvis BB, Wells KM, Lee Y-W et al (1987b) Macrocyclic trichothecene mycotoxins from species of Brazilian *Baccharis*. Phytopathol 77:980–984
Jarvis BB, Kuti J, Bean GA (1988) Phytotoxicity of macrocyclic trichothecenes toward *Baccharis* cell lines. JSM Mycotoxins 1988:199–203
Jarvis BB, Mokhtari-Rejali N, Schenkel EP et al (1991) Trichothecene mycotoxins from Brazilian *Baccharis* species. Phytochemistry 30:789–797
Kingsbury JM (1964) Poisonous plants of the United States and Canada. Prentice Hall, Englewood Cliffs
Kupchan SM, Jarvis BB, Dailey RG Jr et al (1976) Baccharin, a novel potent antileukemic trichothecene triepoxide from *Baccharis megapotamica*. J Am Chem Soc 98:7092–7093
Kupchan SM, Streelman DR, Jarvis BB et al (1977) The isolation of potent new antileukemic trichothecenes from *Baccharis megapotamica*. J Organomet Chem 42:4221–4225
Kuti JO, Jarvis BB, Mokhtari-Rejali N et al (1990) Allelochemical regulation of reproduction and seed germination of two Brazilian *Baccharis* species by phytotoxic trichothecenes. J Chem Ecol 16:3441–3453
Langohr IM, Gava A, Barros CSL (2005) Intoxicação por *Baccharidastrum triplinervium* (Asteraceae) em bovinos. Pesqui Vet Bras 25:235–238
Lértora WJ, Negrette MS (2015) *Baccharis coridifolia* poisoning in water buffalo (*Bubalus bubalis*) in the north of the Province of Corrientes, Argentina. Braz J Vet Pathol 8:1–5
Loukaci A, Kayser O, Bindseil K-U, Siems K et al (2000) New trichothecenes isolated from *Holarrhena floribunda*. J Nat Prod 63:52–56
Manley GD, Edds GT, Sundlof SF (1982) Cattle deaths from poisonous plant. Folia Morphol 11:20
Marsh CD, Clawson AB, Eggleston WW (1920) *Baccharis pteronioides* as a poisonous plant of the Southwest. J Am Vet Med Assoc 57:430–434
McCormick SP, Stanley AM, Stover NA et al (2011) Trichothecenes: from simple to complex mycotoxins. Toxins 3:802–814
Newman DJ, Cragg GM (2015) Endophytic and epiphytic microbes as “sources” of bioactive agents. Front Chem 3:1–13
Oliveira-Filho JC, Carmo PMS, Lucena RB et al (2011) *Baccharis megapotamica* var *weirii* poisoning in water buffalo (*Bubalus bubalis*). J Vet Diagn Investig 23:610–614
Oliveira-Filho JC, Carmo PMS, Iversen A et al (2012) Experimental poisoning by *Baccharis megapotamica* var *weirii* in buffalo. Pesqui Vet Bras 32:383–390
Panziera W, Gonçalves MA, Lorenzett MP et al (2015) Intoxicação natural por *Baccharis megapotamica* var. *weirii* em caprinos. Pesqui Vet Bras 35:360–364
Pedroso PMO, Bandarra PM, Feltrin C et al (2010) Intoxicação por *Baccharis megapotamica* var. *weirii* em ovinos. Pesqui Vet Bras 30:403–405
Richardson AO, Palmer JD (2007) Horizontal gene transfer in plants. J Exp Bot 58:1–9
Rissi DR, Rech RR, Fighera RA et al (2005) Intoxicação espontânea por *Baccharis coridifolia* em bovinos. Pesqui Vet Bras 25:111–114
Rizzo I, Varsavky E, Haidukowski M et al (1997) Macrocyclic trichothecenes in Baccharis coridifolia plants and endophytes and *Baccharis artemisioides* plants. Toxicon 35:753–757
Rodrigues RL, Tokarnia CH. (1995) Fatores que influenciam a toxidez de Baccharis coridifolia (Compositae): um estudo experimental em coelhos. Pesqui Vet Bras 15:51–69
Rozza DB, Raymundo DL, Corrêa et al (2006) Intoxicação espontânea por *Baccharis coridifolia* (Compositae) em ovinos. Pesqui Vet Bras 26:21–25
Schang RJ (1929) Acción tóxica del romerillo o mio-mio (*Baccharis coridifolia*). Algunos conceptos nuevos. Rev Med Vet 11:151–181
Stegelmeier BL, Sani Y, Pfister JA (2009) *Baccharis pteronioides* toxicity in livestock and hamsters. J Vet Diagn Investig 21:208–213
Strobel G, Daisy B (2003) Bioprospecting for microbial endophytes and their natural products. Microbial Mol Biol Rev 67:491–502

Tokarnia CH, Döbereiner J (1975) Intoxicação experimental em bovinos por "mio-mio" *Baccharis coridifolia*. Pesqui Agrop Bras 10:79–97

Tokarnia CH, Dobereiner J (1976) Intoxicação experimental em ovinos por "mio-mio", *Baccharis coridifolia*. Pesqui Agropec Bras 11:19–26

Tokarnia CH, Peixoto PV, Gava A et al (1992) Intoxicação experimental por *Baccharis megapotamica* var *megapotamica* e var *weirii* (Compositae) em bovinos. Pesqui Vet Bras 12:19–31

Tokarnia CH, Brito MF, Barbosa JD et al (2012) Plantas que afetam o trato digestório in Ibid Plantas Tóxicas do Brasil para Animais de Produção, 2nd. Helianthus, Rio de Janeiro, pp 95–146

Trapp SC, Jarvis BB, Hohn TM (1998) Characterization of the macrocyclic trichothecene pathway gene cluster in *Myrothecium roridum*. Mol Gen Genet 257:421–432

Varaschin MS, Alessi AC (2003) Poisoning of mice by *Baccharis coridifolia*: an experimental model. Vet Hum Toxicol 45:42–44

Varaschin MS, Barros CSL, Jarvis BB (1998) Intoxicação experimental por *Baccharis coridifolia* (Compositae) em bovinos. Pesqui Vet Bras 18:69–75

Zhang HJ, Tamez PA, Aydogmus Z et al (2002) Antimalarial agents from plants. III. Trichothecenes from *Ficus fistulosa* and *Rhaphidophora decursiva*. Planta Med 68:1088–1091

Part III
Baccharis: Applications and Innovations

In this part, we bring to light the social and economic relevance that *Baccharis* presents. Many species in this genus are important in the traditions of many countries. Some species have been used mainly as culinary ingredients, utensils, and primarily in popular medicine. Among the most popular and commercialized species for medicinal purposes, *Baccharis trimera* stands out and is therefore widely cultivated. Nevertheless, the applications extend to the cosmetics, food, and pharmaceutical industry. Due to its market potential, many *Baccharis* species have patents about them generated outside their countries of origin.

***Baccharis rhomboidalis*. Illustration by Patrícia Angrisano**

Chapter 16
An Overview of the Cultural and Popular Use of *Baccharis*

Mara Rejane Ritter, Anderson Luiz Christ, Aline de Mello Zevieski, and Marina Fülber

Abstract It is widely known that *Baccharis* is an important genus from the ethnobotanical point of view, but a general revision of their usage was never produced. Through the revision of recent and classic ethnobotanical studies, we made a compilation of data regarding the popular uses of several species of *Baccharis* in South America. Commentaries on the usage of species of *Baccharis* are here presented and grouped by general use. The main categories of use found were as alimentary plants, for the construction of utensils, dye plants, and as medicinal plants. Medicinal uses were classified according to ICD-11. We found that *Baccharis* species are mostly used as medicinal plants, particularly for the treatment of general alignments, external injuries, and digestive diseases. *Baccharis articulata* and *B. trimera* (= *B. crispa*) were the most cited species in the consulted studies. Our results reiterate the importance of this genus in the popular knowledge of communities of South America.

Keywords Ethnobotanical knowledge · Food security · Medicinal plants · Plant popular use

---

M. R. Ritter (✉)
Departamento de Botânica, Instituto de Biociências, Universidade Federal do Rio Grande do Sul, Porto Alegre, Rio Grande do Sul, Brazil

Programa de Pós-Graduação em Botânica, Instituto de Biociências, Universidade Federal do Rio Grande do Sul, Porto Alegre, Rio Grande do Sul, Brazil
e-mail: mara.ritter@ufrgs.br

A. L. Christ · A. de Mello Zevieski
Programa de Pós-Graduação em Botânica, Instituto de Biociências, Universidade Federal do Rio Grande do Sul, Porto Alegre, Rio Grande do Sul, Brazil

M. Fülber
Departamento de Botânica, Instituto de Biociências, Universidade Federal do Rio Grande do Sul, Porto Alegre, Rio Grande do Sul, Brazil

G. W. Fernandes et al. (eds.), *Baccharis*,
https://doi.org/10.1007/978-3-030-83511-8_16

1 Introduction

Many species of *Baccharis* L. are known and used by different traditional communities in many countries around the globe, particularly in Latin America. According to Aguilar et al. (2007) and Molares et al. (2009), the broad usage of *Baccharis* in traditional medicine can be attributed to the phytochemical aspects of the plants, which are rich in flavonoids and terpenes.

Several different uses are reported for *Baccharis* in the traditional literature, including as food, forage for livestock, tools, dye plants, and charcoal, but the ethnobotanical knowledge related to species of this genus is prominently linked to their use as medicine, as noted by Molares et al. (2009). According to Ariza Espinar (1973) and Heiden et al. (2009), many species of *Baccharis* used as medicine have winged stems and are known as "carquejas" (or a variety of names derived from it) and are usually used as chagrin in the form of infusions. Many other species, mostly more robust plants with a shrubby habit, are known as "vassouras" ("brooms") and used for a wide variety of ends, including use as charcoal and crafting of tools (Pio Corrêa 1931; Rodrigues et al. 2002).

In Brazil, there are many sources of information by various naturalists regarding the ethnobotany of *Baccharis*. Brandão (2010) mentions the account of C.J.F. Bunburry, an English naturalist who traveled through Rio de Janeiro and Minas Gerais from 1833 to 1835, regarding the usage of the "carqueja" (*Baccharis crispa* Spreng. or one of its synonyms, such as *B. trimera* DC.):

> "... It is excessively bitter and extensively used in medicine by Brazilian people, particularly for horses."
>
> "The entire plant is extremely bitter and surpasses the quina de Genciana (*Gentiana lutea* L.). It is used against intermittent fevers."

Currently, *Baccharis* species continue to be used in Brazilian popular medicine. *Baccharis trimera* (Less.) DC. (now synonymous with *B. crispa* Spreng.), for example, is listed in the Relação Nacional de Plantas Medicinais de Interesse ao Sistema Único de Saúde -RENISUS (National List of Medicinal Plants of Interest to the Unified Health System) (Brasil 2019) and is indicated by the Formulário de Fitoterápicos da Farmacopéia Brasileira (Pharmacotherapeutic Formulary of the Brazilian Pharmacopoeia) as an antidispeptic (Brasil 2011, 2018). Many species of the genus are commercialized in popular markets, and multiple similar species may be used interchangeably for the treatment of the same diseases (Heiden et al. 2009).

In this context, the objective of the present work was to compile ethnobotanical information of *Baccharis* species.

2 Material and Methods

For the development of this revision, we consulted published articles, monographs, MSc dissertations, and PhD theses. Works on traditional uses of *Baccharis* were accessed through a search on digital databases such as PubMed (https://www.ncbi.nlm.nih.gov/pubmed/), Web of Science (https://clarivate.com/products/web-of-science), Scopus (https://www.scopus.com), Scielo (https://www.SciELO.org/php/index.php), and in University libraries. Searches were conducted without a limiting period. Classical Pharmacognosy and Phytotherapy textbooks were also reviewed. The traditional use was reviewed on recognized ethnobotanical references about Brazilian Flora.

Popular and scientific names of species of *Baccharis* are cited according to their original citations in the consulted literature, disregarding any nomenclatural or taxonomic issues involving these names. This includes original citations of species presently included in *Baccharis*, but previously circumscribed in different genera, such as *Baccharidastrum triplinervium* (Less.) Cabrera (= *Baccharis vulneraria* Baker). Authorship of scientific names follows IPNI (http://www.ipni.org).

The expressions used to describe folk uses found in the ethnobotanical literature were cited as in the original works. Popular therapeutic indications for species of *Baccharis* species found in ethnobotanical surveys were classified according to ICD-11 (WHO 2018).

3 Ethnobotanical Uses

The ethnobotanical data found are presented in the following categories of use.

Alimentary Use

Some species of *Baccharis* are used in the preparation of beverages. *Baccharis articulata* (Lam.) Pers. ("carquejinha") is cited by Kinupp and Lorenzi (2014) as a nonconventional food plant and as sporadically employed in cooking and brewing homemade beers. It is also part of the recipe for a little slimy drink. According to Ferreira (1998), *B. genisteloides* (Lam.) Pers. replaces hops in beer brewing and Hurrell et al. (2011) cite the use of branches to flavor bitter drinks.

Baccharis gilliesii A.Gray, known as "yerba de obeja" ("sheep's weed"), is used as forage for livestock, in general, in some rural communities in Argentina, and noted as being particularly enjoyed by sheep, hence its popular name (Muiño 2010). *Baccharis salicifolia* (Ruiz & Pav.) Pers. ("chilca") is also used as forage in these communities (Muiño 2010). Zardini (1984) cites *B. incarum* Wedd. as being used as forage in Argentina and also as being consumed as food by humans in northern

Chile. According to Loredo-Medina et al. (2002), *Baccharis conferta* Kunth ("escoba") is used as forage in Mexico.

According to Zardini (1984), *B. rufescens* Spreng. is consumed as a salad in Argentina, where there are also records of it being used as a spice in substitution of salt.

Utensils

In the classic work of Pio Corrêa (1931), several uses are cited for species of *Baccharis*, including their usage as material for the constructing of rustic brooms ("vassouras").

Baccharis dracunculifolia DC. is one of the most cited species in the construction of domestic utensils in Brazil. Rodrigues et al. (2002) mention the use of this species (popularly called "alecrim") as firewood and as a broom in the Brazilian state of Minas Gerais. It is also mentioned with these uses in Santa Catarina, denominated by "vassoura-branca" and "vassourão" (Zuchiwschi et al. 2010; Pereira et al. 2016) and in São Paulo (Oliveira Junior et al. 2018). In Rio Grande do Sul, it is known as "vassourão" and used for crafting domestic and agricultural utensils, for structural use, and for firewood (Chaves 2010).

Zardini (1984) mentions ethnobotanical usages of several species of *Baccharis* in Argentina, including a few used as tools: the wood of *Baccharis boliviensis* (Wedd.) Cabrera, *B. dracunculifolia, B. incarum* Wedd., *B. polifolia* Griseb. and *B. tucumanensis* Hook. & Arn. is used as charcoal; *B. notosergila* Griseb. is predominantly used in the crafting of brushes; *B. spartioides* DC. is also used in the crafting of brushes, which are said to be very aromatic and useful to repel insects, and the branches of the plant are also used as brooms.

According to Loredo-Medina et al. (2002), *Baccharis conferta* Kunth ("escoba") is used for the production of handicrafts and for industrial use in Mexico.

Dye Plants

Branches of various species of *Baccharis* are used for dyeing in South America.

According to Ferreira (1998), when the branches are cooked in aluminum and stainless steel pots, the coloration is yellow-gold; when they are cooked in used iron pots or rusty cans, the coloration is yellow-green. *Baccharis calliprinos* Griseb. is used as a dye plant, mostly to obtain the color yellow and *B. notosergila* is used to obtain the color black, which is then used to paint pottery (Zardini 1984).

Baccharis genistelloides, *B. latifolia* (Ruiz & Pav.) Pers., and *B. odorata* Kunth are cited by Usca and Linares (2012) as used in a few communities in Ecuador as dye plants. The usage of *B. latifolia* as a dye plant is also reported by Zardini (1984), this time in Argentina. Some Andean communities still retain the tradition of using some vegetable dyes extracted from shrub species (*Baccharis incarum* (Wedd.) Perkins) such as "thola" (Vidaurre et al. 2006).

Medicinal Use

The therapeutic uses of the species cited as medicinal were compiled and classified according to the international classification of diseases; ICD-11 (WHO 2018) is explained in Table 16.1. A total of 41 species are listed in Table 16.2, which also includes the chapters of ICD-11 used to classify them, their vernacular names found in the consulted literature, and the references of the usages.

Figure 16.1 summarizes the number of species cited for each chapter of ICD-11. Of the 22 categories, 14 could be associated with the medicinal uses found in the consulted literature. No mentions for Chaps. 4, 6, 7, 8, 9, 10, 19, and 20 were found in the consulted literature. Most species of *Baccharis* with known medicinal use could be classified under category 21, a rather unspecific category that mostly covers symptoms not suitable for the other chapters. Among the most cited uses classified hereunder, Chap. 21 covers the usage of *Baccharis* for the treatment of general fevers, inflammations, rheumatism or pain in general, and general malaise. Many species are also commonly used for the treatment of general external injuries (Chap. 22) and for diseases of the digestive system (Chap. 13).

Of the 41 species named in our revision, *B. articulata, B. trimera*, and *B. genisteloides* are the most versatile, with their uses being classified under Chaps. 11, 10, and 9 of ICD-11, respectively. *Baccharis articulata* and *B. trimera* are particularly notable for being the most cited species in our revision, which evidences their statuses as two of the most widely used medicinal plants in southern South America, especially in Argentina and southern Brazil. According to Heiden et al. (2009), however, *B. trimera* is to be treated as a synonym of *B. crispa*, which is also extensively cited in the consulted literature.

The use of several species of *Baccharis* for the digestive system can be explained by the fact that people seek bitter plants to alleviate the symptoms of poor digestion. Olivier and van Wyk (2013) affirm that the use of bitter tonics is an ancient practice believed to have beneficial effects on digestion and high bitterness values presented by these plants may at least be partly ascribed to the bitter tonic effect, that is, the stimulation of gastric juices via the nervus vagus.

Multiple recent sources exist to provide information regarding the usage of *Baccharis* as medicine, particularly traditional studies of ethnobotany. Still, some records of the use of *Baccharis* species as medicinal are found in historical works, such as Pio Corrêa (1931), which in his classic work "Dicionário das Plantas úteis do Brasil" related several medicinal uses for different species of *Baccharis*, such as digestive and as tonics.

Interesting information is provided by Ferreira (1998), who cites the use of *Baccharis genisteloides* in the rural regions of Argentina, where the population believes that the plant is effective in simultaneously combating male impotence and

Table 16.1 Classification of the popular therapeutic indications for *Baccharis* species found in ethnobotanical surveys according to ICD-11 (WHO 2018)

Chapter	International Classification Diseases	Popular uses
01	Certain infectious or parasitic diseases	As anthelmintic, anti-inflammatory, and antibacterial. For infections, hepatitis, leprosy, malaria, nephritis, syphilis, cholera, and Chagas disease
02	Neoplasms	As anticancer
03	Diseases of the blood or blood-forming organs	For anemia and "to the blood"
05	Endocrine, nutritional, or metabolic diseases	To increase the appetite, weakness, anorexia, weight loss, diabetes, hypoglycemic, depurative, menopause, glycosuria. To lower triglycerides and cholesterol
11	Diseases of the circulatory system	For angina and stroke. As hypotensive, restore blood circulation, and varicose veins
12	Diseases of the respiratory system	For asthma, cough, influenza. As anticatarrhal
13	Diseases of the digestive system	For hemorrhoids, constipation, dyspepsia, ulcers, gastritis, nausea, treat heartburn. As antidiarrheic, choleric action, colagoga, tonic, liver, hepatoprotective, digestive, stomach, eupeptic, hepatic stimulant, gastropathy, inflammation of the spleen and gallstones
14	Diseases of the skin	For growth and hair loss, acne treatment, and dermatitis
15	Diseases of the musculoskeletal system or connective tissue	For spine pains
16	Diseases of the genitourinary system	For vaginal lavage, menstrual cramps, metritis, and gynecological disorders. For treating maladies of the kidneys, kidney stone, kidney disease. As a diuretic, urinary tract infection and bladder inflammation
17	Conditions related to sexual health	For male impotence, female sterility, venereal diseases, aphrodisiac, contraceptive, and female fertility regulator
18	Pregnancy, childbirth or the puerperium	To treat postpartum complications, as abortive
21	Symptoms, signs, or clinical findings, not elsewhere classified	For treatment of malaise, inflammation, rheumatism, muscular pains, bones, headaches, cramps. As analgesic, antiseptic, antispasmodic, febrifuge, stimulant action, smooth muscle relaxant (vasodilator action), reduce swelling, jaundice, food poisoning, and fortifier
22	Injury, poisoning, or certain other consequences of external causes	For wounds, burns, ulcers, trauma, contracted diseases of animals, against snakebites and purification and relaxation. As antiseptic and depurative

The categories for which no citations for use of *Baccharis* species were found were excluded

Table 16.2 *Baccharis* species, traditional popular names, chapters of ICD-11, and references

Baccharis species, traditional popular names	Traditional uses according to the Classification of Diseases (ICD-11) (chapters)	References
Baccharidastrum triplinervium (Less.)Cabrera (Erva-santa, erva-santa-maria)	22	Garlet and Irgang (2001) and Soares et al. (2004)
Baccharis altimontana G.Heiden, Baumgratz & R.Esteves	01, 05, 13, 21	Santos et al. (2014)
Baccharis anomala DC. (Parreirinha)	01, 13, 16, 22	Kubo (1997) and Garlet and Irgang (2001)
Baccharis artemisioides Hook. & Arn. (Mío-mío-blanco, pichana-blanca, perkan-kachú, plan-romeriyu, romerillo, romerillo-blanco, romerillo-malo)	21, 22	Zardini (1984)
B. articulata (Lam.) Pers. (Carqueja, carqueja-branca, carqueja-crespa, carqueja-doce-carquejinha, carquejinha-branca, carquejinha-do-campo, carquejinha-miúda, carquejilla, yacaré-ruguai, killá-fosí, kilá-foshí, l'e- tañoni, caá-cambu-y-guazu, cacapeguazú, yaguareté-caá)	01, 03, 05, 11, 12, 13, 15, 16, 17, 21, 22	Pio Corrêa (1931), Mariante (1984), Zardini (1984), Simões et al. (1986), Martins et al. (1994), Alice et al. (1995), Lopes (1997), Kubo (1997), Ferreira (1998), Koch (2000), Pavan-Fruehauf (2000), Garlet and Irgang (2001), Soares et al. (2004), Barbosa (2005), Goleniowski et al. (2006), Barros et al. (2007), Souza (2007), Baldauf et al. (2009), Ceolin et al. (2009), Steffen (2010), Haeffner et al. (2012), Quiroga et al. (2012), Battisti et al. (2013), Santos et al. (2014) and Acosta et al. (2017)
Baccharis calliprinos Griseb. (Chascoma, chilca-dulce, fiamate, palo-blanco)	21	Zardini (1984)
Baccharis conferta Kunth (Escobilla-china)	13	Weimann and Heinrich (1996)
Baccharis crispa Spreng. (Carqueja, carqueja-amargosa, carqueija, carquejilla)	01, 05, 13, 16, 21, 22	Zardini (1984), Paz et al. (1992), Haeffner et al. (2012), Bieski et al. (2015) and Tribess et al. (2015)
Baccharis douglasii DC. (Renegada)	16, 22	Bocek (1984)
Baccharis dracunculifolia DC. (Alecrim-do-campo, carqueja, vassourinha, vassoura-mansa, jatun t'ula, tola)	01, 13, 16, 21, 22	Gavilanes et al. (1981/1982), Garlet and Irgang (2001), Fernandez et al. (2003), Silva et al. (2008) and Quiroga et al. (2012)
Baccharis floribunda Kunth (Waca-ch'illka)	16, 22	Fernandez et al. (2003)

(continued)

Table 16.2 (continued)

Baccharis species, traditional popular names	Traditional uses according to the Classification of Diseases (ICD-11) (chapters)	References
Baccharis gaudichaudiana DC. (Alecrim-do-campo, carqueja, carqueja-doce, vassourinha)	01, 03, 05, 11, 13, 16, 21, 22	Pio Corrêa (1931), Zardini (1984), Pavan-Fruehauf (2000) and Garlet and Irgang (2001)
Baccharis genistelloides (Lam.) Pers. (Carqueja, carqueja-amarga, carqueja-amargosa, carqueja-de-folhas-estreitas, quina-de-condamine, callua-callua, cuchu-cuchu, ischu-tullma, karkeja, kuchu-kuchu, quinsa cuchu, yaja, cucho-cucho, charara, kimsa-kkuchu)	01, 03, 05, 11, 13, 16, 17, 18, 21	Simões et al. (1986), Montes and Wilkomirsky (1988), Cervi et al. (1989), Ferreira (1998), Albuquerque et al. (2005), Osuna et al. (2005), Macía et al. (2005), Aguilar et al. (2007), De-la-Cruz et al. (2007) and Philippi (2012)
Baccharis genistifolia DC. (Wentrú-kulandriya)	21	Zardini (1984)
Baccharis glutinosa Pers. (Batamote, jarilla, hierba-del-pasmo, bachomo, guatamote)	13, 14, 16, 21, 22	Bocek (1984), García-Alvarado et al. (2001), Aguilar et al. (2007) and Moreno-Salazar et al. (2008)
Baccharis grisebachii Hieron. (Quinchamal, romerillo, tancha)	22	Zardini (1984)
Baccharis latifolia (Ruiz & Pav.) Pers. (Chilco, chilca, chilka)	13, 21, 22	Macía et al. (2005), De-la-Cruz et al. (2007) and Sequeda-Castañeda et al. (2015)
Baccharis linearis (Ruiz & Pav.) Pers. (Romerillo)	21	Montes and Wilkomirsky (1988)
Baccharis lundii DC.	13	Pio Corrêa (1931) and Pavan-Fruehauf (2000)
Baccharis macrodonta DC.	21	Pio Corrêa (1931) and Gavilanes et al. (1981/1982)
Baccharis microcephala (Less.) DC. (Carqueja, carqueija)	13	Zardini (1984)
Baccharis microphylla Kunth (Chilca)	11, 13, 21, 22	Morales et al. (2008)
Baccharis notosergila Griseb. (Carqueija, carqueja, koloron-rakté, milgrat, pagueré-lko, oron-rakté, tipisha'i)	1, 18, 21, 22	Pio Corrêa (1931), Zardini (1984) and Pavan-Fruehauf (2000)
Baccharis obovata Hook. & Arn. (Wuatro)	12, 14, 21	Molares et al. (2009)
Baccharis ochracea Spreng. (Erva-santa)	13, 17, 21	Philippi (2012)
Baccharis odorata Kunth (Taya)	21	De-la-Cruz et al. (2007)

(continued)

Table 16.2 (continued)

Baccharis species, traditional popular names	Traditional uses according to the Classification of Diseases (ICD-11) (chapters)	References
Baccharis pedicellata DC. (Chilca-cordillerana)	21	Montes and Wilkomirsky (1988)
Baccharis petiolata DC. (Chilca)	11, 13, 21, 22	Morales et al. (2008)
Baccharis pingraea DC. (Chilca)	01, 02, 21	Goleniowski et al. (2006)
Baccharis punctulata DC. (Erva-santa, ch'illka-saru- saru)	03, 12, 22	Kubo (1997) and De-la-Cruz et al. (2007)
Baccharis rhomboidalis J.Rémy (Chica-cordillerana)	05	Reis et al. (1990)
Baccharis riograndensis Malag. & J.E.Vidal	13, 16, 21	Pio Corrêa (1931), Mariante (1984), Simões et al. (1986), Martins et al. (1994), Alice et al. (1995), Lopes (1997) and Pavan-Fruehauf (2000)
Baccharis salicifolia (Ruiz & Pav.) Pers. (Chilca, chilca-blanca, ckechua-chilca, chillca, chilco, yana-chilca, chilca- negra, chilca-amarga, suncho, caabo-yuqui, ca'gu-si, chilca- dulce, chirca, jarilla, jarilla- del-río, junco, vara-dulce, yuno)	01, 02, 12, 14, 21, 22	Zardini (1984), Scarpa (2004), Goleniowski et al. (2006), Aguilar et al. (2007) and De-la-Cruz et al. (2007)
Baccharis santelicis Phil. (Chilca)	11, 13, 21, 22	Morales et al. (2008)
Baccharis stenocephala Baker	22	Pio Corrêa (1931), Montes and Wilkomirsky (1988) and Pavan-Fruehauf (2000)
Baccharis tridentata Vahl (Carqueja-folhuda)	16	Pio Corrêa (1931)

(continued)

Table 16.2 (continued)

Baccharis species, traditional popular names	Traditional uses according to the Classification of Diseases (ICD-11) (chapters)	References
Baccharis trimera (Less.) DC. (Carqueja, carqueja-folhuda, charruinha, carqueja-doce, bacanta, carqueja-amarga, carqueja-amargosa, capoeira-branca, carqueja-branca, carqueja-crespa, carque, quina-de-condamine, tiririca-de-babado, vassoura, tiririca-de-balaio, yaguaretê- ca'á)	01, 03, 05, 11, 12, 13, 16, 17, 21, 22	Gavilanes et al. (1981/1982), Zardini (1984), Santos et al. (1988), Siqueira (1988), Cervi et al. (1989), Sousa et al. (1991), Paz et al. (1992), Figueiredo et al. (1993), Martins et al. (1994), Figueiredo et al. (1997), Lopes (1997), Piva (1998), Pavan-Fruehauf (2000), Garlet and Irgang (2001), Ladeira (2002) Moreira et al. (2002), Stasi and Hiruma-Lima (2002), Maciel and Cardoso (2003), Albuquerque et al. (2005), Osuna et al. (2005), Brasil (2006), Vendruscolo and Mentz (2006), Albuquerque et al. (2007), Hanazaki et al. (2007), Lorenzi and Matos (2008), Silva et al. (2008), Ceolin et al. (2009), Pereira et al. (2009), Steffen (2010), Castro et al. (2011), Coan and Matias (2013), Löbler et al. (2014), Cercato et al. (2015) and Ribeiro et al. (2017)
Baccharis trinervis (Lam.) DC. (Bejuco-de-valdivia, marucha)	13, 22	Weimann and Heinrich (1996) and Vásquez et al. (2015)
Baccharis triptera Mart. (Carqueja-amarga, cacália)	01, 05, 11,13, 16, 21	Gavilanes et al. (1981/1982), Verardo (1981/1982) and Reis et al. (1990)
Baccharis vulneraria Baker (Mbichini-kaá, yerba-santa)	22	Zardini (1984)
Baccharis sp.	01, 05, 11, 13, 21, 22	Maffei (1969), Souza (2007) and Borges (2010)

female sterility. It also reported that during the cholera epidemic in Brazil in 1849, it was believed that *B. articulata* was fighting anemia and that this species is also used in veterinary medicine to combat cattle diarrhea (Ferreira 1998).

Other Uses

Some other uses have been found for *Baccharis*, including their use as bee forage, as ornamental plants, and their use in religious rituals.

In Southern Brazil, *Baccharis* species are used as bee forage. Reports of this use include the ones of Breyer et al. (2016) and Zuchiwschi et al. (2010) for

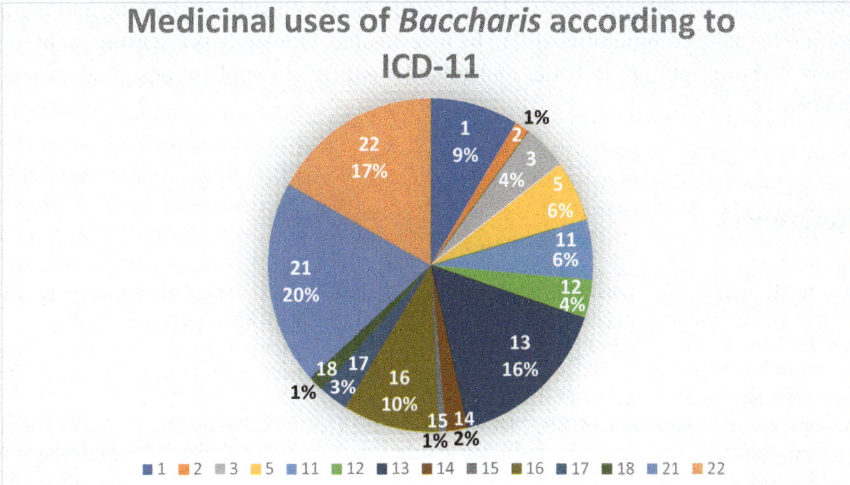

Fig. 16.1 Summary of the amount of species cited for each chapter of medicinal use according to ICD-11

Baccharidastrum triplinervium (Less.) Cabrera ("vassorinha", "erva-desantana") (= *B. vulneria*), and the one of Ferreira (1998) for *B. genisteloides*.

In Brazil, *Baccharis dracunculifolia* is used in liturgical rituals of Candomblé in some communities in Bahia (Pires et al. 2009) and in São Paulo (Oliveira Junior et al. 2018).

The use in personal hygiene is referred in Brazil for *Baccharis dracunculifolia* by Pereira et al. (2016) and for *B. genisteloides*, which has very white wood and can be used to clean the teeth according to Ferreira (1998).

Baccharis milleflora (Less.) DC. and *B. tridentata* Vahl. were pointed by Tognon and Cuquel (2015) as presenting a high ornamental potential for use as cutting foliage with characteristics suitable for use as complements to floral arrangements.

4 Conclusion

Our findings demonstrate the importance of *Baccharis* in the ethnobotany of South America. We found that species of *Baccharis* are mostly used as medicinal plants, particularly for the treatment of general alignments, external conditions, and the treatment of digestive diseases. The latter use is mostly related to the species with winged stems, which are popularly known as "carquejas." However, we also found that there are many other different applications of *Baccharis*, including their use as forage, as dye plants, and their importance in a few religious rituals. The general knowledge in the ethnobotany of *Baccharis* is, however, still incomplete, and there are surely many communities still to be investigated. Our study is a starting point for future revisions aiming to better understand the different applications of species of *Baccharis*.

References

Acosta ME, Ladio AH, Vignale ND (2017) Medicinal plants traded in San Salvador de Jujuy city (Argentina) and its botanical quality. Bol Latinoam Caribe Plant Med Aromat 16:34–52

Aguilar E, Anaya B, Alarcón J et al (2007) Etnobotánica, fitoquímica y farmacologia de especies del género *Baccharis* (Asteraceas) utilizadas como plantas medicinales en el Departamento de Ayacucho. Cien Investig 10:13–19

Albuquerque UP, Almeida CFCBR, Marins JFA (2005) Tópicos em conservação, etnobotânica e etnofarmacologia de plantas medicinais e mágicas. Sociedade Brasileira de Etnobiologia e Etnoecologia. NUPPEA, Recife

Albuquerque UP, Monteiro JM, Ramos MA et al (2007) Medicinal and magic plants from a public market in northeastern Brazil. J Ethnopharmacol 110:76–91

Alice CB, Siqueira NCS, Mentz LA et al (1995) Plantas medicinais de uso popular: atlas farmacognóstico. Editora da ULBRA, Canoas

Ariza Espinar L (1973) Las especies de *Baccharis* (Compositae) de Argentina Central. Bol Acad Nac Cien 50:175–305

Baldauf C, Kubo RR, Silva F, Irgang BE (2009) "Ferveu, queimou o ser da erva": conhecimentos de especialistas locais sobre plantas medicinais na região sul do Brasil. Rev Bras Plant Med 11:282–291

Barbosa JF (2005) Estudo etnobotânico das plantas de uso medicinal e místico utilizadas na comunidade Quilombola de Nova Real, Bom Retiro do Sul, Estado do Rio Grande do Sul. Trabalho de conclusão em Ciências Biológicas, Unisinos, São Leopoldo

Barros FMC, Pereira KN, Zanetti GD et al (2007) Plantas de uso medicinal no município de São Luiz Gonzaga, RS, Brasil. Lat Am J Pharm 26:652–662

Battisti C, Garlet TMB, Essi L et al (2013) Plantas medicinais utilizadas no município e Palmeira das Missões, RS, Brasil. Rev Bras Bioc 11:338–348

Bieski IGC, Leonti M, Arnason JT et al (2015) Ethnobotanical study of medicinal plants by population of Valley of Juruena Region, Legal Amazon, Mato Grosso, Brazil. J Ethnopharmacol 173:383–423

Bocek BR (1984) Ethnobotany of Costanoan Indians, California, based on collections by John P. Harrington. Econ Bot 38:240–255

Borges AM (2010) Plantas medicinais no cuidado em saúde de moradores da Ilha dos Marinheiros: contribuições à enfermagem. Dissertação de Mestrado em Enfermagem, Universidade Federal de Pelotas, Pelotas

Brandão MGL (2010) Plantas úteis de Minas Gerais – na obra dos naturalistas. Código Comunicação, Belo Horizonte

Brasil (2019) Relação Nacional de Plantas Medicinais de Interesse ao SUS – RENISUS. http://portalarquivos2.saude.gov.br/images/pdf/2014/maio/07/renisus.pdf. Accessed 19 Mar 2019

Brasil, Agência Nacional de Vigilância Sanitária (2011) Formulário de Fitoterápicos da Farmacopéia Brasileira/Agência Nacional de Vigilância Sanitária. ANVISA, Brasília

Brasil, Agência Nacional de Vigilância Sanitária (2018) Formulário de Fitoterápicos da Farmacopéia Brasileira/Agência Nacional de Vigilância Sanitária. Primeira edição, Primeiro suplemento. ANVISA, Brasília

Brasil, Ministério da Saúde. Secretaria de Ciências, tecnologia e insumos estratégicos. Departamento de Assistência Farmacêutica (2006) A fitoterapia no SUS e o Programa de Pesquisa de plantas medicinais da central de medicamentos. Ministério da Saúde, Brasília

Breyer HFE, Breyer EDH, Cella I (2016) Produção e beneficiamento da Própolis. Empresa de Pesquisa Agropecuária e Extensão Rural de Santa Catarina, Florianópolis, SC

Castro JA, Brasileiro BP, Lyra DH (2011) Ethnobotanical study of traditional uses of medicinal plants: the flora of Caatinga in the community of Cravolândia-BA, Brazil. J Med Plant Res 5:1905–1917

Ceolin T, Heck RM, Barbieri RL et al (2009) Plantas medicinais utilizadas como calmantes por agricultores ecológicos da região Sul do Rio Grande do Sul, Brasil. Rev Enferm 3:1034–1041

Cercato LM, White PAS, Nampo FK et al (2015) A systematic review of medicinal plants used for weight loss in Brazil: is there potential for obesity treatment? J Ethnopharmacol 176:286–296

Cervi AC, Negrelle RRBN, Sbalchiero D (1989) Espécies vegetais utilizadas na terapêutica popular no município de Curitiba, Paraná, Brasil. Estud Biol 23:5–42

Chaves AS (2010) Etnobotância em comunidades rurais de origem italiana e polonesa do município de Erechim/RS. Dissertação de Mestrado. Programa de Pós-Graduação em Ecologia, Universidade Regional Integrada do Alto Uruguai e das Missões

Coan CM, Matias T (2013) Utilização das plantas medicinais pela comunidade indígena de Ventarra Alta- RS. Rev Educ IDEAU 8:1–14

De la Cruz H, Vilcapoma G, Zevallos PA (2007) Ethnobotanical study of medicinal plants used by the Andean People of Canta, Lima, Peru. J Ethnopharmacol 111:284–294

Fernandez EC, Sandi YE, Kokoska L (2003) Ethnobotanical inventory of medicinal plants used in the Bustillo Province of the Potosi Department, Bolivia. Fitoterapia 74:407–416

Ferreira EL (1998) Corantes Naturais da Flora Brasileira. Guia prático de tingimento com plantas. Optagraf Editora e Gráfica Ltda, Curitiba

Figueiredo GM, Leitão-Filho HF, Begossi A (1993) Ethnobotany of Atlantic Forest coastal communities: diversity of plants uses in Gamboa (Itacuruçá Island, Brazil). Hum Ecol 21:419–430

Figueiredo GM, Leitão-Filho HF, Begossi A (1997) Ethnobotany of Atlantic Forest coastal communities: II. Diversity of plant uses at Sepetiba Bay (SE Brazil). Hum Ecol 25:353–360

García-Alvarado JS, Verde-Star MJ, Heredia NL (2001) Traditional uses and scientific knowledge of medicinal plants from Mexico and Central America. Int J Geogr Inf Syst 8:37–89

Garlet TMB, Irgang BE (2001) Plantas medicinais utilizadas na medicina popular por mulheres trabalhadoras rurais de Cruz Alta, Rio Grande do Sul, Brasil. Rev Bras Plant Med 4:9–18

Gavilanes ML, Brandão M, Cardoso C (1981/1982) Levantamento das plantas daninhas utilizadas como medicinais, de uso popular. Oréades 8:34–47

Goleniowski ME, Bongiovanni GA, Palacio L et al (2006) Medicinal plants from the "Sierra de Comechingones", Argentina. J Ethnopharmacol 107:324–341

Haeffner R, Heck RM, Ceolin T et al (2012) Plantas medicinais utilizadas para o alívio da dor pelos agricultores ecológicos do Sul do Brasil. Rev Eletr Enf 14:596–602

Hanazaki N, Peroni N, Araujo LG et al (2007) Etnobotânica Caiçara no Litoral Paulista. RiMa, São Carlos

Heiden G, Iganci JRV, Macias L (2009) *Baccharis* sect. *Caulopterae* (Asteraceae, Astereae) no Rio Grande do Sul, Brasil, Rodriguésia 60:943–983

Hurrell JA, Ulibarri EA, Arenas PM et al (2011) Plantas de herboristería. Editorial LOLA, Buenos Aires

Kinupp VF, Lorenzi H (2014) Plantas alimentícias não convencionais (PANC) no Brasil: guia de identificação, aspectos nutricionais e receitas ilustradas. Instituto Plantarum de Estudos da Flora, São Paulo

Koch I (2000) Estudo etnobotânico de plantas medicinais na cultura ítalo brasileira, no Rio Grande do Sul. Dissertação de Mestrado Unisinos, São Leopoldo

Kubo RR (1997) Levantamento das plantas de uso medicinal em Coronel Bicaco, RS. Dissertação de Mestrado. Universidade Federal do Rio Grande do Sul, Porto Alegre

Ladeira AM (2002) Plantas medicinais com óleos essenciais. Instituto de Botânica, São Paulo

Löbler L, Santos D, Rodrigues ES et al (2014) Levantamento etnobotânico de plantas medicinais no bairro Três de Outubro, da cidade de São Gabriel, RS, Brasil. Rev Bras Bioc 12:81–89

Lopes AMV (1997) Plantas usadas na medicina popular do Rio Grande do Sul. Infograph, Santa Maria

Loredo-Medina OL, Rodríguez-Chávez JM, Espinosa MGR (2002) Aprovechamiento de recursos vegetales en una localidad de la reserva de la biosfera Mariposa Monarca, Michoacán, México. Etnobiologia 2:1–29

Lorenzi H, Matos FJA (2008) Plantas medicinais no Brasil – nativas e exóticas. Instituto Plantarum, Nova Odessa

Macía MJ, García E, Vidaurre PJ (2005) An ethnobotanical survey of medicinal plants commercialized in the markets of La Paz and El Alto, Bolivia. J Ethnopharmacol 97:337–350

Maciel AC, Cardoso N (2003) Cura, sabor e magia nos quintais de Ilha Grande. UERJ, Rio de Janeiro

Maffei BRA (1969) Plantas medicinales. Ed. Nuestra Tierra, Montevideo

Mariante HM (1984) Medicina campeira e povoeira. Martins Livreiro, Porto Alegre

Martins ER, Castro DM, Castellani DC (1994) Plantas medicinais. UFV Impr Univ, Viçosa

Molares S, González SB, Ladio A et al (2009) Etnobotánica, anatomia y caracterización físico-química del aceite esencial de *Baccharis obovata* Hook. & Arn. (Asteraceae: Astereae). Acta Bot Bras 23:578–589

Montes M, Wilkomirsky T (1988) Medicina tradicional Chilena. Editora de la Universidade De Concepción, Concepción

Morales G, Paredes A, Sierra P et al (2008) Antimicrobial activity of three *Baccharis* species used in the traditional medicine of northern Chile. Molecules 13:790–794

Moreira RCT, Costa LCB, Costa RCS et al (2002) Abordagem etnobotânica acerca do uso de plantas medicinais na Vila Cachoeira, Ilhéus, Bahia, Brasil. Acta Farm Bonaer 21:205–211

Moreno-Salazar SF, Robles-Zepeda RE, Johnson DE (2008) Plant folks medicines for gastrointestinal disorders among the main tribes of Sonora, Mexico. Fitoterapia 79:132–141

Muiño WA (2010) Ethnobotanical study of the rural population of the west of the Pampa plain (Argentina). Ethnobot Res Appl 8:219–223

Olivier DK, van Wyk B-E (2013) Bitterness values for traditional tonic plants of southern Africa. J Ethnopharmacol 147:676–679

Oliveira Júnior CJF, Voigtel SDS, Nicolau AS et al (2018) Sociobiodiversidade e agricultura familiar em Joanópolis, SP, Brasil: potencial econômico da flora local. Hoehnea 45:40–54

Osuna JTA, Rodrigues ACC, Oliveira J et al (2005) Cultivo, beneficiamento e considerações sobre usos de plantas medicinais. Universidade Estadual de Feira de Santana, Feira de Santana

Pavan-fruehauf S (2000) Plantas medicinais da mata atlântica – manejo sustentado e amostragem. Annblume: Fapesp, São Paulo

Paz EA, Bassagoda MAJ, Ferreira F (1992) Uso racional de las plantas medicinales. Editorial Fin de Siglo, Montevideo

Pereira ZV, Mussury RM, Almeida AB et al (2009) Medicinal plants used by Ponta Porã community, Mato Grosso do Sul state. Acta Sci 31:293–299

Pereira GS, Noelli FS, Campos JB et al (2016) Ecologia Histórica Guarani: As plantas utilizadas no Bioma Mata Atlântica do litoral Sul de Santa Catarina, Brasil (Parte 1). Cadernos do LEPARQ 13:197–246

Philippi JMS (2012) Plantas Medicinais – *Baccharis ochracea* Sprengel – Erva Santa. In: Identificação dos elementos histológicos de vegetais consumidos como chás e temperos e assemelhados. Nova Letra, Blumenau

Pio Corrêa M (1931) Diccionario das plantas uteis do Brasil, vol 2. Ministério da Agricultura, Rio de Janeiro

Pires MV, Abreu PP, Soares CS, Souza et al (2009) Etnobotânica de terreiros de candomblé nos municípios de Ilhéus e Itabuna, Bahia, Brasil. Rev Bras Bioc 7:3–8

Piva MG (1998) Estudio Etnobotánico del Municipio de Caxias do Sul/RS- Brasil. Tese de Doutorado. Universidade de León, León

Quiroga R, Meneses L, Bussmann RW (2012) Medicinal ethnobotany in Huacareta (Chuquisaca, Bolivia). J Ethnobiol Ethnomed 8:29–42

Reis GMC, Zurlo MA, Augusto MM (1990) Jardim de cheiros – Plantas medicinais. Governo do Distrito Federal, Secretaria da Agricultura e produção, Fundação Zoobotânica/Jardim Botânico de Brasília

Ribeiro RV, Bieski IGC, Balogun SO et al (2017) Ethnobotanical study of medicinal plants used by Ribeirinhos in the North Araguaia microregion, Mato Grosso, Brazil. J Ethnopharmacol 205:69–102

Rodrigues LA, Carvalho DA, Gomes LJ et al (2002) Espécies vegetais nativas usadas pela população local em Luminárias – MG. Bolm Agropec 52:1–34

Santos CAM, Torres KR, Leonart R (1988) Plantas medicinais (herbarium, flora et scientia). Ícone, São Paulo. Scientia et Labor, Curitiba

Santos MRA, Lima MR, Oliveira CLLG (2014) Medicinal plants used in Rondônia, Western Amazon, Brazil. Rev Bras Plant Med 16:707–720

Scarpa GF (2004) Medicinal plants used by the Criollos of Northwestern Argentine Chaco. J Ethnopharmacol 91:115–135

Sequeda-Castañeda LG, Célis C, Luengas-Caicedo PE (2015) Phytochemical and therapeutic use of *Baccharis latifolia* (Ruiz & Pav.) Pers. (Asteraceae). Pharmacology 2:14–17

Silva F, Lopes RC, Armond C (2008) Folhas de chá. Remédios caseiros e comercialização de plantas medicinais, aromáticas e condimentares. UFV Impr Univ, Viçosa

Simões CMO, Mentz LA, Schenkel EP et al (1986) Plantas da Medicina popular no Rio Grande do Sul. Editora UFRGS, Porto Alegre

SJ Siqueira, J.C (1988) Plantas medicinais – Identificação e uso das espécies dos cerrados. Edições Loyola, São Paulo

Soares ELC, Vendruscolo GS, Eisinger SM, Záchia RA (2004) Estudo etnobotânico do uso dos recursos vegetais em São João do Polêsine, RS, Brasil, no período de outubro 1999 a junho de 2001. I – Origem e fluxo do conhecimento. Rev Bras Plant Med 6:69–95

Sousa MP, Matos MEO, Matos FJA et al (1991) Constituintes químicos ativos de plantas medicinais brasileiras. Edições UFC/Laboratório de produtos naturais, Fortaleza

Souza MB (2007) O uso medicinal e místico de plantas por moradores do bairro Morretes, município de Nova Santa Rita, Rio Grande do Sul. Monografia de Conclusão de Curso. UNILASALLE, Canoas

Stasi LC, Hiruma-Lima CA (2002) Plantas medicinais na Amazônia e na Mata Atlântica. Editora UNESP, São Paulo

Steffen SJCJ (2010) Plantas medicinais – Usos populares tradicionais. Impressos Portão, São Leopoldo

Tognon GB, Cuquel FL (2015) Potencial ornamental de *Baccharis milleflora* e *Baccharis tridentata* como folhagem de corte. Ciência Rural 46:70–75

Tribess B, Pintarelli GM, Bini LA et al (2015) Ethnobotanical study of plants used for therapeutic purposes in the Atlantic Forest region, Southern Brazil. J Ethnopharmacol 164:136–146

Usca YG, Linares LP (2012) Etnobotanica y fitoquimica de plantas tintoreas en las comunidades de Rumira, Chaullacocha y Chupani: provincia de Urubamba – Cusco. Tese, Universidad Nacional de San Antonio Abad del Cusco, Cusco, Peru

Vásquez J, Alarcón JC, Jiménez SL et al (2015) Main plants used in traditional medicine for the treatment of snake bites in the regions of the department of Antioquia, Colombia. J Ethnopharmacol 170:158–166

Vendruscolo GS, Mentz LA (2006) Levantamento etnobotânico das plantas utilizadas como medicinais por moradores do bairro Ponta Grossa, Porto Alegre, Rio Grande do Sul. Iheringia Sér Bot 61:83–103

Verardo SMS (1981/1982) Levantamento preliminar de plantas medicinais em Juiz de Fora e São João Nepomuceno (MG). Oréades 8:92–115

Vidaurre PJ, Paniagua N, Moraes MR (2006) Etnobotánica en los Andes de Bolivia. In: Moraes MR, Øllgaard B, Kvist LP et al (eds) Botánica Económica de Los Andes Centrales. Universidad Mayor de San Andrés, La Paz

Weimann C, Heinrich M (1996) Indigenous medicinal plants in Mexico: the example of the Nahua (Sierra de Zongolica). Bot Acta 110:62–72

WHO (2018) ICD-11 for mortality and morbidity statistics (ICD-11 MMS). https://icd.who.int/. Accessed 12 Dec 2018

Zardini EM (1984) Etnobotánica de Compuestas Argentinas com especial referencia a su uso farmacológico (Primera Parte). Acta Farm Bonaer 3:77–99

Zuchiwschi E, Fantini AC, Alves AC et al (2010) Limitações ao uso de espécies florestais nativas pode contribuir com a erosão do conhecimento ecológico tradicional e local de agricultores familiares. Acta Bot Bras 24:270–282

Chapter 17
CPQBA 1: First Cultivar Registered and Protected from a Brazilian Medicinal Plant

Ilio Montanari Jr

Abstract The production chain of a phytomedicine starts with the production of the raw material. In Brazil, the richest country in biodiversity on the planet, plant resources are rarely sustainably explored. Usually, plant species with some economic importance are exploited directly from nature, leading these species to genetic erosion and extinction risk. Some examples of Brazilian species that are overexploited and suffer ecological damage are *Maytenus ilicifolia, Cephaelis ipecacuanha, Achyrocline satureioides, Pfaffia glomerata, Hebanthe eriantha*, and also *Baccharis trimera*, the theme of this chapter. To alleviate the ecological pressure of extractivism on *B. trimera*, I carried out a process of domestication and breeding, intending to create a cultivar (a cultivable variety) of this species. The aim was to offer the rural producer a new agricultural option and the processing industry a high-quality sustainably produced raw material. I took populations of *B. trimera* to the Pluridisciplinary Center for Chemical, Biological and Agricultural Research of the State University of Campinas (CPQBA-UNICAMP) where, via mass selection with gametic control for five generations, I developed the cultivar CPQBA 1. This variety has low seed dormancy, uniform development, good architecture, uniformity of vegetative cycle, good regrowth capacity, and good productivity. The descriptors of CPQBA 1 were established by the Ministry of Agriculture, Livestock and Supply (MAPA) and the cultivar became the first Brazilian medicinal plant to be registered and protected.

Keywords *Baccharis trimera* · Cultivable variety · Cultivar protection · Domestication processes · Plant breeding

I. Montanari Jr (✉)
Chemical, Biological and Agricultural Pluridisciplinary Research Center. Agrotechnology Division, State University of Campinas, Campinas, Brazil

G. W. Fernandes et al. (eds.), *Baccharis*,
https://doi.org/10.1007/978-3-030-83511-8_17

1 Introduction

The growing demand for natural products has increased the interest in Brazilian plants, because they represent opportunities for the development of several products obtained from our flora, such as medicines, cosmetics, nutraceuticals, and food. Despite its growing economic importance, few native species are produced sustainably with the quality, regularity, and quantity required by the manufacturing industry. Economically important species are being overexploited, endangering their natural populations and the market itself.

Baccharis trimera (Less.) D.C., Asteraceae, popularly known as carqueja, is a dioic shrub, about 1 m high, perennial, branched from the ground level, with winged stems. It is a common species occurring in a great part of South America. Brazil is considered its dispersion center. The species is rich in flavonoids such as apigenin, cirsimaritin, eupatorin, genkvanin, hispidulin, and quercetin (Hänsel et al. 1992). Other chemical constituents found in the species are essential oils with carquejol, carquejilacetat, and β-pinene as main constituents (Brasil 2002). There is a strong economic interest in the plant. In folk medicine, the aerial part of the plant is used to treat the liver, stomach, and intestine problems, besides cases of fever, influenza, common cold, and rheumatism (Correa 1931; Hänsel et al. 1992). The bitter constituents of the plant are also used in the drink industry (Correa 1931).

In 1995, the CPQBA-UNICAMP came up with the idea of developing a carqueja cultivar, that is, a cultivable variety. The reason for the emergence of this idea was due to a confluence of environmental and economic factors. At the time, as the plant was not cultivated, the wild collection was the main way to get the raw material to be used by the industry. This economic interest induced the collection of plants that occur spontaneously in nature, compromising natural populations. By creating a cultivar, the demand for the plant could be supplied by commercial cropping and, consequently, the natural populations would not be further threatened. In addition, it would be possible to supply on a regular basis the companies that transform the raw material into a product with predictable quantities and quality. Thus, the objectives of creating a cultivar were to offer farmers a new agricultural option, improving the quality of the raw material offered to the market and, at the same time, to avoid the ecological problems caused by the wild collection.

2 The Domestication Process

A common idea is that if the plant is born and grows spontaneously, then it is to be much easier to grow under cultivation. This idea is not correct, because the individuals of a wild population are not homogenous and will express individual differences in characteristics such as resistance to pests and diseases, cycle, size, habit, etc. With wild individuals, the commercial cultivation of a species is not feasible. To understand why it is not possible to commercially grow plants that occur spontaneously, we need to keep in mind that individuals who make up the same wild population are almost always genetically different from each other. These differences, attributed to inheritable factors among individuals of the same species, are given the name of genetic variability (Hoyt 1994). This variability is the basis of the dynamics of the evolutionary process and has its origin in the combined action of mutation, genetic recombination, numerical and structural alteration of the chromosomes, genetic drift, migration, reproductive isolation, and adaptation (Allard 1971; Briggs and Knowles 1977). It is through the genetic variability that natural selection acts. Thus, in nature, an individual's chances of leaving offspring increase if the individuals in a population are genetically different from each other.

In nature, genetic variability is very important, but in agriculture causes many difficulties. The term genetic variability includes a large variation in seed dormancy, in the morphology of individuals, in development, in resistance to pests and diseases, in the content of active principles, in plants architecture, vigor, productivity, their response to soil fertility, water conditions, etc., generating technical problems that make them very difficult to grow commercially (Pank 2006). Moreover, in the case of medicinal plants, the raw material obtained from wild populations will be chemically heterogeneous, since it is at the level of the secondary metabolism of plants, governed by the genetic code, that molecules of therapeutic interest, the so-called active principles, are produced (Bu'lock 1969; Man 1987; Khanna & Shukla 1990).

Fig. 17.1 and 17.2 show examples of seed dormancy and big morphological variation in the wild population of *B. trimera*, respectively.

Fig. 17.1 Seedlings of *Baccharis trimera*. These seedlings were sowed at the same time, but the germination was not uniform, showing the consequence of seed dormancy

Fig. 17.2 Morphological differences in same-aged individuals of *Baccharis trimera*

The plants to be grown must be uniform, germinating, and developing without many differences between them so that they can be managed by the farmer in order to have a standardized product. The homogeneity of a population is a prerequisite in agriculture for the success of their cultivation. For this to happen, plants of a species in the wild must, through a selection and breeding work, go through a process of narrowing their genetic base so that techniques and agricultural parameters of the cultivated population can be developed. This process is called domestication. Normally associated with animals and understood day by day with the sense of taming, the word domestication can also be used in relation to plants. Of Latin origin, domestication means to bring to the "domus,",that is to the house. While cultivation is related to human activities in the conduct of the agricultural process, such as fertilization, pruning, soil preparation, irrigation, etc., domestication, in turn, is related to the genetic response of plants (or animals) to the agricultural process (Harlan 1992).

Therefore, the genetic variability in wild populations, if it does not make it impossible, greatly increases the difficulty of their cultivation.

3 The Creation of the Cultivar Cpqba 1

When choosing the species *Baccharis trimera* to create a cultivar, we thought to obtain a plant with good agricultural characteristics. The chemical characteristics were not considered, because it is not yet known which are the molecules responsible for its therapeutic effect. Although there are many chemical and pharmacological studies with the species, there are no studies that link these two fields of knowledge. For this reason, we do not take into account its chemical profile, which would be the ideal situation in the selection of medicinal plants. Thus, we began the selection process aiming at uniform germination, good biomass production, simultaneous flowering of individuals (since the active principles vary according to the physiological stage of each individual and the flowering indicates the moment of harvest), resistance to pests and diseases, standing plants (plant architecture is important in the conduction of the crop and at the time of harvest), and plants with good ability to regrow after harvesting, because the species is perennial.

The selection work began in 1996, with the collection of carqueja seeds from different populations from the states of Minas Gerais, São Paulo, and Paraná. The selection method adopted was the mass selection with gametic control; that is, the male and female plants were selected for the crosses that would form the next generation. This method was chosen because, in addition to being simple, it is efficient in the exploitation of populations with great genetic variance and it is used for both autogamous and allogamous plants, as in the case of carqueja.

The generations were selected in two steps: first at the greenhouse, where the seedlings were grown, and after that at the experimental field. At the greenhouse, from 2000 seedlings, 1000 plants that did not lie down, with rapid germination and vigorous growth, were selected. The selection at the greenhouse was made 4 months after sowing when seedlings were evaluated. Those selected seedlings were then planted in the field at CPQBA-UNICAMP. From those 1000 plants in the field, after 8 months from planting, they were evaluated and selected according to their growth habit and biomass production. Those that did not show an erect habit and/or have a good biomass production were cut out from the field. The selected plants were then cut at 30 cm from the ground. After three months, their regrowth capability was evaluated, and those that did not present good regrowth were rooting out. Then the selected plants completed their reproductive cycle and the seeds were used to compose the next generation to be selected. The total time of each selection cycle was 24 months. The basic statistics, as well the progress achieved in this process, can be observed in Table 17.1.

The data shows that the selections made at the greenhouse improved the height of the seedlings through the generations. This improved height reflects the selection gains for faster-growing plants, as well as an increasing number of plants without seed dormancy, and therefore, germination and development are faster than those with dormancy. Seed dormancy and lying-down plants are very undesirable for the *B. trimera* cultivation. The selection aiming at erect plants proved to be also efficient with a decreasing percentage of lying-down plants inside the more advanced generations. The increase of selected plants in the field reflects the uniformity level obtained by the different generations, showing clear progress in the direction of vigorous and good regrowth capability after harvest. Finally, the gain of biomass obtained was almost twice of the parental population. After 4 generations of selection, the population showed good agricultural characteristics as dormancy loss,

Table 17.1 Basic statistics of the massal selection process conducted at CPQBA-UNICAMP for the creation of the *Baccharis trimera* cultivar CPQBA1

Population	Seedlings height mean (cm)	Seedlings height mean standard deviation	% of lie down plants (greenhouse)	% of plants selected (field)	Dried biomass (kg)/plant
São Paulo	19.165	4.461	15.25	5.83	0.34 (average of the three populations)
Minas Gerais	14.683	3.244	20.08	10.83	
Paraná	13.723	5.963	15.71	2.5	
F_1	16.08	8.645	17.84	26.08	0.32
F_2	19.58	6.876	23.21	37.72	0.39
F_3	26.97	4.653	10.71	57.9	0.45
F_4	40,40	4237	8.36	90.0	0.64

F = filial generation

Fig. 17.3 Uniformity of seedlings without dormancy after 5 selection cycles of *Baccharis trimera*

good vigor, good architecture, good regrowth capability, and good biomass production. Aiming to offer to the farmer seeds of that population, we proceed with the protection of the now on called cultivar "CPQBA 1." Figs. 17.3 and 17.4 show seedlings without dormancy and the field behavior of the cultivar CPQBA 1, respectively.

Fig. 17.4 Cultivated field showing the vigor and uniformity of *Baccharis trimera* cultivar CPQBA 1

4 The Protection of the Cultivar

The protection of a cultivar is equivalent to an industrial patent (Brasil 1997). In order to protect a cultivar, it must be distinguishable from others, be homogeneous (with little variation between individuals), and have stability, that is, maintain its characteristics through generations. Therefore, before being protected, the cultivar undergoes evaluations of distinguishability, homogeneity, and stability. Because it is the first cultivar of the species *Baccharis trimera*, it was necessary to establish which criteria would be used to distinguish it from the wild populations and the possible cultivars that will arise. These criteria are given the names of descriptors, as they describe the characteristics of the cultivar. With the help of the National Service of Protection of Cultivars, agency linked to the Ministry of Agriculture Livestock and Supply, 15 descriptors for the cultivar CPQBA 1 have been established: architecture, height, width, number of chapters, stem thickness, anthocyanin pigmentation in the stem, leaf length, leaf width, green color intensity, number of seeds per plant, cycle until flowering, epicuticle wax, essential oils content, presence of carquejol, and presence of carquejila acetate. With the protection of cultivar CPQBA 1, the technology developed could be licensed by UNICAMP, and today the seeds of that cultivar can be purchased in the seeds market.

5 Conclusion

The domestication of medicinal plants is a way of preserving nature as it eases the ecological pressure caused by extractivism. A new species that enters an agricultural system does not compete with nature for space, for new areas of cultivation; on the contrary, it offers the farmer a new agricultural option. The lack of economically viable agricultural options is one of the reasons that encourage poor agriculture: this exhausts the environment, rendering it unproductive and demanding new areas. The cultivation of medicinal plants guarantees the quality of the raw material, because the cultivation allows controlling the environment, the genetic characteristics of the population under cultivation, the stage of development of the plants at the time of harvest, and the postharvest operations, which are the four factors that influence the pattern of a vegetable raw material. The cultivation guarantees the production of the finished product, since the companies that will transform the raw material into a product can predict the amount, the regularity, and the standard of raw material they will work with.

The cultivar CPQBA 1 has also contributed in the field of agricultural, chemical, pharmacological, and clinical research, as conducting experiments with standardized vegetable raw material helps to ensure that the proposed objectives will be achieved. This occurs because, in science, experiments need to be replicated to come to general conclusions. To obtain replicated results, the same experimental conditions must be maintained. The plant samples used in experiments are part of such experimental conditions. This means that by following several research steps, the plant samples need to be standardized. For this reason, many researchers have worked with this cultivar, since they know that they can count on the same raw material in the different stages of research.

References

Allard RW (1971) Principles of plant breeding. Wiley, New York.

Brasil (1997) Decreto Lei n° 2366, de 05 de novembro de 1997. Regulamenta a Lei n° 9456, de 25 de abril de 1997, que institui a Proteção de Cultivares, dispõe sobre o Serviço Nacional de Proteção de Cultivares SNPC, e dá outras providências. Diário Oficial (da República Federativa do Brasil), Brasília, DF, n.216, p. 25342–25343, 7 nov.1997, Seção 1, p. 242–53

Brasil (2002) Farmacopeia Brasileira. 4th edition. Agência Nacional de Vigilância Sanitária, Brasília

Briggs FN, Knowles PF (1977) Introduction to plant breeding. Reinhold Publishing Corporation, New York

Bu'lock JD (1969) Biossintesis de productos naturales. Urmo, Bilbao

Correa MP (1931) Dicionário das plantas úteis do Brasil e exóticas cultivadas, vol II. Ministério da Agricultura, Rio de Janeiro

Hänsel R, Keller K, Rimpler H, Schneider G (1992) Hagers handbuch für der pharmazeutische praxis. Springer Verlag, Germany

Harlan JR (1992) Crops and man. American Society of Agronomy-Crop Science Society, Madison

Hoyt E (1994) Conservação dos parentes silvestres das plantas cultivadas. Addison-Wesley Iberoamericana

Khanna KR, Shukla S (1990) Genetics of secondary plant products and breeding for their improved content and modified quality. In: Khanna KR (ed) Biochemical aspects of crop improvement. CRC Press, London, pp 283–323

Man J (1987) Secondary metabolism. Clarendon Press, New York

Pank F (2006) Adaptation of medicinal and aromatic plants to contemporary requirements by breeding: aims, methods and trends. Rev Bras Pl Med 8:39–42

Chapter 18
Perspectives of *Baccharis* Secondary Metabolites as Sources for New Anticancer Drug Candidates

Christiane Contigli, Elaine Maria de Souza-Fagundes, Warne Pedro de Andrade, Jacqueline Aparecida Takahashi, Yumi Oki, and G. Wilson Fernandes

Abstract Cancer incidence and mortality are increasing worldwide, and the existing treatment protocols, which most frequently involve chemotherapy, are far from being successful for all kinds of tumors or all patients. Novel chemotherapy strategies are crucial to overcoming this issue, and secondary metabolites from plants and other biological sources are relevant in the search for new anticancer drug candidates. The genus *Baccharis*, mainly represented by the species *B. artemisioides, B. concinna, B. coridifolia, B. dracunculifolia B. gaudichaudiana, B. grisebachii,*

C. Contigli (✉)
Serviço de Biologia Celular, Diretoria de Pesquisa e Desenvolvimento, Fundação Ezequiel Dias, Belo Horizonte, Minas Gerais, Brazil
e-mail: christiane.contigli@funed.mg.gov.br

E. M. de Souza-Fagundes
Laboratório de Biologia Molecular e Celular, Departamento de Fisiologia e Biofísica, Instituto de Ciências Biológicas, Universidade Federal de Minas Gerais, Belo Horizonte, Minas Gerais, Brazil
e-mail: elainefagundes@ufmg.br

W. P. de Andrade
Núcleo de Hematologia e Oncologia, Grupo Oncoclínicas, Belo Horizonte, Minas Gerais, Brazil

Departamento de Obstetrícia e Ginecologia, Escola de Medicina, Universidade Estadual Paulista, São Paulo, Brazil

J. A. Takahashi
Departamento de Química, Instituto de Ciências Exatas, Universidade Federal de Minas Gerais, Belo Horizonte, Minas Gerais, Brazil
e-mail: jat@qui.ufmg.br

Y. Oki · G. W. Fernandes
Laboratório de Ecologia Evolutiva & Biodiversidade, Departamento de Genética, Ecologia & Evolução, Instituto de Ciências Biológicas, Universidade Federal de Minas Gerais, Belo Horizonte, Minas Gerais, Brazil

G. W. Fernandes et al. (eds.), *Baccharis*,
https://doi.org/10.1007/978-3-030-83511-8_18

B. latifolia, B. megapotamica, B. milleflora, B. obtusifolia, B. ochracea, B. penningtonii, B. salicina, B. scandens, and *B. trimera*, has been studied in experimental models, and their chemical extracts have demonstrated toxic effects on a range of tumor cells from different cancer origins, with low or no effects on healthy cells. Green propolis, whose botanical origin is *B. dracunculifolia*, also revealed comparable results. Some phytochemicals isolated from *Baccharis* extracts as well as from green propolis, such as artepillin C, baccharin, curcuphenol, drupanin, gardenin B, quercetin, and spathulenol, demonstrated remarkable anticancer activities in in vitro and in vivo preclinical laboratory studies. Interestingly, through different mechanisms of action, these bioactive compounds are capable of inhibiting tumor cell proliferation, migration and invasion, eliciting death by apoptosis, and affecting the tumor microenvironment. Nevertheless, more preclinical evidence is still required to point out candidates for clinical trials and to drive the development of new drugs to fight cancer.

Keywords Anticancer action · Cancer cells · Drug discovery · Cytotoxicity · Natural products

1 Introduction

Plant secondary metabolites are related to a variety of significant ecological roles, such as plant protection against herbivores and pathogens, the attraction of pollinators and seed dispersers, plant–plant competition as well as in their communication (Harborne 1991a, b; Harborne 1999; Price et al. 1991; Herrera and Pellmyr 2009; Isah 2019), and litter decomposition (review in Chomel et al. 2016), among many other roles (see Yang et al. 2018). These metabolites have been shown to be important sources of novel bioactive compounds with potential application for the discovery of new drugs (e.g., Seca and Pinto 2018; Newman and Cragg 2020). Indeed, around 80% of the world population still uses medicinal plants as the main resource for basic health care, and since 1978, herbal medicines are recognized as important therapeutic resources by the World Health Organization (WHO) (Brandão et al. 2010; WHO 2019). Among the potential chemotherapeutical effects of plant compounds, the anticancer activity has great relevance, since cancer is the second leading cause of death worldwide (WHO 2020).

This chapter presents a review of the scientific literature, along with unpublished data obtained by our research group, regarding the current knowledge about antiproliferative and proapoptotic effects of secondary metabolites and natural products from plants of the genus *Baccharis* against in vitro cancer cells, in addition to in vivo experimental models of cancer. These studies aim to demonstrate the potential of some of these compounds as anticancer drug candidates.

2 Overview About Cancer

Cancer comprises a large group of diseases that are characterized by abnormal cell multiplication and spread from primary lesion, a process known as metastasis. Currently, it is understood as a disease of the genome, arising through the accumulation of genetic mutations and epigenetic modifications that disturb the normal function of signaling pathways responsible for the coordination of cell proliferation, differentiation, and death (Koeffler et al. 1991; Tomasetti et al. 2017).

To acquire a neoplastic phenotype, germinative cells pass through a sequence of changes in their genes, which can be activated or inactivated by genetic mechanisms, such as gene mutations, chromosomal breaks and losses, gene amplifications, and epigenetic events (like DNA methylation or histone acetylation) (Hanahan and Weimberg 2011; Youn and Simon 2011; Roy et al. 2014; Baylin and Jones 2016). Over time, such modifications cause instability in the genome of the affected cells, leading to the phases of initiation and progression of cancer, a process known as carcinogenesis. In fact, to promote tumorigenesis, these modifications must occur in driver genes mainly associated with the regulation of cell growth and differentiation, cell life cycle, and cell death. One important group of driver genes are protooncogenes, responsible for coding transcription factors, growth factors, receptors, and apoptosis regulatory factors, or involved with chromatin remodeling. Another group comprises tumor suppression genes, which encode proteins involved in controlling cell life cycle, damage repair, and apoptosis (Hanahan and Weimberg 2011; Youn and Simon 2011; Collura et al. 2019).

The loss of genomic stability is critical for carcinogenesis, when high rates of mutations, combined with alterations in chromosome number and structure, cause uncontrolled cellular proliferation. Besides these intrinsic factors, the growth of cancer needs an enabling microenvironment around tumor cells, which allows immune system evasion, promotion of angiogenesis, and escape from apoptosis inducers and tumor suppressors (Fig. 18.1; for a review, see Hanahan and Weinberg 2011; Fares et al. 2020).

Some of the changes in the DNA that predispose humans to cancer (5–10% of the events) can be inherited, since they occur in germinative cells – the so-called germline mutations. However, most cases (90–95%) are related to somatic mutations, which are not inherited. Genetic and chromosomic mutations can be the result of a deficient system of DNA repair, commonly caused by aging (Chatterjee and Walker 2017), or can be induced by exposure to external factors and poor lifestyle habits, such as ultraviolet radiation (UV), ionizing radiation, chemical products (called xenobiotics: e.g., alcohol, tobacco, pesticides) (Ali and Bhattacharya 2014), or can also be an effect of biologic carcinogenic agents, like Epstein-Barr virus (EBV), human B and C hepatitis viruses (HBV and HCV), human papillomavirus (HPV), human T-cell lymphotropic virus type 1 (HTLV-1), and the bacteria *Helicobacter pylori* (Bouvard et al. 2009).

Cancer incidence and mortality are rapidly growing worldwide, and according to the Global Cancer Observatory (GLOBOCAN), approximately 18.1 million new

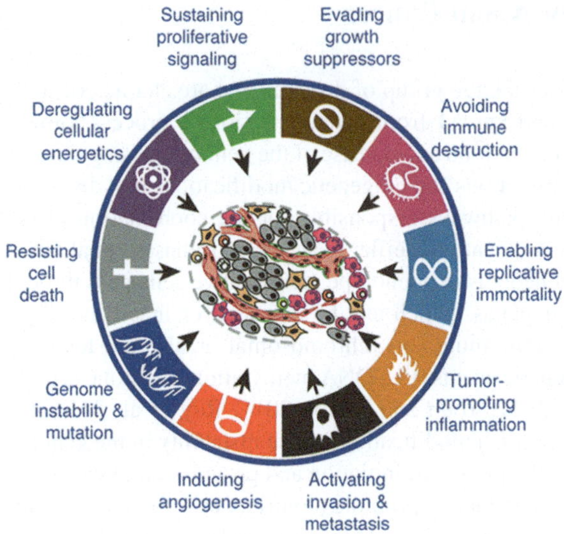

Fig. 18.1 Hallmarks of carcinogenesis. (Modified from Hanahan and Weinberg 2011)

cases and 9.6 million cancer deaths were estimated for 2018 (Bray et al. 2018). For men, lung cancer was responsible for the main estimated incidence (14.5%) and mortality (22.0%), followed by prostate (13.5%), colorectum (10.9%), stomach (7.2%), and liver (6.3%) cancer, concerning incidence. Higher mortalities were expected for liver (10.2%), stomach (9.5%), colorectum (9.0%), and prostate (6.7%) cancer, thus revealing that those types of tumors are among the most frequent and lethal ones. For women, breast cancer had the major incidence (24.2%) and mortality (15.0%), followed by colorectum (9.5%), lung (8.4%), and cervix uteri (6.6%) cancers as the ones showing the highest incidence, while lung (13.8%), colorectum (9.5%), and cervix uteri (7.5%) were the most lethal ones (Fig. 18.2).

In Brazil, for instance, the National Cancer Institute (INCA) has estimated the occurrence of about 625,000 new cases of cancer for the biennium 2020–2021 (Fig. 18.3; INCA 2018). Among the ten types of cancer with the highest incidence, except nonmelanoma skin cancer, the estimates in men are prostate (29.2%), lung (7.9%), intestine (9.1%), stomach (5.9%), and oral cavity (5.2%) cancers. In women, breast (29.7%), intestine (7.4%), cervix (8.1%), lung (5.6%), and thyroid (4.0%) cancers are expected to be the most incident in the population for the current biennium (INCA 2018).

Cancer is one of the most expensive diseases, which demands huge resources, including hospitalization, surgery, chemotherapy, radiotherapy, hormone therapy, and novel approaches as targeted therapy. In the United States, the estimated cost of cancer in the year 2020 reached $157 billion. In addition, up to 73% of cancer survivors in that country face financial difficulties, caused by a series of out-of-pocket costs associated with the disease, such as new supportive needs, aggravated by employment loss and psychological consequences (Lentz et al. 2019).

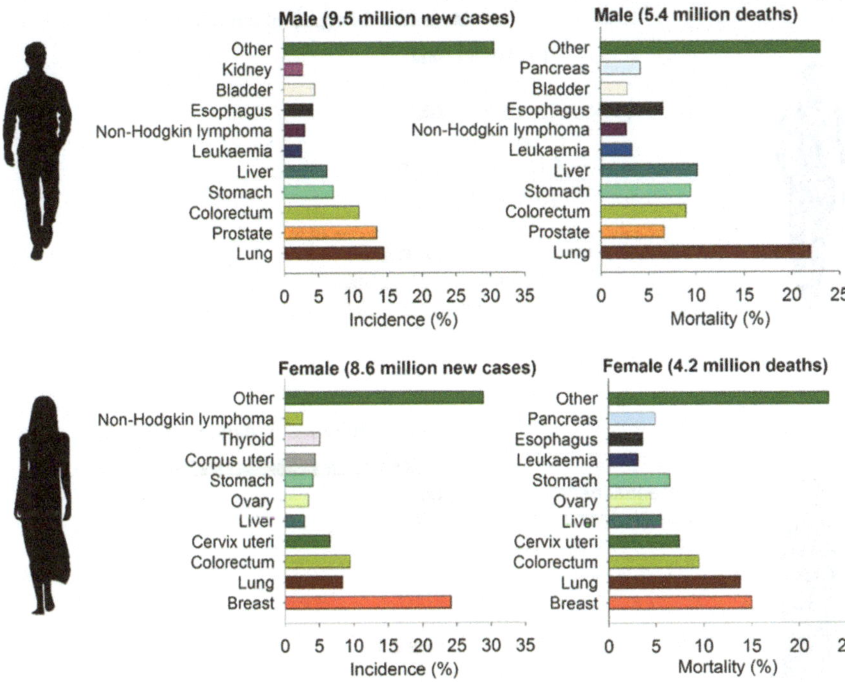

Fig. 18.2 Worldwide distribution of cancer cases and deaths, by gender, in 2018, according to GLOBOCAN report (Bray et al. 2018)

3 Conventional Treatment for Cancer and Drug Resistance

The strategies of treatment for each cancer will depend on its type and stage. As mentioned previously, they may include surgery, radiotherapy, chemotherapy using cytotoxic drugs, hormone therapy, or novel approaches as targeted therapy, which include cancer vaccines, gene therapy, immunotherapy using tumor-specific monoclonal antibodies, or immunomodulation with checkpoint inhibitors. Association or combination of the different methods is used to obtain better effectiveness, such as photodynamic combinational therapy, personalized cancer medicine, conventional therapy, and immunotherapy, among other options (e.g., Jackson and Chester 2015; Zhao 2016; Lee et al. 2018; Wang et al. 2018; Zhang and Li 2018).

Despite all the advances in cancer treatment, chemotherapy remains the most used. Chemotherapy is classified according to the purpose of the treatment, as curative, adjuvant, neoadjuvant, or palliative. Curative chemotherapy has the objective of eradicating the disease, being used for primary tumors as well as on some metastatic tumors. Adjuvant treatment is implemented after surgical removal of the tumor, in order to suppress micrometastasis, eventually not detected by imaging scans. Neoadjuvant treatment, instead, is used with the purpose of reducing the

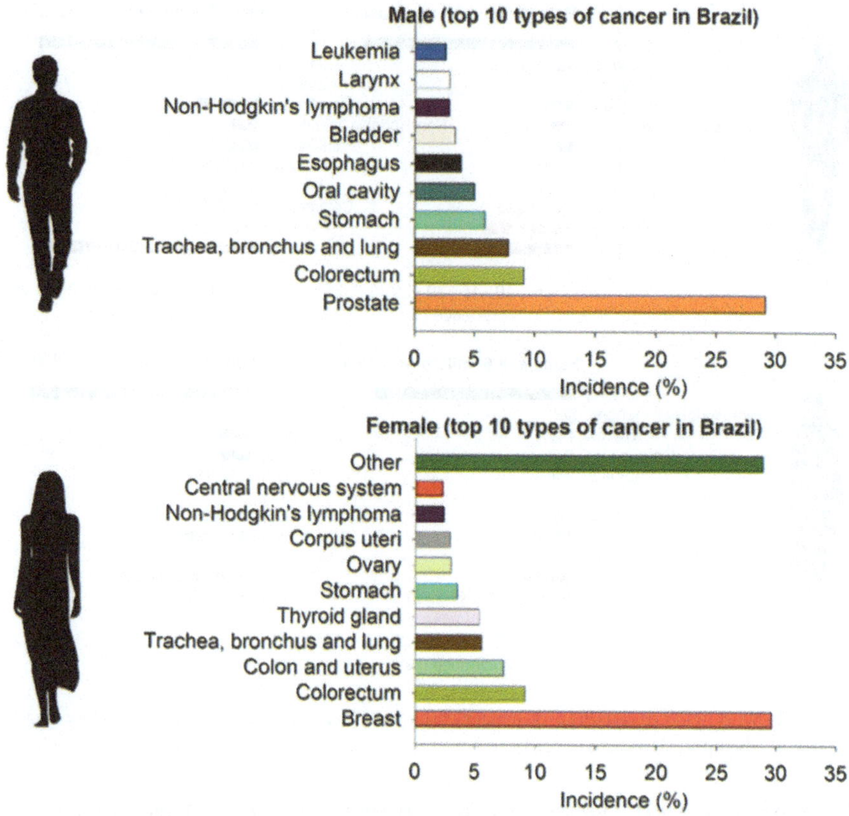

Fig. 18.3 Distribution of the ten more frequent types of cancer in the Brazilian population, by gender, except for nonmelanoma skin cancer, estimated for 2020. For males, 163,910 new cases are expected, and for females, 159,460 new cases. (Data from INCA 2018)

tumor size before surgical removal. Palliative chemotherapy is used to extend survival and to improve the life quality of patients (DeVita et al. 2018).

In clinical practices, the same chemotherapy protocol is not effective to everyone with the same type of cancer, since factors related to the tumor heterogeneity, along with patient health condition, other pre-existing diseases, stage of cancer development, rate of evolution, migration, and metastasis of tumors, among other aspects, may interfere with the success of the chemotherapeutical treatment (Conforti et al. 2018; DeVita et al. 2018; Thallinger et al. 2018; Silva et al. 2019; Fares et al. 2020).

Resistance to chemotherapy represents one of the greatest challenges in oncology (Harder et al. 2017). Some types of tumors, such as melanoma and hepatocellular carcinoma, are intrinsically resistant to chemotherapy, meaning that they do not respond properly since the beginning of conventional drug treatment (Kunjachan et al. 2013). However, other types, such as ovarian, breast, colon, rectum, and lung cancer, may initially be sensitive to treatment and later be resistant to the same drugs, characterizing acquired resistance (Robey et al. 2007). In fact, not all tumors are identical or behave in the same way, although they could receive the same histopathological classification, based on their morphology. The genetic and phenotypic diversity of neoplastic cell populations, among patients and even within the tumor, is responsible for most of the differences observed in response to treatment protocols (DeVita et al. 2018).

Another problem associated with resistance to treatment observed in cancer patients is due to the presence of cancer stem cells (CSCs). These cell types take part in tumor formation, and they have a high capacity for self-renewal, differentiation, and tumorigenicity. They are related to tumor heterogeneity, expression of drug transporters, resistance gene activation, and evasion of cell death. These characteristics lead to chemo- and radioresistance, culminating in tumor recurrence, as well as metastases (Prieto-Vila et al. 2017; Shibue and Weinberg 2017; Smith and Macleod 2019; Fares et al. 2020).

Therefore, considering that cancer is a major problem for public health worldwide, and that novel strategies for prevention and treatment are needed in order to offer better chances of cure, extended survival, and improvement of life quality to all cancer patients, the discovery of more selective and effective anticancer drugs is an urgent need. Cancer also has a high economic cost to the population, and its control could certainly represent a significant input of resources in other sectors of society. These reasons justify the development of scientific studies and the search for new drugs to properly bring this disease to a level in which society can deal with. In this context, biodiversity is a powerful source of potentially bioactive compounds to fight against tumor development.

4 Phytopharmacological Drugs in Use for Cancer Treatment

Until recently, approximately half of the antineoplastic drugs contain, as active principles, chemical compounds from biological sources, or synthetic forms chemically identical or derived from their natural precursors (DeVita et al. 2018; Newman and Cragg 2020). The main classes of phytochemicals used in cancer chemotherapy are taxane diterpenes (paclitaxel and docetaxel), vinca alkaloids (vinblastine, vincristine, and vinorelbine), the lignan podophyllotoxin and its semisynthetic derivatives (etoposide, teniposide, and etoposide phosphate), camptothecin derived alkaloids (topotecan and irinotecan), omacetaxine mepesuccinate, and combretastatins.

Other anticancer compounds from plant origin, still in study, include geniposide, colchicine, artesunate, salvicine, ellipticine, roscovitine, maytanasin, tapsigargin, and bruceantin. Additional substances with anticancer activity from other natural sources, such as psammaplin, didemnin, dolastin, ecteinascidin, and halichondrin, were isolated from microalgae, bacteria, and invertebrates (e.g., DeVita et al. 2018; Lichota and Gwozdzinski 2018; Seca and Pinto 2018).

Table 18.1 presents the more commonly used anticancer phytochemicals, their synthetic/semisynthetic derivatives, as well as their chemotherapeutic targets (see also DeVita et al. 2018; Seca and Pinto 2018; Choudhari et al. 2020; Newman and Cragg 2020; Sabzehzari et al. 2020).

For instance, paclitaxel (brand name: Taxol®), isolated from *Taxus brevifolia* (Taxaceae, common name: Pacific yew), is an antineoplastic drug used as adjuvant, neoadjuvant, and palliative in chemotherapeutic protocols for ovarian, mammary, lung, esophagus, head-and-neck, and bladder cancer, as well as for Kaposi sarcoma related to AIDS (DeVita et al. 2018). Paclitaxel acts specifically in the mitosis phase (M) of the cell cycle, by inhibiting tubulin polymerization, which causes the disruption of microtubules formation, leading to inhibition of cell division and, consequently, cell death (Nobili et al. 2009).

Paclitaxel was approved by the USA Food and Drug Administration (FDA) in 1992, but the first indications of cytotoxic activity on the natural source were reported in 1964. Its active principle was isolated in 1966 and had its chemical structure defined later in 1971 (Wall and Wani 1996). The anticancer activity was established, on animal models, for rat B16 melanoma, lung and breast cancer on animal model LX1 e MX1 (Wall and Wani 1996; Cragg 1998). The yield of paclitaxel from *T. brevifolia* is extremely low to cope with the massive demand for this chemotherapeutic drug. In addition, despite what has already been described, the total synthesis of paclitaxel is not economically viable for industrial purposes, due to its very complex chemical structure, bearing several stereocenters, which demands a long series of chemical reactions. Nowadays, this chemotherapeutic drug is obtained by a semisynthetic process using a taxol-related plant-derived natural diterpene as starting material. Production utilizing endophytes from *T. brevifolia* and other taxol-synthetizing plants and biotechnological production is under study (El-Sayed et al. 2020).

Until today, Taxol® is used as first-line adjuvant therapy, in combination with Platin, for ovarian epithelial carcinoma, and as a second line of treatment for advanced ovarian carcinoma, among other applications (Ye and Bhatia 2012). Therefore, paclitaxel is a good example of a drug extracted from plant sources with successful use in clinical oncologic, with the global market expected to reach $152 million by 2025 (Seca and Pinto 2018; Sabzehzari et al. 2020).

The alkaloids vinblastine and vincristine, both extracted from a plant known as periwinkle (*Catharanthus roseus*: Apocynaceae), act similarly to paclitaxel. Vinblastine (brand name: Velban®) is a chemotherapeutic drug for the treatment of lymphoma, testicular cancer, and Kaposi sarcoma. Vincristine (brand name: Oncovin®) is used for acute lymphoblastic leukemia, lymphomas, multiple myeloma, rhabdomyosarcoma, neuroblastoma, Ewing's sarcoma, Wilm's tumour, chronic leukemia, and trophoblastic neoplasm (DeVita et al. 2018). Vinorelbine (brand name: Navelbine®), derived from vinblastine, is also an antimitotic drug, highly specific for mitotic microtubules and with a low affinity for axonal microtubules. Vinorelbine can additionally inhibit the synthesis of protein, DNA, and RNA. The drug is applied to mammary cancer and small cell lung carcinoma (DeVita et al. 2018).

Table 18.1 Phytochemicals and derivatives in use as cancer chemotherapeutic drugs

Phytochemical or derivative	Original source	Brand name	Chemical class	Cancer target
Paclitaxel	*Taxus brevifolia* (Pacific yew)	Taxol®	Taxane diterpenes	Ovarian epithelial carcinoma, mammary cancer, lung cancer, head-and-neck cancer, esophagus cancer, bladder cancer, Kaposi sarcoma
Docetaxel	Semisynthetic taxane	Taxotere®		Mammary cancer, ovary cancer, lung cancer, prostate cancer, gastric cancer, bladder cancer
Vinblastine	*Catharanthus roseus* (periwinkle)	Velban®	Vinca alkaloids	Lymphoma, testicular cancer, Kaposi sarcoma
Vincristine		Oncovin®		Acute lymphoblastic leukemia, lymphomas, multiple myeloma, rhabdomyosarcoma, neuroblastoma, Ewing's sarcoma, Wilm's tumour, chronic leukemia, trophoblastic neoplasm
Vinorelbine	Synthetic derivative of vinblastine	Navelbine®		Mammary cancer, small cell lung carcinoma
Etoposide	*Podophyllum peltatum* (mayapple)	Etopophos® Vepesid® VP16®	Lignan podophyllotoxins	Germ cell tumors, small cell lung carcinoma, non-Hodgkin lymphoma
Teniposide	Semisynthetic derivatives of etoposide	Vulmon® VM-26®		Germ cell tumors, lung cancer
Etoposide phosphate		Etipophos®		
Irinotecan	*Camptotheca acuminata*	Camptosar® CPT-1®	Camptothecin alkaloids	Colorectal cancer, lung cancer
Topotecan	Semisynthetic derivative of camptothecin	Hycamtin®		Ovarian cancer, cervical cancer, small cell lung carcinoma, acute myeloid leukemia
Omacetaxine mepesuccinate	*Cephalotaxus fortunei*	Synribo®	Alkaloid	Chronic myeloid leukemia
Combrestatin A4 phosphate	*Combretum caffrum*	Fosbretabulin Tromethamine®	Combrestatin	Thyroid cancer and ovarian cancer

Another phytochemotherapeutic drug is etoposide (brand name: Etopophos®), extracted from mayapple (*Podophyllum peltatum*: Berberidaceae) and specific for S and G2 phases of the cell cycle, capable of inhibiting topoisomerase II, stabilizing DNA–topoisomerase II complex, and consequently preventing DNA from unfolding. This drug is active for germinative cell tumors, small cell lung carcinoma, non-Hodgkin lymphoma, and Hodgkin lymphoma recurrence (DeVita et al. 2018).

Topotecan (brand name: Hycamtin®) is a semisynthetic chemotherapeutic agent that causes cell death through the inhibition of topoisomerase II and DNA cleavage. This drug is derived from camptothecin, an alkaloid extracted from the tree *Camptotheca acuminata* (Cornaceae). It is used for ovarian cancer, small cell lung carcinoma, and acute myeloid leukemia (DeVita et al. 2018).

Other examples of plant-derived metabolites approved for anticancer therapy include omacetaxine mepesuccinate, used for chronic myeloid leukemia, and compounds of the combretastatin class, applied for thyroid and ovarian cancers. According to Choudhari et al. (2020), around 30 phytochemicals are currently in preclinical and clinical trials for cancer treatment.

Palliative and preventive cancer therapies also benefit from phytotherapy compounds and traditional herbal preparations. Medicinal plants utilized for cancer treatment by decoction or infusion preparations are well described in the literature. Some of them are under investigation, for instance, *Curcuma longa* (Zingiberaceae), whose roots are rich in curcumin, a potent bioactive polyphenol compound. Pure curcumin and its preparations are on clinical trials for several diseases, including breast, lung, prostate, pancreatic, and colorectal cancers (Salehi et al. 2019; Choudhari et al. 2020). Other plants popularly used against tumors have only indirect evidence of their effectiveness. For example, *Arrabidaea chica* (Bignoniaceae), known in Brazil as pariri, has been related to antioxidant activity because of its phenolic compounds (De Siqueira et al. 2019). However, this is an area of research that should be further explored, since popular knowledge has been of major relevance in many scientific developments. Indeed, ingenol mebutate is a successful case of a bioactive compound applied in preventive cancer therapy. Isolated from *Euphorbia peplus* (Euphorbiaceae), it is indicated for actinic keratosis, a condition that may evolve to squamous cell carcinoma (Choudhari et al. 2020).

Medicinal plants are an important alternative for people from low-income countries, which concentrate higher incidence of some types of cancer. For instance, countries with a human developing index lower than 0.80 concentrated 84% of cases and 88% of deaths due to cervical cancer in 2018 (Wild 2019). In this context, scientific evidence of the biological mechanisms of action, as well as the absence of toxicity, should be encouraged to assure the safety of herbal medicines.

The scientific literature is abundant in studies showing the potential of organisms, from the most diverse taxa, for producing anticancer agents. However, this knowledge rarely reaches the shelves of pharmacies as medicines, a fact that deserves more attention from governments and society, so that the results of basic science will be fully appreciated and supported.

5 Overview of the State-of-the-Art Evidence of Anticancer *Baccharis* Metabolites

As we present in this chapter, there is a considerable number of reports related to the potential anticancer roles of phytochemicals from the *Baccharis* genus, on in vitro and in vivo experimental conditions, as well as descriptions of their chemical nature. Nevertheless, for most scientific studies on phytochemical properties of plant products against cancer, including *Baccharis*, further testing and development are still needed to better understand their mechanisms of action and to foresee their possible applications in clinical oncology.

Among the genus *Baccharis*, which represent more than 440 species from the American continent (see Chap. 2), a small number of species has been used in folk medicine as herbal remedies for a large spectrum of diseases, including few indications for protection or treatment against cancer (see also Chap. 16 – Cultural and popular use of *Baccharis*: drugs, culinary and utensils). Only a few *Baccharis* species have actually been studied and have demonstrated any anticancer effects either in vitro and or in experimental in vivo models. As shown in Table 18.2, 15 species, along with *B. dracunculifolia*-derived green propolis, had their crude extracts or isolated compounds evaluated to determine their potential anticancer properties. *B. dracunculifolia* is the most studied species, followed by *B. trimera*. Notably, the majority of *B. dracunculifolia* studies had comparative evaluations with Brazilian green propolis extracts or analysis of isolated compounds shared by both sources. This is explained by the chemical similarities between plant metabolome and the natural bee product.

The analysis of 36 scientific reports on the strategies carried out to determine the possible anticancer effects of *Baccharis* metabolites and to elucidate their mechanisms of action indicates that the vast majority of studies used in vitro experimental assays with tumor cell lines, and sometimes nontumor cells as controls (Table 18.2). Only 8 articles, out of these 36, report experimental animal models in their studies. Taking the biological effects briefly reported in Table 18.2, which is explored in more details later in this chapter, it is observed that these studies are limited to the early stages of research and development, since they generally performed screening assays to determine the Half Maximum Inhibitory Concentration (IC_{50}) of crude extracts, fractions or pure compounds, while only some of them included additional methodological approaches to investigate the biological mechanisms of action. The number of tumor cell lines tested varied from one, in some articles, to a complete panel, with diverse histopathological origins (see Table 18.2 footprint).

The progress of the studies over time called our attention. Taking as reference the data presented in Table 18.2, after the first studies from 1976 (n = 1) and 1977 (n = 1), there is a gap until the 1990s, with few studies (n = 5), and then a slight increase from 2000 to 2009 (n = 9). In the 2010s, an increased number of scientific articles (n = 20) was finally observed, as well as an improvement in the

Table 18.2 Overview on the anticancer effects of bioactive compounds from *Baccharis* genus and *B. dracunculifolia*–derived green propolis, by in vitro and in vivo experimental models of cancer

	Experimental system			
OriginExtract/molecule	In vitro model/ cell line*	In vivo animal model	Biological effect	References
B. artemisioides				
Ethanol extract	CCRF CEM, CMLK562, CEM/ ADR5000, Lucena 1, and healthy PBMC	–	Cytotoxicity to cancer cells and PBMC	González et al. (2018)
B. concinna				
Hexane and ethyl acetate extracts	HL-60, Jurkat, MCF7, and healthy PBMC	–	Cytotoxicity to cancer cells, no effect on PBMC	See Sect. 7
B. coridifolia				
Dichloromethane extract	KB	–	Cytotoxicity, interaction with DNA, antioxidant activity	Mongelli et al. (1997)
Ethanol and aqueous extracts	HT29, NCI-H460 and U373	–	High cytotoxicity	Monks et al. (2002)
Ethanol extract	CCRF CEM, CMLK562, CEM/ ADR5000, Lucena 1, and healthy PBMC	–	High cytotoxicity to cancer cells and PBMC	González et al. (2018)
B. dracunculifolia				
Methanolic fraction	L-1210	–	High cytotoxicity	Fukuda et al. (2006)
Baccharisketone and *p*-Methoxythymol acetate			Moderate cytotoxicity	
Spathulenol and 3,4,3′,4’-Tetrahydroxy-5,5′-diisopropyl-2,2′-dimethylbiphenyl			High cytotoxicity	
Hexane and ethyl acetate extracts	HL-60, Jurkat, MCF7, and healthy PBMC	–	Cytotoxicity to cancer cells, no effect on PBMC	See Sect. 7
Baccharin	–	Induced colorectal cell carcinogenesis in rat	Chemoprotective, antigenotoxic, reduction of DNA damage, and aberrant crypt foci	Munari et al. (2014)
Artepillin C				
Drupanin				

(continued)

Table 18.2 (continued)

OriginExtract/molecule	Experimental system In vitro model/ cell line*	In vivo animal model	Biological effect	References
B. dracunculifolia–derived green propolis				
Artepillin C	HuH 13	–	DNA fragmentation, apoptosis	Matsuno et al. (1997)
Artepillin C	Leukemic cell lines	Human carcinoma and melanoma xenografts in nude mice	Apoptosis, necrosis, abortive mitosis, suppression of tumor growth, increased CD4/ CD8 T cells ratio	Kimoto et al. (1998)
Propolis extract and artepillin C	–	Induced renal cell carcinogenesis in mice	Prevention of oxidative renal damage and carcinogenesis	Kimoto et al. (2000)
	–	Induced pulmonary cell carcinogenesis in mice	Inhibition of lipid peroxidation, and development of pulmonary cancers from adenomas	Kimoto et al. (2001a)
Artepillin C	Lymphocytic (7 T-cell lines, 5 B-cell lines), myeloid and monocytic leukemia, nonlymphoid nonmyeloid leukemia, and healthy lymphocytes	–	Cytotoxicity, DNA fragmentation, high apoptosis induction on cancer cells, especially T-cell lines; inhibition on activated lymphocytes; limited effect to healthy cells	Kimoto et al. (2001b)
Methanol extract and ethyl acetate fraction	HT-1080 and 26-L5	–	Cytotoxicity	Banskota et al. (1998)
Coniferyl aldehyde, betuletol, kaempferide, ermanin, and others				
Baccharin and drupanin	–	Sarcoma S-180 allograft in mice	Suppression of tumor growth, genotoxicity	Mishima et al. (2005)
Ethanol extract and artepillin C	Angiogenesis model using nontumoral HUVEC	Mouse dorsal air sac assay, transfected with S180 tumor cells	Suppression of angiogenesis	Ahn et al. (2007)

(continued)

Table 18.2 (continued)

OriginExtract/molecule	Experimental system In vitro model/ cell line*	 In vivo animal model	 Biological effect	References
Ethanolic propolis extract	LNCaP	–	TRAIL-induced cytotoxicity and apoptosis	Szliszka et al. (2011)
Artepillin C, quercetin, kempferol and *p*-coumaric acid				Szliszka et al. (2011, 2012)
B. dracunculifolia and green propolis				
Ethanolic plant extract and essential oil	RC-58T/h/ SA#4, DU145, PC-3, and nontumor PrEC	–	Selective inhibition of proliferation, cell cycle arrested at S phase, inhibition of cell cycle–related gene expression on cancer cells	Li et al. (2007)
Ethanolic propolis extract				
Ethanolic plant extract and essential oil	HEp-2	–	Cytotoxicity at high concentration	Bufalo et al. (2009, 2010)
Ethanolic propolis extract				
Caffeic and cinnamic acids				
Artepillin C	HT1080, A549, and U2OS	–	p53 activation, nuclear translocation	Bhargava et al. (2018)
Artepillin C	U343	–	Cytotoxicity	De Oliveira et al. (2014)
Baccharin	B16F10			
Artepillin C	HeLa, SiHa, CaSki, C33A, and nontumor HaCaT	–	Selective apoptosis induction of cancer cells, inhibition of migration and invasion	Souza et al. (2018)
Artepillin C	Hep-2 Langmuir monolayer and GUV* models	–	Dose-dependent cytotoxicity, membrane aggregation and permeabilization, DNA degradation, cell necrosis	Kobal et al. (2020)
B. gaudichaudiana				

(continued)

Table 18.2 (continued)

OriginExtract/molecule	Experimental system In vitro model/ cell line*	In vivo animal model	Biological effect	References
Gaudichaudol C	KB-V1	Lymphocytic leukemia in P-388 mice	Cytotoxicity in vitro and in vivo	Fullas et al. (1994)
Gaudichaudone	–			
Articulin acetate				
Apigenin	BC-1, Col-2, HT-1080, KB, KB-V1, Lu-1, and Mel-2	–		
Hispidulin				
Spathulenol	Mel-2			
B. grisebachii				
Lyophilized decoction	HCT-116, and nontumoral HBL-100	–	Cytotoxicity, also against nontumoral cell line	Gómez et al. (2019)
B. latifolia				
Ethanolic extract	Hep3B, HepG2, PLC/PRF/5, and SNU-182	–	Antiproliferative	Carraz et al. (2015)
	Hep3B		Morphologic modifications	
B. megapotamica				
Alcoholic extract	KB	Lymphocytic leukemia in P-388 mice	Antileukemic in vivo, in vitro cytotoxicity	Kupchan et al. (1976, 1977)
Baccharin				
Baccharinol				
Isobaccharin				
Isobaccharinol				
B. milleflora				
Essential oil	Jurkat, Raji, HL-60, and healthy PBMC	–	Cytotoxicity to tumor cells, inhibited proliferation and DNA replication, G0/G1 arrest of cell cycle, apoptosis and necrosis, selective to cancer cells	Pereira et al. (2017)
B. obtusifolia				
Methanolic extract	RKO and D-384	–	Cytotoxicity	Romero-Benavides et al. (2018)
Genkwanin	RKO			
Apigenin 7,4'-Dimethyl Ether				

(continued)

Table 18.2 (continued)

OriginExtract/molecule	Experimental system: In vitro model/ cell line*	Experimental system: In vivo animal model	Biological effect	References
B. ochracea				
Ethanol and aqueous extracts	HT29, NCI-H460 and U373	–	High cytotoxicity	Monks et al. (2002)
B. penningtonii				
Curcuphenol	Caco-2	–	Inhibited proliferation and DNA replication, increased caspase-3 activity, apoptosis	Rodrigo et al. (2010)
B. salicina				
2-β-(L-Rhamnopyranosyl)-3-α-angeloyloxy-15-acetyloxy-7,13(14)-*E*-dien-*ent*-labdane	A375 and HCT-116 Taq DNA polymerase assay	–	Cytotoxicity	Garro et al. (2020)
2-β-(L-rhamnopyranosyl)-3-α-angeloyloxy-15-hydroxy-7,13(14)-*E*-dien-*ent*-labdane			Cytotoxicity, inhibition of Taq DNA polymerase	
B. scandens				
Quercetin	HL-60 and U-937	–	Cytotoxicity	Cabrera et al. (2016)
Salvigenin				
Xanthomicrol				
Gardenin B			High cytotoxicity, DNA degradation, cell cycle disturbance, activation of multiple caspases, apoptosis	
B. trimera				
Hexane and ethyl acetate extracts	HL-60, Jurkat, MCF7, and healthy PBMC	–	Cytotoxicity to cancer cells, no effect on PBMC	See Sect. 7

(continued)

Table 18.2 (continued)

OriginExtract/molecule	Experimental system			References
	In vitro model/ cell line*	In vivo animal model	Biological effect	
Phenolic compounds	SiHa	–	Cytotoxicity, inhibition of cell colony survival, inhibition of motility, disruption of cell membrane, necrosis	De Oliveira et al. (2013b)
Terpenoid compounds			Cytotoxicity, disruption of cell membrane, apoptosis	
Essential oil	MCF7, HepG2, and nontumoral MCF 10a	–	Cytotoxicity, apoptosis and necrosis, including for nontumoral cells	Moro (2016)
α-Humulene, β-caryophyllene, and caryophyllene oxide				
Eupatorin				
Trimeroside	HT-29, NCI-H460, U-251, KB, HepG2, and nontumoral NIH-3 T3	–	Low cytotoxic effect	Dos Santos et al. (2018)

*Human cancer cell lines: KB – nasopharynx carcinoma, KB-V1 – derived from KB; HEp-2 – laryngeal epidermoid carcinoma; Mel-2, A375 – melanoma; BC-1, MCF7 – breast carcinoma; Col-2, SW480, DLD-1, COLO201, RKO, Caco-2, HT-29 – colon cancer; HCT-116 – colorectal cancer; MKN1, MKN28, MUGC4 – gastric cancer; A549, Lu 1 – lung carcinoma; NCI-H460 – non–small cell lung cancer; HeLa, SiHa, CaSki, C33A – cervical cancer; HuH 13, Hep3B, HepG2, PLC/PRF/5, SNU-182 – hepatocarcinoma; Raji – lymphoblastoid cells derived from Burkitt lymphoma; CCRF-CEM, CMLK562, CEM/ADR5000, Lucena 1, Jurkat – acute lymphoblastic leukemia; HL-60 – acute promyelocytic leukemia; L-1210, NB4, K562– leukemia; U937 – promonocytic myeloid leukemia; D-384 – astrocytoma; U-251, U343 – glioblastoma; HT1080 – fibrosarcoma; U2OS – osteosarcoma; LNCaP, RC-58T/h/SA#4 – prostate cancer; DU145 – brain metastasis of prostate cancer; PC-3 – bone metastasis of prostate cancer

Human nontumoral cells: HBL-100 – epithelial mammary cell line; HaCaT – epithelial; PBMCs – peripheral blood mononuclear cells; HUVEC – umbilical vein endothelial cells; PrEC – prostate epithelial cells; MCF-10a – breast cell line

*Animal cancer cell lines: B16F10 – murine myeloma; 26-L5 – murine colon carcinoma; S180 – murine sarcoma; MDCK – Madin-Darby canine kidney

*Animal nontumoral cell line: NIH-3 T3 – murine embryo fibroblast

**GUV – giant unilamellar vesicles

methodological strategies applied to identify, isolate and purify bioactive compounds, and to understand their mechanisms of action, especially in the case of those molecules shared by *Baccharis* metabolomes as well as by Brazilian green propolis.

Another aspect of the research of novel drug candidates, regardless of their future therapeutic application, is to assure the absence of side effects, especially genotoxicity and mutagenicity to healthy cells. The studies of Munari et al. (2008), Rodrigues et al. (2009), De Oliveira et al. (2013a), and Roberto et al. (2016), among others, have explored these features from *Baccharis* metabolites as well as from *B. dracunculifolia*–derived propolis (discussed later in this chapter).

6 Experimental Evidence of the Anticancer Effects of *Baccharis* Metabolites

A significant number of independent scientific studies, using in vitro and in vivo experimental models, has been calling attention to previous empirical evidence of the potential anticancer effects of *Baccharis*. Indeed, current scientific knowledge strongly supports the development of strategies to use *Baccharis* secondary metabolites on novel approaches to cancer treatment.

Some of the first studies on the anticancer properties of *Baccharis* species were published by Kupchan et al. (1976, 1977). These authors have worked on the bioguided isolation and structural elucidation of new trichothecenes from a *B. megapotamica* alcoholic extract (Fig. 18.4e), which had presented in vitro and in vivo antileukemic activities, respectively to cells derived from human oral epidermoid carcinoma (KB lineage), and P-388 mice leukemia model. Interestingly, those trichothecenes, identified as baccharin, baccharinol, isobaccharinol, and isobaccharin (Fig. 18.4), were described, until then, as fungi secondary metabolites. Trichothecenes are mycotoxins with different chemical structures, including the macrocyclic baccharins, and with diverse biological and toxicity profiles, according to their structural features. Despite their natural toxicity, trichothecenes can either generate immune suppressive or stimulatory responses. Their use as anticancer agents can be accomplished in the form of immunotoxins, which can be prepared by attaching a toxin, as a trichothecene, to an antibody specific to proteins exclusively expressed on cancer cell surface (Wu et al. 2017).

Almost two decades later, Fullas et al. (1994) isolated several diterpenoids from the aerial parts of *B. gaudichaudiana*, in addition to other chemical compounds (Fig. 18.5g). Among all these molecules, the newly described labdane diterpene, named as gaudichaudol C, presented a cytotoxic activity to P-388 murine lymphocytic leukemia (Half-Maximal Effective Dose, ED_{50} = 2.4 μg/mL) and to KB-V1 (ED_{50} = 4.7 μg/mL), a multidrug resistant (including to vinblastine) lineage derived from KB cells. Gaudichaudone, another new compound, as well as the previously described articulin acetate, were both cytotoxic to P-388 cells (respectively, ED_{50}

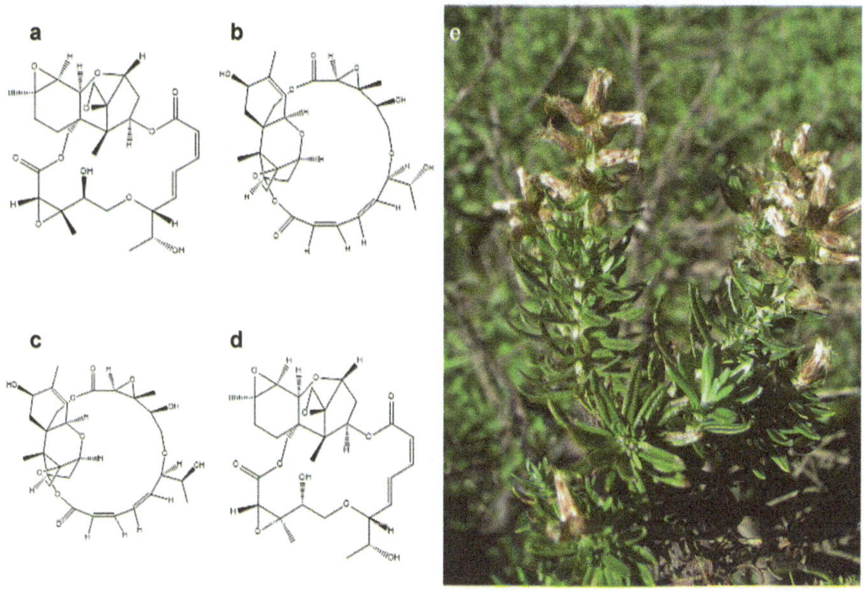

Fig. 18.4 (**a**) Baccharin, (**b**) baccharinol, (**c**) isobaccharinol, and (**d**) isobaccharin, isolated from (**e**) *B. megapotamica*. (Molecules reproduced from NCBI 2020, image of *B. megapotamica* by G Heiden)

= 11.7 μg/mL and 1.7 μg/mL). The flavonoids apigenin and hispidulin were weakly effective to other human tumoral cell lines, such as BC-1 (breast cancer), Col-2 (colon cancer), HT-1080 (fibrosarcoma), KB, KB-V1, Lu-1 (lung cancer), Mel-2 (melanoma), and murine P-388. Sesquiterpene spathulenol was active against Mel-2 (ED_{50} = 6.3 μg/mL).

A few years later, Mongelli et al. (1997), studying the in vitro cytotoxic activity of *B. coridifolia* (Fig. 18.6a), revealed that a dichloromethane extract was also able to inhibit the proliferation of KB human cell line (derived from oral epidermoid carcinoma), presenting an ED_{50} = 4.2 μg/mL. This fraction also elicited a 56% decrease in DNA absorbance at 1000 μg/mL, on DNA-methyl green assay, suggesting the presence of DNA-interacting compounds on its composition. Moreover, measuring the in vitro antioxidant activity, the dichloromethane extract was able to induce oxidative stress in a dose-dependent way.

Studying *B. coridifolia* as well, Monks et al. (2002) reported that the ethanolic and aqueous extracts of that species, and those of *B. ochracea*, were able to elicit pronounced cytotoxicity against cell lines from colon cancer (HT-29), non–small cell lung cancer (NCI-H460), and glioblastoma (U343). In the study of González et al. 2018, a similar ethanolic preparation from *B. coridifolia*, besides *B. artemisioides*, induced potent cytotoxic effects (IC_{50} values from 0.37 to 5.89 *μ*g/mL) on a panel of cell lines representative of acute lymphoblastic leukemia. However, they also affected normal peripheral blood mononuclear cells, used as control. These

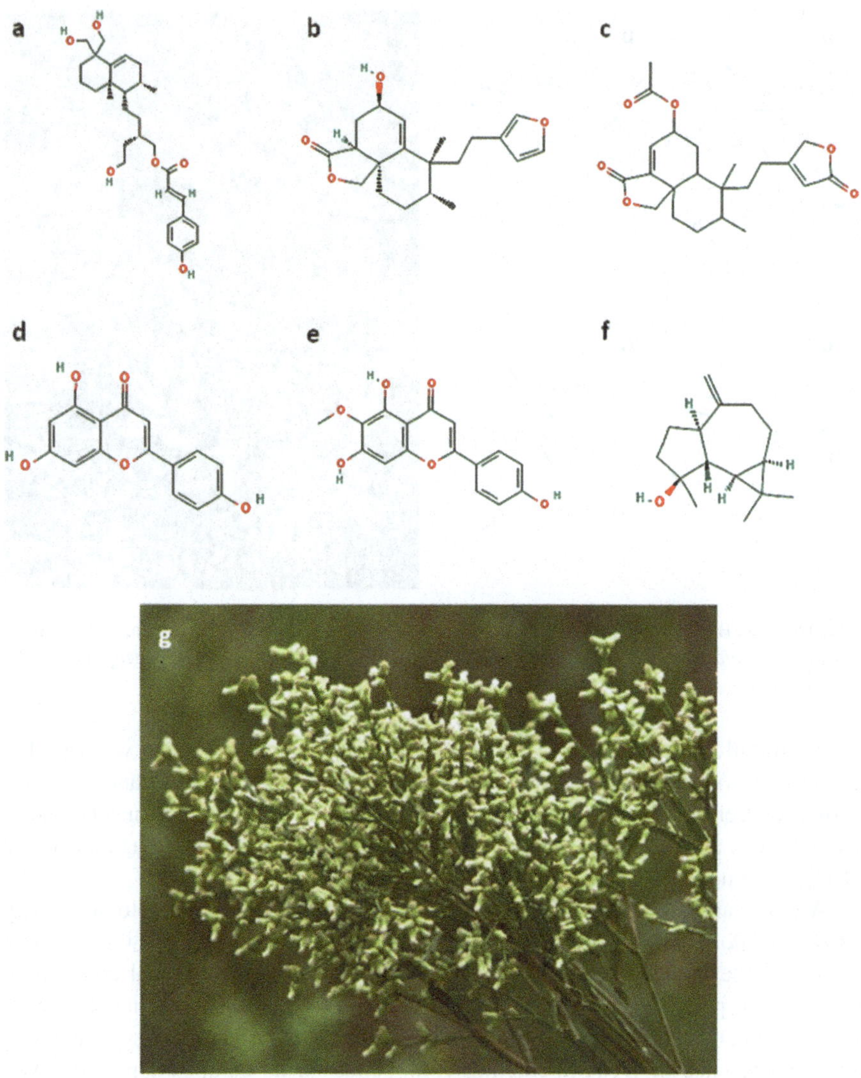

Fig. 18.5 (**a**) Gaudichaudol C, (**b**) gaudichaudone, (**c**) articulin acetate, (**d**) apigenin, (**e**) hispidulin and (**f**) spathulenol, isolated from (**g**) *B. gaudichaudiana*. (Molecules reproduced from NCBI 2020, image of *B. gaudichaudiana* by G Heiden)

in vitro activities observed for *B. coridifolia* compounds might be related to the biological mechanism of action that ultimately leads to cell death. Detailed phytochemical studies of Arisawa et al. (1985) have previously identified schottenol glucoside in this species, which may be one of the active components responsible for the observed results.

Fig. 18.6 (**a**) *B. coridifolia*, (**b**) *B. trimera*, (**c**) *B. latifolia* and (**d**) *B. concinna*. (Images by G Heiden)

B. trimera (Fig. 18.6b) is one species that has been extensively studied to determine its biological properties, including its anticancer potential. In previous studies, De Oliveira et al. (2012) observed that phenolic enriched extracts were able to elicit in vivo anti-inflammatory effects upon carrageenan-induced pleurisy, and antioxidant activity on DPPH and TAR in vitro assays. Subsequently, these authors explored the antitumor properties of *B. trimera* phenolic (PHE) and terpenoid (SAP) compounds by in vitro assay, using SiHa cell line derived from human cervical cancer (De Oliveira et al. 2013b). When exposed to PHE and SAP for 24 h, this cell line presented inhibited viability on MTT (3-(4,5-dimethylthiazol-2-yl)-2,5-diphenyltetrazolium bromide) assays (IC_{50} = 482 μg/mL and 456 μg/mL, respectively). Interestingly, on clonogenic assays and wound-healing migration assays,

PHE was also able to elicit a dose-dependent inhibition of cell colony survival and motility. SAP compounds, instead, increased colony formation and did not affect cell motility. Both compounds were able to affect cell membrane integrity, since a significant dose-dependent increase in extracellular lactate dehydrogenase was observed. Despite that, cell death assays revealed that those *B. trimera* compounds act by different mechanisms: PHE eliciting adverse necrosis, and SAP inducing death via apoptosis. These complementary activities of PHE and SAP indicate these compounds as possible candidates for future anticancer drug development.

B. trimera essential oil and its components, α-humulene, β- caryophyllene, caryophyllene oxide, and eupatorin, were also evaluated in vitro, using MCF7 (breast cancer), HepG2 (hepatocarcinoma), and a nontumoral cell line MCF-10a (Moro 2016). All the samples were cytotoxic, at some level, to tumoral and nontumoral cells, probably through induction of death by apoptosis, but also by the necrosis pathway. The absence of selectivity to tumor cells, besides their high level of elicited necrosis, disqualifies these compounds as chemotherapeutic candidates.

More recently, Dos Santos et al. (2018) isolated a novel normonoterpene glycoside, trimeroside, from *B. trimera* aqueous extract, along with three already known flavonoids, cirsimaritin, luteolin, and quercetin. The antitumor activity of trimeroside was assessed by in vitro viability assays using a panel of tumor cell lines, derived from colon cancer (HT-29), non–small cell lung cancer (NCI H460), nasopharynx carcinoma (KB), hepatocarcinoma (HepG2), glioblastoma (U-251), and nontumor murine embryo fibroblast (NIH-3 T3) as control. Unfortunately, this compound did not show an expressive cytotoxic effect against these cell lines and, as such, it may not represent a good drug candidate at this moment.

Another species that evoked interest was *B. latifolia* (Fig. 18.6c). On an ethnopharmacological survey, Carraz et al. (2015) investigated 51 Peruvian medicinal plant species, evaluating the cytotoxicity of their secondary metabolites against hepatocarcinoma cell lines. The ethanolic extract of *B. latifolia* presented antiproliferative activity on Hep3B, HepG2, PLC/PRF/5, and SNU-182 tumor cell lines (IC_{50} = 10.8 μg/mL, 33.3 μg/mL, 24.3 μg/mL and 20.1 μg/mL, respectively) on ATP-based luminescence assay. Besides, no in vitro cytotoxic effect was shown on healthy human hepatocytes up to 100 μg/mL. *B. latifolia* extract also induced microscopic morphological modifications on Hep3B cells, but no microtubule or mitosis cytoskeleton perturbations were observed on immunofluorescence assay. Although this species is commonly used in Peru for kidney and liver inflammation, the cytotoxic mechanism of action on human hepatocellular carcinoma has not been completely elucidated.

Certainly, one concern about using natural compounds, particularly in cancer protocols, is to prevent possible genotoxic and mutagenic effects on healthy cells. Munari et al. (2008) revealed that the ethyl acetate extract of *B. dracunculifolia* leaves (Fig. 18.7c) presented bioactive metabolites that possibly act as "Janus compounds," with distinct biologic behaviors depending on the concentration. In in vitro assays using Chinese hamster ovary cells, the extract mentioned above at 12.5 μg/mL caused a significant reduction in doxorubicin-induced chromosome damage,

Fig. 18.7 (**a**) Artepillin C and (**b**) drupanin, isolated from (**c**) *B. dracunculifolia* and from green propolis. (Molecules reproduced from NCBI 2020, image by G Heiden)

whereas, at 100 μg/mL, the extract itself was able to increase chromosomal aberrations and abnormal metaphases. The authors have discussed the complexity of components in *B. dracunculifolia* extract, which could be acting alone or synergically to reveal these apparently contrasting results. Additionally, they have remarked that artepillin C (Fig. 18.7), the major flavonoid found in their high-performance liquid chromatography (HPLC) analysis, might be involved in the observed results. Interestingly, artepillin C, also found in green propolis, was previously described as a free radical scavenger (Kumazawa et al. 2003; Nakanishi et al. 2003), an activity that could be correlated to the observed protection from DNA damage.

B. dracunculifolia ethyl acetate extract (Bd-EAE) was evaluated on its chemopreventive potential by in vivo model of experimental colorectal carcinogenesis (Munari et al. 2014). Wistar rats were submitted to a protocol of exposure to the colon carcinogen inducer 1,2-dimethylhydrazine (DMH), or to the ethyl acetate extract before DMH. In vitro comet assays, conducted to analyze DNA degradation, showed that the pretreatment of rat colon cells with Bd-EAE was able to reduce DNA damage elicited by DMH. Moreover, no aberrant crypt foci (ACF) were observed on colon samples from animals that received only Bd-EAE. A significant lower frequency of ACF was encountered on samples from animals treated with DMH plus Bd-EAE, indicating a protective effect against colon carcinogenesis of *B. dracunculifolia* ethyl acetate extract.

Likewise, Rodrigues et al. (2009), investigating the in vivo genotoxic/antigenotoxic and mutagenic effect of *B. trimera*, observed a transient effect of DNA lesions on peripheral blood cells, 3 h after treatment of female adult mice with a single dose of *B. trimera* aqueous extract (500–2000 mg/kg). However, this damage seemed to be reversible or repaired, when analyzed by comet assay, since no other genotoxic effect was observed in blood or liver samples of the animals, 24 h after treatment with three doses of the extract. Additionally, an antigenotoxic effect was observed when blood cells were treated ex vivo with hydrogen peroxide. *B. trimera* extract also presented a dose-dependent antioxidant activity by in vitro DPPH (2,2-diphen yl-1-picryl-hydrazyl-hydrate) and xanthine oxidase assays (IC_{50} = 0.69 mg/mL and 92.39 μg/mL, respectively). In contrast, a higher frequency of micronucleated polychromatic erythrocytes was observed in the bone marrow of treated mice, indicating a possible chromosomal mutagenic activity. These contradictory effects indicate the importance of further studies to better understand the potential and safety of *B. trimera* preparations.

The scientific reports discussed so far, in addition to others listed in Table 18.2, demonstrate that many efforts have been made to assure the safety of *Baccharis* species used as traditional medicines and to isolate bioactive metabolites with a role in in vitro and in vivo models of cancer. Researchers were primarily looking for unique compounds with selective and effective anticancer action. This means that those molecules must be cytotoxic preferentially to tumor cells, with minor side effects on healthy cells and tissues. Furthermore, they need to induce the apoptosis pathway of cellular death, instead of the necrosis pathway, since in vivo immune responses involving inflammation must be avoided. So far, a few candidates fulfill the selection criteria.

7 Potential Anticancer Properties of *Baccharis* Inhabiting the Brazilian Savanna

In this section, we report on our preliminary studies of anticancer in vitro effects of extracts from three *Baccharis* species commonly distributed in the Brazilian Cerrado, a uniquely local savanna: *B. trimera* (Fig. 18.6b), *B. dracunculifolia* (Fig. 18.7c), and *B. concinna* (Fig. 18.6d). The last one is a rare and threatened species that inhabits higher altitudes of Serra do Espinhaço, a mountain range located in the Brazilian Atlantic plateau (see Marques et al. 2002; Gomes et al. 2004; Fernandes et al. 2017a).

To study their potential anticancer properties, the aerial parts of plant samples were separated into different groups, according to branch maturation (young or mature) and sex (female or male, except for *B. trimera*, from which only female plants were available). Each sample was then processed by solvent gradient extraction, generating hexane, ethyl acetate, *n*-butanol, and aqueous fractions. Plant extracts were subsequently tested by in vitro MTT viability/cytotoxic assays (Mosmann 1983), using three human cancer cell lines: Jurkat acute lymphoblastic leukemia, HL-60 acute promyelocytic leukemia, and MCF7 breast adenocarcinoma. Peripheral blood mononuclear cells (PBMC) from healthy donors were used as control. In brief, cells were exposed to 100 μg/mL of each *Baccharis* extract, in triplicate, for 24 h. Cell viability was then determined by spectrophotometric measurement of MTT dye reduction, catalyzed by mitochondrial dehydrogenase of active cells in formazan. Figures 18.8, 18.9, and 18.10 show the cytotoxic profile of those extracts on cancer cell lines and control PBMC.

According to our results, nonpolar extracts (hexane and, especially, ethyl acetate fraction) from all three studied *Baccharis* species contain plant secondary metabolites with selective cytotoxic activity against at least one of the cancer cell lines used (Figs. 18.8, 18.9, and 18.10). These extracts show a discrete to no effect in healthy control cells (cell viability values around 100%, Figs. 18.8 and 18.9), as evidenced by the proportional metabolic reduction of MTT. Among cancer lineages, HL-60 is the one with the highest sensitivity for both nonpolar extracts. MCF7 and Jurkat cells, in contrast, are those that showed less sensitivity to the majority of assayed extracts, although some samples of *B. trimera* ethyl acetate fractions (Fig. 18.8, samples 2 and 6 against Jurkat cells) and *B. dracunculifolia* hexane fraction (Fig. 18.9, sample 13 against Jurkat and MCF7 cells) elicit cytotoxic effects. Hexane extracts of *B. trimera* and *B. dracunculifolia* are also capable of inhibiting the PBMC viability, showing no selectivity of these preparations for cancer cells. In general, extracts obtained with the same solvent give the same cytotoxic effect, regardless of the origin of the plant's sample group, whether from mature or young branches, or from female or male individuals, indicating that there are no apparent differences in the secondary metabolites present in these sample groups (see also Espírito-Santo et al. 2012).

Our results also showed that polar extracts (*n*-butanol and aqueous fractions), particularly from *B. dracunculifolia*, contain metabolites that seem to improve the metabolism of healthy cells (PBMC), as observed by an increase of the number of viable cells (Fig. 18.9). Nevertheless, we also recorded an increase in the metabolism of cancer cells, especially for MCF7 lineage treated with *B. dracunculifolia* and *B. concinna* polar extracts (Figs. 18.9 and 18.10).

The hexane and ethyl acetate extracts, which had caused high-limit cytotoxic effects in screening assays at 100 μg/mL, were then evaluated to determine the IC_{50} for HL-60, since this leukemic cell line was shown as the most sensitive among the

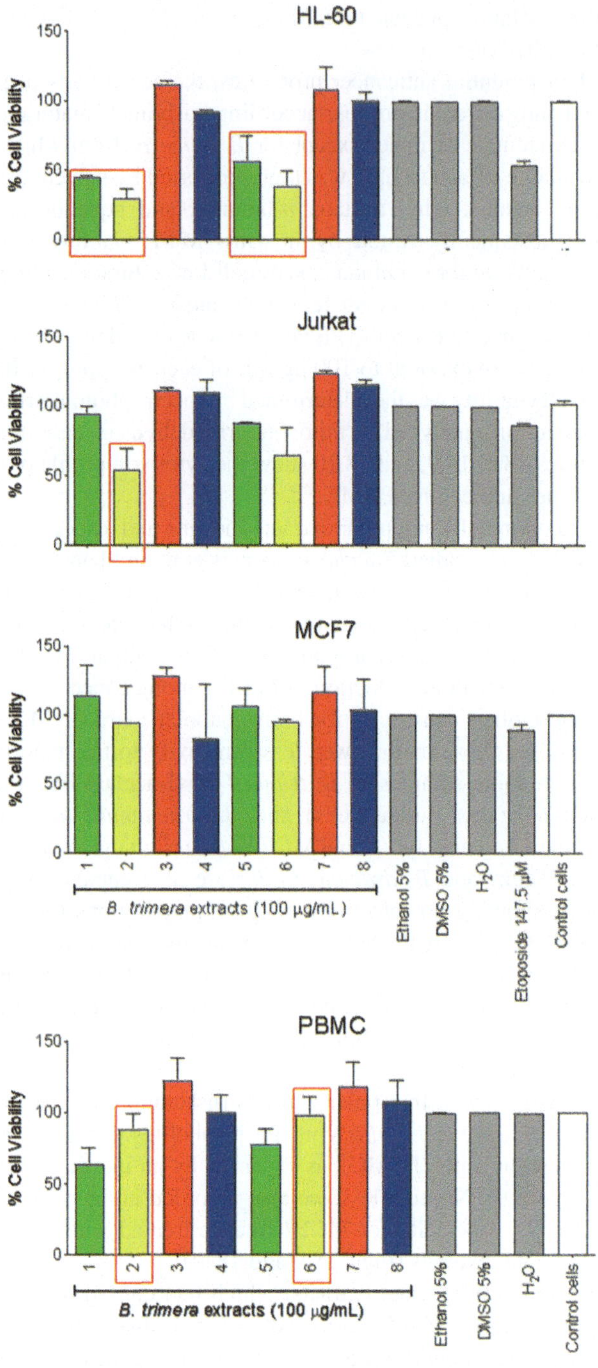

Fig. 18.8 In vitro cytotoxic effect of *B. trimera* metabolites against cancer cell lines of HL-60 acute promyelocytic leukemia, Jurkat acute lymphoblastic leukemia, and MCF7 breast adenocarcinoma, compared to peripheral blood mononuclear cells (PBMC) from healthy donors,

three studied cancer cell lines. Values of IC_{50} ranged from 43.25 to 360.60 μg/mL (Table 18.3), with best results for ethyl acetate fractions from female samples of *B. dracunculifolia* (Bd6, IC_{50} = 47.88 μg/mL) and *B. concinna* (Bc6, IC_{50} = 43.25 μg/mL). Despite the cytotoxicity observed on screening assays ($n = 2$), three fractions (Bd13, Bd14, and Bc5, see Table 18.3) were considered inactive against HL-60 cells up to 100 μg/mL, since IC_{50} could not be determined in the range of assayed concentrations (see Table 18.3 for the nomenclature of plant samples extracts). The loss of activity of these samples in IC_{50} assays might be explained by the additional freeze/thaw cycles, or by poor solubility in ethanol (the solvent used for dissolution of dry extracts) or in culture medium, among other factors well discussed in the literature (Cheng et al. 2003; Di and Kerns 2006).

According to Starlin et al. (2012), phytochemical extracts are considered to have a moderate anticancer effect when IC_{50} values of ≤100 μg/mL are obtained for at least two lineages. Despite the fact that, in our study, the $IC_{50} \leq 100$ μg/mL of non-polar extracts was only determined for HL-60 cell line, their cytotoxic effect seems promising for myeloid leukemia cells. Therefore, further detailed studies are needed to confirm these findings in other cancer lineages.

This study indicates the presence of a variety of secondary metabolites produced by *Baccharis* species, capable of inducing widely divergent biological effects. In addition, it is possible that there are similarities among the metabolome of these three species, since they revealed a common pattern of modulatory effects on tumor and healthy cells. In fact, previous studies have shown that *Baccharis* species share common bioactive compounds in their metabolome (see Table 18.2). Therefore, our findings on the cytotoxic potential of *B. dracunculifolia*, *B. trimera*, and *B. concinna* extracts might be explained by the presence of secondary metabolites with low to medium polarity, such as phenolic compounds like caffeic acids and cinnamic acid derivatives (as *p*-coumaric acid, artepillin C, drupanin, and baccharin), flavonoids (as kaempferol, luteolin, gardenin B, quercetin), and terpenes (as oleanolic acid, clerodane diterpene) already described in similar chemical preparations from some *Baccharis* species (see Table 18.2). Artepillin C, the main bioactive compound of *B. dracunculifolia*, has already demonstrated a selective cytotoxic effect to leukemic cell lines (Kimoto et al. 1998, 2001b). Furthermore, Kabala-Dzik et al. (2018) observed that the flavonoids apigenin and quercetin, commonly present in *Baccharis* species and propolis from different origins, are cytotoxic to MCF7 cells, being able to induce cell cycle arrest and apoptosis. Three metabolites of medium polarity, the flavonoids 3-*O*-methylquercetin, quercetin, and kaempferol, were recently detected in *B. trimera* by High-Resolution Magic Angle Spinning

Fig. 18.8 (continued) elicited after 24 h of exposure. Extracts (100 μg/mL) correspond to hexane (*green*), ethyl acetate (*yellow*), *n*-butanol (*red*), and aqueous (*blue*) fractions of young (*1–4*) or mature branches (*5–8*) from female individuals. Controls correspond to solvents (ethanol 5%, DMSO 5%, or water 5%), cytotoxic etoposide 147.5 μM (*gray*), or culture medium (Control cells, in *white*). The most cytotoxic fractions to cancer cells, with low effect on PBMC, were indicated (*highlighted in red*). Results correspond to mean ± sd (standard deviation) of two independent assays, each one in triplicate, for cancer cell lines, and mean ± sd of five independent assays, in triplicate, for PBMC

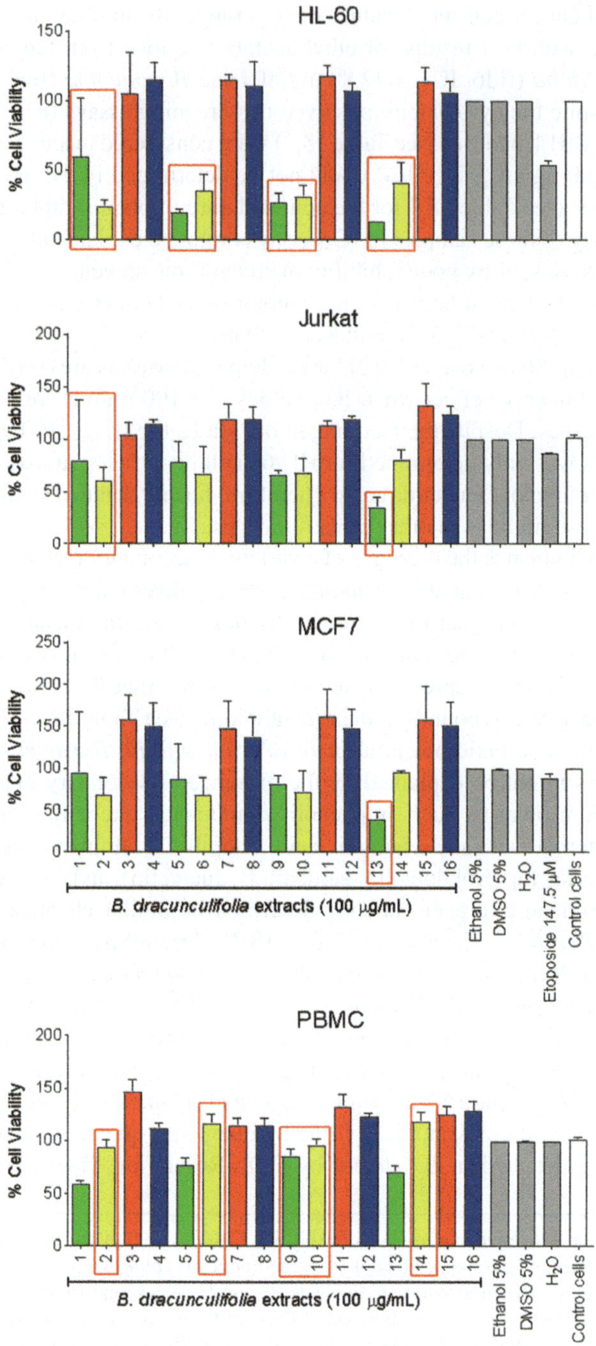

Fig. 18.9 In vitro cytotoxic effect of *B. dracunculifolia* metabolites against cancer cell lines of HL-60 acute promyelocytic leukemia, Jurkat acute lymphoblastic leukemia, and MCF7 breast adenocarcinoma, compared to peripheral blood mononuclear cells (PBMC) from healthy donors,

Nuclear Magnetic Resonance (1H HR-MAS NMR) (Dutra et al. 2020). It is known that flavonoids had been previously associated with the cytotoxic activities of *B. trimera* (Oliveira et al. 2013), specifically, quercetin, lately isolated from *B. scandens* (Cabrera et al. 2016), which has been shown to elicit cytotoxicity to HL-60 (Dutra et al. 2020).

The study presented here comprises the first scientific report of an in vitro anticancer effect of *B. concinna* metabolites, a rare Brazilian *Baccharis* (Marques et al. 2002; Gomes et al. 2004; Fernandes et al. 2017a). Likewise, *B. trimera* and *B. dracunculifolia* had not been previously tested for their cytotoxic effects against HL-60 and Jurkat cell lines, representative from different leukemias, neither *B. dracunculifolia* for breast cancer–derived MCF7 cell line. The HL-60 cell line was previously used for the evaluation of isolated compounds from *B. scandens* (Cabrera et al. 2016, see Table 18.2). This lineage, as well as the Jurkat cell line, is also sensitive to *B. milleflora* essential oil (Pereira et al. 2017, see also Table 18.2). Interestingly, few in vitro *Baccharis* studies have used tumor cells from breast cancer origin. One exception was presented by Cordero et al. (2004), who reported no cytotoxic effect of scopolin, a metabolite isolated from *B. tricuneata*, against four human tumor cell lines, including MCF7. Moro (2016) observed apoptosis and necrosis of MCF7 cells treated with *B. trimera* essential oil or its components, although nontumor cells were also affected. Our present results are the first that indicate an in vitro selective effect of *B. dracunculifolia* nonpolar metabolites on breast cancer (see Fig. 18.9).

Finally, the limited cytotoxic effects of ethyl acetate fractions from *B. trimera* and *B. dracunculifolia* on healthy PBMC mean a selective anticancer activity of their bioactive components. This is a promising result for further investigations of the role of each *Baccharis* metabolite on cancer cell cytotoxicity, looking for more selective and effective antineoplastic drug candidates.

8 Anticancer Mechanisms of Action Elicited by *Baccharis* Secondary Metabolites

The screening of bioactive compounds capable of inducing cytotoxicity in in vitro and in vivo experimental models is only the first step in the way to discover potential antineoplastic drug candidates. It is imperative to understand the cellular and molecular pathways employed by molecules that selectively inhibit tumor growth,

Fig. 18.9 (continued) elicited after 24 h of exposure. Extracts (100 μg/mL) correspond to hexane (*green*), ethyl acetate (*yellow*), *n*-butanol (*red*), and aqueous (*blue*) fractions of young (*1–4*) or mature branches (*5–8*) from female individuals, or young (*9–12*) or mature (*13–16*) branches from male individuals. Controls correspond to solvents (ethanol 5%, DMSO 5%, or water 5%), cytotoxic etoposide 147.5μM (*gray*), or culture medium (Control cells, in *white*). The most cytotoxic fractions to cancer cells, with low effect on PBMC, were indicated (*highlighted in red*). Results correspond to mean ± sd (standard deviation) of two independent assays, each one in triplicate, for cancer cell lines, and mean ± sd of five independent assays, in triplicate, for PBMC

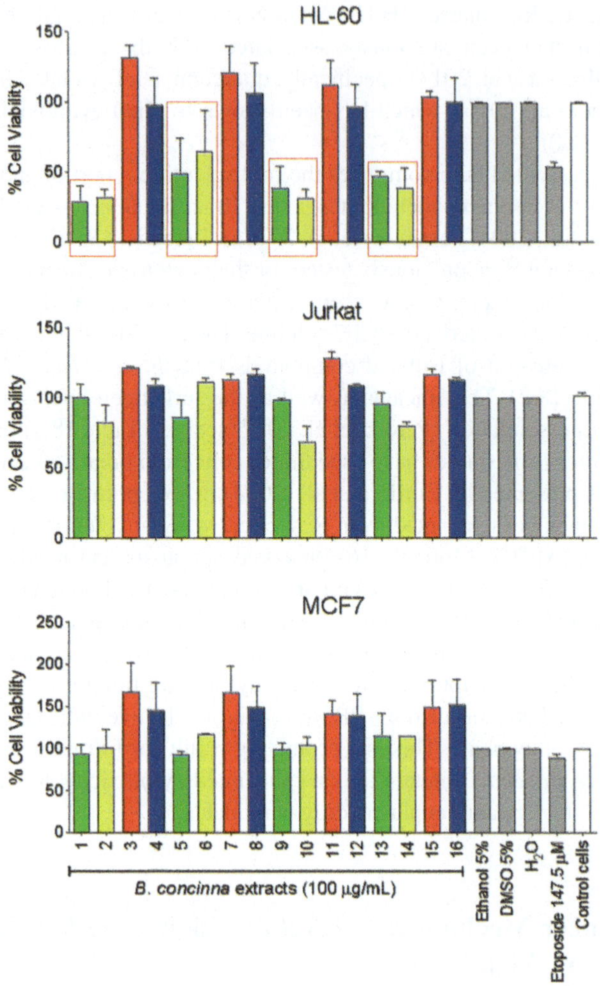

Fig. 18.10 In vitro cytotoxic effect of *B. concinna* metabolites against cancer cell lines of HL-60 acute promyelocytic leukemia, Jurkat acute lymphoblastic leukemia, and MCF7 breast adenocarcinoma, elicited after 24 h of exposure. Extracts (100 µg/mL) correspond to hexane (*green*), ethyl acetate (*yellow*), *n*-butanol (*red*), and aqueous (*blue*) fractions of young (*1–4*) or mature branches (*5–8*) from female individuals, or young (*9–12*) or mature (*13–16*) branches from male individuals. Controls correspond to solvents (ethanol 5%, DMSO 5% or water 5%), cytotoxic etoposide 147.5µM (*gray*), or culture medium (Control cells, in *white*). The most cytotoxic fractions to cancer cells were indicated (*highlighted in red*). Results correspond to mean ± sd (standard deviation) of two independent assays, each one in triplicate

Table 18.3 IC_{50} values of nonpolar extracts from three *Baccharis* species, against HL-60 tumor cell line

Species	Extract[a]	IC_{50} (μg/mL)[b]
B. trimera (Bt)	Bt5	61.32 (32.35–116.2)
	Bt6	77.02 (62.17–95.42)
B. dracunculifolia (Bd)	Bd5	85.28 (67.79–107.3)
	Bd6	**47.88** (38.03–60.28)
	Bd13	>100
	Bd14	>100
B. concinna (Bc)	Bc5	>100
	Bc6	**43.25** (6.396–292.4)

Extracts had their IC_{50} determined after viability assay screening
[a]Extracts: 5 (hexane fraction) and 6 (ethyl acetate fraction), from female mature branches; 13 (hexane fraction) and 14 (ethyl acetate fraction), from male mature branches
[b]IC_{50} (with 95% confidence interval) determined in two independent assays (except for Bt13 and Bt14), each assay in triplicate
Values >100 mean that IC_{50} could not be determined in the concentration range used

migration, metastasis, resistance, or that specifically induce cancer cell apoptosis, as well as the recruitment of a patient's immune system.

It was shown that curcuphenol (Fig. 18.11), a sesquiterpene present in a variety of plants, including *B. penningtonii* and also in marine sponges, has antiproliferative and proapoptotic action against cancer cells (Rodrigo et al. 2010). In an in vitro model, using Caco-2 human colon cancer cell line, these authors observed a dose-dependent effect of curcuphenol (29 to 116 μg/mL) capable of inhibiting cell proliferation (9 to 32%) as well as DNA replication (IC_{50} = 43 μg/mL), and to induce cell death by apoptosis (2.6 to 3.0-fold) with increased caspase-3 activity (6.6- and eightfold at 29 and 58 μg/mL, respectively). Three other polyoxygenated flavonoids, isolated from *B. pentlandii* by Tarqui Tarqui et al. (2012), named 8-methoxycirsilineol, xanthocrimol, and sideritoflavone, had already been associated with cytotoxic and apoptotic activities in a previous study published by Sergeev et al. (2007).

Apigenin, isolated from *B. gaudichaudiana* by Fullas et al. (1994), is a natural dietary flavonoid, abundantly found in many plant species. This flavone has been successfully correlated with in vitro and in vivo anti-inflammatory, antiproliferative, pro-apoptotic, antiangiogenic, and antimetastatic effects on a variety of cancer types, such as head-and-neck, skin, breast, prostate, colorectal, pancreatic, liver, cervical, and ovary tumors (Liu et al. 2005; Ujiki et al. 2006; Vargo et al. 2006; Shukla and Gupta 2007; Bhattacharya et al. 2018; Chang et al. 2018; Hu et al. 2018; Yan et al. 2018; see also the review of Madunic et al. 2018). Recently, Yan et al. (2018) verified that this bioactive compound was responsible for in vitro and in vivo inhibition of human chondrosarcoma cell proliferation by cell cycle arrest at G2/M phase, colony formation impairment, and by eliciting cell apoptosis through the increase of reactive oxygen species (ROS). Likewise, Vargo et al. (2006) had previously observed the effect of apigenin on ROS generation and phosphorylation of

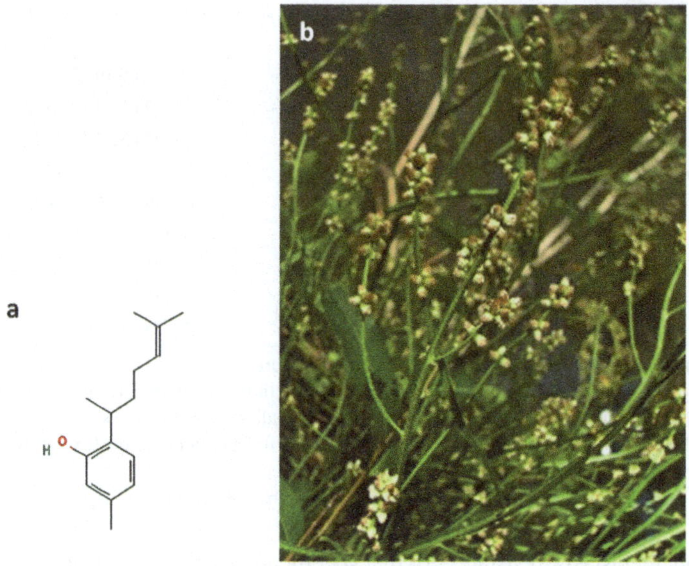

Fig. 18.11 (**a**) Curcuphenol, isolated from (**b**) *B. penningtonii*. (Molecule reproduced from NCBI 2020, image by G Heiden)

MAPK (mitogen-activated protein kinase), p38, and ERK (extracellular signal–regulating kinase) on THP-1 leukemia cells, although they had suggested the existence of an alternative mechanism of apigenin-induced apoptosis, independent of these factors, and related to the activation of caspase-3 and PKCδ (protein kinase Cδ).

Chang et al. (2018) demonstrated the ability of apigenin to reduce proliferation, migration, invasion, and metastasis of non–small cell lung cancer in a study using in vitro assays, as well as an in vivo xenograft model. In another experimental model, Bhattacharya et al. (2018) observed a significant reduction of tumor progression after the administration of apigenin-loaded nanoparticles to rats with chemical-induced hepatocellular carcinoma.

The in vitro and in vivo experimental use of apigenin, combined with conventional chemotherapeutic drugs, has also shown promising results, since this flavone is able to act synergistically, increasing the bioavailability or improving cell intake of some drugs (Madunic et al. 2018). For example, the association of apigenin with cetuximab, an anticancer monoclonal antibody, tested in a nasopharyngeal carcinoma model, revealed a synergistic increase in cell cycle arrest and apoptosis, in addition to inhibition of tumor growth (Hu et al. 2018). All these findings together indicated the relevance of apigenin as a candidate for preclinical trials, as pointed out by Madunic et al. (2018) and Choudhari et al. (2020).

B. scandens is also indicated as a source of bioactive flavonoids, such as salvigenin, gardenin B, xanthomicrol, and quercetin (Cabrera et al. 2016) (Fig. 18.12). Comparing their chemical structures to their cytotoxicity to HL-60 acute myeloid leukemia and U-937 human promonocytic myeloid leukemia cell lines, Cabrera

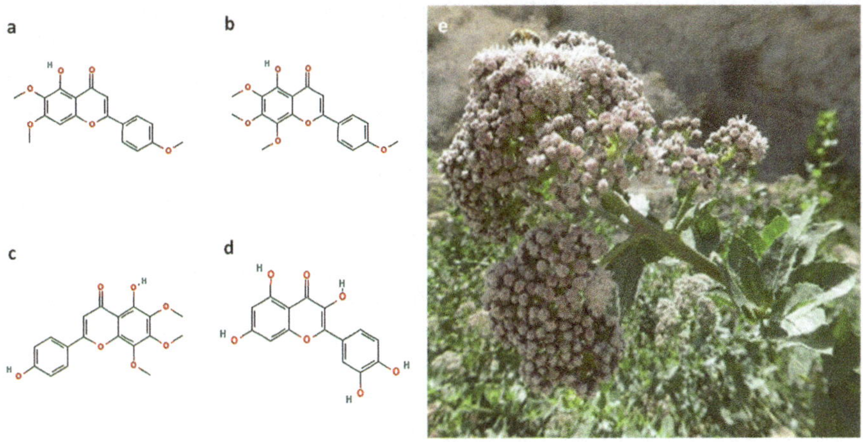

Fig. 18.12 (**a**) Salvigenin, (**b**) gardenin B, (**c**) xanthomicrol, and (**d**) quercetin, isolated from (**e**) *B. scandens*. (Molecules reproduced from NCBI 2020, image by D Montesinos)

et al. (2016) concluded that methoxyl groups at certain positions (C4' and C8) conferred to gardenin B the highest effect among all tested compounds (IC_{50} = 1.6 and 3.0μM, respectively). This activity seemed to be selective to cancer cells, since gardenin B up to 30μM had no significant effect on the viability of quiescent or proliferating (phytohemagglutinin B induced) human peripheral blood mononuclear cells. The cytotoxic effect of this flavonoid on HL-60 and U-937 cells was dose and time dependent and involves death by apoptosis, as evidenced by DNA fragmentation, apoptotic bodies, cell cycle arrest at S and G2/M phases, increase in sub-G1 fraction, and externalization of phosphatidylserine in the cell membrane. Furthermore, these authors demonstrated that gardenin B induces the extrinsic and intrinsic apoptotic pathways by the activation of procaspases-2, −3, −8, and −9, and poly-(ADP-ribose) polymerase, and by the induction of caspase-2, −3, −8, and −9 activities. In the study of Cabrera et al. (2016), similarly to the observations on apigenin by Vargo et al. (2006), the mechanism of cell death elicited by gardenin B was independent of ROS. However, the authors verified that this flavonoid also induced the generation of these oxidants.

Quercetin is also an abundant dietary flavonoid, encountered in many vegetal sources besides *Baccharis*, and its anticancer potential had been previously studied by other authors. In an in vitro model using human oral squamous carcinoma SCC-9 cell line, Haghiac and Walle (2005) demonstrated that quercetin affects DNA synthesis (IC_{50} = 15μM) after 24 h of exposure, strongly inhibiting thymidylate synthase, and elicits an irreversible antiproliferative effect after 48 h (IC_{50} = 94μM) and 72 h (IC_{50} = 50μM). The compound also induces early necrosis at 24 and 48 h of exposure, probably by a direct interaction on the cell membrane and by cell cycle arrest at the S phase. This effect is followed by an increase in apoptosis after 72 h, correlated to caspase-3 activation.

Priyadarsini et al. (2010) and Bishayee et al. (2013), both using the human cervical carcinoma HeLa cell line, confirmed the role of quercetin on cell cycle arrest at G2/M phase, its direct effect on DNA intercalation, cell membrane depolarization, and ROS accumulation, and on apoptosis induction, through p53 upregulation, NF-κB, AKT, and Bcl-2 antiapoptotic proteins' downregulation, and caspase-3 activation.

The antioxidant potential of quercetin from *B. trimera* ethanolic extract was evaluated by Araujo et al. (2017) in SK Hep-1 cells (human liver adenocarcinoma). The authors detected a possible signaling pathway by which *B. trimera* metabolites, probably due to quercetin and other flavones' synergistic action, decrease the expression and the enzymatic activity of protein kinase C (PKC), playing a potential role in ROS inhibition. Although the study did not show this correlation, it is likely that the antioxidant potential of metabolites derived from *Baccharis* may be involved in its anticancer action.

The recent findings of the multiple mechanisms of action of quercetin related to chemoprevention and chemotherapy of cancer (Rather and Bhagat 2020) justify its indication as a potential novel drug candidate. In fact, this flavonoid is in clinical trials not only for prostate cancer treatment, but also for chemoprevention of squamous cell carcinoma and prevention of chemotherapy-induced neuropathic pain in cancer (Choudhari et al. 2020).

In a recent study by Romero-Benavides et al. (2018), the cytotoxic effect of methanol extract from *B. obtusifolia* leaves and two isolated flavonoids, identified as 5,4′-dihydroxy-7-methoxyflavone (known as genkwanin) and 5-hydroxy-7,4′-dimethoxyflavone (known as apigenin 7,4′-dimethyl ether) (Fig. 18.13), was evaluated in in vitro assays against tumor cell lines derived from prostate adenocarcinoma (PC-3), colon carcinoma (RKO), astrocytoma (D-384), and breast adenocarcinoma (MCF7). The authors observed that RKO and D-384 are the cell lines most sensitive to the methanolic extract, their growth being inhibited by 89.2% and 60.6%, respectively. It was also observed that genkwanin and apigenin 7,4′-dimethyl ether are not able to affect PC-3, D-384, or MCF7 cells up to 100 μM. Instead, these flavonoids demonstrate specificity to RKO cell line, with IC_{50} of 68.11 μM (19.48 μg/mL) and 34.30 μM (10.22 μg/mL), respectively. Interestingly, apigenin 7,4′-dimethyl ether presents a methoxy group in the C-4′ position of B ring, similarly to gardenin B (Fig. 18.12b), which suggests that this analogous stereochemistry might be involved in the higher cytotoxicity of these two compounds.

The studies of Kimoto et al. (1998, 2000, 2001a, b) show the efforts to understand the mechanism of action on cancer of the Brazilian green propolis' artepillin C, also found as a secondary metabolite of *B. dracunculifolia* (see Fernandes et al. 2018) (Fig. 18.7). Using in vitro and in vivo cancer experimental models, these researchers observed that artepillin C increases the apoptosis of leukemic cells and inhibits in vitro tumor growth. In addition, Kimoto et al. (1998) had previously shown that xenografts of human carcinoma and melanoma in nude mice are notably affected by artepillin C, which induces apoptosis and necrosis, abortive mitosis, and suppression of tumor growth, and indirectly activates the immune system despite an increase in CD4/CD8 T cells ratio.

Fig. 18.13 (**a**) Genkwanin and (**b**) apigenin 7,4′-dimethyl ether, isolated from (**c**) *B. obtusifolia*. (Molecules reproduced from NCBI 2020, image by G Heiden)

It was shown, in another in vitro study (Kimoto et al. 2001b), that artepillin C is able to inhibit cell proliferation, as measured by the decrease in DNA synthesis, and is also able to induce apoptosis to all tumor cell lines, preferentially to T-cell tumor lines, as evidenced by the frequency of apoptotic bodies and DNA fragmentation. Moreover, this study had shown that artepillin C is cytotoxic to mitogen-stimulated normal blood lymphocytes, but not to nonactivated lymphocytes, used as control, meaning that the compound is probably selective to proliferative transformed cells. The authors concluded that the mechanism of action seems to be partially associated with the higher expression of Fas antigen and with the disturbance of mitochondrial membrane potential.

In experimental renal cell carcinogenesis, the treatment of mice with propolis or artepillin C prevented oxidative renal damage and tumor development (Kimoto et al. 2000). Similarly, on induced pulmonary cell carcinogenesis, artepillin C affected lipid peroxidation and prevented the transformation of adenomas into adenocarcinomas (Kimoto et al. 2001a).

Akao et al. (2003), who studied baccharin (Fig. 18.4), artepillin C, and drupanin (Fig. 18.7), demonstrated that these compounds at 150 μM were able to trigger in vitro cytotoxic effect on a panel of human cell lines derived from gastric and colon cancer, and leukemia (Table 18.2). In a further study using angiogenesis in vitro model with nontumor human umbilical vein endothelial cells (HUVEC), and an in vivo model using the ICR mouse strain transfected with murine Sarcoma S180 tumor cells in dorsal air sac (DAS), propolis ethanolic extract and artepillin C were able to suppress angiogenesis (Ahn et al. 2007).

Complementary studies on possible genotoxic and protective (antigenotoxic) artepillin C activities were presented by De Oliveira et al. (2013a). It was observed in those studies, using in vitro comet and micronucleus assays, that DNA damage and micronucleus frequencies are significantly increased by the treatment of Chinese hamster lung fibroblasts (V79 cells) with artepillin C at 20μM. In antigenotoxicity assays, artepillin C at a low dose (2.5 to 10.0μM) can protect cells from methyl methane sulphonate (MMS) induced genotoxicity, perhaps through its already known antioxidant properties. Roberto et al. (2016) have further confirmed the antigenotoxicity and antimutagenicity effect of ethanolic extracts of green propolis and *B. dracunculifolia* (known by their artepillin C and quercetin contents, among other compounds) on rat hepatoma HTC cell line. DNA damage and micronuclei occurrence were significantly reduced in cells exposed to MMS in the presence of plant extracts (65.32% and 85.00%, respectively) or cells pretreated with propolis extract and then exposed to MMS (62.39% and 86.67%, respectively).

De Oliveira et al. (2014) also investigated the in vitro antiproliferative effect of artepillin C and baccharin, compared to Brazilian green propolis and crude extracts from the source plant *B. dracunculifolia*. The study evaluated a panel of human and murine tumor cell lines, and a control cell line GM07492A from nontumor human lung fibroblasts, in cytotoxic assays using XTT dye [(sodium 2,3,-bis(2-methoxy-4-nitro-5-sulfophenyl)-5-[(phenylamino)-carbonyl]-2H-tetrazolium) salt]. As expected, the isolated compounds were more cytotoxic for tumor cell lines than Brazilian green propolis or *B. dracunculifolia* crude extracts. For example, the lowest IC_{50} value achieved by propolis extract was 41.0 ± 4.5 μg/mL on the human glioblastoma U343 cell line. In contrast, for *B. dracunculifolia* extract, the lowest IC_{50} value achieved was 44.9 ± 7.1 μg/mL on the human hepatocellular carcinoma HepG2 cell line. On the other hand, the lowest observed IC_{50} values were 20.1 ± 2.9 for artepillin C on U343 cells and 13.0 ± 1.5 μg/mL for baccharin on murine melanoma B16F10 cells. The selectivity index (SI) values determined for those compounds indicated a preferential cytotoxic effect on tumor cell lines, compared to control cells, more effective than the chemotherapeutical drugs used as positive controls. For the association of artepillin C plus baccharin, the lowest IC_{50} result was 35.2 ± 0.5 μg/mL for B16F10 cells, although no additive or synergistic effect was observed for most cell lines. Nonetheless, these two compounds from *B.*

dracunculifolia and green propolis origin have been considered promising anticancer candidates.

In a more recent study, Souza et al. (2018) explored the mechanisms of action of artepillin C, evaluating the effect of this isolated compound on a panel of human cervical cancer cell lines, represented by HeLa (human papillomavirus/HPV 18 positive), SiHa (HPV 16 positive), CaSki (HPV 16 and 18 positive), and C33A (HPV negative), compared to the nontumor human epithelial cell line HaCaT. The bioactive compound selectively inhibited viability, migration, and invasion of tumor cells and triggered oxidative stress, inducing apoptosis through its intrinsic pathway. The authors have indicated artepillin C as a promising drug candidate for cervical cancers, including those caused by HPV.

Another aspect of the role of artepillin C on cancer was revealed by molecular docking studies conducted by Bhargava et al. (2018), using the tumor suppressor protein p53 and its inhibitor, the mortalin protein. Through in vitro and in vivo assays, the researchers had compared artepillin C to green propolis-supercritical extract (GPSE) and this extract conjugated with γ cyclodextrin (GPSE-γCD), which contained, respectively, 9.6 and 3.0% of artepillin C. Immunofluorescence staining of treated cells showed that artepillin C and GPSE act by inducing the nuclear translocation and activation of free p53 through its dissociation from the mortalin complex. On viability and clonogenic assays, it was observed that GPSE and GPSE-γCD were more cytotoxic than artepillin C for fibrosarcoma HT1080, lung carcinoma A549, and osteosarcoma U2OS human cell lines. These findings indicate the essential role of other propolis bioactive compounds, as well as the effectiveness of encapsulating propolis with cyclodextrin to improve its solubility in aqueous solvents. Finally, in an in vivo model of subcutaneous tumor xenograft using HT1080-treated nude mice, GPSE and especially GPSE-γCD were able to suppress tumor growth, indicating the potential anticancer use of encapsulated propolis.

Although the complete mechanism of action of many *Baccharis* metabolites against cancer still needs to be elucidated, scientific data have indicated that *B. dracunculifolia*-derived compounds, including Brazilian green propolis, present important characteristics such as (i) the inhibition of tumorigenic signaling pathways that drive cancer cells proliferation, migration, and invasion; (ii) the recruitment of specific anticancer immune responses, instead of an inflammatory immune response, and iii) the impairing of tumor microenvironment, avoiding cancer progression (Fig. 18.14; see also Patel 2016). Therefore, considering the scientific evidence regarding apigenin, quercetin, baccharin, and especially artepillin C mechanisms of action against tumor cells, and their role in preventing cancer development and progression, these bioactive molecules are the most promising antineoplastic candidates among *Baccharis* secondary metabolites and propolis compounds as well.

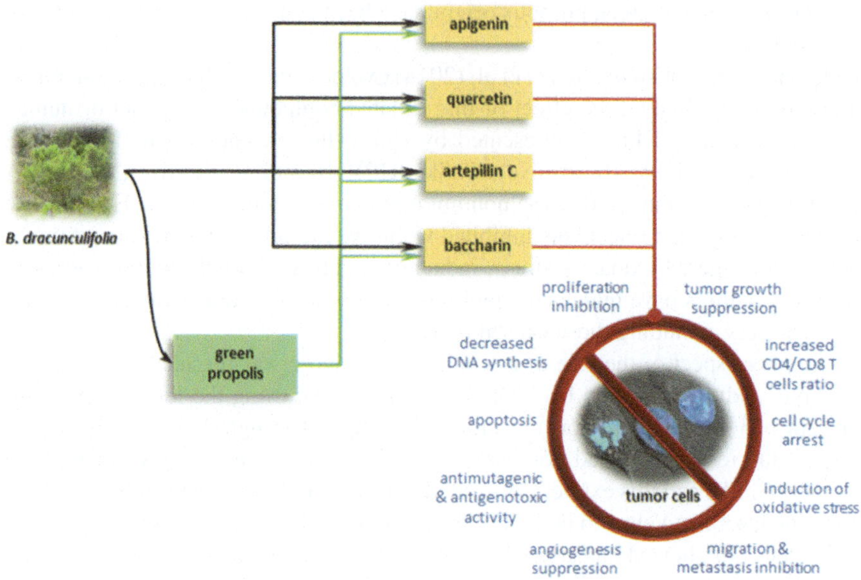

Fig. 18.14 The potential of *B. dracunculifolia* secondary metabolites as anticancer candidates

9 Final Considerations

In the course of the biotechnological roadmap of natural drug discovery, some issues need to be carefully undertaken. First, to consider that long-term studies are needed to go from the ecosystem or biodiversity to drug development. Taking the route of paclitaxel as an example, from the first scientific evidence as an anticancer agent to drug approval, a few more years of research will be necessary to completely elucidate the pathways by which *Baccharis* metabolites exert their biological role against cancer. Therefore, solid scientific observations of the potential chemotherapeutic use of plant species must be strongly supported.

Additionally, metabolite composition can be influenced by genetic, epigenetic, and environmental factors (as discussed by Yang et al. 2018), hence potentially affecting the efficiency of extraction of bioactive compounds from natural sources. Moreover, standard protocols are needed to properly isolate and characterize these bioactive compounds, to better elucidate their biological mechanisms of action, pharmacological applications, side effects, administration routes, formulations, and dosage. As pointed out by Patel (2016), standardized quality controls and properly designed clinical trials are also imperative to allow that *Baccharis* compounds or derivatives are accepted as novel strategies for cancer treatment.

It is equally relevant to consider the economic, environmental, and social viability of exploring natural sources, versus the feasibility of using biotechnological or chemical routes to scale up the production of bioactive molecules present on *Baccharis* metabolome and on propolis composition, such as apigenin, quercetin, baccharin, and artepillin C. Considering the bioavailability of propolis to ingestion, its biosafety acceptation, and low price, as observed by Watanabe et al. (2011) and Patel (2016), the more indicated strategy seems to be optimizing an industrial process to extract these phytochemicals directly from propolis sources.

Finally, but not the least, basic research is fundamental for drug discovery, and current trends of inadequate or decreasing investments must be reverted at all costs. Investments in basic research by the public sector create essential human and intellectual capital and enrich society in unexpected ways, including new treatments and technologies that generate new industries, elevating the global standard of living (GSAS 2020). The expansion of fundamental knowledge about *Baccharis* will result in enormous contributions to solve these global challenges to human health. Most of the *Baccharis* species are solely found in Latin American countries, where financial resources in science have been cut to insufficient levels (e.g., Angelo 2017; Fernandes et al. 2017b; Oliveira et al. 2020). As such, restoration of basic research funding and building a strong, stable, and long-term program for research on biodiversity-derived products represent a wise strategy to elevate the standard of living of the global human population (Joly and Speglich 2003; Joly et al. 2014; Boggio 2019). These challenges need to be met urgently, as the current devastation of the global biodiversity is under acceleration, and many unknown *Baccharis*, for example, *B. concinna*, are in the route of disappearing without being studied for their medicinal properties against cancer and other important worldwide diseases.

Having in mind these considerations, the search for more selective and effective anticancer drugs is justified by the worldwide growth of cancer incidence and mortality, the increase of chemotherapeutical resistance, and the specific lack of drug therapies. Considering this, *Baccharis* phytochemicals have proved to be valuable sources for new anticancer strategies due to the wide mechanisms of action that they trigger against tumor cells, despite being well tolerated by healthy cells, as demonstrated by in vitro and in vivo experimental studies. In addition, some of these bioactive compounds also have chemoprotective properties, opening other possibilities of usage, from primary prevention for healthy individuals to secondary prevention against premalignant lesions, as suggested by López-Romero et al. (2018).

Interestingly, many of the bioactive compounds indicated in this chapter are being isolated from different *Baccharis* species and from diverse *Baccharis*-related origins, such as green propolis, as well as by *Baccharis*-associated endophytic fungi (Fernandes et al. 2018). These multiple sources allow the development of different approaches to scale up the isolation of future drug principles, being eventually the starting point for the development of more effective and selective synthetic analogs for cancer chemotherapy.

As stressed by Alonso-Castro et al. (2011), more studies are essential to deeply understand the mechanisms of action of these bioactive phytochemicals responsible for interfering in cancer progression pathways, in order to move forward to clinical trials with selected anticancer candidates.

Acknowledgments The authors thank Gustavo Heiden and Daniel Montesinos for the images of *Baccharis* species and Elsevier Publisher for their kind permission to use the image of "The hallmarks of cancer" from the original article of Hanahan and Weinberg (2011), reproduced with modifications on Fig. 18.1. Authors also recognize the laboratory help provided by Mércia H Santos, Kahena S Lifsitch, Karina R Nunes, and Luciene AC da Silva of the Fundação Centro Tecnológico de Minas Gerais (CETEC), and Maria das Graças L Brandão, Daniel Negreiros, Carla FD Romeiro, and Lívia C Miranda of the Universidade Federal de Minas Gerais (UFMG). Finally, the authors thank CNPq, CAPES, Planta, and FAPEMIG (Brazil) for several grants.

References

Ahn MR, Kunimasa K, Ohta T, Kumazawa S, Kamihira M, Kaji K, Uto Y, Hori H, Nagasawa H, Nakayama T (2007) Suppression of tumor-induced angiogenesis by Brazilian propolis: major component artepillin C inhibits *in vitro* tube formation and endothelial cell proliferation. Cancer Lett 252:235–243

Akao Y, Maruyama H, Matsumoto K, Ohguchi K, Nishizawa K, Sakamoto T, Araki Y, Mishima S, Nozawa Y (2003) Cell growth inhibitory effect of cinnamic acid derivatives from propolis on human tumor cell lines. Biol Pharm Bull 26:1057–1059

Ali A, Bhattacharya S (2014) DNA binders in clinical trials and chemotherapy. Bioorg Med Chem 22:4506–4521

Alonso-Castro AJ, Villarreal ML, Salazar-Olivo LA, Gomez-Sanchez M, Dominguez F, Garcia-Carranca A (2011) Mexican medicinal plants used for cancer treatment: pharmacological, phytochemical and ethnobotanical studies. J Ethnopharmacol 133:945–972

Angelo C (2017) Scientists plead with Brazilian government to restore funding. Nature 550:166–167

Araujo GR, Rabelo ACS, Meira JS, Rossoni-Júnior JV, Castro-Borges W, Guerra-Sá R, Batista MA, Silveira-Lemos D, Souza GHB, Brandão GC (2017) *Baccharis trimera* inhibits reactive oxygen species production through PKC and down-regulation p47 phox phosphorylation of NADPH oxidase in SK Hep-1 cells. Exp Biol Med 242:333–343

Arisawa M, Kinghorn AD, Cordell GA, Phoebe CH, Farnsworth NR (1985) Plant anticancer agents. XXXVI. Schottenol glucoside from *Baccharis coridifolia* and *Ipomopsis aggregata*. Planta Med 6:544–545

Banskota AH, Tezuka Y, Prasain JK, Matsushige K, Saiki I, Kadota S (1998) Chemical constituents of Brazilian propolis and their cytotoxic activities. J Nat Prod 61:896–900

Baylin SB, Jones PA (2016) Epigenetic determinants of cancer. Cold Spring Harb Perspect Biol 8:a019505

Bhargava P, Grover A, Nigam N, Kaul A, Doi M, Ishida Y, Kakuta H, Kaul SC, Terao K, Wadhwa R (2018) Anticancer activity of the supercritical extract of Brazilian green propolis and its active component, artepillin C: bioinformatics and experimental analyses of its mechanisms of action. Int J Oncol 52:925–932

Bhattacharya S, Mondal L, Mukherjee B, Dutta L, Ehsan I, Debnath MC, Gaonkar RH, Pal MM, Majumdar S (2018) Apigenin loaded nanoparticle delayed development of hepatocellular carcinoma in rats. Nanomedicine 14:1905–1917

Bishayee K, Ghosh S, Mukherjee A, Sadhukhan R, Mondal J, Khuda-Bukhsh AR (2013) Quercetin induces cytochrome-c release and ROS accumulation to promote apoptosis and arrest the cell cycle in G2/M, in cervical carcinoma: signal cascade and drug-DNA interaction. Cell Prolif 46:153–163

Boggio PS (2019) Science and education are essential to Brazil's well-being. Nat Hum Behav 3:648–649

Bouvard V, Baan R, Straif K, Grosse Y, Secretan B, El Ghissassi F, Benbrahim-Tallaa L, Guha N, Freeman C, Galichet L, Cogliano V (2009) WHO International Agency for Research on

Cancer Monograph Working Group. A review of human carcinogens – part B: biological agents. Lancet Oncol 10:321–322

Brandão MGL, Cosenza GP, Stanislau AM, Fernandes GW (2010) Influence of Brazilian herbal regulations on the use and conservation of native medicinal plants. Environ Monit Assess 14:369–377

Bray F, Ferlay J, Soerjomataram I, Siegel RL, Torre LA, Jemal A (2018) Global cancer statistics 2018: GLOBOCAN estimates of incidence and mortality worldwide for 36 cancers in 185 countries. CA Cancer J Clin 68:394–424

Bufalo MC, Candeias JMG, Sforcin JM (2009) *In vitro* cytotoxic effect of Brazilian green propolis on human laryngeal epidermoid carcinoma (HEp-2) cells. Evid-Based Compl Alt Med 6:483–487

Bufalo MC, Candeias JM, Sousa JP, Bastos JK, Sforcin JM (2010) *In vitro* cytotoxic activity of *Baccharis dracunculifolia* and propolis against HEp-2 cells. Nat Prod Res 24:1710–1718

Cabrera J, Saavedra E, Del Rosario H, Perdomo J, Loro JF, Cifuente DA, Tonn CE, García C, Quintana J, Estévez F (2016) Gardenin B-induced cell death in human leukemia cells involves multiple caspases but is independent of the generation of reactive oxygen species. Chem Biol Interact 256:220–227

Carraz M, Lavergne C, Jullian V, Wright M, Gairin JE, Gonzales de la Cruz M, Bourdy G (2015) Antiproliferative activity and phenotypic modification induced by selected Peruvian medicinal plants on human hepatocellular carcinoma Hep3B cells. J Ethnopharmacol 166:185–199

Cheng X, Hochlowski J, Tang H, Hepp D, Beckner C, Kantor S, Schmitt R (2003) Studies on repository compound stability in DMSO under various conditions. J Biomol Screen 8:292–304

Chang JH, Cheng CW, Yang YC, Chen WS, Hung WY, Chow JM, Chen PS, Hsiao M, Lee WJ, Chien MH (2018) Downregulating CD26/DPPIV by apigenin modulates the interplay between Akt and Snail/Slug signaling to restrain metastasis of lung cancer with multiple EGFR statuses. J Exp Clin Cancer Res 37:199

Chatterjee N, Walker GC (2017) Mechanisms of DNA damage, repair, and mutagenesis. Environ Mol Mutagen 58:235–263

Chomel M, Guittonny-Larcheveque M, Fernandez C, Gallet C, Des Rochers A, Pare D, Jackson BG, Baldy V (2016) Plant secondary metabolites: a key driver of litter decomposition and soil nutrient cycling. J Ecol 104:1527–1541

Choudhari AS, Mandave PC, Deshpande M, Ranjekar P, Prakash O (2020) Phytochemicals in cancer treatment: from preclinical studies to clinical practice. Front Pharmacol 10:1614

Collura A, Lefevre JH, Svrcek M, Tougeron D, Zaanan A, Duval A (2019) Instabilité des microsatellites et cancer – De l'instabilité du génome à la médecine personnalisée. Med Sci 35:535–543

Conforti F, Pala L, Bagnardi V, De Pas T, Martinetti M, Viale G, Gelber RD, Goldhirsch A (2018) Cancer immunotherapy efficacy and patients' sex: a systematic review and meta-analysis. Lancet Oncol 19:737–746

Cordero CP, Gómez-González S, León-Acosta CJ, Morantes-Medina SJ, Aristizabal FA (2004) Cytotoxic activity of five compounds isolated from Colombian plants. Fitoterapia 75:225–227

Cragg GM (1998) Paclitaxel (Taxol): a success story with valuable lessons for natural product drug discovery and development. Med Res Rev 18:315–331

De Oliveira CB, Comunello LN, Lunardelli A, Amaral RH, Pires MG, da Silva GL, Manfredini V, Vargas CR, Gnoatto SC, de Oliveira JR, Gosmann G (2012) Phenolic enriched extract of *Baccharis trimera* presents anti-inflammatory and antioxidant activities. Molecules 17:1113–1123

De Oliveira PF, Lima IM, Monteiro Neto M, Bastos JK, da Silva Filho AA, Tavares DC (2013a) Evaluation of genotoxicity and antigenotoxicity of artepillin C in V79 cells by the comet and micronucleus assays. Nutr Cancer 65:1098–1103

De Oliveira CB, Comunello LN, Maciel ES, Giubel SR, Bruno AN, Chiela EC, Lenz G, Gnoatto SC, Buffon A, Gosmann G (2013b) The inhibitory effects of phenolic and terpenoid compounds from *Baccharis trimera* in Siha cells: differences in their activity and mechanism of action. Molecules 18:11022–11032

De Oliveira PF, de Souza Lima IM, Munari CC, Bastos JK, da Silva Filho AA, Tavares DC (2014) Comparative evaluation of antiproliferative effects of Brazilian green propolis, its main source *Baccharis dracunculifolia*, and their major constituents artepillin C and baccharin. Planta Med 80:490–492

De Siqueira FC, Leitão DDSTC, Mercadante AZ, Chisté RC, Lopes AS (2019) Profile of phenolic compounds and carotenoids of *Arrabidaea chica* leaves and the *in vitro* singlet oxygen quenching capacity of their hydrophilic extract. Food Res Int 126:108597

DeVita VT, Rosenberg SA, Lawrence TS (2018) DeVita, Hellman, and Rosenberg's cancer: principles & practice of oncology, 11th edn

Di L, Kerns EH (2006) Biological assay challenges from compound solubility: strategies for bioassay optimization. Drug Discov Today 11:446–451

Dos Santos MS, da Silva J, Menezes APS, Barros FMC, Lemes MLB, Rossatto RR, Feistel C, Almeida ID, Grivicich I, Prado L, Picada JN, Ferraz ABF (2018) Biotoxicological analyses of trimeroside from *Baccharis trimera* using a battery of *in vitro* test systems. Oxidative Med Cell Longev 2018:7804135

Dutra LM, da Conceição Santos AD, Lourenço AVF, Nagata N, Heiden G, Campos FR, Barison A (2020) 1H HR-MAS NMR and chemometric methods for discrimination and classification of *Baccharis* (Asteraceae): A proposal for quality control of *Baccharis trimera*. J Pharmac Biomed Anal 184:113200

El-Sayed ER, Ahmed AS, Hassan IA et al (2020) Semi-continuous production of the anticancer drug taxol by *Aspergillus fumigatus* and *Alternaria tenuissima* immobilized in calcium alginate beads. Bioprocess Biosyst Eng 43:997–1008

Espírito-Santo MM, Neves FS, Fernandes GW, Silva JO (2012) Plant phenology and absence of sex-biased gall attack on three species of *Baccharis*. PLoS One 7:e46896

Fares J, Fares MY, Khachfe HH, Salhab HA, Fares Y (2020) Molecular principles of metastasis: a hallmark of cancer revisited. Signal Transduct Target Ther 5:1–17

Fernandes GW, Duarte HM, Silveira FA, Broetto F, Lüttge U, Rennenberg H (2017a) Effects of sex and altitude on nutrient, and carbon and nitrogen stable isotope composition of the endangered shrub *Baccharis concinna* GM Barroso (Asteraceae). Acta Bot Bras 31:229–240

Fernandes GW, Vale MM, Overbeck GE, Bustamante MM, Grelle CE, Bergallo HG, Magnusson WE, Akama A, Alves SS, Amorim A, Araújo J et al (2017b) Dismantling Brazil's science threatens global biodiversity heritage. Perspect Ecol Conserv 15:239–243

Fernandes GW, Oki Y, Belmiro MS, Resende FM, Junior AC, De Azevedo JL (2018) Multitrophic interactions among fungal endophytes, bees, and *Baccharis dracunculifolia*: resin tapering for propolis production leads to endophyte infection. Arthropod Plant Interact 12:329–337

Fukuda M, Ohkoshi E, Makino M, Fujimoto Y (2006) Studies on the constituents of the leaves of *Baccharis dracunculifolia* (Asteraceae) and their cytotoxic activity. Chem Pharm Bull 54:1465–1468

Fullas F, Hussain RA, Chai HB, Pezzuto JM, Soejarto DD, Kinghorn AD (1994) Citotoxic constituents of *Baccharis gaudichaudiana*. J Nat Prod 57:801–807

Garro HA, Bruna-Haupt E, Cianchino V, Malizia F, Favier S, Menacho-Márquez M, Cifuente D, Fernández CO, Pungitore CR (2020) Verbascoside, synthetic derivatives and other glycosides from Argentinian native plant species as potential antitumoral agents. Nat Prod Res 2020:1–6

Gomes V, Collevatti RG, Silveira FAO, Fernandes GW (2004) The distribution of genetic variability in *Baccharis concinna* (Asteraceae), an endemic, dioecious and threatened shrub of rupestrian fields of Brazil. Conserv Genet 5:157–165

Gómez J, Simirgiotis MJ, Lima B, Paredes JD, Villegas Gabutti CM, Gamarra-Luques C, Bórquez J, Luna L, Wendel GH, Maria AO, Feresin GE, Tapia A (2019) Antioxidant, gastroprotective, cytotoxic activities and UHPLC PDA-Q Orbitrap mass spectrometry identification of metabolites in *Baccharis grisebachii* decoction. Molecules 24:1–19

González ML, Joray MB, Laiolo J, Crespo MI, Palacios SM, Ruiz GM, Carpinella MC (2018) Cytotoxic activity of extracts from plants of central Argentina on sensitive and multidrug-resistant leukemia cells: isolation of an active principle from *Gaillardia megapotamica*. Evid-Based Compl Alt Med 2018:9185935

GSAS, G-Science Academies Statement (2020) Basic research. https://www.academie-sciences.fr/pdf/rapport/20200528_GScience_Basic_Research.pdf. Accessed 12 Aug 2020

Haghiac M, Walle T (2005) Quercetin induces necrosis and apoptosis in SCC-9 oral cancer cells. Nutr Cancer 53:220–231

Hanahan D, Weinberg RA (2011) Hallmarks of cancer: the next generation. Cell 144:646–674

Harborne JB (1991a) The chemical basis of plant defense. In: Palo RT, Robbins CT (eds) Plant defenses against mammalian herbivory. CRC Press, Boca Raton, pp 45–59

Harborne JB (1991b) Flavonoid pigments. In: Rosenthal GA, Berenbaum MR (eds) Herbivores: their interactions with secondary plant metabolites v.1, 2 edn. Academic Press Inc, pp 389–427

Harborne JB (1999) Classes and functions of secondary products from plants. In: Walton NJ, Brown DE (eds) Chemicals from plants – perspectives on secondary compounds. Imperial College Press, London, pp 1–25

Harder B, Tian W, La Clair JJ, Tan AC, Ooi A, Chapman E, Zhang DD (2017) Brusatol overcomes chemoresistance through inhibition of protein translation. Mol Carcinog 56:1493–1500

Herrera CM, Pellmyr O (eds) (2009) Plant animal interactions: an evolutionary approach. Wiley

Hu WJ, Liu J, Zhong LK, Wang J (2018) Apigenin enhances the antitumor effects of cetuximab in nasopharyngeal carcinoma by inhibiting EGFR signaling. Biomed Pharmacother 102:681–688

INCA, Instituto Nacional do Câncer José Alencar Gomes da Silva (2018) Coordenação de prevenção e vigilância. Estimativa 2018: Incidência do Câncer no Brasil. Rio de Janeiro, Ministério da Saúde. http://www.inca.gov.br/estimativa/2018/casos-taxas-brasil. Accessed 01 Apr 2020

Isah T (2019) Stress and defense responses in plant secondary metabolites production. Biol Res 52:39

Jackson SE, Chester JD (2015) Personalized cancer medicine. Int J Cancer 137:262–266

Joly CA, Speglich E (2003) Programa Biota/Fapesp: um novo paradigma no estudo da conservação e do uso sustentável da biodiversidade. Ciên Cult 55:41–43

Joly CA, Metzger JP, Tabarelli M (2014) Experiences from the Brazilian Atlantic Forest: ecological findings and conservation initiatives. New Phytol 204:459–473

Kabała-Dzik A, Rzepecka-Stojko A, Kubina R, Iriti M, Wojtyczka RD, Buszman E, Stojko J (2018) Flavonoids, bioactive components of propolis, exhibit cytotoxic activity and induce cell cycle arrest and apoptosis in human breast cancer cells MDA-MB-231 and MCF-7 – a comparative study. Cell Mol Biol 64:1–10

Kimoto T, Arai S, Kohguchi M, Aga M, Nomura Y, Micallef MJ, Kurimoto M, Mito K (1998) Apoptosis and suppression of tumor growth by artepillin C extracted from Brazilian propolis. Cancer Detect Prev 22:506–515

Kimoto T, Koya S, Hino K, Yamamoto Y, Nomura Y, Micallef MJ, Hanaya T, Arai S, Ikeda M, Kurimoto M (2000) Renal carcinogenesis induced by ferric nitrilotriacetate in mice, and protection from it by Brazilian propolis and artepillin C. Pathol Int 50:679–689

Kimoto T, Koya-Miyata S, Hino K, Micallef MJ, Hanaya T, Arai S, Ikeda M, Kurimoto M (2001a) Pulmonary carcinogenesis induced by ferric nitrilotriacetate in mice and protection from it by Brazilian propolis and artepillin C. Virchows Arch 438:259–270

Kimoto T, Aga M, Hino K, Koya-Miyata S, Yamamoto Y, Micallef MJ, Hanaya T, Arai S, Ikeda M, Kurimoto M (2001b) Apoptosis of human leukemia cells induced by artepillin C, an active ingredient of Brazilian propolis. Anticancer Res 21:221–228

Kobal MB, Pazin WM, Bistaffa MJ, Constantino CJL, Toledo KA, Aoki PHB (2020) Correlating artepillin C cytotoxic activity on HEp-2 cells with bioinspired systems of plasma membranes. Mater Sci Eng C Mater Biol Appl 112:110943

Koeffler HP, Mccormick F, Denny C (1991) Molecular mechanisms of cancer. West J Med 155:505–514

Kumazawa S, Yoneda M, Shibata I, Kanaeda T, Hamasaka T, Nakayama T (2003) Direct evidence for the plant origin of Brazilian propolis by the observation of honeybee behavior and phytochemical analysis. Chem Pharm Bull 51:740–742

Kunjachan S, Rychlik B, Storm G, Kiessling F, Lammers T (2013) Multidrug resistance: physiological principles and nanomedical solutions. Adv Drug Deliv Rev 65:1852–1865

Kupchan SM, Jarvis BB, Dailey RG Jr, Bright W, Bryan RF, Shizuri Y (1976) Baccharin, a novel potent antileukemic trichothecene triepoxide from *Baccharis megapotamica*. J Am Chem Soc 98:7092–7093

Kupchan SM, Streelman DR, Jarvis BB, Dailey RG Jr, Sneden AT (1977) Isolation of potent new antileukemic trichothecenes from *Baccharis megapotamica*. J Organomet Chem 42:4221–4225

Lee YT, Tan YJ, Oon CE (2018) Molecular targeted therapy: treating cancer with specificity. Eur J Pharmacol 834:188–196

Lentz R, Benson AB 3rd, Kircher S (2019) Financial toxicity in cancer care: Prevalence, causes, consequences, and reduction strategies. J Surg Oncol 120:85–92

Li H, Kapur A, Yang JX, Srivastava S, McLeod DG, Paredes-Guzman JF, Daugsch A, Park YK, Rhim JS (2007) Antiproliferation of human prostate cancer cells by ethanolic extracts of Brazilian propolis and its botanical origin. Int J Oncol 31:601–606

Lichota A, Gwozdzinski K (2018) Anticancer activity of natural compounds from plant and marine environment. Int J Mol Sci 19:3533

Liu LZ, Fang J, Zhou Q, Hu X, Shi X, Jiang BH (2005) Apigenin inhibits expression of vascular endothelial growth factor and angiogenesis in human lung cancer cells: implication of chemoprevention of lung cancer. Mol Pharmacol 68:635–643

López-Romero D, Izquierdo-Vega JA, Morales-González JA, Madrigal-Bujaidar E, Chamorro-Cevallos G, Sánchez-Gutiérrez M, Betanzos-Cabrera G, Alvarez-Gonzalez I, Morales-González Á, Madrigal-Santillán E (2018) Evidence of some natural products with antigenotoxic effects. Part 2: plants, vegetables, and natural resin. Nutrients 10:1954

Madunic J, Madunic IV, Gajski G, Popic J, Garaj-Vrhovac V (2018) Apigenin: a dietary flavonoid with diverse anticancer properties. Cancer Lett 413:11–22

Marques AR, Fernandes GW, Reis IA, Assunção RM (2002) Distribution of adult male and female *Baccharis concinna* (Asteraceae) in the rupestrian fields of Serra Do Cipó, Brazil. Plant Biol 4:94–103

Matsuno T, Jung SK, Matsumoto Y, Saito M, Morikawa J (1997) Preferential cytotoxicity to tumor cells of 3, 5-diprenyl-4-hydroxycinnamicacid (artepillin C) isolated from propolis. Anticancer Res 17:3565–3568

Mishima S, Ono Y, Araki Y, Akao Y, Nozawa Y (2005) Two related cinnamic acid derivatives from Brazilian honeybee propolis, baccharin and drupanin, induce growth inhibition in allografted sarcoma S-180 in mice. Biol Pharm Bull 28:1025–1030

Mongelli E, Desmarchelier C, Rodríguez Talou J, Coussio J, Ciccia G (1997) *In vitro* antioxidant and cytotoxic activity of extracts of *Baccharis coridifolia* DC. J Ethnopharmacol 58:157–163

Monks NR, Ferraz A, Bordignon S, Machado KR, Lima MFS, Rocha AB, Schwartsmann G (2002) *In vitro* cytotoxicity of extracts from Brazilian Asteraceae. Pharm Biol 40:494–500

Moro IJ (2016) Avaliação da atividade citotóxica e tipo de morte em linhagens celulares humanas tratadas com derivados vegetais e metabólitos secundários de *Baccharis trimera* (Less.) DC. Dissertation, UNESP

Mosmann T (1983) Rapid colorimetric assay for cellular growth and survival: application to proliferation and cytotoxicity assays. J Immunol Methods 65:55–63

Munari CC, Resende FA, Alves JM, Sousa JPB, Bastos JK, Tavares DC (2008) Mutagenicity and antimutagenicity of *Baccharis dracunculifolia* extract in chromosomal aberration assays in Chinese hamster ovary cells. Planta Med 74:1363–1367

Munari CC, Furtado RA, Santiago ML, Manhas SS, Bastos JK, Tavares DC (2014) Inhibitory effects of *Baccharis dracunculifolia* on 1,2-dimethylhidrazine-induced genotoxicity and preneoplastic lesions in rat colon. Eur J Cancer Prev 23:240–245

Nakanishi I, Uto Y, Ohkubo K, Miyazaki K, Yakumaru H, Urano S, Okuda H, Ueda J, Ozawa T, Fukuhara K, Fukuzumi S, Nagasawa H, Hori H, Ikota N (2003) Efficient radical scavenging ability of artepillin C, a major component of Brazilian propolis, and the mechanism. Org Biomol Chem 1:1452–1454

Newman DJ, Cragg GM (2020) Natural products as sources of new drugs over the nearly four decades from 01/1981 to 09/2019. J Nat Prod 83:770–803

Nobili S, Lippi D, Witort E, Donnini M, Bausi L, Mini E, Capaccioli S (2009) Natural compounds for cancer treatment and prevention. Pharmacol Res 59:365–378

Oliveira CB, Comunello LN, Maciel ES, Giubel SR, Bruno AN, Chiela ECF, Lenz G, Gnoatto SCB, Buffon A, Gosmann G (2013) The inhibitory effects of phenolic and terpenoid compounds from *Baccharis trimera* in SiHa cells: differences in their activity and mechanism of action. Molecules 18:11022–11032

Oliveira EA, Martelli Junior H, Silva ACSE, Martelli DRB, Oliveira MCL (2020) Crise de financiamento da ciência no Brasil e COVID-19: impacto deletério na produção científica. An Acad Bras Cienc 92:e20200700

Patel S (2016) Emerging adjuvant therapy for cancer: propolis and its constituents. J Dietary Suppl 13:245–268

Pereira CB, Kanunfre CC, Farago PV, Borsato DM, Budel JM, Noronha Sales BHL, Campesatto EA, Sartoratto A, Miguel MD, Miguel OG (2017) Cytotoxic mechanism of *Baccharis milleflora* (Less.) DC. essential oil. Toxicol in Vitro 42:214–221

Price PW, Lewinsohn TM, Fernandes GW, Benson WW (1991) Plant–animal interactions: evolutionary ecology in tropical and temperate regions. Wiley, New York

Prieto-Vila M, Takahashi RU, Usuba W, Kohama I, Ochiya T (2017) Drug resistance driven by cancer stem cells and their niche. Int J Mol Sci 18:2574

Priyadarsini RV, Murugan RS, Maitreyi S, Ramalingam K, Karunagaran D, Nagini S (2010) The flavonoid quercetin induces cell cycle arrest and mitochondria-mediated apoptosis in human cervical cancer (HeLa) cells through p53 induction and NF-κB inhibition. Eur J Pharmacol 649:84–91

Rather RA, Bhagat M (2020) Quercetin as an innovative therapeutic tool for cancer chemoprevention: molecular mechanisms and implications in human health. Cancer Med 9:9181–9192

Roberto MM, Matsumoto ST, Jamal CM, Malaspina O, Marin-Morales MA (2016) Evaluation of the genotoxicity/mutagenicity and antigenotoxicity/antimutagenicity induced by propolis and *Baccharis dracunculifolia*, by *in vitro* study with HTC cells. Toxicol in Vitro 33:9–15

Robey RW, Polgar O, Deeken J, To KW, Bates SE (2007) ABCG2: determining its relevance in clinical drug resistance. Cancer Metastasis Rev 26:39–57

Rodrigo G, Almanza GR, Cheng Y, Peng J, Hamann M, Duan RD, Akesson B (2010) Antiproliferative effects of curcuphenol, a sesquiterpene phenol. Fitoterapia 81:762–766

Rodrigues CR, Dias JH, de Mello RN, Richter MF, Picada JN, Ferraz AB (2009) Genotoxic and antigenotoxic properties of *Baccharis trimera* in mice. J Ethnopharmacol 125:97–101

Romero-Benavides JC, Ortega-Torres GC, Villacis J, Vivanco-Jaramillo SL, Galarza-Urgilés KI, Bailon-Moscoso N (2018) Phytochemical study and evaluation of the cytotoxic properties of methanolic extract from *Baccharis obtusifolia*. Int J Med Chem 2018:8908435

Roy DM, Walsh LA, Chan TA (2014) Driver mutations of cancer epigenomes. Protein Cell 5:265–296

Sabzehzari M, Zeinali M, Naghavi MR (2020) Alternative sources and metabolic engineering of Taxol: advances and future perspectives. Biotechnol Adv 43:107569

Salehi B, Stojanović-Radić Z, Matejić J, Sharifi-Rad M, Kumar NVA, Martins N, Sharifi-Rad J (2019) The therapeutic potential of curcumin: a review of clinical trials. Eur J Med Chem 163:527–545

Seca A, Pinto D (2018) Plant secondary metabolites as anticancer agents: successes in clinical trials and therapeutic application. Int J Mol Sci 19:263

Sergeev IN, Ho CT, Li S, Colby J, Dushenkov S (2007) Apoptosis-inducing activity of hydroxylated polymethoxyflavones and polymethoxyflavones from orange peel in human breast cancer cells. Mol Nutr Food Res 51:1478–1484

Shibue T, Weinberg RA (2017) EMT, CSCs, and drug resistance: the mechanistic link and clinical implications. Nat Rev Clin Oncol 14:611–629

Shukla S, Gupta S (2007) Apigenin-induced cell cycle arrest is mediated by modulation of MAPK, PI3K-Akt, and loss of cyclin D1 associated retinoblastoma dephosphorylation in human prostate cancer cells. Cell Cycle 6:1102–1114

Silva C, Perestrelo R, Silva P, Tomás H, Câmara JS (2019) Breast cancer metabolomics: from analytical platforms to multivariate data analysis. A review. Meta 9:102–133

Smith AG, Macleod KF (2019) Autophagy, cancer stem cells and drug resistance. J Pathol 247:708–718

Souza RP, Bonfim-Mendonça PS, Damke GMZF, de Assis Carvalho ARB, Ratti BA, Dembogurski DSO, da-Silva VRS, Silva SO, Da-Silva DB, Bruschi ML, Maria-Engler SS, Consolaro MEL (2018) Artepillin C induces selective oxidative stress and inhibits migration and invasion in a comprehensive panel of human cervical cancer cell lines. Anti Cancer Agents Med Chem 18:1750–1760

Starlin T, Raj AC, Ragavendran P, Gopalakrishnan VK (2012) Phytochemical screening, functional groups and element analysis of *Tylophora pauciflora* wight and arn. Int Res J Pharma 3:180–183

Szliszka E, Zydowicz G, Janoszka B, Dobosz C, Kowalczyk-Ziomek G, Krol W (2011) Ethanolic extract of Brazilian green propolis sensitizes prostate cancer cells to TRAIL-induced apoptosis. Int J Oncol 38:941–953

Szliszka E, Zydowicz G, Mizgala E, Krol W (2012) Artepillin C (3, 5-diprenyl-4-hydroxycinnamic acid) sensitizes LNCaP prostate cancer cells to TRAIL induced apoptosis. Int J Oncol 41:818–828

Tarqui Tarqui S, Flores Segura Y, Almanza Vega GR (2012) Polyoxygenated flavonoids from *Baccharis pentlandii*. Rev Boliv Quím 29:10–14

Thallinger C, Füreder T, Preusser M, Heller G, Müllauer L, Höller C, Prosch H, Frank N, Swierzewski R, Berger W, Jäger U, Zielinski C (2018) Review of cancer treatment with immune checkpoint inhibitors: current concepts, expectations, limitations and pitfalls. Wien Klin Wochenschr 130:85–91

Tomasetti C, Li L, Vogelstein B (2017) Stem cell divisions, somatic mutations, cancer etiology, and cancer prevention. Science 355:1330–1334

Ujiki MB, Ding XZ, Salabat MR, Bentrem DJ, Golkar L, Milam B, Talamonti MS, Bell RH Jr, Iwamura T, Adrian TE (2006) Apigenin inhibits pancreatic cancer cell proliferation through G2/M cell cycle arrest. Mol Cancer 29:1–8

Vargo MA, Voss OH, Poustka F, Cardounel AJ, Grotewold E, Doseff AI (2006) Apigenin-induced-apoptosis is mediated by the activation of PKCdelta and caspases in leukemia cells. Biochem Pharmacol 72:681–692

Wall ME, Wani MC (1996) Camptothecin and taxol: from discovery to clinic. J Ethnopharmacol 51:239–254

Wang JJ, Lei KF, Han F (2018) Tumor microenvironment: recent advances in various cancer treatments. Eur Rev Med Pharmacol Sci 22:3855–3864

Watanabe MA, Amarante MK, Conti BJ, Sforcin JM (2011) Cytotoxic constituents of propolis inducing anticancer effects: a review. J Pharm Pharmacol 63:1378–1386

WHO, World Health Organization (2019) Global report on traditional and complementary medicine 2019. https://www.who.int/publications/i/item/978924151536. Accessed 12 Aug 2020

WHO, World Health Organization (2020) Cancer. https://www.who.int/en/news-room/fact-sheets/detail/cancer. Access 12 Aug 2020

Wild CP (2019) The global cancer burden: necessity is the mother of prevention. Nat Rev Cancer 19:123–124

Wu Q, Wang X, Nepovimova E, Miron A, Liu Q, Wang Y, Su D, Yang H, Kuca K (2017) Trichothecenes: immunomodulatory effects, mechanisms, and anti-cancer potential. Arch Toxicol 91:3737–3785

Yan X-B, Xie T, Wang SD, Wang Z, Li HY, Ye ZM (2018) Apigenin inhibits proliferation of human chondrosarcoma cells via cell cycle arrest and mitochondrial apoptosis induced by ROS generation – an *in vitro* and *in vivo* study. Int J Clin Exp Med 11:1615–1631

Yang L, Wen K-S, Ruan X, Zhao Y-X, Wei F, Wang Q (2018) Response of plant secondary metabolites to environmental factors. Molecules 23:1–26

Ye VM, Bhatia SK (2012) Metabolic engineering for the production of clinically important molecules: Omega-3 fatty acids, artemisinin, and taxol. J Biotechnol 7:20–33

Youn A, Simon R (2011) Identifying cancer driver genes in tumor genome sequencing studies. Bioinformatics 27:175–181

Zhang Q, Li L (2018) Photodynamic combinational therapy in cancer treatment. J BUON 23:561–567

Zhao J (2016) Cancer stem cells and chemoresistance: the smartest survives the raid. Pharmacol Ther 160:145–158

Chapter 19
Innovation and Knowledge of Prospective Studies on the Genus *Baccharis*

Mariana Bonifácio Amancio, Yumi Oki, G. Wilson Fernandes, Carla Rago, Aristóteles Góes-Neto, and Vasco Ariston de Carvalho Azevedo

Abstract *Baccharis* species are well known for their secondary metabolites used in the treatment of several diseases and for having great potential of use in various industries. In order to identify the current status of innovation in which species of this genus are found, an assessment on scientific and technological productions was carried out, as well as on the collaboration of those involved in the generation of scientific knowledge and/or in the production of products and processes. This analysis aimed to measure the contributions of science and technology to society. As of December 2019, about 991 articles and 223 patents on the genus *Baccharis* were registered. Brazil, which harbors a great diversity of species of this genus (178 species), stands out in the number of published scientific articles (45.22%). Moreover, the United States with 21 native *Baccharis* species excels as the country with the highest number of patents for products and processes of this genus (48.02%). Nonetheless, it is worth mentioning that the majority of patents are not for *Baccharis* species from this country. When analyzing the species found in patents and scientific articles, the presence of 21 species of common interest was observed, mostly associated with the biological activity of their secondary metabolites. *Baccharis* patents are focused on the production of drugs for various medicinal applications. Although some scientific articles also deal with this topic, the authors focus on the chemical analysis of different species of *Baccharis*. Among the 401 inventors and 100 authors, only four turned out to be both patent filers and authors of scientific articles. These findings indicate a great difference between the production of articles and patents on *Baccharis*, which consequently, may represent a loss in economic development and performance, and international competitiveness.

Keywords Bioeconomy · Science · Economy · Pharmacology · Innovation · Patents

M. B. Amancio · Y. Oki · G. W. Fernandes (✉) · C. Rago · V. A. de Carvalho Azevedo (✉)
Departamento de Genética, Ecologia & Evolução, Instituto de Ciências Biológicas, Universidade Federal de Minas Gerais, Belo Horizonte, Minas Gerais, Brazil

A. Góes-Neto
Departamento de Microbiologia, Instituto de Ciências Biológicas, Universidade Federal de Minas Gerais, Belo Horizonte, Minas Gerais, Brazil

G. W. Fernandes et al. (eds.), *Baccharis*,
https://doi.org/10.1007/978-3-030-83511-8_19

1 Introduction

Over the last few years, the interest in science-technology relations has increased due to the relevance of knowledge and innovation for economic development and social benefits (Looy et al. 2006). It is an approach worth highlighting, especially in periods of economic recession and questioning of public expenses (Meyer 2000). Works such as those by Narin (1995) and Narin et al. (1997), in which analyses of patents and research activities (articles) were performed, underlined an increasing relationship between science and technology, especially in high-tech areas in the United States. Although technological production is associated with knowledge, this relationship is not always direct and clear. It can vary with the degree of science, and social interest and economic applicability.

Baccharis is one of the genera that aggregates a diversity of economic applications, and, among them, the control of soil erosion (Martinez et al. 2015; Vessella et al. 2015; Woolsey et al. 2018), ornamental use (Tognon and Cuquel 2016), food supplement (Navares et al. 2010), raw material for the development of cosmetics and pharmaceutical products (Sousa et al. 2011; Mejía-Giraldo et al. 2016; Casagrande et al. 2018), source of raw material for bees in propolis production (Souza et al. 2019), and products of botanical origin. Such prominence in the area of natural products is due to its diverse pharmacological properties in the treatment of several diseases such as ulcers, gastrointestinal diseases, diabetes, cancer, inflammations, and bacterial-fungal infections (e.g., Corrêa 1984; Carneiro and Fernandes 1996; Claudino 2013; Silva et al. 2019). These uses have favored studies on the propagation of some species of *Baccharis*. Furthermore, all of these attributions provided by *Baccharis* fall within the paradigm of bioeconomics, which describes the way in which a set of economic activities obtains latent value from biological processes and renewable bioresources to develop and promote better health, growth, and sustainability conditions (Vivien et al. 2019).

Baccharis' scientific knowledge and potential for innovation are high. In this research, we verify the existence of more than 900 published articles and 200 patent applications, as we shall see below. On the other hand, tropical countries such as Brazil, which exhibit a great diversity of species of this genus, do not always present a parallel relationship with scientific production nor are associated with the generation of patents. The challenges of many of these countries involve the ease and success of filing patents (França and Vasconcelos 2018) since the current relationship between science and technology may also help to better diagnose problems and to propose solutions that favor economic development.

2 Scope and the Database

Currently, prospective studies have shown to be a way of stimulus for organizations of innovation systems as it contemplates the interaction between science and technology. The publication of research results is seen as a direct representation of science itself in the form of scientific articles while technology is portrayed through the application of knowledge obtained from the creation of processes and products, represented by patents (Meyer and Batthacharya 2004). Thus, in order to seek comprehension of *Baccharis*' science-technology interaction, this chapter sought to analyze the profile of registered patents and articles on this genus available on virtual databases over the last six decades (1940–2019). To this end, the chapter was divided into five sections. The first section approaches an analysis of the profile of patents and articles on species of the genus *Baccharis*. In this section, we present the chronology, species of interest, and the origin of patents and articles. The second section addresses the processes, products, and international classification of patents that involves the genus. The third section discusses areas of interest and applications of patents and articles. The fourth section deals with the network of authors and inventors involved in patents and articles. Finally, the last section presents remarks on science-technology relations developed with *Baccharis* and their contributions to society.

The database used for patents was Google Patents, United States Patent and Trademark Office (USPTO), Espacenet (developed by the European Patent Office (EPO) together with member states of the European Patent Organization), Instituto Nacional da Propriedade Industrial (INPI), DISCUS, and World Intellectual Property Organization (WIPO). The data searches used for scientific productions was Web of Science.

3 Profile of Patents and Articles of the Genus *Baccharis*

Chronology of Publications and Patent Filings of Species of the Genus Baccharis

The first patent on *Baccharis* registered in the main available patent basis was in 1942. This first order was placed by the private company American District Telegraph Company and entitled "Radiant energy receiving system" (Lindsay and Pearson 1942). The purpose of this patent application was to create an antitheft system equipped with a sensor coupled to the alarm, and the main differential was the high capacity for protection regardless of weather conditions. One of the materials of interest described in this document is the fiber of *Baccharis* sp. (with no species identification), which is presented as one of the materials used in the sensor assembly once this fiber has a high adsorption capacity. Although the first patent

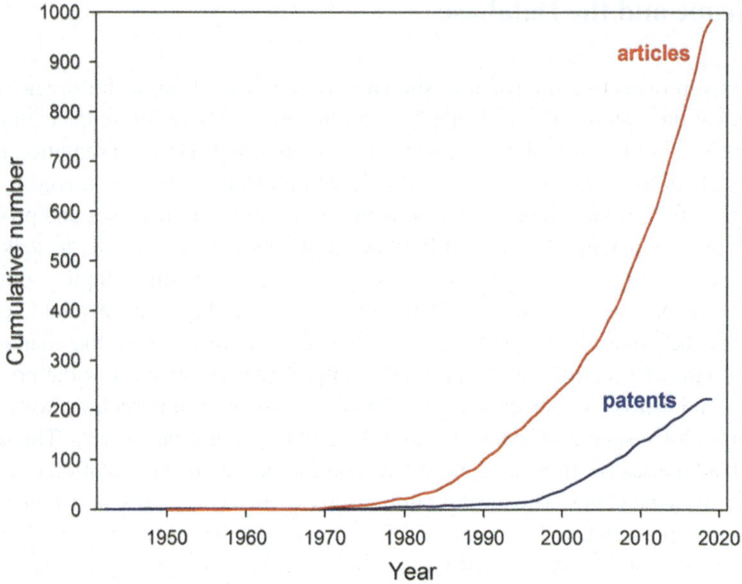

Fig. 19.1 Number of patent filings and publications of scientific articles on species of the genus *Baccharis* until 2019

filing focused on the use of *Baccharis* in the composition of a sensor, no other patent or article with the same purpose was developed afterward.

The first patent registrations occurred in a timid and inconsistent way during 1942 to 1995 (Fig. 19.1). Before that period, the patenting of pharmaceutical products was historically not permitted in many countries. Only in the late twentieth century, mandatory pharmaceutical patent protection was established for all members of the World Trade Organization (WTO) (Shadlen et al. 2019). Since then, registrations have increased rapidly, reaching a total of 223 registrations in 2019. The recent increase is so clear that in the last 10 evaluated years, almost half (98 patents) of *Baccharis* patents were registered.

The first article on *Baccharis* according to the Web of Science (1945–2019) was published in 1961 by Dolejs and collaborators (1961), with the title "Sesquiterpenic compounds of *Baccharis genistelloides* Pers – Structure of palustrol" in the journal *Collection of Czechoslovak Chemical Communications*. This publication analyzed secondary substances in the aerial part of *B. genistelloides*, focusing on the δ-cadinene sesquiterpenes. To date, this phytochemical interest in *Baccharis* species continues, particularly focused on the extraction and identification of chemical compounds (Bettucci et al. 2020).

Unlike patents, the publication of scientific articles on *Baccharis* only displayed a boost almost 10 years after the first scientific article had been published in 1961. Although the publication of articles started slightly later than patents, there has been a substantial growth reaching about 991 articles in 2019. Comparatively, the number of articles represents about 4.5 times more than the number of patents.

One of the factors that influence this variation between numbers of patents and articles is related to differences in the time of evaluation and concession between them. Most articles (67%) take 6 months to 2 years to finalize publication (Noorden 2016). However, even when an author needs to wait a long time for the article to be published, that time is much shorter than one required for a patent to be filed. Moreover, it is also important to highlight that article publication or even a lecture in any type of event before the secrecy period (1–5 years) can make patenting unfeasible. This condition imposes a choice that is not always feasible for the researcher who is being evaluated for his production of articles, and little for the number of patents.

Additionally, a patent application request does not mean that the patent will be granted (França and Vasconcelos 2018). In an analysis of the technological field of herbal medicines in Brazil, where there is the largest number of published articles, from a total of 876 analyzed patent applications, only 12 (1.3%) were granted (Vasconcellos et al. 2004). Therefore, the fact that the patent is granted does not mean that it will be licensed or even exploited in any way. This topic is complex, and the information gathered only outlines some of the many factors present in the context of the comparative and competitive relationship between the basic research of scientific articles and the generation of patent products and/or processes. The incentive to innovation based on basic research represents the entrepreneurial continuity of organizations, taking into account such important issues in the conduction of economic and social growth within companies and universities.

Currently, incentives and guidelines for the effective granting of patent deposits are not uniform across countries. Many developing countries have few incentives, and institutions that assist in the patenting process guarantee sooner success in patent grant (Moreira et al. 2016). Many research institutions validate various pharmacological data every day, but the lack of a culture of protection, or even the lack of knowledge about intellectual property, makes it impossible for these results to promote financial return (Dundas 2012; Fisch et al. 2015; Arqué-Castells et al. 2016).

Species of Interest for Patents and Articles

Baccharis is considered the largest genus that belongs to the Asteraceae family, covering formally 442 species (Fielding 2001; Giuliano 2001; Heiden and Pirani 2016). *Baccharis* has a wide geographic distribution occuring from Canada to the southern tip of Argentina and Chile (Barroso 1976; Fielding 2001; Verdi et al. 2005; Heiden et al. 2006). In patent filing applications, 43 species of this genus were identified. However, patent filing requests without identification at the species level reached 15.7%. In scientific articles, the number of investigated species was greater, 192 studied species (43.4% of the genus). Approximately 10% of published articles present only the genus, without species determination. Those of common interest

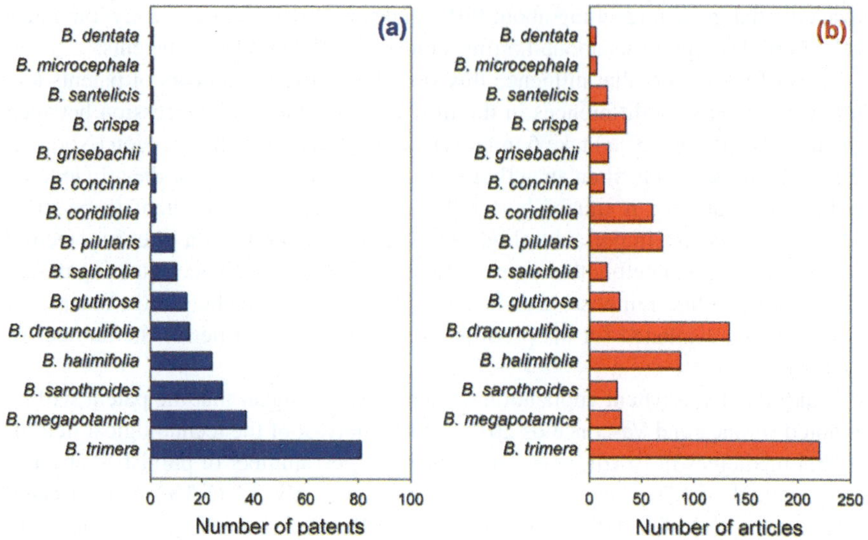

Fig. 19.2 Number of (**a**) patents and (**b**) scientific articles of the most cited species of the genus *Baccharis*

found in articles and patents are: *B. concinna, B. coridifolia, B. cutervensis, B. crispa, B. dentata, B. dracunculifolia, B. erioclada, B. glaziovii, B. glutinosa, B. grisebachii, B. halimifolia, B. megapotamica, B. microcephala, B. pilularis, B. pseudotenuifolia, B. salicifolia, B. santecilis, B. sarothroides, B. stenocephala, and B. trimera* (Fig. 19.2).

The most cited species of the most cited patents are *B. trimera* (36.3%), *B. megapotamica* (16.6%), and *B. sarothroides* (12.6%) (Fig. 19.2). *Baccharis trimera* is considered one of the most studied plant species in botanical, pharmacological, and chemical terms and is present in the list and described in the sixth Edition of the Brazilian Pharmacopoeia (Karam et al. 2013; ANVISA 2019). It has a wide distribution in Brazil, occurring in the Cerrado (Rodrigues and Carvalho 2001), Atlantic Forest (Pavan-Fruehauf 2000; Vieira and Silva 2002), and Pampa (Ritter and Baptista 2005; Caporal and Boldrini 2007) biomes, and can even be found in many ruderal environments (Brandão and Oliveira 1995; Macedo 1995; Vieira and Silva 2002), whereas *B. megapotamica* inhabits wetlands from Brazil and Argentina (Tokarnia et al. 1992; Carneiro and Irgang 2005). This shrub stands out for the presence of a series of potent cytotoxic agents that belong to the trichothecene antibiotics complex. In addition, trichothecenes have exceptionally high toxicity for eukaryotic organisms, including high phytotoxicity (Ueno 1983; Bamburg 1983; Snyder 1986). Among the analyzed chemical compounds, macrocyclic trichothecenes (roridins, verrucarins, and baccharinoids) proved to be among the most potent phytotoxic agents (Cutler and Jarvis 1985). The species *B. sarothroides* has its origin in the desert regions of the United States, and its notoriety is mainly focused on its concentration of flavonoids with proven cytotoxic activity such as centauredin and 3,4′-di-methoxy −3′,5,7-trihydroxy-flavone (Kupchan and Baverschmidt 1971).

The most cited species in articles are *B. trimera* (22.3%), *B. dracunculifolia* (13.5%), and *B. halimifolia* (8.87%). *Baccharis dracunculifolia* is a dioecious shrub whose distribution ranges from southeastern to southern Brazil, reaching Argentina, Uruguay, Paraguay, and Bolivia (Gomes and Fernandes 2002). This species is a facilitator of other species in Cerrado, as it houses a wide network of interactions (Fagundes et al. 2005; Neves et al. 2016; Barbosa et al. 2019; Monteiro et al. 2020; Rodrigues et al. 2020), especially with the Africanized bees *Apis mellifera* that collect their resins for propolis manufacturing (Fernandes et al. 2018; Rodrigues et al. 2020). It is also known for its pharmacological properties and for the treatment of several microbial and fungal diseases, ulcers, leishmania, schistosomiasis, and cancer, among others (e.g., Park et al. 2004; Silva-Filho et al. 2009; Parreira et al. 2010). The shrub species *B. halimifolia*, known for its complex network of interactions, is native to salt marshes and dunes along the Atlantic and North American Gulf Coast (Krischik and Denno 1990; Egerova et al. 2003). It is considered an invasive species in Australia and especially in Europe (Sims-Chilton et al. 2010; Caño et al. 2013, 2016). In the past hundred years, *B halimifolia* has invaded almost all of Europe's estuaries (Caño et al. 2016).

Origin of Patent Filings and Articles

The origin of *Baccharis*' patents is distributed in 53 countries, and the articles have been originated from 174 countries (Fig. 19.3a, b). Despite being a genus of origin in the Americas (Heiden and Pirani 2016), only 5.73% of patent deposits were generated in this region (Fig. 19.3). This number is significantly lower than the frequency of articles published in the same region, which was approximately 67.5%. These results also show that the potential to generate patents with *Baccharis* is not being explored by the countries of origin. Consequently, these countries are not being favored by the economic advantages that patents can bring such as market monopoly for a certain period of time (Moir 2008). This way, these countries through patents could have higher income and profitability indicators.

The country with the highest number of patents (95 patents, 48.02% of patents) for *Baccharis* was the United States (Fig. 19.3a). The country is home to 21 native species (Kupchn and Baverschmidt 1971), of which only six have patents (28.6% of native species in the United States): *B. dioica*, *B. emoryi*, *B. glutinosa*, *B. halimifolia*, *B. pilularis*, and *B. sarothroides*. It is noted; however, that about 50% of the patents in the United States correspond to two non-native species: *B. trimera* (26.3% of the United States patents) and *B. megapotamica* (24.2% of the patents) (Fig. 19.4). *B. trimera* patents are associated with the development of cosmetics (69.23% of patents), pharmacological products (23.08%), and chemical extraction (7.69%), while *B. megapotamica* patents are mostly directed to pharmacology (91.30%). The United States has been known for its mastery in the pharmacological field since the

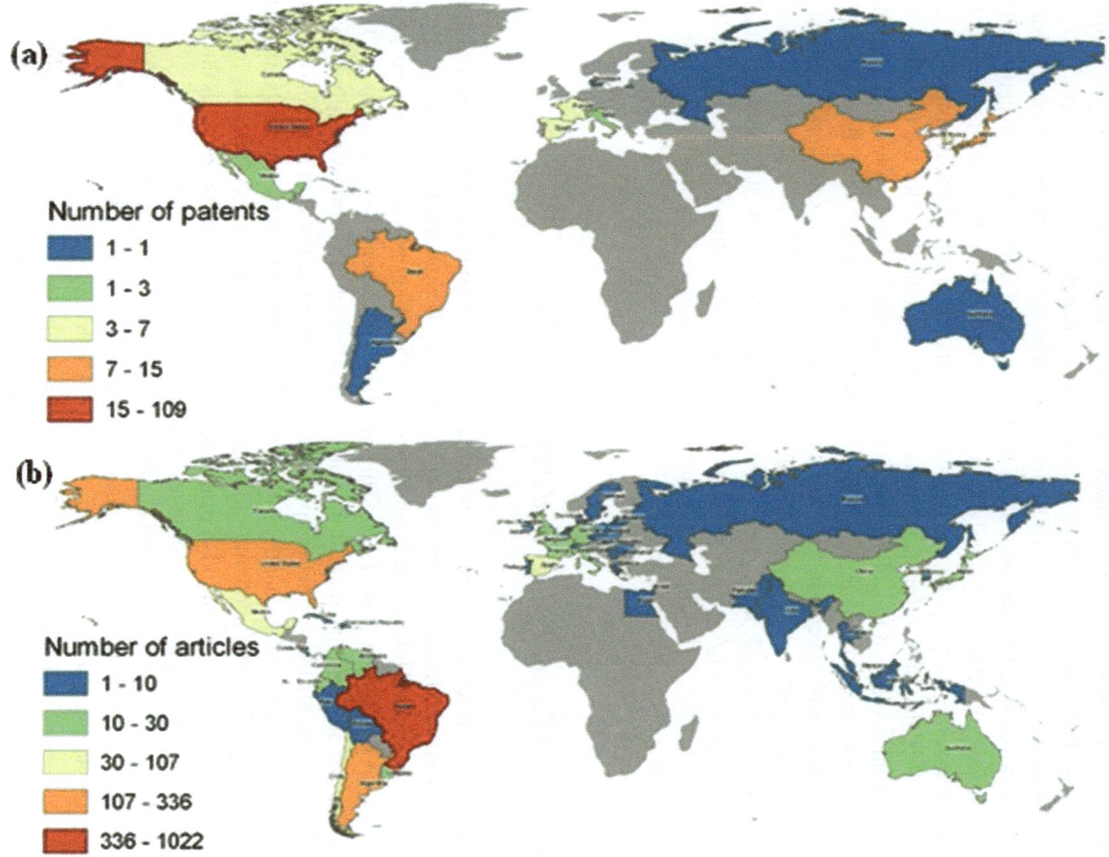

Fig. 19.3 Geographic distribution of the (**a**) number of filed patents and (**b**) published scientific articles related to the genus *Baccharis*

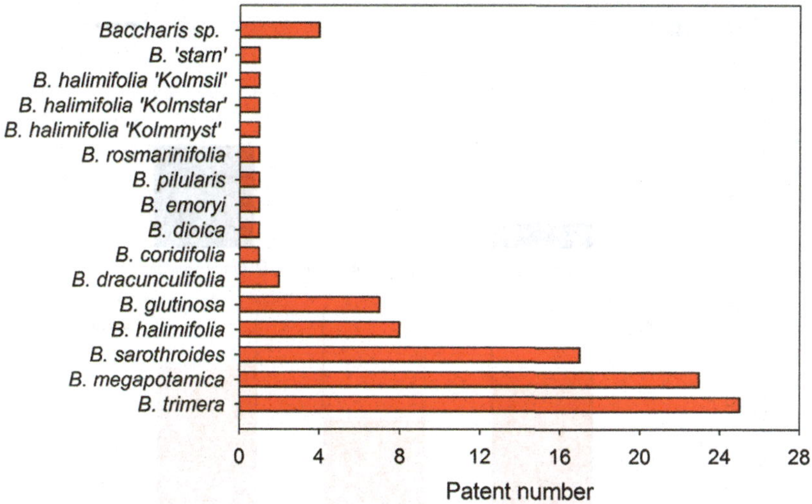

Fig. 19.4 Distribution of the number of patents in the United States (95 patents) per species of *Baccharis*. *Baccharis* sp.* – identification presented in the patent only at gender level

nineteenth century by means of pharmaceutical researches based on the extraction, purification, and isolation of natural products (Radaelli 2007).

The United States' leadership in the number of *Baccharis* patents may be associated with its oldest patent on record as of 1790 and its stronger and facilitative structure in patent generation (Khan 2005). A different profile of countries in South America, such as Brazil, which harbors 178 species of *Baccharis* (about 36% of *Baccharis* species; Heiden et al. 2019), leads the list of countries with the largest number of scientific publications on species of this genus with a frequency of 45.22% (448 articles) among published studies (Fig. 19.3b). Brazil has 12 patents for 30 species of *Baccharis* (30 species out of 175 registered species: 17% of the existing species in the country). Part of this difference may be associated with the country's patent history. Until 1996, pharmaceutical products were excluded from patent protection in the country (Nogués 1990). Currently, unfortunately, the vast majority of patent rights granted in Brazil belong to foreign business groups, especially from the United States (Marques 2000).

Additionally, an analysis on the patents' origin countries of the three most cited *Baccharis* species (*B. trimera*, *B.megapotamica*, *B. dracunculifolia*) indicates that over 50% of the patents were developed outside their countries of origin (Fig. 19.5). About 88% of *B. trimera* patents came from where it is not native: Canada, China, EPO, Spain, United States, France, Japan, United Kingdom, and WIPO (Fig. 19.6). While for *B. megapotamica*, no patent was found for its origin locations. About 53.9% of *B. dracunculifolia* patents were carried out outside their countries of origin: United States, EPO, Japan, and WIPO (Fig. 19.6). In contrast, the origin of studies developed for publication of articles on *B. trimera*, *B. megapotamica*, and *B. dracunculifolia* are, in their majority (over 50%), in the countries where they are native.

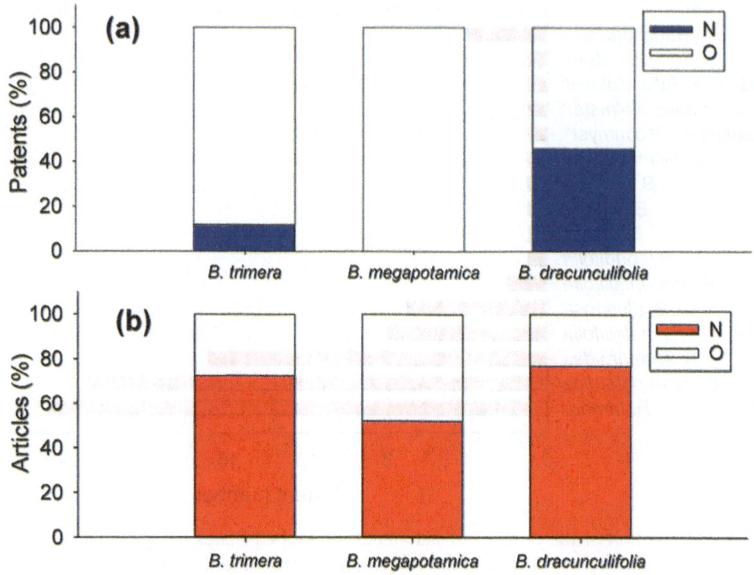

Fig. 19.5 Frequency of patents (**a**) and articles (**b**) found inside (N) and outside (O) their country of origin for *Baccharis trimera*, *Baccharis megapotamica*, and *Baccharis dracunculifolia*

Plant Parts Explored in Patents and Articles

Plants exhibit a wealth of chemical substances that can vary not only among species or plant gender but also according to the plant organ (Ferracini et al. 1995; Muller-Riebau et al. 1997; Perri et al. 1999; Vesela et al. 1999; Lago et al. 2008; Manurung et al. 2017; Feduraev et al. 2019). These chemical differences in composition and contents found among plant organs have been generally considered in studies and scientific innovations. Other environmental factors may influence plant production of chemical compounds such as seasonality, geographical distribution, and the level of stress that they are subjected to (Perri et al. 1999; Carvalho-Filho et al. 2006; Edreva et al. 2008). These factors, regardless of their importance, are hardly explored in patents but practically only in scientific publications.

In patents, pollen represents the part of the plant of greatest interest (37.3%), followed by the use of leaves (25.5%), aerial part as a whole (13.7%), and whole plant (11.8%) (Fig. 19.7). Other organs such as branches, roots, and cladodes are less explored in patents.

The majority of the patents of *Baccharis* pollen is associated with the health benefits it brings, such as nutritional supplementation (Cho et al. 2017), physical vigor, and immune-strengthening (Kroyer and Hegedus 2001). Pollen has been incorporated for the development of patents that involve immunotherapy of allergies and vaccines and administration of immunomodulatory compounds in order to increase immune response or sensitivity of patients with cancer or autoimmune diseases (Graça and Água-Doce 2010; Bartlett et al. 2014; Santos et al. 2014; Zhu

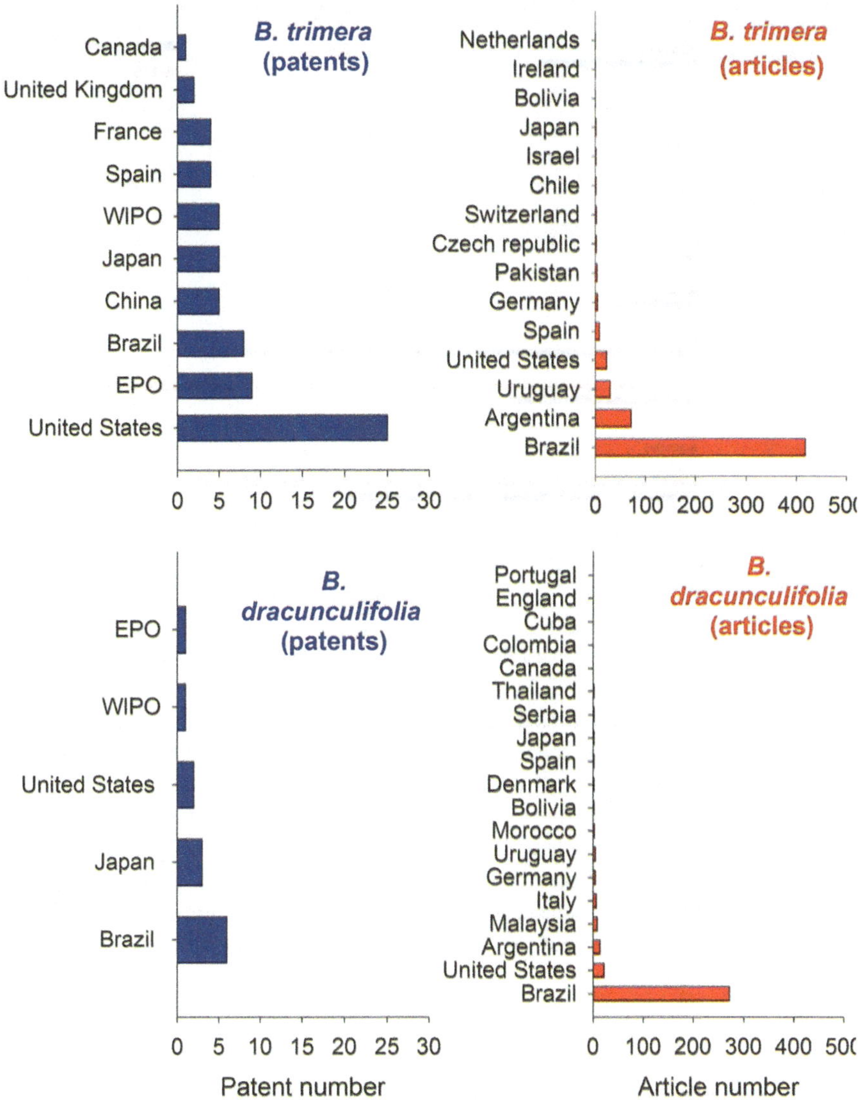

Fig. 19.6 Number of patents and published articles for the species *B. trimera* and *B. dracunculifolia*

2015; Fleisner 2016). Another highlight of the use of pollen in patents, for *B. sarothroides* in this matter, is related to the improvement of immune responses during the treatment of complications derived from transplant and graft rejection. Most of the patents that use pollen as a raw material are for *B. sarothroides* and *B. halimifolia*. *Baccharis halimifolia* is native to Nova Scotia, eastern and southern United States, eastern Mexico, Bahamas, and Cuba (Heering 1907; Hitchcock and Standley 1919; Sánchez and Cano 1983).

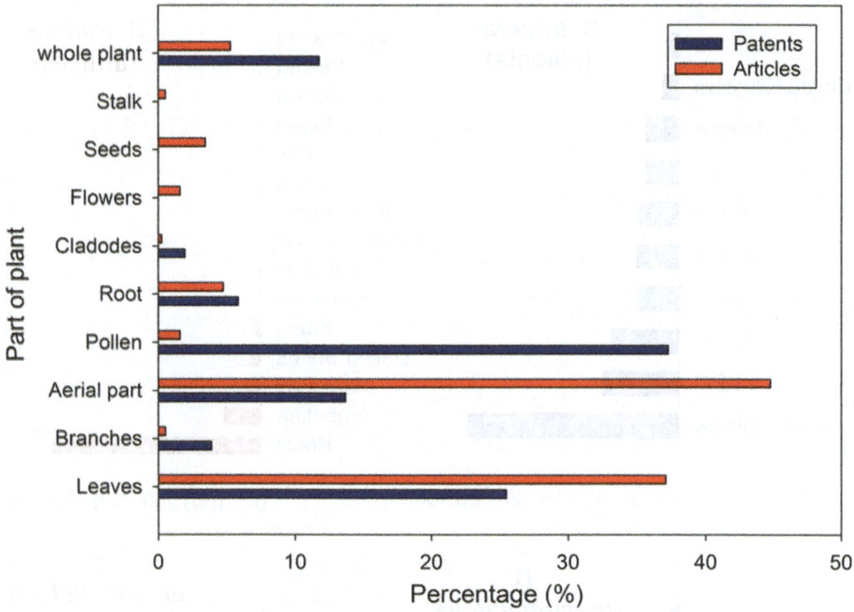

Fig. 19.7 Percentage of patents and articles on plant parts of species of the genus *Baccharis*

In articles, the aerial part as a whole (leaves and branches) was mainly used for studies (44.8%), while the specific use of leaves was found in 37% of the articles (Fig. 19.7). The other parts of research interest were roots, seeds, flowers, pollen, branches, and cladodes, each presenting less than 5% of the articles herein analyzed (Fig. 19.7). *Baccharis* is a genus with a great richness of secondary compounds in leaves, such as flavonoids, phenolics, and, especially, essential oils, which release a distinctive odor that attracts insects (Ferracini et al. 1995; Moreira et al. 2019). Two species of *Baccharis* stand out for the exploration of all plant parts: *B. trimera*, and *B. dracunculifolia*.

4 Processes and Products and the International Classification of Patents

Patents are classified into products, processes, or products and processes (found in a single patent application). The differentiation between product and process can be summarized as follows: the product is directly related to the customer's needs and desires, whereas the process pursues to define how produced items or services can be presented in the most optimized possible way.

As an exemplification of a product patent, there is the first patent application for a body sunscreen that utilized the photo-absorbent chemical compounds of *B.*

sarothroides in the cosmetic composition (Unger and Creery 1998). This application was dated in 1998 by an American company, and the patent was entitled "Sunscreen agents from natural sources". It is known that *B. sarothroides* possesses a high concentration of centaureidin. According to Saeki et al. (2003), centaureidin is able to inhibit the growth of melanocyte dendrites, which reduces epidermal pigmentation. A combination of chemical compounds of natural origin and their proven effectiveness, such as centaureidin, presents an opportunity to create exclusive cosmetic products applied precisely in skin darkening or lightening (Tada et al. 2006).

As to exemplify a process patent, we can reference "Method for obtaining extracts of *Baccharis glutinosa* with antifungal activity" (Rosa-Burgos et al. 2011). This patent, filed in 2015, is the most recent among the reviewed documents in the pharmacology field of plants of the genus *Baccharis* with interest in the development of extraction methods, more specifically of *B. glutinosa* chemical compounds.

Moreover, in product and process patents, the technical formulation characteristics of a given product as well as the methodology used to produce it are described in the same patent application. A good example of this type of patent is "Combination of vegetable extracts modifies immune response and is effective against e.g. multiple sclerosis" (Carreiras 2001). This application fits in the area of oncology and autoimmune diseases. It is a Spanish patent from 2001, which contains in its claims the extraction description of root chemical compounds of several plant species, among them *B. trimera*, followed by the drug formulation and its description as a pharmaceutical product.

In general, about 59.2% of *Baccharis* patents refer to products and processes, 26.3% refer to products, and 14.5% aim at the creation of processes only (Fig. 19.8).

Regardless of the type of patent (product or process), a specific classification is designated. All applications are classified in accordance with the International Patent Classification (IPC) according to their technological area. The assessment of this classification by the users promotes the establishment of yet another effective

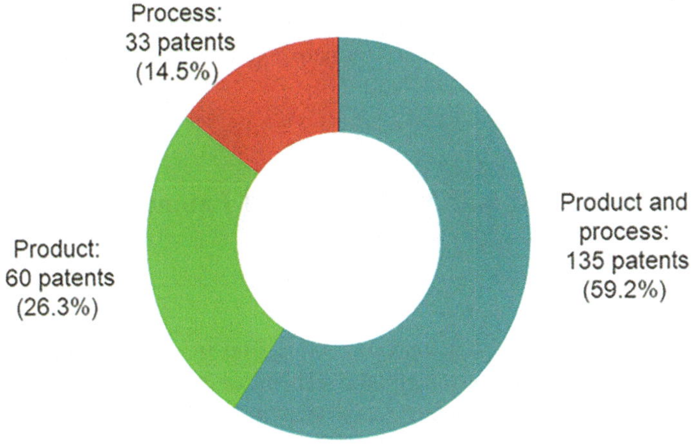

Fig. 19.8 Percentage of *Baccharis* patents classified as product, process, and product and process

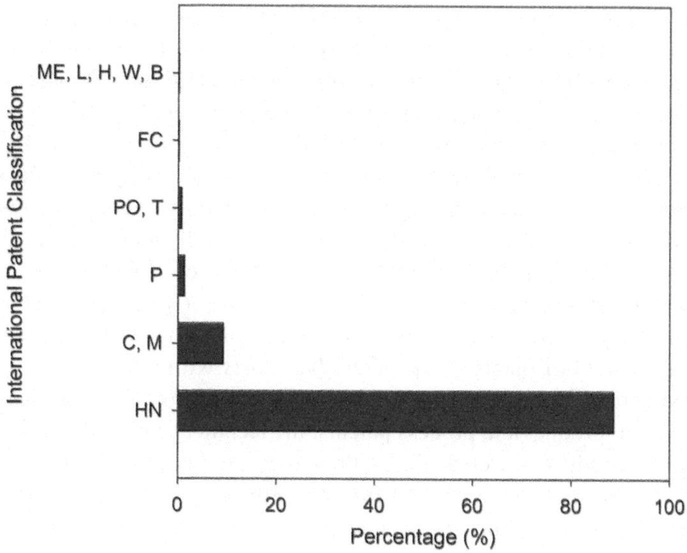

Fig. 19.9 International Patent Classification (IPC) referring to species of the genus *Baccharis*. Abbreviations: *ME* Mechanical Engineering, *L* Lighting, *H* Healting, *W* Weapons, *B* Blasting, *FC* Fixed Constructions, *PO* Performing Operations, *T* Transporting, *P* Physics, *C* Chemistry, *M* Metallurgy, *HN* Human necessities

search tool for the retrieval of patent documents. Statistical analysis of patent filing activity is one of the most used methodologies in technological monitoring (Weid et al. 2018); therefore, IPC can be used as a database for the elaboration of statistics aimed at the industrial property, which allows the assessment of technological development in several areas (OECD 2005). Human needs represent the area of greatest interest for patents of species of the genus *Baccharis* with a frequency of 88.6% (Fig. 19.10). This class of patents encompasses subclasses that involve agriculture, food or food products, articles for personal household use, as well as health and recreation.

5 Areas of Interest and Applications of Patents and Articles

Patents

Baccharis' product and process patents were found in the following application areas: cosmetics, pharmacology, methods of chemical compounds extraction, nutrition, sensors, veterinary, agronomy, environment, and labeling. For *Baccharis*' product patents, the main areas of interest are cosmetics (63.75%) and pharmacology (24.35%) (Fig. 19.9a, Table 19.1). *Baccharis* patents for cosmetic products are

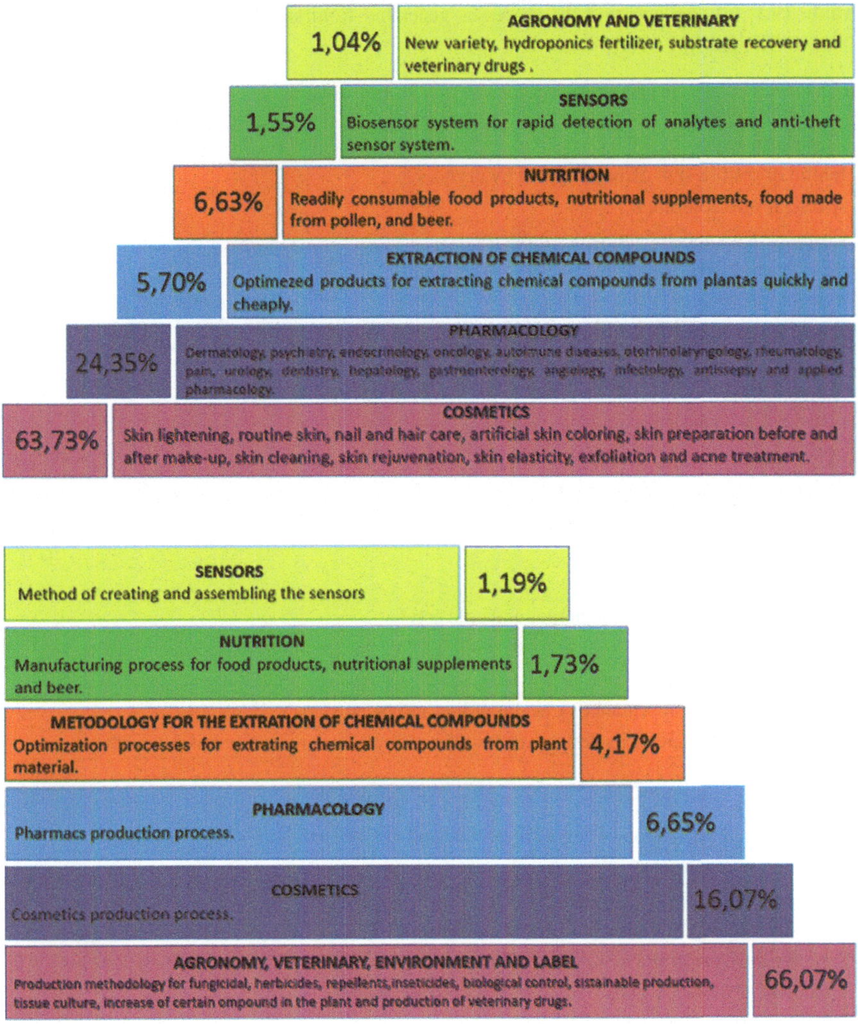

Fig. 19.10 Frequency of patent applications for products (**a**) and processes (**b**) of the genus *Baccharis*, separated by areas of interest

mostly aimed at dermatological treatments, such as hair loss, prevention and treatment of signs caused by age, scar treatment, exfoliating, and others. As for pharmacology, product patents were found for administration and production of drugs in the following areas: psychiatry, endocrinology, oncology, autoimmune diseases, otolaryngology, rheumatology, pain, urology, dentistry, hepatology, gastroenterology, infectology, and applied pharmacology (release of drugs) (Table 19.1). About 18% of patents aimed at oncology draw on the species *B. glutinosa*, *B. megapotamica*, *B. sarothroides*, and also on documents in which the species has not been described at species level, reaching 21% of the analyzed results. Process patents

Table 19.1 (a) Products and (b) processes generated from species of *Baccharis* in the areas of pharmacology and cosmetology

(a) Products
a.1) Dermatology
Skin clarification
Inhibit or restore skin damage caused by dryness
Artificial skin coloration
Cosmetic with anti-inflammatory property
Hair cleaning and hydration
Hair loss treatment
Prevention and cure of gray hair
Sunscreen
Antiperspirant
Routine skin, hair and/or nail care
Skin preparation
Skin protection for diaper wearers
Prevention and treatment for aging
Protection against skin lesions
Artificial skin coloration
Composition for treatment of herpes and cold sores
Diaper rash treatment
Exfoliating
Products for better appearance or health of lips and facial skin
Dehydroascorbic acid or its derivatives for skin coloration; care and/or makeup methods
Wound care
a.2) Psychiatry
Treatment and prevention of cognitive decline and age-related memory impairments
Addiction treatment
a.3) Endocrinology
Sweeteners, promoters, or enhancers of sweeteners
Drug for prevention and treatment of diseases and conditions induced by carbohydrates
Treatment of diabetes, obesity, and other metabolic diseases
Product to promote weight loss
a.4) Oncology and autoimmune diseases
Kit with presegmentation compounds
Treatment or prevention of autoimmune disease, allergic reaction, transplant-related complication, graft rejection
Treatment of infectious diseases, cancer, inflammation, tissue damage, etc.
Induction of immune tolerance
Vaccine to improve immune response
Immunotherapy and vaccine
Immunotherapy and allergy
Treatment of angiogenesis and metastasis
Fc coupled to compositions and methods of its use

(continued)

Table 19.1 (continued)

Product that prevents or limits the effects of chemotherapy
Oncology and autoimmune diseases
a.5) Otorhinolaryngology and allergy
Treatment of one or more nasal or sinus conditions
Composition to prevent and treat type I allergy
Expandable devices for nasal polyps treatment
Treatment for allergies and asthma
Methods and compositions for allergens dosage
Method and device for testing and allergy treatment
a.6) Rheumatology
Treatment for arthritis-associated deformity, including pain and for prevention or treatment of stiffness or inflammation
Prevention, therapy, and osteoporosis treatment
a.7) Ache
Therapeutic inhibitor of vascular smooth muscle cells
Anti-inflammatory (potent COX-2 inhibitors)
a.8) Urology
Treatment of urinary system diseases
a.9) Dentistry
Prevention and treatment of oral diseases
Aroma for oral products
Oral health products
a.10) Hepatology
Compound for hepatitis treatment
a.11) Gastroenterology
Intestinal regulator
Treatment to reverse symptoms of constipation
a.12) Angiology
Treatment of homeostatic instability conditions
Treatment for vascular trauma
Treatment for thrombosis
a.13) Infectology
Malaria treatment
Treatment of diseases related to coronavirus infection (pneumonia and gastroenteritis)
a.14) Applied pharmacology
Product and method for oral administration of nutraceuticals
Promoter of vitamin C transporter production
Drug delivery and release
Antibiotics
a.15) Antisepsis
Paints, coatings, and polymers containing phytochemicals
Paint to apply on surface and inhibit the colonization of organism-painted surfaces
Product with sanitizing, antiseptic, and/or disinfectant function for objects, food, and skin

(continued)

Table 19.1 (continued)

Filtering materials with biocidal phytochemicals
Skin sanitizers (durable)
Hand sanitizer
Antimicrobial film
(b) Processes
b.1) Dermatology
Assessment method of dry skin state
Process for artificial skin coloration
Tanning techniques
Fast and economical process for preparing substituted phenylaldehydes (raw material for the preparation of a large number of aromatic compounds useful in the perfume industry)
b.2) Extraction of compounds
Methodology of secondary metabolites extraction
Identification of compounds found in propolis
Acquisition of propolis extract
Methodology of artepilin C extraction
Process of acquisition of plant extracts and compositions comprising extracellular protease inhibitors
Extraction and use of propolis extracts in products
b.3) Quality control
Methodology for quality control of propolis and extracts of *B. dracunculifolia*
b.4) Antisepsis
Active oxygen scavenger
b.5) Applied pharmacology
Isolation of autoantibodies (autoimmune diseases)
Methodology of product/medicine preparation

focused on pharmacology generally describe the way a certain product was produced or even optimized, the extraction of its compounds, or even the evaluation methods for dermatological problems (Table 19.1).

The greatest highlight in the agronomy, veterinary, and environment fields (about 1% of product patents) is among phytosanitary products formulated with secondary compounds of *Baccharis* species, which have a natural-based formulation and, for this reason, present shorter retention time in the cultivated area and low level of soil and water contamination. Another interesting application is related to hybrid species, that is, the result of natural crossing between genetically similar species, such as the following species: *Baccharis hybris* 'Starn', *Baccharis halimifolia* 'Kolmmyst' (Kolster 2003a), *Baccharis halimifolia* 'Kolmstar' (Kolster 2003b), and *Baccharis halimifolia* 'Kolmsil' (Kolster 2003c). The particularity of these species is their high potential in the recovery of degraded areas due to their morphological characteristics such as deep root system and voluminous aerial part.

Considering the areas of interest of process patents (Fig. 19.9b), cosmetics and pharmacology occupy second and third places with a frequency of 16.07% and 6.55%, respectively. In first place, with 66.07% of frequency, there are processes

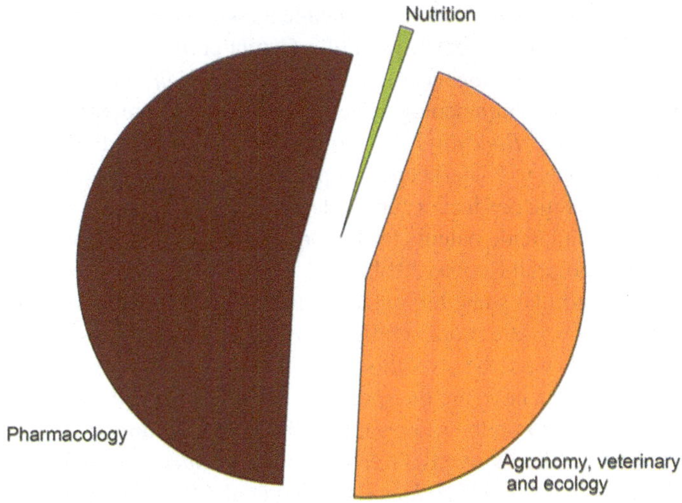

Fig. 19.11 Areas of interest of scientific articles on *Baccharis*

aimed at agronomy and environment, mostly associated with damage reduction caused by herbivores and phytopathogens, but also weed control.

Process patents, when compared to product patents, have been particularly evaluated at a lower economic value, partly due to the difficulty of screening (Kartal 2007). In addition, many product patents are frequently filed with the same methodologic process (Kartal 2007).

Articles

The scientific articles on *Baccharis* can be grouped into three large groups in accordance with similarity of interests: pharmacology (54.5%), agronomy, veterinary and ecology (45.5%), and nutrition (1%) (Fig. 19.11). Within the great area of pharmacology, there are studies focused on the extraction of chemical compounds, taking into account the following aspects: types of solvents, concentration and chemical composition, and according to the plant part, optimization of extraction processes, chromatographic analysis of compounds, and others. In studies that involved tests of medicinal interest, extracted chemical compounds were tested in vivo and in vitro experiments.

Many in vivo and in vitro experiments carried out with *Baccharis* coincide with the testing objectives of patent applicants, such as compounds with carcinogenic, analgesic, anticonvulsant, anti-inflammatory, antihyperglycemic, and antihyperlipidemic functions and antioxidants among others. This similarity suggests an interaction between articles and patents with regard to the knowledge generated on *Baccharis*' biological activities of pharmacological interest, especially those associated with carcinogenic activity of chemical compounds of *Baccharis* (Galvão et al. 2012; González et al. 2018; Jaramillo-García et al. 2018). The species that stood out the most in carcinogenesis studies were *B. coridifolia* and *B. dracunculifolia*. However, within pharmacology scientific articles, 68% refer to biocidal activities found in some *Baccharis*. Some of these articles aim to apply the *Baccharis* compounds to bacteria and fungi responsible for human diseases. In the meantime, no antibiotic product or fungicide with the same activity was observed in patents, only just fungicide patents for the protection of agricultural crops, such as the patent by Burgos and Rocha (2014).

It was also noted that some biological activities, such as the antimalarial action of *Baccharis* species, were more significantly evaluated in articles than in patents. Oliveira et al. (2012) were the inventors of a patent on the use of antimalarial chemical compounds (diterpenes), but the species has not been described at species level. While in articles found on the same subject, the following *Baccharis* species were used in the tests: *B. microdonta*, B. *paucifloculosa*, *B. punctulata*, *B. reticularioides*, *B. stenocephala*, and *B. genisteloides* (Henning et al. 2011; Budel et al. 2018a, b). *Baccharis*' biological activities for Leishmaniasis treatment (Grecco et al. 2010, 2012; Parreira et al. 2010; Passero et al. 2011; Neto et al. 2019) and Chagas disease (Grecco et al. 2010; Vanini et al. 2012; Guerreiro et al. 2018; Ueno et al. 2018) were only addressed in scientific articles. The species *B. uncinella* and *B. dracunculifolia* showed activity to treat Leishmaniasis and Chagas disease. For Chagas disease treatment, the species *B. semiserrata* and *B. retusa* also showed activity. This lack of interaction between science and technology is aimed at curing neglected diseases, although in 2009, they were considered priorities by the World Health Organization (2004).

In the veterinary field, studies focused on species that cause animal poisoning, with emphasis on *B. megapotamica*. Reports on the intoxication of whole herds, induced resistance, experimental intoxication, and animal testing with the purpose to prevent intoxication were found in the studies. Another information only found in articles is related to the antiophidic capacity that the leaves of *B. crispa* present (Costa 2010).

Baccharis friburguensis was the species of interest in only one scientific article; however, this same species has been pointed out to be used as a natural sweetener. In the four remaining scientific articles involving plants of the genus *Baccharis* in the area of nutrition, there are nutritional analyses in search of carbohydrate sources.

In our study, a total of 85 articles involves morpho-anatomical study areas of the species, taxonomic classification, and floristic surveys (involving native, pioneer, endemic, dominant, weed species, etc.). Specifically, inside ecology, studies with *Baccharis* were carried out on the recovery of degraded areas (e.g., Perea et al. 2019), phytoremediation (Haque et al. 2008), and insect–plant interaction (e.g., Fagundes et al. 2005; Fernandes et al. 2014; Monteiro et al. 2020), mainly with gall-inducing insects (e.g., Arduin et al. 2005; Espirito-Santo et al. 2012; Agudelo et al. 2018).

6 Collaborative Network Among Those Involved in the Patents and Articles

Other aspects related to the study of science–technology interaction start from the analysis of coauthorship. One hundred of scientific articles authors were found in this survey, a number far lower than the number of patent applicants for the genus *Baccharis*: four hundred and four. A survey was carried out with ten patent inventors and scientific article authors (Table 19.2). Bastos JK is the author of the largest number of published scientific articles, with 40 publications. Bastos is a Pharmacognosy professor in the Department of Pharmaceutical Sciences at the Faculty of Pharmaceutical Sciences, University of São Paulo. His research is focused on the organic chemistry of natural products, mainly on secondary metabolites of higher plants and propolis, in which species of the genus *Baccharis* lie, with emphasis on *B. dracunculifolia*. The inventor Reno JM stood out in the ranking of the most frequent inventors, with 20 patent applications. All of Reno' patents were applied by the private company Neorx Corporation. The high number of patents invented by Reno draws attention mainly due to the fact that the other inventors present a frequency of 1 to 3 deposits.

Bastos presented another very interesting feature, as he was one of the 4 patent inventors who also appear on the list of scientific article authors, the other three are Jarvis BB, Rosa-Burgos EC, and Cortez-Rocha MO. Bastos and Jarvis are inventors of one patent each; however, Bastos published 40 articles while Jarvis published 21. Rosa-Burgos and Cortez-Rocha filed two patent applications each, yet the first has seven scientific articles published, while the second has five. This amount of inventors–authors demonstrates how weak is the collaboration between professionals involved in the science and technology of species of the genus *Baccharis*. Pravdic and Oluic-Vukovic (1986) stated that more productive authors tend to form partnerships more frequently, and authors who work in multidisciplinary research exhibit the most frequent collaborative behavior with highly productive authors.

Table 19.2 Ranking of inventors and authors who most filed patents and published scientific articles on the genus *Baccharis*

Inventors	Number of patent deposits	Authors	Number of article publications
Reno JM	20	Bastos JK	40
Theodor LJ	13	Fernandes GW	35
Gustavson LM	10	Tonn CE	27
Axworthy DB	9	Giordano OS	24
Leroy Kunz LL	9	Budel JM	23
Portes P	9	Lago JHG	23
Gupta SK	9	Jarvis BB	21
Simonnet JT	8	Heiden G	20
Jia Q	8	Romoff P	19
Laboureau J	6	da Silva AA	18

The future perspective in which society can be favored by the range of benefits provided by species of the genus *Baccharis* focuses on the need of a deep stage of maturity related to interaction and collaboration of professionals interested in patent study and development. According to Moura (2012), authorship in scientific articles has been configured as an opening opportunity, mainly to compete or obtain funding to carry out their research, discussions between counterparts, and the researcher' scientific visibility in their area of expertise. In contrast, coinvention presents criteria of participation for authorship in technological production, different from those used in scientific production.

Collaborations between universities and companies have been growing in recent years (Cohen et al. 2002; Kon 2016). The knowledge generated by universities is extremely important for industrial development, especially high-tech sectors due to their proximity to the scientific knowledge basis (Klevorick et al. 1995). The financial resources derived from companies are pointed out as one of the advantages of partnership for universities (Lehmann and Menter 2016).

7 Final Considerations

Baccharis stands out for its great potential for science and technology, especially in the pharmacological area, with publications and patents that go beyond limits of origin. Patent filings on *Baccharis*; however, have been hampered due to legal reasons, required time for patenting, and difficulty in patent granting after application. These barriers have affected countries such as Brazil, which, although shows the greatest scientific contribution on *Baccharis*, does not have the greatest economic contribution in the form of patents. The largest number of *Baccharis* patent filings is led by the United States, which, throughout its history, has had great incentives and investments in the innovation sector.

Among the species *B. trimera* and *B. dracunculifolia* represent the species with more scientific studies. *Baccharis trimera* also appears at the top of the species of greatest interest to patent inventors, followed by *B. megapotamica*. This great scientific and market interest in *B. trimera* may be related to the use of secondary metabolites in the treatment of various diseases.

Unfortunately; however, *Baccharis*' scientific knowledge has not always contributed to the advance of the generation of patent filings. Published scientific articles could assist in generating patent filings, such as a potential antibiotic and antifungal for human diseases, treatment of various diseases (such as autoimmune diseases, cancers, among others) bringing innovation and economic incentives. The innovative discoveries evidenced in patents of this genus were also not reflected in the scientific deepening and development. There are no reports in the form of articles on the already patent work deposited, *Baccharis* fiber sensors, nor studies that explore the anticarcinogenic activity of the species *B. glutinosa*, *B. megapotamica*, and *B. sarothroides*.

Another aspect that reinforces the poor interaction between the knowledge generated by articles and patents is the low collaboration among authors of scientific articles (mostly researchers from higher education and research institutions) and patent applicants (mostly companies). These findings indicate an urgent need for greater exchange of knowledge generated from articles and patents, and collaborations between *Baccharis* researchers and applicants to promote science and technology. The application of research and technology helps to create new demands and new industries, which drive a future of economic growth and social development.

Acknowledgments We thank Frederico Oliveira for the development of the maps and the funding agencies CAPES, CNPq, Planta Ltda, and FAPEMIG for the scholarships granted to the authors involved.

References

Agência Nacional de Vigilância Sanitária (2019) Farmacopéia Brasília, vol I. Agência Nacional de Vigilância Sanitária, Brasília.

Agudelo I, Cogoi L, Filip R, Kuzmanich N, Wagner ML, Ricco RA (2018) Anatomy, histochemistry, and comparative analysis of hydroxycinnamic derivatives in healthy leaves and galls induced by *Baccharopelma* spp. (Hemiptera: Psyllidae) in *Baccharis spicata* (Lam) Baill (Asteraceae). Biochem Syst Ecol 77:22–30

Arduin M, Fernandes GW, Kraus JE (2005) Morphogenesis of galls induced by *Baccharopelma dracunculifoliae* (Hemiptera: Psyllidae) on *Baccharis dracunculifolia* (Asteraceae) leaves. Braz J Biol 65:559–571

Arqué-Castells P, Cartaxo RM, García-Quevedo J, Godinho MM (2016) Royalty sharing effort and invention in universities: evidence from Portugal and Spain. Res Policy 45:1858–1872

Bamburg JR (1983) Biological and biochemical actions of tricothecene mycotoxins. In: Hahn FE, Kopecko DJ, Müller EG (eds) Progress in Molecular and Subcellular Biology, vol 8. Springer-Verlag, Heidelberg, pp 41–110

Barbosa M, Fernandes GW, Morris RJ (2019) Interaction engineering: non-trophic effects modify interactions in an insect galler community. J Anim Ecol 8:1168–1177

Barroso GM (1976) Compositae – Subtribo Baccharidinae Hoffmann – Estudo das espécies ocorrentes no Brasil. Rodriguésia 28:1–273

Bartlett JB, Muller GW, Schafer PH, Galustian C, Dalgleish AC, Meyer B (2014) Immunologische Verwendungen von immunomodulatorischen Verbindungen zur Impfung und zur Therapie gegen Infektionskrankheiten. Applicant: Celgene Corporation. EP2301535B1. Priority data: 2005-09-01

Bettucci GR, Rodriguez MV, Ferretti M, Campagna MN, Monti J, Carnovale CE, Alvarez MJ, Martínez ML (2020) Choleretic properties of *Baccharis spicata* aerial parts extracts in male Wistar Rats. Rev Bras Farmacogn 30:54–60

Brandão M, Oliveira AK (1995) Plantas ruderais e subespontâneas do Município de Ouro Preto e sua utilização na Medicina Popular. Daphne 5:49–71

Budel JM, Raman V, Monteiro LM, Almeida VP, Bokek VB, Heiden G, Takeda IJM, Khan IA (2018a) Foliar anatomy and microscopy of six Brazilian species of *Baccharis* (Asteraceae). Microsc Res Tech 81:1–11

Budel JM, Wang M, Raman V, Zhao J, Khan SI, Rehman JU, Techen N, Tekwani B, Monteiro LM, Heiden G, Takeda IM, Farago PV, Khan IA (2018b) Essential oils of five *Baccharis* species: investigations on the chemical composition and biological activities. Molecules 23:1–19

Burgos ECR, Rocha MOC (2014) Method for obtaining extracts of *Baccharis glutinosa* with antifungal activity. Applicant: Universidade de Sonora. MX2014000932A. Priority data: 2013-12-13

Caño L, Campos JA, García-Magro D, Herrera M (2013) Replacement of estuarine communities by an exotic shrub: distribution and invasion history of *Baccharis halimifolia* in Europe. Biol Invasions 15:1183–1188

Caño L, Fuertes-Mendizabal T, García-Baquero G, Herrera M, González-Moro MB (2016) Plasticity to salinity and transgenerational effects in the nonnative shrub *Baccharis halimifolia*: insights into an estuarine invasion. Am J Bot 103:808–820

Caporal FJM, Boldrini II (2007) Florística e fitossociologia de um campo manejado na Serra do Sudeste, Rio Grande do Sul. Rev Bras Biocienc 53:37–44

Carneiro MAA, Fernandes GW (1996) Herbivoria. Ciência Hoje 20:35–39

Carneiro AM, Irgang BE (2005) Origem e distribuição geográfica das espécies ruderais da Vila de Santo Amaro, General Câmara, Rio Grande do Sul. Iheringia Ser Bot 60:175–188

Carreiras GJA (2001) Combinacion de extractos vegetales modificadora de la respuesta inmune. Applicants: Carreiras Garcia José Antônio ES2156694A1

Carvalho-Filho JLS, Blank AF, Alves PB, Ehlert PAD, Melo AS, Cavalcanti SCH, Arrigoni-Blank MF, Silva-Mann (2006) Influence of the harvesting time, temperature and drying period on basil (*Ocimum basilicum* L.) essential oil. Rev Bras Farmacogn 16:24–30

Casagrande M, Zanela J, Júnior AW, Busso C, Wouk J, Lurckevicz G, Montanher PF, Yamashita F, Malfatti CRM (2018) Influence of time, temperature, and solvente on the extraction of bioactive compounds of *Baccharis dracunculifolia*: in vitro antioxidante activity, antimicrobial potential, and phenolic compound quantification. Ind Crop Prod 125:207–219

Cho NJ, Jackman JÁ, Mundargi RC, Park JH, Potroz MG (2017) Microemcapsulation compounds into natural spores and pollen grains. Applicant: Nanyang Technological Universitu. WO2017010945A1. Priority data: 2015-07-16

Claudino CC (2013) *Baccharis trimera* (Less) DC. Estudo fitoquímico e avaliação da citotoxicidade com sulforrodamina B. Tese de Mestrado em Ciências Farmacêuticas, Faculdade de Ciências Farmacêuticas, Universidade Estadual Paulista

Cohen WM, Nelson RR, Walsh JP (2002) Links and impacts: the influence of public research on industrial R&D. Manag Sci 48:1–23

Corrêa MP (1984) Dicionário das plantas úteis do Brasil e das exóticas cultivadas. MINAGRI/ IBDF, Rio de Janeiro

Costa PRR (2010) Produtos naturais como ponto de partida para a descoberta de novas substâncias bioativas: Candidatos a fármacos com ação antiofídica, anticâncer e antiparasitária. Rev Virtual Quim 1:58–66

Cutler HC, Jarvis BB (1985) Preliminary observations on the effects of macrocyclic trichothecenes on plant growth. Environ Exp Bot 25:115–128

Dolejs L, Herout V, Sorm F (1961) Sesquiterpenic compounds of *Baccharis genistelloides* Pers: structure of palustrol. Colln Czech Chem Commun Engl Edn 26:811–817

Dundas N (2012) Research intensity and knowledge transfer activity in UK universities. Res Policy 41:262–275

Edreva A, Velikova V, Tsonev T, Dagnon S, Gürel A, Aktaş L, Gesheva E (2008) Stress-protective role of secondary metabolites: diversity of functions and mechanisms. Gen Appl Plant Physiol 34:67–78

Egerova J, Proffitt CE, Travis SE (2003) Facilitation of survival and growth of *Baccharis halimifolia* L. by *Spartina alterniflora* Loisel in a created Louisiana salt marsh. Wetlands 23:250–256

Espirito-Santo MM, Neves FS, Fernandes GW, Silva JO (2012) Plant phenology and absence of sex-biased gall attack on three species of *Baccharis*. PLoS One 10:e46896

Fagundes M, Neves FS, Fernandes GW (2005) Direct and indirect interactions involving ants, insect herbivores, parasitoids, and the host plant *Baccharis dracunculifolia* (Asteraceae). Ecol Entomol 30:28–35

Feduraev P, Chupakhina, Maslennikov P, Skrypnik L (2019) Variation in phenolic compounds content and antioxidant activity of different plant organs from *Rumex crispus* L. and *Rumex obtusifolius* L. at different growth stages. Antioxidants 8:1–15

Fernandes GW, Silva JO, Espírito-Santo MM, Fagundes M, Oki Y, Carneiro MAA (2014) *Baccharis*: a neotropical model system to study insect plant interactions. In: Fernandes GW, Santos JC (eds) Neotropical insect galls. Springer, Dordrecht, pp 193–219

Fernandes GW, Oki Y, Belmiro MS, Resende FM, Junior AC, de Azevedo JL (2018) Multitrophic interactions among fungal endophytes, bees, and *Baccharis dracunculifolia*: resin tapering for propolis production leads to endophyte infection. Arthropod Plant Interact 12:329–337

Ferracini VL, Paraiba LC, Leitao Filho HF, Da Silva AG, Nascimento LR, Marsaioli AJ (1995) Essential oils of seven Brazilian *Baccharis* species. J Essent Oil Res 7:355–367

Fielding RR (2001) *Baccharis*: a genus of the Asteraceae new to Canada. Proc Natl Sci Inst Sci 41:214–215

Fisch CO, Hassel TM, Sandner PG (2015) University patenting: a comparison of 300 leading universities worldwide. J Technol Transfer 40:318–345

Fleisner DC (2016) Methods and devices for allergy testing and treatment. Appicant: Spiriplex, Inc. US2016327567A1. Priority data: 2015-05-07

França E, Vasconcellos AG (2018) Patentes de fitoterápicos no Brasil: uma análise do andamento dos pedidos no período de 1995-2017. Cad Cienc Technol 35:329–359

Galvão LCC, Furletti VF, Bersan SMF, Cunha MG, Ruiz ALTC, Carvalho JE, Santoratto A, Rehder VLG, Figueira GM, Duarte MCT, Ikegaki M, Alencar SM, Rosalen PL (2012) Antimicrobial activity of essential oils against *Streptococcus mutans* and their antiproliferative effects. Evid Based Complement Alternat Med 2012:1–15

Giuliano DA (2001) Classificación infragenérica de las espécies Argentinas de *Baccharis* (Asteraceae, Astereae). Darwin 39:131–154

Gomes V, Fernandes GW (2002) Germination of *Baccharis dracunculifolia* DC (Asteraceae) achene. Acta Bot Bras 16:421–427

González ML, Joray MB, Laiolo J, Crespo MI, Palacios SM, Ruiz GM, Carpinella (2018) Cytotoxic activity of extracts from plants of Central Argentina on sensitive and multidrug-resistant leukemia cells: isolation of an active principle from *Gaillardia megapotamica*. Evid Based Complement Alternat Med 2018:1–13

Graça LRSS, Água-doce AMM (2010) The use of adjuvant to facilitate the induction of immune tolerance. Applicants: Água-doce AMM, Graça LRSS, Instituto de Medicina Molecular, Universidade de Lisboa. WO2010056143A1. Priority data: 2008-11-13

Grecco SS, Reimão JQ, Tempone AG, Sartorelli O, Romoff P, Ferreira MJP, Fávero AO, Lago JHG (2010) Isolation of an antileishmanial and antitrypanosomal flavanone from the leaves of *Baccharis retusa* DC. (Asteraceae). Parasitol Res 106:1245–1248

Grecco SDS, Ferreira MJ, Romoff P, Favero OA, Lago JHG (2012) Phenolic derivatives from *Baccharis retusa* DC (Asteraceae). Biochem Syst Ecol 42:21–24

Guerreiro AC, Cecati FM, Ardánáz CE, Donadel OJ, Tonn CE, Sosa ME (2018) Assessment of the insecticidal potential of the *Eupatorium buniifolium* essential oil against *Triatoma infestans* (Hemiptera: Reduviidae). A chiral recognition approach. Neotrop Entomol 47:418–428

Haque N, Peralta-Videa JR, Jones GL, Gill TE, Gardea-Torresdey JL (2008) Screening the phytoremediation potential of desert broom (*Baccharis sarothroides* Gray) growing on mine tailings in Arizona, USA. Environ Pollut 153:362–368

Heering WCA (1907) Symbolae Antillanae, seu, Fundamenta florae Indiae Occidentalis. editi Ignatius Urban, Germany, 243p

Heiden G, Pirani JR (2016) Taxonomy of *Baccharis* subgen. Tarchonanthoides (Asteraceae: Astereae: Baccharidinae), a group from the southeastern South American grasslands and savannas. Phytotaxa 241:1–70

Heiden G, Barbieri RL, Stumpf ER (2006) Considerações sobre o uso de plantas ornamentais nativas. Ornam Hortic 171:2–7

Heiden G, Antonelli A, Pirani JR (2019) A novel phylogenetic infrageneric classification of *Baccharis* (Asteraceae: Astereae), a highly diversified American genus. Taxon 68:1048–1081

Henning L, Garcia GM, Giannis A, Bussmann RW (2011) New constituents of *Baccharis genistelloides* (Lam.) Pers. ARKIVOC 2011:78–81

Hitchcock AS, Standley PC (1919) Flora of the district of Columbia and Vicinity (Contributions from the United States National Herbarium, vol 21). United States National Museum (Reprinted by Kessinger Publishing, LaVergne, Tennessee, 2010), Washington, pp 329 (+42 plates).

Jaramillo-García V, Trindade C, Lima E, Guecheva TN, Villela I, Martinez-Lopez W, Corrêa DS, Ferraz ABF, Moura S, Sosa MQ, da Silva J, Henriques JAP (2018) Chemical characterization and cytotoxic, genotoxic, and mutagenic properties of *Baccharis trinervis* (Lam. Persoon) from Colombia and Brazil. J Ethnopharmacol 213:210–220

Karam TK, Dalposso LM, Casa DM, de Freitas GBL (2013) Carqueja (*Baccharis trimera*): utilização terapêutica e biossíntese. Rev Bras Plant Med 15:280–286

Kartal M (2007) Intellectual property protection in the natural product drug discovery, traditional herbal medicine and herbal medicinal products. Phytother Res 21:113–119

Khan BZ (2005) The democratization of invention: patents and copyrights in American economic development, 1st edn. Cambridge University Press, New York

Klevorick AK, Levin RC, Nelson RR, Winter SG (1995) On the sources and significance of interindustry differences in technological opportunities. Res Policy 24:185–205

Kolster PR (2003a) *Baccharis* planta named "Kolmmyst". Applicant: Gebr kolster BV. USPP13744P2. Priority data: 1998–01- 23

Kolster PR (2003b) *Baccharis* planta named "Kolmstar". Applicant: Gebr kolster BV. USPP1895P2. Priority data: 2006-03-17

Kolster PR (2003c) *Baccharis* planta named "Kolmsil". Applicant: Gebr kolster BV. USPP1895P2. Priority data: 1998-01-23

Kon A (2016) Ecossistemas de inovação: a natureza da inovação em serviços. RACEF 7:14–27

Krischik VA, Denno RF (1990) Patterns of growth, reproduction, defense, and herbivory in the dioecious shrub *Baccharis halimifolia* (Compositae). Oecologia 83:182–190

Kroyer G, Hegedus N (2001) Evaluation of bioactive properties of pollen extracts as functional dietary food supplement. Innov Food Sci Emer 2:171–174

Kupchan SM, Bauerschmidt E (1971) Cytotoic flavonols fron *Baccharis sarothroides*. Phytochemistry 10:664–666

Lago JHG, Romoff P, Favero OA, Souza FO, Soares MG, Baraldi PT, Corrêa AG (2008) Chemical composition of male and female *Baccharis trimera* (Less.) DC.(Asteraceae) essential oils. Biochem Syst Ecol 36:737–740

Lehmann EE, Menter M (2016) University–industry collaboration and regional wealth. J Technol Transfer 41:1284–1307

Lindsay MHA, Pearson HJ (1942) Radiant energy receiving system. Depositor: American District Telegraph Co. US2278936A. Deposit data: 25-04-1939

Looy BA, Callaert J, Debachere K (2006) Publication and patente behavior of academic researches: conflicting, reinforrcing or merely co-existing? Res Policy 35:596–608

Macedo JF (1995) Fenologia da floração das plantas invasoras no Campus-Pampulha da Universidade Federal de Minas Gerais. Daphne 5:15–27

Manurung H, Kustiawan W, Kusuma IW, Marjenah M (2017) Total flavonoid content and antioxidant activity of tabat Barito (*Ficus deltoidea* Jack) on different plant organs and ages. J Med Plant 5:120–125

Marques MB (2000) Patentes farmacêuticas e acessibilidade aos medicamentos no Brasil. Hist Cienc Saude-Manguinhos 8:7–21

Martinez EGA, Forero LML, Pérez FC (2015) Promoción del crecimiento de *Baccharis macrantha* (Asteraceae) con bacterias solubilizadoras de fosfatos asociadas a su rizosfera. Acta Biol Colomb 20:121–131

Mejía-Giraldo JC, Winkler R, Gallardo C, Sánchez-Zapata AM, Puertas-Mejía MA (2016) Photoprotective potential of *Baccharis antioquensis* (Asteraceae) as natural sunscreen. Photochem Photobiol 92:742–752

Meyer M (2000) Does science push technology? Patents citing science literature. Res Policy 29:409–434

Meyer M, Batthacharya S (2004) Commonalities and differences between scholarly and technical collaboration. Scientometrics 6:443–456

Moir HVJ (2008) Who benefits? an empirical analysis of Australian and US patent ownership. Centre for Governance of Knowledge and Development (CGKD) working paper. https://doi.org/10.2139/ssrn.1409502

Monteiro GF, Macedo-Reis LE, Dáttilo W, Fernandes GW, Siqueira de Castro F, Neves FS (2020) Ecological interactions among insect herbivores, ants and the host plant *Baccharis dracunculifolia* in a Brazilian mountain ecosystem. Austral Ecol 45:158–167

Moreira FGP, Torkomian ALV, Soares TJCC (2016) Exploration and firms' innovative performance. How does this relationship work? Rev Bras Gest Neg 18:392–415

Moreira X, Abdala-Roberts L, Nell CS, Vázquez-González C, Pratt JD, Keefover-Ring K, Mooney KA (2019) Sexual and genotypic variation in terpene quantitative and qualitative profiles in the dioecious shrub *Baccharis salicifolia*. Sci Rep 9:1–10

Moura AMM (2012) Motivação para a pesquisa, determinação de parcerias e divisão da coautoria e coinvenção: principais critérios utilizados pelos pesquisadores da área da Biotecnologia. Em Questão 18:29–45

Muller-Riebau FJ, Berger BM, Yegen O, Cakir C (1997) Seasonal variation in the chemical compositions of essential oils of selected aromatic plants growing in Turkey. J Agric Food Chem 45:4821–4825

Narin F (1995) Patents as indicators for the evaluation of industrial research output. Scientometrics 34:489–496

Narin F, Hamilton KS, Olivastro D (1997) The increasing linkage between US technology and public science. Res Policy 26:317–330

Navares N, Brosh A, Pittrof W (2010) Seasonal dynamics of nutritional quality of California chaparral species. Anim Feed Sci Technol 158:44–56

Neto RNM, Setúbal RFB, Higino TMM, Brelaz-de-Castro MCA., Silva LCN, Aliança ASDS (2019) Asteraceae plants as sources of compounds against leishmaniasis and Chagas disease. Front Pharmacol 10:1–20

Neves FS, Cornelissen T, Coelho MS, Barbosa M, Carneiro MAA, Ribeiro SP, Fernandes GW (2016) Antagonistic interactions in the rupestrian grasslands: new insights and perspectives. In: Fernandes GW (ed) Ecology and conservation of mountaintop grasslands in Brazil. Springer, Switzerland, pp 315–343

Nogués J (1990) Patents and pharmaceutical drugs: understanding the pressures on developing countries. J World Trade 24:81–104

Noorden RV (2016) Snail's pace: nature readers on their longest wait to get published. Nature 19375:1–2

OECD (2005) Annual Report 2005. OECD Publishing, Paris, 143 p

Oliveira AB, Varotti FP, Batista R. (2012) Composições farmacêuticas antimaláricas contendo derivados de diterpenos caurânicos. Applicants: Batista Ronan, Olivera Alaide Braga, Universidade Federal de Minas Gerais, Varotti Fernando de Pilla. WO2012083408A1. Priority data: 2010-12-23

Park YK, Paredes-Guzman JF, Aguiar CL, Alencar SM, Fujiwara FY (2004) Chemical constituents in *Baccharis dracunculifolia* as the main botanical origin of southeastern Brazilian propolis. J Agric Food Chem 52:1100–1103

Parreira NA, Magalhães LG, Morais DR, Caixeta SC, de Sousa JPB, Bastos JK, Cunha WR, Silva MLA, Nanayakkara NPD, Rodrigues V, Filho AAS (2010) Antiprotozoal, schistosomicidal, and antimicrobial activities of the essential oil from the leaves of *Baccharis dracunculifolia*. Chem Biodivers 7:993–1001

Passero LFD, Bomfim-Melo A, Corbett CEP, Laurenti MD, Toyama MH, Toyama DO, Ramoff P, Fávero AO, Grecco SSG, Zalewsky CY, Lago JHG (2011) Anti-leishmanial effects of purified compounds from aerial parts of *Baccharis uncinella* C. DC. (Asteraceae). Parasitol Res 108:529–536

Pavan-Fruehauf S (2000) Plantas medicinais de Mata Atlântica: manejo sustentado e amostragem. Annablume/FAPESP, São Paulo, 216p

Perea R, Cunha JS, Spadeto C, Gomes VM, Moura AL, Rúbia B, Fernandes GW (2019) Nurse shrubs to mitigate plant invasion along roads of montane Neotropics. Ecol Eng 136:193–196

Perri NB, Anderson RE, Brennan NJ, Douglas MH, Heaney AJ, McGimpsey JA, Smallfield BM (1999) Essential oils from dalmatian sage (*Salvia officinalis* L.): variations among individuals, plant parts, seasons, and sites. J Agric Food Chem 47:2048–2054

Pravdic N, Oluic-Vukovic V (1986) Dual approach to multiple authorship in the study of collaborator and scientific output relationship. Scientometrics 10:259–280

Radaelli V (2007) Etapas evolutivas da indústria farmacêutica: da formação à consolidação, expansão e hegemonia das grandes empresas. Pensam Real 20:59–77

Ritter MR, Baptista LRM (2005) Levantamento florístico da família Asteraceae na "Casa Ede Pedra" e áreas adjacentes, Bagé, Rio Grande do Sul. Iheringia Ser Bot 60:5–10

Rodrigues VEG, Carvalho DA (2001) Plantas medicinais no domínio do cerrados. UFLA, Lavras, 180p

Rodrigues DM, De Souza MC, Arruda C, Pereira RAS, Bastos JK (2020) The role of *Baccharis dracunculifolia* and its chemical profile on green propolis production by *Apis mellifera*. J Chem Ecol 46:150–162

Rosas-Burgos ECR, Cortez-Rocha MO, Plascenciajatomea M, Cinco-Moroyoqui FJ, Robles-Zepeda RE, Lópezcervantes J, Sánchez-Machado DI, Lares-Villa F (2011) The effect of *Baccharis glutinosa* extract on the growth of mycotoxigenic fungi and fumonisin B1 and aflatoxin B1 production. World J Microbiol Biotechnol 27:1025–1033

Saeki Y, Kanamaru A, Matsumoto K, Tada A (2003) PP01The inhibitory effects of centaureidin on the outgrowth of dendrites, melanosome transfer and melanogenesis in normal human melanocyte. Pigment Cell Res 16:593–593

Sánchez MS, Cano EFC (1983) Listados Floristicos de México. II. Flora de Quintana Roo. Universidad Nacional Autónoma de México, México

Santos CF, Popa-Mckiver M, McCord AM, Hirschel M (2014) Materials and methods for directing an immune response to an epitope. Applicant: Biovest International Inc. US20140140986A1. Priority data: 2010-11-08

Shadlen KC, Sampat BN, Kapczynsky A (2019) Patents, trade and medicines: past, present and future. Rev Int Polit Econ 27:75–97

Silva Filho AA, Resende DO, Fukui MJ, Santos FF, Pauletti PM, Cunha WR, Silva MLA, Gregório LE, Bastos JK, Nanayakkara NPD (2009) In vitro antileishmanial, antiplasmodial and cytotoxic activities of phenolics and triterpenoids from *Baccharis dracunculifolia* DC (Asteraceae). Fitoterapia 80:478–482

Silva CM, Caetano FH, Pereira FDC, Morales MAM, Sakane KK, Moraes KCM (2019) Cellular and molecular effects of *Baccharis dracunculifolia* D.C. and *Plectranthus barbatus* Andrews medicinal plant extracts on retinoid metabolism in the human hepatic stellate cell LX-2. BMC Complement Altern M 19:1–11

Sims-Chilton NM, Zalucki MP, Buckley YM (2010) Long-term climate effects are confounded with the biological control programme against the invasive weed *Baccharis halimifolia* in Australia. Biol Invasions 12:3145–3155

Snyder AP (1986) Qualitative, quantitative and technological aspects of the trichothecene mycotoxins. J Food Prot 49:544–569

Sousa JPB, Leite MF, Jorge RF, Resende DO, Silva Filho AA, Furtado NA, Soares AEE, Spadaro ACC, Magalhães PM, Bastos JK (2011) Seasonality role on the phenolics from cultivated *Baccharis dracunculifolia*. Evid Based Complement Alternat Med 2011:1–8

Souza ARS, Moraes SZC, Viana-Júnior AB, Araújo ED (2019) Toward a novel pharmacology and therapeutic understanding of Brazilian Propolis: a meta-analytical approach. Pharmacogn Rev 14:1–7

Tada A, Kanamaru A, Ito Y (2006) Control of melanosome transfer by promoting shrinkage or expansion of melanocyte dendrites. Int J Cosmet Sci 28:147–149

Tognon GB, Cuquel FL (2016) Ornamental potential use of *Baccharis milleflora* and *Baccharis tridentata* as ornamental cut foliage. Cien Rural 46:70–75

Tokarnia CH, Peixoto PV, Gava A, Barros CSL (1992) Intoxicação experimental por *Baccharis megapotamica* var. *megapotamica* e var. *weirii* (Compositae) em bovinos. Pesqui Vet Bras 12:19–31

Ueno Y (1983) Tricotecenes-chemical, biological and toxicological aspects. Elsevier, Amsterdam

Ueno AK, Barcellos AF, Costa-Silva TA, Mesquita JT, Ferreira DD, Tempone AG, Romoff P, Antar GM, Lago JHG (2018) Antitrypanosomal activity and evaluation of the mechanism of action of diterpenes from aerial parts of *Baccharis retusa* (Asteraceae). Fitoterapia 12:55–58

Unger EC, Creery TPM (1998) Sunscreen agents from natural sources. Applicant: Imarx Pharmaceutical Corp. US5824312A. Priority data: 1994-03-10

Vanini AB, Fleming AC, Santos TG, Purnhagen L (2012) Chemical characterization and antimicrobial evaluation of the essential oils from *Baccharis uncinella* DC and *Baccharis semiserrata* DC (Asteraceae). J Essent Oil Res 24:547–554

Vasconcellos AG, Esquibel MA, Lage CLS (2004) Proteção patentária de fitoterápicos no Brasil: um estudo sobre os depósitos de patente ao longo da década de 90. Rev Bras Plant Med 7:51–56

Verdi LG, Brighente IMC, Pizzolatti MG (2005) Gênero *Baccharis* (Asteraceae): aspectos químicos, econômicos e biológicos. Quim Nova 28:85–94

Vesela D, Sman D, Valteronová I, Vanek T (1999) Seasonal variations in the contents of taxanes in the bark of *Taxus baccata* L. Phytochem Anal 10:319–321

Vessella F, Sciré M, Salis A, Piovesan G (2015) Natural regeneration and gender-specific spatial pattern of *Taxus baccata* in an old-growth population in Foresta Umbra (Italy). Dendrobiology 73:75–90

Vieira RF, Silva SR (2002) Estratégias para conservação e manejo de recursos genéticos de plantas medicinais e aromáticas. Resultados da 1° reunião técnica. Embrapa/Ibama/MMA/CNPq, Brasília, 184p

Vivien FD, Nieddu M, Befert N, Debref R, Giampietro M (2019) The hijacking of the bioeconomy. Ecol Econ 159:189–197

Weid I von der, Santos, PR, Valadão ABG, Mayerhoff ZDVL (2018) Categorização do setor de biotecnologia baseada na Classificação Internacional de Patentes e análise do panorama de depósito de pedidos de patentes neste Setor, no Brasil (2012–2016). INPI, DIRPA, CEPIT, DIESP, Rio de Janeiro, 28p

Woolsey J, Hanna C, Mceachern K, Anderson S, Hartman BD (2018) Regeneration and expansion of *Quercus tomentella* (Island Oak) groves on Santa Rosa Island. West N Am Nat 78:758–767

World Health Organization (2004) The World Health Organization quality of life (WHOQOL) BREF 2012 revision. World Health Organization, Switzerland

Zhu X (2015) Fc coupled compositions and methods of their use. Applicant: University of Maryland. US2015004161A1. Priority data: 2013-07-01

Part IV
Propolis of *Baccharis*

Green propolis, known as the green treasure, has a high added value and is produced by *Apis mellifera* bees from a special resin collected from *Baccharis dracunculifolia*. The unsurpassed value of green propolis is mainly due to its chemical compounds, such as artepillin C, which is effective in inhibiting the growth of tumor cells. Green propolis is made up of several other chemical constituents of pharmacological importance, such as caffeic, ferulic, p-coumaric and cinnamic acids, kaempferol, and chrysin. Propolis also stands out for its antioxidant properties, immunological potential, and for the treatment of other diseases, such as arthritis, diabetes, ulcer, pneumonia, among others.

Resin collection from *Baccharis dracunculifolia*. Illustration by Patrícia Angrisano

Chapter 20
Chemical Constituents and Antioxidant Properties of Green Propolis

Shigenori Kumazawa

Abstract Propolis is a resinous substance collected by honeybees from various plant sources. Propolis from the southeastern region of Brazil is known to have a greenish-brown or green color because of which it is called "green propolis." The plant origin of Brazilian green propolis is *Baccharis dracunculifolia* DC. (Asteraceae); hence, both Brazilian green propolis and *B. dracunculifolia* contain similar compounds. In this chapter, the chemical constituents and the antioxidant properties of Brazilian green propolis and its plant origin, *B. dracunculifolia*, are reviewed. The major components of the ethanol extracts of Brazilian green propolis are terpenoids and the prenylated derivatives of *p*-coumaric acid. In particular, 3,5-diprenyl-4-hydroxycinnamic acid (artepillin C) is known to be the characteristic compound found in the Brazilian green propolis. The in vitro antioxidant activity of Brazilian green propolis using various methods has been evaluated, and artepillin C and caffeoylquinic acid derivatives have been reported to be its major effective antioxidants. However, few studies have explored the in vivo antioxidant activity of Brazilian green propolis.

Keywords Antioxidant activity · Artepillin C · Cinnamic acid derivatives · Terpenoids · Caffeoylquinic acid derivatives

1 Introduction

Propolis, a natural resinous substance collected by honeybees from buds and exudates of plants, is thought to be used as a protective barrier for the beehive. Propolis has been reported to have many beneficial biological properties, including antibacterial, antiviral, and anticancer properties (Silva-Carvalho et al. 2015); hence, it is

S. Kumazawa (✉)
Department of Food and Nutritional Sciences, University of Shizuoka, Shizuoka, Japan
e-mail: kumazawa@u-shizuoka-ken.ac.jp

G. W. Fernandes et al. (eds.), *Baccharis*,
https://doi.org/10.1007/978-3-030-83511-8_20

extensively used in various health foods and beverages to prevent inflammation, heart disease, diabetes, and cancer (Sforcin 2016; Pasupuleti et al. 2017).

The chemical composition of propolis has been reported to be dependent on the local flora of the site from which it is collected (Bankova 2005; Salatino et al. 2011). Thus, the composition of the plant origin of propolis determines its chemical composition. Various propolis types and their plant origins have been studied. It is generally accepted and chemically demonstrated that the bud exudates of poplar trees are the main source of propolis in the temperate zones (Bankova et al. 2000). Because of this, the European propolis contains the typical "poplar bud" phenolic compounds, such as flavonoid aglycones (flavones and flavanones), phenolic acids, and their esters (Bankova et al. 2002). Poplar trees cannot grow in tropical and subtropical regions; therefore, propolis from these regions has a chemical composition different from that of the "poplar type (poplar bud)" propolis. For example, there are few poplar trees in southeast Brazil; however, honeybees use other plants as their source of propolis (Marcucci and Bankova 1999). A study was previously reported *Baccharis dracunculifolia* DC. (Asteraceae) to be the plant origin of the propolis of the Brazilian green propolis (propolis found in southeast Brazil), by observing the honeybee behavior and doing a phytochemical analysis of propolis (Kumazawa et al. 2003). Brazilian honeybees collect the leaves of this plant and bring them back to their hives to use them as a source of propolis.

The *Baccharis* plant species have been studied for their antifungal (Rahalison et al. 1995), insecticidal (Juan Hikawczuk et al. 2008), and plant growth inhibitory (Céspedes et al. 2002) properties. *B. dracunculifolia* was reported to have cytotoxic (Búfalo et al. 2010), antioxidant (Guimarães et al. 2012), antibacterial (Pereira et al. 2016), and anti-inflammatory (de Figueiredo-Rinhel et al. 2019) properties. The chemical constitution of the Brazilian green propolis is identical to that of *B. dracunculifolia*, which has been shown to be the source of the propolis (Kumazawa et al. 2003; Park et al. 2004). Further, the biological properties of Brazilian green propolis have also been observed in *B. dracunculifolia* (da Silva Leitão et al. 2004; de Funari et al. 2007). In this chapter, the chemical constituents and antioxidant properties of the Brazilian green propolis and its plant origin, *B. dracunculifolia*, are reviewed.

2 Chemical Constituent of Green Propolis

Propolis is a complex resinous mixture, comprising approximately 50% resin, 30% wax, 10% essential oils, 5% pollen, and 5% other organic substances (Toreti et al. 2013). The chemical composition of propolis differs according to the geographic region from where the resins were collected, based on the flora of that region. Additionally, the compositions of the propolis vary according to the season (Bankova et al. 1998).

Many studies have been conducted on the chemical composition and the biological properties of propolis, and it has been shown to contain a variety of chemical compounds, including polyphenols (flavonoids, phenolic acids and their esters), terpenoids, steroids, and amino acids (Salatino et al. 2011). However, because of the geographical differences, propolis from Europe, South America, and Asia have different chemical compositions (Salatino et al. 2011). Propolis from Europe and China comprises many kinds of flavonoids and phenolic acid esters (Bankova et al. 2000). In contrast, the major components in the propolis of Brazilian origin are terpenoids and the prenylated derivatives of *p*-coumaric acid (Marcucci and Bankova 1999; Kumazawa et al. 2003).

According to Park et al. (2002), Brazilian propolis can be divided into 12 classes. Propolis from southeast Brazil is known to have a greenish-brown or green color, because of which it is also called "green propolis" (Park et al. 2002). As described above, because the plant origin of Brazilian green propolis is *B. dracunculifolia* (Kumazawa et al. 2003), the composition of both is similar. Figure 20.1 shows the representative phenolic compounds found in the ethanol extract of Brazilian green propolis. 3,5-Diprenyl-4-hydroxycinnamic acid (artepillin C), known to be the characteristic compound in green propolis and *B. dracunculifolia*, was quantified to be 43.9 mg/g in the ethanol extract of the Brazilian green propolis (Kumazawa et al. 2004), and has been reported to be the extract's main pungent ingredient (Hata et al. 2012). Artepillin C is also known to possess anti-inflammatory and anticancer properties (Paulino et al. 2008; Messerli et al. 2009).

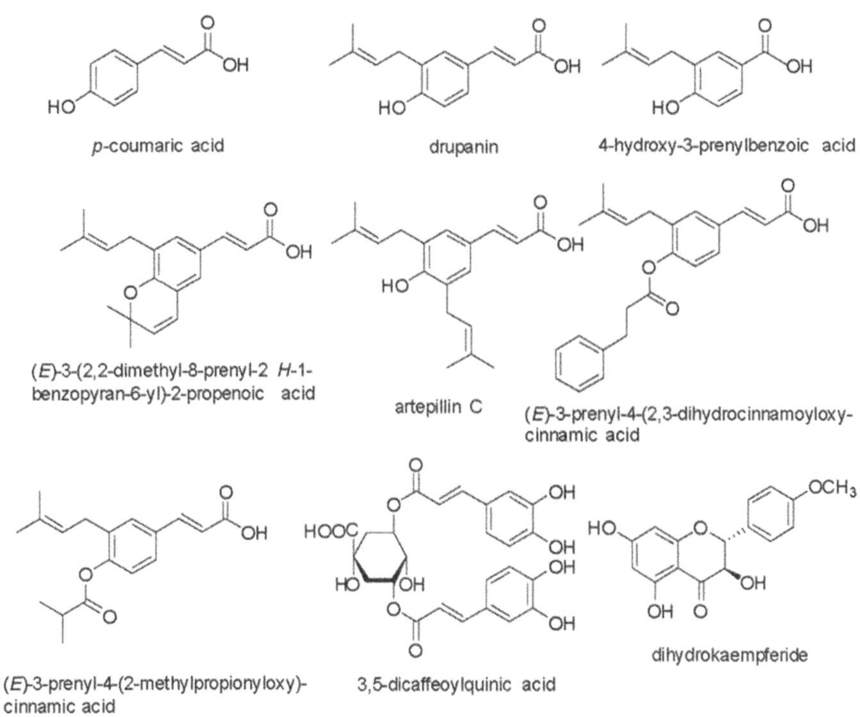

Fig. 20.1 Representative phenolic compounds in Brazilian green propolis

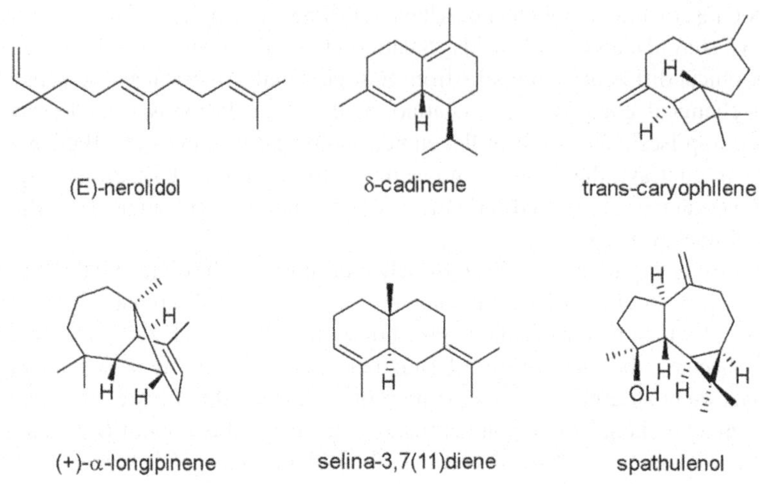

Fig. 20.2 Representative volatile compounds in Brazilian green propolis

The volatile compounds in the Brazilian green propolis have been reviewed (Bankova et al. 2014). The representative volatile compounds are shown in Fig. 20.2. Sesquiterpenes, such as (E)-nerolidol, δ-cadinene, and spathulenol, are the major volatile compounds in the propolis (de Sousa et al. 2009), and these compounds have also been shown to be responsible for its pleasant aroma as well as biological properties (de Sousa et al. 2009; Bankova et al. 2014).

Many analytical methods have been used for the separation and identification of the propolis components. The chemical composition of the propolis also depends on the method of extraction and the solvent used. Commercially sold medicinal products, such as tablets, capsules, and syrups, are usually prepared using the ethanol extracts of propolis. However, there are a few products that use the water extracts of propolis.

3 Antioxidant Properties of Green Propolis

Oxidative stress has been recognized to be involved in the pathology of several human diseases, such as diabetes, neurodegenerative diseases, and aging (Aruoma 1998). An increase in oxidative stress has been associated with many degenerative diseases such as atherosclerosis and cancer, through its effect on DNA mutation, protein oxidation, and lipid peroxidation (Aruoma 1998; Kim et al. 2015). Therefore, many studies have focused on investigating novel antioxidant compounds from natural products. Most propolis extracts are composed of different polyphenols, which have been reported to possess potent antioxidant activity (Toreti et al. 2013; Silva-Carvalho et al. 2015). However, the antioxidant activity of the propolis samples from different geographies varies, based on their chemical constitutions.

There are several methods to study the antioxidant properties of natural products. Because of operational simplicity, radical-scavenging assays such as DPPH (2,2-diphenyl-1-picrylhydrazyl) and ABTS (2,2′-azinobis-(3-ethylbenzothiazoline-6-acid) have been widely used to evaluate antioxidant activity. Other in vitro methods such as FRAP (ferric reducing antioxidant power) and ORAC (oxygen radical absorption capacity) assays are also used to evaluate the in vitro antioxidant activity (Bunaciu et al. 2016). Recently, cell-based in vitro antioxidant assays have been reported (Kellett et al. 2018). Further, the antioxidant activity of propolis from different sources including Brazilian green propolis has been evaluated (Silva-Carvalho et al. 2015).

Many studies have been conducted on the evaluation of the in vitro antioxidant activity of the Brazilian green propolis, using the DPPH assay. For example, Basnet et al. (1997) evaluated the DPPH-free-radical-scavenging activity of the water, methanol, and chloroform extracts of Brazilian propolis, and isolated the antioxidant compound, propol (3-[4-hydroxy-3-(3-oxo-but-1-enyl)-phenyl]-acrylic acid). Propol was found to possess more active antioxidant activity than vitamins C and E (Basnet et al. 1997). Souza et al. also evaluated the ethanol extracts of Brazilian propolis, using the DPPH assay and identified 3-prenyl-4-hydroxycinnamic acid (drupanin) and artepillin C to be the major antioxidants (Souza et al. 2007). Artepillin C has been identified also by Izuta et al. (2009) as the compound with DPPH radical-scavenging activity in Brazilian green propolis. They also reported that the caffeoylquinic acid derivatives (3,4-di-caffeoylquinic acid, 3,5-di-caffeoylquinic acid, and chlorogenic acid) are also the antioxidants found in the Brazilian green propolis. The caffeoylquinic acid derivatives exhibited a DPPH-radical-scavenging activity as strong as that of ascorbic acid and trolox (Izuta et al. 2009). Zhang et al. (2017) have also reported that artepillin C and caffeoylquinic acids were identified as antioxidant components from the Brazilian green propolis. Thus, artepillin C and caffeoylquinic acid derivatives could be the major effective antioxidants with a DPPH-radical-scavenging activity in the Brazilian green propolis. Table 20.1 summarizes some of the studies addressing the antioxidant activity of Brazilian green propolis.

The effects of Brazilian green propolis against UV-irradiation-induced oxidative stress have been evaluated by Fonseca et al. (2011). In the study, oral treatment of commercially available Brazilian green propolis extracts was shown to prevent UV-induced glutathione depletion in hairless mice. Murase et al. (2013) have reported that the ultraviolet A (UVA)-irradiation-induced production of reactive oxygen species is suppressed by pretreatment with the water extracts of Brazilian green propolis (WEP). The same group also reported that WEP acts as an early inducer of HO-1 and a rapid activator of Nrf2 to protect against UVA-induced oxidative stress (Saito et al. 2015).

A study has investigated the potential effects of the Brazilian green propolis on glucose metabolism and its antioxidant function in patients with type-2 diabetes mellitus (T2DM) (Zhao et al. 2016). At the end of the study, serum glutathione and total polyphenols were significantly increased, and serum carbonyls and lactate dehydrogenase activity were significantly reduced in the Brazilian green propolis

Table 20.1 Antioxidant activity of Brazilian green propolis

Type of extract/isolated compound(s)	Methods/activity	References
Water extract/propol	DPPH assay	Basnet et al. (1997)
Ethanol extract/drupanin, artepillin C	DPPH assay	Souza et al. (2007)
Water and ethanol extract/caffeoylquinic acid derivatives, artepillin C	DPPH assay	Izuta et al. (2009) Zhang et al. (2017)
Commercially available extract	UV-induced oxidation in the oral treatment of hairless mice	Fonseca et al. (2011)
Water extract	UV-induced oxidative cell damage	Murase et al. (2013) Saito et al. (2015)
Ethanol extract	Antioxidant function in patients with type 2 diabetes mellitus	Zhao et al. (2016)

group. Zhao et al. (2016) concluded that Brazilian green propolis is effective in improving the antioxidant status of T2DM patients.

4 Summary and Future Directions

Propolis has been commonly used as traditional medicine since ancient times. However, its constitution and biological properties (such as antioxidant activity) differ based on geographical origin. Moreover, the solvents used for extraction (i.e., water or ethanol) also influence the biological properties of propolis. Studies using both, ethanol and water extraction, have explored the antioxidant properties of Brazilian green propolis. While ethanol extracts of the propolis have been shown to contain cinnamic acid derivatives, such as drupanin or artepillin C, its water extracts have been shown to contain primarily caffeoylquinic acid derivatives. All these compounds, derived from ethanol and water extracts, have been shown to possess in vitro antioxidant activity. However, only a few studies have explored the in vivo antioxidant activity of Brazilian green propolis. Furthermore, information on the safe dose needed for clinical use is lacking. Therefore, clinical studies using Brazilian green propolis and its active compounds are needed.

References

Aruoma OI (1998) Free radicals, oxidative stress, and antioxidants in human health and disease. J Am Oil Chem Soc 75:199–212. https://doi.org/10.1007/s11746-998-0032-9

Bankova V (2005) Chemical diversity of propolis and the problem of standardization. J Ethnopharmacol 100:114–117. https://doi.org/10.1016/j.jep.2005.05.004

Bankova V, Boudourova-Krasteva G, Popov S, Sforcin JM, Funari SRC (1998) Seasonal variations of the chemical composition of Brazilian propolis. Apidologie 29:361–367. https://doi.org/10.1051/apido:19980406

Bankova V, de Castro SL, Marcucci MC (2000) Propolis: recent advances in chemistry and plant origin. Apidologie 31:3–15. https://doi.org/10.1051/apido:2000102

Bankova V, Popova M, Bogdanov S, Sabatini AG (2002) Chemical composition of European propolis: expected and unexpected results. Z Naturforschung C J Biosci 57:530–533. https://doi.org/10.1515/znc-2002-5-622

Bankova V, Popova M, Trusheva B (2014) Propolis volatile compounds: chemical diversity and biological activity: a review. Chem Cent J 8:1–8. https://doi.org/10.1186/1752-153X-8-28

Basnet P, Matsuno T, Neidlein R (1997) Potent free radical scavenging activity of propol isolated from Brazilian propolis. Z Naturforsch C J Biosci 52:828–833. https://doi.org/10.1515/znc-1997-11-1217

Búfalo MC, Candeias JM, Sousa JP, Bastos JK, Sforcin JM (2010) In vitro cytotoxic activity of *Baccharis dracunculifolia* and propolis against HEp-2 cells. Nat Prod Res 24:1710–1718. https://doi.org/10.1080/14786410902909300

Bunaciu AA, Danet AF, Fleschin S, Aboul-Enein HY (2016) Recent applications for in vitro antioxidant activity assay. Crit Rev Anal Chem 46:389–399. https://doi.org/10.1080/10408347.2015.1101369

Céspedes CL, Uchoa A, Salazar JR, Perich F, Pardo F (2002) Plant growth inhibitory activity of *p*-hydroxyacetophenones and tremenones from Chilean endemic *Baccharis* species and some analogous: a comparative study. J Agric Food Chem 50:2283–2292. https://doi.org/10.1021/jf011108g

da Silva Leitão DP, da Silva Filho AA, Polizello ACM, Bastos JK, Spadaro ACC (2004) Comparative evaluation of in-vitro effects of Brazilian green propolis and *Baccharis dracunculifolia* extracts on cariogenic factors of *Streptococcus mutans*. Biol Pharm Bull 27:1834–1839. https://doi.org/10.1248/bpb.27.1834

de Figueiredo-Rinhel ASG, de Andrade MF, Landi-Librandi AP, Azzolini AECS, Kabeya LM, Bastos JK, Lucisano-Valim YM (2019) Incorporation of *Baccharis dracunculifolia* DC (Asteraceae) leaf extract into phosphatidylcholine-cholesterol liposomes improves its anti-inflammatory effect in vivo. Nat Prod Res 33:2521–2525. https://doi.org/10.1081/14786419.2018.1448809

de Funari CS, de Oliveira FV, Mathor MB (2007) Analysis of propolis from *Baccharis dracunculifolia* DC. (Compositae) and its effects on mouse fibroblasts. J Ethnopharmacol 111:206–212. https://doi.org/10.1016/j.jep.2006.11.032

de Sousa JPB, Jorge RF, Leite MF, Furtado NAJC, Bastos JK, da Silva Filho AA, Queiroga CL, de Magalhães PM, Soarea AEE (2009) Seasonal variation of the (E)-nerolidol and other volatile compounds within ten different cultivated populations of *Baccharis dracunculifolia* D.C. (Asteraceae). J Essent Oil Res 21:308–314. https://doi.org/10.1080/10412905.2009.9700179

Fonseca YM, Marquele-Oliveira F, Vicentini FT, Furtado NA, Sousa JP, Lucisano-Valim YM, Fonseca MJ (2011) Evaluation of the potential of Brazilian propolis against UV-induced oxidative stress. Evid Based Complement Alternat Med 2011:1–9. https://doi.org/10.1155/2011/863917

Guimarães NS, Mello JC, Paiva JS, Bueno PC, Beretta AA, Torquato RJ, Nantes IL, Rodrigues T (2012) *Baccharis dracunculifolia*, the main source of green propolis, exhibits potent antioxidant activity and prevents oxidative mitochondrial damage. Food Chem Toxicol 50:1091–1097. https://doi.org/10.1016/j.fct.2011.11.014

Hata T, Tazawa S, Ohta S, Rhyu MR, Misaka T, Ichihara K (2012) Artepillin C, a major ingredient of Brazilian propolis, induces a pungent taste by activating TRPA1 channels. PLoS One 7:e48072. https://doi.org/10.1371/journal.pone.9948072

Izuta H, Narahara Y, Shimazawa M, Mishima S, Kondo S, Hara H (2009) 1,1-diphenyl-2-picrylhydrazyl radical scavenging activity of bee products and their constituents determined by ESR. Biol Pharm Bull 32:1947–1951. https://doi.org/10.1248/bpb.32.1947

Juan Hikawczuk VE, Robert Saad J, Giordano OS, García C, Martín T, Martín VS, Sosa ME, Tonn CE (2008) Insect growth regulatory effects of linear diterpenoids and derivatives from *Baccharis thymifolia*. J Nat Prod 71:190–194. https://doi.org/10.1021/np070392o

Kellett ME, Greenspan P, Pegg RB (2018) Modification of the cellular antioxidant activity (CAA) assay to study phenolic antioxidants in a Caco-2 cell line. Food Chem 244:359–363. https://doi.org/10.1016/j.foodchem.2017.10.035

Kim GH, Kim JE, Rhie SJ, Yoon S (2015) The role of oxidative stress in neurodegenerative diseases. Exp Neurobiol 24:325–340. https://doi.org/10.5607/en.2015.24.4.325

Kumazawa S, Yoneda M, Shibata I, Kanaeda J, Hamasaka T, Nakayama T (2003) Direct evidence for the plant origin of Brazilian propolis by the observation of honeybee behavior and phytochemical analysis. Chem Pharm Bull 51:740–742. https://doi.org/10.1248/cpb.51.740

Kumazawa S, Hamasaka T, Nakayama T (2004) Antioxidant activity of propolis of various geographic origins. Food Chem 84:329–339. https://doi.org/10.1016/S0308-8146(03)00216-4

Marcucci MC, Bankova V (1999) Chemical composition, plant origin and biological activity of Brazilian propolis. Curr Top Phytochem 2:115–123

Messerli SM, Ahn MR, Kunimasa K, Yanagihara M, Tatefuji T, Hashimoto K, Mautner V, Uto Y, Hori H, Kumazawa S, Kaji K, Ohta T, Maruta H (2009) Artepillin C (ARC) in Brazilian green propolis selectively blocks oncogenic PAK1 signaling and suppresses the growth of NF tumors in mice. Phytother Res 23:423–427. https://doi.org/10.1002/ptr.2658

Murase H, Shimazawa M, Kakino M, Ichihara K, Tsuruma K, Hara H (2013) The effects of Brazilian green propolis against excessive light-induced cell damage in retina and fibroblast cells. Evid Based Complement Alternat Med 2013:1–12. https://doi.org/10.1155/2013/238279

Park YK, Alencar SM, Aguiar CL (2002) Botanical origin and chemical composition of Brazilian propolis. J Agric Food Chem 50:2502–2506. https://doi.org/10.1021/jf011432b

Park YK, Paredes-Guzman JF, Aguiar CL, Alencar SM, Fujiwara FY (2004) Chemical constituents in *Baccharis dracunculifolia* as the main botanical origin of southeastern Brazilian propolis. J Agric Food Chem 52:1100–1103. https://doi.org/10.1021/jf021060m

Pasupuleti VR, Sammugam L, Ramesh N, Gan SH (2017) Honey, propolis, and royal jelly: a comprehensive review of their biological actions and health benefits. Oxidative Med Cell Longev 2017:1–22. https://doi.org/10.1155/2017/1259510

Paulino N, Abreu SR, Uto Y, Koyama D, Nagasawa H, Hori H, Dirsch VM, Vollmar AM, Scremin A, Bretz WA (2008) Anti-inflammatory effects of a bioavailable compound, Artepillin C, in Brazilian propolis. Eur J Pharmacol 587:296–301. https://doi.org/10.1016/j.ejphar.2008.02067

Pereira CA, Costa AC, Liporoni PC, Rego MA, Jorge AO (2016) Antibacterial activity of *Baccharis dracunculifolia* in planktonic cultures and biofilms of *Streptococcus* mutans. J Infect Public Health 9:324–330. https://doi.org/10.1016/j.jiph.2015.10.012

Rahalison L, Benathan M, Monod M, Frenk E, Gupta MP, Solis PN, Fuzzati N, Hostettmann K (1995) Antifungal principles of *Baccharis pedunculata*. Planta Med 61:360–362. https://doi.org/10.1055/s-2006-958101

Saito Y, Tsuruma K, Ichihara K, Shimazawa M, Hara H (2015) Brazilian green propolis water extract up-regulates the early expression level of HO-1 and accelerates Nrf2 after UVA irradiation. BMC Complement Altern Med 15:1–8. https://doi.org/10.1186/s12906-015-0945-4

Salatino A, Fernandes-Silva CC, Righi AA, Salatino MLF (2011) Propolis research and the chemistry of plant products. Nat Prod Rep 28:925–936. https://doi.org/10.1039/c0np00072h

Sforcin JM (2016) Biological properties and therapeutic applications of propolis. Phytother Res 30:894–905. https://doi.org/10.1002/ptr.5605

Silva-Carvalho R, Baltazar F, Almeida-Aguiar C (2015) Propolis: a complex natural product with a plethora of biological activities that can be explored for drug development. Evid Based Complement Alternat Med 2015:1–30. https://doi.org/10.1155/2015/206439

Souza RM, de Souza MC, Patitucci ML, Silva JF (2007) Evaluation of antioxidant and antimicrobial activities and characterization of bioactive components of two Brazilian propolis samples using a pKa-guided fractionation. Z Naturforsch C J Biosci 62:801–807. https://doi.org/10.1515/znc-2007-11-1205

Toreti VC, Sato HH, Pastore GM, Park YK (2013) Recent progress of propolis for its biological and chemical composition and its botanical origin. Evid Based Complement Alternat Med 2013:1–14. https://doi.org/10.1155/2013/697390

Zhang C, Shen X, Chen J, Jiang X, Hu F (2017) Identification of free radical scavengers from Brazilian green propolis using off-line HPLC-DPPH assay and LC-MS. J Food Sci 82:1602–1607. https://doi.org/10.1111/1750-3841.13730

Zhao L, Pu L, Wei J, Li J, Wu J, Xin Z, Gao W, Guo C (2016) Brazilian green propolis improves antioxidant function in patients with type 2 diabetes mellitus. Int J Environ Res Public Health 13:1–9. https://doi.org/10.3390/ijerph13050498

Chapter 21
Possible Role of Propolis-Derived Components in the Prevention and Treatment of Obesity and Diabetes

Takanori Tsuda

Abstract Propolis consists of the gums, saps, pigments, and essential oils collected by honey bees from particular parts of plants, mainly the sprouts, buds, and bark, and is mixed together with bees' own secretory substances. Recent studies demonstrated that one of the key plant sources in Brazilian green propolis is *Baccharis dracunculifolia*. This chapter presents the possible role of Brazilian green propolis–derived components in the prevention and treatment of obesity and diabetes. Especially, we focus on the regulation of white adipocyte function (upregulating adiponectin expression) and induction of brown-like adipocyte formation by Brazilian green propolis constituents. We demonstrated that artepillin C (ArtC), which is one of the abundant constituents in Brazilian green propolis, significantly inhibited the downregulation of adiponectin expression in white adipocytes. Moreover, we found that ArtC induces brown-like adipocyte formation in white adipose tissues. Finally, we summarize the challenges concerning research on Brazilian green propolis' health benefits. This fundamental research has established a promising new function for Brazilian green propolis constituents, with several exciting possibilities for its application in dietary supplements and the like.

Keywords Adipocyte · Adiponectin · Artepillin C · Bio-economy · Peroxisome proliferator-activated receptor

---

T. Tsuda (✉)
College of Bioscience and Biotechnology and Graduate School of Bioscience and Biotechnology, Chubu University, Kasugai, Aichi, Japan
e-mail: tsudat@isc.chubu.ac.jp

G. W. Fernandes et al. (eds.), *Baccharis*,
https://doi.org/10.1007/978-3-030-83511-8_21

1 Introduction

Humans have been making use of honeybee products since ancient times, most notably in the consumption of honey. Moreover, honeybees are essential to the pollination of crops. Recent concerns over the shortage of honeybees have sparked an interest in studying honeybees and understanding the reasons for their diminishing numbers. In particular, one honeybee product that has been studied extensively over the past decade is Brazilian green propolis, the health benefits of which have been investigated.

Perhaps, the most promising of Brazilian green propolis' health benefits is the prevention of obesity and diabetes. Both obesity and diabetes have become significant global problems in recent decades. In Asia, especially, there has been a notable increase in diabetes in recent years (Scully 2012). This health crisis must be resolved urgently, and toward this end, new means of prevention and treatments using dietary factors are awaited.

To date, various food-derived factors have been investigated for their preventive and curative roles in obesity and diabetes. While there have been some publications on the propolis ethanol extract, data on individual propolis components and their comparative activities are insufficient. In addition, the antiobesity and antidiabetic mechanisms are not yet well understood. To give the current status of research in this area, this chapter presents our progress on the regulation of white adipocyte function (upregulating adiponectin expression) and induction of brown-like adipocyte formation by Brazilian green propolis constituents.

2 Propolis Plant Sources and Health Benefits in Humans

Propolis is a viscous, resinous substance, varying from dark green and brown to dark brown in color. It consists of the gums, saps, pigments, and essential oils collected by honeybees from particular parts of plants, mainly the sprouts, buds, and bark, and is mixed together with bees' own secretory substances. The history of propolis goes as far back as ancient Egypt and Greece, where traditional medicinal uses for propolis were already known (e.g., as an antibacterial treatment) (Kartal et al. 2003). Many physiological functions of propolis have been reported as its chemical constituents have been analyzed over the past 20 years. The use of propolis is expanding both as a popular health food and dietary supplement and as an additive in beauty products (e.g., skin creams) and candies.

Meanwhile, there has also been a considerable amount of false information being spread on the health benefits and chemical composition of propolis. One likely reason for this misinformation involves the premature marketing of propolis on a large scale before the chemistry of its constituents has been adequately elucidated. The chemical composition of propolis varies; more often than not, a constituent present in one propolis is absent in another. The exact composition of propolis depends entirely on its place of origin as well as its extraction method. So, these differences naturally affect the physiological effect and strength of each propolis.

One propolis in particular, Brazilian green propolis, is one of the most studied propolis worldwide. It is characterized by its many cinnamic acid derivatives, such as artepillin C (ArtC). Propolis composition varies by the place of origin because of the different plant sources involved in its creation. From direct behavioral observation of honeybees as well as from chemical analyses, Kumazawa et al. have determined that one of the key plant sources in Brazilian green propolis is *Baccharis dracunculifolia* (Kumazawa et al. 2003).

As much as propolis is a fascinating natural product from the perspective of chemistry, we will focus here on the specific effects that ArtC from *B. dracunculifolia*-derived Brazilian green propolis has on adipose tissues as an example of our research related to the health benefit.

3 Brazilian Green Propolis Constituents and their Physiological Functions

Regulation of White Adipocyte Function

ArtC is a key constituent of Brazilian green propolis, of which *Baccharis dracunculifolia* is one plant source. Besides ArtC, this propolis also contains various other cinnamic acid derivatives, which contain a prenyl group (Fig. 21.1) (Tazawa et al. 1998; Tazawa et al. 1999; Ikeda et al. 2011). In collaboration with the University of Shizuoka, Kitasato University, and Kyoto University, we have demonstrated the ability of Brazilian green propolis–derived compounds to augment adiponectin expression, which in turn affects insulin sensitivity (Ikeda et al. 2011).

Mammals, including humans, have two types of adipose tissues. While brown adipose tissue (BAT) is an organ responsible for fat catabolism and heat generation, white adipose tissue (WAT) exists as subcutaneous and visceral fat. Over the past 20 years, studies have shown that WAT is not just an adipose depot; it is in fact the largest endocrine tissue in the body. It produces and secretes various physiologically active substances, which exert a great influence on the human body as a whole (Matsuzawa 2005). Such bioactive substances secreted by white adipocytes are called adipocytokines. The regulation of adipocytokine expression and secretion is disrupted in hypertrophic adipocytes. To date, numerous adipocytokines and their close relationship with various metabolic disorders and diseases have been identified.

The relationship between obesity and WAT inflammation, and the molecular mechanism thereof, is rapidly being elucidated. After macrophages infiltrate adipose tissue, they secrete an inflammatory cytokine called tumor necrosis factor-α (TNF-α). TNF-α induces chronic inflammatory changes in WAT and disrupts white adipocyte function. Using this white adipocyte inflammatory model as a basis, we developed an evaluation system for food-derived factors that regulate adipocyte

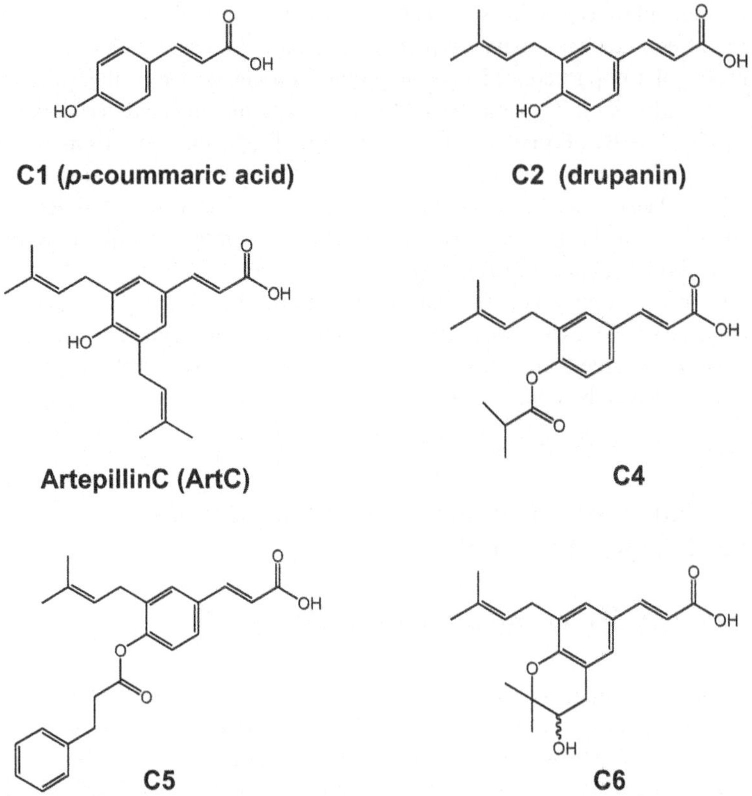

Fig. 21.1 Chemical structure of the propolis-derived components. This figure was reproduced from Ikeda et al. (2011) with permission from Elsevier

function (i.e., adiponectin expression upregulators) (Isa et al. 2008; Ikeda et al. 2011; Yanagisawa et al. 2012). This system revealed that out of the many compounds derived from Brazilian green propolis, ArtC and compound C4 ((*E*)-3-(4-(isobutyryloxy)-3-(3-methylbut-2-enyl) phenyl) acrylic acid) significantly inhibited TNF-α-induced downregulation of adiponectin (Fig. 21.2) (Ikeda et al. 2011). Furthermore, these two compounds were both able to upregulate adiponectin secretions into the culture supernatant (Fig. 21.2) (Ikeda et al. 2011).

Structure–activity relationship analysis revealed that two prenyl groups must be present in order to exert this effect. The presence of a single prenyl group alone did not suffice, with the notable exception of compound C4, which has an isobutylic acid ester at the 4-position in its structure (Ikeda et al. 2011).

Next, we investigated how the two identified compounds inhibit the downregulation of adiponectin. Prior research had shown that adiponectin expression is controlled by peroxisome proliferator–activated receptor γ (PPARγ). A possible mechanism is that our propolis-derived compounds act as PPARγ agonists, thereby driving adiponectin expression. Another possible mechanism by which these

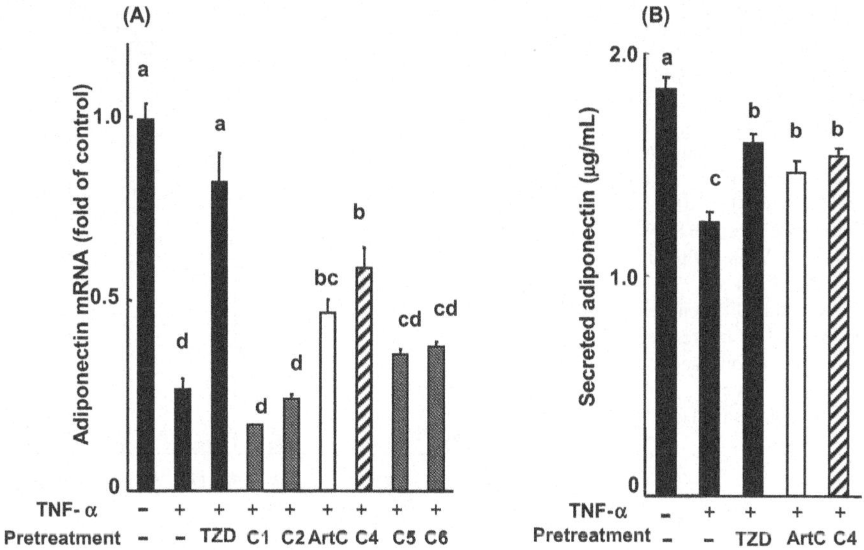

Fig. 21.2 Inhibitory effect of propolis-derived components on the TNF-α-mediated downregulation of adiponectin gene expression (**a**) and the secreted adiponectin concentration in media (**b**) in 3T3-L1 adipocytes. (**a**, **b**) Final concentration of various propolis-derived components was 25 μM. (**a**) Adiponectin gene expression level is expressed relative to the control (=1.0) after normalization using the β-actin gene expression level. Values are means ± SEM, n = 3–6. Values without a common letter are significantly different at $p < 0.05$. This figure was reproduced from Ikeda et al. (2011) with permission from Elsevier

compounds might inhibit the TNF-α-mediated downregulation of adiponectin is through modulation of the c-Jun-NH2-terminal kinase (JNK) signaling pathway. We previously demonstrated that inhibition of TNF-α-mediated JNK activation suppresses the decrease in adiponectin expression (Isa et al. 2008). Based on the above hypothesis, we investigated how the two propolis constituents inhibit the downregulation of adiponectin. Reporter assay results revealed that ArtC was a powerful agonist of PPARγ (Fig. 21.3) (Ikeda et al. 2011). Similar activity was also observed for C4, but the effect was much weaker than that of ArtC, even at a final concentration of 25 μM (Fig. 21.3) (Ikeda et al. 2011). Subsequently, to better understand PPARγ's target gene expression, we tested the effects of ArtC alone or ArtC with PPARγ antagonist pretreatment. The expression levels of both adiponectin and fatty acid–binding protein 4 (aP2) significantly increased after the single ArtC treatment, but this effect was significantly suppressed by the PPARγ antagonist pretreatment. Meanwhile, we conducted the same test with C4 but found that it did not increase the expression of PPARγ target genes (Fig. 21.4) (Ikeda et al. 2011).

Taking these findings together, we concluded that ArtC's effect on upregulation of adiponectin expression is mediated by its PPARγ agonistic effect. However, as this mechanism did not appear to be applicable to C4, we decided to investigate another possible mechanism of action. As a result of this study, C4 significantly

Fig. 21.3 PPARγ transcriptional activity induced by ArtC, C4, or the positive control (TZD). Relative luciferase activities are presented as a fold increase relative to the vehicle control (=1.0). Values are means ± SEM, n = 5–10. Values without a common letter are significantly different at $p < 0.05$. This figure was reproduced from Ikeda et al. (2011) with permission from Elsevier

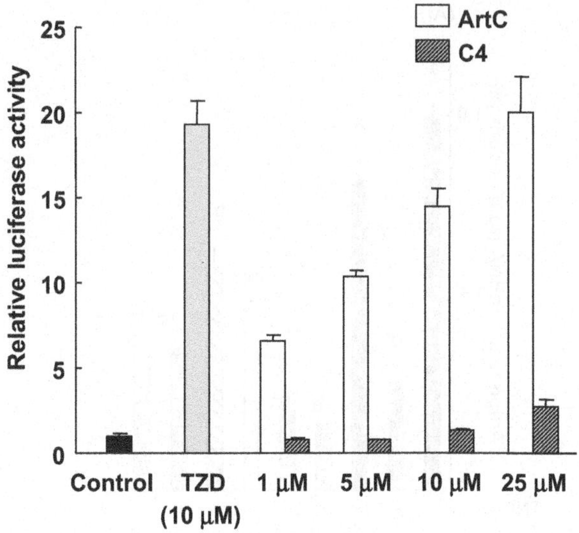

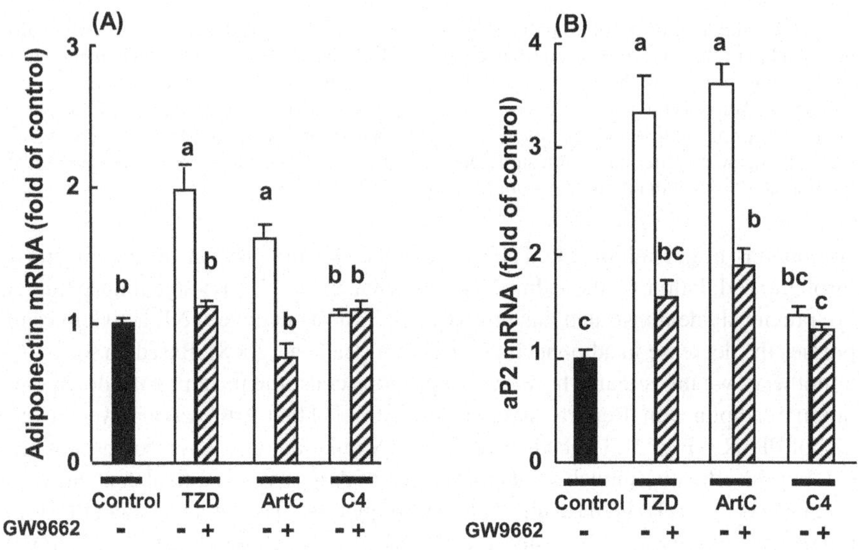

Fig. 21.4 The gene expression levels of adiponectin (**a**) and adipocyte fatty acid–binding protein 4 (aP2) (**b**) in 3T3-L1 adipocytes treated with ArtC, C4 (both 25 μM), or TZD (10 μM) with or without the PPARγ antagonist (GW9662, 20 μM). The gene expression level was expressed relative to the control (=1.0) after normalization using the β-actin gene expression level. Values are means ± SEM, n = 3–6. Values without a common letter are significantly different at $p < 0.05$. This figure was reproduced from Ikeda et al. (2011) with permission from Elsevier

suppressed the TNF-α-induced phosphorylation of JNK. Thus, C4 most likely acted directly on JNK, because it did not inhibit the upstream kinases of JNK. Meanwhile, ArtC, which we showed acts as a PPARγ agonist, did not inhibit the TNF-α-induced activation of JNK (Ikeda et al. 2011).

To further investigate the above results, we performed docking simulations and compared the predicted structure of the complex with that of known ligands and inhibitors. First, we investigated the PPARγ–ArtC complex by selecting the best binding pose (Fig. 21.5) (Ikeda et al. 2011). It became clear that ArtC displays a binding pose very similar to that of a known PPARγ agonist called troglitazone (TZD), which presumably forms hydrogen bonds and hydrophobic interactions with neighboring amino acids. It was also presumed based on binding free energy that ArtC has a strong PPARγ agonistic effect, which was supported by the experimental results thus far. Similarly, we performed docking analyses with JNK. First, we evaluated a known inhibitor of JNK in addition to C4 and determined that both of them bind to the JNK protein via hydrogen bonding and ionic bonding, displaying very similar docking poses to one another. On the other hand, when we evaluated ArtC, it differed in its orientation and a comparison of binding affinity indicated that it does not inhibit JNK (Fig. 21.6) (Ikeda et al. 2011).

These findings could be useful for determining the different applications of individual propolis constituents as well as for designing propolis-containing dietary supplements with specific fortified effects. For instance, since ArtC increases adiponectin expression, it could be given to people whose adiponectin level is naturally low. On the other hand, since C4 counteracts the inflammation-induced reduction of

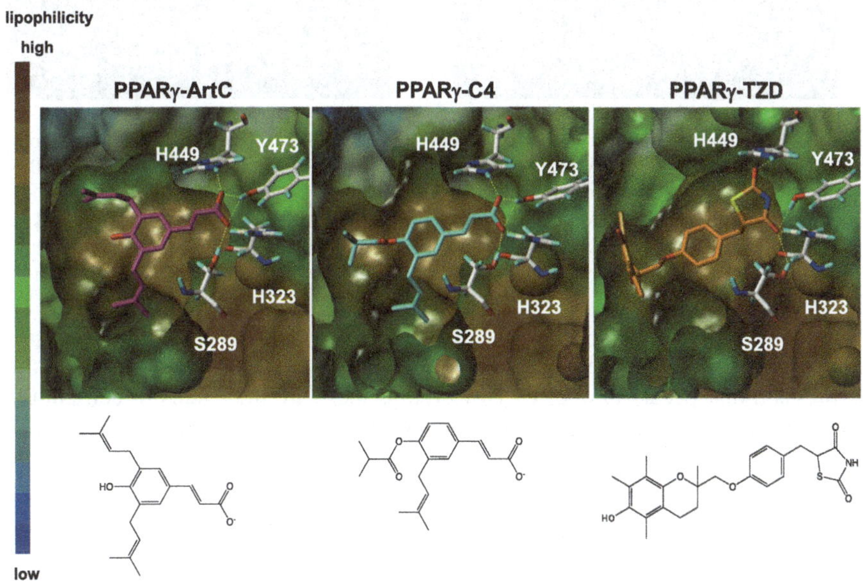

Fig. 21.5 Binding modes of ArtC, C4, or TZD with the PPARγ LBD. This figure was reproduced from Ikeda et al. (2011) with permission from Elsevier

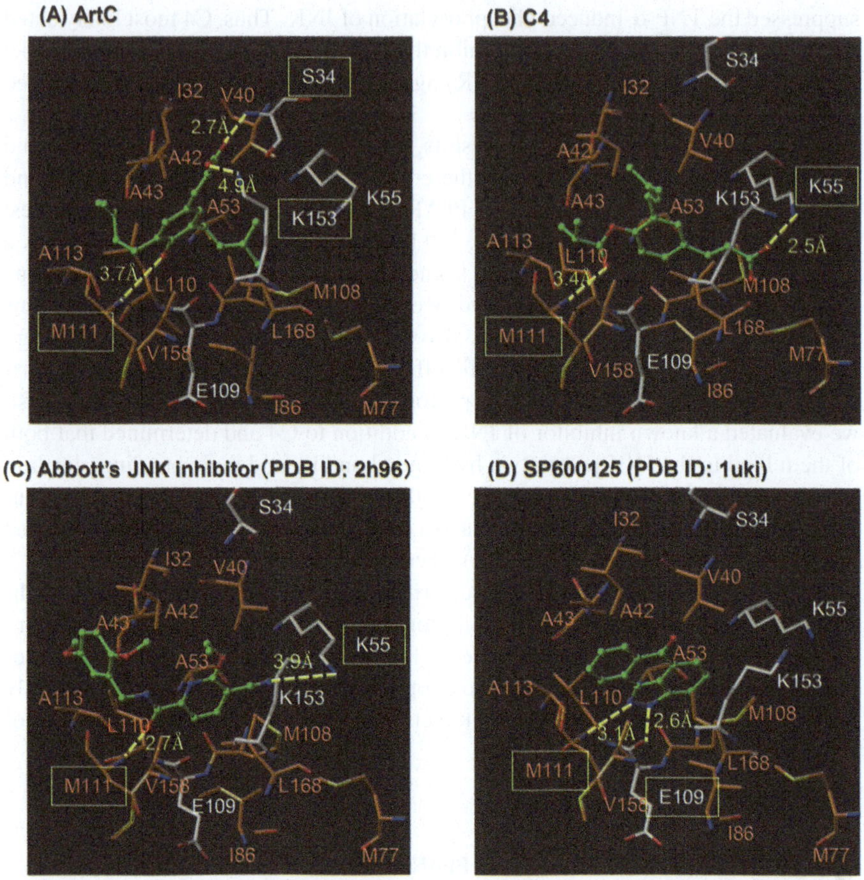

Fig. 21.6 Comparison between docking models and X-ray structures of JNK1-inhibitor complexes. (**a**) Docking model of ArtC, (**b**) docking model of C4, (**c**) X-ray structure of Abbott's JNK inhibitor (PDB identification code: 2 h96), and (**d**) X-ray structure of SP600125 (PDB identification code: 1uki). Inhibitors are represented by ball and stick models. JNK1 is represented by capped stick models and hydrophobic residues are colored orange. Hydrogen bonds and electrostatic interactions are represented by dashed lines. This figure was reproduced from Ikeda et al. (2011) with permission from Elsevier

adiponectin expression in WAT, it may be able to correct the anomalous adiponectin expression that occurs in obesity. However, as these are the results of cell culture experiments, the exact effects in humans must be further investigated (e.g., with clinical trials).

Adipocyte Browning Induction and Mechanism of Action

Recent discoveries have led to a new era of adipocyte research. As mentioned previously, there are two types of mammalian adipose tissues, WAT and BAT. WAT is an energy depot, and one of the body's most significant endocrine organs. It expresses adipocytokines that regulate biological functions. Meanwhile, BAT consumes fat and generates body heat, taking part in thermoregulation (Giralt and Villarroya 2013). Naturally, it follows that an increase in BAT promotes overall energy consumption. Brown adipocytes are rich in mitochondria and contain a brown adipocyte–specific mitochondrial protein called uncoupling protein 1 (UCP1). This protein is involved in heat production in the body. UCP1 is a marker of brown adipocytes and its role is to convert energy to heat by uncoupling oxidative phosphorylation (Cannon and Nedergaard 2004). UCP1 is essential in cold-induced thermoregulation.

BAT has long been deemed physiologically unimportant as it is abundant in newborns but lacking in adults. Recently, however, the presence of BAT has been confirmed in adults as well, and individual differences have also been identified. In addition, BAT quantity decreases with age, which is one of the causes of obesity (Cypess et al. 2009; Saito et al. 2009; van Marken Lichtenbelt et al. 2009; Virtanen et al. 2009; Nedergaard and Cannon 2010; Yoneshiro et al. 2011). Furthermore, there are two distinct populations of brown adipocytes: the classical brown adipocytes that are naturally present, and inducible brown adipocytes (brown-like adipocytes) that differentiate from adipocyte precursor cells but are distinct from the classical ones (Kajimura et al. 2010; Harms and Seale 2013). The brown-like adipocytes are also called beige adipocytes or brite adipocytes and appear in WAT. Their presence leads to a continuous increase in energy consumption due to heat production as well as to a loss of body fat (Yoneshiro et al. 2011). Presently, increasing these energy consumption-promoting brown-like adipocytes is considered an effective and novel strategy to counter obesity (Wu et al. 2012). The induction of brown-like adipocytes formation could be achieved by 1) having a continuous cold stimulus, or 2) inducing the brown adipocyte differentiation of human ES/iPS followed by transplantation. However, method 1) is unrealistic and method 2) is impractical at this time, but it may become feasible in the near future. All in all, a better option seems to be to find a food-derived factor that induces the differentiation of adipocyte precursors into brown-like adipocytes. To do this, the mechanism of action needs to be elucidated in order to form the basis of novel antiobesity and antidiabetic strategies.

Hence, our group decided to investigate food-derived factors that induce brown-like adipocyte formation. In this process, we studied the brown-like adipocyte-inducing ability of ArtC and its mechanism of action (Nishikawa et al. 2016). We knew that thiazolidinediones, such as rosiglitazone, act as PPARγ agonists and that their adiponectin expression–enhancing ability is an effective treatment of diabetes. Furthermore, studies have shown that rosiglitazone can induce brown-like adipocyte formation (Ohno et al. 2012). As earlier results indicated, ArtC also acts as a

PPARγ agonist (Ikeda et al. 2011); thus, it seemed entirely plausible that ArtC could also induce brown-like adipocyte formation.

Therefore, we investigated the brown adipocyte–inducing ability of ArtC using two in vitro evaluation systems using C3H10T1/2 cells (Reznikoff et al. 1973) as well as murine WAT-derived primary cells. C3H1010T1/2 cells are derived from murine mesenchymal stem cells and can differentiate into white and brown adipocytes depending on the conditions. The murine WAT-derived primary cells were prepared from the adipocyte precursor cell–containing fraction of collagenase-treated murine WAT. Brown-like adipocyte induction was evaluated by measuring the expression of brown-like adipocyte marker genes such as UCP1 and cell death–inducing DNA fragmentation factor alpha subunit-like effector A (CideA). From the results of the study, in C3H10T1/2 cells, ArtC strongly increased the mRNA expression of brown-like adipocyte markers in a dose-dependent manner (Fig. 21.7) (Nishikawa et al. 2016). ArtC also significantly increased the expression of brown adipocyte marker mRNA in murine WAT–derived primary cells (Fig. 21.7) (Nishikawa et al. 2016). In addition, ArtC induced the protein expression of UCP1 and PRD1-BF-RIZ1 homologous domain-containing protein-16 (PRDM16) (Fig. 21.8) (Nishikawa et al. 2016), the latter being an essential factor for brown adipocyte differentiation. Previously, we showed that ArtC acts as a PPARγ agonist (Ikeda et al. 2011); so we then investigated the relationship among ArtC, induction of brown-like adipocyte formation, and PPARγ ligand activity. To do this, we treated C3H10T1/2 cells with a combination of ArtC and the PPARγ antagonist. As a result,

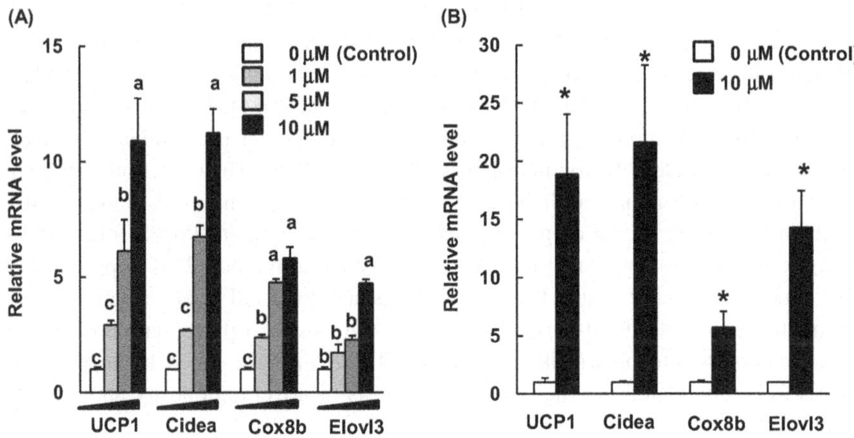

Fig. 21.7 mRNA levels of brown-like adipocyte markers in C3H10T1/2 cells and primary iWAT-derived adipocytes treated with ArtC or vehicle (control). (**a**) C3H10T1/2 cells and (**b**) primary iWAT-derived adipocytes. The mRNA levels are expressed as fold-change relative to control (=1) after normalization using the expression levels of the TBP gene. All data are shown as means ± SEM ($n = 3$). (**a**) Values without a common letter are significantly different at $P < 0.05$; (**b**) * mean values are significantly different from those of the control at $P < 0.05$. Elovl3, elongation of very long-chain fatty acids-like 3. This figure was reproduced from Nishikawa et al. (2016)

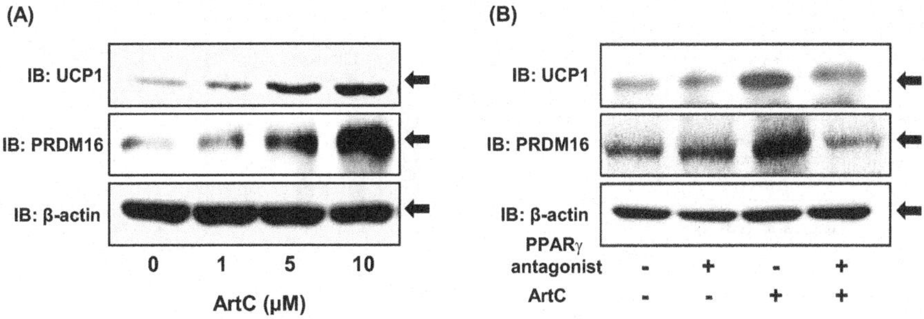

Fig. 21.8 Immunoblot analysis of UCP1, PRDM16, and β-actin proteins in C3H10T1/2 cells differentiated with vehicle or ArtC. (**a**) The expression of UCP1, PRDM16, and β-actin proteins in C3H10T1/2 cells differentiated with ArtC at different doses; (**b**) the effect of the PPARγ antagonist on ArtC-induced UCP1 and PRDM16 protein expression. This figure was reproduced from Nishikawa et al. (2016)

both the UCP1 and PRDM16 protein expression-enhancing effects of ArtC were significantly suppressed by the PPARγ antagonist (Fig. 21.8) (Nishikawa et al. 2016). These results indicated that ArtC induces brown-like adipocyte formation, presumably through its PPARγ-agonist function.

We then looked deeper into the mechanism of brown-like adipocyte induction by ArtC. As shown in previous studies, rosiglitazone induces brown-like adipocyte formation by binding to the complex of the PRDM16 protein and PPARγ as a PPARγ ligand, which stabilizes the PRDM16 protein (i.e., suppressing its degradation) (Ohno et al. 2012). Because of this, it also seemed plausible that ArtC works similarly. To verify this hypothesis, we observed the PRDM16 protein degradation dynamics present in C3H10T1/2 cells by 1) treatment with cycloheximide, a protein synthesis inhibitor, after inducing brown-like adipocyte differentiation with continuous ArtC treatment, and 2) culturing under the control condition and then administering a single dose of ArtC together with cycloheximide. From the results of experiment 1), PRDM16 protein degradation was significantly inhibited in the ArtC group for 4, 8, and 16 h after cycloheximide treatment. The half-life of PRDM16 was calculated as 6.2 h in the control group compared with 19.2 h in the ArtC group (i.e., ArtC more than triple the PRDM16 protein's half-life) (Fig. 21.9) (Nishikawa et al. 2016). Additionally, from experiment 2), PRDM16 protein's degradation was also significantly inhibited for 16 h after cycloheximide treatment (Fig. 21.9) (Nishikawa et al. 2016). PRDM16's half-life for this experiment was 9.4 h in the control group compared with 18.2 h in the ArtC group. These results demonstrated that ArtC acts as a PPARγ ligand and induces brown-like adipocyte formation by inhibiting PRDM16 protein degradation.

Subsequently, we studied the brown-like adipocyte induction of ArtC in vivo. To do so, we administered 5 mg/kg and 10 mg/kg of ArtC orally to C57BL/6 J mice every day for 4 weeks. Hematoxylin and eosin (H&E) staining revealed that induction of brown-like adipocyte formation was dose-dependent in the inguinal WAT (iWAT), visible as multilocular adipocytes. Immunostaining with anti-UCP1 and

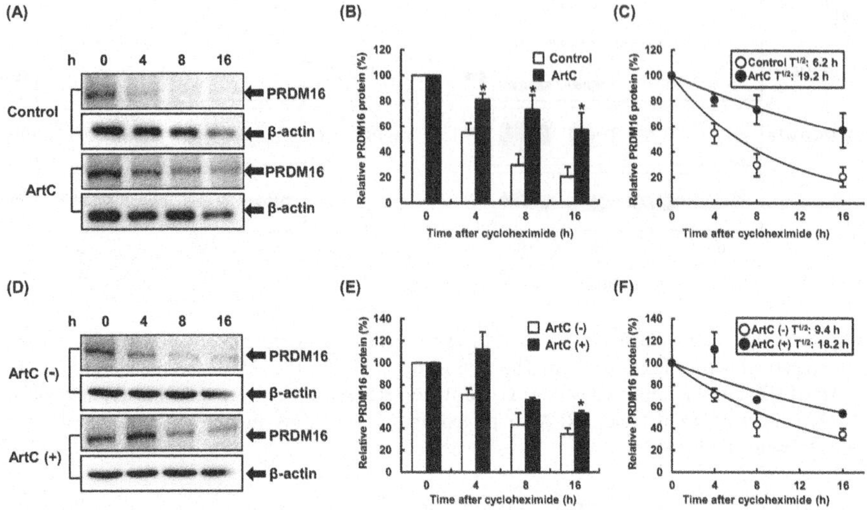

Fig. 21.9 The degradation of PRDM16 protein levels in C3H10T1/2 cells treated with vehicle or ArtC. (**a**, **b**) The degradation of PRDM16 protein in C3H10T1/2 cells differentiated with or without ArtC. C3H10T1/2 cells were differentiated with or without ArtC for 8 days, and then cells were treated with cycloheximide and the degradation of PRDM16 protein was examined using immunoblot analysis at the indicated time point. (**c**) PRDM16 protein stability over time plotted from data in panel (**b**). (**d**, **e**) The degradation of PRDM16 protein in C3H10T1/2 cells differentiated without ArtC. C3H10T1/2 cells were differentiated without ArtC for 8 days, then the cells were treated with vehicle or ArtC in a medium containing cycloheximide. The degradation of PRDM16 protein was examined using immunoblot analysis at the indicated time point. (**f**) PRDM16 protein stability over time plotted from data in panel (**e**). (**b**, **e**) The data are presented as means ± SEM ($n = 3$–5); * mean values are significantly different from those of the control (**b**) or ArtC (−) (**e**) at $P < 0.05$. This figure was reproduced from Nishikawa et al. (2016)

anti-PRDM16 antibodies revealed a strongly positive response in the ArtC-administered group (Fig. 21.10) (Nishikawa et al. 2016). Meanwhile, epididymal WAT (eWAT) tested negative for both multilocular adipocytes formation and immunostaining (Fig. 21.10) (Nishikawa et al. 2016). In addition, there was no difference in H&E or the immunostaining of BAT between the control and ArtC groups; equally, positive staining was observed for both antibodies (Fig. 21.10) (Nishikawa et al. 2016).

Next, we investigated the protein expression of UCP1 and cytochrome c oxidase subunit IV (COXIV), which is a mitochondrial marker in iWAT and BAT. In iWAT, the protein expression of both UCP1 and COXIV was increased in the 10 mg/kg ArtC treatment group, whereas there was no difference among the groups in BAT (Fig. 21.11) (Nishikawa et al. 2016).

Previous research has shown that induction of brown-like adipocyte formation is also induced via the sympathetic nervous system (SNS)-mediated β3-adrenergic signaling pathway. For example, the intake of fish oil significantly increases UCP1 expression in iWAT and BAT and is accompanied by an increase in urinary norepinephrine (NE) excretion (Kim et al. 2015). With this in mind, we decided to

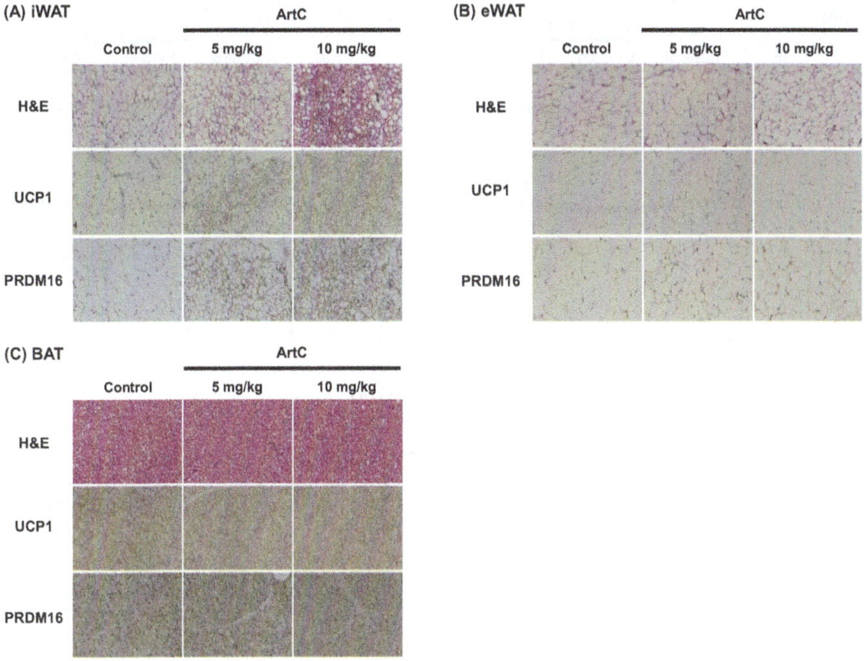

Fig. 21.10 Representative images of H&E staining and immunohistochemical staining of UCP1 and PRDM16 in sections of various adipose tissues from mice treated with vehicle (control) or ArtC (5 mg/kg or 10 mg/kg) for 4 weeks. (**a**) iWAT, (**b**) eWAT, (**c**) BAT. This figure was reproduced from Nishikawa et al. (2016)

investigate the β3-adrenergic signaling pathway response during ArtC treatment. Based on our experiments, we found that there was no difference in plasma NE concentration between the control and ArtC groups. We also determined that there was no difference in the expression levels of β3-adrenergic receptor (β3-AR) mRNA in iWAT and BAT between the control and ArtC groups (Fig. 21.12) (Nishikawa et al. 2016). One may wonder why ArtC did not induce UCP1 expression in BAT. Generally, BAT is more sensitive than iWAT to β3-adrenergic signaling pathway activation (Slavin and Ballard 1978; Trayhurn and Ashwell 1987). Since ArtC does not affect molecules related to the β3-adrenergic signaling pathway, it seems that ArtC-induced brown-like adipocyte formation relies mostly on the aforementioned stabilization of the PRDM16 protein rather than the β3-adrenergic signaling pathway. The exact reason for this (i.e., ArtC did not induce UCP1 expression in BAT) is not yet fully understood.

In sum, our fundamental research has established a promising new function for ArtC, a Brazilian green propolis constituent, with several exciting possibilities for its application in dietary supplements and the like.

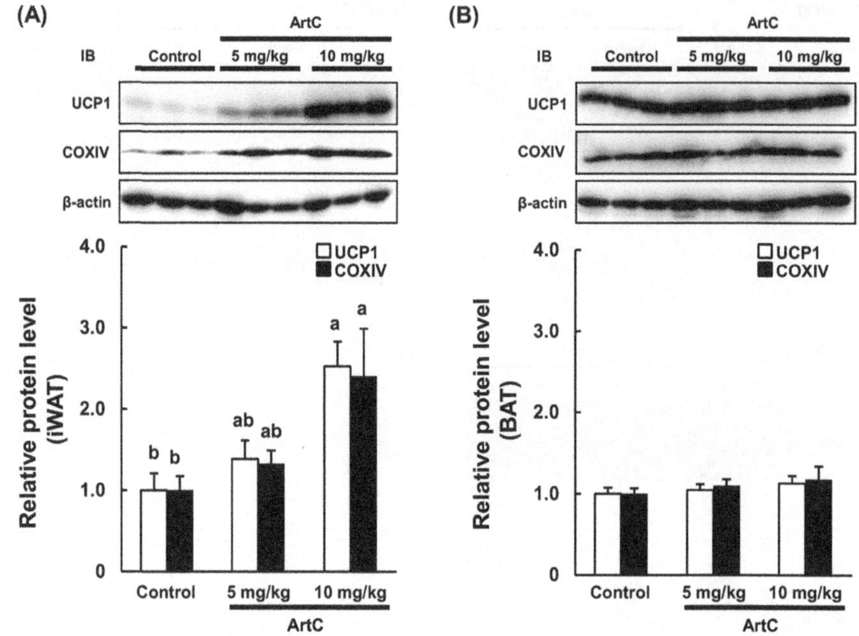

Fig. 21.11 Immunoblot analysis of UCP1, COXIV, and β-actin in the iWAT and BAT of mice treated with vehicle (control) or ArtC (5 mg/kg or 10 mg/kg) for 4 weeks. (**a**) iWAT, (**b**) BAT. Relative protein levels are expressed as fold-change relative to control (=1) after normalization using the β-actin protein level. Data are presented as means ± SEM (n = 9–10). Mean values without a common letter are significantly different at $P < 0.05$. This figure was reproduced from Nishikawa et al. (2016)

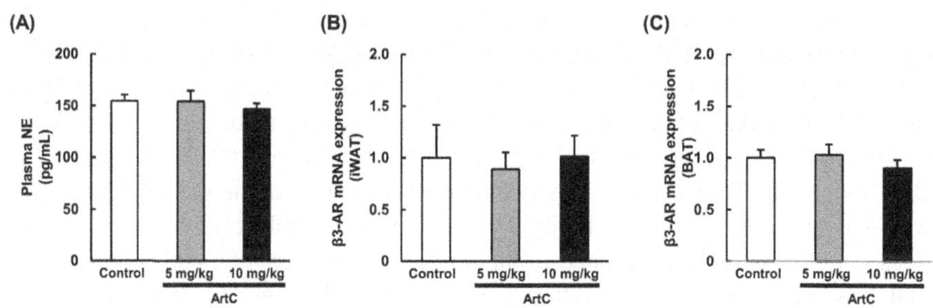

Fig. 21.12 Plasma NE concentration and mRNA levels of β3-AR in iWAT and BAT of mice treated with vehicle (control) or ArtC (5 mg/kg or 10 mg/kg) for 4 weeks. (**a**) Plasma NE concentration (**b**) mRNA levels of β3-AR in iWAT, (**c**) mRNA levels of β3-AR in BAT. The mRNA levels are expressed as fold of control (=1) after normalization using the expression levels of the TATA box–binding protein (TBP) gene. Data are shown as means ± SEM (n = 10). This figure was reproduced from Nishikawa et al. (2016)

4 Health Benefits of Green Propolis: Challenges and Future Prospects

As we better understand the chemical structures and various plant sources of Brazilian green propolis constituents, research regarding their health benefits also advances, such as traditional medicinal use and disease prevention in humans. How should we now proceed further with research into propolis health benefits? The foremost point to consider is the compositional variation of propolis due to both the geographical origin and the extraction method. These greatly influence the presence and strength of its biological activity. This variation is one of the main reasons for the confusion and misunderstanding surrounding *Baccharis*-derived Brazilian green propolis as well as other propolis of different geographical origins and derived from other plant sources. It follows that the amount and stability of constituents must be carefully considered to fully understand the health benefits of *Baccharis*-derived Brazilian green propolis. Below, we have highlighted some key challenges for future propolis health benefits research based on the above points of consideration.

Firstly, more data on the absorption and metabolism of propolis constituents are required. Compared with other food-derived compounds, the absorption and metabolism of Brazilian green propolis constituents are relatively unclear and have not been studied in detail. For example, in the case of ArtC, we need to determine whether it is ArtC itself that acts on tissues or whether it is ArtC's catabolic and metabolic products that are active. Furthermore, there are questions as to whether propolis' activity is affected by a difference in the intestinal flora, and whether propolis intake, in turn, affects our intestinal flora.

Secondly, clinical trials need to be conducted in order to address the lack of efficacy data in humans. Such trials need to be randomized, double-blind, and placebo-controlled. They also need to be cross-sectional with minimized variations in conditions and samples across the research groups.

Finally, when using a propolis extract, we need to determine which constituent is responsible for which effect, or if it is in fact a synergistic effect of multiple constituents. We know that in the case of *Baccharis*-derived Brazilian green propolis, ArtC greatly contributes to its health benefit. We must bear in mind, however, that other constituents are present, and so we must determine whether ArtC requires the presence of other constituents to exert its effects.

Very little is currently known about the health benefits of *Baccharis*-derived Brazilian green propolis. We hope to overcome the issues described in this chapter and realize a future not only in which the advancement of propolis research is contributing greatly to our health but also one in which the advanced application of *Baccharis* plants is being actively promoted.

References

Cannon B, Nedergaard J (2004) Brown adipose tissue: function and physiological significance. Physiol Rev 84:277–359

Cypess AM, Lehman S, Williams G et al (2009) Identification and importance of brown adipose tissue in adult humans. N Engl J Med 360:1509–1517

Giralt M, Villarroya F (2013) White, brown, beige/Brite: different adipose cells for different functions? Endocrinology 154:2992–3000

Harms M, Seale P (2013) Brown and beige fat: development, function and therapeutic potential. Nat Med 19:1252–1263

Ikeda R, Yanagisawa M, Takahashi N et al (2011) Brazilian propolis-derived components inhibit TNF-α-mediated downregulation of adiponectin expression via different mechanisms in 3T3-L1 adipocytes. Biochim Biophys Acta 1810:695–703

Isa Y, Miyakawa Y, Yanagisawa M et al (2008) 6-Shogaol and 6-gingerol, the pungent of ginger, inhibit TNF-α mediated downregulation of adiponectin expression via different mechanisms in 3T3-L1 adipocytes. Biochem Biophys Res Commun 373:429–434

Kajimura S, Seale P, Spiegelman BM (2010) Transcriptional control of brown fat development. Cell Metab 11:257–262

Kartal M, Yildiz S, Kaya S et al (2003) Antimicrobial activity of propolis samples from two different regions of Anatolia. J Ethnopharmacol 86:69–73

Kim M, Goto T, Yu R et al (2015) Fish oil intake induces UCP1 upregulation in brown and white adipose tissue via the sympathetic nervous system. Sci Rep 5:18013

Kumazawa S, Yoneda M, Shibata I et al (2003) Direct evidence for the plant origin of Brazilian propolis by the observation of honeybee behavior and phytochemical analysis. Chem Pharm Bull 51:740–742

Matsuzawa Y (2005) Adipocytokines and metabolic syndrome. Semin Vasc Med 5:34–39

Nedergaard J, Cannon B (2010) The changed metabolic world with human brown adipose tissue: therapeutic visions. Cell Metab 11:268–272

Nishikawa S, Aoyama H, Kamiya M et al (2016) Artepillin C, a typical Brazilian propolis-derived component, induces brown-like adipocyte formation in C3H10T1/2 cells, primary inguinal white adipose tissue-derived adipocytes, and mice. PLoS One 11:e0162512

Ohno H, Shinoda K, Spiegelman BM et al (2012) PPARγ agonists induce a white-to-brown fat conversion through stabilization of PRDM16 protein. Cell Metab 15:395–404

Reznikoff CA, Bertram JS, Brankow DW (1973) Quantitative and qualitative studies of chemical transformation of cloned C3H mouse embryo cells sensitive to postconfluence inhibition of cell division. Cancer Res 33:3239–3249

Saito M, Okamatsu-Ogura Y, Matsushita M et al (2009) High incidence of metabolically active brown adipose tissue in healthy adult humans. Diabetes 58:1526–1531

Scully T (2012) Diabetes in numbers. Nature 485:S2–S3

Slavin BG, Ballard KW (1978) Morphological studies on the adrenergic innervation of white adipose tissue. Anat Rec 191:377–389

Tazawa S, Warashina T, Noro T et al (1998) Studies on the constituents of Brazilian propolis. Chem Pharm Bull 46:1477–1479

Tazawa S, Warashina T, Noro T (1999) Studies on the constituents of Brazilian propolis II. Chem Pharm Bull 47:1388–1392

Trayhurn P, Ashwell M (1987) Control of white and brown adipose tissues by the autonomic nervous system. Proc Nutr Soc 46:135–142

van Marken Lichtenbelt WD, Vanhommerig W, Smulders NM et al (2009) Cold-activated brown adipose tissue in healthy men. N Engl J Med 360:1500–1508

Virtanen KA, Lidell ME, Orava J et al (2009) Functional Brown adipose tissue in healthy adults. N Engl J Med 360:1518–1525

Wu J, Boström P, Sparks LM et al (2012) Beige adipocytes are a distinct type of thermogenic fat cell in mouse and human. Cell 150:366–376

Yanagisawa M, Sugiya M, Iijima H et al (2012) Genistein and daidzein, typical soy isoflavones, inhibit TNF-α-mediated downregulation of adiponectin expression via different mechanisms in 3T3-L1 adipocytes. Mol Nutr Food Res 56:1783–1793

Yoneshiro T, Aita S, Matsushita M et al (2011) Brown adipose tissue, whole-body energy expenditure, and thermogenesis in healthy adult men. Obesity 19:13–16

Chapter 22
Effects of the Green Propolis on the Immune Response

José Maurício Sforcin and Marco Biagi

Abstract Africanized bees have a preference for *Baccharis dracunculifolia* DC. (Asteraceae) as a source of green propolis in Brazil. Propolis has been always mentioned as an immunomodulatory agent, and in vitro and in vivo assays have provided new information concerning its mechanisms of action. Propolis effects on antibody production have been widely investigated, and its impact on different cells of the immune system, involving the innate and adaptive immune response, has been assessed. Propolis increased the microbicidal activity of macrophages, monocytes, and neutrophils, as well as the lytic activity of natural killer cells against tumor cells. Since humans have used propolis for different purposes and propolis-containing products have been marketed, the knowledge of its properties with scientific basis is not only of academic interest but also of those who use propolis as well.

Keywords Antitumor property · *Baccharis dracunculifolia* · Bio-economy · Immune system · Plant chemistry

1 Introduction

Africanized bees have a preference for *Baccharis dracunculifolia* DC. (Asteraceae) as source of propolis in southeast Brazil (Teixeira et al. 2005). Volatile substances, in the resiniferous ducts or gland trichomes, trigger bee attraction. Brazilian "green"

J. M. Sforcin (✉)
Department of Chemical and Biological Sciences, Institute of Biosciences, UNESP, Botucatu, SP, Brazil
e-mail: jose.m.sforcin@unesp.br

M. Biagi
Dipartimento di Scienze Fisiche, della Terra e dell'Ambiente, Siena, Italy
e-mail: biagi4@unisi.it

G. W. Fernandes et al. (eds.), *Baccharis*,
https://doi.org/10.1007/978-3-030-83511-8_22

propolis contains *B. dracunculifolia* resin as its main vegetal source and, as the name indicates, it is greenish in color.

In fact, a detailed investigation of the botanical sources of propolis produced in the Campus of Botucatu – UNESP demonstrated that *B. dracunculifolia* is involved preferentially in propolis production by Africanized honeybees in this region, as well as *Eucalyptus citriodora* Hook and *Araucaria angustifolia* (Bert.) O. Kuntze (Bankova et al. 1999). Parts of the plants preferably visited by bees (exudates from the leaves of *Baccharis* and *Araucaria,* and from the trunk of *Eucalyptus*) were collected and investigated by chromatographic methods. The main components identified in *B. dracunculifolia* and in propolis were almost the same: 3,5-diprenyl-4-hydroxycinnamic acid called artepillin C, other prenylated cinnamic acids, dihydrocinnamic acid, *p*-coumaric acid, prenyl- and diprenyl-*p*-coumaric acids, and flavonoids in similar concentrations. On the other hand, some components were entirely absent in *Baccharis* exudates, and other plants may also contribute as vegetal sources of propolis. The main components of *E. citriodora* were aromatic acids and sugars, and *A. angustifolia* exudates contained only traces of aromatic acids, consisting mainly of diterpenic acids (Bankova et al. 1999). It has been reported that bees do not change its chemical composition in a specific geographic region, because they visit essentially the same vegetal sources (Bankova et al. 2000).

Since clinical and in vivo studies on animal models reported that propolis is well tolerated and nontoxic after administration (Mani et al. 2006; Cornara et al. 2017), we wish to discuss the effects of the green propolis on the immune response.

2 Propolis Immunomodulatory Action

The greatest problem to initiate in vitro and in vivo immunological assays is to design the experimental protocols, since researchers have used different propolis samples from all over the world, different concentrations both in vivo and in vitro, different kinds of extracts, intake periods, and routes of administration (Sforcin 2007; Sforcin and Bankova 2011; Sforcin 2016). Moreover, different assays have been validated to evaluate propolis immunomodulatory action using different immune cells. The improvement of the innate immunity is thought to be the basis of the effectiveness of green propolis in viral, bacterial, and fungal infections; for this reason, our group has analyzed the effects of green propolis mainly on monocytes, macrophages, dendritic cells (DCs), and natural killer (NK) cells, though green propolis has been found to affect also adaptive immunity (Gao et al. 2014).

Monocytes

Most of the authors have investigated the immunomodulatory action of green propolis studying mice or rats and found that this bee product mainly modulates innate immunity (Sforcin 2007; Cornara et al. 2017). We proposed to deeply evaluate the immunomodulatory action of propolis on the initial events of the immune response of human monocytes.

Monocytes develop in the bone marrow and represent the primary type of mononuclear phagocyte found in the blood. When activated in body tissues, monocytes differentiate in macrophages. Monocytes and macrophages are members of the mononuclear phagocyte system, a family of myeloid cells that also comprises dendritic cells (DCs) (Jakubzick et al. 2017). Monocytes were isolated from healthy blood donors and incubated with different concentrations of propolis, up to 20 μg/ml. Propolis exerted immunomodulatory effects on cell receptor expression, cytokine production, and fungicidal activity of human monocytes against *Candida albicans*, without affecting cell viability and depending on concentration. Such actions involved the participation of TLR-2 and TLR-4 (Búfalo et al. 2014).

In order to compare the modulatory effect of propolis samples from three different countries of Latin America, primary human monocytes were incubated with propolis samples (0.2–20 μg/ml) collected in Brazil, Cuba, and Mexico and their effects on pro- and anti-inflammatory cytokine production (TNF-α and IL-10, respectively) were investigated. Data indicated that all samples did not affect monocyte viability. Brazilian green propolis stimulated both TNF-α and IL-10 production by monocytes. Cuban propolis stimulated TNF-α and inhibited IL-10 production, while the Mexican sample exerted the opposite effect, inhibiting TNF-α and stimulating IL-10 production. Cuban propolis sample was originated from the floral resin of the genus *Dalbergia*, especially *D. ecastophyllum* (Leguminosae), while *Populus fremontii* was the main vegetal source of Mexican propolis. The major compounds found in Brazilian, Cuban, and Mexican propolis samples were artepillin C, isoflavonoids, and pinocembrin, respectively. One may conclude that Brazilian, Cuban, and Mexican propolis contained different components that may exert pro- and anti-inflammatory activity depending on concentration, which may provide a novel approach to the development of immunomodulatory drugs containing propolis (Conti et al. 2015).

Recently, our group produced a propolis-based odontological product whose patent was deposited in February 2015 (BR10 2015 003982–4). This product was tested in vivo and beneficial effects were seen in the control of dental plaque in humans. Afterward, we investigated the effects of this odontological product–containing propolis on human monocytes in vitro, since monocytes from the gum bleeding come to the inflammatory site in periodontal diseases. Data showed that this product may favor the recognition of antigens by monocytes, slightly activates the NF-κB signaling pathway, and increases the bactericidal activity of human monocytes against *S. mutans*. Also, this odontological product played a role in anti-inflammatory cytokine production, which can be beneficial in the treatment of periodontal diseases (Santiago et al. 2016).

It has been reported that there may be synergistic effects of several compounds conferring different pharmacological activities to propolis. Since the synergistic effects of propolis constituents are not elucidated, our group evaluated the involvement of phenolic acids (caffeic – Caf, dihydrocinnamic – Cin, and *p*-coumaric – Cou) alone or in combination in the activation of human monocytes, comparing their effects to propolis. Propolis and combinations containing Caf enhanced TNF-α production by resting cells. Propolis, Cin, Cou, and Caf + Cin stimulated IL-6 production. All treatments upregulated IL-10. In LPS-stimulated cells, treatments downregulated IL-6 and maintained TNF-α and IL-10 production. A lower TLR-2 expression was seen than propolis. Caf + Cin enhanced TLR-4 expression. Propolis, Caf, and Caf + Cin stimulated hydrogen peroxide (H_2O_2) production, whereas propolis, Cin, Cou, and Caf + Cin + Cou induced a higher fungicidal activity. Cin and Cin + Cou increased the bactericidal activity of human monocytes. In conclusion, propolis activated human monocytes, and the acids were involved differently in propolis activity (Cardoso et al. 2017). Kujumgiev et al. (1999) have suggested that the biological properties of propolis are due to a natural mixture of its components, and a single propolis constituent does not have an activity greater than that of the total extract.

Macrophages

Macrophages belong to the mononuclear phagocyte system, which initially included circulating monocytes and macrophages and later on included also DCs (Hoeffel and Ginhoux 2018). Activated macrophages display an enhanced microbicidal or tumoricidal capacity and secrete high levels of pro-inflammatory cytokines and mediators such as TNF-α, IL-1β, IL-6, IL-8, and IL-12 (Duque and Descuteux 2014) and oxygen and nitrogen metabolites (Mosser and Edwards 2008).

Regarding reactive oxygen species, NADPH oxidase catalyzes the reduction of molecular oxygen to superoxide anion (O_2^-), which spontaneously recombines with other molecules to produce reactive free radicals, including hydroxyl radical ($OH^·$) and H_2O_2 (Weigert et al. 2018). Neutrophils, macrophages, and monocytes use myeloperoxidase to further combine H_2O_2 with Cl^- to produce hypochlorite, which plays a role in destroying bacteria. Oxidants produced by phagocytes may destroy important biomolecules as well as phagocytosed microorganisms, and are involved in the tissue injury associated with inflammatory diseases (Babior 2000). Antioxidants are present in lower concentrations than biomolecules and may prevent, protect, or reduce the extension of oxidative damage, such as, for example, glutathione peroxidase, catalase, and superoxide dismutase. Other antioxidants, such as ascorbic acid (vitamin C) and tocopherol (vitamin E), are nonenzymatic antioxidants. Thus, there is a delicate balance between the generation and destruction of oxidant agents, which may be beneficial or deleterious to the organism (Novelli 2005).

In vitro, at concentrations <20 μg/ml, green propolis increased H_2O_2 generation in murine peritoneal macrophages (Orsi et al. 2000). On the other hand, Simões et al. (2004) observed an inhibitory effect of propolis and some of its components on superoxide anion production by rabbit neutrophils. These results are interesting, because propolis may exert both pro- and antioxidant activity depending on concentration, type of cell, and experimental protocol. However, propolis mechanism of action on free radical generation by macrophages is unclear (Cuesta et al. 2005).

Nitric oxide (NO) generation is another indicative of macrophage activation. Inducible nitric oxide synthase (iNOS)-derived NO directly or indirectly contributes to the initiation and progression of inflammation and tissue injury. The interaction between NO and O_2 produces the highly cytotoxic oxidant peroxynitrite ONOO (Novelli 2005). NO is an important microbicidal mechanism of macrophages for inhibiting DNA synthesis, mitochondrial respiration, and active transport in fungal and bacterial membranes (Macmicking et al. 1997). Besides, NO is also an important neurotransmitter, vasodilator, and cellular mediator of tissue repair (Chakraborty et al. 2006).

Brazilian green propolis, at low concentrations, inhibited NO generation by peritoneal macrophages (Orsi et al. 2000). Other propolis samples from different geographic regions also affected NO release. Moriyasu et al. (1994) also observed that propolis inhibited NO production by lipopolysaccharide (LPS)-stimulated macrophages, and Krol et al. (1996) linked this effect to flavonoids. Hu et al. (2005) evaluated the action of water and ethanolic extracts of propolis in a murine model of acute inflammation, verifying that both extracts inhibited NO generation.

After propolis treatment of mice for 3 consecutive days (0.2 ml/day of a solution containing 250–6000 μg/ml), peritoneal macrophages were taken and activated in vitro with gamma-interferon (IFN-γ). H_2O_2 and NO production were increased, which suggested that propolis treatment led macrophages to higher responsiveness to stimuli such as IFN-γ (Orsi et al. 2000). However, depending on its concentration, macrophages from propolis-treated animals and stimulated in vitro with IFN-γ showed inhibition in H_2O_2 and NO generation. These findings indicated a modulatory action of propolis on oxygen and nitrogen metabolites production by macrophages.

Propolis effects were analyzed on macrophages of BALB/c mice submitted to immobilization stress for 7 days. Stressed mice showed a higher H_2O_2 generation by peritoneal macrophages, and propolis treatment potentiated H_2O_2 generation and inhibited NO production by these cells (Missima and Sforcin 2007). In acutely stressed mice (3 days), Brazilian green propolis (200 mg/kg/day) restored TLR-2 and TLR-4 expression, contributing to the recognition of microorganisms during stressful conditions (Pagliarone et al. 2009a).

The effect of three main vegetal sources of propolis in our apiary (*Baccharis, Araucaria* and *Eucalyptus*) on H_2O_2 and NO generation by macrophages was analyzed. Data revealed no effects related to such extracts on these metabolites production (Lopes et al. 2003). Propolis action results from plant-derived products, and isolated extracts of its plant sources did not have the same effect in this assay. There may be synergistic effects involved in the different pharmacological activities of propolis.

Since *B. dracunculifolia* is the main propolis source in our region, the effect of *B. dracunculifolia* extracts and some purified compounds on H_2O_2 production by peritoneal macrophages of male BALB/c mice were also analyzed. Data revealed that the leaf, leaf rinse, and root extracts increased H_2O_2 release by such cells. Among the isolated compounds, baccharis oxide and friedelanol increased H_2O_2 production. These results suggest a stimulant action of extracts and isolated compounds of *B. dracunculifolia* on macrophages (Missima et al. 2007).

In order to evaluate propolis effect on macrophage's microbicidal action, our group carried out some works, challenging macrophages with different microorganisms. Macrophages were incubated with Brazilian propolis (5–20 μg/ml) and subsequently challenged with *Paracoccidioides brasiliensis* – a dimorphic fungus and the causative agent of the disease paracoccidioidomycosis. Propolis increased the fungicidal activity of macrophages nonsignificantly (Murad et al. 2002). Brazilian propolis increased the bactericidal activity of macrophages against *Salmonella typhimurium*, depending on concentration, which was associated with oxygen and nitrogen intermediates (Orsi et al. 2005).

Natural Killer Cells

The cytotoxic activity of natural killer cells (NK) after in vivo propolis treatment of rats was also evaluated by our group. Resistance to spontaneous tumor development has been associated with the cytotoxic activity of NK cells, found both in man and experimental animals. NK cells exert lytic activity toward several types of tumor and virus-infected cells, and we evaluated their activity by the ^{51}Cr-release cytotoxicity assay against Yac-1 cells (Sforcin et al. 2002a).

In order to investigate the seasonal effect on propolis samples, hydroalcoholic solutions were prepared with samples from spring, summer, autumn, and winter and administered to rats by gavage over three days (0.4 ml twice a day). Data indicated that NK activity increased in propolis-treated animals. There were no significant differences related to the seasonal effect on the immunomodulatory action of propolis (Sforcin et al. 2002a).

Recently, Takeda et al. (2018) also reported that propolis enhanced the NK cytotoxicity in mice, and artepillin C and *p*-coumaric acid may be involved in propolis effects.

Oral intake of WPP enhances NK cell cytotoxic activity, but not cell number, in a manner dependent on IFN-γ and independent of acquired immune responses. Further, artepillin C and p-coumaric acid, but not drupanin, appear to be the critical components of WPP, which augment NK cell cytotoxicity. These findings suggest the possible utility of WPP for use as a therapeutic to prevent cancer development and virus infection through NK cell activation.

Dendritic Cells

Dendritic cells (DCs) are found in the blood, epithelia, and lymphoid tissues. DCs are professional antigen-presenting cells that recognize pathogens and initiate an immune response by internalizing extracellular antigens and presenting peptides to naïve T cells. Previous works of our laboratory assayed the effects of propolis on the maturation and function of human DCs. Propolis, in the range 5–40 μg/ml, activated human DCs, inducing the NF-κB signaling pathway and TNF-α, IL-6, and IL-10 production. The inhibition of hsa-miR-148a and hsa-miR-148b abolished the inhibitory effects on HLA-DR and pro-inflammatory cytokines. The increased expression of hsa-miR-155 was correlated to the increase in TLR-4 and CD86 expression, maintaining LPS-induced expression of HLA-DR and CD40. Such parameters may be involved in the increased bactericidal activity of DCs against *Streptococcus mutans* (Conti et al. 2016).

Antibody Production

Concerning the humoral immune response, the effects of propolis and its constituents on antibody production have been mentioned since a long time ago, increasing the antibody production using different models and antigens (Scheller et al. 1988; Park et al. 2004; Sforcin et al. 2005; Chu 2006; Fischer et al. 2007; Gao et al. 2014; Yang et al. 2015).

Propolis effect on antibody production has been associated with macrophages activation, which leads to cytokine production, regulating the functions of B and T cells. Scheller et al. (1988) observed higher antibody levels when propolis was administered in the short term to the animals. Orsolic and Basic (2003) suggested that the increased IL-1β production by macrophages from propolis-treated mice might be associated with enhanced T and B cell proliferation.

Still, concerning adaptive immunity, our group evaluated the effect of green propolis on antibody production by rats. In this work, it was also analyzed the seasonal effect on propolis action, the action of some active compounds, and the effects of *Baccharis* extract on antibody production. Propolis administration (0.4 ml of a hydroalcoholic propolis solution 120 mg/ml for three days) to rats increased antibody production after 15 days of immunization (Sforcin et al. 2005). Propolis ability to modulate antibody synthesis is a part of its adjuvant activity, since it has been shown recently that propolis has a potent effect on different cells of innate immune response (Sforcin et al. 2002a; Orsi et al. 2005). No differences were seen between the samples from each season, which is in accordance with previous works of our laboratory (Sforcin et al. 2000, 2001, 2002a, b). Caffeic acid and quercetin had no effects on antibody production (Sforcin et al. 2005). *Baccharis* extract did not increase antibody production significantly when compared to control, but efficiently when compared to propolis-treated rats (Sforcin et al. 2005).

The use of propolis in combination with some vaccines has been evaluated. Mice inoculated with the inactivated Suid herpesvirus type 1 (SuHV-1) vaccine plus aluminum hydroxide and propolis presented higher levels of antibodies. The use of SuHV-1 vaccine plus propolis alone did not induce significant levels of antibodies; however, the combination was able to increase the cellular immune response, evidenced by the increase in the expression of mRNA to IFN-γ. Besides, propolis increased the percentage of protected animals challenged with a lethal dose of SuHV-1, suggesting its use in vaccines as an adjuvant (Fischer et al. 2007).

Propolis in combination with the inactivated vaccine against *Aeromonas hydrophila* was analyzed in carps. The phagocytic activity of these fishes and their serum antibody levels against *A. hydrophila* were higher comparing to nonadjuvant vaccinated fishes (Chu 2006). Immunostimulants could activate antigen-presenting cells (e.g. macrophages) and stimulate these cells to produce cytokines, which in turn activate T and B lymphocytes. These data strongly suggested the adjuvant capacity of propolis in association with vaccines.

The effects of propolis on immobilization stress-challenged animals were also investigated. Brazilian green propolis increased IL-4 production, favoring humoral immune response (Pagliarone et al. 2009b).

Finally, it is important to mention that ethanol 70% (propolis solvent) had no effects in all immunological assays of our group.

Figure 22.1 shows the main outcomes regarding propolis immunomodulatory action in some cells.

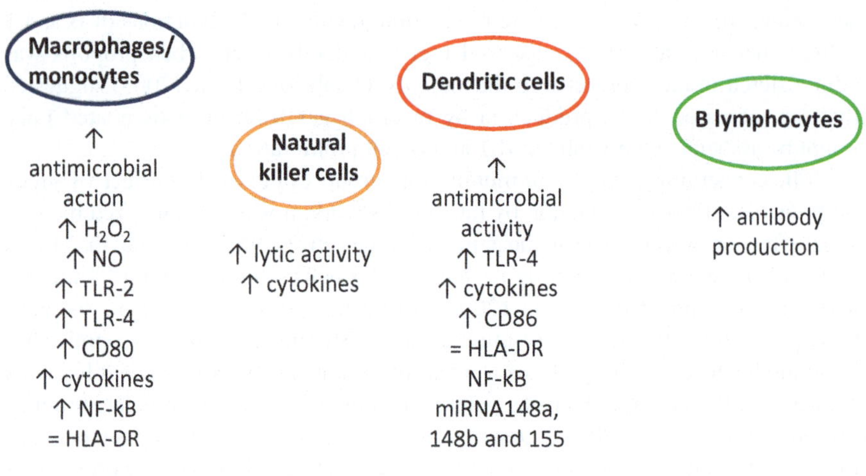

Fig. 22.1 Main outcomes regarding propolis immunomodulatory action in some cells

3 Concluding Remarks

Propolis chemical composition as well as the identification of its vegetal sources enables us to carry out the assays with chemically characterized samples. Propolis action may occur due to a synergistic effect of its constituents (Cardoso et al. 2017). The fact that no seasonal effect was seen on Brazilian green propolis composition suggests the use of samples collected in the same place all over the year (Bankova et al. 1999, 1998a, b), although in some regions, such as the temperate zone of the Northern Hemisphere, bees collect propolis mainly in the summer. Propolis shows antimicrobial activities, and its effects may occur through a direct action on microorganisms, as well indirectly, via stimulation of the immune cells and further microorganism killing.

Propolis biological properties have been intensely investigated in the last years, attracting a great interest of consumers in propolis-containing products marketed by health-food stores and pointing out propolis potential for the development of new drugs. However, in order to establish minimum requirements or setting standards to start the investigation of new drugs, immunomodulatory assays must include tests with positive controls, such as LPS, concanavalin A (Con A), phorbol myristate acetate (PMA), cytokines (IFN-γ), or others to compare propolis efficiency (Sforcin and Bankova 2011).

The knowledge of propolis mechanisms of action on the immune system has advanced in the last years. In vitro and in vivo assays demonstrated that propolis may activate macrophages, monocytes, and DCs, increasing their microbicidal activity. Propolis enhances the lytic activity of NK cells against tumor cells. It also stimulates higher antibody production, suggesting its use in vaccines, as an adjuvant. Although related articles provide new information to postulate some hypothesis and explanations, propolis' mechanisms of action are not fully elucidated, and further investigation will help to a better understanding of its effects on the immune system.

The number of formulations containing propolis and patents has increased. Since humans have been using propolis for a long time, scientific-based information brings an important contribution, evidencing the necessity of basic researches in this field and opening perspectives for new works.

Although the published evidence to date supports propolis safety and effectiveness, its importance to human health is not known with sufficient detail, which opens a new perspective for further studies. Several in vitro assays and preclinical studies have been carried out revealing important targets and mechanisms of action of propolis or its constituents. However, clinical trials are still needed to recommend propolis for veterinary and human purposes and to obtain adequate and standardized outcomes (Sforcin 2016).

References

Babior BM (2000) Phagocytes and oxidative stress. Am J Med 109:33–44

Bankova V, Boudourova-Krasteva G, Popov S, Sforcin JM, Funari SRC (1998a) Seasonal variations in essential oil from Brazilian propolis. J Essent Oil Res 10:693–696

Bankova V, Boudourova-Krasteva G, Popov S, Sforcin JM, Funari SRC (1998b) Seasonal variations of the chemical composition of Brazilian propolis. Apidologie 29:361–367

Bankova V, Boudourova-Krasteva G, Popov S, Sforcin JM, Frete X, Kujumgiev A, Maimoni-Rodella R, Popov S (1999) Phytochemical evidence for the plant origin of Brazilian propolis from São Paulo state. Z Naturforsch C J Biosci 54:401–405

Bankova V, Castro SL, Marcucci MC (2000) Propolis: recent advances in chemistry and plant origin. Apidologie 31:3–15

Búfalo MC, Bordon-Graciani AP, Conti BJ, Golim MA, Sforcin JM (2014) The immunomodulatory effect of propolis on receptors expression, cytokine production and fungicidal activity of human monocytes. J Pharm Pharmacol 66:1497–1504

Cardoso EO, Conti BJ, Santiago KB, Conte FL, Oliveira LPG, Hernandes RT, Golim MA, Sforcin JM (2017) Phenolic compounds alone or in combination may be involved in propolis effects on human monocytes. J Pharm Pharmacol 69:99–108

Chakraborty PD, Bhattacharyya D, Pal S, Ali N (2006) *In vitro* induction of nitric oxide by mouse peritoneal macrophages treated with human placental extracts. Int Immunopharmacol 6:100–107

Chu WH (2006) Adjuvant effect of propolis on immunization by inactivated *Aeromonas hydrophila* in carp (*Carassius auratus gibelio*). Fish Shellfish Immunol 21:113–117

Conti BJ, Santiago KB, Búfalo MC, Herrera FH, Alday E, Velazquez C, Hernandez J, Sforcin JM (2015) Modulatory effects of propolis samples from Latin America (Brazil, Cuba and Mexico) on cytokine production by human monocytes. J Pharm Pharmacol 67:1431–1348

Conti BJ, Santiago KB, Cardoso EO, Freire PP, Carvalho RF, Golim MA, Sforcin JM (2016) Propolis modulates miRNAs involved in TLR-4 pathway, NF-κB activation, cytokine production and in the bactericidal activity of human dendritic cells. J Pharm Pharmacol 68:1604–1612

Cornara L, Biagi M, Xiao J, Burlando B (2017) Therapeutic properties of bioactive compounds from different honeybee products. Front Pharmacol 8:1–20. https://doi.org/10.3389/fphar.2017.00412

Cuesta A, Rodríguez A, Esteban MA, Meseguer J (2005) *In vivo* effects of propolis, a honeybee product, on gilthead seabream innate immune responses. Fish Shellfish Immunol 18:71–80

Duque GA, Descoteaux A (2014) Macrophage cytokines: involvement in immunity and infectious diseases. Front Immunol 5:1–12. https://doi.org/10.3389/fimmu.2014.00491

Fischer G, Conceição FR, Leite FPL, Dummer LA, Vargas GD, Hübner SO, Dellagostin OA, Paulino N, Paulino AS, Vidor T (2007) Immunomodulation produced by a green propolis extract on humoral and cellular responses of mice immunized with SuHV-1. Vaccine 25:1250–1256

Gao W, Wu J, Wei J, Pu L, Guo C, Yang J, Yang M, Luo H (2014) Brazilian green propolis improves immune function in aged mice. J Clin Biochem Nutr 55:7–10

Hoeffe G, Ginhoux F (2018) Fetal monocytes and the origins of tissue-resident macrophages. Cell Immunol330:5-15

Hu F, Hepburn HR, Li Y, Chen M, Radloff SE, Daya S (2005) Effects of ethanol and water extracts of propolis (bee glue) on acute inflammatory animal models. J Ethnopharmacol 100:276–283

Jakubzick CV, Randolph GJ, Henson PM (2017) Monocyte differentiation and antigen-presenting functions. Nat Rev Immunol 17:349–362

Krol W, Scheller S, Czuba Z, Matsuno T, Zydowicz G, Shani J, Mos M (1996) Inhibition of neutrophils' chemiluminescence by ethanol extract of propolis (EEP) and its phenolic components. J Ethnopharmacol 55:19–25

Kujumgiev A, Tsvetkova I, Serkedjieva Y, Bankova V, Christov R, Popov S (1999) Antibacterial, antifungal and antiviral activity of propolis of different geographic origin. J Ethnopharmacol 64:235–240

Lopes FC, Bankova V, Sforcin JM (2003) Effect of three vegetal sources of propolis on macrophages activation. Phytomedicine 10:343
MacMicking JD, North RJ, LaCourse R, Mudgett JS, Shah SK, Nathan CS (1997) Identification of nitric oxide synthase as a protective locus against tuberculosis. Proc Natl Acad Sci U S A. 94:5243–5248
Mani F, Damasceno HCR, Novelli ELB, Martins EAM, Sforcin JM (2006) Propolis: effect of different concentrations, extracts and intake period on seric biochemical variables. J Ethnopharmacol 105:95–98
Missima F, Sforcin JM (2007) Green Brazilian propolis action on macrophages and lymphoid organs of chronically stressed mice. Evid Based Complement Alternat Med 5:71–75
Missima F, Silva Filho AA, Nunes GA, Bueno PCP, De Sousa JPB, Bastos JK, Sforcin JM (2007) Effects of *Baccharis dracunculifolia* D.C. (Asteraceae) extracts and its isolated compounds on macrophage activation. J Pharm Pharmacol 59:463–468
Moriyasu J, Arai S, Motoda R, Kurimoto M (1994) *In vitro* activation of mouse macrophage by propolis extract powder. Biotherapy 8:364–365
Mosser DM, Edwards JP (2008) Exploring the full spectrum of macrophage activation. Nat Rev Immunol 8:958–969
Murad JM, Calvi SA, Soares AMVC, Bankova V, Sforcin JM (2002) Effects of propolis from Brazil and Bulgaria on fungicidal activity of macrophages against *Paracoccidioides brasiliensis*. J Ethnopharmacol 79:331–334
Novelli ELB (2005) Nutrição e vida saudável. Estresse oxidativo e metabolismo energético. Tecmedd, Brazil, 288p
Orsi RO, Funari SRC, Soares AMVC, Calvi SA, Oliveira SL, Sforcin JM, Bankova V (2000) Immunomodulatory action of propolis on macrophage activation. J Venom Anim Toxins incl Trop Dis 6:205–219
Orsi RO, Sforcin JM, Funari SRC, Bankova V (2005) Effects of Brazilian and Bulgarian propolis on bactericidal activity of macrophages against *Salmonella typhimurium*. Int Immunopharmacol 5:359–368
Orsolic N, Basic I (2003) Immunomodulation by water-soluble derivative of propolis: a factor of antitumor reactivity. J Ethnopharmacol 84:265–273
Pagliarone AC, Orsatti CL, Búfalo MC, Missima F, Bachiega TF, Araújo JP Jr, Sforcin JM (2009a) Propolis effects on pro-inflammatory cytokine production and toll-like receptor 2 and 4 expression in stressed mice. Int Immunopharmacol 9:1352–1356
Pagliarone AC, Missima F, Orsatti CL, Bachiega TF, Sforcin JM (2009b) Propolis effect on Th1/Th2 cytokines production by acutely stressed mice. J Ethnopharmacol 125:230–233
Park JH, Lee JK, Kim HS, Chung ST, Eom JH, Kim KA, Chung SJ, Paik SY, Oh HY (2004) Immunomodulatory effect of caffeic acid phenethyl ester in BALB/c mice. Int Immunopharmacol 4:429–436
Santiago KB, Conti BJ, Cardoso EO, Golim MA, Sforcin JM (2016) Immunomodulatory/anti-inflammatory effects of a propolis-containing mouthwash on human monocytes. Pathog Dis 74:ftw081. https://doi.org/10.1093/femspd/ftw081
Scheller S, Gazda G, Pietsz G, Gabrys J, Szumlas J, Eckert L, Shani J (1988) The ability of ethanol extract of propolis to stimulate plaque formation in immunized mouse spleen cells. Pharmacol Res Commun 20:323–328
Sforcin JM (2007) Propolis and the immune system: a review. J Ethnopharmacol 113:1–14
Sforcin JM (2016) Biological properties and therapeutic applications of propolis. Phytother Res 30:894–905
Sforcin JM, Bankova V (2011) Propolis: is there a potential for the development of new drugs? J Ethnopharmacol 133:253–260
Sforcin JM, Fernandes A Jr, Lopes CAM, Bankova V, Funari SRC (2000) Seasonal effect on Brazilian propolis antibacterial activity. J Ethnopharmacol 73:243–249

Sforcin JM, Fernandes A Jr, Lopes CAM, Funari SRC, Bankova V (2001) Seasonal effect of Brazilian propolis on *Candida albicans* and *Candida tropicalis*. J Venom Anim Toxins incl Trop Dis 7:139–144
Sforcin JM, Kaneno R, Funari SRC (2002a) Absence of seasonal effect on the immunomodulatory action of Brazilian propolis on natural killer activity. J Venom Anim Toxins incl Trop Dis 8:19–29
Sforcin JM, Novelli ELB, Funari SRC (2002b) Seasonal effect of Brazilian propolis on seric biochemical variables. J Venom Anim Toxins incl Trop Dis 8:244–254
Sforcin JM, Orsi RO, Bankova V (2005) Effects of propolis, some isolated compounds and its source plant on antibody production. J Ethnopharmacol 98:301–305
Simões LMC, Gregório LE, Silva Filho AA, Souza ML, Azzolini AECS, Bastos JK, Lucisano-Valim YM (2004) Effect of Brazilian green propolis on the production of reactive oxygen species by stimulated neutrophils. J Ethnopharmacol 94:59–65
Takeda K, Nagamatsu K, Okumura K (2018) A water-soluble derivative of propolis augments the cytotoxic activity of natural killer cells. J Ethnopharmacol 218:51–58
Teixeira EW, Negri G, Meira RMSA, Message D, Salatino A (2005) Plant origin of green propolis: bee behavior, plant anatomy and chemistry. Evid Based Complement Alternat Med 2:85–92
Weigert A, von Knethen A, Fuhrmann D, Dehne N, Brüne B (2018) Redox-signals and macrophage biology. Mol Asp Med 63:70–87
Yang Y, Wei K, Yang S, Li B, Zhang Y, Zhu F, Wang D, Chi S, Jiang X, Zhu R (2015) Co-adjuvant effects of plant polysaccharide and propolis on chickens inoculated with *Bordetella avium* inactivated vaccine. Avian Pathol 44:248–253

Chapter 23
From Innovation to Market: An Analysis of the Propolis Production Chain

Frederico Ivair Santiago de Oliveira, Yumi Oki, Fernando M. Resende, Patrícia Angrisano, Dian Carlos Pinheiro Rosa, Lucas Arantes-Garcia, and G. Wilson Fernandes

Abstract Propolis is a bee product that has wide pharmacological potential and has been used in several industrial sectors. Green propolis, a propolis derived from *Baccharis dracunculifolia*, has been highlighted for being a product with relevant pharmacological properties and high commercial value. To contribute to the formulation of policies for the regulation, production, and/or commercialization of propolis, this chapter searched for technologies and products involving propolis with emphasis on green propolis. The main countries, periods, and technological profiles of patents were identified, as well as manufacturers, types of products, and market value of propolis (all types) and green propolis. Patent registrations for propolis have been found on all continents, but especially in China where most are concentrated. Most patents are associated with health care, although most products are from the food and cosmetics industries. The value of products with propolis ranged from US$ 0.10 to US$ 1005.18 and a kilogram of raw propolis from US$ 26.42 to US$ 499.30. Japan presented the largest technological and industrial domain over green propolis. Even though Brazil is the exclusive producer of green propolis, the country is not a major technology developer. Brazil produces the greatest amount of products and the best raw materials in the world but stands out internationally only in the primary phase of the propolis production chain. Analysis of patents and products with propolis identified aspects of different stages of production and increased the understanding of the processes of development and production of items with this bee product.

Keywords Bio-economy · Marketplaces · Patents · Types of products

F. I. S. de Oliveira · Y. Oki · F. M. Resende · P. Angrisano · L. Arantes-Garcia ·
G. W. Fernandes (✉)
Ecologia Evolutiva and Biodiversidade/DGEE, CP 486, ICB, Universidade Federal de Minas Gerais, Belo Horizonte, MG, Brazil

D. C. P. Rosa
Instituto de Ciências e Tecnologia das Águas/ICTA, Universidade Federal do Oeste do Pará, Santarém, PA, Brazil

G. W. Fernandes et al. (eds.), *Baccharis*,
https://doi.org/10.1007/978-3-030-83511-8_23

1 Introduction

The past few years have seen a progressive increase in new patents and natural products based on propolis due to high consumer demand (Mintel 2017, 2018). Currently, there are more than 4500 registered propolis patents from all continents throughout the world (reviewed in this chapter). This is quite a large number of patents for a natural product, especially considering that incentives for such patents at the global level only started in the late 1990s (Shadlen et al. 2019). The explanation for the high number of patents involving propolis is not only a consequence of growing market interest but also due to increased knowledge and production and technological advances related to this natural product. Investments in the development of new propolis products have been extensive and distributed among several sectors including food, pharmaceuticals, cosmetics, and veterinary medicine (Barros et al. 2019; Santos et al. 2019). It is estimated that the current global supply of propolis is between 700 and 800 tons per year (Nothenberg 1997; Pinto et al. 2011), representing an estimated market value of US$ 700 million per year, with a forecast to reach US$ 829 million by 2027 (Openpr 2020). Furthermore, the global propolis market is a sector considered to be invulnerable to economic crises as it is directly associated with human needs (i.e., health market). Indeed, during the present COVID-19 pandemic, there has been a marked increase in demand for natural products with antiviral properties and immune system strengthening capacity such as propolis. In Brazil, for example, a private company (i.e., "Bio Mundo – Produtos Naturais") announced an increase in propolis sales of 835%, with 680% specifically for green propolis consumption (Ingizza 2020). This is just one example among many other potential companies and such a marked increase was found comparing just the month prior to and immediately after the quarantine was officially decreed in Brazil.

Part of this significant propolis demand originates from ancient traditional knowledge. There are records of propolis use since Ancient Egypt at 1700 BCE, when it was used in, among other purposes, the process to mummify pharaohs (Pereira et al. 2002). With time, some of this traditional knowledge was passed through different civilizations and cultures. There are also records of propolis being used as a remedy in the Old Testament and during the Roman wars to heal wounds from arrows and spears (Castaldo and Capasso 2002; Pereira et al. 2002). Persian, Arabian, and Incan civilizations have also been reported to have used propolis for a wide range of purposes. In Persian ancient manuscripts, propolis is described as a product that is used against eczema, myalgia, and rheumatism (Kuropatnicki et al. 2013). The Arabs used propolis since the Middle Ages due to its antiseptic and healing actions, while the Incas used it for its anti-inflammatory and antipyretic properties (Castaldo and Capasso 2002).

Another reason for the great interest in propolis is that there are many types of this bee product, each associated with a highly diversified range of chemical compounds. For instance, more than 3000 compounds (Huang et al. 2014; Anjum et al. 2019) have been identified in propolis with proven efficacy in strengthening the immune system and other health benefits. Some of these are the treatment of diseases, whether they are caused by pathogens (e.g., bacteria, fungi, viruses, and protozoa) or other of origins, such as respiratory diseases, sclerosis, arthritis, and cancer (Marcucci 1995; Sforcin 2009; Veiga et al. 2017; Bhargava et al. 2018). This high diversity of propolis types fosters opportunities for a wide range of technological and market innovations. The uniqueness of each propolis derives from the fact of it being manufactured by a specific bee species utilizing a set of bioactive compounds from a specific plant species and in a specific geographical region (Bankova 2005). In fact, the same bee species can produce different types of propolis depending on the associated plant species (Oliveira and Bastos 1999; Salatino and Salatino 2018). The multiple combination of these characteristics is what determines the pharmacological potential of a type of propolis, which is further reflected in demand and market value. In Brazil, the Africanized honeybee (*Apis mellifera*) produces red propolis through the collection of sap resin from *Dalbergia ecastaphyllum* (Fabaceae), and green propolis when collecting from the vegetative apices of *Baccharis dracunculifolia* (Oliveira and Bastos 1999; Park et al. 2002; Sforcin 2012; Salatino and Salatino 2018).

Among the various types of propolis, green propolis (also known as Brazilian propolis) has received major attention in the international market since 1980 (Berretta et al. 2017). This high level of interest is partially due to the presence of over 150 different secondary metabolites in the composition of green propolis (Chang et al. 2008) with numerous therapeutic properties of high effectiveness such as antibacterial, antiviral, antitumor, and antioxidant activities as well as benefits against diabetes and respiratory, cardiac, and degenerative diseases. Among these chemical compounds, great interest has been associated with artepillin C (3.5-diprenyl-4-hydroxycinnamic acid), a compound with pronounced cytotoxic activity against several tumor strains (Messerli et al. 2009; Carvalho et al. 2011; Souza et al. 2018; Beserra et al. 2020). For instance, a milligram of artepillin C may cost from US$ 168.00 to US$ 297.00 in the North American market.

In the international market, green propolis is particularly renowned and consumed in Japan (Berretta et al. 2017). In addition to its pharmacological properties, the great demand for green propolis by the Japanese market is also attributed to the organoleptic characteristics of the natural product and the relatively lower content of heavy metals and other environmental pollutants (Pereira et al. 2002). Furthermore, the Africanized bees, *A. mellifera*, that produce green propolis are characterized as being more resistant to diseases, which reduces the need to subject the bees to chemical treatments, as is common in other countries. Ultimately, the absence of these chemical treatments guarantees a propolis of greater quality and free from potential chemical residue contamination (Berretta et al. 2017).

The combination of these positive characteristics is attributed to the increased export of green propolis to the rest of the world. Brazil is fortunate to be the exclusive producer of this natural product; it has exported ca. 140 tons of crude propolis for medicinal purposes over the past few years to countries in Asia and Europe and to the United States (Berretta et al. 2017). This great international interest has served as an economic incentive for Brazilian beekeepers, as it comprises a market responsible for more than US$ 500 million per year and with great potential for growth (Openpr 2020).

Regardless of the type, the high diversity of chemical and pharmacological properties of propolis enables a wide range of applications, which favors the manufacture of several products for direct consumption (capsules, extracts, sprays) and derivatives (candies, shampoos, soaps, ointments, hand and body cream, facial cleanser, toothpaste, among others) (Lihong et al. 2009). However, the main challenge in the propolis market has been the lack of standardization among the companies that commercialize it (Bankova 2005; Tagliacollo and Orsi 2011; Katekhaye et al. 2019). For instance, many brands report the propolis concentration used in the composition of their products, but not the propolis type and, even less often, the quality of the natural product (Bankova 2005; Tagliacollo and Orsi 2011; OpenPR 2020). Such information is of extreme importance considering that the core market interest in propolis comes from its natural properties and health benefits (e.g., nutritional and pharmacological). Thus, this lack of information and transparency may restrict the market growth of propolis worldwide (Tagliacollo and Orsi 2011; OpenPR 2020).

In turn, the distribution and commercialization of products based on propolis are usually done in local shops, supermarket chains, pharmacies, or marketplaces. Marketplaces are websites that bring together buyers and smaller traders and are increasingly relevant in global negotiations (Bakos 1991, 1998). The revenue from this virtual market is expected to increase from US$ 18.7 billion in 2017 to US$ 40.1 billion in 2022 (Coresight Research 2018; Thredup 2018). Despite the economic relevance of marketplaces, the market potential that it may provide for propolis commercialization is still poorly understood.

2 Scope

This chapter sought to analyze the profiles of patents involving propolis that were registered over the last four decades (1980–2017), as well as the international propolis market. Due to its great importance, more emphasis was placed on green propolis, which is produced from *Baccharis dracunculifolia*. To this end, the chapter was divided into three sections. The first section presents an analysis of patent filings for all types of propolis including green propolis. The second focused on the propolis and green propolis market, specifically on products advertised on marketplaces. Finally, the third section discusses the relationship between these patents and the market.

3 Patents

Global Distribution of Propolis Patent Registration

A search of the main patent databases available online[1] revealed that the number of patent registrations with the term propolis has increased significantly between 1980 and 2017 (Fig. 23.1a). During this period, 4989 registered patents were associated

[1] Four databases were consulted: 1) European Patent Office – EPO, which gathers data on more than 100 million patent documents from about 70 countries (EPO 2018); 2) World Intellectual

with all types of propolis, which represents an average increase of 131 patents per year. The registration of patents associated with green propolis also increased between 1980 and 2017 (Fig. 23.1a). The first green propolis patent was registered in the early 1990s; since then, the number has increased rapidly, reaching a total of 70 registrations in 2017, which represents 1.3% of all patents with propolis. A further analysis clearly indicates the recent increase in patent registrations associated with propolis. Specifically, around 50% (2519 patents) of all propolis patents and 40% (28) of green propolis patents were registered in the last five years (2013–2017). Furthermore, the last year evaluated (2017) represented a peak in propolis patent registrations, whether for all types (890) or just green propolis (12).

Most of the countries responsible for the increase in propolis patent registration in the last four decades are in Asia (Fig. 23.1b). Specifically, Japan and Russia were the countries that registered the most propolis patents in the 1990s (Fig. 23.1b), but this changed after a steady growth in propolis patent registration by China in the 1990s and 2000s. In the last five years evaluated, China has registered more than 150 patents per year, making it the absolute leader for all types of propolis. In 2017, China had 3132 patents associated with propolis after registering 819 in a single year. China currently has 63% of all the registered propolis patents. Surpassing Japan and Russia, South Korea became the second country in the number of patents due to a steady increase over the last five years.

Asian countries also stand out for the number of registered patents associated with the green propolis (Fig. 23.1c). Japan was the only country with these specific patents until around 2000, and it is still currently the leader in registered green propolis patents with 46% of the total number (32) (Fig. 23.1c). Throughout the 2000s, Brazil also registered some green propolis patents, but did not maintain steady growth over subsequent years. Since 2010, Taiwan, China, and France contributed to the increase in the number of patents associated with green propolis. Currently, China and Taiwan are, respectively, second and third in the number of green propolis patents with 19% (13) and 10% (7) of all the existing patents associated with green propolis.

Considering the global distribution of patent applications, a total of 43 countries located in different regions of the world registered patents associated with some type of propolis between 1980 and 2017 (Fig. 23.2a). As previously presented, Asian countries (i.e., mainly China, South Korea, and Japan) have the largest number of registered patents considering all types of propolis (Fig. 23.1b). Despite less participation, European countries have also played an important role in the number of patents associated with propolis. There are currently five European countries

Property Organization – WIPO, an organization associated with the United Nations (UN) that contains more than 60 million documents from around 50 countries (WIPO 2018a, b); 3) Japanese Patent Office – JPO, which allows access to documents deposited in Japan through the Japanese Patent Information Platform (J-PlatPat 2010); and 4) National Institute of Industrial Property – INPI, the Brazilian federal government agency responsible for patent registrations in the country (INPI 2017, 2018). These databases are commonly used in technological prospecting studies (e.g., Mayerhoff 2008; Soares et al. 2010; Souza et al. 2012; Castro 2014; INPI 2015; Ghesti et al. 2016).

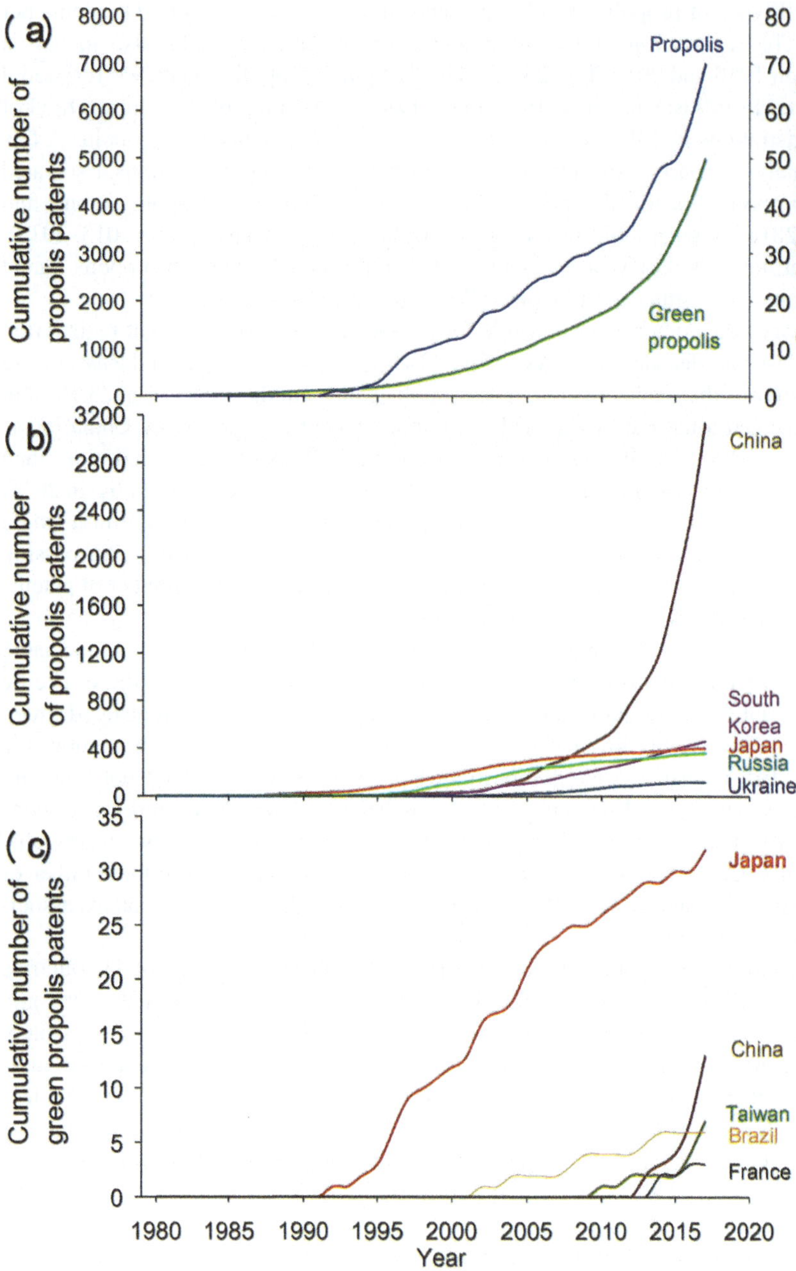

Fig. 23.1 (**a**) Temporal trend of the cumulative number of patents for all types of propolis and green propolis. Note that the left vertical axis represents the number of patents accumulated with all types of propolis and the right vertical axis refers to green propolis. (**b**) Temporal trend of the cumulative number of patents for all types of propolis and (**c**) for green propolis among the five countries with the highest number of patents

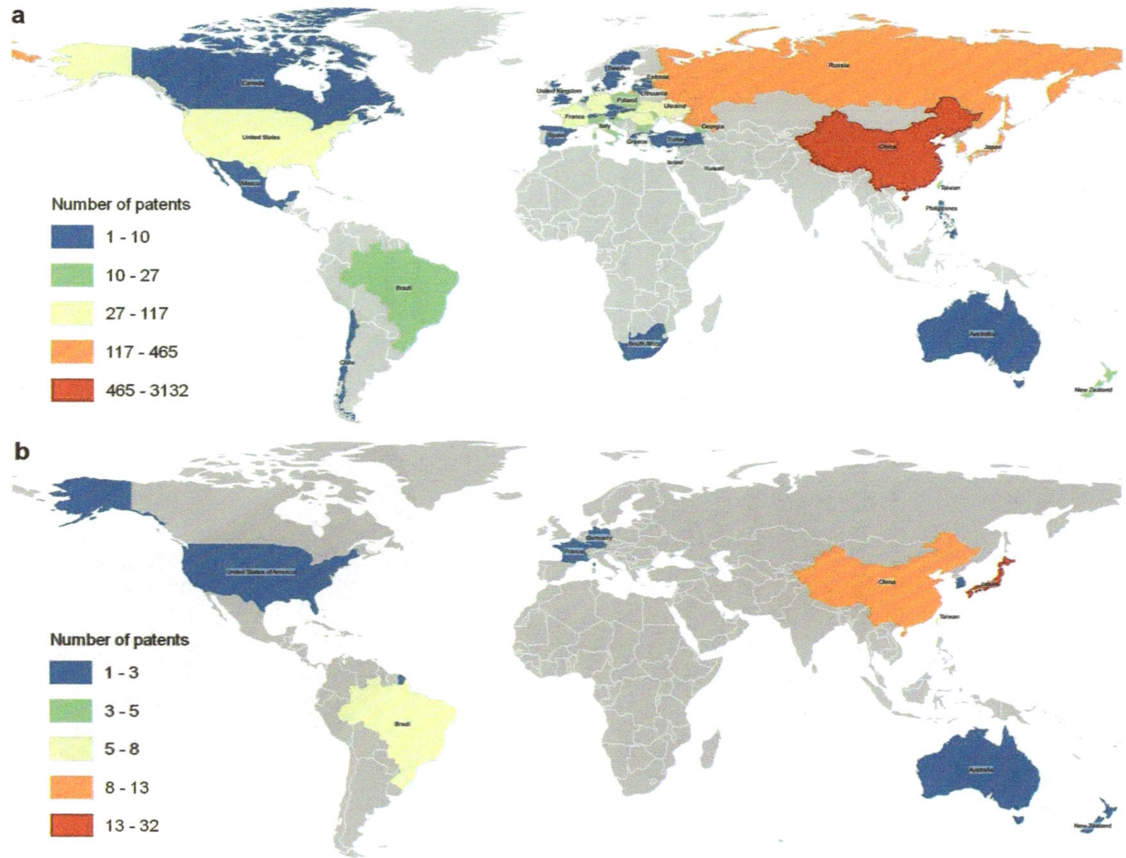

Fig. 23.2 Distribution of patent registrations associated with propolis per country between 1980 and 2017, considering: **(a)** all types of propolis and **(b)** green propolis. Records of 4989 patents associated with all types of propolis are distributed among 43 countries, while 70 green propolis patents are distributed among 10 countries

among the top ten in number of propolis patents (i.e., Ukraine, Romania, France, Germany, and Hungary). In turn, patents associated with green propolis are distributed among just 10 countries (Fig. 23.2b), in which Asian countries again concentrate the largest number of these patents (i.e., Japan, China, and Taiwan; Fig. 23.1c). These are followed by South America (Brazil), Europe (France and Germany), and Oceania (New Zealand and Australia). Although Brazil is fourth in the number of green propolis patents (8.6% of the total; 6 patents), its participation is still low, considering that green propolis is produced exclusively in the country.

Global Players and Technological Profiles Associated with the Registration of Green Propolis Patents

The global players responsible for the research and technology behind the development of green propolis patents between 1980 and 2017 comprise a group of 52 companies, universities, institutes, and individuals (Fig. 23.3). Among these, companies and partnerships between universities and companies are responsible for a great part of the registration of green propolis patents with 61.4% (43) and 2.9% (2) of all patents, respectively. The high interest from the private sector in the commercialization of green propolis reinforces its great potential in the development of medicines and other items, as presented below (see sect. 23.4 Products and market). Universities (10 patents, 14% of the total number) and individuals (14%; 10 patents) also represent important players.

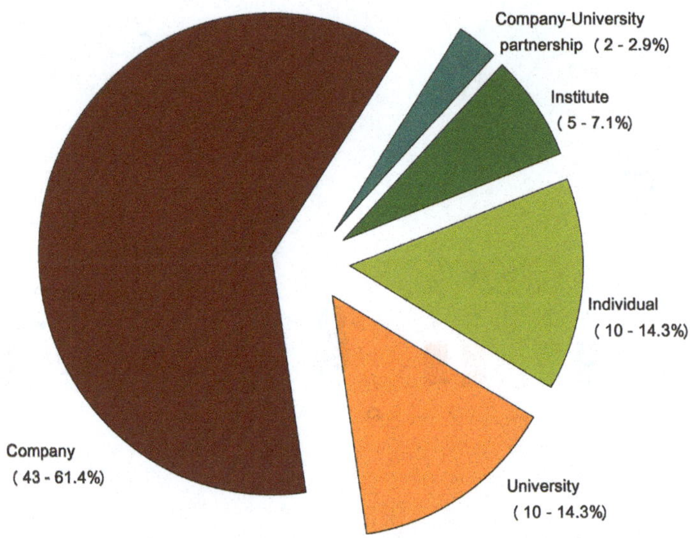

Fig. 23.3 Distribution of green propolis patents according to applicants. A total of 52 players registered 70 green propolis patents between 1980 and 2017

Analysis of the technological profiles associated with green propolis patents according to International Patent Classification (IPC) and Cooperative Patent Classification (CPC) revealed that most of them are linked with human needs (section A; 58.2% of all classifications; 57 patents), as well as chemical or metallurgical purposes (section C; 9.2% of the total; 9 patents; Fig. 23.4a). Patents related to human needs mainly involve applications in the field of health and hygiene (subsection A61; 64.5% of the patents in human needs section) and food or food products (subsection A23; 32.9% of the patents; Fig. 23.4b). Patents related to chemistry or metallurgy are concentrated mainly in the areas of organic chemistry (subsection C07; 64% of the patents in chemistry or metallurgy section) and biochemistry, beer, alcohol, wine, vinegar, microbiology, enzymology, genetic engineering, or mutation (subsection C12; 24% of the patents; Fig. 23.4c).

4 Products and Market

An analysis of products containing propolis that are commercialized in five marketplaces with global relevance (i.e., Mercado Livre, Rakuten, Ebay, Amazon, and Walmart) showed that 47 countries manufacture such products (Fig. 23.5a), whereas products that contain exclusively green propolis are sold in eight countries (Fig. 23.5b). In total, analyzed marketplaces sell 3065 products containing propolis, 8% (244 units) of which specifically contain green propolis. The countries that produce the largest number of products containing some type of propolis are Brazil (24% of the total; 721 products), Japan (20%; 623), and the United States (16%; 482). In turn, Brazil and Japan are the countries with the largest number of products with green propolis with 60% (145 products) and 25% (60) of all the total products, respectively. However, as described in detail hereafter, Brazil has the lowest added value for commercialized green propolis products in the world, while Japan has the highest followed by South Korea and Canada, both with 5% of all products (12 products).

Several types of products that contain propolis are commercialized on marketplaces with a wide range of purposes including raw materials, capsules, lozenges, liquid extracts, creams, and lotions. The vast majority of propolis products can be classified as food (49% of all products; 1498) or cosmetics (45%; 1379; Fig. 23.6). These two categories are also the most common for products with green propolis, in which 65% (159 products) and 18% (45) of all products are food and cosmetics, respectively (Fig. 23.6). Products with raw propolis and therapeutic and veterinary use were also found, albeit in small proportions.

The evaluated products containing propolis that are commercialized in marketplaces varied widely in price. Products with all types of propolis are sold for values between US$ 0.10 and US$ 1005.18 per unit (average value of US$ 29.40), while products with green propolis are sold for values between US$ 1.59 and US$ 641.49 (average value of US$ 39.55). Product prices follow a similar pattern among the ten

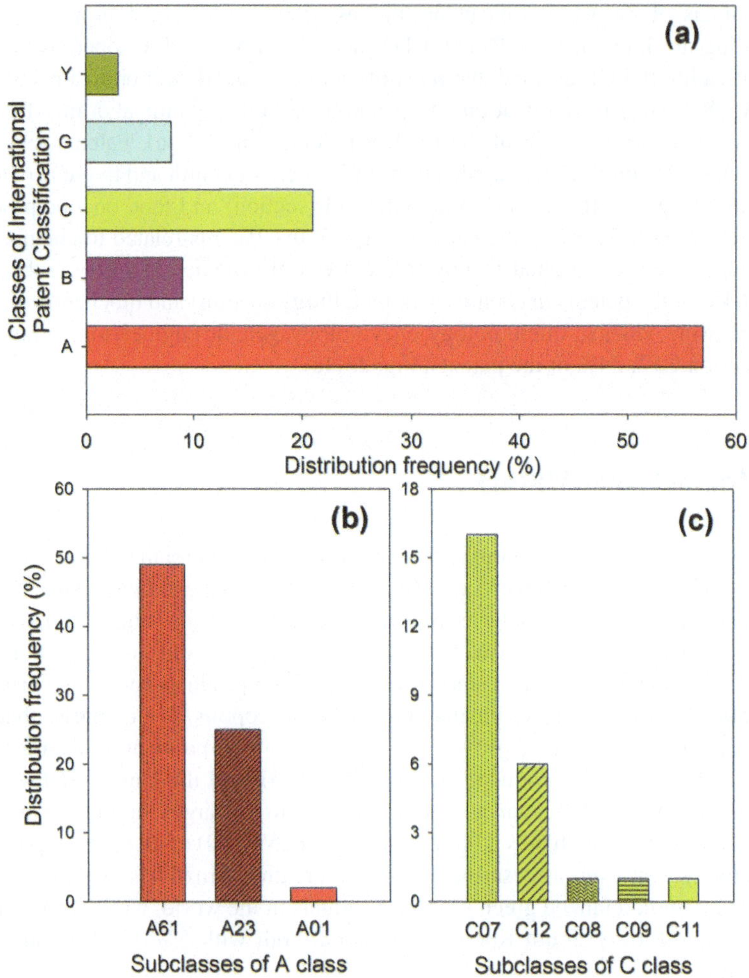

Fig. 23.4 (**a**) Distribution of green propolis patents according to the International Patent Classification (IPC) and its more refined version, the Cooperative Patent Classification (CPC). Patents in section "A" are associated with products and processes related to human necessities; those in section "B" are related to carrying out operations, transportation, separation, or mixing of substances or materials; those in section "C" are related to chemistry or metallurgy; those in are section "G" related to physics; and those in section "Y" are related to the general marking of new technological developments and overall identification of transversal technologies. (**b**) Distribution of green propolis patents classified in section "A". Class A61: veterinary science, medicine, and hygiene; class A23: food, food products; class A01: agriculture, forestry, livestock, hunting, trapping, fishing; (**c**) Distribution of green propolis patents classified in section "C". Class C07: organic chemistry; class C12: biochemistry, beer, alcohol, wine, vinegar, microbiology, enzymology, genetic engineering; class C08: organic macromolecular compounds; class C09: dyes, paints, polishers, natural resins, adhesives; class C11: animal or vegetable oils, fats, fatty substances or waxes, detergents, candles

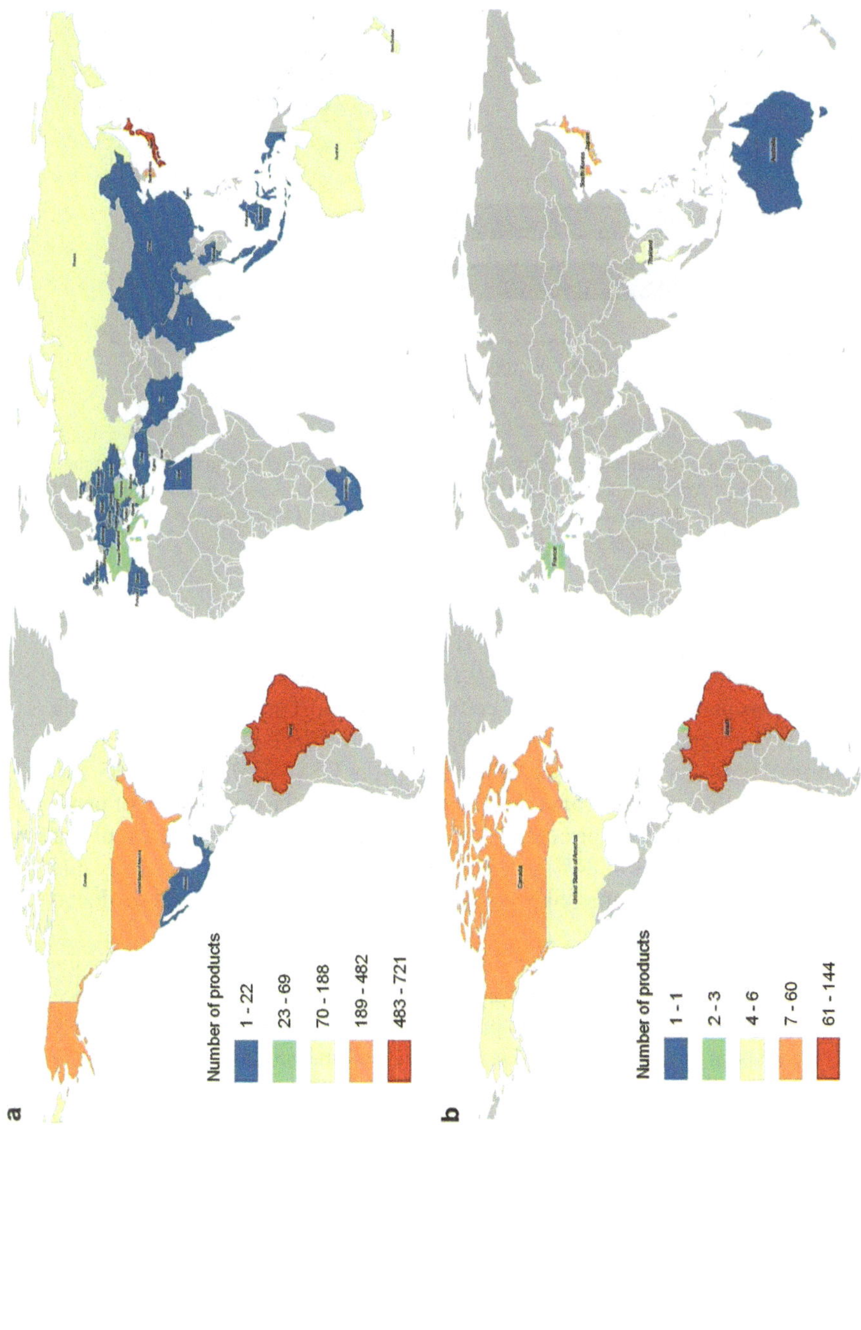

Fig. 23.5 Distribution of commercialized products containing propolis of the evaluated marketplaces (i.e., Mercado Livre, Rakuten, Ebay, Amazon, and Walmart) according to the country of the manufacturer. (**a**) Products with all types of propolis and (**b**) with green propolis. The total quantity of products is 3065 for all types of propolis and 244 for green propolis

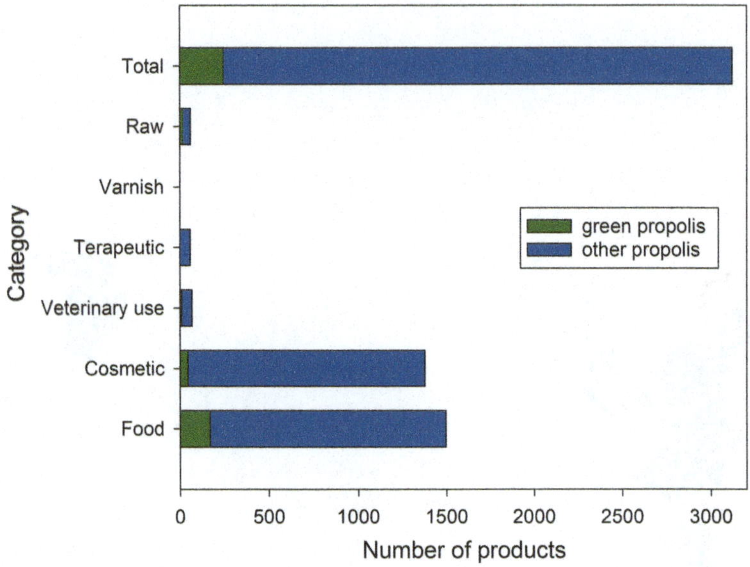

Fig. 23.6 Distribution of commercialized products of the evaluated marketplaces (i.e., Mercado Livre, Rakuten, Ebay, Amazon, and Walmart) by category for products with green propolis and with all types of propolis. The total quantity of products is 3065 and 244 for products with all types of propolis and green propolis, respectively

countries that sell more propolis products (Fig. 23.7), with most of the items being sold at intermediate prices (between US$ 9.14 and US$ 59.60 per unit). Brazil is the exception to this pattern as it is the only country where most products are sold at the lowest prices (between US$ 0.10 and US$ 9.14). Japan is the country with the highest quantity of products sold at the highest prices (from US$ 19.90 to US$ 1005.18). Green propolis had a similar profile, but the asymmetry in the distribution of products per country did not allow price analysis per country to be performed. Brazil and Japan concentrate the sales of 84% (204 products) of green propolis products, while the remaining items are commercialized by the other countries (16%; 40 products) (see Fig. 23.5).

Part of the commercialized products (2% of the all products; 60) contain propolis in its raw state. Brazil stands out in the production of propolis in this category, since the majority of products (87% of all products; 52) are produced in the country (Table 23.1). About a third (32% of all products; 19) of the products in the raw state are green propolis. Other countries that sell raw propolis are the United States (4 products), Russia (2), Turkey (1), and Bulgaria (1). The prices for raw propolis range from US$ 26.42 (mix of Brazilian propolis produced by "Uruçu", "Mandaçaia", and "Manduri" bees) to US$ 499.30 per kilogram (propolis produced in Turkey). Raw green propolis showed an intermediate price (US$ 114.75 per kilogram) when compared to other types of raw propolis.

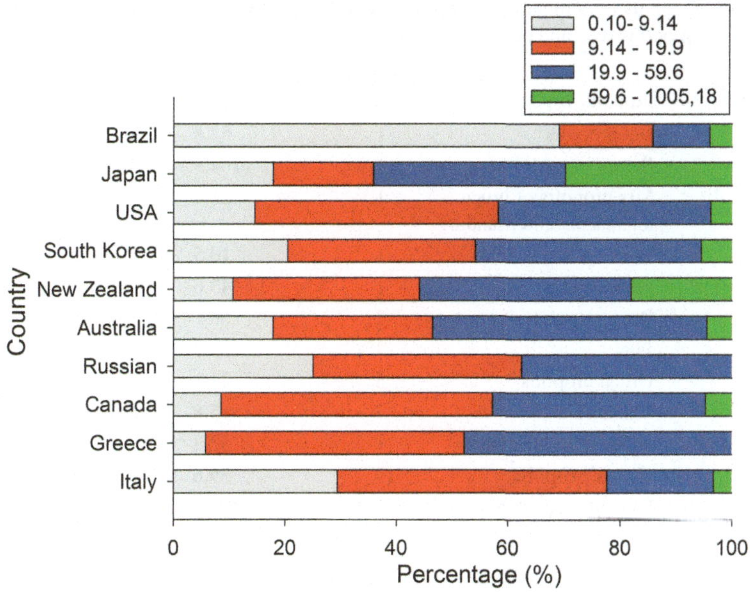

Fig. 23.7 Distribution of commercialized products of the evaluated marketplaces (i.e., Mercado Livre, Rakuten, Ebay, Amazon, and Walmart) containing all types of propolis and their price range (US dollar). Each price range comprises 30% of the marketed products, except for the last price range, which contains 10% of the products with the highest values. The 10 countries that sell the most products on the marketplaces are presented; these countries are responsible for the manufacture of 88% (2712 products) of the sold products containing any types of propolis

5 Relationship between Patents and the Market

Propolis patent registrations have grown significantly over the past four decades. The strong interest in propolis products and the associated technology observed in Asian countries is remarkable. This is partially explained by the long tradition held by Asian countries in general of using these natural products (Pereira et al. 2002). Brazil, in turn, has a modest position among countries with registered patents, even though it is one of the largest manufacturers of propolis-based products and the exclusive producer of green propolis. The large number of countries involved in the manufacture of products containing propolis indicates the widespread interest throughout the globe for the production and commercialization of these items. The overall profile of products that contain propolis is mainly associated with applications in food and cosmetics markets. Brazil and Japan are the countries that manufacture the largest number of products with propolis, but the price range by which these products are sold in Brazil is much lower. On the other hand, the Japanese market sells the most expensive propolis-based products.

The profiles of patents and products associated with propolis described in this chapter will be useful to decision makers involved in propolis regulation,

Table 23.1 Quantity of commercialized products of the marketplaces containing raw propolis by country with propolis type and prices (US dollar)

Country	Propolis Type	Number of Raw Products	Minimum Price (USD)	Maximum Price (USD)
Brazil	Geopropolis	1	52.8	52.8
	Brown propolis	1	104.5	104.5
	Mix propolis of uruçu, mandaçaia and Manduri	1	26.4	26.4
	Mombucão própolis	1	10.5	10.6
	Tubuna propolis	1	43.6	43.6
	Borá propolis	2	66.1	264.2
	Jataí	3	63.4	158.5
	Uruçu propolis	5	21.1	52.8
	Mandaçaia	7	15.8	52.8
	Unidentified propolis	11	52.3	111.7
	Green propolis	**19**	**39.6**	**166.2**
Turkey	Unidentified propolis	1	499.3	499.3
Bulgaria		1	466.0	466.0
Russia		2	49.9	244.0
United states		4	13.2	167.7
Total (Brazil)		**52**		
Total (all countries)		**60**		

production, and marketing policies (Roessing 2002). Specifically, such information can guide strategic actions aimed at boosting the production chain based on integrating research and production (Dagnino 2004). Technology plays an increasingly important role in the competitiveness of industries, highlighting the need to effectively integrate technological development and production processes, especially in countries that produce items with low added value (Batalha and Silva 2013).

Historically, Brazil has stood out in the international market as a major exporter of raw materials and commodities (Furtado 2005), as is the case of iron ore and coffee, soybeans, and other agricultural products (Dagnino 2004; Furtado 2005; Hausmann et al. 2011; Coelho et al. 2013; Conceição et al. 2017). The present study demonstrated that this pattern is also observed in the propolis market, since Brazilian propolis represents more than 80% of the raw propolis found in marketplaces. Despite the fact that Brazil is the country that commercializes the largest number of products containing propolis in marketplaces, it is also the country with the lowest price ranges for these products when compared to other countries that produce and sell them. Specifically, a 30 ml bottle of propolis extract is sold in Brazil for ca. US$ 7.00, while in Japan, it is sold for ca. US$ 53.00. In this context, cooperation between the scientific community and the productive sector can contribute to the production of national items with greater added value. The integration of other sectors is also essential for market expansion. The efficiency of the regulatory sectors in Brazil should be a priority in the case of propolis, which is identified as a drug in many countries but is treated as a conventional food in Brazil, which hampers exportation efforts (Apex 2014; Berretta et al. 2017). Marketing actions are also fundamental for the differentiation of agroindustrial products (Conceição et al. 2017) and should integrate the strategies of companies promoting propolis products.

While China concentrates the most propolis patents, it does not manufacture a lot of propolis-based products. This discrepancy suggests that the number of patents may not be a good indicator of the production and commercialization intensity of a country. The reasons for the great growth in China's patent registrations may lie beyond the increase in the level of technological dynamism or the development of new products. For example, patent registration can be used by large corporations to inhibit the production of similar products by competitors and/or to increase a companies' market value (Bagattolli and Dagnino 2013). In turn, the high cost of the patent registration process favors large companies (Bagattolli and Dagnino 2013) and explains the dominance of these players among the main patent applicants. Thus, the results found in this study question the effectiveness of the patent registration policy in relation to the expected stimulus for the development and availability of new products and technologies for society.

Japan is the main importer of green propolis (Berreta et al. 2017) and is distinguished from other countries by its production of a great number of products containing propolis as well as in the number of patents and products associated with green propolis. Products from Japan showed high added value, indicating a potentially high level of specialization in the industrialization of apiarian products in the country. In fact, Japan is traditionally known for its innovations and technological production, as well as for its highly complex and diverse economy (Hausmann et al. 2011).

The results found in the present study reinforce the association between the increasing number of patents associated with propolis and the growing demand for it, mainly motivated by health benefits (Sforcin and Bankova 2011; Apex 2014; Berretta et al. 2017; Mintel 2017). This demand is particularly accentuated during pandemics, such as the present with COVID-19. Since the beginning of this pandemic, green propolis exports by Brazil have increased by 30%, mainly destined to Asian countries. Concomitantly, Brazil's domestic consumption of propolis-based products also increased. As a consequence, the gross value of green propolis in the Brazilian domestic market went from US$ 32.00 to US$ 75.00 dollars per kilogram (Faemg 2014). Yet, green propolis supplies in Brazil have not been able to meet the increased demand. In addition, the growing consumption and release of new propolis-based products (see Mintel 2017), as well as natural products in general, is a trend observed worldwide in recent years, especially among food and beverages (Mintel 2018).

Several studies emphasize the importance of integrating innovation and productive sectors as a relevant competitive factor within production chains (e.g., Roessing 2002; Dagnino 2004; Batalha and Silva 2013). The analysis of patents and products with propolis presented here made it possible to broaden the understanding of the innovation and production processes of items containing this natural product. Some discrepancies were found between the sectors of innovation and production, demonstrating that the interaction between these sectors is more complex than initially expected and suggesting the need for future detailed analyses addressing this relationship. Another fundamental aspect that still needs to be analyzed is the productive chain of apiarian products around the world, bearing in mind the potential importance for biodiversity conservation, income generation, and human well-being.

References

Anjum SI, Ullah A, Khan KA, Attaullah M, Khan H, Ali H, Bashir MA, Tahir M, Ansari MJ, Ghramh HA, Adgaba N, Dash CK (2019) Composition and functional properties of propolis (bee glue): A review. Saudi J Biol Sci 27:1695–1703. https://doi.org/10.1016/j.sjbs.2018.08.013

Apex - Brasil. Associação Brasileira de Promoção de Exportações e Investimentos (2014) China: Perfil e Oportunidades Comerciais em Alimentos, Bebidas e Agronegócio. Apex, Brasília

Bagattolli C, Dagnino RP (2013) Política de estímulo às patentes no Brasil: Avançando na contramão? Rev Econ Tecnol 9:73–86

Bakos Y (1991) A strategic analysis of electronic marketplaces. MIS Quarterly 15:295–310. https://doi.org/10.2307/249641

Bakos Y (1998) The emerging role of electronic marketplaces on the internet. Commun ACM 41:35–42

Bankova V (2005) Chemical diversity of propolis and the problem of standardization. J Ethnopharmacol 100:114–117

Barros KBNT, Neto EMR, Fonteles MMF (2019) Propolis and its cosmetic applications: a technological prospection. J Young Pharm 11:350–352

Batalha MO, Silva AL (2013) Gerenciamento de sistemas agroindustriais: definições, especificidades e correntes metodológicas. In: Batalha MO (ed) Gestão Agroindustrial, 3rd edn. Atlas, São Paulo, pp 1–62

Berretta AA, Arruda C, Miguel FG, Baptista N, Nascimento AP, Marquele-Oliveira F, Hori JI, Barud HS, Damaso B, Ramos C, Ferreira R, Bastos JK (2017) Functional properties of Brazilian propolis: From chemical composition until the market. In: Waisundara V, Shiomi N (eds) Superfood and functional food - An overview of their processing and utilization. IntechOpen Limited, London, pp 55–98. https://doi.org/10.5772/65932

Beserra FP, Gushiken LFS, Hussni MF, Ribeiro VP, Bonamin F, Jackson CJ, Pellizon CH, Bastos JK (2020) Artepillin C as an outstanding phenolic compound of Brazilian green propolis for disease treatment: a review on pharmacological aspects. Phytother Res 35:2274–2286. https://doi.org/10.1002/ptr.6875

Bhargava P, Grover A, Nigam N (2018) Anticancer activity of the supercritical extract of Brazilian green propolis and its active component, artepillin C: bioinformatics and experimental analyses of its mechanisms of action. Int J Oncol 52:925–932

Ingizza C (2020) Própolis e vitamina D: venda de produtos para imunidade sobe com Covid-19. https://exame.com/pme/propolis-e-vitamina-d-venda-de-produtos-para-imunidade-sobe-com-covid-19/. Acessed 30 Mar 2020

Carvalho AA, Finger D, Machado CS, Schmidt EM, Costa PM, Alves AP, Morais TM, Queiroz MG, Quináia SP, Rosa MR, Santos JM (2011) In vivo antitumoural activity and composition of an oil extract of Brazilian propolis. Food Chem 126:1239–1245

Castaldo S, Capasso F (2002) Propolis, an old remedy used in modern medicine. Fitoterapia 73:1–6

Castro AMG (2014) Prospecção de cadeias produtivas e gestão da informação. Transinformação 13:55–72

Chang R, Piló-Veloso D, Morais SA, Nascimento EA (2008) Analysis of a Brazilian green propolis from *Baccharis dracunculifolia* by HPLC-APCI-MS and GC-MS. Rev Bras Farmacogn 18:549–556

Coelho MS, Resende FM, Fernandes GW (2013) Chinese economic growth: implications for Brazilian conservation policies. Nat Conserv 11:88–91. http://doi.org/10.4322/natcon.2013.014
Conceição JCPR, Junior RGE, Conceição PHZ (2017) Cadeia agroindustrial do café no Brasil: uma análise do período recente. Radar 53:25–30
Coresight Research (2018) Quick take: the boom in marketplace platforms. https://coresight.com/research/quick-take-the-boom-in-marketplace-platform/. Accessed 20 May 2020
Dagnino R (2004) A relação pesquisa-produção: em busca de um enfoque alternativo. In: Santos LW (ed) Ciência, tecnologia e sociedade: o desafio da interação, 2nd edn. Iapar, Londrina, pp 101–151
EPO (2018) European Patent Office (EPO). https://www.epo.org/index.html. Accessed 21 May 2019
Faemg (2014) Preço da própolis em alta no mercado. Federação da Agricultura e Pecuária do Estado de Minas Gerais. http://www.sistemafaemg.org.br/Noticia.aspx?Code=5620&Portal=1&PortalNews=1&ParentCode=139&ParentPath=None&ContentVersion=R. Acessed 20 May 2020
Furtado C (2005) Formação econômica do Brasil. 32 rd ed., Companhia Editora Nacional, São Paulo
Ghesti GF, Faria BS, Dias FR, Junior GCC, Amaral HL, Lima LA, Pires LGS, Araújo LP, Tenório LXS, Silva ML, Fernandes TL (2016) Tutorial de busca nos principais bancos de patentes. Centro de Apoio ao Desenvolvimento Tecnológico da Universidade de Brasília - CDT/UnB, Brasília
Hausmann R, Hidalgo CA, Bustos S, Coscia M, Simões A, Yildirim MA (2011) The atlas of economic complexity: Mapping paths to prosperity. Center for International Development Harvard University, Cambridge
Huang S, Zhang CP, Wang K, Li GQ, Hu FL (2014) Recent advances in the chemical composition of propolis. Molecules 19:19610–19632
INPI (2015) Manual para o depositante de patentes. Ministério do Desenvolvimento, Indústria e Comércio Exterior - Mdic, Instituto Nacional da Propriedade Industrial - INPI. http://www.INPI.gov.br/menu-servicos/patente/arquivos/manual-para-o-depositante-de-patentes.pdf. Accessed 21 May 2019
INPI (2017) Introdução à classficação cooperativa de patentes (CCP). http://www.ufpb.br/inova/contents/documentos/tutorial-cpc-INPI.pdf. Accessed 21 May 2019
INPI (2018) INPI — Instituto Nacional da Propriedade Industrial. http://www.INPI.gov.br/. Accessed 21 May 2019
J-PlatPat (2010) Japanese platform for patent information. https://www.j-platpat.inpit.go.jp/. Accessed 21 May 2019
Katekhaye S, Fearnley H, Fearnley J, Paradkar A (2019) Gaps in propolis research: challenges posed to commercialization and the need for an holistic approach. J Apic Res 58:604–616
Kuropatnicki AK, Szliszka E, Krol W (2013) Historical aspects of propolis research in modern times. Evid Based Complement Alternat Med 2013:1–12. https://doi.org/10.1155/2013/964149
Lihong C, Sanguandeekul R, Fuxing Z, Deowanish S, Thapa R, Wongsiri S (2009) Advances in propolis research and propolis industry in China. J. Royal Inst Thailand 1:136–151
Marcucci MC (1995) Propolis: chemical composition, biological properties and therapeutic activity. Apidologie 26:83–99
Mayerhoff ZDVL (2008) Uma análise sobre os estudos de prospecção tecnológica. Cad Prospeção 1:7–9. https://doi.org/10.9771/cp.v1i1.3538
Messerli SM, Ahn MR, Kunimasa K, Yanagihara M, Tatefuji T, Hashimoto K, Mautner V, Uto Y, Hori H, Kumazawa S, Kaji K (2009) Artepillin C (ARC) in Brazilian green propolis selectively blocks oncogenic PAK1 signaling and suppresses the growth of NF tumors in mice. Phytother Res 23:423–427
Mintel (2017) New products containing propolis. Global analysis report. Commodity innovation series. http://www.agr.gc.ca/resources/prod/Internet-Internet/MISB-DGSIM/ATS-SEA/PDF/6861-eng.pdf. Accessed 21 May 2019

Mintel (2018) Tendências globais em alimentos & bebidas. https://downloads.mintel.com/private/9pRqD/files/650337/. Accessed 21 May 2019

Nothenberg M (1997) Propolis overcomes the challenge of researchers. Quim Deriv 1:24–28

Oliveira VC, Bastos EM (1999) Aspectos morfo-anatômicos da folha de *Baccharis dracunculifolia* DC. (Asteraceae) visando a identificação da origem botânica da própolis. Acta Bot Bras 12:431–439

OpenPR (2020) Propolis Market 2020 Global share, growth, size, opportunities, trends, regional overview, leading company analysis, and key country forecast to 2027. https://www.openpr.com/news/2042327/propolis-market-2020-global-share-growth-size. Accessed 20 May 2020

Park YK, Alencar SM, Aguiar CL (2002) Botanical origin and chemical composition of Brazilian propolis. J Agric Food Chem 50:2502–2506

Pereira AS, Seixas FRMS, Neto FRA (2002) Própolis: 100 anos de pesquisa e suas perspectivas futuras. Quim Nova 25:321–326

Pinto LDMA, Prado NRT, Carvalho LB (2011) Propriedades, usos e aplicações da própolis. Rev Eletr Farm 8:25–25. https://doi.org/10.5216/ref.v8i3.15805

Roessing AC (2002) Cadeias produtivas: roteiro para estudo de sistemas agroalimentares, 1rd ed. Empresa Brasileira de Pesquisa Agropecuária - Embrapa, Londrina

Salatino A, Salatino MLF (2018) Brazilian red propolis: legitimate name of the plant resin source. MOJ Food Process Technol 6:21–22

Santos LM, Fonseca MS, Sokolonski AR, Deegan KR, Araújo RP, Umsza-Guez MA, Barbosa JD, Portela RD, Machado BA (2019) Propolis: types, composition, biological activities and veterinary product patent prospecting. J Sci Food Agr 100:1369–1382

Sforcin JM (2009) Própolis e imunidade: comprovações científicas. UNESP, São Paulo

Sforcin JM (2012) Interação entre as abelhas e a *Baccharis dracunculifolia* para a elaboração da própolis. In: Sforcin JM, Sousa JPB, Filho AADS, Bastos JK, Búfalo MC, Tonuci LRDS (eds) *Baccharis dracunculifolia* - Uma das principais fontes vegetais da própolis brasileira. UNESP, São Paulo, pp 9–15

Sforcin JM, Bankova V (2011) Propolis: is there a potential for the development of new drugs? J Ethnopharmacol 133:253–260

Shadlen KC, Sampat BN, Kapczynski A (2019) Patents, trade and medicines: past, present and future. Rev Int Political Econ 27:75–97. https://doi.org/10.1080/09692290.2019.1624295

Soares JM, Dias MC, Correa V (2010) Como pesquisar o perfil patentário de um fármaco: o caso Efavirenz. Quim Nov 33:1216–1219

Souza IC, Souza MP, Pedon JS (2012) Estudos prospectivos em cadeias produtivas - Revisão teórico - metodológica. Anais do I Encontro de ensino e pesquisa em administração da Amazônia, EnEPA, Porto Velho, pp. 1–15.

Souza RP, Mendonça PSB, Damke GM, Carvalho ARA, Ratti BA, Dembogurski DS, Silva VR, Silva SO, Silva DB, Bruschi ML, Engler SSM (2018) Artepillin C induces selective oxidative stress and inhibits migration and invasion in a comprehensive panel of human cervical cancer cell lines. Anti-Cancer Agent Med 18:1750–1760. https://doi.org/10.2174/1871520618666180604092930

Tagliacollo VA, Orsi RDO (2011) Quality of propolis commercialized in the informal market. Food Sci Technol 31:752–757

Thredup (2018) report of Resale report. https://www.thredup.com/resale/2018?tswc_redir=true. Accessed 20 May 2020

Veiga RS, Mendonça S, Mendes PB, Paulino N, Mimica MJ, Netto AAL, Lira IS, López BCG, Negrão V, Marcucci MC (2017) Artepillin C and phenolic compounds responsible for antimicrobial and antioxidant activity of green propolis and *Baccharis dracunculifolia* DC. J Appl Microbiol 122:911–920

WIPO (2018a) WIPO - World Intellectual Property Organization. https://www.WIPO.int/portal/en/. Accessed 21 May 2019

WIPO (2018b) Guide to the International Patent Classification. https://www.WIPO.int/export/sites/www/classifications/ipc/en/guide/guide_ipc.pdf. Accessed 21 May 2019

Index

G. W. Fernandes et al. (eds.), *Baccharis*,
https://doi.org/10.1007/978-3-030-83511-8

C

D

The manufacturer's authorised representative in the EU is Springer Nature Customer Service Centre GmbH, Europaplatz 3, 69115 Heidelberg, Germany. If you have any concerns regarding our products, please contact ProductSafety@springernature.com

Printed and bound by CPI Group (UK) Ltd, Croydon, CR0 4YY
30/09/2026
02225333-0016